Revolutionizing Tropical Medicine

Revolutionizing Tropical Medicine

Point-of-Care Tests, New Imaging Technologies and Digital Health

Edited by

Kerry Atkinson
University of Queensland Centre for Clinical Research,
Brisbane, Australia and the University of Technology/ Institute of Health
and Biomedical Innovation, Brisbane, Australia

David Mabey
London School of Hygiene and Tropical Medicine, London, UK

The right of Kerry Atkinson and David Mabey to be identified as the authors of the editorial material in this work has been asserted in accordance with law.

Registered Office(s)
John Wiley & Sons, Inc., 111 River Street, Hoboken, NJ 07030, USA

Editorial Office
The Atrium, Southern Gate, Chichester, West Sussex, PO19 8SQ, UK

For details of our global editorial offices, customer services, and more information about Wiley products visit us at www.wiley.com.

Wiley also publishes its books in a variety of electronic formats and by print-on-demand. Some content that appears in standard print versions of this book may not be available in other formats.

Library of Congress Cataloging-in-Publication Data has been applied for:

ISBN: 9781119282648 (Hardback)

Cover Design: Wiley
Cover Images: © notbad/Shutterstock, © piick/Shutterstock, © Wikimedia Commons

Set in 10/12pt WarnockPro by SPi Global, Chennai, India
Printed and bound in Singapore by Markono Print Media Pte Ltd

10 9 8 7 6 5 4 3 2 1

"The wounded surgeon plies the steel
That questions the distempered part;
Beneath the bleeding hands we feel
The sharp compassion of the healer's art
Resolving the enigma of the fever chart."

Four Quartets. IV. East Coker
T.S. Eliot, Nobel Laureate in Literature, 1948.

Table of Contents

Editor's Preface

I first became interested in tropical medicine 50 years ago when I did an elective period as a medical student at Makerere University and Mulago Hospital in Kampala, Uganda. Searing medical images from that visit remain vivid to me to this day. One of the most enduring was the sight of six men in a male inpatient ward each having a tuberculous empyema drained. Another event was memorable: I was waiting to pay my respects to the Dean of the Medical School on the day I arrived when a man came down the corridor and sat down next to me. He asked what I was doing and I told him. I asked him what he was doing and he told me he was going "on a cancer safari." I asked him, embarrassingly, if it was "anything to do with that Burkitt's thing?" He turned and looked at me and said "I am Burkitt."

Serendipity, however, played its role and I subsequently found myself in medical oncology and through that into hematopoietic stem cell transplantation. A number of coincidences occurred. One of my pathology lecturers at the Middlesex Hospital Medical School in London was Tony Epstein, who co-discovered the Epstein-Barr virus that Denis Burkitt was tracking and relating its location in Africa with that of Burkitt's lymphoma. Later when I was at the Royal Marsden Hospital in London, Peter Clifford was an ENT surgeon there. In fact he removed our youngest daughter's tonsils while my wife Pauline and I were working there. Peter had previously worked for many years in East Africa and realized that the only therapeutic approach to most cases of Burkitt's lymphoma was the use of high-dose chemotherapy. To apply this, however, he knew that he had to harvest and cryopreserve the patient's marrow cells prior to the chemotherapy in order to infuse them after the chemotherapy in order to reconstitute the patient's hematopoietic system previously destroyed by the chemotherapy. To perform this Peter hired Reginald Clift, who had been working in pathology in England. E. Donnall Thomas, who pioneered bone marrow transplantation in Seattle and who won the Nobel Prize in Medicine in 1989 for it, heard of this work and promptly hired Reg to work with him in Seattle. After the Marsden I spent five years on Don's unit in Seattle and got to know Reg well. I'm not sure how many degrees of separation this amounts but it is certainly less than six.

Fast forward four decades: I left oncology and stem cell transplantation and went back to my original interest in tropical medicine. I enrolled in the spectacular East Africa course in Tropical Medicine and Hygiene organized each year by Dr Philip Gothard from the London School of Hygiene and Tropical Medicine and held over three months in Tanzania and Uganda. I had been working on a mesenchymal stem cell book with John Wiley Publishers over the preceding two years and during the African course I received an email from them asking if there was any other area of medicine that

I would be interested in writing about. At this time we were learning on the course about rapid diagnostic tests and innovative cheap medical imaging technologies so I asked them if they would be interested on a book on these new and other exciting innovative developments in tropical medicine. It was on the course that I met David who agreed to edit the book with me and without whom this manuscript would never have materialized.

Sydney, 2019

Kerry Atkinson

David Mabey

After medical school and core medical training in the UK I went to work as a clinician at the Medical Research Council Unit in The Gambia, where I spent eight years. For the last four years I was in charge of the clinical services at the unit, which comprised a 40 bed medical and pediatric ward and an outpatient department which saw about 12 000 patients per year. In 1986 I moved back to the UK as a Clinical Senior Lecturer at the London School of Hygiene & Tropical Medicine (LSHTM) with an honorary Consultant Physician post at the Hospital for Tropical Diseases. I ran the DTM&H course at LSHTM for a number of years, and later helped Phil Gothard to set up the East African DTM&H course. My research has focused on sexually transmitted infections (STIs) and neglected tropical diseases (NTDs) in Africa and Asia, in particular on diseases caused by *Chlamydia trachomatis* and *Treponema pallidum*, which cause both STIs and NTDs.

London, 2019

David Mabey

Acknowledgments

We would both like to acknowledge the tremendous expertise, knowledge and willingness to contribute demonstrated by all the authors and co-authors in this book. Any book is only as good as its content and we feel very lucky to have been able to tap the content that has been shared with us.

We would also like to thank the staff at Wiley for their capability in shepherding the book through each of its phases – Antony Sami, Metilda Shummy, Shalisha Sukunya, Ramprasad Jayakumar and especially Mindy Okura-Marszycki who commissioned the book three and a half years ago.

List of Contributors

Helen Allott
The Liverpool School of Tropical
Medicine
Liverpool
UK

Odinaka Anyanwu
Ross University School of Medicine and
University of Texas Southwestern
Dallas
TX
USA

Kerry Atkinson
University of Queensland Centre for
Clinical Research
Brisbane
Queensland
Australia;
The University of Technology/Institute of
Health and Biomedical Innovation
Brisbane
Queensland
Australia

Gemma Bale
Department of Medical Physics and
Biomedical Engineering
University College
London
UK

I.G. Barr
WHO Collaborating Centre for Reference
and Research on Influenza
Victorian Infectious Disease Reference
Laboratory (VIDRL)
Doherty Institute
Melbourne, Victoria, Australia;
Department of Microbiology
and Immunology
Faculty of Medicine, Dentistry
and Health Sciences
University of Melbourne
Parkville
Victoria
Australia;
Faculty of Science and Technology
Federation University
Churchill
Victoria
Australia

Sophia M. Bartels
Dartmouth College
Hanover
NH
USA

Andrew Bastawrous
Faculty of Infectious & Tropical Diseases
Clinical Research Department
London School of Hygiene and
Tropical Medicine
International Centre for Eye Health
(ICEH)
London
UK

Sabine Bélard
Department of Pediatric Pneumology and
Immunology
Charité-Universitätsmedizin Berlin
Germany;
Berlin Institute of Health
Berlin
Germany

James Berkley
Centre for Tropical Medicine and Global
Health
University of Oxford
Oxford
UK;
KEMRI/Wellcome Trust Research
Programme
Kilifi
Kenya;
The Childhood Acute Illness and
Nutrition Network (CHAIN)
Nairobi
Kenya

Marleen Boelaert
Institute of Tropical Medicine
Antwerp
Belgium;
DNDi
Geneva
Switzerland;
Geneva University Hospitals
Geneva
Switzerland

Debrah I. Boeras
International Diagnostics Centre
Clinical Research Department
London School of Hygiene & Tropical
Medicine
London, UK;
Global Health Impact Group
Atlanta
GA
USA

Nigel M. Bolster
University of Strathclyde
Biomedical Engineering
Glasgow
UK

André Briend
University of Tampere School of
Medicine and Tampere University
Hospital
Tampere
Finland;
Department of Nutrition,
Exercise and Sports
Faculty of Science
University of Copenhagen
Copenhagen
Denmark

Colin Brown
King's Sierra Leone Partnership, King's
Centre for Global Health, King's Health
Partners
King's College London
London
UK;
National Infection Service, Public Health
England
London
UK;
Department of Infection
Royal Free Hampstead NHS Trust
London
UK

Philippe Büscher
Department of Biomedical Sciences
Institute of Tropical Medicine
Antwerp
Belgium

Liam J. Caffery
Centre for Online Health
Faculty of Medicine
The University of Queensland
Brisbane
Queensland
Australia

François Chappuis
Institute of Tropical Medicine
Antwerp
Belgium;
DNDi
Geneva
Switzerland;
Geneva University Hospitals
Geneva
Switzerland

Blanche C. Collins
Centers for Disease Control and
Prevention
Center for Surveillance, Epidemiology
and Laboratory Services
Atlanta
GA
USA

Jane Crawley
Centre for Tropical Medicine and Global
Health
University of Oxford
Oxford
UK

Jane Cunningham
Global Malaria Program, World Health
Organization
Geneva
Switzerland

Ishita Desai
Department of Radiology
University of California, Los Angeles
Los Angeles
CA
USA

Lauren Duckworth
Flinders University
Adelaide
South Australia
Australia

Tyler Evans
AIDS HealthCare Foundation
Los Angeles
CA
USA

Farhad Fatehi
Centre for Online Health
Faculty of Medicine
The University of Queensland
Brisbane
Queensland
Australia

Dunia Faulx
PATH (www.path.org)

Chuanping Feng
School of Water Resources and
Environment
China University of Geosciences (Beijing)
Beijing
China

Cai Fong
Thoracic Research Centre
Faculty of Medicine
The University of Queensland
Brisbane
Queensland
Australia

Hamish Graham
Centre for International Child Health
University of Melbourne
Murdoch Research Children's Institute
The Royal Children's Hospital
Melbourne
Victoria
Australia

Heiner Grosskurth
London School of Hygiene and Tropical
Medicine
Department of Infectious Disease
Epidemiology, based at the Mwanza
Intervention Trials Unit (MITU)
National Institute for Medical Research
(NIMR)
Dar es Salaam
Tanzania

Heather Halls
Flinders University
Adelaide
South Australia
Australia

Michael Harrison
Sullivan Nicolaides Pathology
Brisbane
Queensland
Australia

Epco Hasker
Department of Public Health
Institute of Tropical Medicine
Antwerp
Belgium

Tom Heller
Lighthouse Clinic
Kamuzu Central Hospital
Lilongwe
Malawi

Heidi Hopkins
Faculty of Infectious and Tropical
Diseases
London School of Hygiene and Tropical
Medicine
London
UK

Catherine Houlihan
Department of Infection and Immunity
University College London
London
UK

A.C. Hurt
WHO Collaborating Centre for
Reference and Research on Influenza
Victorian Infectious Disease Reference
Laboratory (VIDRL)
Doherty Institute
Melbourne
Victoria
Australia;
Department of Microbiology and
Immunology
Faculty of Medicine, Dentistry, and
Health Sciences
University of Melbourne
Parkville
Victoria
Australia

Michaëla A.M. Huson
Center of Tropical Medicine and Travel
Medicine
Division of Infectious Diseases
Academic Medical Center
University of Amsterdam
Amsterdam
The Netherlands

Meghan L. Jardon
Department of Radiology
University of California, Los Angeles
Los Angeles
CA
USA

Joseph N. Jarvis
Department of Clinical Research
Faculty of Infectious and Tropical
Diseases
London School of Hygiene and Tropical
Medicine
London
UK;
Botswana-UPenn Partnership
Gaborone
Botswana;
University of Botswana
Gaborone
Botswana;
Division of Infectious Diseases
Perelman School of Medicine
University of Pennsylvania
Philadelphia
PA
USA

Elizabeth Joekes
Department of Radiology
Royal Liverpool University Hospitals
Liverpool
UK

Dan Kaminstein
Department of Emergency Medicine
and Hospitalist Service
Medical College of Georgia at Augusta
University
Augusta
GA
USA

Carina King
Institute for Global Health
University College London
London
UK

Carrie Kovarik
Dermatology, Dermatopathology, and
Infectious Diseases
University of Pennsylvania
Philadelphia
PA
USA

Veerle Lejon
Unité Mixte de Recherche IRD-CIRAD
177 INTERTRYP
Institut de Recherche pour le
Développement
Montpellier
France

Richard Lessells
KwaZulu-Natal Research Innovation and
Sequencing Platform
Nelson R Mandela School of Medicine
University of KwaZulu-Natal
Durban
South Africa;
Department of Clinical Research
London School of Hygiene and Tropical
Medicine
London
UK

David Mabey
Clinical Research Department
London School of Hygiene and
Tropical Medicine
London
UK

Eric D. McCollum
Department of Pediatrics, Eudowood
Division of Pediatric Respiratory Sciences
Johns Hopkins School of Medicine
Baltimore
MD
USA;
Department of International Health
Johns Hopkins Bloomberg School of
Public Health
Baltimore
MD
USA

Michael Marks
Clinical Research Department
London School of Hygiene &
Tropical Medicine
London
UK

Lisa A. Marsch
Dartmouth College
Hanover
NH
USA

Hannah K. Mitchell
Botswana-UPenn Partnership
Gaborone
Botswana

Lara Motta
Flinders University
Adelaide
South Australia
Australia

Claire Mullender
St George's Healthcare Trust
London
UK

Martha Mwangome
KEMRI/Wellcome Trust Research
Programme
Kilifi
Kenya

Amina Nardo-Marino
Department of Haematology, Herlev and
Gentofte Hospital
University of Copenhagen
Herlev
Denmark

John A. Naslund
Harvard Medical School
Boston
USA

Benjamin Y.C. Ng
Centre for Personalized NanoMedicine
Australian Institute for Bioengineering
and Nanotechnology
The University of Queensland
Brisbane
Queensland
Australia;

School of Chemistry and Molecular
Biosciences
The University of Queensland
Brisbane
Queensland
Australia

Roger B. Peck
PATH (www.path.org)

Rosanna W. Peeling
International Diagnostics Centre
Clinical Research Department
London School of Hygiene & Tropical
Medicine
London
UK

David Peiris
George Institute for Global Health
Sydney
Australia;

and

University of New South Wales Sydney
Australia

Kelsey L. Pomykala
Department of Radiology
University of California, Los Angeles
Los Angeles
CA
USA

Kara-Lee Pool
Department of Radiology
University of California, Los Angeles
Los Angeles
CA
USA

Giselle Prado
Orange Park Medical Center
Jacksonville
FL
USA

Suman Rijal
Institute of Tropical Medicine
Antwerp
Belgium;
DNDi
Geneva
Switzerland;
Geneva University Hospitals
Geneva
Switzerland

Tala de los Santos
PATH (www.path.org)

Martin Seneviratne
Department of Biomedical Informatics
Stanford School of Medicine
Stanford
CA
USA;

and

The Royal Prince Alfred Hospital
Sydney
NSW
Australia

Janet G. Shaw
Department of Thoracic Medicine
The Prince Charles Hospital
Brisbane
Queensland
Australia;
Thoracic Research Centre
Faculty of Medicine
The University of Queensland
Brisbane
Queensland
Australia

Mark Shephard
Flinders University
Adelaide
South Australia
Australia

Anthony C. Smith
Centre for Online Health
Faculty of Medicine
The University of Queensland
Brisbane
Queensland
Australia

Emma Smith
Department of Thoracic Medicine
The Prince Charles Hospital
Brisbane
Queensland
Australia;
Thoracic Research Centre
Faculty of Medicine
The University of Queensland
Brisbane
Queensland
Australia

Brooke Spaeth
Flinders University
Adelaide
South Australia
Australia

Svetlana Stevanovic
International Laboratory for Air Quality
and Health
Queensland University of Technology
Brisbane
Queensland
Australia

Ilias Tachtsidis
Department of Medical Physics and
Biomedical Engineering
University College
London
UK

Monica Taylor
Centre for Online Health
Faculty of Medicine
The University of Queensland
Brisbane
Queensland
Australia

Mark W. Tenforde
Division of Allergy and Infectious
Diseases
University of Washington School of
Medicine
Seattle
WA
USA;
Department of Epidemiology
University of Washington School of
Public Health
Seattle
WA
USA

Matt Trau
Centre for Personalized NanoMedicine
Australian Institute for Bioengineering
and Nanotechnology
The University of Queensland
Brisbane
Queensland
Australia;
School of Chemistry and Molecular
Biosciences
The University of Queensland
Brisbane
Queensland
Australia

Annalicia Vaughan
Department of Thoracic Medicine
The Prince Charles Hospital
Brisbane
Queensland
Australia;
Thoracic Research Centre
Faculty of Medicine
The University of Queensland
Brisbane
Queensland
Australia

Catherine J. Wedderburn
International Diagnostics Centre
Clinical Research Department
London School of Hygiene & Tropical
Medicine
London
UK

Eugene J.H. Wee
Centre for Personalized NanoMedicine
Australian Institute for Bioengineering
and Nanotechnology
The University of Queensland
Brisbane
Queensland
Australia

Nicholas P. West
School of Chemistry and Molecular
Biosciences
The University of Queensland
Brisbane
Queensland
Australia

James Whitehorn
University College
London
UK

Tom N. Williams
The KEMRI/Wellcome Trust Research
Programme
Kilifi
Kenya;
Imperial College London
London
UK

Ian A. Yang
Department of Thoracic Medicine
The Prince Charles Hospital
Brisbane
Queensland
Australia;

Thoracic Research Centre, Faculty of
Medicine
The University of Queensland
Brisbane
Queensland
Australia

Chengzhong Yu
Australian Institute for Bioengineering
and Nanotechnology
The University of Queensland
Brisbane
Queensland
Australia

Jing Zhang
Australian Institute for Bioengineering
and Nanotechnology
The University of Queensland
Brisbane
Queensland
Australia;
School of Water Resources and
Environment
China University of Geosciences (Beijing)
Beijing
China

Part I

The Health of Low- and Middle-Income Countries Today

1

The Burden of Communicable Diseases in Low- and Middle-Income Countries

Kerry Atkinson[1, 2] and David Mabey[3]

[1] *University of Queensland Centre for Clinical Research, Brisbane, Queensland, Australia*
[2] *The University of Technology/Institute of Health and Biomedical Innovation, Brisbane, Queensland, Australia*
[3] *London School of Hygiene and Tropical Medicine, London, UK*

Revolutionizing Tropical Medicine: Point-of-Care Tests, New Imaging Technologies and Digital Health,
First Edition. Edited by Kerry Atkinson and David Mabey.
© 2019 John Wiley & Sons, Inc. Published 2019 by John Wiley & Sons, Inc.

1.1   Introduction

The Global Burden of Disease 2015 Study provided a comprehensive assessment of all-cause and cause-specific mortality for 249 causes in 195 countries and territories from 1980 to 2015 (GBD 2015 Mortality and Causes of Death Collaborators 2016). Among its key findings were Global life expectancy from birth increased from 61·7 years in 1980 to 71·8 years in 2015.

Several countries in sub-Saharan Africa had very large gains in life expectancy from 2005 to 2015, rebounding from an era of exceedingly high loss of life due to HIV/AIDS.

At the same time, many areas saw life expectancy stagnate or decline, particularly for men and in countries with rising mortality from war or interpersonal violence.

Total deaths due infectious diseases declined significantly from 2005 to 2015, largely attributable to decreases in mortality due to HIV/AIDS, malaria, and acute respiratory infections in children under five years, although neonatal mortality (first month of life) has fallen very little.

In 2015 rotaviral enteritis was the leading cause of under five years deaths due to diarrhea, and pneumococcal pneumonia was the leading cause of under five years deaths due to lower respiratory infections. Deaths due to acute lower respiratory illnesses have fallen overall due to the roll-out of vaccines against measles, *Streptococcus pneumoniae* and *Haemophilus influenzae* b.

Malaria, HIV/AIDS, and tuberculosis (TB) remained leading causes of death in the LMICs.

This chapter describes the global burden of communicable diseases while Chapter 2 describes the global burden of non-communicable diseases. The majority of the data in this chapter was sourced from the World Health Organization (WHO) or the Global Burden of Disease Project (GBD).

Many of the references in this chapter are taken from documents, such as those produced by the WHO, which are only published online. Such documents are referenced alphabetically by the last name of the first author/s or organization in the Bibliography and by their Uniform Resource Locator (URL) in the Webliography (www = World Wide Web; http or https = Hypertext Transfer Protocol/s).

1.2   Definition of a Communicable Disease

A communicable disease, also known as an infectious disease or a transmissible disease, is an illness resulting from an infection. Infectious agents include viruses, bacteria, fungi, protozoa, nematodes and prions.

1.3   Definition of Low- and Middle-Income Countries

In the past, and still currently, the phrases "developing world, developing countries and developing nations" have been used to describe countries with low individual incomes and poor health and poor education systems. However, there is a marked disparity in these parameters between different countries in the "developing world." For this reason we have elected to use the term "low- and -middle income countries" (LMICs) in this book, because there is an objective classification for this term (see below). In some cases, the term "resource-limited settings" has also been used.

The commonest measure for assessing a nation's developmental status is to use the gross domestic product (GDP) per head of population (per capita) as determined by the Organisation of Economic Development (OECD). The OECD defines GDP as "an aggregate measure of production equal to the sum of the gross values added of all resident and institutional units engaged in production plus any taxes, and minus any subsidies, on products not included in the value of their outputs" (Organization for Economic Development definition of GDP 1993; http://esa.un.org/unsd/sna1993/introduction.asp).

Figure 1.1 shows the GDP (PPP) per capita in US dollars for all countries in 2015.

From this map it is clear that countries with a GDP per capita of US$35 000-US$50 000 or more comprise the USA, Canada, some countries in South America, some countries in Western Europe, Saudi Arabia, the Gulf States, Brunei, Taiwan, Japan, South Korea, Australia and New Zealand. In contrast, many nations in Sub-Saharan Africa have a GDP per capita of US$2000 or less.

The World Bank classifies countries into four income groups using the gross national income (GNI) per capita (http://data.worldbank.org/indicator/NY.GNP.PCAP.CD?order=wbapi_data_value_2014+wbapi_data_value+wbapi_data_value-last&sort=desc). The GNI per capita is the US dollar value of a country's final income in a year, divided by its population. It reflects the average income of a country's citizens.

Knowing a country's GNI per capita is a useful first step toward understanding the country's economic strengths and needs, as well as the general standard of living enjoyed by the average citizen. A country's GNI per capita tends to be closely linked with other indicators that measure the social, economic, and environmental well-being of the country and its people. For example, generally people living in countries with higher GNI per capita tend to have longer life expectancies, higher literacy rates, better access to safe water, and lower infant mortality rates. In general, people in LMICs have a lower life expectancy, less education and less money than people in developed nations.

The World Bank's GNI per capita figures are set each year on 1 July. Economies were divided according to 2016 GNI per capita using the following ranges of income:

Low income countries had a GNI per capita of US$1025 or less.

Lower middle-income countries had a GNI per capita between US$1026 and US$4035.

Upper middle-income countries had a GNI per capita between US$4036 and US$12 475.

High income countries had a GNI per capita above US$12 476.

1.4 Definition of Burden of Disease

1.4.1 Disability-Adjusted Life Years

A commonly used metric for assessing the burden of disease is the Disability-Adjusted Life Year (DALY) developed by the WHO, the World Bank and the Harvard School of Public Health (WHO | Metrics: Disability-Adjusted Life Year [DALY]).

The DALY is a time-based measurement that combines years of life lost due to premature mortality and years of life lost due to time lived in health states less than ideal health. One DALY is defined as one lost year of "healthy" life, and the burden of disease

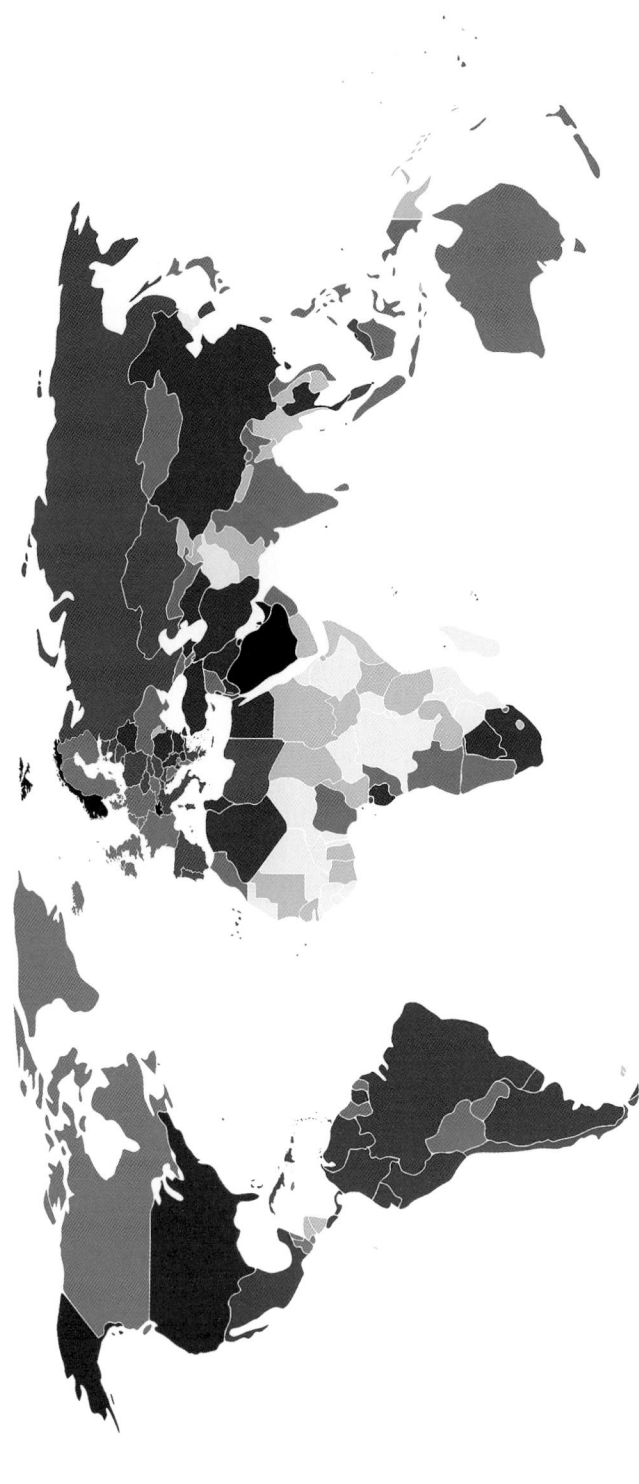

Figure 1.1 World map of GDP per Capita by Country. Gross domestic product (GDP) is converted to international dollars using purchasing power parity rates (PPP). World Bank, International Comparison Program database (2011–2014). GDP per capita - PPP in international US $ ▮ 50000 or more ▮ 35000–50000 ▮ ▮ less than 2000 ▮ no data available. Source: This image, reproduced here from Wikimedia Commons, is in the public domain because its creator, Rfassbind, has placed it there. https://commons.wikimedia.org/wiki/File:World_map_GDP_per_capita.svg. ▮ 20000–35000 ▮ 10000–20000 ▮ 5000–10000 ▮ 2000–5000 *(See color plate section for the color representation of this figure.)*

is a measurement of the gap between current health status and an ideal situation where everyone lives into old age, free of disease and disability.

Additional impacts are due to the cost of treatment, the cost of prevention, loss of income, loss of agricultural productivity, and loss of education (Hotez et al. 2009).

1.5 Definition of Disease Elimination

Elimination of a disease refers to the reduction to zero (or a very low defined target rate) of new cases in a defined geographical area. Elimination requires continued measures to prevent re-establishment of disease transmission.

1.6 Definition of Disease Eradication

Eradication of a disease refers to the complete and permanent worldwide reduction to zero new cases of the disease through deliberate efforts. If a disease has been eradicated, no further control measures are required.

1.7 Definition of the Primary Point-of-Care

The primary point-of-care is the first site that a sick person visits to seek treatment. This may be a traditional healer, an allied health professional such as a pharmacist, or a nurse or a doctor. It is currently rare for it to be a doctor in the LMICs. Because of low income and a long distance to medical facilities, or both, access to diagnostic and treatment clinics is difficult for many people in the LMICs. Primary point-of-care assessment and testing has the potential to make a significant impact on the management many diseases by accelerating diagnosis and instigating immediate treatment or by accelerating triage to an appropriate health facility.

1.8 The 2000 Millennium Development Goals (MDGs) and Their Outcomes

In September 2000 189 heads of state adopted the United Nations (UN) Millennium Development Goals and endorsed a framework for their implementation. The plan was for countries and development partners to work together to reduce poverty and hunger and to tackle ill health, lack of education, gender inequality, lack of access to clean water and environmental degradation (United Nations Millennium Development Goals http://www.un.org/millenniumgoals/pdf/MDG_Report_2009_ENG.pdf).

The Eight Millennium Development Goals were

1) Eradication of extreme poverty and hunger
2) Achievement of universal primary education
3) Promotion of gender equality and the empowerment of women
4) Reduction in child mortality
5) Improvement of maternal health

6) Combatting HIV/AIDS, malaria and other diseases
7) Ensuring environmental sustainability
8) Development of a global partnership for development.

The MDGs were considered interdependent. They had targets of being achieved by 2015. Indicators to monitor progress were prescribed for each Goal. Summary reports from the specific health-related MDGs are as follows and are sourced from: WHO. The Millennium Development Goals Report (2015), p. 4: http://www.un.org/millenniumgoals/2015_MDG_Report/pdf/MDG%202015%20Summary%20web_english.pdf.

MDG Goal 4: Reduce childhood mortality rate

- The global under-five mortality rate has declined by more than half, dropping from 90 to 43 deaths per 1000 live births between 1990 and 2015.
- Despite population growth in the developing regions, the number of deaths of children under five has declined from 12.7 million in 1990 to almost 6 million in 2015 globally.
- Since the early 1990s, the rate of reduction of under-five mortality has more than tripled globally.
- In sub-Saharan Africa, the annual rate of reduction of under-five mortality was over five times faster during 2005–2013 than it was during 1990–1995.
- Measles vaccination helped prevent nearly 15.6 million deaths between 2000 and 2013. The number of globally reported measles cases declined by 67% for the same period.
- About 84% of children worldwide received at least one dose of measles-containing vaccine in 2013, up from 73% in 2000.

MDG Goal 5: Improve maternal health

Since 1990 the maternal mortality ratio has declined by 45% worldwide, and most of the reduction has occurred since 2000.

- In Southern Asia the maternal mortality ratio declined by 64% between 1990 and 2013, and in sub-Saharan Africa it fell by 49%.
- More than 71% of births were assisted by skilled health personnel globally in 2014, an increase from 59% in 1990.
- In Northern Africa the proportion of pregnant women who received four or more antenatal visits increased from 50 to 89% between 1990 and 2014.
- Contraceptive prevalence among women aged 15–49, married or in a union, increased from 55% in 1990 worldwide to 64% in 2015.

MDG Goal 6: Combat HIV/AIDS, malaria and other diseases

- New HIV infections fell by approximately 40% between 2000 and 2013, from an estimated 3.5 million cases to 2.1 million.
- By June 2014, 13.6 million people living with HIV were receiving antiretroviral therapy (ART) globally, an immense increase from just 800 000 in 2003. ART averted 7.6 million deaths from AIDS between 1995 and 2013.
- Over 6.2 million malaria deaths have been averted between 2000 and 2015, primarily of children under five years of age in sub-Saharan Africa. The global malaria incidence rate has fallen by an estimated 37% and the mortality rate by 58%.

- More than 900 million insecticide-treated mosquito nets were delivered to malaria-endemic countries in sub-Saharan Africa between 2004 and 2014.
- Between 2000 and 2013, TB prevention, diagnosis and treatment interventions saved an estimated 37 million lives. The mortality rate fell by 45% and the prevalence rate by 41% between 1990 and 2013.

1.9 Major Individual Diseases in the LMICs: The Big Three - Malaria, HIV/AIDS and Tuberculosis

The three most important infectious diseases in the LMICs remain malaria, HIV/AIDS and tuberculosis. They are associated with poverty. It is of interest that these three diseases are the most prevalent in LMICs as geographically distant from each other as, for example, Tanzania and Papua New Guinea.

1.9.1 Malaria

Almost half of the world's population is at risk of malaria (WHO – 10 facts on malaria 2016). In 2015 90% of malaria cases and 92% of malaria deaths were in sub-Saharan Africa.

Further details on malaria are shown in Table 1.1.

Table 1.1 The global annual incidence, the global prevalence, the number of global all-age annual deaths, the number of global under five years annual deaths, the number of global all-age DALYs lost due to malaria in 2005 and 2015 and the % change from 2005 to 2015.

Year/% change	Global annual incidence	Global prevalence	Global all-age deaths	Global under five years deaths	Global all-age DALYS lost
2005	350–500 million[a]. No figure stated by GBD Collaborators	No data from WHO. 2005 number not stated by GBD Collaborators	More than 1 million[a]. 1.167 million[b]	No data from WHO[a]. 1.167 million[b]	No data from WHO. 90.4 million[c]
2015	212 million[d]. 286.8 million[e]	No data from WHO. 295.7 million[e]	429 000[a]. 730 500[b]	303 000[d]. 474 100[b]	No data from WHO. 55.8 million[c]
Change from 2005 to 2015	A decrease of 15.4%[e]	An increase of 29.9%[e]	A decrease of 37.4%[b]	A decrease of 42.8%[b]	A decrease of 62%[c]

a) WHO World Malaria Report WHO (2005). http://www.who.int/malaria/publications/atoz/9241593199/en
b) GBD 2015 Mortality and Causes of Death Collaborators (2016).
c) GBD 2015 DALYs and HALE Collaborators (2016).
d) World Malaria Report (2016). http://www.who.int/malaria/publications/world-malaria-report-2016/report/en
e) GBD 2015 Disease and Injury Incidence and Prevalence Collaborators (2016).

Table 1.2 The global annual incidence, the global prevalence, the number of global all-age annual deaths, the number of global under five years annual deaths, the number of global all-age DALYs lost due to HIV/AIDS in 2005 and 2015 and the % change from 2005 to 2015.

Year/% change	Global annual incidence	Global prevalence	Global all-age deaths	Global under five years deaths	Global all-age DALYs lost
2005	4.9 million[a]. Children less than 15 years 2.3 million[a]. No figure stated by GBD Collaborators	40.3 million[a]. No figure given by GBD Collaborators[b]	3.1 million[a]. 1.79 million[c]	570 000 children under 15 years[d]; No figure given in GBD Collaborators[c]	98.9 million[e]
2015	2.0 million[d] Children less than 15 years 150,000[d]	36.7 million[a]. 37.2 million[d]	1.1 million[d]. 1.19 million[c]	No figure given in UNAIDS Fact sheet. 88 9004 -	66.7 million[e]
Change from 2005 to 2015	An all-age decrease of 41%[a),d]	An increase of 21%[b]	A decrease of 33%[c]	A decrease of 51%[c]	A decrease of 39.9%[e]

a) WHO. AIDS epidemic update, December (2005). http://www.who.int/hiv/epi-update2005_en.pdf?ua=1
b) GBD 2015 Disease and Injury Incidence and Prevalence Collaborators (2016).
c) GBD 2015 Disease and Injury Incidence and Prevalence Collaborators (2016).
d) UNAIDS Fact Sheet - latest statistics on the status of the AIDS epidemic. http://www.unaids.org/en/resources/fact-sheet.
e) GBD 2015 DALYs and HALE Collaborators (2016).
The UNAIDS AIDS Epidemic Update (2005) lists 570 000 deaths from HIV/AIDS during 2005 but this was for children under 15 years of age rather than those under 5 years of age[d].

In Tables 1.1 and 1.12 it should be noted that WHO did not routinely publish disease incidence data in 2005 or 2015. The GBD Collaborators did not always publish numbers for 2005, although they routinely calculated the % difference between 2005 and 2015.

Thus, increased prevention, treatment and control measures have led to a 37.4% reduction in the malaria mortality rate globally between 2005 and 2015, a 42.8% reduction in the malaria under-fives mortality rates globally between 2005 and 2015 and a 38% decrease in DALYs lost globally due to malaria between 2005 and 2015.

The most important contributors to these improvements are the increased use of insecticide-impregnated bed nets and access to artemisinin-based combination therapy (ACT).

Approximately half the countries with ongoing malaria transmission are on track to meet the World Health Assembly and Roll Back Malaria targets to achieve a 75% reduction in malaria cases by 2015 as compared to those in 2000.

1.9.2 HIV/AIDS

More than 65 million people have been infected with HIV and 30 million people have died due to AIDS-related causes since the emergence of AIDS in 1981. In 2015 there were 36.3 million people living with HIV including 1.8 million children less than 15 years of age. HIV/AIDS has an uneven geographical distribution with Sub-Saharan Africa bearing more than two-thirds of the global burden (Figures 1.2 and 1.3), followed by India and parts of Southeast Asia. The estimated number of people *living* with HIVin 2016 by WHO region is shown in Figure 1.2 and the number of people *dying* of HIV in 2016 by WHO region is shown in Figure 1.3.

Further details on HIV/AIDS are shown in Table 1.2.

These improvements are due to the roll out of ART treatments. As an example, ART access in 2013 averted an estimated 1 051 354 deaths in South Africa and an estimated 422 448 deaths in Nigeria (Granich et al. 2015).

Additionally, 77% of pregnant women living with HIV were able to access antiretroviral medications in 2014 compared with 37% in 2009.

This has halved the mother-to-child transmission rate of HIV across the Global Plan priority countries from 28 to 14%.

These countries are Angola, Botswana, Burundi, Cameroon, Chad, Côte d'Ivoire, the Democratic Republic of the Congo, Ethiopia, Ghana, India, Kenya, Lesotho, Malawi, Mozambique, Namibia, Nigeria, South Africa, Swaziland, Uganda, the United Republic of Tanzania, Zambia, and Zimbabwe. Together, these countries accounted for 90% of the total number of pregnant women living with HIV that needed services to prevent mother-to-child transmission of HIV in 2009 (UNAIDS 2015 Progress Report 2015).

The life expectancy of people living with HIV has dramatically increased since effective ART has been available and still continues to improve (Nakagawa et al. 2013).

1.9.3 Tuberculosis (TB)

The world is continuing to experience a high burden of disease due to TB. One third of the world's population is infected with Mycobacterium tuberculosis. Tuberculosis is the second most lethal communicable disease worldwide after HIV/AIDS with 1.19 million and 1.1 million deaths respectively in 2015. In 2015 there were an estimated 10.4 million new TB cases worldwide, of which 5.9 million (56%) were in men, 3.5 million (34%) in women and 1.0 million (10%) in children (WHO Global Tuberculosis Report 2016). People living with HIV accounted for 1.2 million (11%) of all new TB cases.

The estimated incidence of TB in 2015 is shown in Figure 1.4.

In the United States the annual incidence was <4 per 100 000 population, but in some countries in sub-Saharan Africa and Asia the annual incidence was several hundred per 100 000. Africa had the most severe burden relative to population. India, Indonesia and China had the largest number of cases by country with 23, 10 and 10% respectively. Six countries accounted for 60% of the new cases: India, Indonesia, China, Nigeria, Pakistan, and South Africa (WHO Global Tuberculosis Report 2016).

Over 95% of TB deaths occurred in low- and middle-income countries with the highest burden in Asia and Africa.

TB is a leading cause of death in people who are HIV-infected.

Further details on TB are shown in Table 1.3.

The changes between 2005 and 2015 for TB are not as striking as those for malaria and HIV/AIDS, although during this decade there was a

17.4% reduction in global all-age deaths
20.5% reduction in global under-fives mortality rate
19% reduction in global all-age DALYs lost

Estimated number of people living with HIV, 2016
By WHO region

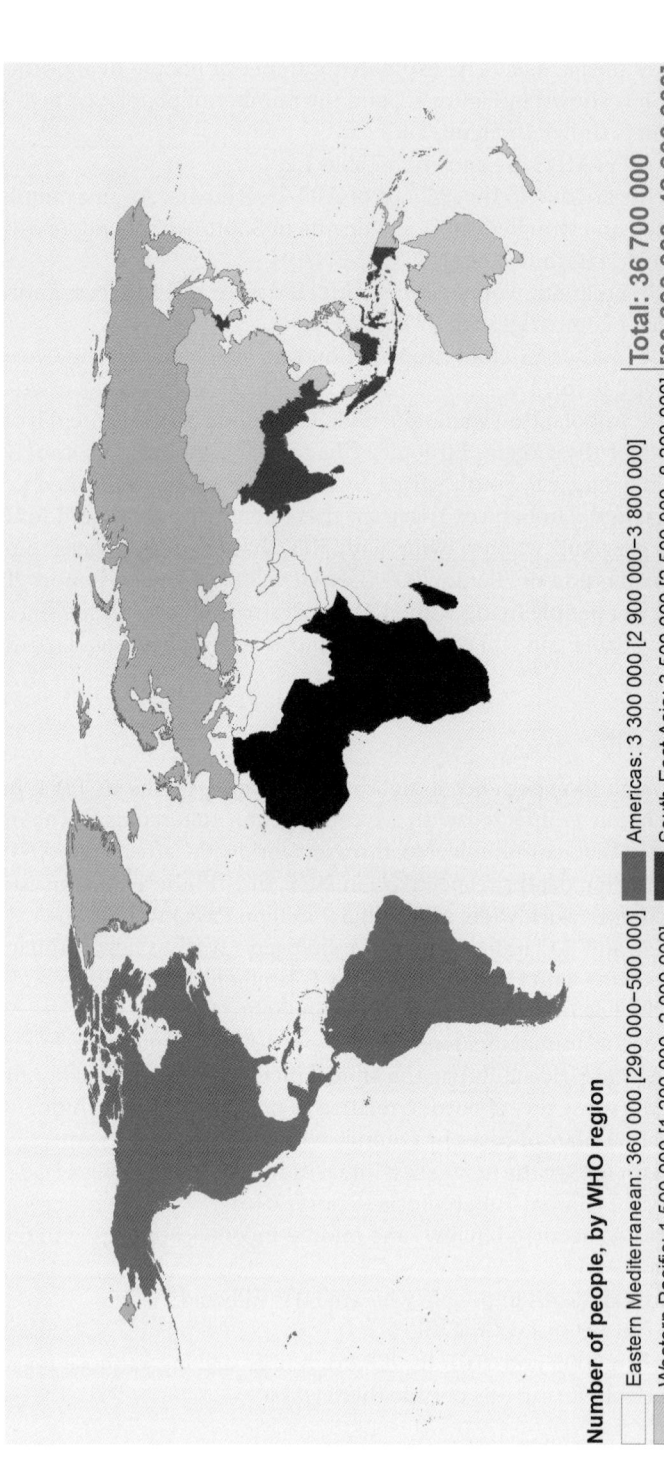

Number of people, by WHO region

Eastern Mediterranean: 360 000 [290 000–500 000]

Western Pacific: 1 500 000 [1 200 000–2 000 000]

Europe: 2 400 000 [2 300 000–2 600 000]

Americas: 3 300 000 [2 900 000–3 800 000]

South-East Asia: 3 500 000 [2 500 000–8 200 000]

Africa: 25 600 000 [22 900 000–28 600 000]

Total: 36 700 000

[30 800 000–42 900 000]

The boundaries and names shown and the designations used on this map do not imply the expression of any opinion whatsoever on the part of the World Health Organization concerning the legal status of any country, territory, city or area or of its authorities, or concerning the delimitation of its frontiers or boundaries. Dotted and dashed lines on maps represent approximate border lines for which there may not yet be full agreement.

Data Source: World Health Organization
Map Production: Information Evidence and Research (IER)
World Health Organization

0 875 1,750 3,500 Kilometers

World Health Organization

Figure 1.2 Estimated number of people *living* with HIV, 2016 – By WHO region. *Source:* WHO/UNAIDS/UNICEF. Copyright World Health Organization. Reproduced with permission. (*See color plate section for the color representation of this figure.*)

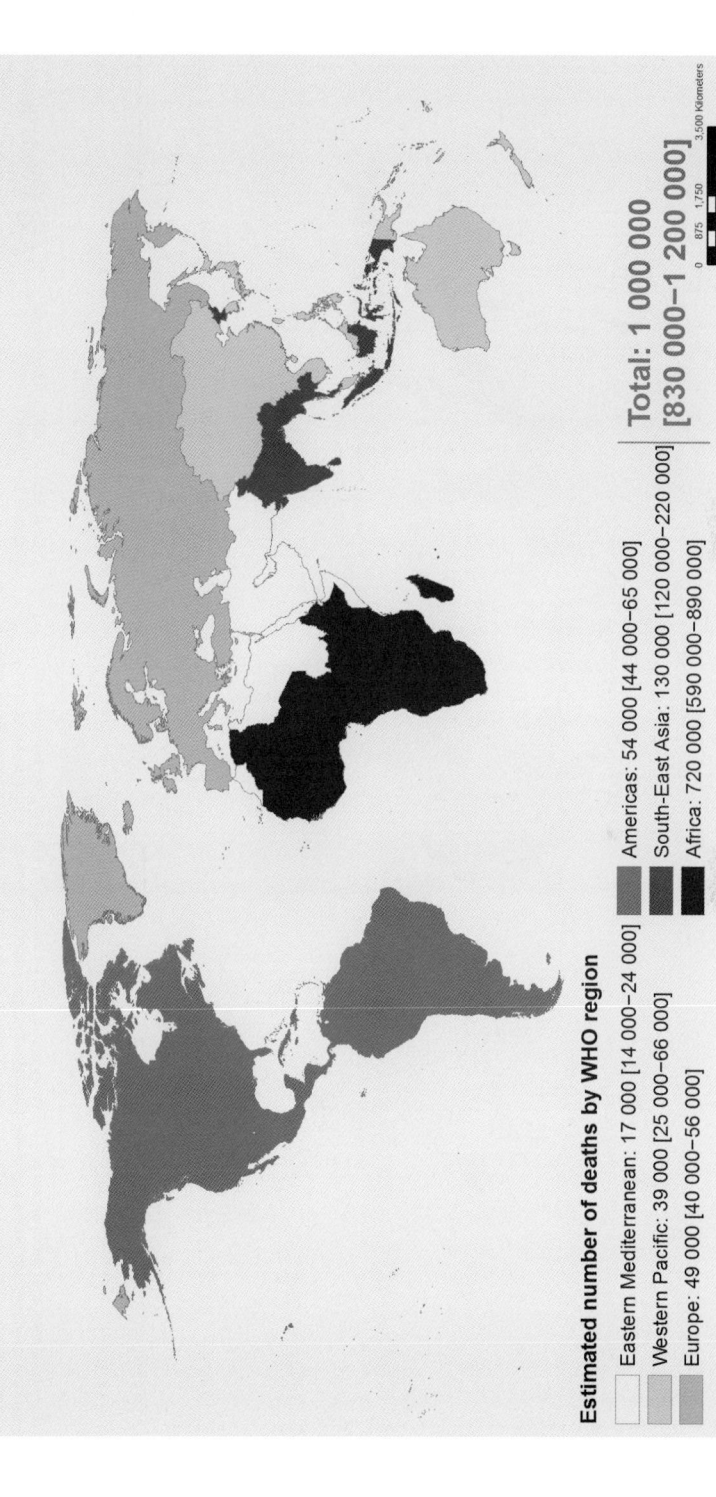

Estimated number of people dying from HIV-related causes, 2016
By WHO region

Estimated number of deaths by WHO region

Eastern Mediterranean: 17 000 [14 000–24 000]

Western Pacific: 39 000 [25 000–66 000]

Europe: 49 000 [40 000–56 000]

Americas: 54 000 [44 000–65 000]

South-East Asia: 130 000 [120 000–220 000]

Africa: 720 000 [590 000–890 000]

Total: 1 000 000
[830 000–1 200 000]

0 875 1,750 3,500 Kilometers

Data Source: World Health Organization
Map Production: Information Evidence and Research (IER)
World Health Organization

World Health Organization

Figure 1.3 Estimated number of people *dying* from HIV-related causes with HIV, 2016 – By WHO region. *Source:* WHO/UNAIDS/UNICEF. Copyright World Health Organization. Reproduced with permission. (*See color plate section for the color representation of this figure.*)

Estimated TB incidence rates, 2016

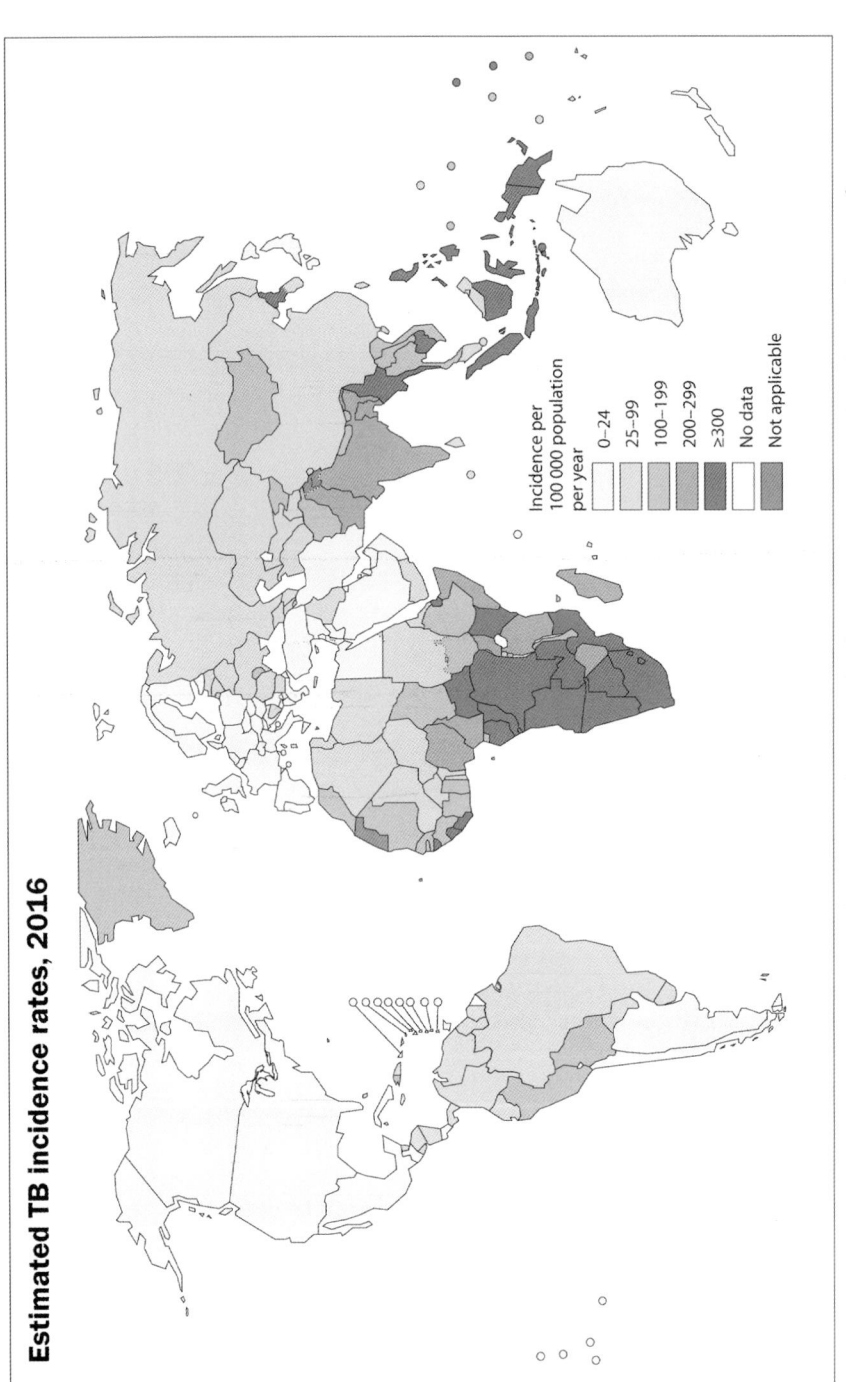

Incidence per
100 000 population
per year

- 0–24
- 25–99
- 100–199
- 200–299
- ≥300
- No data
- Not applicable

The boundaries and names shown and the designations used on this map do not imply the expression of any opinion whatsoever on the part of the World Health Organization concerning the legal status of any country, territory, city or area or of its authorities, or concerning the delimitation of its frontiers or boundaries. Dotted and dashed lines on maps represent approximate border lines for which there may not yet be full agreement.

Data Source: *Global Tuberculosis Report 2017.* WHO, 2017.

© WHO 2017. All rights reserved.

Figure 1.4 Estimated TB incidence rates, 2016. *Source:* Global Tuberculosis Report 2017. WHO (2017). Copyright WHO (2017). All rights reserved. Reproduced with permission. (*See color plate section for the color representation of this figure.*)

Table 1.3 The annual global incidence, the global prevalence, the number of global all-age annual deaths, the number of global under five years annual deaths, the number of global all-age DALYs lost due to tuberculosis in 2005 and 2015 and the % change from 2005 to 2015.

Year/% change	Global annual incidence (new cases)	Global prevalence	Global all-age deaths	Global under five years deaths	Global all-age DALYs lost
2005	8.8 million in 2003[a] No figure stated by GBD Collaborators	2 billion people infected[a] No figure stated by GBD Collaborators	2 million[a] No figure stated by GBD Collaborators	No WHO found. 250 000[b]	49.8 million[b]
2015	10.4 million[c,d] No figure stated by GBD Collaborators	No figure stated by WHO. 8.9 million[d]	1.4 million[c] 1.1 million[e]	200 000[b]	40.3 million[b]
Change from 2005 to 2015	An increase of 18%[c,d]	An increase of 9.2%[d]	A decrease of 17.4%[e]	A decrease of 20.5%[b]	A decrease of 19%[b]

a) WHO Global Factsheet (2005). http://who.int/tb/publications/tb_global_facts_sep05/en
b) GBD 2015 DALYs and HALE Collaborators (2016).
c) WHO Tuberculosis Fact sheet updated in March (2017). http://www.who.int/mediacentre/factsheets/fs104/en.
d) GBD 2015 Disease and Injury Incidence and Prevalence Collaborators (2016).
e) GBD 2015 Mortality and Causes of Death Collaborators (2016).

However, multi-drug resistant TB and extensively drug-resistant TB have emerged over the last two decades.

1.9.3.1 Multidrug-Resistant Tuberculosis

Multi-drug resistant TB (MDR-TB) and extensively drug-resistant (XDR-TB) TB are increasing and pose potentially disastrous public health crises especially in India, China and Russia (Zignol et al. 2016). The incidence of MDR-TB is rising and reached 60 000 cases in the 27 high MDR-TB burden countries worldwide in 2011. MDR-TB is resistant to the two most effective drugs for TB - isoniazid and rifampicin. It poses a major threat to the control of TB worldwide.

An estimated 480 000 new cases of MDR-TB were diagnosed in 2015. Some countries have proportions of MDR-TB as high as 20%. Globally an estimated 3.3% of new cases and 20% of previously treated cases had MDR-TB. These levels have remained virtually unchanged in recent years. An additional 100 000 people had rifampicin-resistant TB (RR-TB). The percentages of new cases of MDR-TB/rifampicin-resistant-TB is shown in Figure 1.5.

Drugs for treating MDR-TB include the fluoroquinolones (e.g. levofloxacin, moxifloxacin), second-line injectable agents (e.g. amikacin, capreomycin) and other core second-line agents (e.g. ethionamide/prothionamide, cycloserine, linezolid and clofazimine [Caminero et al. 2010]). Treatment is lengthy, disruptive and has significant side-effects including depression, psychosis, hearing loss and impairment of renal and hepatic function.

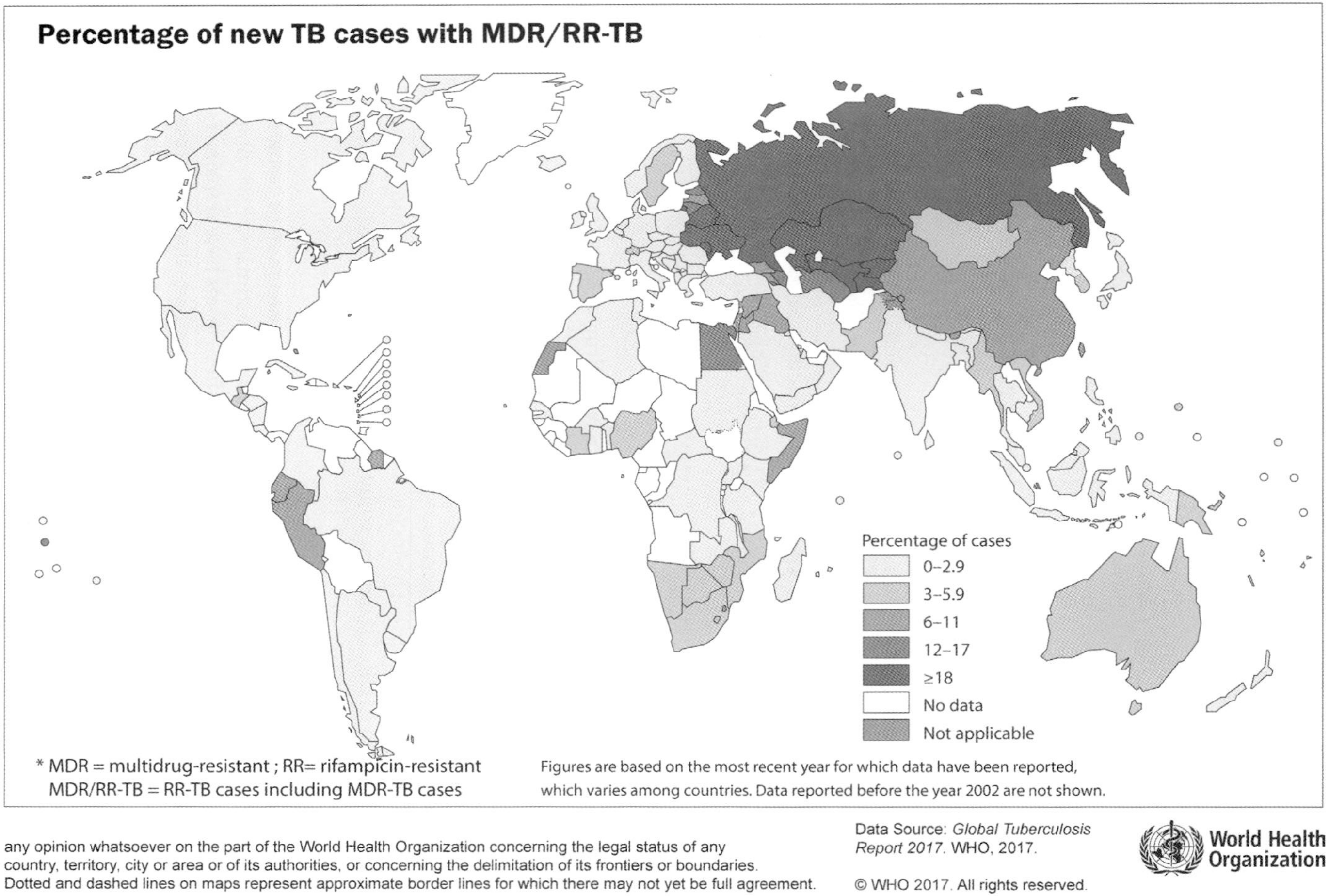

Percentage of new TB cases with MDR/RR-TB

Percentage of cases

- 0–2.9
- 3–5.9
- 6–11
- 12–17
- ≥18
- No data
- Not applicable

Figures are based on the most recent year for which data have been reported, which varies among countries. Data reported before the year 2002 are not shown.

Data Source: *Global Tuberculosis Report 2017*. WHO, 2017.

© WHO 2017. All rights reserved.

World Health Organization

* MDR = multidrug-resistant ; RR= rifampicin-resistant
MDR/RR-TB = RR-TB cases including MDR-TB cases

any opinion whatsoever on the part of the World Health Organization concerning the legal status of any country, territory, city or area or of its authorities, or concerning the delimitation of its frontiers or boundaries. Dotted and dashed lines on maps represent approximate border lines for which there may not yet be full agreement.

Figure 1.5 Percentage of new TB cases with MDR/RR-TB. *Source:* Global Tuberculosis Report 2017. WHO (2017). Copyright WHO (2017). All rights reserved. Reproduced with permission. (*See color plate section for the color representation of this figure.*)

Countries with the highest percentages of new MDR-TB cases included those in Eastern Europe and the Russian Federation. The percentage of previously treated TB cases with MDR/RR-TB is shown in Figure 1.6. Among patients with previously treated TB who were notified in 2014, an estimated 300 000 had MDR-TB. More than half of these patients were in India, China and the Russian Federation.

1.9.3.2 Extensively Drug-Resistant TB (XDR-TB)

Extensively drug-resistant TB is defined as any case of TB resistant to isoniazid and rifampicin plus any fluoroquinolone and at least one of three injectable second-line drugs (amikacin, kanamycin, or capreomycin). XDR-TB has been reported by 105 countries. On average an estimated 9.7% of people with MDR-TB have XDR-TB. The prognosis is extremely poor.

Countries that had used bedalaquine or delamadid for the treatment of M/XDR-TB by the end of June 2017 are shown in Figures 1.7 and 1.8 respectively.

1.9.3.3 The Co-epidemics of TB and HIV/AIDS

The global disease burden of TB remains high with resurgence in many areas due to co-infection with HIV/AIDS. An estimated 13% of TB cases in 2011 were co-infected with HIV and 430 000 deaths were in HIV-positive people.

In 2014 an estimated 1.2 million (12%) of the 9.6 million people who developed TB worldwide were HIV-positive. The African Region accounted for 74% of the estimated number of HIV-positive TB cases. The number of people dying from HIV-associated TB peaked at 570 000 in 2004 and has since fallen to 390 000 in 2014. In 2014 HIV-associated TB deaths accounted for 25% of all TB deaths and one third of the estimated 1.2 million deaths from HIV/AIDS.

1.10 Other Important Communicable Diseases in the LMICs

These include acute lower respiratory tract infections, diarrheal diseases, meningitis, syphilis, chlamydial infection, gonorrhea, trichomoniasis, hepatitis, measles, whooping cough, tetanus and yellow fever.

Between 2005 and 2015 significant progress has been made in a number of these diseases, although others are unchanged and some are becoming more problematic. An overview is given below and in greater detail for each disease category in Tables 1.4–1.12

- A significant reduction has been made in the under-five mortality rate from **lower respiratory tract infections** due to the roll out of measles, pneumococcal and *H. influenzae* b vaccination.
- It is anticipated that rotavirus vaccination will prevent 15–34% of **severe diarrhea** in the LMICs (Soares-Weiser et al. 2012).
- Since 2010 countries in the Sahel region of Africa (the meningitis belt stretching from Senegal in the west to Ethiopia in the east) introduced a new serogroup A meningococcal conjugate vaccine which confers both individual protection and herd immunity. As a result of this, epidemics of **serogroup A meningococcal disease** are disappearing (Gabutti et al. 2015) although other serogroups still cause epidemics, albeit at a lower frequency and of a smaller size.

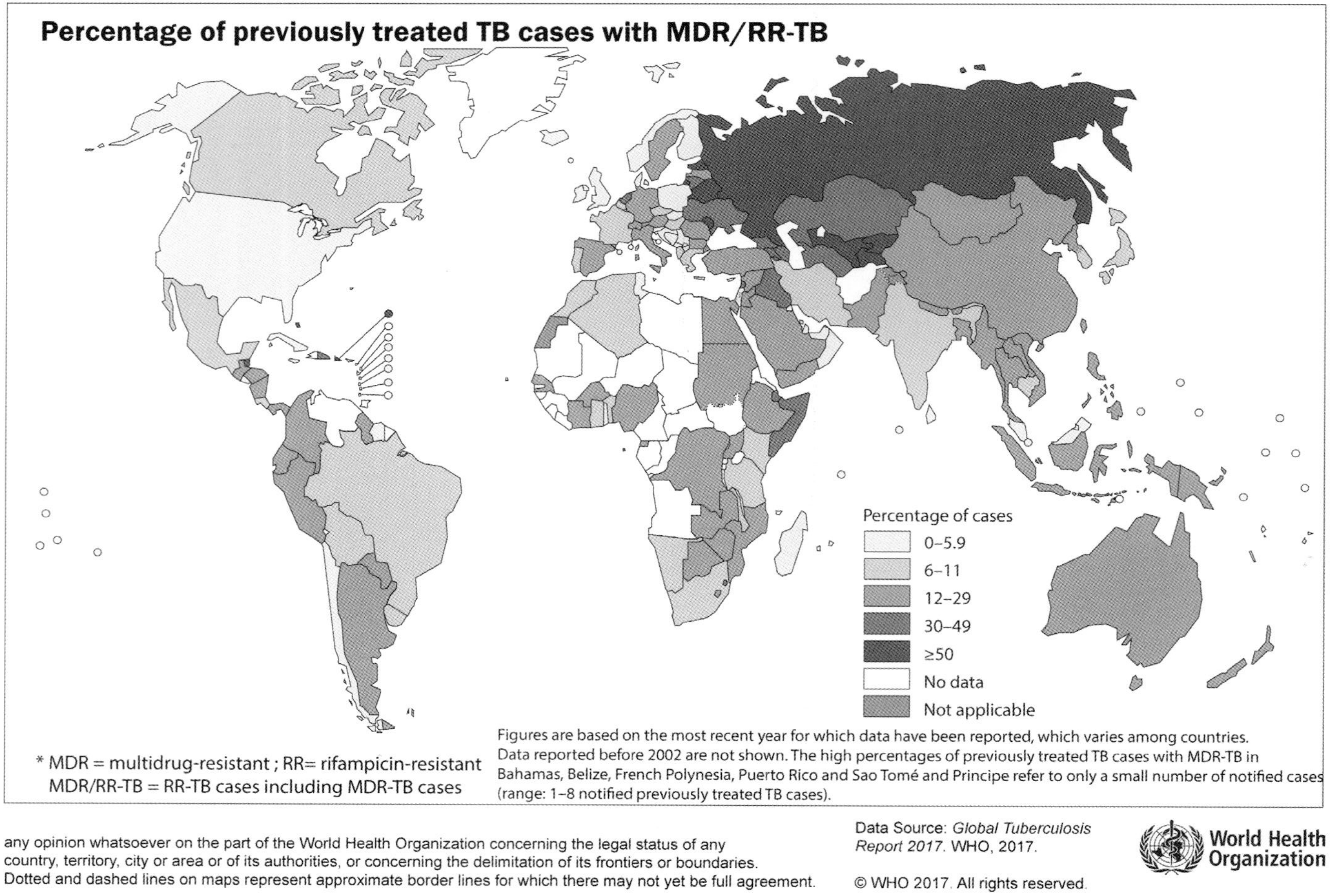

Percentage of previously treated TB cases with MDR/RR-TB

Percentage of cases

- 0–5.9
- 6–11
- 12–29
- 30–49
- ≥50
- No data
- Not applicable

Figures are based on the most recent year for which data have been reported, which varies among countries.
Data reported before 2002 are not shown. The high percentages of previously treated TB cases with MDR-TB in
Bahamas, Belize, French Polynesia, Puerto Rico and Sao Tomé and Principe refer to only a small number of notified cases
(range: 1–8 notified previously treated TB cases).

Data Source: *Global Tuberculosis Report 2017*. WHO, 2017.

© WHO 2017. All rights reserved

* MDR = multidrug-resistant ; RR= rifampicin-resistant
MDR/RR-TB = RR-TB cases including MDR-TB cases

Figure 1.6 Percentage of previously treated TB cases with MDR/RR-TB. *Source:* Global Tuberculosis Report 2017. WHO (2017). Copyright WHO (2017). All rights reserved. Reproduced with permission. (*See color plate section for the color representation of this figure.*)

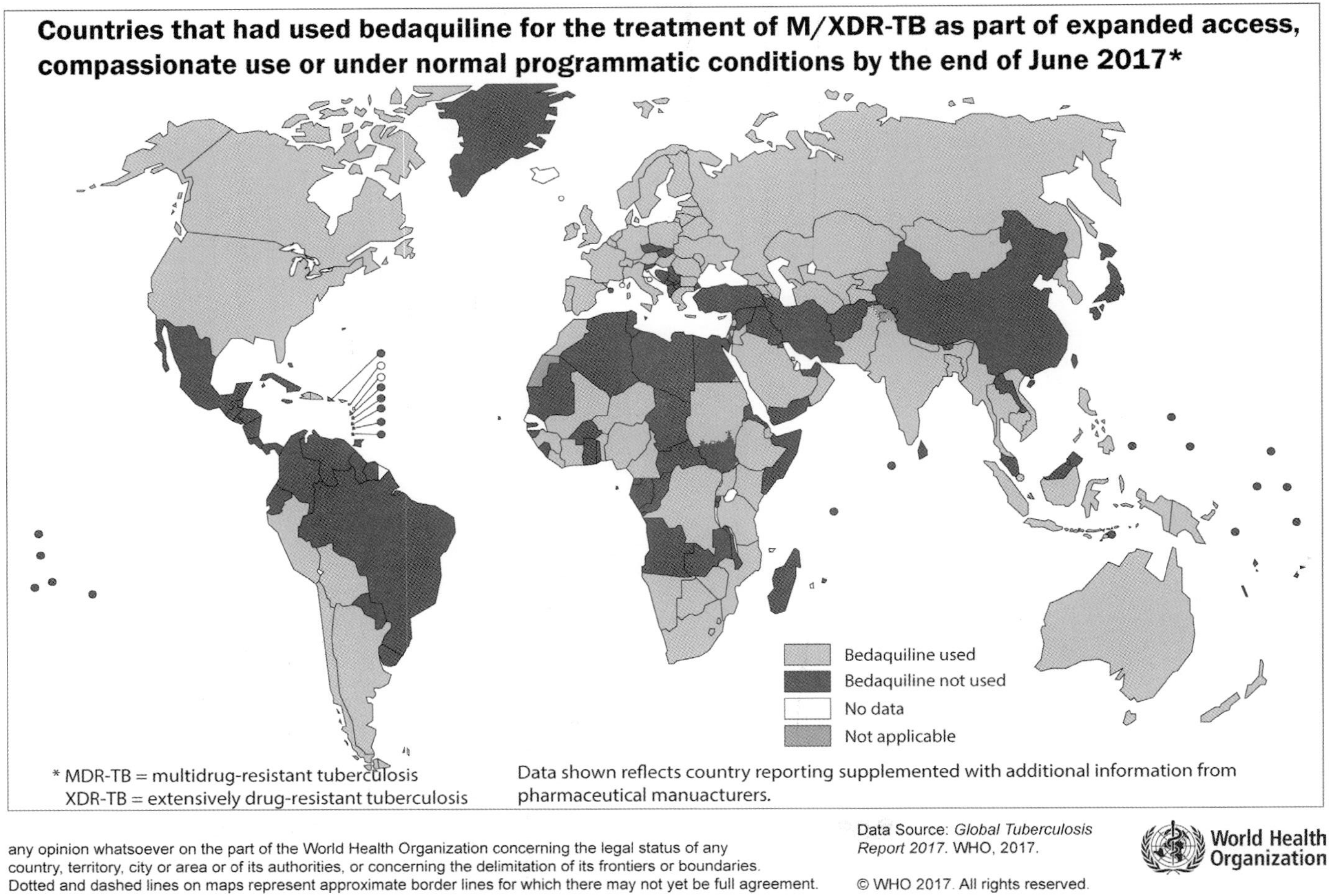

Countries that had used bedaquiline for the treatment of M/XDR-TB as part of expanded access, compassionate use or under normal programmatic conditions by the end of June 2017*

Bedaquiline used

Bedaquiline not used

No data

Not applicable

* MDR-TB = multidrug-resistant tuberculosis
XDR-TB = extensively drug-resistant tuberculosis

Data shown reflects country reporting supplemented with additional information from pharmaceutical manuacturers.

any opinion whatsoever on the part of the World Health Organization concerning the legal status of any country, territory, city or area or of its authorities, or concerning the delimitation of its frontiers or boundaries. Dotted and dashed lines on maps represent approximate border lines for which there may not yet be full agreement.

Data Source: *Global Tuberculosis Report 2017.* WHO, 2017.

© WHO 2017. All rights reserved.

World Health Organization

Figure 1.7 Countries that had used the new agent bedaquiline for the treatment of M/XDR-TB as part of expanded access, compassionate use, or under normal programmatic conditions by the end of June 2017. *Source:* Global Tuberculosis Report 2017. WHO (2017). Copyright WHO (2017). All rights reserved. Reproduced with permission. (*See color plate section for the color representation of this figure.*)

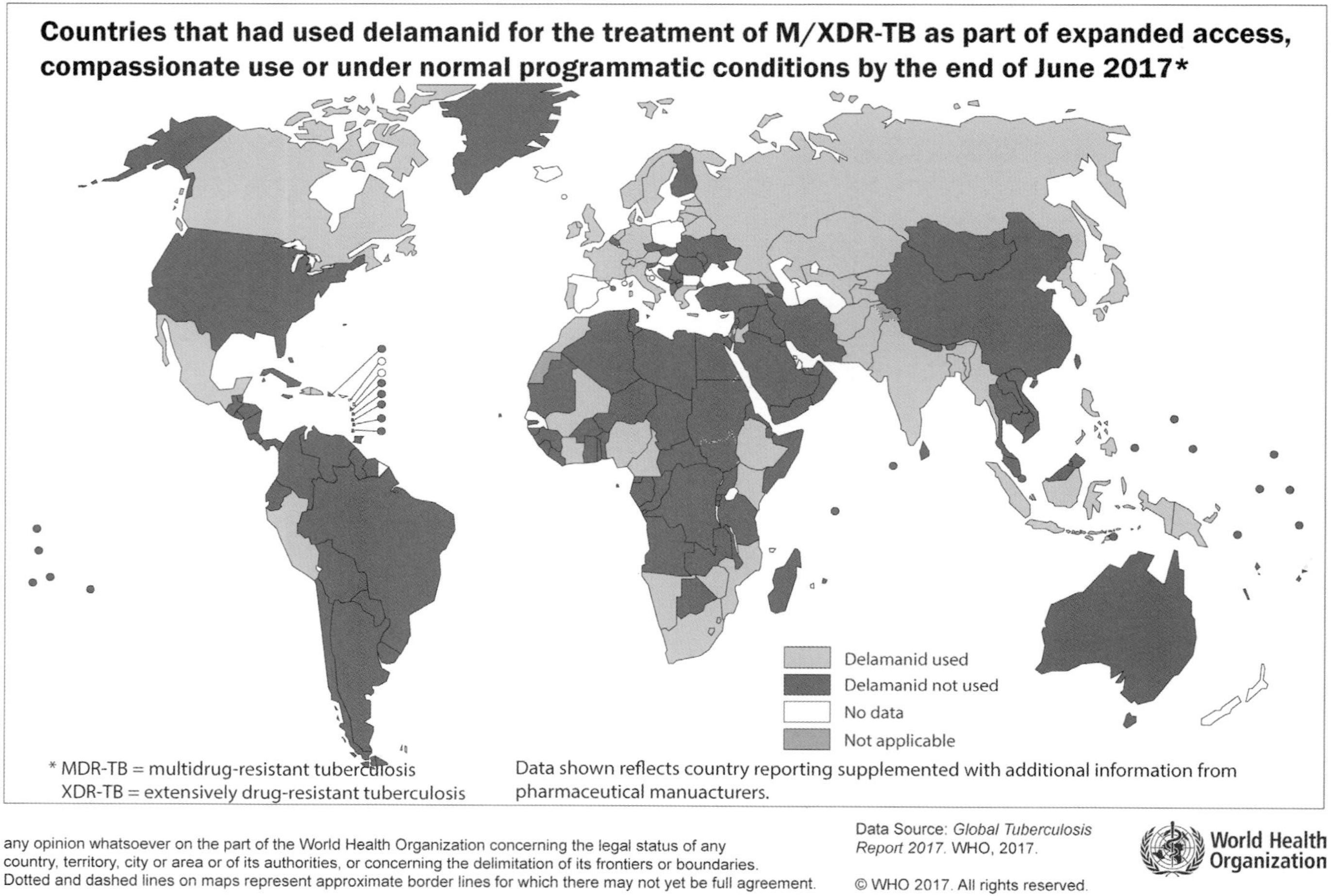

Countries that had used delamanid for the treatment of M/XDR-TB as part of expanded access, compassionate use or under normal programmatic conditions by the end of June 2017*

Delamanid used

Delamanid not used

No data

Not applicable

Data shown reflects country reporting supplemented with additional information from pharmaceutical manuactures.

 **World Health Organization**

* MDR-TB = multidrug-resistant tuberculosis
XDR-TB = extensively drug-resistant tuberculosis

Data Source: *Global Tuberculosis Report 2017.* WHO, 2017.

Figure 1.8 Countries that had used the new agent delamanid for the treatment of M/XDR-TB as part of expanded access, compassionate use, or under normal programmatic conditions by the end of June 2017. *Source:* Global Tuberculosis Report 2017. WHO (2017). Copyright WHO (2017). All rights reserved. Reproduced with permission. (*See color plate section for the color representation of this figure.*)

Table 1.4 The total number of new cases globally in 2015, the global prevalence, the number of global all-age annual deaths, the number of global under five years annual deaths, the number of global all-age DALYs lost due to lower respiratory tract infections in 2015 and the % change from 2005 to 2015.

Year/% change	Global annual incidence[a),b)]	Global prevalence[b)]	Global all-age deaths[c)]	Global under five years deaths[c)]	Global all-age DALYS lost[d)]
Lower respiratory tract infections in 2015	3.2 million in 2015[a)]. 29.2 million[b)]	8.99 million	2.8 million	703 900	135.3 million
Change 2005 and 2015	An increase of 6.8%[b)]	An increase of 5.1%	A decrease of 3.2%	A decrease of 20%	A decrease of 28.8%

a) WHO Top 10 causes of death in 2015. Fact sheet updated in January (2017). http://www.who.int/mediacentre/factsheets/fs310/en.
b) GBD 2015 Disease and Injury Incidence and Prevalence Collaborators (2016).
c) GBD 2015 Mortality and Causes of Death Collaborators (2016).
d) GBD 2015 DALYs and HALE Collaborators (2016).

Table 1.5 The number of new cases globally in 2015, the global prevalence, the number of global all-age annual deaths, the number of global under five years annual deaths, the number of global all-age DALYs lost due to diarrheal diseases in 2015 and the % change from 2005 to 2010.

Disease/% change	Global annual incidence	Global prevalence[a)]	Global all-age deaths[b),c)]	Global under five years deaths[c),d)]	Global all-age DALYS lost[e)]
Diarrhoeal diseases	1.7 billion cases in children each year[d)]. 239 million[a)]	35.8 million	1.4 million[b)]. 1.66 million[c)]	525 000[d)] 498 900[c)]	71.6 million
Change between 2000 or 2005 and 2015	Almost a 50% decrease[a)]	An increase of 6.4%	Almost a 50% decrease between 2000 and 2015[b)] A decrease of 20.8%[c)]	A decrease of 34.3%[c)]	A decrease of 27.2%

a) GBD 2015 Disease and Injury Incidence and Prevalence Collaborators (2016).
b) WHO Top 10 causes of death in 2015. Fact sheet updated in January (2017). http://www.who.int/mediacentre/factsheets/fs310/en.
c) GBD 2015 Mortality and Causes of Death Collaborators (2016).
d) WHO Fact sheet on diarrheal disease updated in May (2017). http://www.who.int/mediacentre/factsheets/fs330/en.
e) GBD 2015 DALYs and HALE Collaborators (2016).

Table 1.6 The number of new cases globally in 2015, the global prevalence, the number of global all-age annual deaths, the number of global under five years annual deaths, the number of global all-age DALYs lost due to meningitis in 2015 and the % change from 2005 to 2015.

Disease	Global annual incidence	Global prevalence in 2015 and per cent change from 2005	Global all-age deaths in 2015 and per cent change from 2005	Global under five years deaths in 2015 and percent change from 2005	Global all-age DALYS lost in 2015 and percent change from 2005
Pneumococcal meningitis	No found	7.3 million/ an increase of 22%[a]	112 900/ an increase of 0.7%[b]	No data found	7.8 million/ a decrease of 0.7%[c]
Haemophilus influenzae b meningitis	No found	2.45 million/ a decrease 8.3%[a]	71 500/ a decrease of 35.4%[b]	No data found	5.3 million/ a decrease of 37%[c]
Meningococcal meningitis	During the 2014 epidemic season 19 African countries implementing enhanced surveill-ance reported 11 908 suspected cases[d]	1.7 million/ an increase of 22%[a]	1,1461 73 300/ a decrease of 1.3%[b]	No data found	5.1 million/ a decrease of 3.3%[c]
All meningitis	Every year bacterial meningitis epidemics affect more than 400 million people living in the 26 countries of the extended African meningitis belt from Senegal to Ethiopia[e]	8.7 million/ an increase of 16.9%[a]	379 200/ a decrease of 7.6%[b]	25 800/ a decrease of 15.6%[b]	25.4 million/ a decrease of 10.6%[c]

a) GBD 2015 Disease and Injury Incidence and Prevalence Collaborators (2016).
b) GBD 2015 Mortality and Causes of Death Collaborators (2016).
c) GBD 2015 DALYs and HALE Collaborators (2016).
d) Meningococcal meningitis WHO Fact sheet Number 141 updated in November 2015.
e) WHO Global Health Observatory Data http://www.who.int/gho/epidemic_diseases/meningitis/en.

- There has also been a dramatic decrease in *Haemophilus influenzae b* **meningitis** in children due to the roll out of *H. influenzae b* vaccination.
- **Syphilis** is by far the most lethal sexually transmitted disease (excluding HIV/AIDS). According to WHO estimates it caused 143 000 early fetal deaths and stillbirths, 62 000 neonatal deaths and 44 000 preterm or low birth weight births in 2012 (Wijesooriya et al. 2016).
- Sexually transmitted diseases remain an important cause of infertility, ectopic pregnancy, and blindness in infancy.

Table 1.7 The global prevalence, the number of global age annual deaths, the number of global under five years annual deaths, the number of global all-age DALYs lost due to sexually transmitted diseases (excluding HIV/AIDS) in 2015 and the % change from 2005 to 2015.

Disease	Global annual incidence	Global prevalence in 2015 and per cent change from 2005	Global all-age deaths and per cent change from 2005	Global under five years deaths and percent change from 2005	Global all-age DALYS lost and percent change from 2005
Syphilis	5.6 million[a]	43.6 million /an increase of 3.7%[b]	106 800/ a decrease of 20.3%[c]	90 500/a decrease of 21.1%[c]	8.96 million/ a decrease of 20%[d]
Chlamydia	131 million[a]	82.8 million/ an increase of 8.3%[b]	A decrease of 5.3%[c]	No data found	369 800/an increase of 9.7%[d]
Gonorrhea	78 million[a]	47.5 million/ an increase of 25.3%[b]	700/ A decrease of 15.7%[c]	No data found	469 800/an increase of 22.7%[d]
Trichomoniasis	143 million[a]	167.6 million/ an increase of 16.2%[b]	300/ a decrease of 16.8% (includes all other sexually transmitted diseases)[c]	No data found	194 300/an increase of 16.1%[d]

a) Newman et al. (2015).
b) GBD 2015 Disease and Injury Incidence and Prevalence Collaborators (2016).
c) GBD 2015 Mortality and Causes of Death Collaborators (2016).
d) GBD 2015 DALYs and HALE Collaborators (2016).

- The prevalence of **viral hepatitis (A, B, C, and E)** is increasing. Hepatitis B and C are risk factors for both hepatic cirrhosis and hepatocellular cancer. The all-age global death number in 2015 for viral hepatitis was 1.23 million, which was higher that of HIV/AIDS (1.19 million), TB (1.1 million) and malaria (730 000). Unlike viral hepatitis, the mortality rates of the latter three diseases are declining. Vaccination for hepatitis B is effective and safe, and vaccination at birth is recommended since mother-to-child transmission is common. Anti-viral medications are available to treat hepatitis B and hepatitis C infections but are expensive and current access is poor in the LMICs.
- There has been a massive decrease in the prevalence, global death number, under five death number and global DALYs lost due to **measles** because of the success of measles vaccination.
- There has been a marked improvement in outcomes for **whooping cough** (pertussis) due to vaccination.
- Diseases well controlled by vaccination programs, such as **yellow fever**, can re-occur if vaccination programs break down.
- A very severe outbreak of **cholera** occurred in Yemen in 2017 and an outbreak of **plague** occurred in Madagascar, also in 2017.

Table 1.8 The global incidence, the global prevalence, the number of global all-age annual deaths, the number of global under five years annual deaths, the number of global all-age DALYs lost due to hepatitis in 2015 and the % change from 2005 to 2015.

Disease	Global incidence in 2015	Global prevalence in 2015 and percent change from 2005	Global all-age deaths and percent change from 2005[a]	Global all-age DALYS lost and percent change from 2005
Acute hepatitis A	1.5 million[b] 114 million[c]	8.8 million/ an increase of 4.2%[c]	169 000/a decrease of 34%[d]	1.03 million /a decrease of 30.7%[e]
Hepatitis B	111 million[c]	356 million/ an increase of 16.7%[c] 257 million HBsAg positive[f]	714 000/a decrease of 8%[d] 887,000[f]	2.7 million/ a decrease of 9.1%[e]
Hepatitis C	No data found	142 million/an increase of 18%[c] 71 million[g]	28 000/a decrease of 9.8%[d] 399,0004	89 000/a decrease of 9.5%[e]
Acute hepatitis E	1.95 million[b] 20 million[h]	1.5 million/an increase of 3.5%[c]	319 000/a decrease of 16%[d] 44,000[h]	1.4 million/a decrease of 23%[e]

a) No data found for under-fives mortality.
b) Franco and colleagues reported in 2012 that there were approximately 1.5 million clinical cases of hepatitis A diagnosed annually, but acknowledged that the rate of infection was probably as much as 10 times higher.
c) GBD 2015 Disease and Injury Incidence and Prevalence Collaborators (2016).
d) GBD 2015 Mortality and Causes of Death Collaborators (2016).
e) GBD 2015 DALYs and HALE Collaborators (2016).
f) WHO Hepatitis B Fact sheet updated in April 2017.
g) WHO Hepatitis C Fact sheet updated in April 2017.
h) WHO Hepatitis E Fact sheet updated in July 2016.
i) The WHO Global Hepatitis Report (2017).

1.10.1 Lower Respiratory Tract Infections

Details are given in Table 1.4.

There is a striking disparity (almost 1 log) for incidence figures of lower respiratory tract infections between the WHO figure and the figure stated by the GBD Collaborators. This must be due either to different methods of data capture or different methods of data analysis, or both. Details of data capture and data analysis from each of these two sources are detailed below.

1.10.1.1 WHO

1.10.1.2 Data Sources and Analysis

Every year WHO analyses data from its 194 Member States and produces burden of disease and mortality health estimates, which are published in the "World Health

Table 1.9 The global incidence, the global prevalence, the number of global all-age annual deaths, the number of global under five years annual deaths, the number of global all-age DALYs lost due to measles in 2015 and the % change from 2005 to 2015.

Year/% change	Annual global incidence	Global prevalence	Global all-age deaths	Global under five years deaths[f]	Global all-age DALYs lost
Measles in 2015	195 762[a,b]	127 000[c]	134 000[a,b] 73 400[d]	Most of the global deaths were in children less than 5 years of age[b] 62,600[d]	6.1 million[e]
% change between 2005 and 2015		A decrease of 70.2%[c]	A decrease of 75%[d]	A decrease of 75%[d]	A decrease of 75%[e]

a) WHO Measles Fact sheet reviewed in March 2017. http://www.who.int/mediacentre/factsheets/fs286/en.
b) WHO Measles http://www.who.int/immunization/monitoring_surveillance/burden/vpd/surveillance_type/active/measles/en.
c) GBD 2015 Disease and Injury Incidence and Prevalence Collaborators (2016).
d) GBD 2015 Mortality and Causes of Death Collaborators (2016).
e) GBD 2015 DALYs and HALE Collaborators (2016).
f) Measles global all age deaths in 2000 were 733 000 versus 164 000 in 2008. CDC MMWR (December 2009). Global measles mortality 2000–2008. https://www.cdc.gov/mmwr/preview/mmwrhtml/mm5847a2.htm.

Table 1.10 The global incidence, the global prevalence, the number of global all-age annual deaths, the number of global under five years annual deaths, the number of global all-age DALYs lost due to whooping cough (pertussis) in 2015 and the % change from 2005 to 2015.

Year/% change	Annual global incidence[a]	Global prevalence	Global all-age deaths	Global under five years deaths[a]	Global all-age DALYs lost
2015	142 512[b]	2.2 million[c]	89 000[b] 54 500[d]	54 500[d]	5.1 Million[e]
Change between 2005 and 2015		A decrease of 27.4%	A decrease of 41%[d]		A decrease of 48%

a) The CDC reports currently that there are 16 million cases of whooping cough in children each year, resulting in 195 000 deaths. https://www.cdc.gov/pertussis/countries/index.html.
b) WHO Pertussis. http://www.who.int/immunization/monitoring_surveillance/burden/vpd/surveillance_type/passive/pertussis/en.
c) GBD 2015 Disease and Injury Incidence and Prevalence Collaborators (2016).
d) GBD 2015 Mortality and Causes of Death Collaborators (2016).
e) GBD 2015 DALYs and HALE Collaborators (2016).

Table 1.11 The global incidence, the global prevalence, the number of global all-age annual deaths, the number of global under five years annual deaths, the number of global all-age DALYs lost due tetanus in 2015 and the % change from 2005 to 2015.

Year/% change[a]	Annual global incidence	Global prevalence	Global all-age deaths	Global under five years deaths	Global all-age DALYs lost
2015	10 337[a]	209 300[b]	56 700[c]	25 500[c]	351 000[d]
Change between 2005 and 2015		An increase of 8.4%[b]	A decrease of 47.5%[c]	A decrease of 57%[c]	A decrease of 53%[d]

a) WHO. Tetanus. Immunization, vaccines and biologicals. http://www.who.int/immunization/ monitoring_surveillance/burden/vpd/surveillance_type/passive/tetanus/en. WHO estimated that there were 72 600 deaths in children less than 5 years old in 2011. WHO estimated that neonatal tetanus killed approximately 49 000 newborn children in 2013, a 94% reduction from the situation in 1988 when an estimated 787 000 newborn babies died of tetanus within their first month of life.
b) GBD 2015 Disease and Injury Incidence and Prevalence Collaborators (2016).
c) GBD 2015 Mortality and Causes of Death Collaborators (2016).
d) GBD 2015 DALYs and HALE Collaborators (2016).

Table 1.12 The global prevalence, the number of global all-age annual deaths, number of global all-age DALYs lost due to yellow fever.

Disease	Global prevalence	Global all-age deaths	Global all-age DALYs lost
Yellow fever	In 2016, 84 000 – 170 0001	In 2016 29 000–60 000[a]	In 2003, 38 000–1.78 million[b]

a) WHO Fact sheet on yellow fever updated in May (2016). http://www.who.int/mediacentre/factsheets/ fs100/en.
b) LaBeaud (2011).

Statistics." WHO statistics are generated from multiple sources using a variety of data collection methods, including household surveys, routine reporting by health services, civil registration and censuses and disease surveillance systems.

Currently WHO receives cause-of-death statistics regularly from only about 100 Member States. Globally, two-thirds (38 million) of 56 million annual deaths are not registered.

WHO makes its data and analyses accessible through the Global Health Observatory portal as well as databases that provide statistics on a wide range of diseases and health indicators.

An example of WHO's data capture for TB has been published: four main methods were used to derive incidence. These were

1) Case notification data combined with expert opinion about case detection gaps (74 countries representing 22% of global incidence in 2015).
2) Results from TB prevalence surveys (20 countries, 62% of global incidence).
3) Notifications in high-income countries adjusted by a standard factor to account for under-reporting and under-diagnosis (118 countries, 15.5% of global incidence).
4) Capture-recapture modeling (five countries, 0.5% of global incidence).

Mortality was obtained from national registration systems of mortality surveys in 128 countries (52% of global HIV-negative TB mortality). In other countries mortality was derived indirectly from incidence and case fatality ratio numbers (Glaziou et al. 2016).

1.10.1.3 Global Burden of Disease Collaborators, 2016

Data Sources   Data sources were as quoted directly: "...we updated data searches through systematic data and literature reviews for 85 causes published up to October 31st, 2015. For other causes, input from GBD collaborators resulted in the identification and inclusion of a small number of additional studies published after January 2013. Data were systematically screened from household surveys archived in the Global Health Data Exchange, sources suggested to us by in-country experts and surveys identified in major multinational survey data catalogues and Ministry of Health and Central Statistical Office websites. Case notifications reported to WHO were updated up to and including 2015."

Data Analysis   "For GBD 2015, the computational engine of DisMod-MR 2.1 remained unchanged, but we substantially rewrote the code that organises the flow of data and settings at each level of the analytical cascade. The sequence of estimation occurs at five levels: global, super-region, region, country, and where applicable, subnational locations (Global Burden of Disease 2015, Appendix pp. 611–624). At each level of the cascade, the DisMod-MR 2.1 computational engine enforces consistency between all disease parameters." "For GBD 2015, we generated fits for the years 1990, 1995, 2000, 2005, 2010 and 2015. We log-linearly interpolated estimates for the intervening years in each 5-year period. Greater detail on DisMod-MR 2.1 is available at Global Health Data Exchange.

We estimated incidence and prevalence by age, sex, cause, year and geography with a wide range of updated and standardized analytical procedures. Improvements from GBD 2013 included the addition of new data sources, updates to literature reviews for 85 causes, and the identification and inclusion of additional studies published up to November 2015, to expand the database used for estimation of non-fatal outcomes. Prevalence and incidence by cause and sequelae were determined with DisMod-MR 2.1, an improved version of the DisMod-MR Bayesian meta-regression tool first developed for GBD 2010 and GBD 2013. For some causes, we used alternative modelling strategies where the complexity of the disease was not suited to DisMod-MR 2.1 or where incidence and prevalence needed to be determined from other data."

GBD data analysis was performed by the Institute for Health Metrics and Evaluation, Seattle, USA.

Commenting on these approaches is beyond the scope of this book, but clearly an attempt to reconcile statistical differences between these two major sources is required.

Of note is the publication in 2016 of a set of guidelines for maximizing the accuracy and transparency in the estimation of health metrics (The GATHER Working Group 2016). The GATHER working group was convened by the WHO in 2014 with the aim of defining and promoting good practice in reporting health estimates. The working group consisted of experts on global health estimates from academia, the WHO, journal editors, members of the EQUATOR network (Enhancing the QUAlity and Transparency Of health Research), and members of existing guideline steering groups. The GATHER working group reviewed existing reporting guidelines for relevance to health estimates and concluded that available reporting guidelines would not ensure adequate reporting of global health estimates.

Based on the review of existing guidance and reporting guidelines and on input from working group members, the group generated a comprehensive list of potential reporting items. They subsequently sought feedback from a broader community of researchers and users of the estimates via an online survey between January and February 2015. The responses were compiled, summarized, and presented at a two-day consensus meeting held in London, United Kingdom, in February 2015.

The primary objective of the working group consensus meeting was to agree on the list of items that should be reported whenever health estimates are published. This list was refined over the course of 2015 and published as part of the GATHER Statement in 2016 (The GATHER Working Group 2016).

1.10.2   Diarrheal Diseases

The total number of new cases globally in 2015, the global prevalence, the number of global all-age annual deaths, the number of global under five years annual deaths, the number of global all-age DALYs lost due to diarrheal diseases in 2015 and the % change from 2005 to 2015 are shown in Table 1.5.

1.10.3   Meningitis

The number of new cases globally in 2015, the global prevalence, the number of global all-age annual deaths, the number of global under five years annual deaths, the number of global all-age DALYs lost due to meningitis in 2015 and the % change from 2005 to 2015 are shown in Table 1.6.

1.10.4   Sexually Transmitted Diseases (Excluding HIV/AIDS)

The global prevalence, the number of global age annual deaths, the number of global under five years annual deaths, the number of global all-age DALYs lost due to sexually transmitted diseases (excluding HIV/AIDS) in 2015 and the % change from 2005 to 2015 are shown in Table 1.7

1.10.5   Hepatitis

Details are shown in Table 1.8. Again, there were extreme differences for many parameters of hepatitis between data from the GBD Collaborators and those from the WHO. The same comments apply as those stated for the figures for parameters of lower respiratory tract infections above.

1.10.6   Measles

Details are shown in Table 1.9.

1.10.7   Whooping Cough (Pertussis)

Details are shown in Table 1.10.

1.10.8 Tetanus

Details are shown in Table 1.11.

1.10.9 Yellow Fever

Details are shown in Table 1.12.

Although yellow fever has been largely prevented by vaccination, outbreaks can still occur when vaccination programs break down: an example is the outbreak in Angola and the Democratic Republic of Congo in 2015–2016, when 7300 suspected cases were reported and in which yellow fever was confirmed in 962 cases. The outbreak was subsequently controlled by an extensive vaccination program.

1.11 Neglected Tropical Disease (NTDs) Prioritized by the World Health Organisation

A number of neglected tropical diseases have been prioritized by the WHO for research, treatment, and prevention. These, together with their management, are shown in Table 1.13.

Table 1.13 Measures to control the neglected tropical diseases.

Disease	Control strategy	Elimination target
Schistosomiasis	MDA, health education, sanitation, snail control	Yes (in China)
Onchocerciasis	MDA, vector control	Yes (the Americas)
Lymphatic filariasis	MDA, vector control	Yes
Trachoma	MDA, water and sanitation, health education	Yes
Yaws	MDA	Yes
Soil-transmitted helminths	MDA	No
Guinea worm	Safe water, health education	Yes
African trypanosomiasis	Case finding and treatment, vector control	Yes (*T. b. gambiense*)
Visceral leishmaniasis	Case finding and treatment	Yes (subcontinent)
Leprosy	Case finding and treatment	Yes
Cysticercosis	Sanitation, meat inspection, vaccination of pigs	No
Echinococcosis	Abbatoir control, treatment of dogs,	No
Fascioliasis	Treatment of sheep, health education	No
Chagas' disease	Vector control, blood screening	Yes (some countries)
Buruli ulcer	Case finding and treatment	No
Rabies	Vaccination of dogs, health education	No
Dengue	Vector control	No

Abbreviation used: MDA, mass drug administration.

These diseases are called neglected because they have been largely eliminated or eradicated in the more developed parts of the World and persist only in the poorest, most marginalized communities and in areas of armed conflict.

They are common in 149 countries, affect more than 1.4 billion people, including more than 500 million children, and cost developing economies billions of dollars every year.

Guinea-worm disease and yaws have been targeted for eradication, and the number of cases of guinea worm has been reduced from more than 3 million in 1986 to less than 30 in 2016. A major campaign to eradicate yaws was led by WHO from 1952 until 1964 but, in spite of treating more than 50 million people with injectable penicillin and reducing the global prevalence by more than 95%, it was ultimately unsuccessful. Yaws has again been targeted for eradication by 2020, using MDA with azithromycin, but little progress has been made. Trachoma, African trypanosomiasis, leprosy and lymphatic filariasis are targeted for elimination as a public health problem by 2020.

The commonest NTDs are the soil-transmitted helminthic (STH) infections which affect one third of the almost three billion people living on less than US$2 per day in developing regions of Sub-Saharan Africa, Asia, and the Americas. STHs are the

Table 1.14 The global all-age DALYs for neglected tropical diseases.

Disease	Global all-age DALYs in 2015
Buruli ulcer	No data found
Chagas' disease	236 000[a]
Dengue fever	1.8 million[a]
Chikungunya fever	No data found
Guinea worm disease	Not reported
Cystic echinococcosis	172 000[a]
Food-borne trematodiases	1.7 million[a]
African trypanosomiasis	202 000[a]
Leischmaniasis	1.4 million[a]
Leprosy	31 000[a]
Lymphatic filariasis	2.07 million[a]
Onchocerciasis	1.13 million[a]
Rabies	932 000[a]
Schistosomiasis	2.6 million[a]
Hookworm	1.75 million[a]
Ascariasis	1.07 million[a]
Trichuriasis	544 000[a]
Taeniasis/ Cysticercosis	303 000[a]
Trachoma	2.7 million[a]
Yaws	1.6 million estimated for 2015– 2050 if disease not eradicated[b]

a) GBD 2015 DALYs and HALE Collaborators (2016).
b) Mitjà et al. (2015).

roundworm (*Ascaris lumbricoides*), whipworm (*Trichuris trichiura*) and hookworms (*Necator americanus* and *Ancylostoma duodenale*).

Integrated control of these neglected tropical diseases relies on health education to promote hygiene and sanitation, supplemented by mass administration of donated drugs, vector control or case finding and treatment, as shown in the Table 1.13.

The global all-age DALYs for the NTDs in 2015 are shown in Table 1.14.

1.12 A Comparison of Health Metrics in an LMIC (Papua New Guinea) and a Developed Country (Australia) with a 7 km Distance Between them

Australia is a developed country. Papua New Guinea is an LMIC. Australia owns almost all the islands in the Torres Strait. The most northerly Australian island is 7 km from the main island of Papua New Guinea.

Table 1.15 Comparison of health metrics between Papua New Guinea and Australia.

Metric	Papua New Guinea	Australia
Population[a]	7.6 million	23.1 million
GDP per capita[a],[b]	US$1217[b]	More than US$50000
People living in rural area1	87%	13%
Life expectancy (years)	63[a]	82[a]
Maternal mortality per 10000 live births	215[a]	6[a]
Infant mortality per 1000 live births	47[a]	3[a]
Commonest causes of death and morbidity	HIV, TB and malaria; acute respiratory infections; other infectious diseases; maternal, neonatal and nutritional diseases; cardiovascular diseases and diabetes; cancer; chronic respiratory diseases and other non-communicable diseases; suicide, homicide and conflict; unintentional injuries. All data from[c]	Coronary artery disease; other cardiovascular diseases; Alzheimer's and other dementias; lung cancer; chronic obstructive pulmonary disease; colorectal cancer; prostate cancer; diabetes mellitus; breast cancer. Data from[d],[e]
Physicians per 1000 people	0.05[f]	2.5[f]

a) Australian Doctors International Annual Report (2015–2016); www.adi.org.au/wp-content/uploads/2016/.../ADI-Annual-Report-2015-6_WEB.pdf.
b) World Economic Outlook Database (2016) https://www.imf.org/external/pubs/ft/weo/2016/01/weodata/index.aspx.
c) WHO Papua New Guinea statistical profile: http://www.who.int/gho/countries/png.pdf?ua=1.
d) Australian Institute of Health and Welfare (2012). A working guide to international comparisons of health. Cat. no. PHE 159. Canberra: AIHW; www.aihw.gov.au/publication-detail/?id=10737421561.
e) WHO Australia statistical profile: http://www.who.int/gho/countries/aus.pdf?ua=1.
f) NationMaster: Health > Physicians > per 1000 people: countries compared. http://www.nationmaster.com/country-info/stats/Health/Physicians/Per-1,000-people.

Table 1.15 illustrates the huge gap in health-related metrics between these two countries.

1.13 Conclusions

Much progress has been made the first 15 years of the twenty-first century in malaria, HIV/AIDS, TB, neglected tropical diseases and other communicable diseases. This has been due to many factors including new vaccination programs, new medications particularly the HAART drugs for HIV/AIDS, improved sanitation, access to clean drinking water and increased public health education. Much remains to be done. New rapid diagnostic tests, the use of telemedicine, and the role of the (ubiquitous) mobile phone represent platforms for achieving further improvement.

The WHO Millennium Development Goals have been replaced by the WHO Sustainable Development Goals for the period 2016 to 2030. These are much more complex and have a much wider remit than health goals alone. Some, if achieved, such as the end of global poverty, will produce a striking improvement in health in the LMICs and are addressed in Chapter 38.

Bibliography

Australian Doctors International Annual Report 2015–2016. www.adi.org.au/

Australian Institute of Health and Welfare. 2012. A working guide to international comparisons of health. Cat. no. PHE 159. Canberra: www.aihw.gov.au/publication-detail/?id=10737421561

Bhutta, Z.A., Sommerfeld, J., Zohra, S. et al. (2014). Global burden, distribution, and interventions for infectious diseases of poverty. *Infect. Dis. Poverty* 3: 21. https://doi.org/10.1186/2049-9957-3-21.

Brice, J.B., Boschi-Pinto, C., Shibuya, K. et al. (2005). WHO estimates of the causes of death in children. *Lancet* 365: 1147–1152.

Caminero, J.A., Sotgiu, G., Zumla, A. et al. (2010). Best drug treatment for multidrug-resistant and extensively drug-resistant tuberculosis. *Lancet Infect. Dis.* 10: 621–629. https://doi.org/10.1016/S1473-3099(10)70139-0.

CDC MMWR. (December 2009). Global measles mortality 2000–2008. https://www.cdc.gov/mmwr/preview/mmwrhtml/mm5847a2.htm

CDC report on pertussis in countries other than the USA. https://www.cdc.gov/pertussis/countries/index.html

Centers for Disease Control and Prevention, USA. https://www.cdc.gov/globalhealth/ntd/diseases/index.html - Neglected tropical diseases

Disease and Injury Incidence and Prevalence Collaborators (2015). Global, regional, and national incidence, prevalence, and years lived with disability for 310 diseases and injuries, 1990–2015: A systematic analysis for the Global Burden of Disease Study 2015. *Lancet* 388: 545–1602; Appendix, pp. 611–624.

Franco, E., Meleleo, C., and Serino, L. (2012). Hepatitis A: epidemiology and prevention in developing countries. *World J. Hepatol.* 4: 68–73.

Gabutti, G., Stefanati, A., and Kuhdari, P. (2015). Epidemiology of Neisseria meningitidis infections: case distribution by age and relevance of carriage. *J. Prev. Med. Hyg.* 56: E116–E120.

GBD 2015 DALYs and HALE Collaborators (2016). Global, regional, and national disability-adjusted life-years (DALYs) for 315 diseases and injuries and healthy life expectancy (HALE), 1990–2015: a systematic analysis for the Global Burden of Disease Study 2015. *Lancet* 388: 1603–1658.

GBD 2015 Disease and Injury Incidence and Prevalence Collaborators (2016). Global, regional, and national incidence, prevalence, and years lived with disability for 310 diseases and injuries, 1990–2015: a systematic analysis for the Global Burden of Disease Study 2015. *Lancet* 388: 1545–1602.

GBD 2015 Mortality and Causes of Death Collaborators (2016). Global, regional, and national life expectancy, all-cause mortality, and cause-specific mortality for 249 causes of death, 1980–2015: a systematic analysis for the Global Burden of Disease Study 2015. *Lancet* 388: 1459–1544.

Glaziou, P., Sismandis, C., Zignol, M. and Floyd, K. (2016). Methods used by WHO to estimate the global burden of TB disease. http://www.who.int/tb/publications/ global_report/gtbr2016_online_technical_appendix_global_disease_burden_ estimation.pdf?ua=1

Granich, R., Gupta, S., Hersh, B. et al. (2015). Trends in AIDS deaths, new infections and ART coverage in the top 30 countries with the highest AIDS mortality burden; 1990–2013. *PLoS One* 10: e0131353. https://doi.org/10.1371/journal.pone.0131353.

Hotez, P.J., Fenwick, A., Savioli, L., and Molyneux, D.H. (2009). Rescuing the bottom billion through control of neglected tropical diseases. *Lancet* 373: 1570–1575.

LaBeaud, A.D., Bashir, F., and King, C.H. (2011). Measuring the burden of arboviral diseases: the spectrum of morbidity and mortality from four prevalent infections. *Popul. Health Metrics* 9: 1. https://doi.org/10.1186/1478-7954-9-1.

Mitjà, O., Marks, M., Konan, D.J. et al. (2015). Global epidemiology of yaws: a systematic review. *Lancet Glob. Health* 3: e324–e331. https://doi.org/10.1016/ S2214-109X(15)00011-X.

Nakagawa, F., May, M., and Phillips, A. (2013). Life expectancy living with HIV: recent estimates and future implications. *Curr. Opin. Infect. Dis.* 26: 17–25. https://doi. org/10.1097/QCO.0b013e32835ba6b1.

NationMaster. n.d. Health > Physicians > Per 1,000 people: Countries Compared. http:// www.nationmaster.com/country-info/stats/Health/Physicians/Per-1,000-people

Newman, L., Rowley, J., Vander Hoorn, S. et al. (2015). Global estimates of the prevalence and incidence of four curable sexually transmitted infections in 2012 based on systematic review and global reporting. *PLoS One* 10 (12): e0143304. https://doi. org/10.1371/journal.pone.0143304.

Organization for Economic Development definition of GDP. (1993). https://data.oecd.org/ gdp/gross-domestic-product-gdp.htm

Soares-Weiser, K., Maclehose, H., Bergman, H. et al. (2012). Vaccines for preventing rotavirus diarrhoea: vaccines in use. *Cochrane Database Syst. Rev.* 11: CD008521. https://doi.org/10.1002/14651858.CD008521.pub3.

The GATHER Working Group (2016). Guidelines for accurate and transparent health estimates reporting: the GATHER statement. *Lancet* published online June 28. doi:https://doi.org/10.1016/S0140-6736(16)30388-9.

The World Bank Data – Gross National Income. n.d. http://data.worldbank.org/indicator/
  NY.GNP.PCAP.CD?order=wbapi_data_value_2014+wbapi_data_value+
  wbapi_data_value-last&sort=desc

UNAIDS AIDS Epidemic Update. (2005). http://data.unaids.org/publications/irc-pub06/
  epi_update2005_en.pdf

UNAIDS Fact Sheet - latest statistics on the status of the AIDS epidemic. n.d. http://www.
  unaids.org/en/resources/fact-sheet

UNAIDS Progress Report. (2015). http://www.unaids.org/sites/default/files/media_asset/
  jc2774_2015progressreport_globalplan_en.pdf

United Nations Millennium Development Goals. n.d. http://www.un.org/millenniumgoals/
  pdf/MDG_Report_2009_ENG.pdf

WHO. AIDS epidemic update. (December 2005). http://www.who.int/hiv/epi-update2005_en.
  pdf?ua=1

WHO. n.d. Accelerating progress towards the health-related Millennium Development
  Goals. http://www.who.int/topics/millennium_development_goals/MDG-NHPS_
  brochure_2010.pdf?ua=1

WHO – 10 facts on malaria. (2016). http://www.who.int/features/factfiles/malaria/en

WHO | Metrics: Disability-Adjusted Life Year [DALY]. n.d. http://www.who.int/
  healthinfo/global_burden_disease/metrics_daly/en

WHO Australia statistical profile. n.d. http://www.who.int/gho/countries/aus.pdf?ua=1

WHO Fact sheet on diarrhoeal disease updated in May. 2017. http://www.who.int/
  mediacentre/factsheets/fs330/en

WHO Fact sheet on hepatitis C updated in April. n.d.

WHO Fact sheet on yellow fever updated in May (2016). http://www.who.int/mediacentre/
  factsheets/fs100/en

WHO Global Hepatitis Report. (2017). http://www.who.int/hepatitis/publications/
  global-hepatitis-report2017/en

WHO Global Tuberculosis Report. (2016). http://www.who.int/tb/publications/global_
  report/en

WHO MDGs: Progress made in health. (2015). http://www.who.int/news-room/fact-
  sheets/detail/millennium-development-goals-(mdgs)

WHO Measles. n.d. http://www.who.int/immunization/monitoring_surveillance/burden/
  vpd/surveillance_type/active/measles/en

WHO Measles Fact sheet reviewed in March (2017). http://www.who.int/mediacentre/
  factsheets/fs286/en

WHO Millennial Development Goals. (2010). http://www.who.int/topics/millennium_
  development_goals/MDG-NHPS_brochure_2010.pdf?ua=1

WHO Papua New Guinea statistical profile. n.d. http://www.who.int/gho/countries/png.
  pdf?ua=1

WHO Ten Fact on Malaria updated in December 2016. http://www.who.int/features/
  factfiles/malaria/en

WHO Tetanus. Immunization, vaccines and biologicals. n.d. http://www.who.int/
  immunization/monitoring_surveillance/burden/vpd/surveillance_type/passive/tetanus/en

WHO Top 10 causes of death in 2015. Fact sheet updated in January. 2017. http://www.
  who.int/mediacentre/factsheets/fs310/en

WHO: The Millennium Development Goals Report. (2015). http://www.un.org/
  millenniumgoals/2015_MDG_Report/pdf/MDG%202015%20Summary%20web_english.pdf

WHO: World Malaria Report. (2016). http://www.who.int/malaria/publications/world-malaria-report-2016/report/en

Wijesooriya, N.S., Rochat, R.W., Kamb, M.L. et al. (2016). Global burden of maternal and congenital syphilis in 2008 and 2012: a health systems modelling study. *Lancet Glob. Health* 4: e527–e533.

World Economic Outlook Database. (2016). International Monetary Fund.

Zignol, M., Dean, A.S., Falzon, D. et al. (2016). Twenty years of global surveillance of antituberculosis-drug resistance. *N. Engl. J. Med.* 375: 1081–1089.

Webliography

www.aihw.gov.au/publication-detail/?id=10737421561 - Australian Institute of Health and Welfare 2012. A working guide to international comparisons of health. Cat. no. PHE 159. Canberra: AIHW.

http://www.who.int/tb/publications/global_report/gtbr2016_online_technical_appendix_global_disease_burden_estimation.pdf?ua=1 - Glaziou, P., Sismandis, C., Zignol, M. and Floyd, K. Methods used by WHO to estimate the global burden of TB disease.

https://doi.org/10.1371/journal.pone.0131353 - Granich, R., Gupta, S., Hersh, B., et al. 2015. Trends in AIDS Deaths, New Infections and ART Coverage in the Top 30 Countries with the Highest AIDS Mortality Burden; 1990–2013. PLoS One, 2015

http://data.unaids.org/publications/irc-pub06/epi_update2005_en.pdf - UNAIDS AIDS Epidemic Update 2005.

http://data.worldbank.org/indicator/NY.GNP.PCAP.CD?order=wbapi_data_value_2014+wbapi_data_value+wbapi_data_value-last&sort=desc - World Bank definition of Gross National Income (GNI) per capita

http://esa.un.org/unsd/sna1993/introduction.asp - OECD definition of GDP.

http://www.nationmaster.com/country - NationMaster: health physicians per 1,000 people: countries compared.

http://www.un.org/millenniumgoals/2015_MDG_Report/pdf/MDG%202015%20Summary%20web_english.pdf - WHO: The Millennium Development Goals Report 2015.

http://www.un.org/millenniumgoals/pdf/MDG_Report_2009_ENG.pdf - United Nations Millennium Development Goals.

http://www.unaids.org/en/resources/fact-sheet - UNAIDS Fact Sheet - latest statistics on the status of the AIDS epidemic.

http://www.who.int/features/factfiles/malaria/en - WHO: 10 facts on malaria, updated December 2016.

http://www.who.int/gho/countries/aus.pdf?ua=1 - WHO Australia statistical profile

http://www.who.int/gho/map_gallery/en - The World Health Organization disease distribution maps.

http://www.who.int/healthinfo/global_burden_disease/metrics_daly/en - WHO Metrics: Disability-Adjusted Life Year (DALY).

http://www.who.int/hepatitis/publications/global-hepatitis-report2017/en - WHO Global Hepatitis Report, 2017.

http://www.who.int/hiv/epi-update2005_en.pdf?ua=1 - WHO. AIDS epidemic update, December 2005.

http://www.who.int/immunization/monitoring_surveillance/burden/vpd/surveillance_type/active/measles/en - WHO Measles.

http://www.who.int/immunization/monitoring_surveillance/burden/vpd/surveillance_type/active/measles/en - Measles.

http://www.who.int/immunization/monitoring_surveillance/burden/vpd/surveillance_type/passive/tetanus/en WHO - Tetanus. Immunization, vaccines and biologicals.

http://www.who.int/malaria/publications/world-malaria-report-2016/report/en - WHO World Malaria Report 2016.

http://www.who.int/mediacentre/factsheets/fs100/en - WHO Fact sheet on yellow fever updated in May 2016.

http://www.who.int/mediacentre/factsheets/fs164/en - WHO Fact sheet on hepatitis C updated in April 2017.

http://www.who.int/mediacentre/factsheets/fs286/en

http://www.who.int/mediacentre/factsheets/fs286/en - WHO Measles Fact sheet reviewed in March 2017.

http://www.who.int/mediacentre/factsheets/fs310/en - WHO Top 10 causes of death in 2015. Fact sheet updated in January 2017.

http://www.who.int/mediacentre/factsheets/fs330/en - WHO Fact sheet on diarrhoeal disease updated in May 2017.

http://www.who.int/topics/millennium_development_goals/MDG-NHPS_brochure_2010.pdf?ua=1 - WHO: Accelerating progress towards the health-related Millennium Development Goals.

http://www.who.int/topics/millennium_development_goals/post2015/en - WHO MDGs: Progress made in health.

https://www.cdc.gov/mmwr/preview/mmwrhtml/mm5847a2.htm - CDC MMWR December 2009. Global measles mortality 2000–2008.

https://www.cdc.gov/pertussis/countries/index.html - CDC report on pertussis in countries other than the USA.

https://www.imf.org/external/pubs/ft/weo/2016/01/weodata/index.aspx - World Economic Outlook Data Base, 2016.

http://www.who.int/en/news-room/fact-sheets/detail/measles - WHO Measles Fact sheet reviewed in March 2017.

http://www.who.int/tb/publications/global_report/en - WHO Global Tuberculosis Report, 2016.

www.adi.org.au/wp-content/uploads/2016/.../ADI-Annual-Report-2015-6_WEB.pdf - Australian Doctors International Annual Report, 2015–2016.

2

The Burden of Non-communicable Diseases in Low- and Middle-Income Countries

Heiner Grosskurth

London School of Hygiene and Tropical Medicine, Department of Infectious Disease Epidemiology, based at the Mwanza Intervention Trials Unit (MITU), National Institute for Medical Research (NIMR), Dar es Salaam, Tanzania

---

CHAPTER MENU

Revolutionizing Tropical Medicine: Point-of-Care Tests, New Imaging Technologies and Digital Health, First Edition. Edited by Kerry Atkinson and David Mabey.

2.1   Introduction

This Chapter begins with a summary of the most common non-communicable diseases (NCDs) in low- and middle-income countries (LMICs). It then looks at the epidemiology of a selection of common NCDs in LMICs and of risk factors associated with NCDs, as well as the overlap between NCDs and communicable diseases. This is followed by data on the burden that NCDs represent for health services, and the economic consequences associated with NCDs at patient, health system and macroeconomic levels. Next it examines the response to the growing NCD epidemic at international and national levels, and within the private sector. The Chapter then explores how health services in LMICs currently cope with the growing NCD burden, identifying the challenges they encounter. The subsequent section provides practical guidance on how to improve the performance of NCD services, based on the experience from a control program implemented in East Africa and ends with guidance on how to evaluate programs that aim to improve NCD services at the primary healthcare level. Policy advisers, health program managers and NCD control project staff may wish to find practically relevant additional information. They can find links to such material in the list of references which whenever possible refers to open access publications.

2.2   Common Non-communicable Diseases in Low- and Middle-Income Countries

Non-communicable diseases include a large and varied group of health conditions, all of which are not or not directly caused by an infectious pathogen. In public health (and in this chapter), the term "NCDs" has a much more restricted meaning and refers to a group of common chronic diseases that mainly comprise four types: cardiovascular diseases including their consequences (such as ischemic heart disease and stroke), diabetes mellitus, chronic respiratory diseases (in particular chronic obstructive pulmonary disease – or "COPD," and asthma) and cancers. These four disease groups are responsible for about 80% of NCD deaths worldwide (WHO 2014a).

2.3   NCD Epidemiology

NCDs have been highly prevalent in industrialized countries for a long time, but in LMICs they occurred much less frequently until about the last two decades of the twentieth century. The burden of NCDs has since been rising in LMICs, repeating to some extent a shift of morbidity and mortality that high income countries (HICs) had experienced much earlier. In HICs this has given rise to the theory of epidemiological transition (Omran 1971), and there is strong evidence that this transition has reached LMICs. For example, according to the Global Burden of Disease Study, the total number of deaths per year due to NCDs rose by approximately 14% in 2005 to approximately 40 million in 2015 (GBD 2015, 2016a), of which more than 80% have been recorded in LMICs (WHO 2014a). During the same period, the death toll attributable to communicable, maternal, neonatal, and nutritional conditions significantly declined, mostly because of advances with respect to HIV disease, malaria and birth complications (GBD 2015, 2016a). Interestingly, *age-adjusted* mortality rates for both NCDs and other conditions have declined to some extent. However, because populations increase particularly in LMICs and because overall life expectancy is increasing, the number of deaths due to NCDs is also increasing (GBD 2015, 2016a). Modeling studies predict that the proportion of deaths caused by NCDs will rise to about 70% of all deaths by about 2030 (Mathers and Loncar 2006).

Globally, men and women are almost equally affected, and NCDs are not just a problem among older age groups. According to WHO estimates derived from the Global Burden of Disease study, more that 40% of all deaths due to NCDs occur below the age of 70, and this proportion is higher in LMICs than in high income countries (WHO 2014a).

2.3.1 Arterial Hypertension and Cardiovascular Diseases

Hypertension is the most important risk factor for cardiovascular diseases. Hypertension is characterized by an elevated usual resting blood pressure with a systolic pressure of 140 mmHg or more and/or a diastolic pressure of 90 mmHg or more (Joint National Committee 2004). The term pre-hypertension is used if the systolic and diastolic blood pressure are 120–139 and/or 80–89 mmHg, respectively. These definitions are somewhat arbitrary as the mortality associated with elevated blood pressure increases for all age bands from about 115 mmHg (systolic) and from about 75 mmHg (diastolic) upwards. For example, for the age group of 40–69 years, each increase of 20 mmHg of systolic or 10 mmHg of diastolic blood pressure leads to a twofold increase in the death rate due to stroke (Lewington et al. 2002). Hypertension can be a symptom of other diseases such as a tumor or dysfunction of the adrenal glands, but in most hypertensive patients no distinct cause can be identified.

Only few data have been available from LMICs for the time before the year 2000, but hypertension seems to have been an uncommon problem there until recently. For example, a survey of about 1700 men from a rural community in Kenya conducted in 1983 found no individual with hypertension, and blood pressure did not increase until approximately the age of 60 (Poulter et al. 1984). In contrast, a survey from 2012 among 2100 adults (40% men) from another rural community in Kenya, the overall prevalence of hypertension was 21%. The prevalence rose rapidly with increasing age and reached 47% in those aged 55 years and above. Additionally, 4% of young adults aged 18–24 years were affected (Hendriks et al. 2012).

A high prevalence of hypertension has been reported by various other recent population studies in African countries including Tanzania and Uganda (Kavishe et al. 2015), Namibia and Nigeria (Hendriks et al. 2012), South Africa (Rheeder et al. 2017), as well as from other LMICs outside Africa (Irazola et al. 2016). The prevalence in some groups exceeded 50%. While earlier studies showed that hypertension was more a problem of urban populations, rural areas are now equally affected.

Several risk factors predispose to hypertension, including male gender, obesity, alcohol use, lack of physical exercise, increased levels of blood cholesterol, diabetes, and high salt consumption. The latter is of particular concern as a high sodium intake has become very common in LMICs (Powles et al. 2013).

More than 80% of patients with hypertension found in population studies were unaware of their condition, and of those who are, few receive treatment and even fewer are controlled (Hendriks et al. 2012; Kavishe et al. 2015).

2.3.2 Diabetes Mellitus

Diabetes mellitus type 1 is caused by an autoimmune process that results in the failure of the pancreas to produce sufficient insulin. Usually the disease begins at an early age but onset can occur at all age groups. All patients with type 1 diabetes require insulin therapy. In contrast, type 2 diabetes occurs when endogenously produced insulin is not sufficiently effective in controlling blood glucose levels. Type 2 diabetes usually begins at a higher age, but may occur in adolescents and young adults, particularly in the presence of lifestyle-related risk factors such as obesity or lack of exercise. In general, type 2 diabetes responds well to reduction of these risk factors, but in some cases, it may require oral antidiabetic medication or even insulin. Both types of diabetes lead to complications most of which result from damage to blood vessels. Complications include diabetic nephropathy, neuropathy, retinopathy and diabetic ulcers of the feet, as well as heart disease, stroke and ischemic limb disease. Diabetes thus contributes significantly to the pathogenesis of cardiovascular diseases (Alberti and Zimmet 1998), and diabetes is often also associated with hypertension. In LMICs where specialized laboratory services are scarce, the diagnostic distinction between the two types of diabetes cannot always be accurately made, but in terms of absolute numbers, the burden of diabetes type 2 is far higher than that of diabetes type 1.

Worldwide, according to WHO estimates, the prevalence of diabetes has almost doubled between 1980 and 2014. This increase has been much more prominent in LMICs than in high-income countries. Globally, more than 400 million adults are affected by diabetes, and about 1.5 million deaths occur annually due to it (WHO 2016a). In population-based studies from various LMICs around the world, the prevalence of diabetes is usually between 5 and 10%, but ranges from 1 to 30% (Kavishe et al. 2015; Kengne et al. 2013; Nanditha et al. 2016), and has been reported to exceed nearly 40% in some Pacific island populations (Nanditha et al. 2016). Diabetes is generally more prevalent in urban populations, but some studies report a high prevalence from rural areas as well (Hwang et al. 2012). Men and women are generally equally affected.

The main risk factors for diabetes type 2 are overweight and obesity in combination with a lack of physical activity (GBD 2013, 2015). Other factors include a high consumption of saturated fatty acids (Ley et al. 2014), low birth weight (Whincup et al. 2008) and smoking (Willi et al. 2007). Of particular importance for the epidemiology of diabetes in LMIC is the observation that poor fetal growth and low birth weight seem to increase the risk of developing metabolic diseases later in life (Whincup et al. 2008). Genetic factors also play a role: diabetes occurs more frequently in individuals with a family history of the disease (Dagenais et al. 2016), and people from South East Asia have a higher risk of developing diabetes than other populations, even in the absence of obesity (Ramachandran et al. 2010).

Diabetes type 2 often develops slowly so that patients remain unaware of their condition for a long time and therefore are not treated, thus losing valuable time for the prevention of complications (Beagley et al. 2014).

2.3.3 The Metabolic Syndrome

Certain risk factors for cardiovascular diseases and diabetes often appear in combination, and this occurs more often than expected by chance. These factors include elevated fasting blood glucose, high triglycerides, reduced high density cholesterol, elevated blood pressure and being overweight. This combination of factors is called the metabolic syndrome. Different definitions of the metabolic syndrome can be found in the literature. More recently it has been agreed that for patients to qualify for the diagnosis of the syndrome, they should have at least three of these five factors, and because a higher than normal waist circumference as a marker for central obesity is frequently present, it can be used as a screening tool to detect the metabolic syndrome (Alberti et al. 2009). The presence of the metabolic syndrome is a strong predictor for developing diabetes and cardiovascular diseases.

The prevalence of the metabolic syndrome in LMIC varies across countries, seems to have increased over recent years and reaches very high levels in some populations, for example, nearly 50% in some groups from Pakistan (Ranasinghe et al. 2017). From a public health perspective, it may be a useful marker to monitor the risk for cardiovascular diseases and diabetes within populations (Ofori-Asenso et al. 2017).

2.3.4 Chronic Kidney Disease

Chronic kidney disease (CKD) is highly prevalent in LMICs. It is clinically defined as a disturbed kidney structure or function lasting longer than three months. Criteria for the diagnosis of CKD include a glomerular filtration rate (GFR) of less than $60 \, \text{mL/min/1.73 m}^2$, and/or markers of kidney malfunction such as albuminuria (ISN 2013).

CKD is often a chronic complication of other NCDs, mainly diabetes and hypertension, or of specific renal diseases such as glomerulonephritis. However, there are also many CKD patients in LMICs for which an underlying cause cannot be established (Lunyera et al. 2016). Prevalence data from LMICs are scarce, ranging from 5 to 15% in general population studies (Jha et al. 2013; Stanifer et al. 2014), but exceed 20% in some populations. CKD can result in end-stage kidney failure, cardiovascular diseases, anemia (due to inadequate renal production of erythropoietin) and bone disorders (Jha et al. 2013) (the latter being due to the inability of damaged kidneys to balance phosphorus and calcium levels in the blood). Risk factors other than diabetes and hypertension include schistosomiasis, inflammatory diseases of the kidney, environmental toxins and poverty. CKD has greater public health importance in LMICs than is currently reflected in the NCD-related literature.

2.3.5 Chronic Obstructive Pulmonary Disease (COPD)

COPD is a chronic lung disease marked by inflammatory changes of small airways and a parenchymal damage of alveolar tissue (GOLD 2017a). COPD patients suffer from progressive dyspnea, and often also from chronic cough and sputum production. The disease is characterized by a reduction in airflow that cannot be fully reversed using a bronchodilator, and this is an important criterion in the differential diagnosis of COPD and asthma. The diagnosis of COPD is made by a forced vital capacity test using respirometry after applying an inhaled bronchodilator: the maximal forced expiratory volume in one second (FEV1) is reduced relative to the forced vital capacity (FVC) that a person can maximally exhale. In individuals with COPD, the FEV1/FVC ratio is less than 0.7 (GOLD 2017b).

In a population-based study from 2007 conducted in 12 sites around the world with a total sample of more than 9000 men and women, the prevalence of manifest COPD was about 10% overall, 12% among men and 9% among women. The risk of having COPD increased with age by an odds ratio of about 2 per 10-year age increment (Buist et al. 2007). Studies from sub-Saharan Africa showed a prevalence ranging from 2 to 25% in different populations (Finney et al. 2013; Kavishe et al. 2015), with the highest levels seen in urban South Africa. It has been estimated that globally in 2010 the number of COPD cases was more than 380 million, and that about 3 million people died from COPD per year (GOLD 2017a). Due to aging and an increase in smoking in LMICs, the expected number of deaths is predicted to rise to more than 4.5 million deaths by 2030 (Lopez et al. 2006).

Key risk factors for developing COPD in LMICs include the exposure to indoor air pollution caused by cooking with biomass fuel (see Chapter 20), and in some countries tobacco smoking which is one of the main causes of COPD in high income countries. Generalized air pollution due to industry and traffic are important too. Industrial workers in LMICs are often also exposed to occupational dusts and chemicals that are associated with COPD (Mannino and Buist 2007). Low birth weight and respiratory infections during childhood increase the risk for the development of COPD in adulthood. Genetic factors such as deficiency of alpha-1 antitrypsin may play a role (de Serres and Blanco 2014).

2.3.6 Asthma

In contrast to COPD, asthma is a chronic disease characterized by episodes of reversible constriction of the airways. Typical symptoms include wheezing, breathlessness and sometimes cough. In susceptible patients, asthma attacks can be triggered by allergens such as house dust mites, pollens, but also by tobacco smoke, polluted air, respiratory tract infections and physical exercise and exposure to cold water (Vernon et al. 2012; Marks et al. 2014). In many cases an allergic cause cannot be found.

Asthma often begins during childhood, but most children who have asthma become free of symptoms during adolescence. On the other hand, asthma can also newly develop later in life. Asthma is less common than COPD, but WHO estimates that worldwide, more than 230 million people have asthma (Marks et al. 2014).

Reliable data on the prevalence and incidence of asthma from LMICs are scarce. For a long time, it was thought that asthma was a problem of northern populations, but there is some indication that the prevalence is rising fast and that nowadays most asthmatic patients live in LMICs (Global Asthma Report 2014). The mortality of asthma is low, causing not more than 1% of all deaths worldwide. Nevertheless, it is believed that about 340 000 asthma patients die per year (Marks et al. 2014). Proportionally, mortality rates are higher among older persons than among children. However, in older age groups asthma patients may also have COPD, making a clear allocation of mortality rates difficult. South Africa has the highest age-standardized asthma mortality rate worldwide with about 290 deaths per million (Global Asthma Report 2014).

2.3.7 Cancer

Data on cancer epidemiology from LMICs are less reliable than those from high-income countries due to the lack of systematic cancer registries in many countries (Bray et al. 2015). However, it has been estimated that more than 60% of all cancers and about 70% of cancer deaths occur in LMICs (Ferlay et al. 2015).

There are wide variations in the distribution pattern of specific types of cancer between and within different regions (Forman and Ferlay 2014). Across all world regions LMICs have the highest burden of stomach, liver, esophageal and cervical cancer. As in industrialized countries, prostate, breast, lung, and colorectal cancers are also frequent (Torre et al. 2016).

In line with the size of its population, Asia bears the highest burden of cancers of all world regions, and about 30% of all cancers in the world occur in China and India. The most common forms of cancer among men in Asia are lung, stomach, liver, colon and esophageal cancers, each with an age-standardized incidence of 10 or more per 100 000 per year, and all of these cancers have high mortality rates. Although lung cancer has the highest incidence and mortality rate among men in the region (35 and 32% per 100 000 respectively), it is still substantially less frequent in Asia than in Europe and North America. Rates of liver and stomach cancer in Asian men are higher than those in other world regions. In women, as in other regions, breast cancer is the most prevalent cancer, but lung cancer is the most frequent cause of cancer-related deaths (Forman and Ferlay 2014).

The epidemiology of cancers in sub-Saharan Africa (SSA) differs from other world regions in some important aspects: among women, the incidence of cervical cancer is as high as that of breast cancer, and its incidence and mortality (35/100 000 and 23/100 000 respectively) are the highest among all world regions. In men prostate cancer is the most common: prostate cancer has an incidence of 28/100 000 and the mortality rate is nearly as high as the incidence, in contrast to industrialized countries where prostate cancer mortality tends to be only about a sixth to a tenth of the incidence, largely due to more advanced options for case detection and treatment. Liver cancer too has a high incidence and mortality, each exceeding 10/100 000 in both men and women. Kaposi sarcoma[1] is a common complication of advanced HIV infection, and because the burden of HIV infection is higher in SSA than other regions, the incidence of this tumor is particularly high there (about 7/100 000) (Forman and Ferlay 2014). Currently the rate of Kaposi sarcoma is falling due to the increasing use of antiretroviral therapy (ART) (Bohlius et al. 2014).

In Latin America, including the Caribbean subregion, breast cancer and prostate cancer are the most frequent forms of cancer among women and men respectively, similar to the situation in Europe and North America. However, the incidence and mortality of cervical cancer (21/100 000) is much higher than in industrialized regions, albeit not as high as in SSA. Lung, colon and stomach cancers are the second, third and fourth frequent forms of cancer among both women and men.

Common risk factors for the most frequent cancers include some that are also associated with other major NCDs: tobacco smoking, abuse of alcohol, low fruit and vegetable intake and obesity (GBD 2015, 2016b). Some cancers are directly caused by viral infec-

---

[1] Moritz Kaposi (1857–1902) was a Hungarian physician and dermatologist who described this disease in 1872.

tions: hepatitis B (hepatocellular cancer), hepatitis C (hepatocellular cancer), human papilloma virus infection (cervical cancer), human herpes virus 8 (Kaposi's sarcoma), and Epstein–Barr virus (Burkitt's lymphoma (Plummer et al. 2016), whilst for some other forms of cancer, infections play a strong contributing role, in particular HIV infection. Increasing age is associated with various forms of cancer, probably because carcinogenic factors have a longer time to exert their effects and to lead to cell mutations.

2.4 Prevention of Non-communicable Diseases

As mentioned above, individual NCDs share a number of life-style related risk factors, notably tobacco use, being overweight or obese, harmful use of alcohol, an unhealthy diet that is high in calories, saturated fatty acids or salt but low in fiber and vitamins, and insufficient physical work or exercise (WHO 2014a). About three-quarters of all cases of cardiovascular diseases and diabetes can be attributed to these risk factors, and thus could be prevented. Smoking and obesity are of particular concern. Smoking alone caused about 5 million deaths in 2010, and this is expected to rise to 10 million or more over the next few decades (Jha and Peto 2014), which would then account for about 10% of all deaths. Approximately 80% of these deaths are expected to occur in LMICs.

The prevalence of being overweight and obesity is surprisingly high in LMICs. In a study of more than 31 000 adults from 9 LMICs, the prevalence of central obesity (i.e. obesity determined by measuring waist circumference) ranged from 19 to 79% (Patel et al. 2016). Women seem to be more affected than men (Kavishe et al. 2015). Obesity is strongly associated with diabetes and hypertension (Patel et al. 2016). Obesity often begins in childhood. It has been estimated that worldwide in 2015 there were 42 million overweight children (UNICEF, WHO, World Bank 2015), and this is a rapidly growing problem in LMICs also (WHO 2016a). Being overweight or obese are also associated with various types of cancer, including cancers of the breast, colon and liver (Steele 2017).

Consequently, there is a need for interventions to reduce these risk factors, both in high-income countries and LMICs. It has been estimated that, if the key NCD risk factors could be eliminated, most cases of heart disease, stroke and diabetes type 2 and about 40% of cancers would be prevented (WHO 2013a). Higher taxation on tobacco products, a ban of smoking at work and public places, a ban on alcohol advertisements and laws on the reduction of salt and saturated fatty acids in processed food could make a major contribution to this goal. Many countries, including about 40 LMICs have implemented such legislation with respect to smoking (WHO 2013a), but much less progress has been made on the other factors. Obviously, the interests of large and powerful industrial companies are at odds with those of public health (Buse et al. 2017).

2.5 The Relationship Between Communicable and Non-communicable Diseases

For a number of reasons, the dichotomy between communicable diseases and NCDs is not as clear-cut as much of the current literature seems to suggest. Firstly, the distinction is blurred because some NCDs, in particular cancers, have a viral infectious origin, for example, hepatocellular cancer (viral hepatitis B and C) cervical cancer (human papilloma virus), and lymphatic tissue (Epstein–Barr virus for non-Hodgkin lymphoma and human herpes virus 8 for Kaposi's sarcoma). Bacterial infections can also play a role in the etiology of cancers. For example, the risk of developing gastric cancer is doubled among individuals infected with *Helicobacter pylori* (Eslick 2006).

Secondly, some infectious diseases are associated with NCDs. For example, HIV-infected patients, particularly those on anti-retroviral therapy, seem to have a substantially higher risk of acquiring cardiovascular diseases and diabetes mellitus than HIV-negative individuals (Dillon et al. 2013; Mathabire Rücker et al. 2017). Vice versa, some NCDs increase the risk for certain infectious diseases, for example, patients with diabetes mellitus type 2 have a threefold higher risk to developing tuberculosis (Riza et al. 2014) and there is some evidence that diabetes may increase the risk of malaria (Danquah et al. 2010).

Thirdly, certain infectious diseases and some NCDs have shared risk factors (Oni and Unwin 2015) (Figure 2.1).

Lastly, from a health system perspective, the requirements for the chronic care of HIV and NCD patients are similar, and it has been suggested that NCD control programs in LMICs should adopt care strategies that have worked well for HIV-infected patients (Duffy et al. 2017). Both groups of patients need regular follow-up to monitor effectiveness of treatment, provide drug refills and reinforce health messages on adherence to medication and on risk reduction. Obviously, the growing number of patients with HIV-NCD comorbidity is in itself a convincing reason to integrate both kinds of services (Oni and Unwin 2015).

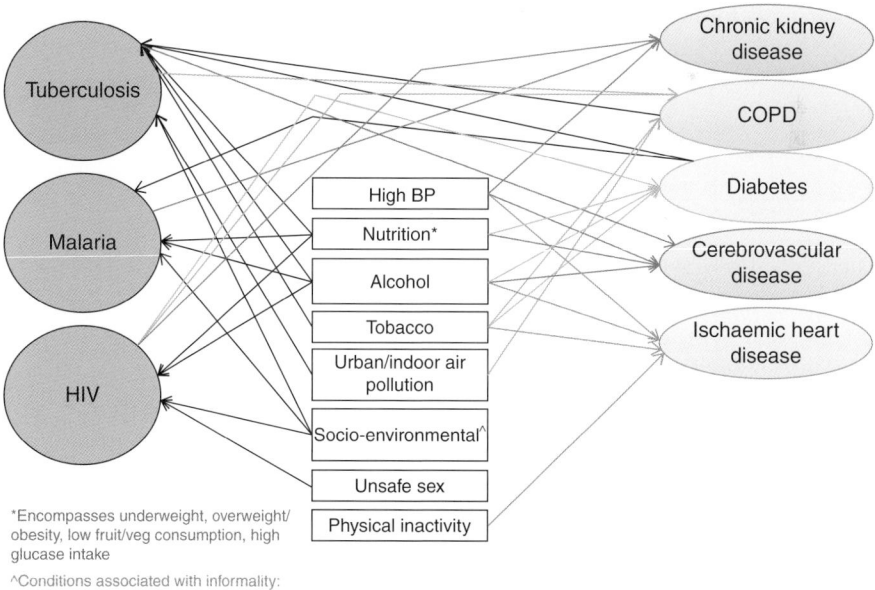

Figure 2.1 Interaction between tuberculosis, malaria and HIV disease with risk factors/disease precursors and non-communicable diseases. BP: blood pressure; COPD: chronic obstructive pulmonary disease. *Source:* This figure has been reproduced from the publication by Oni and Unwin (2015), which is an open access article distributed under the terms of CC BY-NC-ND. (*See color plate section for the color representation of this figure.*)

2.6   The Health System Burden of NCDs

Data on the burden of NCDs on specific health services are surprisingly limited, and only a few studies seem to have systematically investigated this important question. Given the high and growing prevalence of NCDs as observed in population-based studies, one would expect that primary care facilities would face a huge NCD patient burden at all levels, from tertiary and district hospitals down to health centers and even to communal health stations. However, in reality there are substantial and problematic discrepancies between the patient load observed at hospitals and that seen at lower level facilities.

For example, in the context of an NCD intervention programs conducted in East Africa, the health system burden of selected NCDs was studied in the outpatient departments (OPDs) of 52 health facilities in Uganda and Tanzania (8 hospitals, 20 health centers and 24 small primary care dispensaries). The observations across countries were very similar (Katende et al. 2015; Peck et al. 2014). Data are shown here for the study in Tanzania (Figure 2.2).

As in many other LMICs, public health centers in East Africa serve a subdistrict with a population of about 20 000 people, and together with dispensaries, are the main point of call for patients with any health problem. Health centers are staffed with individuals with some clinical training and occasionally with doctors or assistant medical officers. Dispensaries have clinicians and nurses and serve a population of about 5000 individuals. In rural areas most patients with NCDs would be expected to present at their local health facility, because this would save them time and money. However, this study from a representative sample of health facilities in northwestern Tanzania found that the mean number of outpatient visits related to chronic

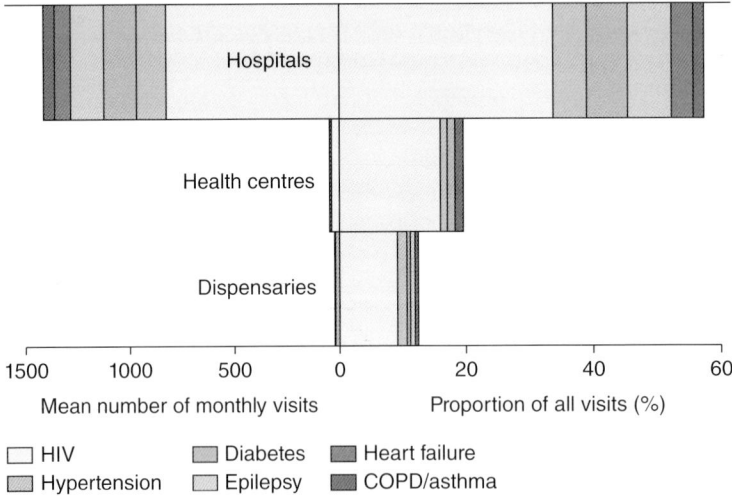

Figure 2.2 Burden of chronic diseases at 24 health facilities in northwest Tanzania. The mean number of chronic disease visits per month per facility is displayed to the left of the midline and the proportion of all outpatient visits due to chronic diseases is on the right. Data were collected from the adult outpatient departments of these 24 health facilities. COPD, chronic obstructive pulmonary disease. *Source:* This figure has been reproduced from the publication by Peck et al. (2014), which is an open access article distributed under the terms of CC BY-NC-ND. (*See color plate section for the color representation of this figure.*)

diseases per month (including chronic HIV infection) was 1411 for hospitals, but only 44 for health centers, and 22 for dispensaries (Figure 2.2). Most outpatient services for NCDs were provided in hospitals instead of health centers and dispensaries. In fact, most dispensaries and half of health centers reported not providing services for common NCDs such as hypertension and diabetes, despite being expected to do so according to current policies, and instead routinely referred such patients to hospital. As a result, some hospital OPDs were overburdened by routine NCD cases whilst peripheral health facilities were underutilized. Obviously, targeted interventions are required to enable and encourage peripheral health facilities to provide routine outpatient NCD services.

The few available data on the utilization of NCD services from other LMICs show similar results. For example, a study from southern China showed that 75% of NCD patients presented at district or even tertiary hospitals (Yang et al. 2014), and hospital-based diabetes clinics reported a heavy NCD patient burden, as also shown in studies from Cameroon, Mali, Tanzania and South Africa (Brown et al. 2014).

2.7   The Economic Impact of NCDs

2.7.1   The Patient's Perspective

The essential medicines for the basic treatment of uncomplicated NCDs are comparatively inexpensive. The costs of a monthly supply of drugs for the treatment of hypertension, diabetes or COPD are much lower than those, for example, of antiretroviral drugs. However, in LMICs where patients are often asked to pay for the medicines received or to share some of the costs, the expenditures for NCD medication overstretch the budget of the poor, whose average monthly cash income is at the poverty line of around US$1.25 per day, and this is particularly devastating in the elderly who often suffer from more than one NCD (Pati et al. 2014). Households respond to this pressure by reducing spending on food and education, and thus NCDs contribute to increasing impoverishment (Engelgau et al. 2011). When NCDs eventually lead to disabling or fatal complications, patients and their families face catastrophic healthcare expenditures (CHEs). CHEs are more common among NCD patients than among patients with non-chronic diseases (Si et al. 2017). CHEs due to NCDs have been observed in numerous LMICs and at various income levels, and for different groups has been reported by 6-84% of the households studied (Jaspers et al. 2015).

2.7.2   The Provider's Perspective

From a healthcare provider point of view, expenditures for NCD case management comprise capital costs such as building use, equipment and staff training, and recurrent costs such as medicines and diagnostic consumables, staff time, general supplies and utilities (Settumba et al. 2015). Estimates of the costs per patient per year incurred for specific NCD care activities vary widely, depending on the country, medical condition, intensity of health service utilization, and on the underlying estimation methodology used (Brouwer et al. 2015). For example, in a systematic review of provider costs the annual expenditures per patient for basic hypertension outpatient care ranged from US$38 (Tanzania) to $566 (China), and those for basic diabetic care including oral metformin treatment from $77 (India) to $989 (China). The treatment costs for complications are substantially higher. For example, for stroke the costs per patient varied from $160 (Tanzania) to $1851 (Pakistan) and even reached $16993 (South Africa). Those for conservative diabetic foot care varied from $86 (India) to $1619 (Nigeria) (Brouwer et al. 2015). Given the high prevalence of NCDs in the general population of LMICs, healthcare expenditures are expected to rise massively, putting a huge and growing burden on health systems that continue to struggle with a high burden of communicable diseases as well (WHO 2011).

2.7.3 Macroeconomic Effects

In 2011 at the request of WHO, in preparation for the United Nations Head of States' meeting on NCDs, the Harvard School of Public Health and the World Economic Forum issued a report on the estimated macroeconomic impact of NCDs. The report stated that "NCDs already pose a substantial economic burden and this burden will evolve into a staggering one over the next two decades." For example, with respect to cardiovascular disease, chronic respiratory disease, cancer, diabetes and mental health, the macroeconomic simulations suggested that there would be a cumulative output loss of US$47 trillion over the next two decades. This loss represented 75% of the global gross domestic product in 2010 (US$63 trillion) (Bloom et al. 2011). For LMICs this loss was calculated at US$21 trillion over the same period.

Estimating the economic impact of diseases is not straight forward, and other authors have come to somewhat lower estimates (Abegunde et al. 2007; WHO 2014a). However, all agree that the economic loss is already huge, is expected to grow and will substantially damage national economies. This is understandable because at the individual level over the coming years, as described above, NCDs are likely to push many more people below the poverty threshold and reduce their economic outlook. Because the burden of NCDs seems is rising quickly, it is expected that the economic loss will rise exponentially (Bloom et al. 2011).

It has also been estimated that compared to this economic impact, the costs of even a highly intensive worldwide NCD intervention program is small. The costs of such a program are thought to be in the order of US $10–12 billion per year (WHO 2014a). In other words, the costs of not intervening are expected to outweigh by far the costs for a thorough intervention program.

2.8 The Response to the NCD Epidemic in LMICs

2.8.1 Response at the International Level

An increase in the prevalence of certain NCDs, in particular cardiovascular diseases and diabetes mellitus type 2, in low income countries was noted for some time before the end of the last century (Beevers and Prince 1991; King and Rewers 1993). However, only recently did the international community realize the size and importance of the growing NCD epidemic (Daar et al. 2007). In 2011 the issue was eventually brought to the attention of the United Nations General Assembly, and Heads of States recognized that NCDs presented a world-wide "challenge of epidemic proportions" that was expected to grow particularly quickly in low and midlevel economies. A declaration was signed by all countries (UN 2011) in which Heads of States committed themselves to five action areas which included efforts to: (i) reduce NCD risk factors within populations; (ii) launch national NCD control policies and strengthen health systems in affected countries; (iii) collaborate internationally to better control NCDs; (iv) promote relevant research; and (v) to set and monitor measurable targets. WHO was tasked with developing detailed recommendations and providing technical guidance, overseen by a WHO Assistant Director-General for NCDs and Mental Health. Subsequently in 2013 the UN formed the United Nations Interagency Task Force on NCDs (UNIATF). UNIATF's role is to bring the different UN agencies together to address NCDs and to assist individual governments in the development of effective responses for the prevention and control of NCDs (UNIATF 2017). UNIATF's work is coordinated by WHO. Also, in 2013 WHO formulated a Global NCD Action Plan for the period which set nine targets including the aim of reducing NCD-related mortality by 25% until the year 2025, and to ensure that at least 50% of NCD patients who require drug therapy should receive it (WHO 2013a). As part of this process and to strengthen primary healthcare services, a package of essential NCD control interventions (PEN) was published. This package is a useful tool for anybody responsible for the practical planning and implementation of NCD control activities in LMICs, and is available online (WHO 2013b).

In 2014 the UN World Assembly reviewed the progress made, and noted that this has been rather limited and uneven across member states and that NCD control efforts should be intensified substantially (UN 2014). Member states then committed themselves to meet a list of specific time-bound aims which were to be monitored by WHO (WHO 2017a). Further progress was to be reported to the General Assembly in 2018. Ten indicators were agreed. These stipulated that by 2018 at the latest, each member state should have established a multisectoral NCD control strategy, conducted surveillance of NCD risk factors using the STEPwise approach to Surveillance (STEPS) instrument (WHO 2012), enforced comprehensive bans on tobacco and alcohol advertising and introduced increased taxation, launched activities to improve dietary habits at the population level including a reduction in sodium consumption, and introduced primary care services for major NCDs using evidence-based treatment protocols (WHO 2017a).

2.8.2 Response at the National Level

Some limited progress has been made over recent years. Most LMICs adopted a national NCD control strategy, often combined with a written action plan. These policy documents describe the national control strategy and often provide useful information on the importance of NCDs, the local epidemiology, and the various sectors that should play a role in the prevention and control of NCDs. For examples see Ministry of Health (MoH) Cambodia (2013), MoH Ghana (2012), MoH Trinidad and Tobago (2017).

A few countries have introduced regular monitoring of the prevalence of NCD risk factors at the population level using serially conducted WHO STEPS surveys (WHO 2017b). The results of STEP surveys have been published as fact sheets, and these confirm that in the participating LMICs the prevalence of elevated blood pressure and blood glucose levels and of various NCD risk factors is indeed high. In countries that conducted serial STEPS surveys, an increase of risk factor prevalence over time has been documented.

Some countries also evaluated the readiness of their healthcare systems to provide NCD care (for example, Nyarko et al. 2016), or started training programs to improve the awareness and skills of their primary care health workers (for example, Davila et al. 2015).

However, overall little progress had been made on the ground: in 2015, based on self-reports from 177 national programs, including 80 from LIMCs, WHO stated that there was still a striking discrepancy between NCD control policies and plans, and their actual implementation, particularly in low income countries (WHO 2015). In most LMICs the response to NCDs was still limited to the health sector (rather than ensuring a multisectoral approach), there were few functioning NCD case finding activities, drug stocks for NCD treatments remained insufficient and budgets for NCD prevention and care were far too small to cope with the huge disease burden. A complicating factor is that the existing drug supply systems are administratively independent of NCD control programs and have not adapted swiftly to the increase in demand. Very few countries have managed to introduce the agreed ground-breaking changes in tax legislation and advertising on alcohol and processed food.

2.8.3 Response by the Private Sector

There have been encouraging developments in the private sector worldwide, both internationally and nationally, to address the NCD challenge. In 2009 several international agencies (including the International Diabetes Federation, the World Heart Federation and organizations engaged in responding to respiratory disease and cancer) formed a global partnership: the NCD Alliance. Over time, about 2000 local civil society-based organizations joined this network. Activities include advocacy at international and national levels, and provision of assistance to local partners in terms of training and capacity building. In LMICs some Alliance members launched well-coordinated programs assisting ministries of health with regards to training and capacity building. A good example is the Tanzanian Diabetes Foundation (TDA) which helped to establish integrated diabetes clinics in regional hospitals across the country. These clinics provide not just diabetes care but also NCD services in general. In collaboration with the Ministry of Health, the Foundation published case management and training guidelines that can be freely accessed on the internet (for example, TDA 2013, 2014).

Similarly, the Global Initiative for Obstructive Lung Diseases (GOLD) was formed in 1998. This initiative aims to increase awareness and to monitor the epidemiology of COPD, and has issued carefully validated treatment guidelines (e.g. GOLD 2017b).

2.9 The Readiness of Primary Healthcare Services in LMICs to Cope with the NCD Burden

In 2007 in an attempt to facilitate a common approach to the understanding and strengthening of health systems, WHO published a framework of six essential health systems' building blocks (WHO 2007), all of which should function well in order to enable a health system to meet expectations. The six blocks are

1) health service delivery
2) the health workforce
3) health information systems

4) equitable access to medical products and technologies
5) health financing
6) governance

Subsequently WHO issued a set of indicators to determine health systems functions, and a handbook describing how to collect data on these (WHO 2010a, b). To further facilitate data collection on these indicators, WHO published the Service Availability and Readiness Assessment (SARA) questionnaire, a set of comprehensive tools to systematically assess and monitor health services (WHO 2013c).

The building block framework, the corresponding indicators and the SARA questionnaires are helpful tools for assessing the functionality of NCD control services, and these instruments are available online (WHO 2007, 2010a, b, 2013c). Furthermore, the SARA tool has been adapted to specifically evaluate NCD services at primary care facilities including hospital outpatient services, and has been amended with an instrument to assess health workers' knowledge on selected NCDs and, for comparison, on HIV infection (Peck et al. 2014) including supplementary online material. Together these resources represent a useful arsenal of methods to assess NCD services and to plan interventions for their improvement.

These validation methods have been applied to assess the functionality of NCD care at primary level in an NCD research program from East Africa (Katende et al. 2015; Peck et al. 2014), and similar assessments have been conducted in a variety of other LMICs around the world (Mannava et al. 2015; Nyarko et al. 2016; O'Neill et al. 2013; Pakhare et al. 2015; Siddharthan et al. 2015; Van Minh et al. 2014; Wangchuk et al. 2014; Yassoub et al. 2014).

The sobering overall result is that primary care services in most LMICs are not yet able to cope with the growing NCD burden: hospital outpatient departments (OPDs) of district and even tertiary hospitals are overwhelmed with an ever growing number of NCD patients, while health centers and other peripheral health facilities are bypassed and under-utilized, mainly because they are not yet ready to manage routine NCD conditions.

Most NCD patients could in principle be easily managed at smaller facilities: for example, uncomplicated cases of diabetes mellitus type 2 or chronic hypertension do not require specialist care. Strengthening peripheral facilities would go a long way to alleviate the burden of NCD care at hospitals, whilst saving time and travel costs incurred by patients. However, this is not happening for a number of reasons:

- Historically, all levels of healthcare in LMICs had a focus on the management of acute infectious diseases such as malaria, diarrhea and respiratory tract infections, and the prevention and care of NCDs seemed of little importance. Health services in LMICs were not designed to deal with chronic conditions (Beaglehole et al. 2008).
- Most health workers are not trained to manage NCDs, the local logistics required to ensure a regular drug supply have not been developed, and the systems necessary to monitor patients with chronic diseases are not in place. This could have changed after the wide-scale introduction of ART, which turned the previously fatal HIV/AIDS disease into a chronic treatable illness. From an organizational point of view, the primary care management of NCDs is similar to that of HIV infection in many regards, and so NCD services could have learned from those effective for HIV care (Harries et al. 2009). Unfortunately, the influence HIV care programs have had on routine outpatient services has been limited because the response to the HIV epidemic was, for understandable reasons, driven by vertical programs, and so routine NCD services did not

benefit from this development. Still, there is an opportunity here to learn from HIV care services, and both kind of services would benefit, given that many HIV patients face a high risk of developing NCDs (Duffy et al. 2017).

There are other problems too: reliable data to assess the preparedness, quality and effectiveness of current NCD services are largely missing, and so it remains difficult to plan for effective NCD control measures (Ali et al. 2013). This resulted in a vicious cycle in which the absence of planning data and an inadequate response at the primary care level reinforced each other, thus perpetuating the problem.

Whilst many countries have national NCD control policies by now, information about them has usually not reached the health services on the ground (Peck et al. 2013; Van Minh et al. 2014).

Finally, simple case management guidelines are usually not in place. A multitude of NCD drugs are mentioned in some national essential drug lists, but health workers are not used to most of them, and drugs are frequently out of stock. Standardized case management algorithms currently in use (for example, control programs for sexually transmitted diseases (Garcia et al. 2012), have yet to be introduced for NCDs. Even the most essential diagnostic equipment such as sphygmomanometers, weighing scales, or glucometers are often not available (Katende et al. 2015; Peck et al. 2014).

2.10 Introducing Effective NCD Control at Primary Care Services: A Practical Approach

Whilst NCD control in LMICs requires a multi-sector response, its success will hinge hugely on the effective and widespread introduction of NCD services at the primary healthcare level. Given the complex challenges described above, a combination of coordinated interventions is required to overcome them. Any program aiming to improve NCD prevention, detection and care will need to strengthen primary care services using a holistic approach that should include the 10 components described below.

2.10.1 Raising NCD Awareness Within the Health Services

Many health program managers at provincial and district levels are not yet aware of the rapidly growing NCD problem, and consequently do not feel that NCDs have become a priority. NCD programs would therefore benefit from awareness and advocacy activities. These could consist of brief workshops organized for health service leaders, organized, for example, at a provincial level. There are useful brochures available that could be shared on such occasions such as those produced by WHO (2017c). Similarly, fact sheets from national STEPS surveys could help provide information about the situation within a specific country. Finally, both health program managers and all staff who provide OPD NCD care should be made aware of their country's national NCD control policy, if this already exists.

2.10.2 Training Healthcare Workers

A variety of approaches have been used, differing with regards to what topics should be covered, how many days such training should take and which health worker cadres should be included. In our experience from an NCD program in East Africa, it has been helpful to initially focus on the diagnosis and management of the most common NCDs rather than to strive for completeness. A five day training course is sufficient to teach the most important principles of diagnostic and case management procedures. Classroom-based teaching should be combined with some practical training delivered by experienced trainers and clinicians utilizing the opportunity of a hospital OPD NCD clinic.

The training should be as interactive as possible and cover at least the following topics:

- the growing burden of NCDs in LMICs
- the most common clinical conditions
- typical long-term complications
- NCD diagnosis
- standard treatment algorithms
- referral guidelines
- case detection strategies
- simple record keeping
- practical exercises on health education for both individual patients and in the outpatient waiting area

An example of a basic NCD training course timetable is available on the Internet and is shown in Figure 2.3.

An example of a Clinic-held NCD patient file is shown in Figure 2.4.

Naturally, medical doctors and experienced clinical officers at hospital OPD services have found it helpful to expose such staff members to the same training as provided to lower cadres or those based at peripheral health centers. The aim is to make hospital doctors aware of newly introduced standardized NCD case management procedures and to ensure that they understand the knowledge and skills now available at lower levels of healthcare providers, which would still refer to them any complicated cases, and to which they should refer patients back once stabilized.

It is important during the training to emphasize that NCDs such as diabetes, hypertension or COPD can often be successfully treated without drugs if patients can be persuaded to follow advice on life style and nutrition. Health workers have a strong tendency to immediately prescribe medicines and to spend little time on health education. Learning to give convincing life style advice requires special attention during the training course, and should be practiced, for example, using role playing.

2.10.3 Conducting Support Supervision

The most important component of health workers' NCD training actually happens after completion of the initial course. It is essential that within days after the course, health workers will be visited within their own working environment by a supervisor in order to reinforce what they have learned. Ideally this should be done by one of the course trainers. Diagnostic and therapeutic procedures are rehearsed, difficult cases discussed, and clinical records jointly reviewed. If possible, some NCD patients should be jointly attended to. Without this support health workers are likely to forget what they learned during the course. Note that just one such support visit is not sufficient; we recommend repeating it initially at monthly intervals, and later at least four times per year. A visit roster should be agreed and adhered to.

Figure 2.3 below illustrates the following:

1. NCD training course timetable
2. Checklist for support supervision
3. Clinic-held NCD patient file form
4. Patient-held NCD card
5. Scoring sheets for the evaluation of NCD services at primary care facilities

1. NCD training course for primary care health workers

NCD Course Timetable

Adapted from training courses held by the East African Chronic Disease Programme 2013 – 2017

Time	Content
Day 1	
08.30 – 09.00	Session 1: Aims, objectives and course schedule
09.00 – 09.30	Session 2: Assessment of students' current knowledge on chronic diseases
09.30 – 10.00	Session 3: How to use the training materials provided
10.00 – 10.30	Session 4: Introduction to non-communicable diseases (part 1)
10.30 – 11.00	**Refreshments**
11.00 – 11.30	Session 5: Introduction to non-communicable diseases (part 2)
11.30 – 12.15	Session 6: What is diabetes mellitus? What are its complications?
12.15 – 13.00	Session 7: How can diabetes mellitus type-2 be prevented?
13.00 – 14.00	**Lunch Break**
14.00 – 14.30	Session 8: Diagnosis of diabetes mellitus
14.30 – 15.30	Session 9: Non-pharmacological management of diabetes type 2 (How to educate patients and family members; with role playing)
15.30 – 15.50	**Refreshments**
15.50 – 16.20	Session 10: Pharmacologic management of non-insulin dependent diabetes mellitus
16.20 – 17.00	Session 11: 1. Summary and review of day 1 material: key issues to remember 2. Review of day 1 impressions and questions
Day 2	
08:30 – 09.15	Session 12: Review of day 1: what did we learn?
09.15 – 09.45	Session 13: Case management flow chart for diabetes mellitus
09.45 – 10.30	Session 14: What is hypertension? What are its complications?
10.30 – 11.00	**Refreshments**
11.00 – 11.30	Session 15: How can hypertension be prevented?
11:30 – 12:00	Session 16: Diagnosis of hypertension
12:00 – 13:00	Session 17: Non-pharmacological management of hypertension (How to educate patients and family members; with role playing)

Figure 2.3 Non-communicable disease (NCD) training course

(Continued)

13.00 – 14.00	**Lunch Break**
14.00 – 14.45	Session 18: Pharmacological management of hypertension
14.45 – 15.30	Session 19: Case management flow chart for hypertension
15.30 – 15.50	**Refreshments**
15.50 – 16.20	Session 20: Follow-up of patients with hypertension and diabetes mellitus
16.20 – 17.00	Session 21: 1. Summary and review of day 2 material: key issues to remember 2. Review of day 2 impressions and questions
Day 3	
08:30 – 09.00	Session 22: Review of day 2: what did we learn?
09.00 – 09.45	Session 23: Recording and reporting: how to use clinic-held patient files, patient cards and referral forms (part 1)
09.45 – 10.30	Session 24: Recording and reporting: how to use clinic-held patient files, patient cards and referral forms (part 2), with practical exercises
10:30 – 11.00	**Refreshments**
11.00 – 12.00	Session 25: How to educate patients waiting in the OPD area (with role playing)
12.00 – 12.30	Session 26: Case detection of hypertension (HT) and diabetes (DM) patients and of patients with HT/DM risk factors: how to screen patients
12.30 – 13.00	Session 27: Review of hypertension and diabetes: diagnosis and treatment flow charts
13.00 – 14.00	**Lunch break**
14.00 – 14.45	Session 28: What is chronic obstructive pulmonary disease (COPD)? What are its complications?
14.45 – 15.30	Session 29: What is asthma? How to distinguish it from COPD? What are its complications?
15.30 – 15.50	**Refreshments**
15.50 – 16.40	Session 30: Diagnosis of COPD and asthma
16.40 – 17.00	Session 31: 1. Summary and review of day 3 material: key issues to remember 2. Review of day 3 impressions and questions
Day 4	
08:30 – 09.00	Session 32: Review of day 3: what did we learn?
09.00 – 10.00	Session 33: Non-pharmacological management of COPD (How to educate patients and family members; with role playing)
10.00 – 10.30	Session 34: Pharmacological management of COPD
10:30 – 11.00	**Refreshments**
11.00 – 11.30	Session 35: Pharmacological management of asthma
11.30 – 12.00	Session 36: Case management flowchart for COPD and asthma
12.00 – 12.30	Session 37: Follow-up of patients with COPD and asthma
12.30 – 13.00	Session 38: Common cancers: overview

Figure 2.3 (Continued)

13.00 – 1400	Lunch break
14.00 – 14.45	Session 39: The role of primary care health facilities in cancer detection
14.45 – 15.30	Session 40: The role of primary care health facilities in cancer prevention
15.30 – 15.50	**Refreshments**
15.50 – 16.30	Session 41: How to promote linkage to care, retention and adherence
16.30 – 17.00	Session 42: 1. Summary and review of day 4 material: key issues to remember 2. Review of day 3 impressions and questions
Day 5	
08:30 – 09.00	Session 43: Review of day 4: what did we learn?
09.00 – 09.30	Session 44: Review of COPD/asthma flow chart
09.30 – 10.00	Session 45: Review of NCD diagnosis and management
10.00 – 10.30	Session 46: Review of NCD case detection
10:30 – 11.00	**Refreshments**
11.00 – 11.45	Session 47: How to integrate NCD education and case detection into mobile outreach activities
11.45 – 12.30	Session 48: Post test
12.30 – 13.30	**Lunch break**
13.30 – 16.00	Practical: Visit OPD services at selected primary care centres
Day 6	
08:30 – 13.00	Practical: Visit OPD services at selected primary care centres
13.00 – 14.00	**Lunch break**
14.00 – 15.30	Session 49: Discussion of cases seen
Day 7	
08:30 – 13.00	Practical: Visit OPD services at selected primary care centres
13.00 – 14.00	**Lunch break**
14.00 – 15.30	Session 50: Discussion of cases seen
15.30 – 15.50	**Refreshments**
15.50 – 16.20	Session 51: Preparation of 1st supervisory support visit (to be conducted within 2 weeks after the end of the training course): timing, logistics, expectations
16.20 – 16.50	Session 52: Course evaluation: participants' feedback (in writing and by discussion)
16.50 – 17.00	Closing of training course

Figure 2.3 (Continued)

2.   Checklist for support supervision visits

Health Facility_____Health Facility Code_____

Names of Supervisor(s)

_____

Date _____

Date of last supervision visit _____

1.   **Staff**

     • Number trained in CD   ☐

     • Number who still work at this facility   ☐

     • Number available/present at HF at time of visit   ☐

1   • Describe any additional staffing problems +/- suggested solutions

_____

_____

_____

_____

2.   **Treatment guidelines and tools**

     • Is the desk guide for management of hypertension and diabetes available and in use at the HF?   ☐

(1 = available and in use, 2 = available but not in use, 3 = not available)

     • Is the algorithm for management of diabetes available and in use at at the HF?   ☐

(1 = available and in use, 2 = available but not in use, 3 = not available)

     • Is the algorithm for management of hypertension available and in use at the HF?   ☐

(1 = available and in use, 2 = available but not in use, 3 = not available)

     • Is the chart for interpreting body mass index (BMI) available and in use at the HF?   ☐

(1 = available and in use, 2 = available but not in use, 3 = not available)

     • Is the patient education information leaflet available and in use at the HF?   ☐

(1 = available and in use, 2 = available but not in use, 3 = not available)

Figure 2.3 (Continued)

- Use the space below to describe any additional problems with treatment guidelines

_____

_____

_____

_____

3. **Case detection of NCDs**
    - Is screening for HT/DM being done routinely at the OPD? (Y/N)
    - Is the screening register being used appropriately (Y/N)
    - How many patients have been screened since the last visit?
    - Describe below any additional problems with screening activities

_____

_____

_____

_____

4. **Management of NCDs**
    - Does the treatment for diabetes follow the guidelines? (Y/N)
    - Does the treatment for hypertension follow the guidelines? (Y/N)
    - Please describe below any additional problems with the treatment

_____

_____

_____

_____

5. **Referral of patients**

    Examine the referral documents to ascertain the following points:
    - How many patients have been referred since the last visit?
    - List below the health facilities where the patients were referred to.

_____

_____

    - How many patients have been back-referred?
    - List below the facilities which referred patients back to this facility.

_____

Figure 2.3 (Continued)

- How many referrals did this facility receive from elsewhere? ☐

- List the health facilities from which patients were referred.

_____

_____

- How many patients were referred back from here? ☐

- Provide any additional information on problems related to referrals

_____

_____

6.   **Essential medicines**

- Availability and supply

Medicine	Physically in stock today? (Y/N)	Usually in stock? (Y/N)	Date of last supply	Source of last supply: 1=NMS, 2=Project, 3=Other
Metformin				
Glibenclamide				
Insulin				
Bendrofluazide				
Nifedipine				
Atenolol				

- Describe any problems and proposed solutions:

_____

_____

_____

_____

_____

_____

_____

Figure 2.3 (Continued)

7. **Essential equipment**
    - Availability and use of equipment

Equipment	Is equipment from project still present? (1=available and in use, 2=available but not in use, 3=not available)	Does the facility have alternative equipment? (1=available and in use, 2=available but not in use, 3=not available)
Measuring tape		
Weighing scale		
Stadiometer (for measuring height)		
Glucometer		
Blood pressure machine		

- Describe any problems and proposed solutions:

_____

_____

_____

_____

_____

8. **Documentation and Monitoring**
    - Are tools/forms available and in use? Are they being used appropriately?

Tool/Form	Available and in use? (1=available and in use, 2=available but not in use, 3=not available) If 2, or 3 discuss with staff and summarise discussion (problems and solutions) below	Being used appropriately? (Y/N) If No, discuss with staff and summarise problems and solutions below
Screening register		
Patient-held cards		
Facility-held patient files		
NCD register		
Appointment book		
Referral form		

Figure 2.3 (Continued)

- Describe any problems and proposed solutions:

_____

_____

_____

_____

_____

_____

_____

9. **Community outreach**
    - Does this facility conduct any community outreach activities? (Y/N)

    ┌─────────┐
    │   │
    └─────────┘

    - Please describe any activities conducted since last support visit:

_____

_____

_____

_____

_____

    - How frequently have these activities been performed?

_____

_____

10. **Utilization of facility**

    - Number of new patients registered with diabetes

    - Number of diabetes patients seen for follow-up (FU)

    - Number of new patients registered with hypertension

    - Number of hypertension patients seen for FU

    - Number of other NCD patients newly registered

    - Number of NCD health education sessions held

11. **Patient-held card**

Front page

Figure 2.3 (Continued)

NCD PATIENT ID CARD for presentation at health facilities

Health facility name: …………......................…

Patient number: ……….................................…

Patient name: ……………...…

Age: ……..…...years   Sex: M / F

Address: …………...…

Telephone: ……..…

Diagnoses: (please underline)

  Diabetes | Hypertension | COPD | Other

  If other, specify

  Overleaf

Date of appointment	Current treatment

Figure 2.3 (Continued)

These supervisory visits also provide an excellent opportunity to monitor the operational performance of NCD services, and to address problems that are likely to emerge over time. For this purpose, supervisors should use standardized checklists that allow a regular review of key functions and collection of essential data. As indicated above, this information should include:

- the presence and functionality of equipment
- the availability of essential drugs
- health workers' adherence to treatment guidelines and algorithms
- the number and diagnoses of NCD patients registered
- the conduct of case detection activities
- the number of patients referred and back-referred
- the number of patients reached through health education sessions in OPD waiting areas.

Any outreach activities should also be monitored with respect to NCD services provided, new cases detected and educational sessions conducted. The information collected can be compiled at a central level and used for the evaluation of the program.

Figure 2.4 shows an example of a clinic-held NCD patient file.

Patient name: .... **Patient number:** .... **Page:** ....

Health facility:		
Patient address:	Sex:	DOB:
	Clinic number:	Date of 1st visit:
Ward:	Phone:	Treatment supporter:
District:	Diagnoses:	Relationship:
		Phone:

Height:

Date Attended	Symptoms / Complications	Wt (kg)	BMI	Waist	Hip	BP	Blood glucose Fasting	Blood glucose Random	Urine Glucose	Urine Proteins	Urine Ketones	Education Given? Disease Education	Education Given? Lifestyle advice	Education Given? Education leaflet used?	TREATMENT SPECIFICS (Current medication; side effects if any)	LIFE STYLE ADVICE GIVEN (Details)	Date agreed for next visit

Complications: A, none; **B,** blood glucose > 10mmol/L; **C,** hyperosmolar coma; **D,** ketoacidosis; **E,** hypoglycemia; **F,** foot infection / ulcers; **G,** neuropathy, extremity numbness/pain; **H,** erectile dysfunction; **I,** cataract; **J,** retinopathy; **K,** nephropathy / proteinuria

Figure 2.4 Clinic-held NCD patient file

Who should perform these support supervision visits? This is ideally done by district level staff, although hospital-based supervision services also exist in some places. There are similarities here with the district tuberculosis control or the district AIDS control programs, which are in place in many countries. District medical officers should nominate NCD coordinators who will be responsible for training, supervision and monitoring of NCD control activities. It is important that the right staff members are identified for such tasks. NCD coordinators should typically be experienced clinical officers who have managerial skills and an enthusiastic interest in public health.

2.10.4 Providing Essential Diagnostic Equipment and Tests

These should focus on the most common NCDs. The minimum equipment required at peripheral health facilities include a sphygmomanometer with different cuff sizes to measure blood pressure, a glucometer with a reliable supply of strips to determine blood sugar levels, a stethoscope for the auscultation of the heart and lungs, and a tape measure to measure waist circumference for the documentation and monitoring of overweight patients, and of those with metabolic syndrome. Instead of the tape measure, a weighing scale, stadiometer and a body mass index (BMI) calculation chart are often preferred. The diagnosis of COPD can be made using a simple handheld respirometer after application of a short-acting bronchodilator. Visual acuity can be tested in diabetes patients with suspected retinopathy using a Snellen chart. Standard dipstick strips can be used to detect proteinuria when diabetic kidney disease is suspected. At district hospital OPDs, an HbA1c blood test to determine antidiabetic treatment efficacy, a urine test for microalbuminuria, and a hand-held ophthalmoscope for the diagnosis of retinopathy should be available.

2.10.5 Ensuring a Reliable Supply of Essential NCD Drugs

A huge number of drugs can be found in national 'essential' drug lists in some LMICs, often reflecting the preference that different specialists may historically have had when these lists were developed. For the routine care of most NCD patients only a few types of medicines are needed, and low-cost generic drugs should be obtained for this purpose.

The first line drug treatment of hypertension requires a thiazide (for example, bendrofluazide or hydrochlorothiazide). If an additional drug is required, a calcium channel blocker (for example, nifedipine), an acetyl choline esterase (ACE) inhibitor (for example, lisinopril) or a β-blocker (for example, atenolol) can be added.

For diabetic patients a biguanide (metformin) is required, which can if necessary be combined with a drug from the sulfonylurea group (for example, glibenclamide). Patients that require insulin injections should initially be managed at a hospital or a diabetes clinic, but once stable may be monitored at peripheral health facilities close to their home.

The list of medicines for COPD at the primary care level should include a short-acting inhaled β_2-agonist (for example, salbutamol) and/or an inhaled acetylcholine blocker (for example, ipratropium), plus an oral methylxanthine (for example, aminophylline) if necessary. Patients with acute exacerbations of COPD may require an inhaled β_2-agonist and short-term oral glucocorticosteroids (for example, prednisolone), and possibly a broad-spectrum antibiotic treatment (Ait-Khaled et al. 2001; GOLD 2017a, b). Treatment with bronchodilators in nebulizers may be given at hospitals.

2.10.6 Introducing Standardized NCD Case Management Algorithms

Most cases of uncomplicated NCDs can be managed with life style adaptation and a small selection of inexpensive drugs that are applied according to simple standardized treatment guidelines. These guidelines can be condensed into treatment algorithms that should be available in each OPD consultation room. Such standardized algorithms are based on easy-to-follow diagnostic steps and treatment principles, and are an essential tool for any NCD control program (Figure 2.5). Treatment algorithms should be accompanied by a more detailed desk guide so that health workers can look for additional guidance if needed. Examples of NCD desk guides can be found on the internet, for example, TDA (2013).

2.10.7 Treating Patients Close to Their Residence: Effective Referral and Back-referral

A major challenge for NCD patients is the frequent lack of treatment opportunities within their community, particularly in rural areas, and this is a major problem for both linkage and adherence to care. Consequently, patients remain either untreated or often

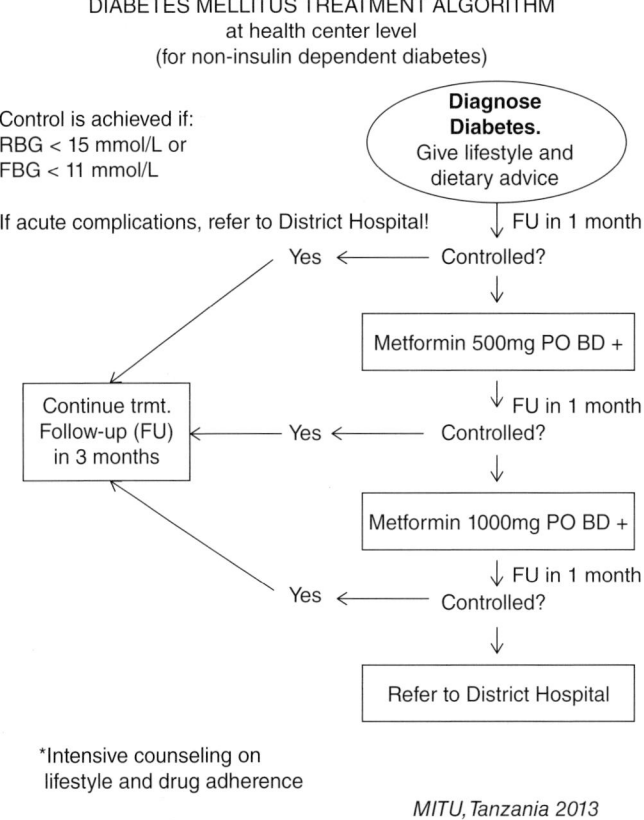

Figure 2.5 An example of a treatment algorithm for diabetes mellitus used in primary care facilities in an NCD control project from a resource-restricted setting in East Africa.

interrupt their medication, and in any event have a high risk of long-term complications. Those that do link to care face travel costs and loss of income whilst away. A major objective of NCD control programs is therefore to strengthen peripheral health facilities so that they can provide essential services close to patients' area of residence. The referral of patients to higher levels of care should be reserved for those with complications or who are difficult to treat. Importantly, care providers at hospitals must back-refer patients once they are stabilized. This in turn requires some cooperation between the facilities involved, including communication by brief but comprehensive referral and back-referral notes, ideally using a standard form. This need for an effective collaboration between smaller peripheral health facilities and the hospital NCD clinic is a good reason to include staff from both levels of healthcare in the same NCD training program.

An alternative strategy to ensuring user-friendly NCD services is the allocation of dedicated community nurses who regularly contact patients at or near their homes to monitor their health, provide drug refills and conduct a brief examination. Observations are recorded using a brief standardized checklist. If anything unusual is observed, the nurse will contact a facility-based clinician by mobile phone to seek advice. If necessary, the patient is sent to hospital for further examination. This approach has been successfully implemented, for example, in a NCD control program in South Africa (Coleman et al. 1998). A similar strategy has been used for the chronic care of HIV patients on ART in Uganda. Interestingly, such models compare favorably with purely facility-based care both in terms of clinical effectiveness and costs (Jaffar et al. 2009).

2.10.8 Introducing Health Education and NCD Case Detection

The majority of NCD patients are not aware of their condition (Beagley et al. 2014; Hendriks et al. 2012; Kavishe et al. 2015), and are usually first seen at health services once they develop complications which are often lethal. For example, a recent analysis of inpatient records from a large tertiary hospital in northern Tanzania revealed that hypertension has become the second most common cause of mortality among patients admitted to a medical ward, surpassed only by HIV disease (Peck et al. 2013). Yet almost all such cases could in principle be prevented, provided that they were detected early. It is of great public health importance that routine primary care services make a major effort to identify patients with undetected NCDs as early as possible and ensure that they are monitored and receive adequate treatment.

Our own experience with the introduction of screening services at general OPD clinics both at hospitals and smaller health facilities has been encouraging. Patients waiting for consultation for any kind of ailment were invited to listen to a health talk on common NCDs. Afterwards all were offered to have their blood pressure measured. Individuals aged 40 years and above, those who were obviously overweight and anybody with a family history of diabetes were screened for an elevated random blood glucose level. Patients with suspected hypertension or diabetes were then asked to return for a confirmatory investigation on a subsequent day. For people with suspected diabetes this included a fasting blood glucose test. A similar strategy was applied at antenatal clinics. With his approach large numbers of individuals were screened, and many unknown NCD cases detected and subsequently followed up.

2.10.9   Establishing an Effective Recording System

Keeping carefully completed patient records is essential for all patients attending a health facility, but even more so for NCD patients, as they will be in contact with the health system many times, possibly lifelong, and previous records are needed to see whether a chronic condition improves or worsens. Unfortunately, health workers are already burdened with much paper work, so keeping records brief and simple is important. The widespread system in LMIC primary care facilities of recording subsequent patients in large ledger books is not really helpful as with this system it is always difficult to trace previous entries.

The problem can be solved by learning from HIV care programs which introduced individual clinic-held patient files, marked with the name of the patient or a unique individual number. The minimum set of data to be monitored and recorded at the consultation visit should include the name, gender, age, diagnosis, date of last visit, clinical information at each visit (blood pressure, blood sugar level, etc.), current symptoms and current treatment. Patients themselves can be provided with a small card with identifying information, diagnoses and current treatment. The card enables rapid identification of the respective clinic-held patient file and is also helpful if the patient presents to a different clinic. Examples of a standard clinic-held patient NCD file and of a patient-held NCD card are available online (see Figures 2.3 and 2.4)

2.10.10   Integrating NCD Control Measures into Outreach Activities

Many health facilities regularly reach out to the community, for example, for vaccination or antenatal services. On these occasions they often collaborate with community health workers or local village health teams. Such outreach activities offer a unique opportunity to also conduct public information campaigns on NCDs. The objective is to raise awareness on NCDs in the general population and to address erroneous beliefs and myths, but also to integrate case detection and treatment into other ongoing services. The concept of NCDs is still alien to many people, including the need to make life style changes for primary or secondary NCD prevention, or for long-term treatment. Misconceptions are frequent and include beliefs that NCDs result from ill-intended supernatural influences, or that modern medicines are ineffective or even harmful. Not uncommonly NCD patients even experience stigmatization (Nnko et al. 2015). It is therefore important to use such outreach activities to educate the general public on NCDs.

2.11   The Role of Primary Healthcare Services in Cancer Prevention and Care

The effective management of cancers requires specialist skills and is costly. Most LMICs have only a few national centers that provide such services. Unfortunately, cancer patients often present at a late stage, and the mortality of cancer in LMICs is high. For patients and their families, most cancer diagnoses represent a disaster, resulting in catastrophic healthcare expenditures and poverty (ASEAN Cancer Action Study Group 2017).

Yet for some common cancers, primary healthcare services can play an important role in prevention or early detection.

Among women one of the most frequent tumors is cancer of the uterine cervix. In high income countries screening for the detection of early cancerous lesions is performed using cytology or human papilloma virus (HPV) tests. In LMICs these are not routinely available. However, a simple and reasonably effective alternative screening strategy is the visual inspection with acetic acid (VIA) or Lugol's iodine (VILI) applied during a speculum examination of the cervix. The technique is inexpensive and does not require much training (Fokom-Domgue et al. 2015). It can be applied by nurse midwives at primary care clinics provided women are made aware that such service is locally available. Women with a positive screening result are then referred for gynecological treatment. It is incomprehensible that this simple and low-cost screening strategy has not yet been widely introduced in LMICs.

Cervical cancer can be reliably prevented by vaccination against infection with high-risk strains of the HPV (WHO 2014b). HPV vaccination is currently being introduced in many LMICs (Gallagher et al. 2017), and once this is well established it can be expected that primary care services will play a pivotal role in achieving wide coverage by vaccinating adolescent girls. Two doses are recommended, given six months apart, but a three-dose schedule is also in use. Similarly, a large proportion of liver cancers can be prevented through vaccination against hepatitis B virus infection (Nelson, Easterboork and McMahorn 2016). This vaccine is given at birth, and three further doses are provided jointly with routine DPT vaccinations.

Since the onset of the AIDS epidemic, Kaposi's sarcoma has become another very common cancer in LMICs, affecting men and women at a young age (Forman and Ferlay 2014). Facility or home-based HIV counseling and testing provided through primary care services can help to diagnose HIV infection early, and to initiate ART before the patient's immune system is seriously damaged (Ruzagira et al. 2017). Effective ART treatment prevents HIV disease progression and reliably reduces the risk of developing Kaposi's sarcoma (Bohlius et al. 2014).

In addition, once primary care health workers have adopted the basic principles of care for common non-cancerous NCDs, they can also be trained in the recognition of symptoms and signs that are suggestive of cancers of the breast, skin, prostate and the intestinal tract. The health workers' role will then be to refer patients with such symptoms to specialist care as early as possible. Promising research is underway toward the development of technologies for the diagnosis and treatment of cancers in LMICs, including early cancer diagnosis at primary care facilities (Pearlman et al. 2016).

Figure 2.6 (a) Evaluation of NCD services at 75 primary care facilities in Uganda and Tanzania that participated in an NCD intervention trial. The figure shows facility performance scores (0–100) averaged over intervention and control units, by country. The score is a composite measure of NCD service availability and readiness including health workers' knowledge. (b) Evaluation of NCD services at 75 primary care facilities in Uganda and Tanzania that participated in an NCD intervention trial. The figure shows the proportion of NCD patients adequately managed, by country and health facility type. The assessment was based on a composite measure to determine the quality of NCD care received by a total of 300 individual patients.

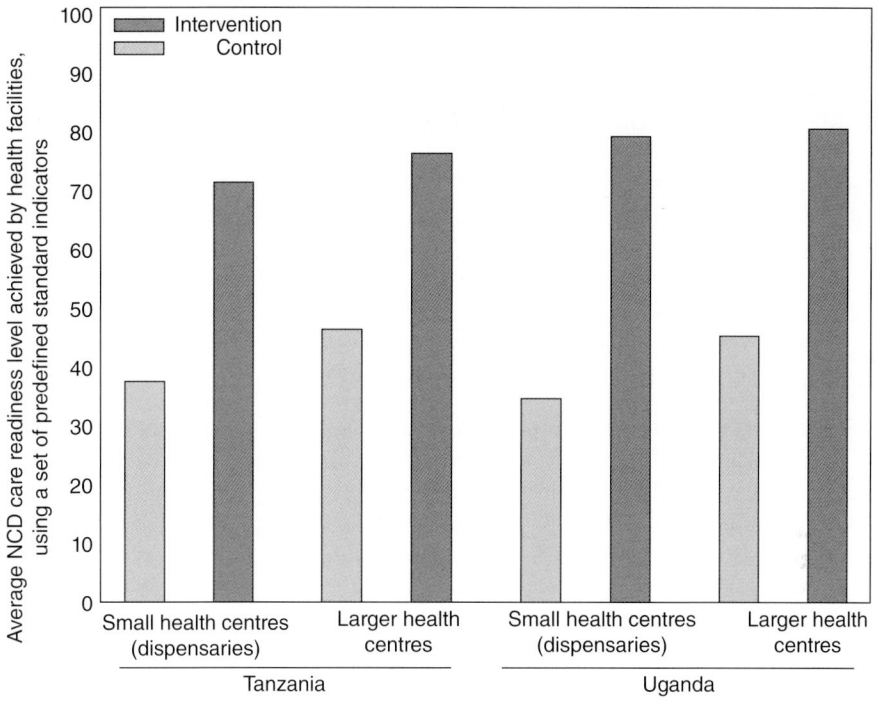

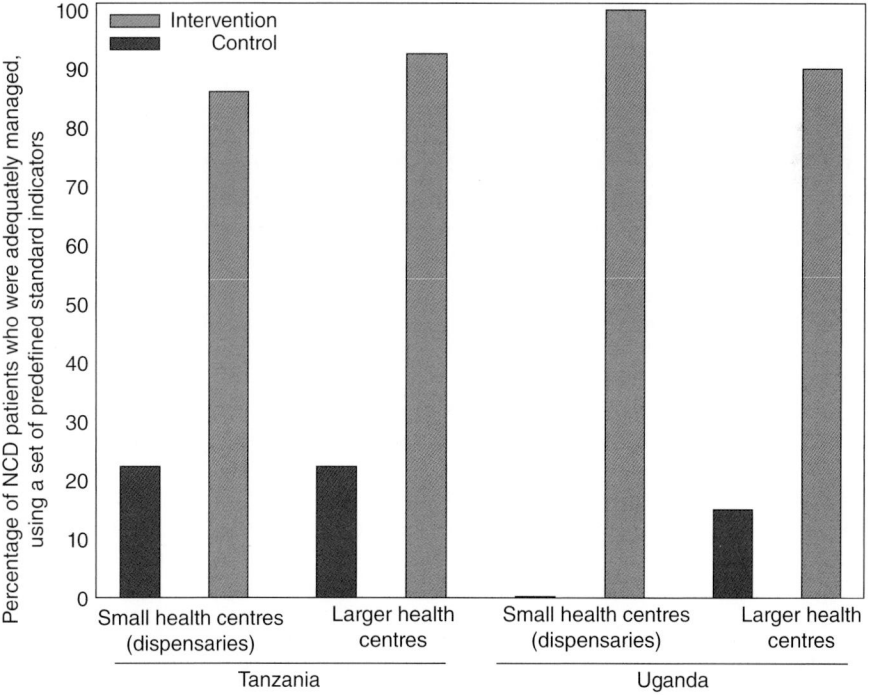

2.12 Evaluating Programmes to Strengthen NCD Services at Primary Care Level

There are different strategies to evaluate the effectiveness of programs that aim to improve NCD services at primary care level. One option is to apply the WHO Service Availability and Readiness (SARA) tool (WHO 2013c). However, it would be ideal to measure the effectiveness of an intervention program in a quantifiable way that would also determine the quality of care actually provided to a selection of individual NCD patients.

In an NCD intervention trial conducted at primary care facilities in East Africa with funding from the Medical Research Council UK and with material support from the TDA (Trial registration number ISRCTN27340385), the investigators designed a system to evaluate two important aspects of NCD services:

1) Service readiness assessed by physical inspection, including health workers' knowledge on NCDs assessed through a knowledge test
2) Quality of care received by randomly selected patients, assessed through interviews, examination and evaluation of patient records. For each of these components, a scoring system was an a priori definition against which health facilities were evaluated. Such assessment may be applied before and after the intervention, or, as in the case of this trial, through a randomized controlled design. Examples of the scoring sheets used in this NCD control project are available online (see Figures 2.3 and 2.4)

The trial demonstrated that an affordable and sustainable intervention comprising the 10 intervention components described above can be highly effective in improving the response to NCDs at peripheral health facilities (manuscript submitted) (Figure 2.6a and b).

The overall scores achieved in the intervention group of health facilities as compared to the control group, based on the physical readiness of services with respect to the availability of equipment, medicines, policies and guidelines and utilization of services and health workers' knowledge on NCDs, increased from approximately 40/100 to more than 75/100, and the proportion of patients with hypertension or diabetes who were adequately managed more than quadrupled from approximately 20 to about 90%. The intervention was also cost-effective when analyzing the costs per case adequately treated. Based on these encouraging results, the intervention has subsequently been extended to the control facilities.

2.13 Conclusions

Non-communicable diseases represent a major burden of disease in the LMICs. This is slowly being realized and methods are being put in place to diagnose and treat them early.

Bibliography

Abegunde, D.O., Mathers, C.D., Adam, T. et al. (2007). The burden and costs of chronic diseases in low-income and middle-income countries. *Lancet* 370 (9603): 1929–1938.

Addo, J., Agyemang, C., Smeeth, L. et al. (2012). A review of population-based studies on hypertension in Ghana. *Ghana Med. J.* 46 (2 Suppl): 4–11.

Ait-Khaled, N., Enarson, D., and Bousquet, J. (2001). Chronic respiratory diseases in developing countries: the burden and strategies for prevention and management. *Bull. World Health Organ.* 79: 971–979.

Alberti, K.G. and Zimmet, P.Z. (1998). Definition, diagnosis and classification of diabetes mellitus and its complications. Part 1: diagnosis and classification of diabetes mellitus provisional report of a WHO consultation. *Diabet. Med.* 15 (7): 539–553.

Alberti, K.G., Eckel, R.H., Grundy, S.M. et al. (2009). Harmonizing the metabolic syndrome: a joint interim statement of the International Diabetes Federation Task Force on Epidemiology and Prevention; National Heart, Lung, and Blood Institute; American Heart Association; World Heart Federation; International Atherosclerosis Society; and International Association for the Study of Obesity. *Circulation* 120 (16): 1640.

Ali, M.K., Rabadán-Diehl, C., Flanigan, J. et al. (2013). Systems and capacity to address noncommunicable diseases in low- and middle-income countries. *Sci. Transl. Med.* 5: 181.

ASEAN Cancer Action Study Group (2017). Policy and priorities for national cancer control planning in low- and middle-income countries: lessons from the Association of Southeast Asian Nations (ASEAN) costs in oncology prospective cohort study. *Eur. J. Cancer* 74: 26–37.

Beaglehole, R., Epping-Jordan, J., and Patel, V. (2008). Improving the prevention and management of chronic disease in low-income and middle-income countries: a priority for primary health care. *Lancet* 372: 940–949.

Beagley, J., Guariguata, L., Weil, C., and Motala, A. (2014). Global estimates of undiagnosed diabetes in adults. *Diabetes Res. Clin. Pract.* 103 (2): 150–160.

Beevers, D.G. and Prince, J.S. (1991). Some recent advances in non-communicable diseases in the tropics. 1. Hypertension: an emerging problem in tropical countries. *Trans. R. Soc. Trop. Med. Hyg.* 85 (3): 324–326.

Bloom, D.E., Cafiero, E.T., Jané-Llopis, E. et al. (2011). *The Global Economic Burden of Non-communicable Diseases*. Geneva: World Economic Forum http://apps.who.int/medicinedocs/documents/s18806en/s18806en.pdf.

Bohlius, J., Valeri, F., Maskew, M. et al. (2014). Kaposi's sarcoma in HIV-infected patients in South Africa: multicohort study in the antiretroviral therapy era. *Int. J. Cancer* 135 (11): 2644–2652.

Bray, F., Ferlay, J., Laversanne, M. et al. (2015). Cancer incidence in five continents: inclusion criteria, highlights from volume X and the global status of cancer registration. *Int. J. Cancer* 137 (9): 2060–2071.

Brouwer, E.D., Watkins, D., Olson, Z. et al. (2015). Provider costs for prevention and treatment of cardiovascular and related conditions in low- and middle-income countries: a systematic review. *BMC Public Health* 15: 1183.

Brown, J.B., Ramaiya, K., Besançon, S. et al. (2014). Use of medical services and medicines attributable to diabetes in Sub-Saharan Africa. *PLoS One* 9 (9): e106716.

Buist, A.S., McBurnie, M.A., Vollmer, W.M. et al. (2007). International variation in the prevalence of COPD (the BOLD Study): a population-based prevalence study. *Lancet* 370 (9589): 741–750.

Buse, K., Tanaka, S., and Hawkes, S. (2017). Healthy people and healthy profits? Elaborating a conceptual framework for governing the commercial determinants of non-communicable diseases and identifying options for reducing risk exposure. *Glob. Health* 13 (1): 34.

Coleman, R., Gill, G., and Wilkinson, D. (1998). Noncommunicable disease management in resource-poor settings: a primary care model from rural South Africa. *Bull. World Health Organ.* 76 (6): 633–640.

Daar, A.S., Singer, P.A., Persad, D.L. et al. (2007). Grand challenges in chronic non-communicable diseases. *Nature* 450 (7169): 494–496.

Dagenais, G., Gerstein, H.C., Zhang, X. et al. (2016). Variations in diabetes prevalence in low-, middle-, and high-income countries: results from the prospective urban and rural epidemiological study. *Diabetes Care.* 39 (5): 780–787.

Danquah, I., Bedu-Addo, G., and Mockenhaupt, F.P. (2010). Type 2 diabetes mellitus and increased risk for malaria infection. *Emerg. Infect. Dis.* 16: 1601–1604.

Davila, E.P., Suleiman, Z., Mghamba, J. et al. (2015). Non-communicable disease training for public health workers in low- and middle-income countries: lessons learned from a pilot training in Tanzania. *Int. Health* 7 (5): 339–347.

Dillon, D.G., Gurdasani, D., Riha, J. et al. (2013). Association of HIV and ART with cardiometabolic traits in sub-Saharan Africa: a systematic review and meta-analysis. *Int. J. Epidemiol.* 42 (6): 1754–1771.

Duffy, M., Ojikutu, B., Andrian, S. et al. (2017). Non-communicable diseases and HIV care and treatment: models of integrated service delivery. *Trop. Med. Int. Health* 22 (8): 926–937.

Engelgau, M., Rosenhouse, S., El-Saharty, S., and Mahal, A. (2011). The economic effect of noncommunicable diseases on households and nations: a review of existing evidence. *J. Health Commun.* 16 (Suppl 2): 75–81.

Eslick, G.D. (2006). Helicobacter pylori infection causes gastric cancer a review of the epidemiological, meta-analytic, and experimental evidence. *World J. Gastroenterol.* 12: 2991–2999.

Ferlay, J., Soerjomataram, I., Dikshit, R. et al. (2015). Cancer incidence and mortality worldwide: sources, methods and major patterns in GLOBOCAN 2012. *Int. J. Cancer* 136 (5): E359–E386.

Finney, L.J., Feary, J.R., Leonardi-Bee, J. et al. (2013). Chronic obstructive pulmonary disease in sub-Saharan Africa: a systematic review. *Int. J. Tuberc. Lung Dis.* 17 (5): 583–589.

Fokom-Domgue, J., Combescure, C., Fokom-Defo, V. et al. (2015). Performance of alternative strategies for primary cervical cancer screening in sub-Saharan Africa: systematic review and meta-analysis of diagnostic test accuracy studies. *BMJ* 351: h3084.

Forman, D. and Ferlay, J. (2014). The global and regional burden of cancer. In: *World Cancer Report* (ed. B. Stewart and C. Wild), 16–27. Lyon, France: International Agency for Research on Cancer Available at http://publications.iarc.fr/Non-Series-Publications/ World-Cancer-Reports/World-Cancer-Report-2014.

Gallagher, K., Howard, N., Kabakama, S. et al. (2017). Lessons learnt from human papillomavirus (HPV) vaccination in 45 low- and middle-income countries. *PLoS One* 12 (6): e0177773.

Garcia, P., Carcamo, C., Garnett, G. et al. (2012). Improved STD syndrome management by a network of clinicians and pharmacy workers in Peru: the PREVEN network. *PLoS One* 9 (10): e47750.

GBD 2013 (2015). Global burden of diseases risk factors collaborators. Global, regional, and national comparative risk assessment of 79 behavioural, environmental and occupational, and metabolic risks or clusters of risks in 188 countries, 1990–2013: a systematic analysis for the global burden of disease study 2013. *Lancet* 386 (10010): 2287–2323.

GBD 2015 (2016a). Mortality and causes of death collaborators. Global, regional, and national life expectancy, all-cause mortality, and cause-specific mortality for 249 causes of death, 1980–2015: a systematic analysis for the global burden of disease study 2015. *Lancet* 388: 1459–1544.

GBD 2015 (2016b). Risk factors collaborators. Global, regional, and national comparative risk assessment of 79 behavioural, environmental and occupational, and metabolic risks or clusters of risks, 1990–2015: a systematic analysis for the global burden of disease study 2015. *Lancet* 388 (10053): 1659–1724.

Global Asthma Report (2014). Auckland, New Zealand: Global Asthma Network. ISBN: 978-0-473-29125-9 (PRINT) | 978-0-473-29126-6 (ELECTRONIC) http://www.globalasthmareport.org/burden/burden.php

GOLD (2017a). Global Initiative for Chronic Obstructive Lung Disease. Global strategy for the diagnosis, management and prevention of chronic obstructive lung disease. 2017 report. http://goldcopd.org/gold-2017-global-strategy-diagnosis-management-prevention-copd

GOLD (2017b). Global Initiative for Chronic Obstructive Lung Disease. Pocket guide to COPD diagnosis, management and prevention. A guide for health care professionals. 2017 Edition. http://goldcopd.org/wp-content/uploads/2016/12/wms-GOLD-2017-Pocket-Guide.pdf

Harries, A.D., Zachariah, R., Jahn, A. et al. (2009). Scaling up antiretroviral therapy in Malawi-implications for managing other chronic diseases in resource-limited countries. *JAIDS* 52: S14–S16.

Hendriks, M.E., Wit, F.W., Roos, M.T. et al. (2012). Hypertension in Sub-Saharan Africa: cross-sectional surveys in four rural and urban communities. *PLoS One* 7 (3): e32638.

Hwang, C.K., Han, P.V., Zabetian, A. et al. (2012). Rural diabetes prevalence quintuples over twenty-five years in low- and middle-income countries: a systematic review and meta-analysis. *Diabetes Res. Clin. Pract.* 96 (3): 271–285.

Irazola, V.E., Gutierrez, L., Bloomfield, G. et al. (2016). Hypertension prevalence, awareness, treatment, and control in selected LMIC communities: results from the NHLBI/UHG network of centers of excellence for chronic diseases. *Glob. Heart* 11 (1): 47–59.

ISN (2013). International Society of Nephrology. KDIGO 2012 Clinical Practice Guideline for the Evaluation and Management of Chronic Kidney Disease. 2013: 1: 5–8 http://www.kdigo.org/clinical_practice_guidelines/pdf/CKD/KDIGO_2012_CKD_GL.pdf

Jaffar, S., Amuron, B., Foster, S. et al. (2009). Rates of virological failure in patients treated in a home-based versus a facility-based HIV-care model in Jinja, SE Uganda: a cluster-randomised equivalence trial. *Lancet* 374: 2080–2089.

Jaspers, L., Colpani, V., Chaker, L. et al. (2015). The global impact of non-communicable diseases on households and impoverishment: a systematic review. *Eur. J. Epidemiol.* 30 (3): 163–188.

Jha, V., Garcia-Garcia, G., Iseki, K. et al. (2013). Chronic kidney disease: global dimension and perspectives. *Lancet* 382 (9888): 260–272.

Jha, P. and Peto, R. (2014). Global effects of smoking, of quitting, and of taxing tobacco. *N. Engl. J. Med.* 370: 60–68.

Joint National Committee. (2004). National high blood pressure education program. The seventh report of the Joint National Committee on prevention, detection, evaluation, and treatment of high blood pressure. 2004. U.S. Department of Health and Human

Services. National Institutes of Health. National Heart, Lung and Blood Institute. http://www.nhlbi.nih.gov/files/docs/guidelines/jnc7full.pdf

Katende, D., Mutungi, G., Baisley, K. et al. (2015). Readiness of Ugandan health services for the management of outpatients with chronic diseases. *Trop. Med. Int. Health* 20: 1385–1395.

Kavishe, B., Biraro, S., Baisley, K. et al. (2015). High prevalence of hypertension and of risk factors for non-communicable diseases (NCDs): a population based cross-sectional survey of NCDs and HIV infection in Northwestern Tanzania and Southern Uganda. *BMC Med.* 13 (1): 126.

Kengne, A.P., Echouffo-Tcheugui, J.B., Sobngwi, E., and Mbanya, J.C. (2013). New insights on diabetes mellitus and obesity in Africa-part 1: prevalence, pathogenesis and comorbidities. *Heart* 99 (14): 979–983.

King, H. and Rewers, M. (1993). Diabetes in adults is now a third world problem. World Health Organization Ad Hoc Diabetes Reporting Group. *Ethn. Dis.* 3 (Suppl): S67–S74.

Lewington, S., Clarke, R., Qizilbash, N. et al. (2002). Age-specific relevance of usual blood pressure to vascular mortality: a meta-analysis of individual data for one million adults in 61 prospective studies. *Lancet* 360 (9349): 1903.

Ley, S.H., Hamdy, O., Mohan, V., and Hu, F.B. (2014). Prevention and management of type 2 diabetes: dietary components and nutritional strategies. *Lancet* 383 (9933): 1999–2007.

Lopez, A.D., Shibuya, K., Rao, C. et al. (2006). Chronic obstructive pulmonary disease: current burden and future projections. *Eur. Respir. J.* 27 (2): 397–412.

Lunyera, J., Mohottige, D., Von Isenburg, M. et al. (2016). CKD of uncertain etiology: a systematic review. *Clin. J. Am. Soc. Nephrol.* 11 (3): 379–385.

Mannava, P., Abdullah, A., James, C. et al. (2015). Health systems and noncommunicable diseases in the Asia-Pacific region: a review of the published literature. *Asia Pac. J. Public Health* 27 (2): NP1–NP19.

Mannino, D.M. and Buist, A.S. (2007). Global burden of COPD: risk factors, prevalence, and future trends. *Lancet* 370 (9589): 765–773.

Marks, G., Pearce, N., Strachan, D. and Asher, I. 2014. Global Burden of Disease due to Asthma; in: Global Asthma Report, chapter 2. ISBN: 978-0-473-29125-9 (PRINT) | 978-0-473-29126-6 http://www.globalasthmareport.org/burden/burden.php

Mathabire Rücker, S.C., Tayea, A., Bitilinyu-Bangoh, J. et al. (2017). High rates of hypertension, diabetes, elevated low-density lipoprotein cholesterol, and cardiovascular disease risk factors in HIV-infected patients: results from a cross-sectional survey in Malawi. *AIDS* https://doi.org/10.1097/QAD.0000000000001700.

Mathers, C.D. and Loncar, D. (2006). Projections of global mortality and burden of disease from 2002 to 2030. *PLoS Med.* 3 (11): e442.

MoH Cambodia. (2013). National Strategic Plan for the Prevention and Control of Noncommunicable Diseases 2013–2020. Ministry of Health, Royal Government of Cambodia. National Strategic Plan for the Prevention and Control of Noncommunicable Diseases 2013. http://www.iccp-portal.org/system/files/plans/KHM_B3_NSP-NCD %202013-2020_Final%20approved.pdf

MoH Ghana. (2012). Republic of Ghana. Ministry of Health. National policy for the prevention and control of chronic non-communicable diseases in Ghana. 2012. http://www.pascar.org/uploads/files/Ghana_-_National_Policy-_Prevention_and_control_of_chronic_non-communicable_diseases.pdf

MoH Trinidad and Tobago. (2017). Ministry of Health. National strategic plan for the prevention and control of non communicable diseases. Trinidad and Tobago 2017–2021 https://www.psi.org/publication/national-strategic-plan-for-the-prevention-and-control-of-non-communicable-diseases-Trinidad-and-Tobago-2017-2021/

Nanditha, A., Ma, R.C., Ramachandran, A. et al. (2016). Diabetes in Asia and the Pacific: implications for the global epidemic. *Diabetes Care* 39 (3): 472–485.

Nelson, N., Easterbrook, P., and McMahorn, B. (2016). Epidemiology of hepatitis B virus infection and impact of vaccination on disease. *Clin. Liver Dis.* 20 (4): 607–628.

Nnko, S., Bukenya, D., Kavishe, B. et al. (2015). Chronic diseases in north-west Tanzania and southern Uganda. Public perceptions of terminologies, aetiologies, symptoms and preferred management. *PLoS One* 10: e0142194.

Nyarko, K.M., Ameme, D.K., Ocansey, D. et al. (2016). Capacity assessment of selected health care facilities for the pilot implementation of a package for essential non-communicable diseases (PEN) intervention in Ghana. *Pan Afr. Med. J.* 25 (Suppl 1): 16.

Ofori-Asenso, R., Agyeman, A.A., and Laar, A. (2017). Metabolic syndrome in apparently "healthy" Ghanaian adults: a systematic review and meta-analysis. *Int. J. Chronic Dis.* 2017: 2562374.

Omran, A. (1971). The epidemiologic transition. A theory of the epidemiology of population change. *Milbank Mem. Fund Q.* 49 (4): 509–538.

O'Neill, K., Takane, M., Sheffel, A. et al. (2013). Monitoring service delivery for universal health coverage: the service availability and readiness assessment. *Bull. World Health Organ.* 91 (12): 923–931.

Oni, T. and Unwin, N. (2015). Why the communicable/non-communicable disease dichotomy is problematic for public health control strategies: implications of multimorbidity for health systems in an era of health transition. *Int. Health.* 7 (6): 390–399.

Pakhare, A., Kumar, S., Goyal, S., and Joshi, R. (2015). Assessment of primary care facilities for cardiovascular disease preparedness in Madhya Pradesh, India. *BMC Health Serv. Res.* 15: 408.

Patel, S., Ali, M., Alam, D. et al. (2016). Obesity and its relation with diabetes and hypertension: a cross-sectional study across four low- and middle-income country regions. *Glob. Heart* 11 (1): 71–79.e4.

Pati, S., Agrawal, S., Swain, S. et al. (2014). Non communicable disease multimorbidity and associated health care utilization and expenditures in India: cross-sectional study. *BMC Health Serv. Res.* 14: 451.

Pearlman, P., Divi, R., Gwede, M. et al. (2016). The National Institutes of Health Affordable Cancer Technologies Program: improving access to resource-appropriate technologies for cancer detection, diagnosis, monitoring, and treatment in low- and middle-income countries. *IEEE J. Transl. Eng. Health Med.* 4: 2800708.

Peck, R.N., Green, E., Mtabaji, J. et al. (2013). Hypertension-related diseases as a common cause of hospital mortality in Tanzania: a 3-year prospective study. *J. Hypertens.* 31 (9): 1806–1811.

Peck, R., Mghamba, J., Vanobberghen, F. et al. (2014). Preparedness of Tanzanian health facilities for outpatient primary care of hypertension and diabetes: a cross-sectional survey. *Lancet Glob. Health* 2 (5): e285–e292. Including access to supplementary online material.

Plummer, M., de Martel, C., Vignat, J. et al. (2016). Global burden of cancers attributable to infections in 2012: a synthetic analysis. *Lancet Glob. Health* 4 (9): e609–e616.

Poulter, N., Kaytee, K., Hopwood, B. et al. (1984). Blood pressure and associated factors in a rural Kenyan community. *Hypertension* 6: 810–813.

Powles, J., Fahimi, S., Micha, R. et al. (2013). Global, regional and national sodium intakes in 1990 and 2010: a systematic analysis of 24 h urinary sodium excretion and dietary surveys worldwide. *BMJ Open* 3 (12): e003733.

Ramachandran, A., Ma, R.C., and Snehalatha, C. (2010). Diabetes in Asia. *Lancet* 375 (9712): 408–418.

Ranasinghe, P., Mathangasinghe, Y., Jayawardena, R. et al. (2017). Prevalence and trends of metabolic syndrome among adults in the Asia-pacific region: a systematic review. *BMC Public Health* 17 (1): 101.

Rheeder, P., Morris-Paxton, A., Ewing, R., and Woods, D. (2017). The noncommunicable disease outcomes of primary healthcare screening in two rural subdistricts of the Eastern Cape Province, South Africa. *Afr. J. Prim. Health Care Fam. Med.* 9 (1): 1466.

Riza, A.L., Pearson, F., Ugarte-Gil, C. et al. (2014). Clinical management of concurrent diabetes and tuberculosis and the implications for patient services. *Lancet Diabetes Endocrinol.* 2: 740–753.

Ruzagira, E., Grosskurth, H., Kamali, A., and Baisley, K. (2017 Oct). Brief counselling after home-based HIV counselling and testing strongly increases linkage to care: a cluster-randomised trial in Uganda. *J. Int. AIDS Soc.* 20 (2): https://doi.org/10.1002/jia2.25014.

de Serres, F. and Blanco, I. (2014). Role of alpha-1 antitrypsin in human health and disease. *J. Intern. Med.* 276 (4): 311–335.

Settumba, S., Sweeney, S., Seeley, J. et al. (2015). The health system burden of chronic disease care: an estimation of provider costs of selected chronic diseases in Uganda. *Trop. Med. Int. Health* 20 (6): 781–790.

Si, Y., Zhou, Z., Su, M. et al. (2017). Catastrophic healthcare expenditure and its inequality for households with hypertension: evidence from the rural areas of Shaanxi Province in China. *Int. J. Equity Health* 16 (1): 27.

Siddharthan, T., Ramaiya, K., Yonga, G. et al. (2015). Noncommunicable diseases in East Africa: assessing the gaps in care and identifying opportunities for improvement. *Health Aff. (Millwood).* 34 (9): 1506–1513.

Stanifer, J.W., Jing, B., Tolan, S. et al. (2014). The epidemiology of chronic kidney disease in sub-Saharan Africa: a systematic review and meta-analysis. *Lancet Glob. Health* 14: 1–8.

Steele, B., Thomas, C., Henley, J. et al. (2017). Vital signs: trends in incidence of cancers associated with overweight and obesity — United States, 2005–2014. *CDC Morb. Mortal. Wkly. Rep. (MMWR)* 66 (39): 1052–1058.

TDA. (2013). Tanzania Diabetes Association; World Diabetes Foundation; United Republic of Tanzania, Ministry of Health and Social Welfare. NCD – Case Management desk guide. 2013 https://www.worlddiabetesfoundation.org/sites/default/files/NCD%20Desk%20Guide%20Tanzania.pdf

TDA. (2014). Tanzania Diabetes Association; United Republic of Tanzania. Ministry of Health and Social Welfare. - Cardiovascular Disease, Type 2 Diabetes, Obesity, Cancer, COPD and Hyperlipidaemia Care. - Case management training modules. 2014 https://www.worlddiabetesfoundation.org/sites/default/files/Tanzania_NCD%20Training%20Manuals%202014.pdf

Torre, L., Siegel, R., Ward, E., and Jemal, A. (2016). Global cancer incidence and mortality rates and trends - an update. *Cancer Epidemiol. Biomark. Prev.* 25 (1): 16–27.

UN. (2011). Political Declaration of the High-level Meeting of the General Assembly on the Prevention and Control of Non-communicable Diseases http://www.who.int/nmh/events/un_ncd_summit2011/political_declaration_en.pdf?ua=1

UN. (2014). Outcome document of the high-level meeting of the General Assembly on the comprehensive review and assessment of the progress achieved in the prevention and control of non-communicable diseases. http://www.who.int/nmh/events/2014/a-res-68-300.pdf

UNIATF. (2017). The United Nations Interagency Task Force on the Prevention and Control of Non-communicable Diseases. Working together for health and development. http://www.who.int/ncds/un-task-force/working-together-adaptation.pdf?ua=1

UNICEF, WHO, World Bank (2015). *Levels and Trends in Child Malnutrition: UNICEF-WHO-World Bank Joint Child Malnutrition Estimates*. New York: UNICEF WHO, Geneva; World Bank, Washington, DC. http://www.who.int/nutgrowthdb/jme_brochure2016.pdf.

Van Minh, H., Do, Y.K., Bautista, M.A., and Tuan Anh, T. (2014). Describing the primary care system capacity for the prevention and management of non-communicable diseases in rural Vietnam. *Int. J. Health Plann. Manag.* 29 (2): e159–e173.

Vernon, M.K., Wiklund, I., Bell, J.A. et al. (2012). What do we know about asthma triggers? A review of the literature. *J. Asthma* 49 (10): 991–998.

Wangchuk, D., Virdi, N.K., Garg, R. et al. (2014). Package of essential noncommunicable disease (PEN) interventions in primary health-care settings of Bhutan: a performance assessment study. *WHO South East Asia J. Public Health* 3 (2): 154–160.

Whincup, P.H., Kaye, S.J., Owen, C.G. et al. (2008). Birth weight and risk of type 2 diabetes: a systematic review. *J. Am. Med. Assoc.* 300: 2886–2897.

WHO. (2007). Everybody's business: strengthening health systems to improve health outcomes. WHO's Framework for Action. World Health Organization 2007. ISBN 978 92 4 159607 7 http://apps.who.int/iris/bitstream/10665/43918/1/9789241596077_eng.pdf

WHO. (2010a). Monitoring the building blocks of health systems: a handbook of indicators and their measurement strategies. World Health Organization 2010. ISBN 978 92 4 156405 2 http://www.who.int/healthinfo/systems/WHO_MBHSS_2010_full_web.pdf

WHO. (2010b). Package of essential noncommunicable (PEN) disease interventions for primary health care in low-resource settings. World Health Organization 2010. ISBN 978 92 4 159899 6 http://www.who.int/nmh/publications/essential_ncd_interventions_lr_settings.pdf

WHO. (2011). Scaling up action against noncommunicable diseases: how much will it cost? World Health Organization 2011. ISBN 978 92 4 150231 3 http://apps.who.int/iris/bitstream/10665/44706/1/9789241502313_eng.pdf

WHO. (2012). Noncommunicable diseases and their risk factors - Monitoring and surveillance of noncommunicable diseases. - STEPwise approach to surveillance (STEPS)- The STEPS Instrument and Support Materials. World Health Organization 2012. http://www.who.int/ncds/surveillance/steps/instrument/en and http://www.who.int/chp/steps/instrument/STEPS_Instrument_V3.2.pdf?ua=1

WHO. (2013a). Global action plan for the prevention and control of NCDs 2013–2020. World Health Organization 2013. http://apps.who.int/iris/bitstream/10665/94384/1/9789241506236_eng.pdf?ua=1

WHO. (2013b). Implementation tools: package of essential noncommunicable (PEN) disease interventions for primary health care in low-resource settings. World Health Organization 2013. ISBN 978 92 4 150655 7. http://apps.who.int/iris/bitstream/10665/133525/1/9789241506557_eng.pdf

WHO (2013c). *Service Availability and Readiness Assessment (SARA)*. Geneva: World Health Organization http://www.who.int/healthinfo/systems/SARA_Reference_Manual_Full.pdf.

WHO. (2014a). Global status report on noncommunicable diseases 2014. World Health Organization 2014. ISBN 978 92 4 156485 4. http://apps.who.int/iris/bitstream/10665/148114/1/9789241564854_eng.pdf

WHO (2014b). Human papillomavirus vaccines: WHO position paper. *Wkly Epidemiol. Rec.* 89 (43): 465–491.

WHO. (2015). Assessing national capacity for the prevention and control of noncommunicable diseases: report of the 2015 global survey. World Health Organization 2015. ISBN 978 92 4 156536 3. http://apps.who.int/iris/bitstream/10665/246223/1/9789241565363-eng.pdf

WHO. (2016a). Global report on diabetes. World Health Organization 2016. ISBN 978 92 4 156525 7. http://apps.who.int/iris/bitstream/10665/204871/1/9789241565257_eng.pdf

WHO. (2016b). Report of the commission on ending childhood obesity. World Health Organization 2016. ISBN 978 92 4 151006 6.

WHO. (2017a). Technical note. How WHO will report in 2017 to the United Nations General Assembly on the progress achieved in the implementation of commitments included in the 2011 UN Political Declaration and 2014 UN Outcome Document on NCDs. http://www.who.int/nmh/events/2015/Updated-WHO-Technical-Note-NCD-Progress-Monitor-September-2017.pdf?ua=1

WHO. (2017b. STEPS Country Reports. http://www.who.int/chp/steps/reports/en

WHO (2017c). *Noncommunicable Diseases – Fact Sheet.* Geneva: World Health Organization http://www.who.int/mediacentre/factsheets/fs355/en.

Willi, C., Bodenmann, P., Ghali, W.A. et al. (2007). Active smoking and the risk of type 2 diabetes: a systematic review and meta-analysis. *J. Am. Med. Assoc.* 298 (22): 2654–2664.

Yang, H., Huang, X., Zhou, Z. et al. (2014). Determinants of initial utilization of community healthcare services among patients with major non-communicable chronic diseases in South China. *PLoS One.* 9 (12): e116051.

Yassoub, R., Hashimi, S., Awada, S., and El-Jardali, F. (2014). Responsiveness of Lebanon's primary healthcare centers to non-communicable diseases and related healthcare needs. *Int. J. Health Plann. Manag.* 29 (4): 407–421.

Webliography

http://apps.who.int/medicinedocs/documents/s18806en/s18806en.pdf. Bloom, D.E., Cafiero, E.T., Jané-Llopis, E., et al. 2011. The Global Economic Burden of Non-communicable Diseases. Geneva: World Economic Forum.

http://publications.iarc.fr/Non-Series-Publications/World-Cancer-Reports/World-Cancer-Report-2014. Forman, D., Ferlay, J. The global and regional burden of cancer. Stewart, B., Wild, C. World Cancer Report. International Agency for Research on Cancer, Lyon, France, 16 – 27, 978-92-832-0443-5.

http://www.globalasthmareport.org/burden/burden.php Global Asthma Report 2014. Auckland, New Zealand: Global Asthma Network, 2014. ISBN: 978-0-473-29125-9 (PRINT) | 978-0-473-29126-6 (ELECTRONIC).

http://goldcopd.org/gold-2017-global-strategy-diagnosis-management-prevention-copd GOLD 2017a. Global Initiative for Chronic Obstructive Lung Disease. Global strategy for the diagnosis, management and prevention of chronic obstructive lung disease. 2017 report.

http://goldcopd.org/wp-content/uploads/2016/12/wms-GOLD-2017-Pocket-Guide.pdf GOLD 2017b. Global Initiative for Chronic Obstructive Lung Disease. Pocket guide to

COPD diagnosis, management and prevention. A guide for health care professionals. 2017 Edition.

http://www.kdigo.org/clinical_practice_guidelines/pdf/CKD/KDIGO_2012_CKD_GL.pdf ISN 2013. International Society of Nephrology. KDIGO 2012 Clinical Practice Guideline for the Evaluation and Management of Chronic Kidney Disease. 2013 1: 5–8.

http://www.nhlbi.nih.gov/files/docs/guidelines/jnc7full.pdf Joint National Committee 2004. National high blood pressure education program. The seventh report of the Joint National Committee on prevention, detection, evaluation, and treatment of high blood pressure. 2004. U.S. Department of Health and Human Services. National Institutes of Health. National Heart, Lung and Blood Institute.

http://www.globalasthmareport.org/burden/burden.php Marks, G., Pearce, N., Strachan, D., Asher, I. Global Burden of Disease due to Asthma; in: Global Asthma Report 2014, chapter 2. ISBN: 978-0-473-29125-9 (PRINT) | 978-0-473-29126-6.

http://www.iccp-portal.org/system/files/plans/KHM_B3_NSP-NCD%202013-2020_Final%20approved.pdf MoH Cambodia 2013. National Strategic Plan for the Prevention and Control of Noncommunicable Diseases 2013–2020. Ministry of Health, Royal Government of Cambodia. National Strategic Plan for the Prevention and Control of Noncommunicable Diseases 2013.

http://www.pascar.org/uploads/files/Ghana_-_National_Policy-_Prevention_and_control_of_chronic_non-communicable_diseases.pdf MoH Ghana 2012. Republic of Ghana. Ministry of Health. National policy for the prevention and control of chronic non-communicable diseases in Ghana. 2012.

https://www.psi.org/publication/national-strategic-plan-for-the-prevention-and-control-of-non-communicable-diseases-Trinidad-and-Tobago-2017-2021/ MoH Trinidad and Tobago 2017. Ministry of Health. National strategic plan for the prevention and control of non-communicable diseases. Trinidad and Tobago 2017–2021.

https://www.worlddiabetesfoundation.org/sites/default/files/NCD%20Desk%20Guide%20Tanzania.pdf TDA 2013. Tanzania Diabetes Association; World Diabetes Foundation; United Republic of Tanzania, Ministry of Health and Social Welfare. NCD - Case Management desk guide. 2013.

https://www.worlddiabetesfoundation.org/sites/default/files/Tanzania_NCD%20Training%20Manuals%202014.pdf TDA 2014. Tanzania Diabetes Association; United Republic of Tanzania. Ministry of Health and Social Welfare. - Cardiovascular Disease, Type 2 Diabetes, Obesity, Cancer, COPD and Hyperlipidaemia Care. - Case management training modules. 2014.

http://www.who.int/nmh/events/un_ncd_summit2011/political_declaration_en.pdf?ua=1 UN 2011. Political Declaration of the High-level Meeting of the General Assembly on the Prevention and Control of Non-communicable Diseases.

http://www.who.int/nmh/events/2014/a-res-68-300.pdf UN 2014. Outcome document of the high-level meeting of the General Assembly on the comprehensive review and assessment of the progress achieved in the prevention and control of non-communicable diseases.

http://www.who.int/ncds/un-task-force/working-together-adaptation.pdf?ua=1 UNIATF 2017. The United Nations Interagency Task Force on the Prevention and Control of Non-communicable Diseases. Working together for health and development.

http://www.who.int/nutgrowthdb/jme_brochure2016.pdf. UNICEF, WHO, World Bank 2016. Levels and Trends in Child Malnutrition: UNICEF-WHO-World Bank Joint Child Malnutrition Estimates. UNICEF, New York; WHO, Geneva; World Bank, Washington DC: 2015.

http://apps.who.int/iris/bitstream/10665/43918/1/9789241596077_eng.pdf WHO 2007. Everybody's business: strengthening health systems to improve health outcomes. WHO's Framework for Action. World Health Organization 2007. ISBN 978 92 4 159607 7.

http://www.who.int/healthinfo/systems/WHO_MBHSS_2010_full_web.pdf WHO 2010a. Monitoring the building blocks of health systems: a handbook of indicators and their measurement strategies. World Health Organization 2010. ISBN 978 92 4 156405 2.

http://www.who.int/nmh/publications/essential_ncd_interventions_lr_settings.pdf WHO 2010b. Package of essential noncommunicable (PEN) disease interventions for primary health care in low-resource settings. World Health Organization 2010. ISBN 978 92 4 159899 6.

http://apps.who.int/iris/bitstream/10665/44706/1/9789241502313_eng.pdf WHO 2011. Scaling up action against noncommunicable diseanses: how much will it cost? World Health Organization 2011. ISBN 978 92 4 150231 3.

http://www.who.int/ncds/surveillance/steps/instrument/en and http://www.who.int/chp/steps/instrument/STEPS_Instrument_V3.2.pdf?ua=1 WHO 2012. Noncommunicable diseases and their risk factors - Monitoring and surveillance of noncommunicable diseases. - STEPwise approach to surveillance (STEPS)- The STEPS Instrument and Support Materials. World Health Organization 2012.

http://apps.who.int/iris/bitstream/10665/94384/1/9789241506236_eng.pdf?ua=1 WHO 2013a. Global action plan for the prevention and control of NCDs 2013–2020. World Health Organization 2013.

http://apps.who.int/iris/bitstream/10665/133525/1/9789241506557_eng.pdf WHO 2013b. Implementation tools: package of essential noncommunicable (PEN) disease interventions for primary health care in low-resource settings. World Health Organization 2013. ISBN 978 92 4 150655 7.

http://www.who.int/healthinfo/systems/SARA_Reference_Manual_Full.pdf WHO 2013c. Service Availability and Readiness Assessment (SARA). Geneva: World Health Organization; 2013.

http://apps.who.int/iris/bitstream/10665/148114/1/9789241564854_eng.pdf WHO 2014. Global status report on noncommunicable diseases 2014. World Health Organization 2014. ISBN 978 92 4 156485 4.

http://apps.who.int/iris/bitstream/10665/246223/1/9789241565363-eng.pdf WHO 2015. Assessing national capacity for the prevention and control of noncommunicable diseases: report of the 2015 global survey. World Health Organization 2015. ISBN 978 92 4 156536 3.

http://apps.who.int/iris/bitstream/10665/204871/1/9789241565257_eng.pdf WHO 2016. Global report on diabetes. World Health Organization 2016. ISBN 978 92 4 156525 7.

http://www.who.int/nmh/events/2015/Updated-WHO-Technical-Note-NCD-Progress-Monitor-September-2017.pdf?ua=1 WHO 2017a. Technical note. How WHO will report in 2017 to the United Nations General Assembly on the progress achieved in the implementation of commitments included in the 2011 UN Political Declaration and 2014 UN Outcome Document on NCDs.

http://www.who.int/chp/steps/reports/en. WHO 2017b. STEPS Country Reports.

http://www.who.int/mediacentre/factsheets/fs355/en. WHO 2017c. Noncommunicable Diseases – Fact Sheet. World Health Organization, Geneva. 2017.

Part II

How to Improve Healthcare in Low- and Middle-Income Countries by Primary Point-of-Care Rapid Diagnostic Testing

"And my best friend, my doctor
Won't even say what it is I've got"

From the song Just like Tom Thumb's Blues by Bob Dylan, Nobel Laureate in Literature 2016, from his album Highway 61 Revisited issued in 1965.

3

The Optimal Features of a Rapid Point-of-Care Diagnostic Test

David Mabey and Rosanna Peeling

London School of Hygiene and Tropical Medicine, London, UK

CHAPTER MENU

3.1   Introduction

Lack of access to quality diagnostic tests remains a major contributor to the health burden in resource-limited settings. Point-of-care (POC) tests have the potential to reduce this burden by making high quality diagnostic tests available to populations that do not have access to laboratory tests. The optimal features of a POC diagnostic test can be summarized by the acronym ASSURED (Mabey et al. 2004). It should be Affordable, Sensitive, Specific, User-friendly, Rapid and Robust, Equipment-free and Deliverable to end-users (Table 3.1).

Although technological advances have led to the development of many excellent POC tests in recent years which fulfill these criteria, there is inevitably a trade-off between them. A test which costs less than US$0.5, requires no equipment and gives a result in less than 30 minutes is likely to be less sensitive than a more expensive test requiring equipment and electricity, which gives a result in more than two to three hours.

3.2   Accuracy Versus Accessibility

In deciding which is the optimal test to use in a given situation, it is important to bear in mind that a less sensitive POC test which is widely accessible may prevent more adverse health outcomes than a highly sensitive laboratory-based test available only to a privileged few (Smit et al. 2013a).

Revolutionizing Tropical Medicine: Point-of-Care Tests, New Imaging Technologies and Digital Health, First Edition. Edited by Kerry Atkinson and David Mabey.
© 2019 John Wiley & Sons, Inc. Published 2019 by John Wiley & Sons, Inc.

Table 3.1 The ASSURED criteria grouped into the 3 A's of diagnostics.

	Criteria	Description	Category
A	Affordable	Affordability to end users and health systems	Affordability
S	Sensitive	Avoid false negatives	Accuracy
S	Specific	Avoid false positives	
U	User-friendly	Simple to perform in a few steps, minimal training, non-invasive specimens	Access
R	Rapid and robust	Results allow treatment at same visit, tests remain robust through varying supply chain and storage conditions	
E	Equipment free	Least use of expensive equipment	
D	Deliverable to end-users	Tests must be available and accessible to those who need them	

Gift et al. (1999) have drawn attention to what they call the rapid test paradox, whereby a less sensitive POC test for *Chlamydia trachomatis*, which enables same-day treatment, results in more infected people receiving treatment than a sensitive, laboratory-based test which requires patients to make a second visit to receive their result and treatment. It is important for regulatory agencies, and those responsible for drafting target product profiles for diagnostic tests, to bear this in mind when approving or licensing new POC tests, and not to let the best be the enemy of the good. The World Health Organization (WHO) has taken this into account in developing target product profiles for POC syphilis tests, in which a minimal sensitivity of 80% compared to a laboratory-based assay is considered acceptable. This would enable countries to make syphilis screening available to all women attending antenatal clinics (ANCs) and increase the proportion of pregnant women with syphilis identified in sub-Saharan Africa, where approximately 90% of pregnant women attend an ANC at least once per pregnancy, from less than 50% to more than 70% (Figure 3.1)(WHO 2016; Wijesooriya et al. 2016).

3.3 Quality Assurance

Ensuring the quality of both POC tests and that of testing performed at hundreds or thousands of different sites by healthcare workers, who are not skilled in reading test

	Sensitivity			
Access	**100**	**90**	**80**	**70**
100	100	90	80	70
90	90	81	72	63
80	80	72	64	56
70	70	63	56	49
60	60	54	48	42
50	50	45	40	35
40	40	36	32	28
30	30	27	24	21
20	20	18	16	14
10	10	9	8	7

Figure 3.1 The trade-off between access and sensitivity.

results, is a major challenge (Nkengasong et al. 2016). Approaches to external quality assessment (EQA) for POC tests include the use of proficiency panels consisting of dried positive and negative samples provided to the POC testing sites as blinded panels, in order to determine if the healthcare workers can obtain the correct results. These dried tube samples do not require refrigeration and have been used in many remote settings including the Amazon region of Brazil (Benzaken et al. 2014). Ideally the results of EQA results from POC testing sites should be read electronically and reported automatically to a central database (Cheng et al. 2016). Another method of assuring quality consists of collection of a second sample at the point-of-care for re-testing, using a "gold standard" test, at a central laboratory. Dry blood spots have been used in this way for EQA of syphilis testing in rural Tanzania (Smit et al. 2013b).

3.4 The Importance of Connectivity

Electronic readers and smartphones have the potential to standardize the interpretation of POC tests. Many POC test manufacturers now incorporate connectivity into POC and near-POC instruments. Alternatively, lateral flow tests can be scanned by a reader or a mobile phone, making it possible to link data from POC test readers and devices to Ministries of Health and thus provide critical information on testing coverage and disease trends (Wedderburn et al. 2015). The Ebola crisis demonstrated the importance of integrating digital technology with a new generation of POC molecular tests that are highly sensitive and specific. Digital technology helped facilitate and improve patient record management, clinical treatment of patients and quality monitoring. These advances can also form the basis of early warning systems to identify disease outbreaks and optimize control and elimination interventions, and can be used to support quality assurance programs. With the addition of barcodes, 2D codes, or electronic storage to the tests, other data can be collected by smartphones, including manufacturing information (for example, lot numbers) and dates, expiry dates, stock availability and possibly even environmental conditions such as temperature and humidity under which the tests have been manufactured, transported, stored and used (Figure 3.2).

Figure 3.2 Connectivity not only turns data into intelligence but also enables improvement of the quality of testing, supply chain management and patient care.

3.5 Environmental Friendliness

As more and more diagnostic tests are used in both urban and rural areas, there should be more effort devoted to mitigating the environmental impact of these tests. Although individual tests do not pose a significant risk, many thousands of tests performed at each site pose environmental risks. The plastics used in current rapid tests cannot be recycled and produce toxic fumes if burnt. Focus needs to be placed on the materials used for these tests, including the housing, substrate matrix materials and reagents used in the test. Paper is an obvious choice of substrate material and has many inherent advantages over standard plastic materials. There is currently a lot of activity in the field of micropaper-based analytical devices (Cate et al. 2015; Martinez et al. 2010). Similar materials could be used for storage, making tests biodegradable or easily disposable. Additional factors that also need to be considered include materials used for active components such as electronic tracks and electrodes, and the disposal of samples.

References

Benzaken, A.S., Bazzo, M.L., Galban, E. et al. (2014). External quality assurance with dried tube specimens (DTS) for point-of-care syphilis and HIV tests: experience in an indigenous populations screening programme in the Brazilian Amazon. *Sex. Transm. Infect.* 90 (1): 14–18.

Cate, D.M., Adkins, J.A., Mettakoonpitak, J., and Henry, C.S. (2015). Recent developments in paper-based microfluidic devices. *Anal. Chem.* 87: 19–41.

Cheng, B., Cunningham, B., Boeras, D.I. et al. (2016). Data connectivity: a critical tool for external quality assessment. *Afr. J. Lab. Med.* 5 (2): 535. https://doi.org/10.4102/ajlm.v5i2.535.

Gift, T.L., Pate, M.S., Hook, E.W. et al. (1999). The rapid test paradox: when fewer cases detected lead to more cases treated: a decision analysis of tests for *Chlamydia trachomatis. Sex. Transm. Dis.* 26: 241–242.

Mabey, D., Peeling, R., Ustianowski, A., and Perkins, M. (2004). Diagnostics for the developing world. *Nat. Rev. Microbiol.* 2: 231–240.

Martinez, A.W., Phillips, S.T., Whitesides, G.M., and Carrilho, E. (2010). Diagnostics for the developing world: microfluidic paper-based analytical devices. *Anal. Chem.* 82: 3–10.

Nkengasong, J., Boeras, D.I., Abimiku, A., and Peeling, R.W. (2016). Assuring the quality of diagnostic testing: the future is now. *Afr. J. Lab. Med.* 5 (2): a558. https://doi.org/10.4102/ajlm.v5i2.558.

Smit, P.W., Mabey, D., Changalucha, J. et al. (2013a). The trade-off between accuracy and accessibility of syphilis screening assays. *PLoS One* 8 (9): e75327. https://doi.org/10.1371/journal.pone.0075327.

Smit, P.W., van der Vlis, T., Mabey, D. et al. (2013b). The development and validation of dried blood spots for external quality assurance of syphilis serology. *BMC Infect. Dis.* 13 (1): 102.

Wedderburn, C.J., Murtagh, M., Toskin, I., and Peeling, R.W. (2015). Using electronic readers to monitor progress toward elimination of mother-to-child transmission of HIV and syphilis: an opinion piece. *Int. J. Gynaecol. Obstet.* 130 (Suppl 1): S81–S83. https://doi.org/10.1016/j.ijgo.

Wijesooriya, N.S., Rochat, R.W., Kam, M.L. et al. (2016). Global burden of maternal and congenital syphilis in 2008 and 2012: A health systems modelling study. *Lancet Global Health.* 4 (8): e525–e533. https://doi.org/10.1016/s2214-109x(16)30135-8.

World Health Organisation. (2014). Sexually transmitted infections point-of-care testing, 06-08 May 2014, Annecy, http://www.who.int/reproductivehealth/potc-tpps-2-16.pdf

Webliography

http://doi.org/10.4102/ajlm.v5i2.558. Nkengasong, J., Boeras, D.I., Abimiku, A., and Peeling, R.W. Assuring the quality of diagnostic testing: The future is now. Afr J Lab Med. 2016 5(2) a558

http://who.int/reproductivehealth/POTC-TPPs-2016.pdf?ua=1

4

Revolutionizing HIV Healthcare Delivery Through Rapid and Point-of-Care Testing

Catherine J. Wedderburn, Debrah I. Boeras, and Rosanna W. Peeling

International Diagnostics Centre, Clinical Research Department, London School of Hygiene & Tropical Medicine, London, UK

CHAPTER MENU

4.1 Synopsis

A range of innovative rapid diagnostic tests are commercially available for increasing access to the serological diagnosis of human immunodeficiency virus (HIV) disease HIV-1 and -2 in adults, virological diagnosis of HIV in infants born to seropositive mothers, monitoring HIV viral load in patients on treatment, and CD4$^+$T cell

Revolutionizing Tropical Medicine: Point-of-Care Tests, New Imaging Technologies and Digital Health,
First Edition. Edited by Kerry Atkinson and David Mabey.
© 2019 John Wiley & Sons, Inc. Published 2019 by John Wiley & Sons, Inc.

enumeration for the management of advanced HIV disease. Studies have shown that point-of-care (POC) testing has the potential to expand testing coverage, enhance linkage to care, minimize loss to follow up and increase the efficiency of the healthcare system by reducing the number of patient visits and healthcare workload. Challenges include increased demands on training, quality assurance, and supply chain management to cover all POC sites. Advances in POC diagnostic technologies can be leveraged to revolutionize healthcare delivery for HIV if they are implemented within a connected diagnostic system that consists of a system of laboratories linked to a network of POC testing sites, with clinical pathways that are optimized to take full advantage of the test results being available within a single patient visit.

4.2 Introduction

HIV is a global epidemic which has crossed cultures and countries, races and ages. There has been huge progress in the diagnosis and treatment of HIV over the past decade, but the infection continues to spread and HIV incidence remains high. In 2016 there were an estimated 1.8 million new HIV infections and 1 million deaths from acquired immune deficiency syndrome (AIDS), most of which were in resource-limited settings, with the majority in sub-Saharan Africa (UNAIDS 2017). In 2014 the UN launched the 90-90-90 goals with the aim that by 2020, 90% of all people living with HIV will know their status, 90% of people diagnosed with HIV infection will receive sustained antiretroviral treatment (ART) and 90% of all people receiving ART will have viral suppression (UNAIDS 2014). Accessible high quality diagnostics are critical to achieving these global targets and rapid tests that can be performed at the POC allow HIV programs to extend access to testing outside of laboratory settings. With most countries committing to attaining the United Nations' Sustainable Development Goals (SDGs), empowering communities by increasing access to testing so that no one is left behind is an important priority.

In this Chapter we will discuss

1) The challenges of delivering HIV care in resource-limited settings and the diagnostic tests needed to achieve the 90-90-90 goals
2) Recent advances in POC tests for HIV diagnosis and monitoring, the impact they have and the remaining challenges, and
3) What is needed to revolutionize healthcare delivery to help countries achieve their 90-90-90 targets toward universal access to healthcare as part of the SDGs.

4.3 Diagnostic Tests in Resource-Limited Settings

4.3.1 Challenges of Healthcare Delivery in Resource-Limited Settings

High quality tests are commercially available for the diagnosis of most infectious diseases but they are neither affordable nor accessible to patients in resource-limited settings. Conventional testing for HIV has routinely involved laboratory-based tests performed using large expensive equipment that requires a steady source of electricity, sophisticated laboratory infrastructure and highly trained laboratory technicians.

It was recognized early in the HIV epidemic that diagnostic tests must be made more accessible and affordable so that they can be performed closer to the patient (for example, at the POC) with the potential to:

- Expand testing coverage by decentralizing testing to health facilities at different levels of the healthcare system.
- Allow same-day testing and results to ensure earlier initiation of treatment and linkage to care.
- Allow treatment monitoring where needed.
- Reduce loss to follow up across the testing-treatment cascade.
- Empower health workers to deliver immediate care at every level.

Two key aspects for healthcare delivery in resource-limited settings are that

- The tests need to be easy to use so that healthcare workers at all levels of the healthcare system can perform these tests with minimal training.
- They can be stored without refrigeration for months, and preferably for one–two years, without impacting their accuracy and quality.

These qualities are captured by the ASSURED criteria which stands for Affordable, Sensitive and specific, User-friendly, Rapid and robust, Equipment free and Deliverable to end-users, and guides target features of POC diagnostic tests (Mabey et al., 2004).

The diagnosis of HIV may be made by either detecting the virus in the blood following infection using a molecular assay or waiting for the body's immune system to produce antibodies in response to the infection and using serological assays to detect HIV antigen or antibody. Since most infected individuals are not aware of when they become infected and remain asymptomatic for many years, diagnostic tests to detect antibodies to HIV are a highly effective way of diagnosing HIV (Zhang and Versalovic 2002).

The first generation of POC tests for HIV were rapid diagnostics tests (RDTs) that detected antibodies to HIV-1 and -2 in 15 minutes using finger-pricked whole blood samples. The tests were designed as immunochromatographic assays in a lateral flow assay format and required very minimal training to perform three simple steps – collecting the specimen, performing the testing and reading the result. The WHO has pre-qualified 15 of these RDTs as fulfilling the criteria set by WHO that will ensure accurate and timely results (WHO updated 2018).

These RDTs worked well for identifying infected individuals but other diagnostic tests are needed for delivery of a complete package of care for HIV patients.

4.3.2 Diagnostic Tests Needed to Reach the 90-90-90 Targets

All of the 90-90-90 targets require quality-assured diagnostic tests to accurately identify HIV-infected individuals, place them on appropriate treatment, properly manage their care and monitor their viral load to ensure viral suppression. Currently, we are only at 70% of the first UN 90 target with a huge gap and substantial challenges in ensuring diagnostic tests are available to all who need them (Table 4.1) (UNAIDS 2017, 2018).

4.3.2.1 The First 90

The first 90 requires all people at risk to be tested and given a diagnosis. However, testing is not available in all settings across the world. Results from laboratory-based testing may take a long time to reach the patient, leading to significant losses to follow up.

Table 4.1 Diagnostic tests needed for each of the 90-90-90 targets and challenges.

Cohort	Targets	Laboratory tests	POC tests	Challenges
1st 90	90% of people living with HIV know their status	Immunoassays	Rapid tests for adults widely available	Adults: stigma, risk perception. Infants: limited access to virologic diagnosis
2nd 90	90% of people diagnosed with HIV are on ART	No test required if test and treat; otherwise, use CD4 tests to determine eligibility	POC CD4 test implementation stalled in many countries as test and treat is being scaled up	Access to POC CD4 testing critical for patients with advanced disease
3rd 90	90% of those on ART are virally suppressed	Viral load (VL) testing only available in national reference laboratories	Limited availability for POC VL tests	Limited access to VL monitoring lack of awareness of ongoing transmission

Many at-risk individuals may not come forward for testing because of stigma or the perception that they are not at risk. There are ongoing efforts to allow self-testing in order to overcome these challenges.

Lack of access to timely diagnosis and treatment for infants born to HIV-positive women is a critical barrier in reaching the first 90 for many low- and middle-income countries (LMICs). As maternal antibodies persist in the blood of infants for around 12–18 months, the diagnosis of HIV in the infant must involve detection of the virus itself by a molecular assay such as polymerase chain reaction (PCR). RDTs used in adults may incorrectly diagnose the child as HIV-positive when maternal antibodies are detected. Previously, early infant diagnosis (EID) involved conventional laboratory molecular assays, which require large expensive equipment and trained personnel. In many LMICs where there is limited access to laboratories, access to infant testing entails sending blood samples to centralized laboratories with very long turnaround times from testing to diagnosis and initiation on ART. This results in high rates of loss to follow up (Sibanda et al. 2013). Because studies have shown that 50% of HIV-infected children will die before their 2nd birthday without treatment, with the highest peak of mortality being in the first few months of life (Newell et al. 2004), this delay in diagnosis dramatically affects mortality. In 2015 it was estimated that only 51% of HIV-exposed infants were tested for HIV within the first two months of life (UNAIDS 2015), resulting in long delays in treatment initiation despite the evidence that early ART in the first few months of life reduces mortality by 75% (Violari et al. 2008).

Overall, there is a critical need for new and novel ways of early infant HIV testing that can accurately and rapidly identify infected infants and link them immediately to care and treatment (Ferrand 2017). In 2014 WHO called for elimination of mother-to-child transmission of HIV (EMTCT) (WHO 2014) requiring diagnosis and treatment of all HIV-infected pregnant women. However, many countries have not yet been able to meet these targets and ongoing transmissions from mother-to-child still continues. There were 150 000 new infants diagnosed with HIV in 2015, and although this number has decreased over time, the detrimental effects of HIV in children if not diagnosed early and started on treatment have been well-described (UNICEF 2016c). POC EID technology that allows for faster delivery of results, potentially on the same day of sample collection, and therefore earlier ART initiation, is needed to reduce loss to follow up across the diagnostic and treatment cascade and to prevent mortality from delayed treatment.

4.3.2.2 The Second 90

Although the WHO guidelines changed in 2014 to advise that all people with HIV should be initiated on ART ("Test and Start" or "Universal Test and Treat"), this is not possible in most LMICs with limited access to affordable HIV drugs. These countries still need to rely on CD4 testing to prioritize eligibility for treatment (Peeling et al. 2015). Implementation of POC CD4 testing has stalled in some countries as the Test and Start option is being scaled up, often leaving patients in rural areas without access to CD4 testing.

Worldwide it is estimated that 30–50% of HIV patients already have advanced disease at presentation (WHO 2017d; Waldrop et al. 2016). For these patients, POC CD4 testing is needed to determine if they need prophylaxis against opportunistic infections (WHO 2017a).

4.3.2.3 The Third 90

The third 90 requires lifelong monitoring to ensure the ongoing efficacy of treatment and to immediately identify treatment failure in order to initiate further treatment. A positive viral load can trigger investigation into treatment compliance, and where there is true treatment failure, the patient should be switched to a different drug regimen (WHO 2017b). Viral load testing is traditionally performed using quantitative molecular assays that require sophisticated laboratory equipment and highly trained laboratory technicians. They are expensive and usually only available in national reference laboratories. In many settings, HIV patients are either not monitored for viral load or dried blood spots are collected and sent to central laboratories for viral load testing. This process results in long delays in return of results to rural health centers. Patients often stay on ineffective drug regimens, allowing onward transmission, or become lost to follow-up across the treatment cascade. Having viral load assays that can be performed at district level hospitals or health facilities is a priority (UNAIDS 2017).

4.4 Challenges of Using Rapid and Point-of-Care Testing Within the Context of the Healthcare System

The diagnostic tests identified above need to be made available where they are needed and fit within, and complement, a complex and often fragmented healthcare system, with many other obstacles beyond testing. The most concentrated number of people living with HIV reside in sub-Saharan Africa (UNAIDS 2017), where the healthcare system is often overburdened and laboratory infrastructure is limited and underfunded. Having new technologies alone cannot succeed when the health system is stressed because of an already weak infrastructure. Innovation in delivery must accompany technological innovation so that the introduction and scale up of new diagnostic tests can improve patient outcomes while strengthening healthcare systems.

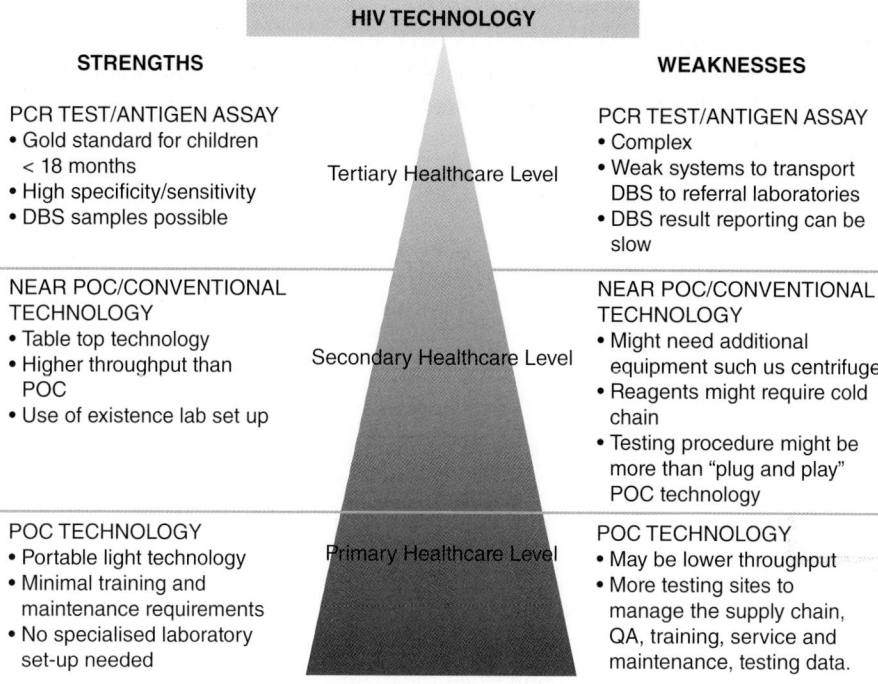

Figure 4.1 A schematic diagram of types of HIV testing within each level of the healthcare system. Source: *Taken from* UNICEF: Accelerate access to innovative point-of-care (POC) HIV diagnostics: CD4, EID and VL 2016 (UNICEF 2016d).

While RDTs for the serological diagnosis of HIV can be used at all levels of the healthcare system, the majority of device-based POC/near-POC assays such as CD4, EID and viral load monitoring are too expensive to be made available everywhere. Although these near-POC assays are sample-in-answer-out or "plug and play" devices, they still require more extensive training and maintenance than RDTs. For optimal usage, cost-effectiveness and revolutionary impact on patient care, POC tests are best placed at district level health facilities and hospitals, in clinics or hospitals in urban areas to complement existing laboratory infrastructure, or at central testing hubs each providing testing at a group of sites (Diallo et al. 2017). Countries need to strategically place devices taking into account patient load, geographical locations, device characteristics and existing laboratory infrastructure (Diallo et al. 2017) and, in parallel, determine how this will strengthen the laboratory–clinic interface (UNICEF 2016d) (Figure 4.1).

4.5 Recent Advances in HIV Diagnosis and Monitoring and Their Impact

4.5.1 HIV Self-Testing and Multiplex Testing

HIV self-testing (HIVST) using either oral or blood-based testing is the latest advancement in bringing testing closer to the patient. With self-testing, the person performs the HIV diagnostic test and interprets the results in private. In 2015 WHO released guidance to support the implementation and scale-up of ethical, effective, acceptable and evidence-based approaches to HIVST (WHO, 2016a).

This approach will still require appropriate quality assurance and monitoring to determine whether this intervention will contribute to closing the testing gap and achieving the United Nations' 90–90–90 global goals.

Multiplex testing allows for HIV testing to be combined with other testing that is appropriate for HIV-infected patients, including, for example, tuberculosis, syphilis, hepatitis B and hepatitis C. This approach would presumably provide a larger catchment of testing and faster results, and would perhaps be less costly for the healthcare system. Studies are ongoing to determine if multiplex testing can increase overall diagnosis and early detection and increase efficiency of the healthcare system.

4.5.2 Early Infant Diagnosis of HIV

In recent years there has been an increase in EID testing and expansion of the laboratory network with the use of dried blood spots (DBS). Filter paper is used to collect blood specimens in remote areas and then transport them dried at room temperature to the laboratory for testing. However, in 2015 only 51% of at-risk infants received an HIV test before two months of age (UNAIDS 2016), highlighting the need for a new strategy to revolutionize the healthcare of infants. Alongside the WHO call for EMTCT of HIV, over the past few years there has been a push for the development of POC diagnostics for EID. In 2015 UNITAID published the diagnostic pipeline of EID technologies (UNITAID 2015) (Figure 4.2).

These devices may be used near or at the POC and are able to process specimens and provide results the same day. These POC nucleic acid testing devices are able to perform qualitative EID molecular testing, traditionally performed in centralized laboratories, in almost any setting by trained healthcare workers using portable or semi-portable devices.

The quality of EID POC testing has been evaluated in both laboratory and field settings. Four POC EID technologies have received CE-IVD (Conformité Européene *In Vitro* Diagnostics) approval and two POC EID technologies have met WHO prequalification requirements: Alere q HIV 1/2 Detect and Cepheid Xpert HIV-1 qual (WHO 2017d). In 2015 the EID Consortium was established to bring together global expertise in the fields of public health, HIV diagnostics and infant health in order to overcome the challenges of diagnosing and treating infant HIV (www.eidconsortium. org). The first priority for the EID Consortium was to determine the field performance of POC EID devices through a pooled analysis of existing data for the first two products ready for evaluation: the Alere q HIV-1/2 Detect and the Cepheid GeneXpert HIV-1 qual. Data from nine independent field evaluations from six countries were pooled (Carmona et al. 2016). The results showed that both the Alere q HIV-1/2 Detect and the Cepheid Xpert HIV-1 qual performed well in field testing with high sensitivity and specificity, comparable to laboratory testing and thus supported the use of these POC EID devices (WHO 2017c).

The impact on infant case finding and treatment initiation has been investigated in pilot studies from Mozambique and Malawi (Jani et al. 2016; Mwenda 2016). Both studies found that POC EID significantly reduced the turnaround time of results when compared with standard of care conventional testing. The majority received same-day results, and consequently POC EID allowed earlier initiation of ART. Additionally, the

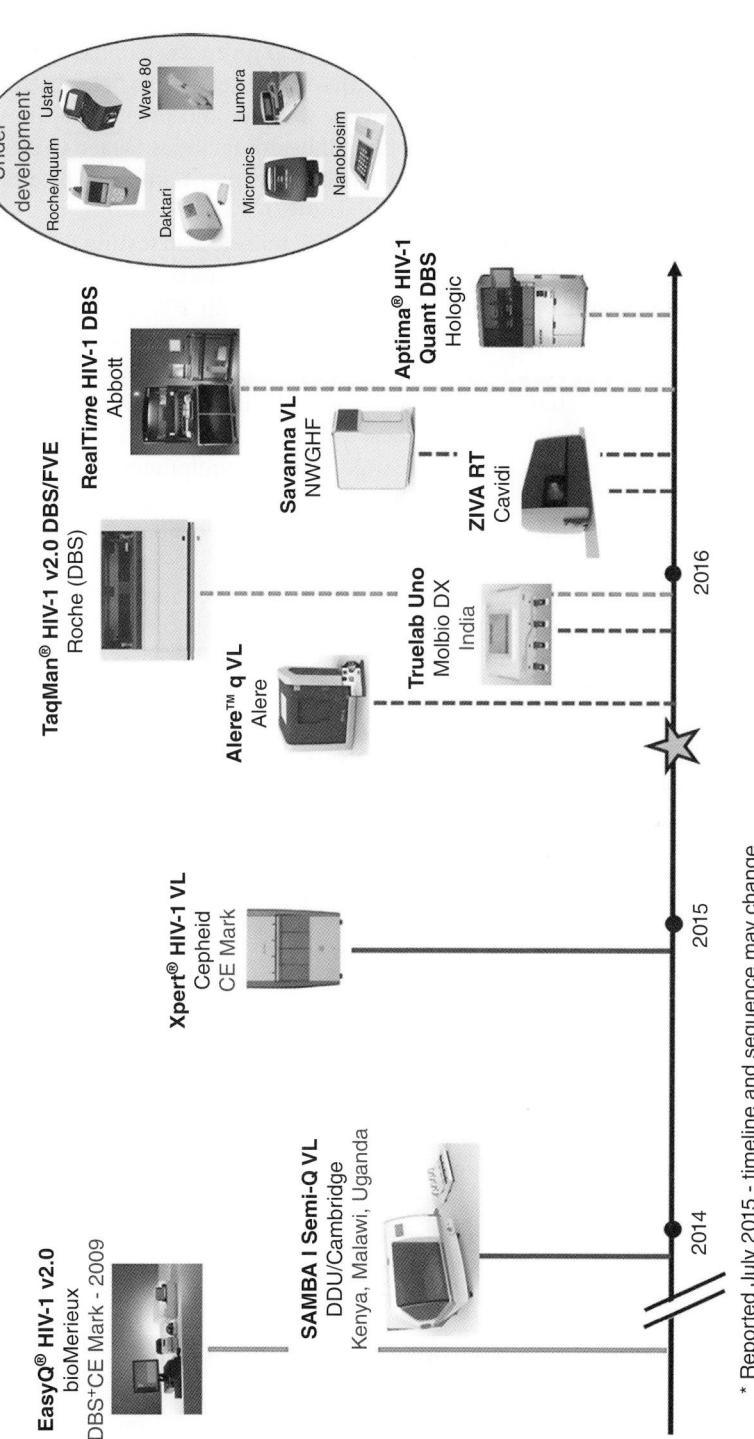

Figure 4.2 HIV/AIDS point-of-care viral load technologies in the pipeline. *Source:* UNITAID. HIV/AIDS Diagnostics Technology Landscape, October 2015. http://www.unitaid.org/assets/UNITAID_HIV_Nov_2015_Dx_Landscape-1.pdf Accessed on 18th February 2018. Abbreviations used: DBS, dried blood spot, CE, Conformité Européenne, meaning European Conformity. (CE marking is a certification mark that indicates conformity with health, safety, and environmental protection standards for products sold within the European Economic Area.)

study in Malawi showed an overall increase in ART initiation (Mwenda 2016) and the results from Mozambique demonstrated improved ART retention rates, reducing loss to follow up (Jani et al. 2016), and addressing a critical gap given the extensive loss to follow up across the EID cascade(Siyanda et al. 2013). Mwenda et al. also demonstrated that POC EID devices were acceptable to healthcare workers and patients, although some of the near POC devices could not be used in all health facilities (Maenad 2016). Provider acceptability of using POC EID devices has been further reinforced by a field implementation study in South Africa (Dunning et al. 2017).

The WHO have supported the use of POC or near POC nucleic acid testing technologies as described in the 2016 WHO Consolidated Guidelines on the use of antiretroviral drugs for treating and preventing HIV infection (WHO 2013). In 2017 the WHO recommended the rapid national regulatory approval and initiation of scale up of these devices and incorporation of POC EID into National HIV Care and Treatment Guidelines (WHO 2017c). However, there are concerns about the high cost and shelf-life of cartridges, the supply chain and availability of local technical support for repairs. There are also challenges to sustainability and to having the POC equipment accessible to every infant (birth testing). The different POC machines are not all for use in primary healthcare facilities and district hospitals may serve as the best location for balancing distance and the cost of running the test.

Individual strategies for different countries should be developed for prioritizing the POC devices and need to be based on patient throughput and maximization of POC EID utilization. In Mozambique a study showed that a prototype POC p24 Ag detection test (the LYNX), despite its low sensitivity of 72%, had the potential to provide test results for up to 81% more patients compared to sending dried blood spots to laboratories for testing (Maggi et al. 2017). This prototype POC p24 assay is feasible for use in primary healthcare settings and is more affordable than near-POC molecular tests for EID.

4.5.3 POC CD4 Testing for Managing HIV Patient Care

4.5.3.1 Before the Initiation of ART

The use of POC diagnostic technologies significantly improves retention in the pre-ART care cascade. POC CD4 testing can reduce time to result from ten days to less than one day on average (Vojnov et al. 2016). In Mozambique POC CD4 testing reduced loss to follow-up before ART initiation by 50% (Jani et al. 2011). POC CD4 testing is still being used to initiate treatment in many countries. A qualitative study in Uganda found that diagnostic tests, such as POC testing, combined with interventions to change provider behaviors enhanced uptake of innovations targeting the HIV cascade in clinical settings (Semitala et al. 2017). In Namibia POC CD4 testing was feasible and effective when task-shifted to lay health workers. The rollout of POC CD4 testing improved access to CD4 testing and retention in care between HIV diagnosis and ART initiation in LMICs (Kaindjee-Tjituka et al. 2017).

4.5.3.2 The Use of the CD4 Test for Early HIV Disease

The landscape for diagnostic tests, and particularly those performed at the primary POC is continually rapidly changing. In 2010 WHO updated its recommendations for monitoring treatment effectiveness in HIV patients on antiretroviral therapy. Until that year the international recommendation was a combination of clinical monitoring and the CD4$^+$ T cell (CD4) count. However, other research showed that monitoring viral load was an earlier and more sensitive way to identify treatment failure (WHO 2013).

In 2016 the WHO Consolidated Guidelines on the use of Antiretroviral Drugs for Treating and Preventing HIV infection recommended that all patients who tested positive for HIV initiate antiretroviral therapy, regardless of the CD4 count, and emphasized that viral load monitoring should be used to determine treatment failure (WHO 2013). While CD4 monitoring is being phased out in many places in favor of annual viral load monitoring, the WHO treatment guidelines still recommend the use of CD4 testing for clinical management (WHO 2017a). CD4 count is the best predictor for disease status and immediate risk of death and thus should be used to identify those who have advanced HIV disease. Patients who are unstable or have advanced HIV disease should have a CD4 test every six months until stable.

4.5.3.3 The CD4 Test for Advanced HIV Disease

The WHO Guidelines for Managing Advanced HIV Disease and Rapid Initiation of Antiretroviral Therapy defines a role for CD4 testing in identifying and managing patients with advanced HIV disease through a package of services (WHO 2017a). Performing a CD4 test at baseline is important because, even though many countries have adopted the test and start approach, the percentage of patients starting ART with advanced disease has not changed and new data suggest that a high proportion of patients with advanced disease are asymptomatic. Data from the International Epidemiology Databases to Evaluate AIDS (IeDEA) cohort suggest that in some settings more than 50% of people are starting ART with advanced HIV disease (Avila et al. 2014).

The WHO Guidelines for Managing Advanced HIV Disease and Rapid Initiation of Antiretroviral Therapy uses a CD4$^+$ T cell threshold of <200 cells/mm^3 to identify people with advanced HIV disease who are eligible for elements of a package of care (WHO 2017a). This new role for CD4 can leverage existing capacity for CD4 testing, including instrument-based and POC testing. A simple to use, instrument-free, disposable semi-quantitative lateral flow strip test has been approved for sale in Europe (CE mark obtained) for CD4 enumeration at thresholds <300 cells/mm^3 (Omega Diagnostics Group, accessed March 2018). The results, which can be read by eye after 40 minutes, can be digitized and the data transmitted to a central database.

Policy makers should consider mapping existing resources within the healthcare system and re-engineering patient flow so that viral load testing is scaled up while CD4 testing is phased back from its role in initiation and monitoring of HIV. It is important to consider the remaining life of the existing instruments and ongoing barriers to receiving testing cartridges and supplies to prevent critical stockouts.

4.5.4 POC HIV Viral Load Monitoring

The 2016 WHO Consolidated Guidelines on the Use of Antiretroviral Drugs for Treating and Preventing HIV Infection Recommendations for a Public Health Approach state that viral load testing is the preferred monitoring approach to diagnose and confirm treatment failure (WHO 2013). Routine viral load testing should be carried out at 6 and 12 months after initiating ART and, if the patient is stable on ART, every 12 months thereafter. In 2015 UNITAID published the diagnostic pipeline of POC viral load technologies (UNITAID 2015) (Figure 4.2). Many countries are implementing viral load testing as the preferred monitoring technology for people on ART and POC viral load testing technologies have the potential to expand access to viral load testing.

In July of 2017 the WHO released an Information note on What's New in Treatment Monitoring: Viral Load and CD4 Testing (WHO 2017b). The note reinforced the importance of routine HIV viral load monitoring to diagnose and confirm treatment failure. The note once again allowed for the use of dried blood spot specimens using venous or capillary whole blood (in the event that plasma was unavailable) to determine the HIV viral load using a treatment failure threshold of 1000 copies/mL. As previously mentioned, the scale up for viral load testing has been slow despite tremendous support from The President's Emergency Plan For AIDS Relief (PEPFAR) and partners, with less than 40% of patients having access to routine viral load testing (WHO 2015b).

4.6 WHO Recommendations: POC Diagnostics for Achieving the 90-90-90 Goals

- First 90: The WHO recommends initiating treatment in all HIV-infected people and rapid HIV testing and EID POC can provide same day testing and results and allows healthcare workers to initiate treatment on the same day as testing (WHO 2017c). This has been shown to

  1) increase testing uptake in areas of need, and
  2) increase linkage to care (Jani et al. 2016; Mwenda 2016).

- Second 90: The WHO recommends that all HIV-infected people be screened for advanced disease to manage their care through a package of services. Rapid and POC CD4 testing is needed to screen for advanced disease, which can then be followed by rapid cryptococcal antigen screening and Xpert MTB/RIF POC. a nucleic acid based molecular test for diagnosing tuberculosis in those with $CD4^+$ T cell count ≤100 (WHO 2017a).
- Third 90: The WHO recommends that all HIV-infected people on treatment be monitored for virological failure using HIV viral load testing. Viral load POC testing and dried blood spots can provide viral load monitoring for populations that are otherwise hard to reach and for those in remote areas (WHO 2017b).

The WHO list of prequalified *in vitro* diagnostic products for HIV that are currently available is summarized in Figure 4.2.

4.7 Remaining Challenges – Human Resources, Quality Assurance, and Test Selection and Placement

The field of HIV diagnostics has made huge advances in recent years. However, with 36.7 million people in the world living with HIV, and with ongoing transmission, there remains an urgent need to improve diagnostic techniques to meet the needs of the people and curb the epidemic. The introduction of every new technology puts more demands on the healthcare system for human resources, capacity building, and support services.

Diagnostic tests must be selected (Kosack et al. 2017) and implemented to fit within healthcare systems in three important aspects:

- where they are most cost effective and sustainable
- where quality of tests and testing can be assured
- where linkage to care across the treatment continuum is ensured.

Apart from technological innovation, the cascade of HIV testing and care from diagnosis to treatment initiation to monitoring and maintenance has steadily increased the complexity of service delivery under current healthcare systems. Country capacity to sustain programs continues to be an issue and will be critical to optimize the full potential of POC testing as part of a strengthened healthcare system.

De Cock et al. (2011) reflected on 30 years of HIV/AIDS in Africa and advocated that HIV testing policies need to be reconsidered within a human rights-based approach that specifically addresses the role of HIV testing as a public health emergency and as a measure of social justice. Further research is needed to integrate measures to address socioeconomic and cultural barriers preventing access to HIV diagnostic tests.

4.8 Moving Forward

To truly revolutionize healthcare delivery across resource-limited settings, both innovative POC technologies and innovative strategies for delivery need to be coordinated into a model that ultimately increases the efficiency and effectiveness of a healthcare system.

There is a need to focus not just on innovative diagnostics, such as POC tests, but to develop in each country a comprehensive diagnostic system that comprises of a central reference laboratory connected to a system of regional laboratories and a network of POC testing sites. The central reference laboratory acts as a command center to select appropriate diagnostics for each level of the healthcare system and provides training and quality assurance across the entire diagnostic system (Boeras et al. 2017). It is critical to avoid adding stress to fragile health systems. Rather, studies have shown that POC diagnostic tests can be used to strengthen the health system by improving the quality of care through empowering providers to use POC diagnostic test results to make evidence-based management decisions. This also reduces the frequency of patient visits and health provider work-load as well as loss to follow-up across the HIV treatment continuum (Garcia et al. 2013; Jani and Peter 2013; Mabey et al. 2012).

Careful planning needs to be carried out before implementation to ensure appropriate devices are placed in locations that best fit patient needs and will help countries reach the 90-90-90 targets: these can be conventional technologies or near POC or POC instruments, depending on the local infrastructure. Rapid and POC testing can support expansion of testing by increasing coverage and improving access but selection of the right devices to be placed in the right places and then properly monitored, supplied and maintained is critically needed. Increased access to testing will only have its intended impact if the quality of the POC tests and of testing are assured (Boeras et al. 2017; Fonjungo et al. 2016).

Devices that are multipurpose and dual or multiplex tests that may diagnose multiple diseases will allow for increased testing capacity and efficiency, bridge the laboratory-clinic interface and improve integration of healthcare programs (UNICEF 2016).

Countries can leverage investments in epidemic preparedness to build such diagnostic systems.

To optimize the full potential of rapid and POC testing, it is critical that connectivity solutions can be used to integrate testing, quality assurance and supply chain management. Investing in connectivity solutions across the entire diagnostic system allows for real time monitoring of testing. Connectivity is a powerful tool for turning data into intelligence for patient management and disease control and prevention (Cheng et al. 2016). Further research is needed to determine how countries can utilize current capacity in their setting to develop connectivity to allow real-time results reporting and monitoring.

This model of healthcare delivery will require a cadre of trained healthcare workers capable of providing a diagnostic solution instead of just trained to perform the testing. Healthcare workers, at every level of the system, should be empowered to act upon the test results and work with clinicians to enroll patients into the care pathway. Since almost all of the POC and near-POC test results can be digitized and transmitted, new models of training need to include training on information and communications technology and quality assurance.

Lastly, one of the most radical and revolutionizing innovations in healthcare delivery has been the use of Unmanned Aerial Vehicles (UAVs) or drones to improve supply chains in resource-limited settings (see Chapter 35). Drones can be used to deliver HIV diagnostic test kits, consumables, and supplies to hard-to-reach areas. In Malawi UNICEF has tested the use of drones to deliver HIV test kits to rural areas (McNeish 2016).

4.9 Conclusions

Innovative rapid diagnostic technologies, such as POC testing, have the potential to revolutionize HIV healthcare delivery across the patient care pathway from testing to treatment initiation to monitoring and helping countries meet their 90-90-90 targets.

While POC diagnostics in themselves can be revolutionary, their strategic placement within the healthcare system to disrupt stalled approaches can also revolutionize programs and streamline patient flow. Multipurpose tests will bring programs closer together, while POC testing will introduce a new cadre of healthcare providers empowered to link patients to care. Some of the more revolutionary diagnostics available right now allow testing at birth for EID of HIV and HIV self-testing at home, thus, bringing diagnostics, the patient and the healthcare provider ever closer together.

In order to fully leverage these advances to improve HIV care, countries need to develop a fully connected diagnostic system comprising a system of laboratories linked to a network of POC testing sites offering patient-centered treatment services, while continuing to address socioeconomic and cultural challenges to uptake of testing. Understanding the available technologies and what works in different contexts is critical to ensuring the best possible care for patients with HIV.

Finally, it is important to take the lessons learned from HIV testing with all the resources, technology and investment over the years to other diseases and programs, and ensure integrated care across the healthcare system. Innovations to create a well-connected patient-centered diagnostic system with intimate links to service delivery must become government policies in order to truly revolutionize healthcare in resource-limited settings across the world.

Bibliography

Boeras, D.I., Nkengasong, J.N., and Peeling, R.W. (2017). Implementation science: the laboratory as a command Centre. *Curr. Opin. HIV AIDS* 12 (2): 171–174.

Boeras, D.I. and Peeling, R.W. (2016). External quality assurance for HIV point-of-care testing in Africa: a collaborative country-partner approach to strengthen diagnostic services. *Afr. J. Lab. Med.* 5 (2): 556.

Carmona, S., Wedderburn, C., Macleod, W., et al. (2016). *Field performance of point-of-care HIV testing for early infant diagnosis: Pooled analysis from six countries from the EID Consortium.* IAS: AIDS 2016; Durban, South Africa.

Cheng, B., Cunningham, B., Boeras, D.I. et al. (2016). Data connectivity: a critical tool for external quality assessment. *Afr. J. Lab. Med.* 5 (2): 535.

De Cock, K.M., Jaffe, H.W., and Curran, J.W. (2011). Reflections on 30 years of AIDS. *Emerg. Infect. Dis.* 17 (6): 1044–1048.

Diallo, K., Modi, S., Hurlston, M. et al. (2017). A proposed framework for the implementation of early infant diagnosis point-of-care. *AIDS Res. Hum. Retrovir.* 33 (3): 203–210.

Dunning, L., Kroon, M., Hsiao, N.Y., and Myer, L. (2017). Field evaluation of HIV point-of-care testing for early infant diagnosis in Cape Town, South Africa. *PLoS One* 12 (12): e0189226.

Ferrand, R.A. (2017). Gaps in the early infant diagnosis cascade in a high HIV prevalence setting. *Public Health Action* 7 (2): 78.

Fonjungo, P.N., Osmanov, S., Kuritsky, J. et al. (2016). Ensuring quality: a key consideration in scaling-up HIV-related point-of-care testing programs. *AIDS* 30 (8): 1317–1323.

Garcia, P.J., Carcamo, C.P., Chiappe, M. et al. (2013). Rapid syphilis tests as catalysts for health systems strengthening: a case study from Peru. *PLoS One* 8 (6): e66905.

IeDea, Collaborations ARTC, Avila, D., Althoff, K.N. et al. (2014). Immunodeficiency at the start of combination antiretroviral therapy in low-, middle-, and high-income countries. *J. Acquir. Immune Defic. Syndr.* 65 (1): e8–e16.

Jani, I.V., Meggi, B., Mabunda, N., et al. (2016). *Effect of point-of-care early infant diagnosis on retention of patients and rates of antiretroviral therapy initiation on primary healthcare clinics: a cluster-randomized trial in Mozambique.* International AIDS Society: AIDS; Durban, South Africa.

Jani, I.V. and Peter, T.F. (2013). How point-of-care testing could drive innovation in global health. *N. Engl. J. Med.* 368 (2324): 2319–2324.

Jani, I.V., Sitoe, N.E., Alfai, E.R. et al. (2011). Effect of point-of-care CD4 cell count tests on retention of patients and rates of antiretroviral therapy initiation in primary health clinics: an observational cohort study. *Lancet* 378 (9802): 1572–1579.

Kaindjee-Tjituka, F., Sawadogo, S., Mutandi, G. et al. (2017). Task-shifting point-of-care CD4+ testing to lay health workers in HIV care and treatment services in Namibia. *Afr. J. Lab. Med.* 6 (1): 643.

Kosack, C.S., Page, A.L., and Klatser, P.R. (2017). A guide to aid the selection of diagnostic tests. *Bull. World Health Org.* 95 (9): 639–645.

Mabey, D., Peeling, R.W., Ustianowski, A., and Perkins, M.D. (2004). Diagnostics for the developing world. *Nat. Rev. Microbiol.* 2 (3): 231–240.

Mabey, D.C., Sollis, K.A., Kelly, H.A. et al. (2012). Point-of-care tests to strengthen health systems and save newborn lives: the case of syphilis. *PLoS Med.* 9 (6): e1001233.

McNeish, H. (2016). The First HIV-Fighting Drones Have Been Deployed in Africa: Vice. https://www.vice.com/en_ca/article/8gkpbk/unicef-just-launched-the-first-hiv-fighting-drones-in-africa

Meggi, B., Bollinger, T., Mabunda, N. et al. (2017). Point-of-care p24 infant testing for HIV may increase patient identification despite low sensitivity. *PLoS One* 12 (1): e0169497.

Mwenda, R. (2016). *Impact of Point-of-Care EID Testing into the National EID Program: Pilot Experiences from Malawi*. International AIDS Society: AIDS 2016; Durban, South Africa.

Newell, M.L., Coovadia, H., Cortina-Borja, M. et al. (2004). Mortality of infected and uninfected infants born to HIV-infected mothers in Africa: a pooled analysis. *Lancet* 364 (9441): 1236–1243.

Omega Diagnostics Group. (2018). Available from http://www.omegadiagnostics.com/Products/Infectious-Diseases/HIV/CD4. Accessed March 2018.

Peeling, R.W., Sollis, K.A., Glover, S. et al. (2015). CD4 enumeration technologies: a systematic review of test performance for determining eligibility for antiretroviral therapy. *PLoS One* 10 (3): e0115019.

Semitala, F.C., Camlin, C.S., Wallenta, J. et al. (2017). Understanding uptake of an intervention to accelerate antiretroviral therapy initiation in Uganda via qualitative inquiry. *J. Int. AIDS Soc.* 20 (4): https://doi.org/10.1002/jia2.25033.

Sibanda, E.L., Weller, I.V., Hakim, J.G., and Cowan, F.M. (2013). The magnitude of loss to follow-up of HIV-exposed infants along the prevention of mother-to-child HIV transmission continuum of care: a systematic review and meta-analysis. *AIDS* 27 (17): 2787–2797.

UNAIDS. (2014). 90-90-90 An ambitious treatment target to help end the AIDS epidemic.

UNAIDS. (2015). Children and HIV:n Fact Sheets. http://www.unaids.org/sites/default/files/media_asset/FactSheet_Children_en.pdf

UNAIDS. (2016). On the fast-track to an AIDS free generation: The incredible journey of the Global Plan towards the elimination of new HIV infections among children by 2015 and keeping their mothers alive. http://www.unaids.org/sites/default/files/media_asset/GlobalPlan2016_en.pdf

UNAIDS. (2017). Ending AIDS: Progress towards the 90-90-90 targets. http://www.unaids.org/en/resources/documents/2017/20170720_Global_AIDS_update_2017

UNAIDS. Fact sheet - Latest statistics on the status of the AIDS epidemic 2017. http://www.unaids.org/en/resources/fact-sheet

UNAIDS. AIDSinfo. (2018) http://aidsinfo.unaids.org

UNICEF. (2016a). For every child, end AIDS: Seventh Stocktaking Report.

UNICEF. (2016b). Key Considerations for Introducing Point of Care Diagnostic Technologies in National Laboratory Programmes: A Forthcoming Resource.

UNICEF. (2016c). Women: At the heart of the HIV response for children. https://www.unicef.de/blob/171450/3867f9482f89ea120e73c77e3c390046/unicef-report-women-at-the-heart-of-the-hiv-response-for-children-data.pdf

UNICEF. (2016d) Accelerate access to innovative point of care (POC) HIV diagnostics: CD4, EID and VL 2016. https://www.unicef.org/innovation/innovation_82102.html

UNITAID. (2015). HIV/AIDS Diagnostics Technology Landscape. 2015 October.

Violari, A., Cotton, M.F., Gibb, D.M. et al. (2008). Early antiretroviral therapy and mortality among HIV-infected infants. *N. Engl. J. Med.* 359 (21): 2233–2244.

Vojnov, L., Markby, J., Boeke, C. et al. (2016). POC CD4 testing improves linkage to HIV care and timeliness of ART initiation in a public health approach: a systematic review and meta-analysis. *PLoS One* 11 (5): e0155256.

Waldrop, G., Doherty, M., Vitoria, M., and Ford, N. (2016). Stable patients and patients with advanced disease: consensus definitions to support sustained scale up of antiretroviral therapy. *Trop. Med. Int. Health* 21 (9): 1124–1130.

WHO (2013). *Consolidated guidelines on the use of antiretroviral drugs for treating and preventing HIV infection*. Geneva, Switzerland: World Health Organisation.

WHO (2014). *Elimination of mother-to-child transmission (EMTCT) of HIV and syphilis: Global guidance on criteria and processes for validation*. Geneva, Switzerland: World Health Organisation.

WHO (2015a). *Consolidated guidelines on HIV testing services*. Geneva, Switzerland: World Health Organisation.

WHO (2015b). *HIV diagnostic tests in low- and middle-income countries: forecasts of global demand for 2014–2018*. Geneva, Switzerland: World Health Organisation.

WHO. (2016a). Guidelines on HIV self-testing and partner notification.

WHO (2016b). *Consolidated Guidelines on the Use of Antiretroviral Drugs for Treating and Preventing HIV Infection*. Geneva, Switzerland: World Health Organisation.

WHO (2017a). *Guidelines for managing advanced HIV disease and rapid initiation of antiretroviral therapy*. Geneva, Switzerland: World Health Organisation.

WHO (2017b). *What's new in treatment monitoring: Viral Load and CD4 testing*. Geneva, Switzerland: World Health Organisation.

WHO (2017c). *Novel point-of-care tools for early infant diagnosis of HIV*. Geneva, Switzerland: World Health Organisation.

WHO. (2017d). The evolving role of CD4 cell counts in HIV care. https://www.who.int/hiv/pub/journal_articles/The_evolving_role_of_CD4_cell_counts_in_HIV_care.pdf

WHO. (2018). WHO HIV test kit evaluations. https://www.who.int/diagnostics_laboratory/evaluations/hiv/en/ Accessed 2018.

WHO. In vitro diagnostics and laboratory technology. http://www.who.int/diagnostics_laboratory/en

Zhang, M. and Versalovic, J. (2002). HIV update. Diagnostic tests and markers of disease progression and response to therapy. *Am. J. Clin. Pathol.* 118 (Suppl): S26–S32.

Webliography

http://aidsinfo.unaids.org UNAIDS. AIDSinfo.

http://www.omegadiagnostics.com/Products/Infectious-Diseases/HIV/CD4 Accessed March 2018. Omega Diagnostics Group.

http://www.unaids.org/en/resources/documents/2017/20170720_Global_AIDS_update_2017 UNAIDS. Ending AIDS: Progress towards the 90-90-90 targets. 2017.

http://www.unaids.org/en/resources/fact-sheet UNAIDS. Fact sheet - Latest statistics on the status of the AIDS epidemic 2017.

http://www.unaids.org/sites/default/files/media_asset/GlobalPlan2016_en.pdf 2016. UNAIDS. On the fast-track to an AIDS free generation: The incredible journey of the

Global Plan towards the elimination of new HIV infections among children by 2015 and keeping their mothers alive.

http://www.who.int/diagnostics_laboratory/en WHO. *In vitro* diagnostics and laboratory technology.

https://www.unicef.org/innovation/innovation_82102.html UNICEF. Accelerate access to innovative point of care (POC) HIV diagnostics: CD4, EID and VL 2016.

https://www.vice.com/en_ca/article/8gkpbk/unicef-just-launched-the-first-hiv-fighting-drones-in-africa McNeish H. The First HIV-Fighting Drones Have Been Deployed in Africa: Vice; 2016.

www.eidconsortium.org - early infant death consortium.

5

Rapid Point-of-Care Diagnostic Tests for Tuberculosis

Richard Lessells

KwaZulu-Natal Research Innovation and Sequencing Platform, Nelson R Mandela School of Medicine, University of KwaZulu-Natal, Durban, South Africa
Department of Clinical Research, London School of Hygiene and Tropical Medicine, London, UK

5.1 Introduction

Despite effective antimicrobial chemotherapy that can cure most cases of TB, around 28 000 people develop TB disease and 5000 people die from TB every day (World Health Organization 2016b). The World Health Organization (WHO) declared TB a global public health emergency in 1993, yet two decades later it is the leading cause of death from infectious diseases (WHO 1993, 2016b).

Early diagnosis of TB with universal drug susceptibility testing (DST) is one of the core components of the WHO End TB Strategy (WHO 2015a). Early case detection and initiation of appropriate anti-TB therapy is important not only to reduce TB-related morbidity and mortality but also to interrupt transmission. As *Mycobacterium tuberculosis* is transmitted primarily by people with active pulmonary disease, detection of pulmonary TB is the priority for TB programs. Universal DST is important because around one in 20 TB cases globally are caused by rifampicin-resistant or multidrug-resistant *M. tuberculosis* strains. This chapter therefore concentrates on diagnostics for active TB disease and for detection of drug resistance.

Although there is no accepted universal definition of point-of-care (POC) testing, it generally implies the ability to perform a test at or near the patient to enable the rapid initiation of appropriate treatment. Point-of-care testing for TB encompasses testing of people admitted to hospital, testing of people attending a peripheral clinic, or testing people in a community setting or at home. This chapter includes discussion on the diagnostic technologies that are suitable for use in these contexts and appraises the evidence around their use in POC testing programs. While the focus is on technologies endorsed by the WHO, it is also important to look to the future and discuss products in development.

5.2 The Need for Rapid Point-of-Care TB Diagnostic Tests

5.2.1 The Diagnostic Gap

In 2015 there were an estimated 10.4 million new TB cases worldwide. Only 6.1 million TB cases were notified, suggesting that around two in every five people with TB went undiagnosed or unreported (WHO 2016b). Of the notified cases, only around half were bacteriologically confirmed (i.e. had a positive diagnostic test for TB).

The situation was worse for drug-resistant TB, where only around one in four of the estimated 580 000 cases of rifampicin-resistant or multidrug-resistant TB was detected and notified (WHO 2016b). This is partly explained by the widespread lack of DST. Even among the bacteriologically confirmed TB cases in 2015, less than one in three had any form of DST for rifampicin.

There is also a huge diagnostic gap for children. Estimates suggest that around two-thirds of all children with active TB disease are not notified, and that the vast majority of those with drug-resistant TB disease are not diagnosed (Dodd et al. 2014, 2016).

5.3 Weaknesses in the TB Diagnostic Cascade

Even when people with TB present to health services, not all will be diagnosed and start treatment in a timely fashion. Substantial health system delays have been documented, which are partly attributable to the poor performance of diagnostic tests and broader weaknesses in the systems in which diagnostics are deployed (Sreeramareddy et al. 2009; Storla et al. 2008). In addition to delays, people can also drop out at several points in the diagnostic cascade. Weaknesses in the diagnostic cascade are highlighted by the finding that in Africa about one in five people with a confirmed TB diagnosis do not start treatment (MacPherson et al. 2014). These delays and loss to follow-up are accentuated for drug-resistant TB (DR-TB), even in the era of rapid molecular diagnostic tests. In a nationwide retrospective cohort study from South Africa, around 40% of people with rifampicin-resistant TB detected by Xpert (see Section 5.7) in 2013 had not started treatment within six months (Cox et al. 2017).

5.4   Potential Impact of Rapid Point-of-Care Diagnostic Tests

Mathematical modeling has highlighted the potential impact not only of improved sensitivity of TB diagnostics but also of improving access through decentralized deployment. One mathematical model suggested that a new highly accurate rapid diagnostic test could prevent around one quarter of all TB deaths if implemented within existing healthcare infrastructure. If access was improved and it was deployed at a more peripheral level then the impact could be even greater with over one-third of all TB deaths being averted (Keeler et al. 2006). Findings from this model also emphasized the impact of loss to follow-up within the diagnostic process, with a reduction in the lives saved by almost half with 20% loss to follow-up (failure to provide all specimens or to return for results).

5.5   Defining the Diagnostic Needs

Historically TB diagnostic development has often taken place without the end user in mind. Given the importance of improved diagnostics for achieving the goals of the End TB strategy, efforts have been made to identify the highest priority needs to guide diagnostic developers and innovators (Denkinger et al. 2015; WHO 2014). The high-priority diagnostic needs with a summary of the target product profiles (TPP) agreed during a WHO-led consensus meeting in 2014 are shown in Table 5.1.

5.6   Smear Microscopy

Direct observation of *M. tuberculosis* organisms by microscopy (smear microscopy) remains the primary diagnostic method for TB in much of the world (Cazabon et al. 2017). Smear microscopy is relatively simple, rapid, and inexpensive. Microscopy identifies the most infectious cases but performs poorly at detecting paucibacillary disease. The sensitivity of conventional microscopy on two sputum smears, compared to a reference standard of mycobacterial culture, is around 50–70% in well-conducted clinical studies (Cuevas et al. 2011; Davis et al. 2013). However, the actual sensitivity of smear microscopy in routine practice may be as low as 20%, particularly in high HIV prevalence settings (Perkins and Cunningham 2007; Steingart et al. 2006). Other limitations of microscopy include the inability to discriminate between *M. tuberculosis* and non-tuberculous mycobacteria, to determine the viability of organisms and to detect drug resistance.

Table 5.1 The four target product profiles identified by stakeholders at a WHO-convened meeting in 2014 considered to be of high priority.

1. A point-of-care non-sputum-based test capable of detecting all forms of TB by identifying characteristic biomarkers or biosignatures (known as the biomarker test)
2. A point-of-care triage test, which should be a simple, low-cost test that can be used by first-contact healthcare providers to identify those who need further testing (the triage test)
3. A point-of-care sputum-based test to replace smear microscopy for detecting pulmonary TB (the smear-replacement test)
4. A rapid drug-susceptibility test that can be used at the microscopy-centre level of the healthcare system to select first-line regimen-based therapy (the rapid DST test).

5.6.1 Improving the Performance of Smear Microscopy

The performance of smear microscopy can be improved by relatively simple physical and chemical processing methods, such as treating the specimen with bleach or sodium hydroxide followed by centrifugation (Steingart et al. 2006). Fluorescence microscopy also improves sensitivity by around 10% compared to conventional microscopy, without loss of specificity (Steingart et al. 2006). Light-emitting diode (LED) microscopes have lower cost, power and maintenance requirements than conventional fluorescent microscopes and are more suited for use at peripheral levels. In 2011 the WHO recommended that conventional fluorescence microscopy be replaced by LED microscopy, and that LED microscopy be phased in as an alternative for conventional light microscopy (WHO 2011a).

Another potential incremental improvement to microscopy could be the use of automated digital systems that can evaluate microscopic images to identify TB-like organisms. Such technology could reduce the need for skilled microscopists in high burden settings. Initial studies of the TBDx system (Applied Visual Sciences, Herndon, VA) in South Africa, Peru and Vietnam have demonstrated suboptimal performance as a stand-alone test compared to conventional microscopy (Ismail et al. 2015; Nabeta et al. 2017). However, algorithms in which automated digital microscopy is followed by confirmatory testing for low-positive results could be a cost-effective alternative to molecular diagnostics (Jha et al. 2016).

5.6.2 Point-of-Care Smear Microscopy

Although smear microscopy can be used at peripheral levels of the health system, POC testing programs providing same-day diagnosis and treatment are surprisingly rare (Davis et al. 2012; WHO 2011c). Modeling has suggested that same-day microscopy could have a similar effect on TB incidence as the introduction of a rapid molecular diagnostic test such as Xpert (Dowdy et al. 2013).

In one of the randomized controlled trials evaluating the impact of Xpert MTB/RIF (the TB-NEAT study), the control arm involved POC microscopy at primary healthcare clinics in southern Africa. In this study where 60% of the study population were HIV positive, the sensitivity of smear microscopy (fluorescent microscopy in most sites) was around 50% and around one in three people with culture-positive TB received same-day treatment after POC microscopy (Theron et al. 2014). These findings suggest that, although point-of-care deployment of smear microscopy could translate to same-day treatment for some of the most infectious people with TB, the potential for greater impact is limited primarily by the poor sensitivity of the test.

5.7 Molecular Diagnostic Tests

5.7.1 Xpert MTB/RIF

Xpert® MTB/RIF (Cepheid, Sunnyvale, CA) is a molecular test for the detection of *M. tuberculosis* complex and rifampicin resistance (Boehme et al. 2010, 2011; Helb et al. 2010). Through targeting the *rpo*B gene, the assay detects the DNA sequence specific to the *M. tuberculosis* complex and identifies mutations associated with rifampicin resistance. Rifampicin resistance is considered a reasonable proxy for multidrug-resistant TB (MDR-TB), as approximately 90% of rifampicin-resistant *M. tuberculosis* isolates worldwide are also resistant to isoniazid (Kurbatova et al. 2012; Smith et al. 2012).

Xpert has good accuracy for the detection of *M. tuberculosis* (Steingart et al. 2014). A single Xpert test on sputum will detect around 80% of culture-confirmed cases. Sensitivity is somewhat lower in people with HIV. In comparison to smear microscopy, Xpert increases TB detection among culture-confirmed cases by 23% (Steingart et al. 2014). Xpert also provides accurate results for rifampicin resistance, although assessment of accuracy depends on the reference standard employed. Discordance between Xpert and other genotypic and phenotypic DST methods can occur for a number of reasons (Van Rie et al. 2012; Zetola et al. 2014).

In 2011 the WHO first recommended the use of Xpert for people living with HIV and people at high risk of MDR-TB (WHO 2011b). In 2013 the policy was updated to recommend Xpert as the initial diagnostic test for all people requiring investigation for TB (World Health Organization 2013a). Scale-up of Xpert in most countries has been relatively slow, and by 2015 only one in three countries with a high burden of TB and/or drug-resistant TB were using Xpert as the initial diagnostic test for all individuals (Albert et al. 2016; Cazabon et al. 2017; Pai and Furin 2017). In the 22 high burden countries, on average there are still about 10 smear microscopy tests done for every Xpert test (Cazabon et al. 2017). The exception has been South Africa, which adopted Xpert as the initial diagnostic test for all people in 2011 and achieved full coverage through the national laboratory system by 2013 (Albert et al. 2016).

Whilst cost may be main driver of the slow uptake, another factor may be the lack of compelling evidence of benefit over smear microscopy. The introduction of Xpert stimulated a range of robust independent studies to evaluate the effectiveness and impact in real-world settings. The TB-NEAT study was a randomized controlled trial (RCT) in which 1502 individuals with symptoms suggestive of pulmonary TB disease at primary healthcare (PHC) clinics were randomly allocated to have sputum smear microscopy or Xpert MTB/RIF diagnostic testing (Theron et al. 2014). In all but one site, diagnostic testing was performed on site. There was no difference between the microscopy and Xpert groups in the primary outcome of TB morbidity at two months or six months in culture-positive cases that started treatment. Despite this, there were some clear benefits to Xpert testing: Xpert detected a greater proportion of culture-positive cases than microscopy (83% versus 50%) and a greater proportion of culture-positive cases received same-day TB treatment in the Xpert arm (60% versus 37%).

The XTEND study was a pragmatic, cluster randomized trial nested within the South African national roll-out of Xpert (Churchyard et al. 2015). Twenty clusters (each cluster consisted of two PHC clinics and an off-site laboratory) were allocated to

Xpert or microscopy. Analysis included 4656 individuals investigated for TB. There was no difference in the primary outcome of mortality at six months: 4% in the Xpert group versus 5% in the microscopy group. There was a modest increase in diagnostic yield from the initial sputum specimen but pre-treatment loss to follow-up at day 28 was similar in both groups (17% in Xpert group versus 15% in microscopy group). Therefore, the improved yield from Xpert may have been offset somewhat by the testing not being performed at the point-of-care.

5.7.2   Point-of-Care Testing with Xpert

Although initially developed as a near-patient test suitable for POC use, its deployment so far has largely been through established centralized laboratory networks (Albert et al. 2016; Weyer et al. 2013). The two RCTs described above were both designed to evaluate the impact of Xpert compared to microscopy, but neither were designed to test specifically the impact of POC testing. TB-NEAT employed POC testing in both arms, whilst XTEND used off-site laboratory testing for both arms. A cluster randomized trial in rural South Africa addressed the impact of POC placement by directly comparing a nurse-delivered POC Xpert testing program at clinic level to standard laboratory-based Xpert testing (Figure 5.1) (Lessells et al. 2013, 2017).

In this study there was no evidence that the POC Xpert testing program increased the number of culture-confirmed cases on appropriate treatment by 30 days. However, there was evidence that POC testing strengthened the diagnostic cascade and accelerated initiation of appropriate treatment. In this and other studies of POC Xpert testing programs, 50–70% of people with a positive Xpert test received same-day treatment (Hanrahan et al. 2013, 2015; Lessells et al. 2017; Theron et al. 2014). This is an important reminder that in the real world POC testing does not automatically translate to

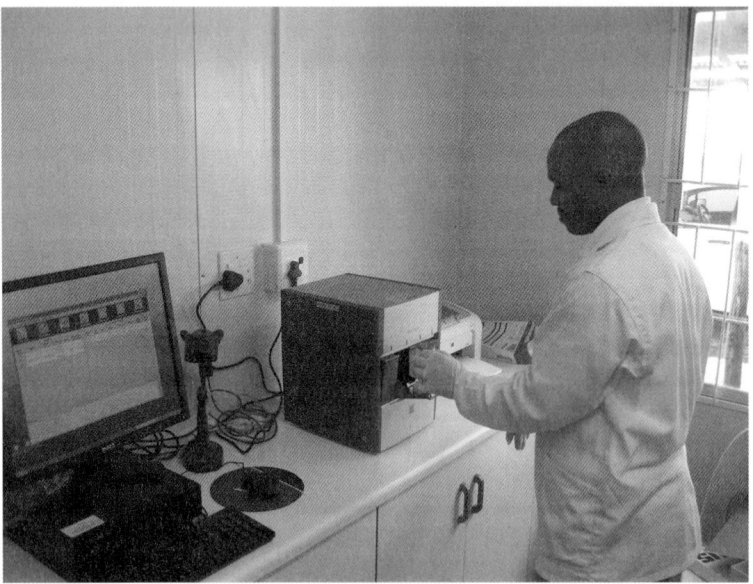

Figure 5.1 Nurse conducting Xpert MTB/RIF testing at a primary healthcare clinic in South Africa (Lessells et al. 2013).

same-day treatment, and that better technologies and testing strategies are needed to achieve same-day treatment for all.

As TB case finding moves out of centralized health systems into communities, it is also important to evaluate diagnostic tests in this context. Xpert has been compared with smear microscopy as a POC testing system for community-based intensified case finding. In the XACT study more people with culture-confirmed TB started treatment within 60 days (86% versus 56% in the microscopy arm); around half received same-day treatment in the Xpert arm (Calligaro et al. 2017).

5.7.3 Impact of Xpert for Drug-Resistant TB

In South Africa the detection of rifampicin resistance with Xpert has helped to reduce time to initiation of drug-resistant TB treatment. Under culture-based DST systems there was usually two to three months between sputum specimen submission and initiation of multidrug-resistant TB treatment (Heller et al. 2010; Jacobson et al. 2013; Loveday et al. 2012). Xpert MTB/RIF has reduced this to around three weeks (Cox et al. 2017; Dlamini-Mvelase et al. 2014; Iruedo et al. 2017; Naidoo et al. 2014) and to under a week in some settings linked to decentralized treatment programs (Cox et al. 2015). Similar reductions in time to MDR-TB treatment from other settings in Europe and Asia have been observed (Nair et al. 2016; Stagg et al. 2016; van Kampen et al. 2015a, 2015b). However, there are as yet no reports of a POC testing program achieving same-day treatment for drug-resistant TB. This will require substantial parallel capacity building for delivery of drug-resistant TB treatment at primary healthcare clinics.

5.7.4 New Developments with Xpert MTB/RIF

Xpert MTB/RIF Ultra has been developed as a next generation assay to overcome some of the limitations of the original Xpert MTB/RIF test (Chakravorty et al. 2017b). The new assay uses different technology (automated post-PCR melt curve analysis) in order to improve sensitivity for detection of *M. tuberculosis* and to improve accuracy of resistance mutation detection. Evidence from an initial multicenter diagnostic accuracy study comparing a single Ultra assay to a single Xpert MTB/RIF, with culture as the reference standard, has suggested a modest gain in overall sensitivity (88% versus 83%) at the expense of a loss of specificity (95% versus 98%). The superior sensitivity seemed to be largely due to improved detection of *M. tuberculosis* in people with HIV-associated TB. Accuracy for the detection of rifampicin resistance was similar between the two assays (Chakravorty et al. 2017b; Foundation for Innovative New Diagnostics 2017; WHO 2017).

The first generation Xpert MTB/RIF assay will now be phased out and replaced by the Ultra assay and the WHO and the Global Laboratory Initiative (GLI) have released guidance on adoption of the new assay (Global Laboratory Initiative 2017; WHO 2017). While the improved sensitivity is encouraging, the implications of the reduced specificity are not completely clear. There have already been concerns about false positive Xpert results in people with prior TB treatment (Theron et al. 2016). This problem may become more pronounced with the Ultra assay and diagnostic algorithms may become more complicated as a result (Global Laboratory Initiative 2017). The impact of the reduced specificity on treatment decisions and patient-relevant outcomes will need to be studied carefully as the assay is introduced in high burden settings.

Currently in development is a new assay for the GeneXpert system that detects mutations associated with resistance to isoniazid, fluoroquinolones, and injectable agents, through the identification of mutations in six genes and promoter regions (*inh*A promoter, *kat*G, *gyr*A, *gyr*B, *eis* promoter and *rrs*) (Chakravorty et al. 2017a; Xie et al. 2017). While initial evaluation of the assay looks promising (Xie et al. 2017), results of independent evaluation studies are awaited and this has yet to undergo formal evaluation by the WHO.

Also, in development is a rechargeable battery-powered system (GeneXpert Omni) with POC use in home and community-based settings in mind (Cepheid 2015). This is a small portable unit weighing just 1 kg that is designed to have 16 hours of power without mains electricity. As yet there are no published data on the performance of the Xpert assays using this platform. While it may make POC testing more feasible, each unit is single use and the assay time will not be substantially shortened. There is one ongoing RCT (XACT-2) exploring the impact of Xpert testing using the Omni system in a community-based intensified case finding (ICF) study (University of Cape Town 2017).

5.8 Loop-Mediated Isothermal Amplification (LAMP)

Loop-mediated isothermal amplification (LAMP) is a simple, low cost molecular assay which has been applied to several infectious diseases (Mori and Notomi 2009). It is a temperature-independent technique for amplifying DNA that does not require complex laboratory infrastructure and can therefore be used at peripheral levels of the health system. A commercial assay for TB, the Loopamp™ MTBC Detection Kit (Eiken Chemical, Tokyo, Japan), was developed as a potential replacement for smear microscopy. TB-LAMP is a relatively high-throughput, low complexity manual assay that requires less than one hour to perform and can be read with the naked eye under ultraviolet light.

When the WHO first considered the TB-LAMP assay in 2012, an expert group felt that there was insufficient evidence to recommend its use as a replacement for smear microscopy. In particular, there were concerns over suboptimal specificity, extensive training requirements, and cost (WHO 2013b). In 2016 the WHO reappraised the evidence surrounding the TB-LAMP assay, following modifications to the technology and a number of independent evaluations (WHO 2016c). The evidence that was appraised included results of 13 diagnostic accuracy studies, with a total of 4760 participants. Overall the evidence suggested that sensitivity of TB-LAMP was marginally better than smear microscopy, but below that of Xpert MTB/RIF. This time the WHO recommended that TB-LAMP could be used, as a replacement to smear microscopy or as a follow-on test, at peripheral levels of the health system where infrastructure would not allow use of Xpert (WHO 2016c). As expected, sensitivity of TB-LAMP was lower in people living with HIV, and no recommendation was given about its use in this group. Beyond the diagnostic accuracy studies, there is no robust evidence around the effectiveness and impact of TB-LAMP. At best this might provide a small incremental benefit to smear microscopy, but the limitations of not detecting drug resistance and of being operator-dependent mean that it is unlikely to be transformative.

5.9 Line Probe Assays

Line probe assays (LPA) are rapid molecular diagnostic tests that can detect *M. tuberculosis* and drug resistance. The WHO-recommended commercially available tests include GenoType MTBDR*plus* and MTBDR*sl* (Hain Lifescience, Nehren, Germany), and the Nipro NTM + MDRTB detection kit 2 (Nipro, Osaka, Japan) (WHO 2016d). LPAs can be performed on sputum specimens (direct testing) or on cultured isolates of *M. tuberculosis* (indirect testing). The first versions of the Hain assays had suboptimal sensitivity for the detection of *M. tuberculosis* on direct testing. However, with the version 2 MTBDR*plus* assay, the sensitivity for the detection of *M. tuberculosis* in direct testing is approximately equivalent to that of Xpert (Barnard et al. 2012; Crudu et al. 2012).

The Genotype MTBDR*plus* assay detects resistance to rifampicin (*rpo*B) and isoniazid (*inh*A promoter and *kat*G). The MTBDRsl assay detects resistance to fluoroquinolones (*gyr*A and *gyr*B) and second-line injectable agents (*eis* promoter and *rrs*). The NTM+MDRTB detection kit 2 detects resistance to rifampicin (*rpo*B gene), isoniazid (*inh*A promoter, *kat*G, *fab*G1, *fur*A), pyrazinamide (*pnc*A) and fluoroquinolones (*gyr*A).

Although these tests can be performed within one day, the complexity of the laboratory processes and interpretation means that they are not suitable for point-of-care use at peripheral levels (WHO 2016d). Even when deployed at higher levels they are rarely processed to allow same-day treatment decisions.

5.10 Other Molecular Tests

Following the adoption of Xpert, it was expected that several competing automated, modular technologies (fast followers) would come to the market. In reality progress has been slow and, although the first wave of products has now reached the market, none have yet been endorsed by the WHO.

Truenat™ MTB (Molbio Diagnostics, Goa, India) is a chip-based real-time polymerase chain reaction (PCR) test for the detection of TB (Nikam et al. 2013, 2014). The instrument is battery-powered and portable so it is potentially suitable for use in peripheral laboratories, but is single use, thus suitable only for low throughput scenarios. The test is not fully integrated but involves a two-step process of DNA extraction followed by amplification and detection, and is therefore more complicated than Xpert. Overall turnaround time is about an hour. An additional assay taking an hour is required to detect rifampicin resistance. There is some evidence that the performance of the test for detection of *M. tuberculosis* is similar to that of Xpert (Nikam et al. 2014). Results are awaited from a multicenter study coordinated by the Indian Council of Medical Research and additional evaluation studies conducted by the Foundation for Innovative New Diagnostics (FIND). The cost of the instrument and assay for the public sector has not yet been agreed (UNITAID 2017).

Genedrive® MTB/RIF (Epistem Ltd., Manchester, UK) is a real-time PCR assay for the detection of *M. tuberculosis* and rifampicin resistance (Castan et al. 2014). This also involves a two-step process with a paper-based DNA extraction method followed by a cartridge-based PCR amplification and detection using a small, single use, battery-powered system. Overall turnaround time is around one hour. Initial reports from the manufacturers seemed promising, with performance similar to Xpert using spiked sputum specimens (Castan et al. 2014). However, a recent independent evaluation showed extremely poor performance, with sensitivity for the detection of TB below that of smear microscopy (Shenai et al. 2016). In addition, there was evidence of cross-reactivity with some non-tuberculous mycobacteria (NTM) and evidence that specimen processing did not make the specimen biosafe, an important issue for decentralized deployment. The low sensitivity was postulated to be due to the small volume of specimen placed into the assay cartridge. Despite this, the product is still being marketed in India.

5.11 Antigen Tests

5.11.1 Urine Lipoarabinomannan Assay

Lipoarabinomannan (LAM) is a glycolipid component of the cell wall that is released from actively replicating and dying mycobacteria. Tests based on the detection of LAM in the urine have been developed for POC diagnosis of active TB disease in people living with HIV (Lawn 2012). The simple, low-cost lateral-flow (LF) assay (Determine TB LAM Ag, Alere Inc., Waltham, MA) can be performed at the bedside (Figure 5.2). A small amount of urine is added to the test strip and the result is then determined after 25 minutes by comparing visible bands on the test strip to a reference card.

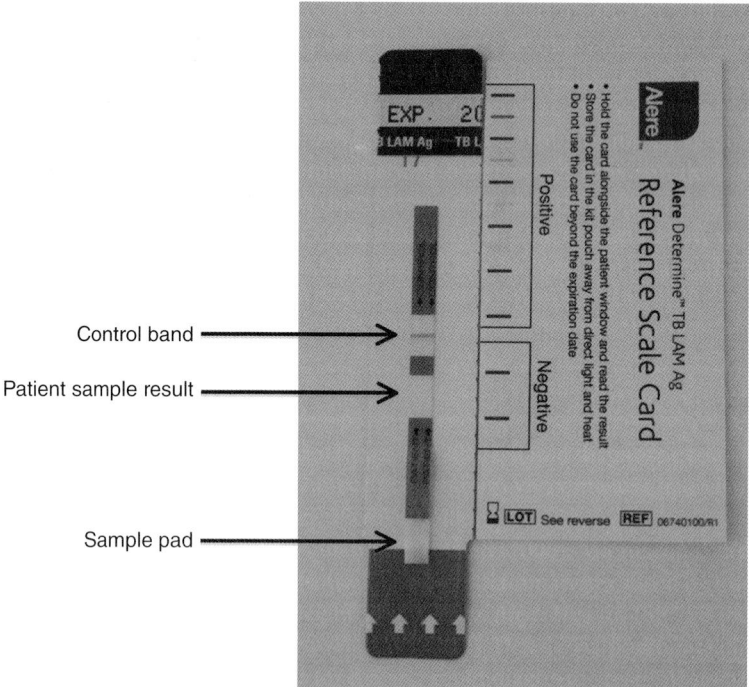

Figure 5.2 Lateral flow lipoarabinomannan test strip (Lawn 2012).

It was initially presumed that detection of LAM in urine resulted from simple renal filtration of LAM circulating in the bloodstream. However, this has recently been disputed and it has been argued that LAM appears in the urine in most cases from direct involvement of the renal tract (Cox et al. 2015; Lawn and Gupta-Wright 2016). The LF-LAM cannot distinguish *M. tuberculosis* complex from non-tuberculous mycobacteria, but the clinical significance and impact of this is uncertain (Nel et al. 2017).

A Cochrane review used data from studies evaluating the urine LAM assay for detection of TB in HIV-positive adults with symptoms suggestive of TB (Shah et al. 2016). Compared to a microbiological reference standard, the pooled sensitivity and specificity of a single urine LAM assay was 45 and 92% respectively. Pooled sensitivity was higher as the CD4$^+$ cell count decreased, reaching 62% in people with CD4$^+$ cell counts <50 cells per µL (Figure 5.3).

A multicenter randomized controlled trial in four African countries provided evidence of the impact of the urine LAM in HIV-positive adults admitted to hospital with symptoms suggestive of TB (Peter et al. 2016). Addition of urine LAM testing to routine TB diagnostics (smear microscopy, Xpert, and culture) led to a modest reduction in all-cause mortality at eight weeks - 21% versus 25%. The impact of the urine LAM test was driven predominantly by a difference in mortality for people with CD4$^+$ cell counts <50 cells per µL. The reduction in mortality occurred despite only a small increase in the overall proportion starting TB treatment. However, there was some evidence that LAM testing accelerated treatment initiation, particularly in sites with more scarce diagnostic resources (Peter et al. 2016).

At present the WHO recommends that the urine LAM may be used to assist in the diagnosis of TB in HIV-positive adult inpatients with signs and symptoms of TB (pulmonary and/or extrapulmonary) who have a CD4$^+$ cell count of ≤100 cells/µL or who are seriously ill regardless of the CD4$^+$ cell count (WHO 2015b). The RCT evidence may lead to strengthening of this policy recommendation but there is no evidence to suggest a role for POC LAM testing at more peripheral levels of the health system.

Despite the evidence and WHO recommendations, there has been poor uptake of LAM testing within national TB programs (Lessem 2017). As of September 2017, no country had adopted the LAM test into routine diagnostic algorithms. The FIND is currently working with Fujifilm (Tokyo, Japan) to develop a new, more sensitive next generation urine LAM assay.

5.12 Combination Diagnostic Packages

Given the strengths and weaknesses of individual diagnostic tests, additional impact could be achieved with combination diagnostic packages. This may be particularly the case for testing HIV-positive adults with advanced immunodeficiency. In this context the combination of sputum Xpert and urine testing (either with LAM or Xpert) offers a potential for better yield. One study in South Africa evaluated different combinations against a composite reference standard of TB diagnosis from any respiratory or non-respiratory specimen (Lawn et al. 2015, 2017). Sputum Xpert alone only detected around one in four TB cases. Sputum Xpert plus urine LAM detected 53% of cases, and sputum Xpert plus urine Xpert detected 78% of cases. The combinations had a higher

(a) Sensitivity

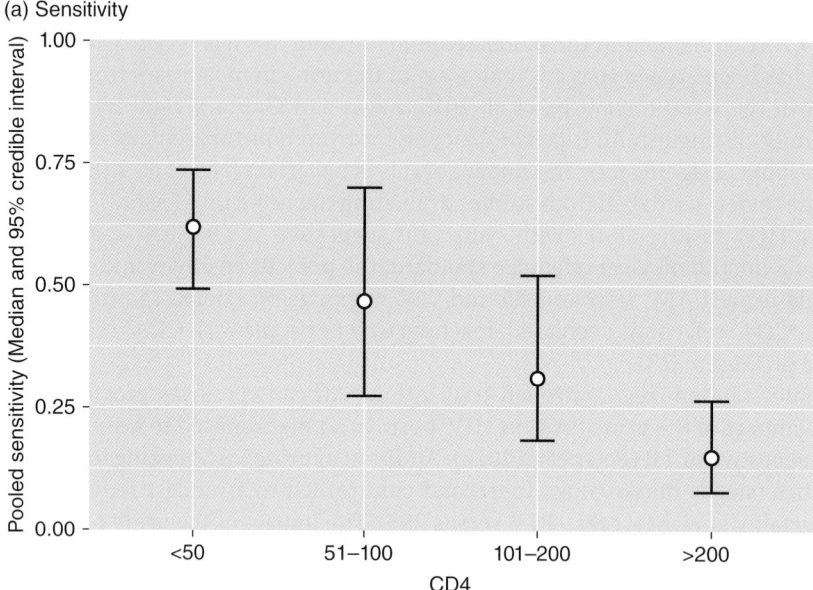

(b) Specificity

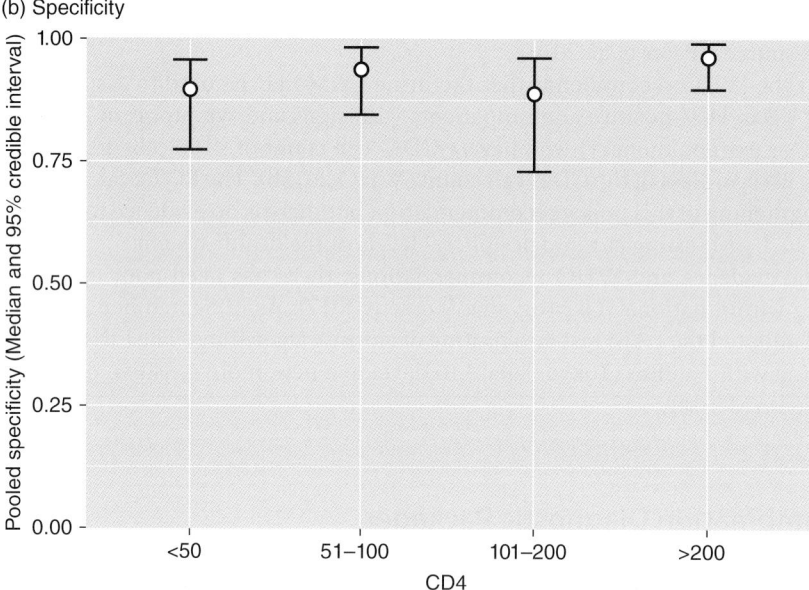

Figure 5.3 Plots of (a) sensitivity and (b) specificity of LF-LAM for TB diagnosis stratified by CD4[+] cell count (Shah et al. 2016).

yield in those with a CD4[+] cell count of <100 cells per µL: 72% for sputum Xpert plus urine LAM and 86% for sputum Xpert plus urine Xpert. The impact of these combination diagnostic packages on mortality is now being evaluated in a randomized controlled trial in South Africa and Malawi (Gupta-Wright et al. 2016).

5.13   Next Generation Sequencing

Next generation sequencing technologies have the potential to transform the fields of clinical infectious diseases and public health. Whole genome sequencing (WGS) can provide rapid identification of pathogens and associated drug resistance mutations. Furthermore, analysis of WGS data can deliver real-time surveillance to inform public health interventions (Didelot et al. 2012; Grad and Lipsitch 2014), something which holds promise for TB. This has been illustrated recently by the use of a portable nanopore DNA sequencing instrument for real-time surveillance of Ebola and Zika virus outbreaks (Grubaugh et al. 2017; Quick et al. 2016).

WGS of *M. tuberculosis* offers the potential for a more comprehensive prediction of drug susceptibility. One of the main technical challenges until now has been the need for culture as a first step, as sputum specimens contain *M. tuberculosis* mixed with large amounts of other bacterial and human material. However, novel DNA extraction and enrichment techniques have been developed that may allow sequencing directly from clinical specimens (Brown et al. 2015; Doughty et al. 2014; Votintseva et al. 2015, 2017). Until now there have only been retrospective and prospective validation studies of WGS against phenotypic DST as a predictor of treatment response (Pankhurst et al. 2016; Walker et al. 2015) and case reports or small case series from low burden settings of the application of WGS to guide DR-TB treatment (Koser et al. 2013; Witney et al. 2015). There is now a need for prospective studies to explore the impact and cost-effectiveness of sequencing within DR-TB diagnostic algorithms. A randomized controlled trial of WGS versus standard genotypic and phenotypic DST commenced in South Africa in 2017 (Centre for the Aids Programme of Research in South Africa 2017).

5.14   Diagnostic Imaging

Chest X-ray (CXR) imaging has good sensitivity but poor specificity for the detection of active TB. Although CXRs are widely used for the diagnosis of TB in high burden countries, there is little consistency in how they are used in diagnostic algorithms and X-ray systems are almost exclusively located at hospital level (Pande et al. 2015). There is now renewed interest in positioning X-ray systems at more peripheral levels or within mobile units to support community-based active case finding. This has been prompted partly by findings from recent national TB prevalence surveys in which 20–40% of people with bacteriologically confirmed TB reported no symptoms yet had detectable abnormalities on CXR (Kapata et al. 2016; Senkoro et al. 2016). This raises the question as to whether CXR has a role as a triage tool within diagnostic algorithms for TB.

Digital x-ray systems can provide high quality images rapidly at relatively low cost and can be implemented within primary healthcare clinics (Muyoyeta et al. 2014; Zachary et al. 2012). Digital imaging also allows for computer-aided detection (CAD), where software can be programmed to analyze digital images for abnormalities consistent with TB disease. There is currently one proprietary system (CAD4TB) developed by Delft Imaging Systems (Veenendaal, the Netherlands). As the software can be modified, systems can achieve high sensitivity but usually at the expense of low specificity

(Muyoyeta et al. 2014). There is a recognized need for more coordinated research to define the role of digital CXR and CAD in screening and diagnostic algorithms for TB (Ahmad Khan et al. 2017; WHO 2016a).

5.15 Other Diagnostics

The development of new non-sputum biomarker tests for active TB is a priority (Table 5.1). There is interest in using the host response to diagnose TB, using various "omics" approaches (Haas et al. 2016). The first challenge here is to identify a biosignature that consistently differentiates TB disease from latent TB infection and other diseases. The second challenge is to translate the technology into a rapid diagnostic test suitable for use at peripheral health centers and in community settings.

Another area of interest is in the use of volatile organic compounds (VOC) in breath. The product furthest along in development is the Aeonose™ for TB (The eNose Company, Zutphen, The Netherlands), which is currently undergoing validation studies (Bruins et al. 2013).

5.16 Conclusions

This chapter has included discussion of some of the TB diagnostic tests suitable for POC use that are currently in use or in the late stages of development (Figure 5.4).

Whilst the past few years has seen incremental improvements in TB diagnostics, there is still a need for transformative tools that will allow a healthcare worker in any location to detect TB and drug resistance within a few minutes at low cost. Whilst there are many obstacles along the path from conception to deployment, there are

	Non-sputum-based biomarker test for all forms of tuberculosis	Simple, low cost triage test (rule-out test)	Sputum-based replacement test for smear microscopy	Rapid drug susceptibility test
Setting	Health posts without laboratories	Community level	Primary healthcare centres	Microscopy centre
Target population	Adults and children with possible PTB & EPTB	Adults and children with signs and symptoms of TB	People with possible TB able to produce sputum	People with possible TB (especially if high risk DRTB)
Target user	HCWs with minimal training	Community health workers	HCWs with minimal training	HCWs with minimal training
Sample type	Urine, blood, saliva, exhaled air	Urine, blood, saliva, exhaled air	Sputum	Sputum
Time to result	<20 minutes	<5 minutes	<20 minutes	<30 minutes
Price of test	<USD 4	<USD 1	<USD 4	<USD 10

Figure 5.4 Summary of different tests for diagnosing pulmonary tuberculosis.

now several competing technologies in the TB diagnostic pipeline (Lessem 2017; UNITAID 2017). This gives hope that we will get the tools we need to achieve a world free of TB.

References

Ahmad Khan, F., Pande, T., Tessema, B. et al. (2017). Computer-aided reading of tuberculosis chest radiography: moving the research agenda forward to inform policy. *Eur. Respir. J.* 50: https://doi.org/10.1183/13993003.00953-2017.

Albert, H., Nathavitharana, R.R., Isaacs, C. et al. (2016). Development, roll-out and impact of Xpert MTB/RIF for tuberculosis: what lessons have we learnt and how can we do better? *Eur. Respir. J.* 48: 516–525.

Barnard, M., Gey Van Pittius, N.C., Van Helden, P.D. et al. (2012). The diagnostic performance of the GenoType MTBDRplus version 2 line probe assay is equivalent to that of the Xpert MTB/RIF assay. *J. Clin. Microbiol.* 50: 3712–3716.

Boehme, C.C., Nabeta, P., Hillemann, D. et al. (2010). Rapid molecular detection of tuberculosis and rifampin resistance. *N. Engl. J. Med.* 363: 1005–1015.

Boehme, C.C., Nicol, M.P., Nabeta, P. et al. (2011). Feasibility, diagnostic accuracy, and effectiveness of decentralised use of the Xpert MTB/RIF test for diagnosis of tuberculosis and multidrug resistance: a multicentre implementation study. *Lancet* 377: 1495–1505.

Brown, A.C., Bryant, J.M., Einer-Jensen, K. et al. (2015). Rapid whole-genome sequencing of mycobacterium tuberculosis isolates directly from clinical samples. *J. Clin. Microbiol.* 53: 2230–2237.

Bruins, M., Rahim, Z., Bos, A. et al. (2013). Diagnosis of active tuberculosis by e-nose analysis of exhaled air. *Tuberculosis (Edinb.)* 93: 232–238.

Calligaro, G.L., Zijenah, L.S., Peter, J.G. et al. (2017). Effect of new tuberculosis diagnostic technologies on community-based intensified case finding: a multicentre randomised controlled trial. *Lancet Infect. Dis.* 17: 441–450.

Castan, P., De Pablo, A., Fernandez-Romero, N. et al. (2014). Point-of-care system for detection of mycobacterium tuberculosis and rifampin resistance in sputum samples. *J. Clin. Microbiol.* 52: 502–507.

Cazabon, D., Suresh, A., Oghor, C. et al. (2017). Implementation of Xpert MTB/RIF in 22 high tuberculosis burden countries: are we making progress? *Eur. Respir. J.* 50: https://doi.org/10.1183/13993003.00918-201.

Centre for the Aids Programme of Research in South Africa. (2017). The Individualized M(X) Drug-resistant TB Treatment Strategy Study (InDEX) [Online]. Available: https://ClinicalTrials.gov/show/NCT03237182 [Accessed Sep 1 2017].

Cepheid. (2015). World's Most Portable Molecular Diagnostics System Unveiled at AACC [Online]. Available from https://www.prnewswire.com/news-releases/worlds-most-portable-molecular-diagnostics-system-unveiled-at-aacc-300119213.html Accessed Sep 1 2017.

Chakravorty, S., Roh, S.S., Glass, J. et al. (2017a). Detection of isoniazid-, fluoroquinolone-, amikacin-, and kanamycin-resistant tuberculosis in an automated, multiplexed 10-color assay suitable for point-of-care use. *J. Clin. Microbiol.* 55: 183–198.

Chakravorty, S., Simmons, A.M., Rowneki, M. et al. (2017b). The New Xpert MTB/RIF ultra: improving detection of mycobacterium tuberculosis and resistance to rifampin in an assay suitable for point-of-care testing. *MBio* 8: https://doi.org/10.1128/mBio.00812-17.

Churchyard, G.J., Stevens, W.S., Mametja, L.D. et al. (2015). Xpert MTB/RIF versus sputum microscopy as the initial diagnostic test for tuberculosis: a cluster-randomised trial embedded in South African roll-out of Xpert MTB/RIF. *Lancet Glob. Health* 3: e450–e457.

Cox, H.S., Daniels, J.F., Muller, O. et al. (2015). Impact of decentralized care and the Xpert MTB/RIF test on rifampicin-resistant tuberculosis treatment initiation in Khayelitsha, South Africa. *Open Forum Infect. Dis.* 2: ofv014.

Cox, H., Dickson-Hall, L., Ndjeka, N. et al. (2017). Delays and loss to follow-up before treatment of drug-resistant tuberculosis following implementation of Xpert MTB/RIF in South Africa: a retrospective cohort study. *PLoS Med.* 14: e1002238.

Cox, J.A., Lukande, R.L., Kalungi, S. et al. (2015). Is urinary Lipoarabinomannan the result of renal tuberculosis? Assessment of the renal histology in an autopsy cohort of Ugandan HIV-infected adults. *PLoS One* 10: e0123323.

Crudu, V., Stratan, E., Romancenco, E. et al. (2012). First evaluation of an improved assay for molecular genetic detection of tuberculosis as well as rifampin and isoniazid resistances. *J. Clin. Microbiol.* 50: 1264–1269.

Cuevas, L.E., Yassin, M.A., Al-Sonboli, N. et al. (2011). A multi-country non-inferiority cluster randomized trial of frontloaded smear microscopy for the diagnosis of pulmonary tuberculosis. *PLoS Med.* 8: e1000443.

Davis, J.L., Cattamanchi, A., Cuevas, L.E. et al. (2013). Diagnostic accuracy of same-day microscopy versus standard microscopy for pulmonary tuberculosis: a systematic review and meta-analysis. *Lancet Infect. Dis.* 13: 147–154.

Davis, J.L., Dowdy, D.W., Den Boon, S. et al. (2012). Test and treat: a new standard for smear-positive tuberculosis. *J. Acquir. Immune Defic. Syndr.* 61: e6–e8.

Denkinger, C.M., Kik, S.V., Cirillo, D.M. et al. (2015). Defining the needs for next generation assays for tuberculosis. *J. Infect. Dis.* 211 (Suppl 2): S29–S38.

Didelot, X., Bowden, R., Wilson, D.J. et al. (2012). Transforming clinical microbiology with bacterial genome sequencing. *Nat. Rev. Genet.* 13: 601–612.

Dlamini-Mvelase, N.R., Werner, L., Phili, R. et al. (2014). Effects of introducing Xpert MTB/RIF test on multi-drug resistant tuberculosis diagnosis in KwaZulu-Natal South Africa. *BMC Infect. Dis.* 14: 442.

Dodd, P.J., Gardiner, E., Coghlan, R., and Seddon, J.A. (2014). Burden of childhood tuberculosis in 22 high-burden countries: a mathematical modelling study. *Lancet Glob. Health* 2: e453–e459.

Dodd, P.J., Sismanidis, C., and Seddon, J.A. (2016). Global burden of drug-resistant tuberculosis in children: a mathematical modelling study. *Lancet Infect. Dis.* 16: 1193–1201.

Doughty, E.L., Sergeant, M.J., Adetifa, I. et al. (2014). Culture-independent detection and characterisation of Mycobacterium tuberculosis and M. africanum in sputum samples using shotgun metagenomics on a benchtop sequencer. *Peer J.* 2: e585.

Dowdy, D.W., Davis, J.L., Den Boon, S. et al. (2013). Population-level impact of same-day microscopy and Xpert MTB/RIF for tuberculosis diagnosis in Africa. *PLoS One* 8: e70485.

Foundation for Innovative New Diagnostics. (2017). A multicentre non-inferiority diagnostic accuracy study of the Ultra assay compared to the Xpert MTB/RIF assay. Geneva.

Global Laboratory Initiative (2017). *Planning for Country Transition to Xpert MTB/RIF Ultra Cartridges.* Geneva: Global Laboratory Initiative.

Grad, Y.H. and Lipsitch, M. (2014). Epidemiologic data and pathogen genome sequences: a powerful synergy for public health. *Genome Biol.* 15: 538.

Grubaugh, N.D., Ladner, J.T., Kraemer, M.U.G. et al. (2017). Genomic epidemiology reveals multiple introductions of Zika virus into the United States. *Nature* 546: 401–405.

Gupta-Wright, A., Fielding, K.L., Van Oosterhout, J.J. et al. (2016). Rapid urine-based screening for tuberculosis to reduce AIDS-related mortality in hospitalized patients in Africa (the STAMP trial): study protocol for a randomised controlled trial. *BMC Infect. Dis.* 16: 501.

Haas, C.T., Roe, J.K., Pollara, G. et al. (2016). Diagnostic 'omics' for active tuberculosis. *BMC Med.* 14: 37.

Hanrahan, C.F., Clouse, K., Bassett, J. et al. (2015). The patient impact of point-of-care versus laboratory placement of Xpert(R) MTB/RIF. *Int. J. Tuberc. Lung Dis.* 19: 811–816.

Hanrahan, C.F., Selibas, K., Deery, C.B. et al. (2013). Time to treatment and patient outcomes among TB suspects screened by a single point-of-care Xpert MTB/RIF at a primary care clinic in Johannesburg, South Africa. *PLoS One* 8: e65421.

Helb, D., Jones, M., Story, E. et al. (2010). Rapid detection of Mycobacterium tuberculosis and rifampin resistance by use of on-demand, near-patient technology. *J. Clin. Microbiol.* 48: 229–237.

Heller, T., Lessells, R.J., Wallrauch, C.G. et al. (2010). Community-based treatment for multidrug-resistant tuberculosis in rural KwaZulu-Natal, South Africa. *Int. J. Tuberc. Lung Dis.* 14: 420–426.

Iruedo, J., O'mahony, D., Mabunda, S. et al. (2017). The effect of the Xpert MTB/RIF test on the time to MDR-TB treatment initiation in a rural setting: a cohort study in South Africa's Eastern Cape Province. *BMC Infect. Dis.* 17: 91.

Ismail, N.A., Omar, S.V., Lewis, J.J. et al. (2015). Performance of a novel algorithm using automated digital microscopy for diagnosing tuberculosis. *Am. J. Respir. Crit. Care Med.* 191: 1443–1449.

Jacobson, K.R., Theron, D., Kendall, E.A. et al. (2013). Implementation of genotype MTBDRplus reduces time to multidrug-resistant tuberculosis therapy initiation in South Africa. *Clin. Infect. Dis.* 56: 503–508.

Jha, S., Ismail, N., Clark, D. et al. (2016). Cost-effectiveness of automated digital microscopy for diagnosis of active tuberculosis. *PLoS One* 11: e0157554.

Kapata, N., Chanda-Kapata, P., Ngosa, W. et al. (2016). The prevalence of tuberculosis in Zambia: results from the first national TB prevalence survey, 2013-2014. *PLoS One* 11: e0146392.

Keeler, E., Perkins, M.D., Small, P. et al. (2006). Reducing the global burden of tuberculosis: the contribution of improved diagnostics. *Nature* 444 (Suppl 1): 49–57.

Koser, C.U., Bryant, J.M., Becq, J. et al. (2013). Whole-genome sequencing for rapid susceptibility testing of M. tuberculosis. *N. Engl. J. Med.* 369: 290–292.

Kurbatova, E.V., Cavanaugh, J.S., Shah, N.S. et al. (2012). Rifampicin-resistant Mycobacterium tuberculosis: susceptibility to isoniazid and other anti-tuberculosis drugs. *Int. J. Tuberc. Lung Dis.* 16: 355–357.

Lawn, S.D. (2012). Point-of-care detection of lipoarabinomannan (LAM) in urine for diagnosis of HIV-associated tuberculosis: a state of the art review. *BMC Infect. Dis.* 12: 103.

Lawn, S.D. and Gupta-Wright, A. (2016). Detection of lipoarabinomannan (LAM) in urine is indicative of disseminated TB with renal involvement in patients living with HIV and advanced immunodeficiency: evidence and implications. *Trans. R. Soc. Trop. Med. Hyg.* 110: 180–185.

Lawn, S.D., Kerkhoff, A.D., Burton, R. et al. (2017). Diagnostic accuracy, incremental yield and prognostic value of determine TB-LAM for routine diagnostic testing for tuberculosis in HIV-infected patients requiring acute hospital admission in South Africa: a prospective cohort. *BMC Med.* 15: 67.

Lawn, S.D., Kerkhoff, A.D., Burton, R. et al. (2015). Rapid microbiological screening for tuberculosis in HIV-positive patients on the first day of acute hospital admission by systematic testing of urine samples using Xpert MTB/RIF: a prospective cohort in South Africa. *BMC Med.* 13: 192.

Lessells, R.J., Cooke, G.S., Mcgrath, N. et al. (2013). Impact of a novel molecular TB diagnostic system in patients at high risk of TB mortality in rural South Africa (Uchwepheshe): study protocol for a cluster randomised trial. *Trials* 14: 170.

Lessells, R.J., Cooke, G., Mcgrath, N. et al. (2017). Impact of point-of-care Xpert MTB/RIF on tuberculosis treatment initiation: a cluster randomised trial. *Am. J. Respir. Crit. Care Med.* 196 (7): 901–910. https://doi.org/10.1164/rccm.201702-0278OC.

Lessem, E. (2017). The tuberculosis diagnostics pipeline. In: *2017 Pipeline Report HIV, TB & HCV: Drugs, Diagnostics, Vaccines, Preventive Technologies, Cure Research, and Immune-Based and Gene Therapies in Development* (ed. M. Frick, A. Gaudino, M. Harrington, et al.), 129–142. New York: Treatment Action Group.

Loveday, M., Wallengren, K., Voce, A. et al. (2012). Comparing early treatment outcomes of MDR-TB in decentralised and centralised settings in KwaZulu-Natal, South Africa. *Int. J. Tuberc. Lung. Dis.* 16: 209–215.

Macpherson, P., Houben, R.M., Glynn, J.R. et al. (2014). Pre-treatment loss to follow-up in tuberculosis patients in low- and lower-middle-income countries and high-burden countries: a systematic review and meta-analysis. *Bull. World Health Organ.* 92: 126–138.

Mori, Y. and Notomi, T. (2009). Loop-mediated isothermal amplification (LAMP): a rapid, accurate, and cost-effective diagnostic method for infectious diseases. *J. Infect. Chemother.* 15: 62–69.

Muyoyeta, M., Maduskar, P., Moyo, M. et al. (2014). The sensitivity and specificity of using a computer aided diagnosis program for automatically scoring chest X-rays of presumptive TB patients compared with Xpert MTB/RIF in Lusaka Zambia. *PLoS One* 9: e93757.

Nabeta, P., Havumaki, J., Ha, D.T. et al. (2017). Feasibility of the TBDx automated digital microscopy system for the diagnosis of pulmonary tuberculosis. *PLoS One* 12: e0173092.

Naidoo, P., Du Toit, E., Dunbar, R. et al. (2014). A comparison of multidrug-resistant tuberculosis treatment commencement times in MDRTBPlus line probe assay and Xpert(R) MTB/RIF-based algorithms in a routine operational setting in Cape Town. *PLoS One* 9: e103328.

Nair, D., Navneethapandian, P.D., Tripathy, J.P. et al. (2016). Impact of rapid molecular diagnostic tests on time to treatment initiation and outcomes in patients with multidrug-resistant tuberculosis, Tamil Nadu, India. *Trans. R. Soc. Trop. Med. Hyg.* 110: 534–541.

Nel, J.S., Lippincott, C.K., Berhanu, R. et al. (2017). Does disseminated nontuberculous mycobacterial disease cause false-positive determine B-LAM lateral flow assay results? A retrospective review. *Clin. Infect. Dis.* 65 (7): 1226–1228.

Nikam, C., Jagannath, M., Narayanan, M.M. et al. (2013). Rapid diagnosis of Mycobacterium tuberculosis with Truenat MTB: a near-care approach. *PLoS One* 8: e51121.

Nikam, C., Kazi, M., Nair, C. et al. (2014). Evaluation of the Indian TrueNAT micro RT-PCR device with GeneXpert for case detection of pulmonary tuberculosis. *Int. J. Mycobacteriol.* 3: 205–210.

Pai, M. and Furin, J. (2017). Tuberculosis innovations mean little if they cannot save lives. *elife* 6: https://doi.org/10.7554/eLife.25956.

Pande, T., Pai, M., Khan, F.A., and Denkinger, C.M. (2015). Use of chest radiography in the 22 highest tuberculosis burden countries. *Eur. Respir. J.* 46: 1816–1819.

Pankhurst, L.J., Del Ojo Elias, C., Votintseva, A.A. et al. (2016). Rapid, comprehensive, and affordable mycobacterial diagnosis with whole-genome sequencing: a prospective study. *Lancet Respir. Med.* 4: 49–58.

Perkins, M.D. and Cunningham, J. (2007). Facing the crisis: improving the diagnosis of tuberculosis in the HIV era. *J. Infect. Dis.* 196 (Suppl 1): S15–S27.

Peter, J.G., Zijenah, L.S., Chanda, D. et al. (2016). Effect on mortality of point-of-care, urine-based lipoarabinomannan testing to guide tuberculosis treatment initiation in HIV-positive hospital inpatients: a pragmatic, parallel-group, multicountry, open-label, randomised controlled trial. *Lancet* 387: 1187–1197.

Quick, J., Loman, N.J., Duraffour, S. et al. (2016). Real-time, portable genome sequencing for Ebola surveillance. *Nature* 530: 228–232.

Senkoro, M., Mfinanga, S., Egwaga, S. et al. (2016). Prevalence of pulmonary tuberculosis in adult population of Tanzania: a national survey, 2012. *Int. J. Tuberc. Lung. Dis.* 20: 1014–1021.

Shah, M., Hanrahan, C., Wang, Z.Y. et al. (2016). Lateral flow urine lipoarabinomannan assay for detecting active tuberculosis in HIV-positive adults. *Cochrane Database Syst. Rev.* CD011420. https://doi.org/10.1002/14651858.CD011420.pub2.

Shenai, S., Armstrong, D.T., Valli, E. et al. (2016). Analytical and clinical evaluation of the Epistem Genedrive assay for detection of Mycobacterium tuberculosis. *J. Clin. Microbiol.* 54: 1051–1057.

Smith, S.E., Kurbatova, E.V., Cavanaugh, J.S., and Cegielski, J.P. (2012). Global isoniazid resistance patterns in rifampin-resistant and rifampin-susceptible tuberculosis. *Int. J. Tuberc. Lung. Dis.* 16: 203–205.

Sreeramareddy, C.T., Panduru, K.V., Menten, J., and Van Den Ende, J. (2009). Time delays in diagnosis of pulmonary tuberculosis: a systematic review of literature. *BMC Infect. Dis.* 9: 91.

Stagg, H.R., White, P.J., Riekstina, V. et al. (2016). Decreased time to treatment initiation for multidrug-resistant tuberculosis patients after use of Xpert MTB/RIF test, Latvia. *Emerg. Infect. Dis.* 22: 482–490.

Steingart, K.R., Henry, M., Ng, V. et al. (2006). Fluorescence versus conventional sputum smear microscopy for tuberculosis: a systematic review. *Lancet Infect. Dis.* 6: 570–581.

Steingart, K.R., Ng, V., Henry, M. et al. (2006). Sputum processing methods to improve the sensitivity of smear microscopy for tuberculosis: a systematic review. *Lancet Infect. Dis.* 6: 664–674.

Steingart, K.R., Schiller, I., Horne, D.J. et al. (2014). Xpert(R) MTB/RIF assay for pulmonary tuberculosis and rifampicin resistance in adults. *Cochrane Database Syst. Rev.* 1: CD009593.

Storla, D.G., Yimer, S., and Bjune, G.A. (2008). A systematic review of delay in the diagnosis and treatment of tuberculosis. *BMC Public Health* 8: 15.

Theron, G., Venter, R., Calligaro, G. et al. (2016). Xpert MTB/RIF results in patients with previous tuberculosis: can we distinguish true from false positive results? *Clin. Infect. Dis.* 62: 995–1001.

Theron, G., Zijenah, L., Chanda, D. et al. (2014). Feasibility, accuracy, and clinical effect of point-of-care Xpert MTB/RIF testing for tuberculosis in primary-care settings in Africa: a multicentre, randomised, controlled trial. *Lancet* 383: 424–435.

UNITAID (2017). Tuberculosis Diagnostics Technology Landscape. 5th Edition.

University of Cape Town. Xpert Active Case-finding Trial 2: Community-based Active Case-finding for Tuberculosis in South Africa (XACT-2) [Online]. Available: https://clinicaltrials.gov/ct2/show/NCT03168945?term=dheda&rank=1 [Accessed July 4 2017].

Van Kampen, S.C., Susanto, N.H., Simon, S. et al. (2015). Effects of introducing Xpert MTB/RIF on diagnosis and treatment of drug-resistant tuberculosis patients in Indonesia: a pre-post intervention study. *PLoS One* 10: e0123536.

Van Kampen, S.C., Tursynbayeva, A., Koptleuova, A. et al. (2015). Effect of introducing Xpert MTB/RIF to test and treat individuals at risk of multidrug-resistant tuberculosis in Kazakhstan: a prospective cohort study. *PLoS One* 10: e0132514.

Van Rie, A., Mellet, K., John, M.A. et al. (2012). False-positive rifampicin resistance on Xpert(R) MTB/RIF: case report and clinical implications. *Int. J. Tuberc. Lung. Dis.* 16: 206–208.

Votintseva, A.A., Bradley, P., Pankhurst, L. et al. (2017). Same-day diagnostic and surveillance data for tuberculosis via whole-genome sequencing of direct respiratory samples. *J. Clin. Microbiol.* 55: 1285–1298.

Votintseva, A.A., Pankhurst, L.J., Anson, L.W. et al. (2015). Mycobacterial DNA extraction for whole-genome sequencing from early positive liquid (MGIT) cultures. *J. Clin. Microbiol.* 53: 1137–1143.

Walker, T.M., Kohl, T.A., Omar, S.V. et al. (2015). Whole-genome sequencing for prediction of Mycobacterium tuberculosis drug susceptibility and resistance: a retrospective cohort study. *Lancet Infect. Dis.* 15: 1193–1202.

Weyer, K., Mirzayev, F., Migliori, G.B. et al. (2013). Rapid molecular TB diagnosis: evidence, policy making and global implementation of Xpert MTB/RIF. *Eur. Respir. J.* 42: 252–271.

Witney, A.A., Gould, K.A., Arnold, A. et al. (2015). Clinical application of whole genome sequencing to inform treatment for multi-drug resistant tuberculosis cases. *J. Clin. Microbiol.* 53: 1473–1483.

World Health Organization (1993). *WHO Declares Tuberculosis a Global Emergency [Press Release]*. Geneva: World Health Organization.

World Health Organization (2011a). *Fluorescent Light-Emitting Diode (LED) Microscopy for Diagnosis of Tuberculosis*. Geneva: World Health Organization.

World Health Organization (2011b). *Policy statement: automated real-time nucleic acid amplification technology for rapid and simultaneous detection of tuberculosis and rifampicin resistance: Xpert MTB/RIF system*. Geneva: World Health Organization.

World Health Organization (2011c). Same-day diagnosis of tuberculosis by microscopy: policy statement. Geneva: World Health Organization.

World Health Organization (2013a). *Automated real-time nucleic acid amplification technology for rapid and simultaneous detection of tuberculosis and rifampicin* resistance: Xpert MTB/RIF system for the diagnosis of pulmonary and extrapulmonary TB in adults and children: policy update. Geneva: Switzerland.

World Health Organization (2013b). *The use of a commercial loop-mediated isothermal amplification assay (TB-LAMP) for the detection of tuberculosis.* Geneva: World Health Organization.

World Health Organization (2014). *High-priority target product profiles for new tuberculosis diagnostics: report of a consensus meeting.* Geneva: Switzerland.

World Health Organization (2015a). *Gear up to end TB: Introducing the End TB Strategy.* Geneva, Switzerland: World Health Organization.

World Health Organization (2015b). *The use of lateral flow urine lipoarabinomannan assay (LF-LAM) for the diagnosis and screening of active tuberculosis in people living with HIV. Policy guidance.* Geneva: World Health Organization.

World Health Organization (2016a). *Chest radiography in tuberculosis detection – summary of current WHO recommendations and guidance on programmatic approaches.* Geneva: World Health Organization.

World Health Organization (2016b). *Global tuberculosis report 2016 [Online].* Geneva: World Health Organization. Available: http://apps.who.int/iris/bitstream/ 10665/250441/1/9789241565394-eng.pdf?ua=1 [Accessed].

World Health Organization (2016c). *The use of loop-mediated isothermal amplification (TB-LAMP) for the diagnosis of pulmonary tuberculosis: policy guidance.* Geneva: World Health Organization.

World Health Organization (2016d). *The use of molecular line probe assay for the detection of resistance to isoniazid and rifampicin: policy update.* Geneva: World Health Organization.

World Health Organization (2017). *WHO meeting report of a technical expert consultation: non-inferiority analysis of Xpert MTF/RIF Ultra compared to Xpert MTB/RIF.* Geneva: World Health Organization.

Xie, Y.L., Chakravorty, S., Armstrong, D.T. et al. (2017). Evaluation of a rapid molecular drug-susceptibility test for tuberculosis. *N. Engl. J. Med.* 377: 1043–1054.

Zachary, D., Schaap, A., Muyoyeta, M. et al. (2012). Changes in tuberculosis notifications and treatment delay in Zambia when introducing a digital X-ray service. *Public Health Action* 2: 56–60.

Zetola, N.M., Shin, S.S., Tumedi, K.A. et al. (2014). Mixed Mycobacterium tuberculosis complex infections and false-negative results for rifampin resistance by GeneXpert MTB/RIF are associated with poor clinical outcomes. *J. Clin. Microbiol.* 52: 2422–2429.

6

Rapid Diagnostic Tests for Syphilis

David Mabey[1], Michael Marks[1], and Rosanna W. Peeling[2]

[1] Clinical Research Department, London School of Hygiene & Tropical Medicine, London, UK
[2] International Diagnostics Centre, Clinical Research Department, London School of Hygiene & Tropical Medicine, London, UK

CHAPTER MENU

6.1   Introduction

Syphilis is caused by the bacterium *Treponema pallidum* and can be cured with a single dose of long acting penicillin. It is transmitted between partners during sexual intercourse, and from infected pregnant women across the placenta to the developing fetus.

The clinical manifestations of syphilis can be divided into three stages. The primary chancre is an ulcer, usually painless, at the site of inoculation (Figure 6.1).

It heals after a few weeks even in the absence of treatment. Some weeks later the secondary stage occurs, usually with a generalized rash which affects the palms and soles, and symptoms involving many body systems (Figure 6.2).

This also resolves, usually over weeks or months, in the absence of treatment, and is followed by the latent stage, in which the patient has no symptoms or clinical signs. Most women who transmit syphilis to their infant have latent syphilis, and can only be identified by screening programs. The tertiary stage may occur many years later, affecting the skin, central nervous, or cardiovascular systems.

The WHO estimates that there are approximately six million new cases of syphilis per year, with the highest burden in sub-Saharan Africa, where as many as 5% of women attending antenatal clinics may be infected (Chico et al. 2012; Newman et al. 2015) (Figure 6.3).

Revolutionizing Tropical Medicine: Point-of-Care Tests, New Imaging Technologies and Digital Health,
First Edition. Edited by Kerry Atkinson and David Mabey.

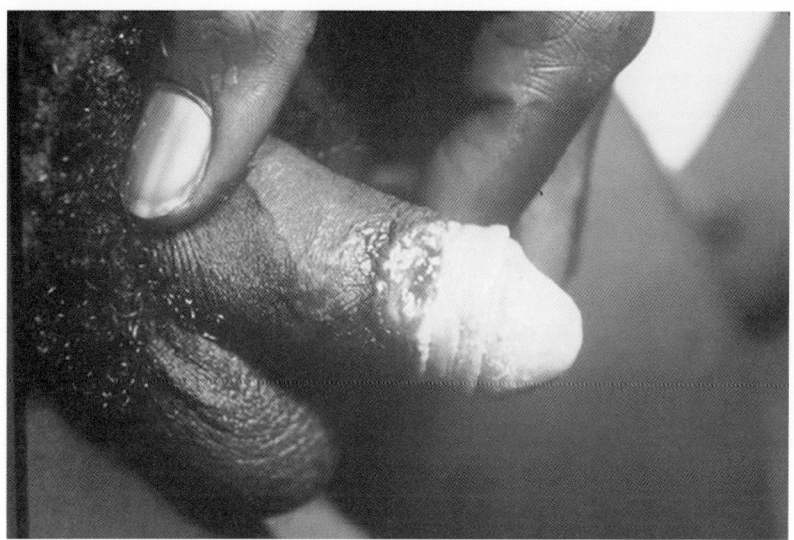

Figure 6.1 A primary syphilis chancre on the penis.

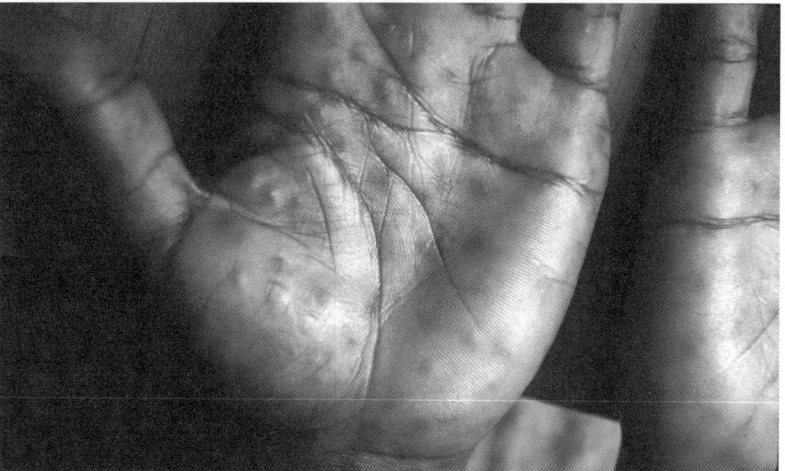

Figure 6.2 A generalized rash which affects the palms and soles in secondary syphilis.

In 2012 about one million pregnant women globally were estimated to have active syphilis but, in sub-Saharan Africa, less than 50% of pregnant women attending antenatal clinic were screened for syphilis (Wijesooriya et al. 2016). A study in Tanzania showed that, among women with syphilis who were not treated, 25% delivered a stillborn baby, and 33% a low birth weight baby (Watson-Jones et al. 2002). These adverse pregnancy outcomes can be prevented by a single dose of long acting benzathine penicillin given before 28 weeks gestation (Watson-Jones et al. 2002).

Appropriate diagnostic tests and early treatment are crucial in preventing long-term complications of syphilis and mother-to-child transmission (MTCT). Screening pregnant women for syphilis is one of the most cost-effective health interventions. In Mwanza, Tanzania, where the prevalence of active syphilis in pregnant women was

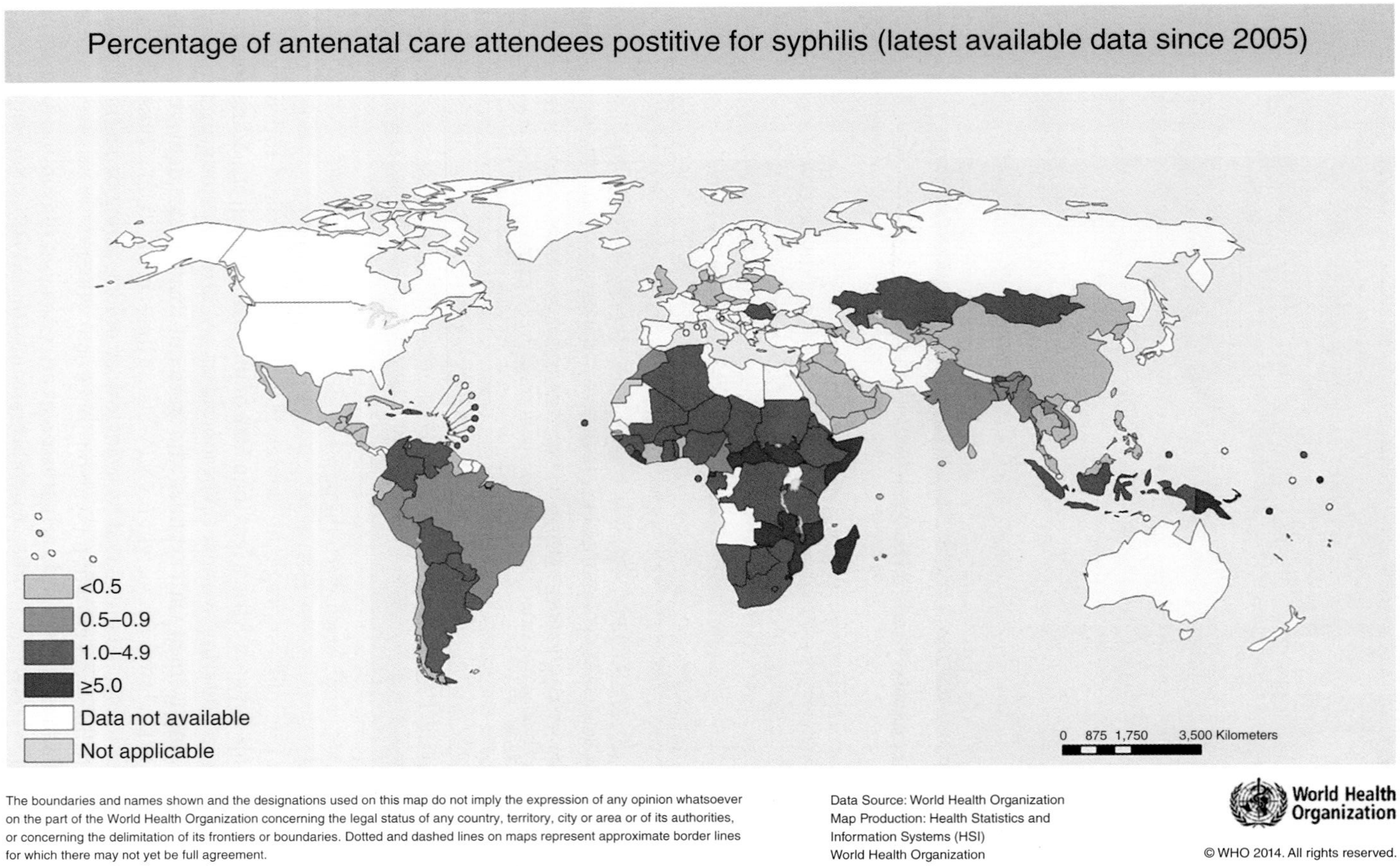

Percentage of antenatal care attendees postitive for syphilis (latest available data since 2005)

<0.5
0.5–0.9
1.0–4.9
≥5.0
Data not available
Not applicable

0   875 1,750   3,500 Kilometers

Data Source: World Health Organization
Map Production: Health Statistics and
Information Systems (HSI)
World Health Organization

World Health
Organization

Figure 6.3 A WHO map showing the percentage of antenatal attendees positive for syphilis. (*See color plate section for the color representation of this figure.*)

6%, the cost per disability-adjusted life year (DALY) saved was US$10.56 if the impact on stillbirths averted was included (Terris-Prestholt et al. 2003). Congenital syphilis and adverse pregnancy outcomes due to untreated syphilis represent a missed opportunity for disease prevention.

6.2 The Diagnosis of Syphilis

Appropriate diagnostic and screening tests are vital for the control of syphilis, since the symptoms and signs are non-specific, and many people with syphilis have no symptoms.

6.2.1 Laboratory Based Tests

In patients with primary or secondary syphilis, *T. pallidum* can be identified in swabs from moist lesions by dark field microscopy or by nucleic acid amplification tests (NAAT) such as polymerase chain reaction (PCR). Microscopy requires a skilled observer and lacks sensitivity, and NAATs are not widely available, so syphilis is usually diagnosed serologically, and serology is all that is available for the diagnosis of latent syphilis.

There are two types of antibody test: treponemal and non-treponemal. Treponemal tests are specific and detect antibodies to *T. pallidum*. They include the *T. pallidum* hemagglutination assay (TPHA), *T. pallidum* particle agglutination assay (TPPA), fluorescent treponemal antibody-absorbed (FTA-Abs), treponemal enzyme immunoassay (ELISA) and a number of POC tests. Once positive, treponemal tests usually remain positive for life.

Non-treponemal tests include the venereal diseases research laboratory (VDRL) test and the rapid plasma reagin (RPR) test, both of which are agglutination tests, requiring electricity, laboratory equipment and a skilled observer to identify positive samples.

Non-treponemal tests are less specific than treponemal tests, and false positive results are more common in pregnant women. Unlike the treponemal tests, they usually revert to negative some months after successful treatment, and falling titers can be used to monitor the response to treatment. Neither treponemal nor non-treponemal tests can distinguish between syphilis and the non-venereal treponematoses (yaws, pinta and endemic syphilis, the latter also known as bejel).

6.2.2 Point-Of-Care Tests

Laboratory-based tests require technical expertise, equipment and electricity. Fortunately, POC tests are now available for the serological diagnosis of syphilis which do not require equipment or electricity, can be stored at room temperature, require only one drop of blood and can give a result in 15 minutes. Using these tests syphilis can be diagnosed at the POC, and treatment given immediately, so that patients do not have to return for their results: "Same-day Testing and Treatment" (STAT). Almost every country has a policy for antenatal syphilis screening and, now that these tests are available, all pregnant women who attend antenatal clinic can be screened and treated at their first visit.

A number of syphilis POC tests are available, many of which fulfill the ASSURED criteria (Peeling and Mabey 2010) (Tables 6.1 and 6.2).

These are mostly treponemal tests, which remain positive for life, but POC tests have recently become available which identify both treponemal and non-treponemal antibodies. They are mostly immuno-chromatographic strip (ICS)-based assays which can be used with whole blood, serum, or plasma. Antigen–antibody reactions appear as a colored line or spot on the membrane. The majority of the tests on the market fulfill the ASSURED criteria, though they vary in their sensitivity, and sensitivity is reduced when using whole blood rather than plasma or serum (Mabey et al. 2006). A systematic review by Tucker et al. (2010) identified 15 studies evaluating syphilis POC tests, representing 23 055 individual test results. Thirteen of the studies were from low- or middle-income countries, and all were performed in an antenatal clinic (ANC) or a sexually transmitted infection (STI) clinic setting. The median ICS syphilis test sensitivity was 86% (interquartile range [IQR] 75-94%). The specificity of syphilis ICS tests ranged from 90.9–100.0% with a median of 99%.

Jafari et al. (2013) carried out a systematic review of POC tests for syphilis using serum and whole blood samples. Only POC tests that met the ASSURED criteria were included, of which 14 were identified (Table 6.2).

The majority were ICS-based assays. The meta-analysis showed that the Determine rapid test, using a serum sample, had the best estimate for sensitivity (92.03%, 95% CI 87.22–95.77) and Syphicheck had the best specificity (99.44%, 95% CI 98.96–99.81), compared to a laboratory-based treponemal test. In all comparisons the estimated sensitivity was higher using serum than using whole blood.

Combined treponemal and non-treponemal dual ICS POC tests are now available which enable healthcare workers to distinguish between active syphilis and past, treated infection (Table 6.2). The Dual Path Platform (DPP) test was the first of these (Figure 6.4).

It has been shown to have good sensitivity and specificity in a variety of settings, although the sensitivity is decreased at low RPR titres (Marks et al. 2016). In theory these tests could be used to monitor the response to treatment, since the non-treponemal test should revert to negative, but they have not yet been evaluated as a test of cure.

Table 6.1 The ASSURED criteria.

ASSURED
Affordable by those at risk of infection
Sensitive (few false-negatives)
Specific (few false-positives)
User-friendly (simple to perform and requiring minimal training)
Rapid (to enable treatment at first visit) and **R**obust (does not require refrigerated storage)
Equipment-free
Delivered to those who need it

Table 6.2 Point-of-care tests on the market with available sensitivities and specificities.

Test	Sample type	Sensitivity (%)	Specificity (%)
Alere Determine Syphilis TP	Whole blood/serum/plasma	59.6–100	95.7–100
Omega VisiTect Syphilis	Whole blood/serum/plasma	72.7–98.2	98.1–100
Qualpro Syphicheck-WB	Whole blood/serum/plasma	64–97.6	98.4–99.7
SD Bioline Syphilis 3.0	Whole blood/serum/plasma	85.7–100	95.5–99.4
Span Diagnostics Crystal TP Syphilis Test	Whole blood/serum/plasma	Not available	Not available
CTK Biotech OnSite™ Syphilis Ab combo Rapid	Whole blood	Not available	Not available
Diagnostics Direct Syphilis Health Check™	Whole blood/serum/plasma	Not available	Not available
Uni-Gold™ syphilis Treponemal	Whole blood/serum/plasma	Not available	Not available
Dual Path Platform (DDP®) Syphilis Test (Chemio Diagnostic Systems, Inc)	Treponemal antibody	90.1–98.2	91.2–98.0
	Non-Treponemal	80.6–98.2	89.4
SD Bioline HIV/Syphilis Duo Rapid Test (Alere/Standard Diagnostics, Inc)	HIV	97.9–99.0	99.0–100
	Treponema pallidum	93.0–99.6	99.1–100
DPP® HIV-Syphilis Assay (Chembio Diagnostic Systems, Inc)	HIV	98.9	97.9–99.6
	T. pallidum	95.3	97.0–99.6
Multiplo Rapid TP/HIV Antibody Test (MedMira, Inc)	HIV	97.9	94.2–99.5
	T. pallidum	94.1	94.2–99.1
INSTI™ HIV/Syphilis Multiplex Test (Biolytical Laboratories, Inc)	HIV	Not available	Not available
	T. pallidum	Not available	Not available
OnSite HIV/Syphilis Ab Combo Rapid Test (CTK Biotech)	HIV	Not available	Not available
	T. pallidum	Not available	Not available

6.3 The Impact of POC Testing for Syphilis

POC tests have been shown to increase the uptake of screening for syphilis in a variety of resource-limited settings. Treponemal POC tests have been rolled out, and their impact evaluated in rural antenatal care clinics in Tanzania, Uganda and China; in both rural and urban clinics in Peru and Zambia; and in remote indigenous communities in Brazil (Mabey et al. 2012). The introduction of POC tests increased the proportion of antenatal care attenders screened for syphilis to 90%, and the proportion of pregnant women with syphilis who were treated the same day exceeded 90% in all countries. Modeling from this study has shown that POC tests are more cost-effective in screening and treating syphilis than laboratory-based tests such as the RPR (Terris-Prestholt et al. 2015). POC tests for syphilis were rolled out nationally in China in 2010 after these studies were completed. Between 2012 and 2014 the number of people reported to the

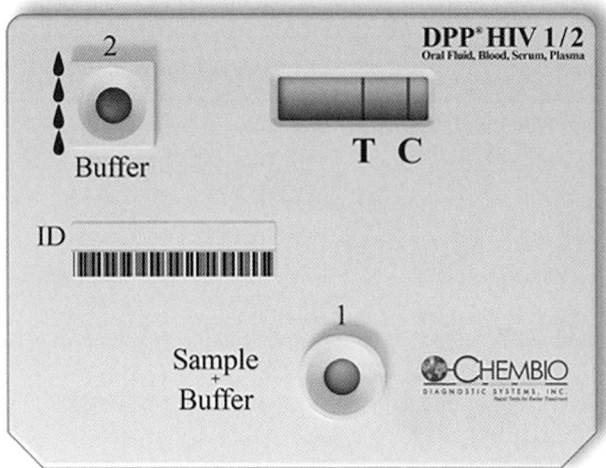

Figure 6.4 The Dual Path Platform (DPP) test was the first immuno-chromatographic strip (ICS)-based assay POC test which enabled healthcare workers to distinguish between active syphilis and past, treated infection.

National Centre for STD Control as having been tested for syphilis increased from 4.2 million to 29.6 million per year (Chen X, personal communication).

Strasser et al. (2012) conducted a field acceptability and feasibility study of including treponemal POC syphilis testing within prevention of mother-to-child transmission (PMTCT) HIV programs in Uganda and Zambia (Strasser et al. 2012). Significant increases in syphilis testing and treatment using a POC test were demonstrated, especially in Uganda, where access to syphilis testing was previously limited, with only 4% of women who attended an ANC being tested for syphilis in rural Jinja District. Post intervention a huge increase was seen: 99.7% of women were tested (p < 0.0001), and 97.8% received STAT. There was no deterioration in HIV services with the addition of the rapid syphilis test. Bronzan et al. (2007) found similar results in rural antenatal clinics in South Africa, where the onsite POC test resulted in 89% of pregnant women being correctly diagnosed and treated for syphilis, compared with 61% at standard practice clinics (Bronzan et al. 2007).

The World Health Organization has acknowledged the value of POC testing for syphilis, and the opportunity it provides for universal screening of pregnant women. In 2007 it launched the Initiative for the Global Elimination of Mother to Child Transmission of both HIV and syphilis (WHO 2007). The goal was for at least 90% of pregnant women to be tested for HIV and syphilis, and for at least 90% of seropositive pregnant women to receive treatment. The major investment in HIV PMTCT programs in many low- and middle-income countries in recent years offers an important opportunity to increase the uptake of syphilis screening. The London School of Hygiene and Tropical Medicine provides a toolkit, available online, on how two such programs can be integrated (London School of Hygiene & Tropical Medicine 2011). Considerable progress has been made since 2007, with five countries so far validated as having achieved the dual elimination targets, but it will be important for those countries to continue to maintain their coverage targets (WHO 2015). The new dual syphilis and HIV POC tests have the potential to accelerate this process (Kiarie et al. 2015).

Point-of-care tests have also been used to screen hard-to-reach populations. In Brazil healthcare workers in remote communities succeeded in screening 55% of the sexually active population (defined as ≥10 years of age) for syphilis, exceeding the 30–40% target originally set (Mabey et al. 2012). Modeling studies have estimated the impact of using rapid tests to screen female sex workers for syphilis and have shown that screening with POC tests could dramatically reduce syphilis prevalence among this hard-to-reach group. However, strategies to reduce re-infection from regular non-commercial partners are needed to maximize impact (Mitchell et al. 2013).

6.4 Challenges in the Implementation of POC Testing

6.4.1 Training and Logistics

Introducing POC tests can be challenging in resource-limited settings, where a small cadre of often overworked healthcare workers needs to be trained, and provided with incentives to undertake additional work. Ensuring a regular supply of POC tests can also be challenging. The roll out of POC testing for syphilis in antenatal clinics in Ghana made it possible to screen women in clinics where screening had not previously been available but, paradoxically, coverage actually decreased in larger clinics that had previously had access to laboratory testing. This was attributed to frequent stockouts of POC tests in the region (Dassah et al. 2015a). A second study in Ghana found that pregnant women seeking antenatal care in the private sector were not screened for syphilis, since the roll out of POC tests had only included public sector health facilities (Dassah et al. 2015b).

6.4.2 Quality Assurance

As countries begin to implement POC testing, adequate external quality assurance (QA) programs must be developed in parallel. These are routinely implemented in most laboratories, but have been largely neglected in the case of POC tests. Montoya et al. (2006) noted the accuracy of POC testing was greater when used by laboratory staff with higher levels of training and in a setting with better infrastructure and supervision than when carried out by staff at health facilities. It is clearly important that health workers using the tests are properly trained and well supervised, but it is also important to monitor their performance regularly and offer remedial training to those who do not meet the required standard.

Two approaches to external QA have been piloted in the case of syphilis POC tests: the use of a proficiency panel consisting of dried positive and negative serum samples that can be reconstituted in the clinic has been evaluated in the Amazonas region of Brazil (Benzaken et al. 2014) and the collection of dried blood spots (DBS) at the POC, which can be re-tested in the laboratory, has been piloted in rural Tanzania (Smit et al. 2013). In Brazil 268 healthcare workers tested five samples each: 9.3% of all samples tested were diagnosed incorrectly and 4.1% of healthcare workers reported difficulties with the POC tests. This method enabled remedial action to be taken when an incorrect result was obtained. The Tanzanian study was based in six villages in northern Tanzania. The conclusion was that DBS samples can be recommended for use with TPPA, and may be of value for external QA of POC syphilis tests.

6.5 The Future

POC tests make it possible to screen all pregnant women who attend an ANC for syphilis, and to treat them immediately. What is needed now is advocacy to convince governments, funding agencies and health program managers that this is an important and worthwhile objective, and that syphilis screening should be a political and healthcare priority. The tools are available; the next step is to make them available to all.

A recent study in Peru has shown that the roll out of POC tests offers an opportunity to improve other aspects of health systems as well as increasing coverage of syphilis screening (García et al. 2013). Widespread adoption and use of POC tests depends on the following: engaging the authorities; dissipating tensions between providers and identifying champions; training according to the needs of healthcare workers; providing monitoring, supervision, support and recognition; sharing results and discussing actions together; consulting and obtaining feedback from clients and from those tasked with performing POC tests; and integrating syphilis screening with other services such as rapid HIV testing.

POC tests can be used for self-testing outside of a clinical setting through community-based organizations, in pharmacies, or at home. Home-based testing for HIV has been shown to reach wide sections of communities in a diverse range of contexts and settings, providing early access to treatment (Sabapathy et al. 2012). Decentralizing testing for curable STIs may increase access to testing and awareness of STIs, but linkage to clinical care will be essential for diagnostic confirmation, treatment, counseling and follow-up (Tucker et al. 2013).

References

Benzaken, A.S., Bazzo, M.L., Galban, E. et al. (2014). External quality assurance with dried tube specimens (DTS) for point of care syphilis and HIV tests: experience in an indigenous populations screening programme in the Brazilian Amazon. *Sex. Transm. Infect.* 90: 14–18.

Bronzan, R.N., Mwesigwa-Kayongo, D.C., Narkunas, D. et al. (2007). Onsite rapid antenatal syphilis screening with an immunochromatographic strip improves case detection and treatment in rural South African clinics. *Sex. Transm. Dis.* 34 (7): S55–S60.

Chico, R.M., Mayaud, P., Ariti, C. et al. (2012). Prevalence of malaria and sexually transmitted and reproductive tract infections in pregnancy in sub-Saharan Africa. *JAMA* 307: 2079–2086.

Dassah, E.T., Adu-Sarkodie, Y., and Mayaud, P. (2015a). Estimating the uptake of maternal syphilis screening and other antenatal interventions before and after national rollout of syphilis point-of-care testing in Ghana. *Int. J. Obstet. Gynecol.* 130: S63–S69.

Dassah, E.T., Adu-Sarkodie, Y., and Mayaud, P. (2015b). Factors associated with failure to screen for syphilis during antenatal care in Ghana: a case control study. *BMC Infect. Dis.* 15: 125.

García, P.J., Cácamo, C.P., Chaippe, M. et al. (2013). Rapid syphilis tests as catalysts for health systems strengthening: a case study from Peru. *PLoS One* 8 (e1001351).

Jafari, Y., Peeling, R.W., Shivkumar, S. et al. (2013). Are *Treponema pallidum* specific rapid and point-of-care tests for syphilis accurate enough for screening in resource limited settings? Evidence from a meta-analysis. *PLoS One* 8 (2): e54695.

Kiarie, J., Mishra, C.K., Temmerman, M., and Newman, L. (2015). Accelerating the dual elimination of mother-to-child transmission of syphilis and HIV: why now? *Int. J. Gynaecol. Obstet.* 130: S1–S3.

London School of Hygiene and Tropical Medicine. (2011). The London School of Hygiene publishes a Rapid Syphilis Test Toolkit. http://www.globe-network.org/en/london-school-hygiene-publishes-rapid-syphilis-test-toolkit

Mabey, D., Peeling, R.W., Ballard, R. et al. (2006). Prospective, multi-Centre clinic-based evaluation of four rapid diagnostic tests for syphilis. *Sex. Transm. Infect.* 82 (5): v13–v16.

Mabey, D.C., Sollis, K.A., Kelly, H.A. et al. (2012). Point-of-care tests to strengthen health systems and save newborn lives: the case of syphilis. *PLoS Med.* 9: 8.

Marks, M., Yin, Y.P., Chen, X.S. et al. (2016). Meta Analysis of the Performance of a Combined Treponemal and Nontreponemal Rapid Diagnostic Test for Syphilis and Yaws. *Clin. Infect. Dis.* 63: 627.

Mitchell, K.M., Cox, A.P., Mabey, D. et al. (2013). The impact of syphilis screening among female sex workers in China: a modelling study. *PLoS One* 8: e55622.

Montoya, P.J., Lukehart, S.A., Brentlinger, P.E. et al. (2006). Comparison of the diagnostic accuracy of a rapid immunochromatographic test and the rapid plasma regain test for antenatal syphilis screening in Mozambique. *Bull. World Health Organ.* 84 (2): 97–104.

Newman, L., Rowley, J., Vander Hoorn, S. et al. (2015). Global estimates of the prevalence and incidence of four curable sexually transmitted infections in 2012 based on systematic review and global reporting. *PLoS One* 10: e0143304.

Peeling, R.W. and Mabey, D. (2010). Point-of-care tests for diagnosing infections in the developing world. *Clin. Microbiol. Infect.* 16: 1062–1069.

Sabapathy, K., Van den Bergh, R., Fidler, S. et al. (2012). Uptake of home-based voluntary HIV testing in sub-Saharan Africa: a systematic review and meta-analysis. *PLoS Med.* 9: e1001351.

Smit, P.W., van der Vlis, T. et al. (2013). The development and validation of dried blood spots for external quality assurance of syphilis serology. *BMC Infect. Dis.* 13: 102.

Strasser, S., Bitarakwate, E., Gill, M. et al. (2012). Introduction of rapid syphilis testing within prevention of mother-to-child transmission of HIV programs in Uganda and Zambia: a field acceptability and feasibility study. *J. Acquir. Immune Defic. Syndr.* 61 (3): 340–346.

Terris-Prestholt, F., Watson-Jones, D., Mugeye, K. et al. (2003). Is antenatal syphilis screening still cost effective in sub-Saharan Africa. *Sex. Transm. Infect.* 79: 375–381.

Terris-Prestholt, F., Vickerman, P., Torres-Rueda, S. et al. (2015). The cost-effectiveness of 10 antenatal syphilis screening and treatment approaches in Peru, Tanzania, and Zambia. *Int. J. Gynaecol. Obstet.* 130: S73–S80.

Tucker, J.D., Bien, C.H., and Peeling, R.W. (2013). Point-of-care testing for sexually transmitted infections: recent advances and implications for disease control. *Curr. Opin. Infect. Dis.* 26: 73–79.

Tucker, J.D., Bu, J., Brown, L.B. et al. (2010). Accelerating worldwide syphilis screening through rapid testing: a systematic review. *Lancet Infect. Dis.* 10: 381–386.

Watson-Jones, D., Changalucha, J., Gumodoka, B. et al. (2002). Syphilis in pregnancy in Tanzania. I. Impact of maternal syphilis on outcome of pregnancy. *J. Infect. Dis.* 186: 940–947.

Watson-Jones, D., Gumodoka, B., Weiss, H. et al. (2002). Syphilis in pregnancy in Tanzania II. The effectiveness of antenatal syphilis screening and single dose benzathine penicillin treatment for the prevention of adverse pregnancy outcomes. *J. Infect. Dis.* 186: 948–957.

WHO. (2016). validates the elimination of mother-to-child transmission of HIV and syphilis in Thailand, Armenia, Belarus and the Republic of Moldova. June 8 2016. http://www.who.int/reproductivehealth/topics/rtis/en

Wijesooriya, N.S., Rochat, R.W., Kamb, M.L. et al. (2016). Maternal and congenital syphilis in 2008 and 2012: a health systems modelling study. *Lancet Glob. Health.* 4 (8): e525–e533.

World Health Organisation. (2007). The Global Elimination of Congenital Syphilis: rationale and strategy for action. http://www.who.int/reproductivehealth/publications/rtis/9789241595858/en/index.html.

World Health Organization. (2015). WHO validates the elimination of mother-to-child transmission of HIV and syphilis in Cuba. June 30 2015. http://www.who.int/reproductivehealth/topics/rtis/en

7

Point-of-Care and Near-Point-of-Care Diagnostic Tests for Malaria: Light Microscopy, Rapid Antigen-Detecting Tests and Nucleic Acid Amplification Assays

Heidi Hopkins[1], and Jane Cunningham[2]

[1] *Faculty of Infectious and Tropical Diseases, London School of Hygiene and Tropical Medicine, London, UK*
[2] *Global Malaria Programme, World Health Organization, Geneva, Switzerland*

7.1   Introduction

Malaria is caused by parasites of the *Plasmodium* genus, transmitted by *Anopheles* mosquitoes to humans, who may be infected with *Plasmodium (P) falciparum, Plasmodium vivax, Plasmodium ovale, Plasmodium malariae* and *Plasmodium knowlesi*. Malaria is endemic to tropical and subtropical regions. Sub-Saharan Africa, where *P. falciparum* predominates, accounts for the vast majority of global malaria cases and deaths, while *P. vivax* is more common in endemic areas of the Americas and Asia. The World Health Organization (WHO) estimates that in 2016 there were 216 million cases of malaria in 91 countries (WHO 2016). After an encouraging period of improving malaria control, progress appears to have slowed and even reversed recently in some regions (WHO 2016).

The clinical manifestations of malaria may be categorized as uncomplicated or severe. Typical symptoms and signs of uncomplicated malaria include fever, malaise, myalgia, arthralgia and headache. Untreated *P. falciparum* (and in some cases *P. vivax and P. knowlesi*) malaria may progress within hours to days to severe malaria, with severe

Revolutionizing Tropical Medicine: Point-of-Care Tests, New Imaging Technologies and Digital Health,
First Edition. Edited by Kerry Atkinson and David Mabey.

anemia, metabolic acidosis, respiratory distress, cerebral malaria, renal failure and death. Of note, on the basis of symptoms and signs alone, malaria cannot be distinguished from other many other causes of illness including self-limited illnesses as well as other infections that require specific treatment. Even in endemic areas a clinical diagnosis of malaria is frequently incorrect (Chandramohan et al. 2001, 2002; Luxemburger et al. 1998; Mwangi et al. 2005; Ndyomugyenyi et al. 2007; Olivar et al. 1991; Sowunmi and Akindele 1993; Vinnemeier et al. 2012). An accurate diagnosis requires the detection of malaria parasites.

Providing prompt and appropriate care for patients with malaria has been a long-standing challenge. For decades the vast majority of patients have lacked access to reliable, affordable diagnostic services; treatment therefore has been presumptive based on symptoms (Hopkins et al. 2009; Whitty et al. 2008). However, rapid and accurate diagnosis of malaria is critical for effective treatment and for reduction in morbidity and mortality. Accurate detection is also necessary for epidemiological screening and surveillance in order to inform malaria control strategies (WHO 2012a, 2013).

In order to improve case management and the rational use of artemisinin-based combination therapies (ACTs), the WHO recommended in 2010 that all suspected cases of malaria should have parasitological confirmation before treatment (WHO 2010a, 2011a). This Chapter outlines the diagnostic tools that are available, and in development, to support implementation of this policy.

7.2 Diagnosis of Malaria

Malaria should be suspected in a patient with fever and epidemiologic exposure including residence in, or travel to, an area where malaria transmission occurs (WHO 2010a, 2015a). A diagnosis of malaria is made in a patient who has symptoms and signs consistent with malaria and a positive malaria diagnostic test.

Qualities of a useful malaria diagnostic test include the ability to:

1) confirm the presence or absence of infection,
2) identify the species of malaria present,
3) quantify the level of parasitemia,
4) detect low-level parasitemia, and
5) allow monitoring of response to antimalarial therapy.

Of note, individuals who acquire partial immunity to malaria due to repeated exposures in endemic settings may carry parasites in their bloodstream without showing symptoms (asymptomatic parasitemia). There is no diagnostic test that can distinguish between parasitemia causing clinical malaria and febrile illness of a different etiology in a patient who also has asymptomatic parasitemia.

The relative importance of various test characteristics depends on malaria epidemiology, laboratory infrastructure and goals of detection in the context where the test is used. Tests for parasite-based diagnosis in routine clinical settings include microscopy and malaria rapid diagnostic tests (mRDTs). Molecular techniques for detection of parasite genetic material are primarily used in research settings and in non-endemic clinical settings to investigate febrile illnesses among returning travelers.

7.3 Light Microscopy of Blood Smears

Examination of Giemsa-stained blood smears under light microscopy has been the standard malaria diagnostic tool for decades (Figure 7.1) (Warhurst and Williams 1996).

Microscopy permits quantification of parasitemia and identification of *Plasmodium* species, and can be used both for initial diagnosis and for following a patient's response to treatment (Figures 7.2).

In addition, skilled microscopists can diagnose hematologic abnormalities and other comorbid infectious diseases such as filariasis, trypanosomiasis and babesiosis. With expert personnel and good quality materials, microscopy can be very sensitive, detecting malaria parasites at densities as low as 4–20 parasites/μL of blood (approximately 0.0001–0.0005% parasitemia) (Bruce-Chwatt 1984; Dowling and Shute 1966; Moody 2002; Payne 1988).

The disadvantages of microscopy are that it is labor- intensive and requires substantial training to develop and maintain expertise (Wongsrichanalai et al. 2007). Variation in techniques used for preparation and interpretation can influence results (Kettelhut et al. 2003; Milne et al. 1994; Thomson et al. 2000; Warhurst and Williams 1996). Prior antimalarial therapy can alter the appearance of parasites and affect their identification in blood smears (Beaudoin and Aikawa 1968; Jiang et al. 1985; Moody 2002). Microscopy cannot reliably detect very low parasitemia (<5 to 10 parasites/μL) or cases where most of the parasite biomass is sequestered, such as placental sequestration in pregnant women (Murray et al. 2008; Nosten et al. 2007). The sensitivity and specificity of malaria microscopy in resource-limited settings is often below levels achievable in reference or

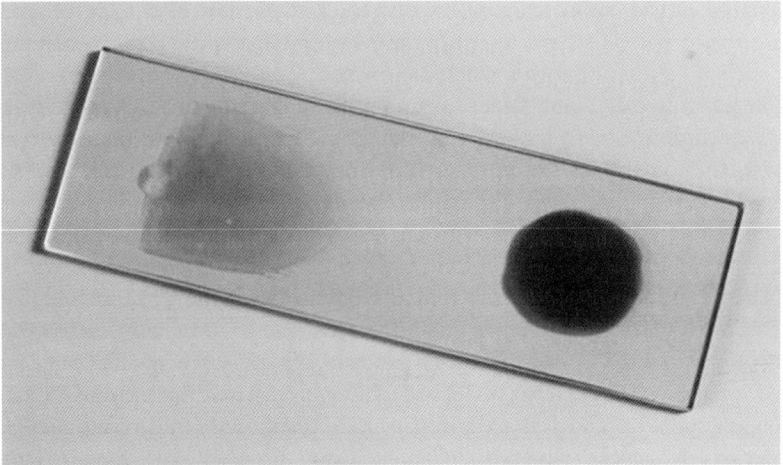

Figure 7.1 This prepared slide offers a good example of a thin (left) and thick (right) blood smear to be examined under the microscope. This image, reproduced here from Wikimedia Commons, is in the public domain because it is a work of the Centers for Disease Control and Prevention, part of the United States Department of Health and Human Services, taken or made as part of an employee's official duties. The source is the CDC – Public Health Image Library (PHIL), #5901. The author is CDC/ Steven Glenn, Laboratory and Consultation Division. By CDC/ Steven Glenn, Laboratory and Consultation Division – CDC – Public Health Image Library (PHIL), #5901, Public Domain, https:// commons.wikimedia.org/w/index.php?curid=4878350

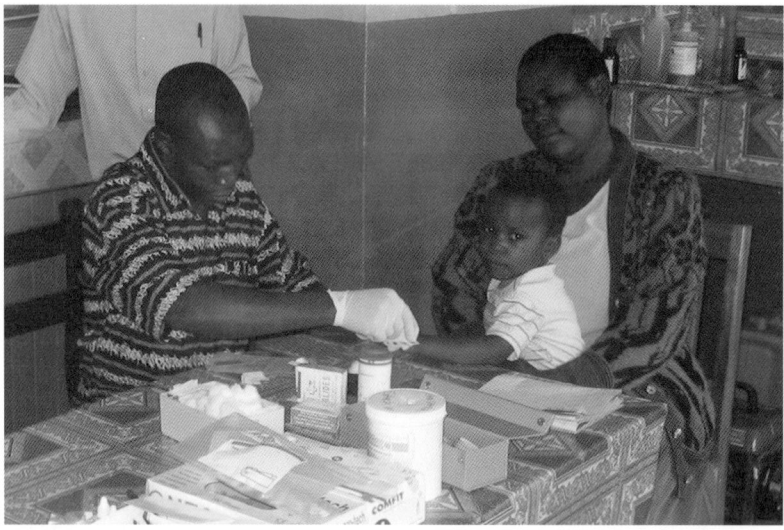

Figure 7.2 Laboratory technician performing a malaria rapid diagnostic test (mRDT) for a child in western Uganda. *Source credit:* Heidi Hopkins.

research laboratories (Bulletin of WHO 1988; Coleman et al. 2002; Kahama-Maro et al. 2011; Kilian et al. 2000; Reyburn et al. 2006).

For malaria microscopy a drop of blood (fingerprick, heelstick or venous) is placed on a glass microscopy slide, then dried, stained and examined optimally under 100x magnification with a microscope's oil immersion objective lens. Thick and thin blood smears are typically prepared on the same or separate slides: thick smears are used to screen for presence or absence of parasites and to estimate parasite density, while thin smears allow for easier identification of parasite stage and/or species.

A number of authoritative detailed guides for preparation, staining and interpretation of blood smears, including bench aids, are available in print and electronic formats (Centers for Disease Control and Prevention 2017a, b; WHO 2009, 2010b, 2015a, 2017a,b,c).

7.4 Rapid Diagnostic Tests for Malaria (mRDTs)

Point-of-care (POC) tests that detect malaria parasite antigens are widely used in resource-limited endemic settings due to their accuracy and ease of use. mRDTs are relatively simple, inexpensive and reliable point-of-care tests that can be used where high-quality microscopy services are not available (Mabey et al. 2004). They do not require electricity or laboratory infrastructure, can be performed by health workers with minimal training and give results within 15–20 minutes.

Most antigen-detecting mRDTs are immunochromatographic lateral flow tests, consisting of a nitrocellulose strip packaged alone as a dipstick or housed within a plastic cassette or paper card (Bell et al. 2017; Moody 2002; Wilson 2012).

One end of the test strip contains labeled, unbound antibodies. A 5–10 μL drop of blood is placed in the sample well, buffer is added and the sample migrates along the strip by capillary action together with the labeled antibodies. Along the strip the migrating liquid encounters a test line (bound antibody specific to a parasite antigen) and a control line (bound antibody specific to the labeled antibody, to confirm adequate migration). A positive test shows a colored band at the test line and at the control line; a negative test shows a colored band only at the control line; absence of a control line indicates an invalid test result.

mRDTs provide a qualitative binary result (positive versus negative), but do not give information about parasite density. Depending on the antigen/s targeted, an mRDT may identify *Plasmodium* genus only or may distinguish *P. falciparum* and *non-falciparum* or *P. vivax* infections. mRDTs that detect antibodies produced by an individual exposed to malaria infection are also available, but are not useful for diagnosing acute infection. The most widely used antigen-detecting mRDTs detect one or more of the following:

1) histidine-rich protein 2 (HRP2)
2) pan- and/or species-specific *Plasmodium* lactate dehydrogenase (pLDH).

HRP2 is one of a family of *P. falciparum* histidine-rich proteins (Howard et al. 1986). It is only produced by *P. falciparum*. HRP2-based mRDTs can detect lower levels of parasitemia than RDTs based on other target antigens (Hendriksen et al. 2011; Heutmekers et al. 2012; Hopkins et al. 2007a, b; Rakotonirina et al. 2008; WHO 2017c). However, detectable HRP2 antigen may persist in the bloodstream for between a few days to several weeks after parasitemia has been cleared (Houze et al. 2009; Mayxay et al. 2001; Singh and Shukla 2002; Swarthout et al. 2007; Tjitra et al. 2001). In the absence of an alternative confirmatory assay, there is currently no way to distinguish HRP2 antigenemia due to a new or persistent infection (i.e. as in treatment failure) from antigenemia persisting after a recently treated infection. There is some risk, therefore, of clinically false-positive HRP2-based test results in which positive results may occur as a result of prior infection. HRP2-based assays are not useful for post-treatment monitoring (Houze et al. 2009; Mayxay et al. 2001; Singh and Shukla 2002; Swarthout et al. 2007; Tjitra et al. 2001). Variability in the gene sequence for *P. falciparum* histidine-rich protein 2 (pfhrp2) may affect test performance in some circumstances (Kumar et al. 2012; Lee et al. 2006).

There has long been evidence of *P. falciparum* parasites lacking the pfhrp2 gene in regions of South America (Gamboa et al. 2010), and public health authorities have always cautioned against using mHRP2-based RDTs in the Amazon River basin. More recently, in an investigation in Eritrea, 80% of microscopy-confirmed positive samples were negative by HRP2-based RDT, a finding attributed to deletion of the pfhrp2 gene (Berhane et al. 2017, 2018). Thus, HRP2-based tests are not reliable for infections contracted in these geographic areas (see Section 7.8).

Plasmodium lactic dehydrogenase (pLDH) is the terminal enzyme in the malaria parasite's glycolytic pathway and is produced by asexual and sexual forms of all *Plasmodium* species (Makler et al. 1998). Two types of pLDH-based mRDTs are available: those targeting a conserved pLDH element in all human malaria species, and those targeting species-specific regions that distinguish *P. falciparum* from *P. vivax*. pLDH antibodies specific for *P. ovale* and *P. malariae* have been described but are not yet commercially available (Bell and Peeling 2006). pLDH-based RDTs are less sensitive than HRP2-based tests for detection of *P. falciparum* infection, especially at relatively low parasite densities (< 500 parasites/µL) (Craig et al. 2002; Hendriksen et al. 2011; Hopkins et al. 2007a; WHO 2017c; WHO/FIND/CDC 2017). Serum pLDH levels become undetectable at approximately the same time that blood smears become microscopy-negative after antimalarial therapy (Houze et al. 2009; Makler et al. 1998;

Moody et al. 2000; Oduola et al. 1997). Antigenic variation does not appear to significantly affect pLDH-based detection (Talman et al. 2007). The reported accuracy of mRDTs varies widely, and is affected by comparison standard/s used, malaria epidemiology in the study area, study population, personnel performing the RDTs, parasite antigen/s targeted, type of antibody used (monoclonal versus polyclonal), source of antigen used to induce the test antibodies and the epitope/s targeted by test antibodies (Abba et al. 2014; Bell and Peeling 2006; Hopkins et al. 2007a; Murray et al. 2008). The most systematic and up-to-date data on accuracy of commercially available mRDTs, assessed in controlled laboratory settings against a standardized global panel of parasite antigens, are available on-line from the WHO-FIND mRDT evaluation program (WHO 2017c) (see also Section 7.7.2).

The choice of mRDT for a given context depends on the malaria epidemiology and objectives in the area (Bell and Peeling 2006; WHO 2011a). In regions where *P. falciparum* predominates, such as much of sub-Saharan Africa, use of HRP2-based tests, which detect *P. falciparum* only, is often considered appropriate and cost-effective. In regions where both *falciparum* and *non-falciparum* parasites are endemic, or where *P. vivax* predominates, use of a combination mRDT that can detect and distinguish between these species is optimal. The performance of current mRDTs for detection of *P. malariae*, *P. ovale* and *P. knowlesi* is much less studied but published reports suggest that sensitivity is reduced compared to that of *P. falciparum* and *P. vivax* (Barber et al. 2013; Foster et al. 2014; Heutemakers et al. 2012a, b; Maltha et al. 2011; Williams et al. 2016a).

7.5 Nucleic Acid Amplification-Based Tests (NAATs) for Malaria

NAATs allow sensitive detection of low-density malaria infections (< 1 parasite/µL), of value primarily for research and epidemiologic purposes (Proux et al. 2011). Due to the infrastructure, costs and expertise required, NAAT performance is generally limited to reference laboratories. In recent years, however, more affordable, field-adapted approaches have been developed. It is well established that low-density malaria infections, below the detection threshold of microscopy or mRDTs, occur and can contribute to transmission (Manjurano et al. 2011; Okell et al. 2009; Ouedraogo et al. 2009; Schneider et al. 2007).

However, more research is needed to determine if detection of these low-density infections through surveillance or specifically targeted interventions (for example, mass screening and treatment (MSAT), can help reach malaria elimination goals (Feachem et al. 2010; Hsiang et al. 2012; Moonen et al. 2010; Mueller et al. 2011).

Loop-mediated isothermal amplification (LAMP) for detection of malaria parasite DNA has been developed as a field assay to facilitate use of molecular technology in endemic areas (Hopkins et al. 2013; Notomi et al. 2000; Poon et al. 2006). LAMP amplification of DNA occurs at a steady temperature so does not require the thermocycler equipment needed for standard polymerase chain reaction (PCR) testing. LAMP results can be available within an hour, and can be detected visually under fluorescence or by turbidimetry. Sensitivity and specificity of LAMP assays depend on genetic target sequences and processing methods but limits of detection of ≤ 1 parasite/µL can be achieved (Han et al. 2007; Lucchi et al. 2010; Paris et al. 2007; Polley et al. 2010, 2013; Ponce et al. 2017; Poschl et al. 2010; Sirichaisinthop et al. 2011).

7.6 Impact of Point-Of-Care Testing for Malaria

Appropriate diagnosis and treatment of malaria has been a public health challenge for decades. The WHO's 2010 recommendation that all suspected cases of malaria receive parasitological confirmation before treatment (WHO 2010c, 2011a,b, 2015a,b) represented a landmark policy shift, and most endemic countries have now adopted this policy at least in public health facilities.

mRDTs have been key to implementing this policy change, with a dramatic increase in access to, and use of, malaria diagnostics since the introduction mRDTs. Global mRDT procurement has surged from 45 million tests in 2008 to over 300 million annually in recent years (UNITAID 2016b; WHO 2017d). In the public healthcare sector, the proportion of suspected malaria cases receiving a parasitological test has increased in most endemic regions since 2010 (Figure 7.3).

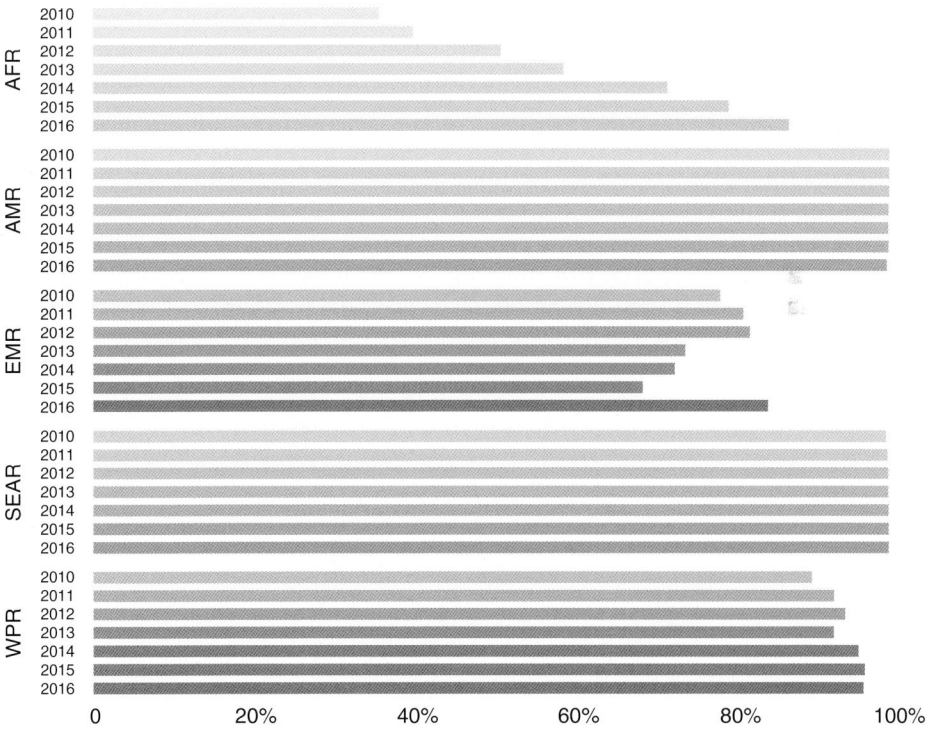

Proportion of suspected cases who were tested in public health facilities

Figure 7.3 Proportion of suspected malaria cases attending public health facilities who received a diagnostic test by WHO region 2010–2016. *Source:* National malaria control programmes. WHO (2017d, p. 25). Abbreviations used: AFR, WHO African region; AMR, WHO Region of the Americas; EMR, WHO Eastern Mediterranean Region; SEAR, WHO South-East Asia Region; WPR, WHO Western Pacific Region.

The greatest estimated rise, in Africa, was from 36 to 87% between 2010 and 2016, mostly due to an increase in mRDT use (WHO 2017d). Before the widespread adoption of mRDTs, clinical trials and early pilot projects encouraged their implementation, though with some heterogeneity of results (Odaga et al. 2014). Compared to presumptive symptom-based treatment, case management based on mRDTs generally reduced antimalarial prescription, especially in settings with low malaria prevalence where health workers' prescribing closely adhered to test results (Bastiaens et al. 2011; D'Acremont et al. 2011; Kyabayinze et al. 2010; Msellem et al. 2009; Mubi et al. 2011; Thiam et al. 2011; Yukich et al. 2012). However, although provider adherence to negative mRDT results was high in some studies (Msellem et al. 2009; Mubi et al. 2011; Skarbinski et al. 2009; Yeboah-Antwi et al. 2010), it was low in others (Bisoffi et al. 2009; Chinkhumba et al. 2010; Reyburn et al. 2007).

A recent synthesis of data from 10 studies of mRDT introduction, representing more than half a million patient consultations, found that mRDTs were associated with significantly lower artemisinin-based combination therapy (ACT) prescription (range 8–69% versus 20–100%), but that prescribing did not always adhere to malaria test results. In several settings ACTs were prescribed to more than 30% of test-negative patients or to fewer than 80% of test-positive patients (Bruxvoort et al. 2017). In addition, recent analyses show that mRDT introduction, and a shift away from presumptive treatment of fever as malaria, has uncovered a broader healthcare challenge – how to effectively diagnose and treat "non-malaria febrile illness," i.e. patients with fever whose malaria test is negative. Across many tropical and subtropical zones, acute febrile illness is one of the most common reasons for seeking healthcare (Crump and Kirk 2015), while the etiologies of these fevers remain largely unknown and undiagnosed (Prasad et al. 2015). In many settings where mRDT introduction has led to a desired reduction in the empirical use of antimalarials, this has been accompanied by an unintended shift toward increased empirical use of antibiotics, especially for patients with negative malaria test results (D'Acremont et al. 2011; Hopkins et al. 2017; Ndhlovu et al. 2015). The potential for this prescribing shift was recognized in the early days of mRDT scale-up (Baiden et al. 2011), and these concerns now appear to be justified in many settings.

7.7 Challenges in Implementation of POC Testing for Malaria

7.7.1 Training and Logistics

In many under-resourced or remote settings, malaria diagnostics, and especially mRDTs, may be the only biomedical diagnostic technology available. To establish and maintain malaria diagnostic services in these contexts, systems for regular and appropriate transport and storage of supplies, and appropriate personnel training, must be developed where none existed before. A number of international and national guidelines are available from the WHO, national ministries of health, and other agencies, on training for malaria microscopy and mRDT use, and for procurement and supply chain management (WHO 2011a, 2015, 2018).

7.7.2 Quality Assurance and Quality Control (QA/QC)

A reliable health service includes regular quality assurance of diagnostic test results. Quality assurance of malaria microscopy is labor- and logistics-intensive, and in many endemic areas tends to be performed intermittently if at all. Again, guidance is available from the WHO and others, as are published reports on quality assurance programs (Ashraf et al. 2012; Khan et al. 2011; Maguire et al. 2006; Sakande et al. 2014; Wafula et al. 2014; Wanja et al. 2017; WHO 2016).

Reference sets or banks of standardized microscopy slides may be used to assess and improve the competence and performance of malaria microscopists (WHO 2016). Efforts are also underway to develop digital or automated slide reading systems for training and QA purposes.

There is a recognized need for quality assurance of mRDTs, especially in hot and humid conditions typical along supply chains in malaria-endemic regions, where lot-to-lot variation and susceptibility to deterioration have been documented (Albertini et al. 2012; Bell and Peeling 2006; Chiodini et al. 2007; Jorgensen et al. 2006; WHO/FIND/CDC 2015). Over the past decade a three-tiered international QA/QC scheme has been developed. First, the WHO malaria RDT product testing program conducts regular laboratory evaluations of commercial mRDT products, including testing assessment against well-characterized panels of parasites. Reports are available on-line (WHO 2017a; WHO/FIND/CDC 2017).

Second, any national program or organization procuring mRDTs may request pre- or post-purchase lot testing from the WHO-FIND program (WHO 2016, 2017d). Positive control wells (PCWs) have been proposed as POC QC tools, as a third component of a tiered QA program (Aidoo et al. 2012; Lon et al. 2005; Tamiru et al. 2015; Versteeg and Mens 2009). Prototype PCWs have undergone field testing but development of universal control materials has not been achieved yet due to the variable reactivity of RDT products with recombinant antigens (Bell et al. 2017).

The reproducibility of NAAT results for malaria is variable (Alemayehu et al. 2013). An international external quality assurance program has been established with the aim of improving NAAT performance standards, although this is unlikely to have POC applications in the near term.

7.7.3 Special Populations

In some areas the safety and acceptability of withholding antimalarial treatment from patients with negative malaria test results remains a concern, especially for vulnerable populations like young children (D'Acremont et al. 2009; English et al. 2009; WHO Global Malaria Program 2015). Favorable outcomes for mRDT-based management of uncomplicated febrile illness in children have been documented in endemic settings, including Benin (Faucher et al. 2010), Ghana (Ansah et al. 2010), Papua New Guinea (Senn et al. 2012), Tanzania (D'Acremont et al. 2010; Mtove et al. 2011a, b), and Zambia (Chanda et al. 2011). However, mRDTs do not have adequate negative predictive value to justify withholding treatment in the setting of severe illness (Hendriksen et al. 2011; Manning et al. 2012; Mtove et al. 2011a, b).

While microscopy and mRDTs are generally accurate for the diagnosis of clinical malaria in pregnant women, microscopy in particular fails to detect a large proportion of asymptomatic placental malaria (Mockenhaupt et al. 2002). Submicroscopic malaria infections in pregnancy are associated with adverse outcomes (Adegnika et al. 2006; Mockenhaupt et al. 2000, 2006). Use of mRDTs, and of other methods with lower thresholds of detection, for the diagnosis of asymptomatic malaria in pregnancy is the subject of ongoing study (Williams et al. 2016a, b).

7.8 The Future

On-going research and development efforts aim to develop more sensitive antigen-detecting mRDTs and automated reader systems in order to enhance the performance of current mRDTs. Highly sensitive mRDTs hold potential for use in surveillance and elimination programs, and possibly for detection of placental malaria, with better portability, ease of use, and speed than current POC or near-POC NAAT options. The first commercially available highly sensitive HRP2 mRDT for detection of *P. falciparum* was launched in 2017 (Alere 2017). As mentioned earlier, HRP2 gene deletions leading to false-negative RDT results have been identified as prevalent in *P. falciparum* parasites from the Peruvian Amazon and from Eritrea, and HRP2-based tests are not reliable for infections contracted in these areas (Berhane et al. 2018; Gamboa et al. 2010; Maltha et al. 2012; Menegon et al. 2017). Several reports of pfhrp2 and pfhrp3 mutations or deletions have emerged from countries in Africa and from India leading in some (Berhane et al. 2018; Menegon et al. 2017), but not all to false-negative HRP2-based RDT results (Beshir et al. 2017; Bharti et al. 2016; Koita et al. 2012; Kozycki et al. 2017; Parr et al. 2017). Although the scale and global distribution of these parasite mutants remains to be elucidated, it is clear that their presence is not limited to the Amazon basin, where the first clinical cases were reported in 2010 (Gamboa et al. 2010). The WHO has issued epidemiological and laboratory guidance for assessment of mutations (Cheng et al. 2014) and a global action plan (WHO 2017d). The need for research into new antigen targets as alternatives to HRP2 is a high priority for the future.

In order to meet global goals for malaria control and elimination, the implementation of malaria POC tests will need to be expanded and improved. Following the successful introduction of mRDTs into public sector care in many endemic countries, mRDTs are also being introduced by private retail and community health providers (Ansah et al. 2015; Aung et al. 2015; Awor et al. 2014; Cohen et al. 2015; Hamer et al. 2012; Mukanga et al. 2012; Ruizendaal et al. 2014; UNITAID 2016b; Yeung et al. 2011). Future implementation of POC tests for malaria and other conditions should benefit from lessons already learned with malaria diagnostics – in particular, attention to technical details must be accompanied by an understanding of patient and health worker expectations and behaviors in a given context, and of the factors that shape these (Burchett et al. 2017; Ochodo et al. 2016).

Bibliography

Abba, K., Kirkham, A.J., Olliaro, P.L. et al. (2014). Rapid diagnostic tests for diagnosing uncomplicated non-falciparum or Plasmodium vivax malaria in endemic countries. *Cochrane Database Syst. Rev.* 12: CD011431.

Adegnika, A.A., Verweij, J.J., Agnandji, S.T. et al. (2006). Microscopic and sub-microscopic Plasmodium falciparum infection, but not inflammation caused by infection, is associated with low birth weight. *Am. J. Trop. Med. Hyg.* 75 (5): 798–803.

Aidoo, M., Patel, J.C., and Barnwell, J.W. (2012). Dried Plasmodium falciparum-infected samples as positive controls for malaria rapid diagnostic tests. *Malar. J.* 11: 239.

Albertini, A., Lee, E., Coulibaly, S.O. et al. (2012). Malaria rapid diagnostic test transport and storage conditions in Burkina Faso, Senegal, Ethiopia and the Philippines. *Malar. J.* 11: 406.

Alemayehu, S., Feghali, K.C., Cowden, J. et al. (2013). Comparative evaluation of published real-time PCR assays for the detection of malaria following MIQE guidelines. *Malar. J.* 12: 277.

Alere. Alere Malaria Ag P.f. (2017). https://www.alere.com/en/home/product-details/alere-malaria-ag-pf.html?utm_source=offline0marketing&utm_medium=infographic&utm_campaign=world_malaria_day_2017 (Accessed December 2017).

Ansah, E.K., Narh-Bana, S., Affran-Bonful, H. et al. (2015). The impact of providing rapid diagnostic malaria tests on fever management in the private retail sector in Ghana: a cluster randomized trial. *BMJ* 350: h1019.

Ansah, E.K., Narh-Bana, S., Epokor, M. et al. (2010). Rapid testing for malaria in settings where microscopy is available and peripheral clinics where only presumptive treatment is available: a randomised controlled trial in Ghana. *BMJ* 340: c930.

Ashraf, S., Kao, A., Hugo, C. et al. (2012). Developing standards for malaria microscopy: external competency assessment for malaria microscopists in the Asia-Pacific. *Malar. J.* 11: 352.

Aung, T., White, C., Montagu, D. et al. (2015). Improving uptake and use of malaria rapid diagnostic tests in the context of artemisinin drug resistance containment in eastern Myanmar: an evaluation of incentive schemes among informal private healthcare providers. *Malar. J.* 14: 105.

Awor, P., Wamani, H., Tylleskar, T. et al. (2014). Increased access to care and appropriateness of treatment at private sector drug shops with integrated management of malaria, pneumonia and diarrhoea: a quasi-experimental study in Uganda. *PLoS One* 9 (12): e115440.

Baiden, F., Webster, J., Owusu-Agyei, S., and Chandramohan, D. (2011). Would rational use of antibiotics be compromised in the era of test-based management of malaria? *Trop. Med. Int. Health* 16 (2): 142–144.

Barber, B.E., William, T., Grigg, M.J. et al. (2013). Evaluation of the sensitivity of a pLDH-based and an aldolase-based rapid diagnostic test for diagnosis of uncomplicated and severe malaria caused by PCR-confirmed Plasmodium knowlesi, Plasmodium falciparum, and Plasmodium vivax. *J. Clin. Microbiol.* 51 (4): 1118–1123.

Bastiaens, G.J., Schaftenaar, E., Ndaro, A. et al. (2011). Malaria diagnostic testing and treatment practices in three different Plasmodium falciparum transmission settings in Tanzania: before and after a government policy change. *Malar. J.* 10: 76.

Beaudoin, R.L. and Aikawa, M. (1968). Primaquine-induced changes in morphology of exoerythrocytic stages of malaria. *Science* 160 (3833): 1233–1234.

Bell, D., Bwanika, J.B., Cunningham, J. et al. (2017). Prototype positive control wells for malaria rapid diagnostic tests: prospective evaluation of implementation among health workers in Lao People's Democratic Republic and Uganda. *Am. J. Trop. Med. Hyg.* 96 (2): 319–329.

Bell, D. and Peeling, R.W. (2006). Evaluation of rapid diagnostic tests: malaria. *Nat. Rev. Microbiol.* 4 (9 Suppl): S34–S38.

Bell, D., Wongsrichanalai, C., and Barnwell, J.W. (2006). Ensuring quality and access for malaria diagnosis: how can it be achieved? *Nat. Rev. Microbiol.* 4 (9): 682–695.

Berhane, A., Mihreteab, S., Gresty, K. et al. (2018). Threat to malaria control programs by Plasmodium falciparum lacking histidine-rich protein 2, Eritrea. *Emerg. Infect. Dis.* 24 (3): 462–470.

Berhane, A., Russom, M., Bahta, I. et al. (2017). Rapid diagnostic tests failing to detect Plasmodium falciparum infections in Eritrea: an investigation of reported false negative RDT results. *Malar. J.* 16 (1): 105.

Beshir, K.B., Sepulveda, N., Bharmal, J. et al. (2017). Plasmodium falciparum parasites with histidine-rich protein 2 (pfhrp2) and pfhrp3 gene deletions in two endemic regions of Kenya. *Sci. Rep.* 7 (1): 14718.

Bharti, P.K., Chandel, H.S., Ahmad, A. et al. (2016). Prevalence of pfhrp2 and/or pfhrp3 gene deletion in Plasmodium falciparum population in eight highly endemic states in India. *PLoS One* 11 (8): e0157949.

Bisoffi, Z., Sirima, B.S., Angheben, A. et al. (2009). Rapid malaria diagnostic tests vs. clinical management of malaria in rural Burkina Faso: safety and effect on clinical decisions. A randomized trial. *Trop. Med. Int. Health* 14 (5): 491–498.

Bruce-Chwatt, L.J. (1984). DNA probes for malaria diagnosis. *Lancet* 1 (8380): 795.

Bruxvoort, K.J., Leurent, B., Chandler, C.I.R. et al. (2017). The impact of introducing malaria rapid diagnostic tests on fever case management: a synthesis of ten studies from the ACT consortium. *Am. J. Trop. Med. Hyg.* 97 (4): 1170–1179.

Burchett, H.E., Leurent, B., Baiden, F. et al. (2017). Improving prescribing practices with rapid diagnostic tests (RDTs): synthesis of 10 studies to explore reasons for variation in malaria RDT uptake and adherence. *BMJ Open* 7 (3): e012973.

Centers for Disease Control & Prevention (CDC). (2017a). DPDx – Laboratory identification of parasitic diseases of public health concern. 5 December 2017. https://www.cdc.gov/dpdx/2017

Centers for Disease Control & Prevention (CDC). (2017b). DPDx: Diagnostic procedures. 18 Sept 2017. https://www.cdc.gov/dpdx/diagnosticProcedures/index.html2017

Chanda, P., Hamainza, B., Moonga, H.B. et al. (2011). Community case management of malaria using ACT and RDT in two districts in Zambia: achieving high adherence to test results using community health workers. *Malar. J.* 10: 158.

Chandramohan, D., Carneiro, I., Kavishwar, A. et al. (2001). A clinical algorithm for the diagnosis of malaria: results of an evaluation in an area of low endemicity. *Trop. Med. Int. Health* 6 (7): 505–510.

Chandramohan, D., Jaffar, S., and Greenwood, B. (2002). Use of clinical algorithms for diagnosing malaria. *Trop. Med. Int. Health* 7 (1): 45–52.

Cheng, Q., Gatton, M.L., Barnwell, J. et al. (2014). Plasmodium falciparum parasites lacking histidine-rich protein 2 and 3: a review and recommendations for accurate reporting. *Malar. J.* 13: 283.

Chinkhumba, J., Skarbinski, J., Chilima, B. et al. (2010). Comparative field performance and adherence to test results of four malaria rapid diagnostic tests among febrile patients more than five years of age in Blantyre, Malawi. *Malar. J.* 9: 209.

Chiodini, P.L., Bowers, K., Jorgensen, P. et al. (2007). The heat stability of Plasmodium lactate dehydrogenase-based and histidine-rich protein 2-based malaria rapid diagnostic tests. *Trans. R. Soc. Trop. Med. Hyg.* 101 (4): 331–337.

Cohen, J., Fink, G., Maloney, K. et al. (2015). Introducing rapid diagnostic tests for malaria to drug shops in Uganda: a cluster-randomized controlled trial. *Bull. World Health Organ.* 93: 142–151.

Coleman, R.E., Maneechai, N., Rachaphaew, N. et al. (2002). Comparison of field and expert laboratory microscopy for active surveillance for asymptomatic Plasmodium falciparum and Plasmodium vivax in western Thailand. *Am. J. Trop. Med. Hyg.* 67 (2): 141–144.

Craig, M.H., Bredenkamp, B.L., Williams, C.H. et al. (2002). Field and laboratory comparative evaluation of ten rapid malaria diagnostic tests. *Trans. R. Soc. Trop. Med. Hyg.* 96 (3): 258–265.

Crump, J.A. and Kirk, M.D. (2015). Estimating the burden of febrile illnesses. *PLoS Negl. Trop. Dis.* 9 (12): e0004040.

D'Acremont, V., Kahama-Maro, J., Swai, N. et al. (2011). Reduction of anti-malarial consumption after rapid diagnostic tests implementation in Dares Salaam: a before-after and cluster randomized controlled study. *Malar. J.* 10: 107.

D'Acremont, V., Lengeler, C., Mshinda, H. et al. (2009). Time to move from presumptive malaria treatment to laboratory-confirmed diagnosis and treatment in African children with fever. *PLoS Med.* 6 (1): e252.

D'Acremont, V., Malila, A., Swai, N. et al. (2010). Withholding antimalarials in febrile children who have a negative result for a rapid diagnostic test. *Clin. Infect. Dis.* 51 (5): 506–511.

Dowling, M.A. and Shute, G.T. (1966). A comparative study of thick and thin blood films in the diagnosis of scanty malaria parasitaemia. *Bull. World Health Organ.* 34 (2): 249–267.

English, M., Reyburn, H., Goodman, C., and Snow, R.W. (2009). Abandoning presumptive antimalarial treatment for febrile children aged less than five years–a case of running before we can walk? *PLoS Med.* 6 (1): e1000015.

Faucher, J.F., Makoutode, P., Abiou, G. et al. (2010). Can treatment of malaria be restricted to parasitologically confirmed malaria? A school-based study in Benin in children with and without fever. *Malar. J.* 9: 104.

Feachem, R.G., Phillips, A.A., Hwang, J. et al. (2010). Shrinking the malaria map: progress and prospects. *Lancet* 376 (9752): 1566–1578.

Foster, D., Cox-Singh, J., Mohamad, D.S. et al. (2014). Evaluation of three rapid diagnostic tests for the detection of human infections with Plasmodium knowlesi. *Malar. J.* 13: 60.

Gamboa, D., Ho, M.F., Bendezu, J. et al. (2010). A large proportion of P. falciparum isolates in the Amazon region of Peru lack pfhrp2 and pfhrp3: implications for malaria rapid diagnostic tests. *PLoS One* 5 (1): e8091.

Hamer, D.H., Brooks, E.T., Semrau, K. et al. (2012). Quality and safety of integrated community case management of malaria using rapid diagnostic tests and pneumonia by community health workers. *Pathog Glob Health* 106 (1): 32–39.

Han, E.T., Watanabe, R., Sattabongkot, J. et al. (2007). Detection of four Plasmodium species by genus-and species-specific loop-mediated isothermal amplification for clinical diagnosis. *J. Clin. Microbiol.* 45 (8): 2521–2528.

Hendriksen, I.C., Mtove, G., Pedro, A.J. et al. (2011). Evaluation of a PfHRP2 and a pLDH-based rapid diagnostic test for the diagnosis of severe malaria in 2 populations of African children. *Clin. Infect. Dis.* 52 (9): 1100–1107.

Heutmekers, M., Gillet, P., Cnops, L. et al. (2012). Evaluation of the malaria rapid diagnostic test SDFK90: detection of both PfHRP2 and Pf-pLDH. *Malar. J.* 11: 359.

Hopkins, H., Asiimwe, C., and Bell, D. (2009). Access to antimalarial therapy: accurate diagnosis is essential to achieving long term goals. *BMJ* 339: b2606.

Hopkins, H., Bebell, L., Kambale, W. et al. (2007b). Rapid diagnostic tests for malaria at sites of varied transmission intensity in Uganda. *J. Infect. Dis.* 197 (4): 510–518. https://doi.org/10.1086/526502.

Hopkins, H., Bruxvoort, K.J., Cairns, M.E. et al. (2017). Impact of introduction of rapid diagnostic tests for malaria on antibiotic prescribing: analysis of observational and randomised studies in public and private healthcare settings. *BMJ* 356: j1054.

Hopkins, H., Gonzalez, I.J., Polley, S.D. et al. (2013). Highly sensitive detection of malaria parasitemia in a malaria-endemic setting: performance of a new loop-mediated isothermal amplification kit in a remote clinic in Uganda. *J. Infect. Dis.* 208 (4): 645–652.

Hopkins, H., Kambale, W., Kamya, M.R. et al. (2007a). Comparison of HRP2- and pLDH-based rapid diagnostic tests for malaria with longitudinal follow-up in Kampala, Uganda. *Am. J. Trop. Med. Hyg.* 76 (6): 1092–1097.

Houze, S., Boly, M.D., Le Bras, J. et al. (2009). PfHRP2 and PfLDH antigen detection for monitoring the efficacy of artemisinin-based combination therapy (ACT) in the treatment of uncomplicated falciparum malaria. *Malar. J.* 8: 211.

Howard, R.J., Uni, S., Aikawa, M. et al. (1986). Secretion of a malarial histidine-rich protein (Pf HRP II) from Plasmodium falciparum-infected erythrocytes. *J. Cell Biol.* 103 (4): 1269–1277.

Hsiang, M.S., Hwang, J., Kunene, S. et al. (2012). Surveillance for malaria elimination in Swaziland: a national cross-sectional study using pooled PCR and serology. *PLoS One* 7 (1): e29550.

Jiang, J.B., Jacobs, G., Liang, D.S., and Aikawa, M. (1985). Qinghaosu-induced changes in the morphology of Plasmodium inui. *Am. J. Trop. Med. Hyg.* 34 (3): 424–428.

Jorgensen, P., Chanthap, L., Rebueno, A. et al. (2006). Malaria rapid diagnostic tests in tropical climates: the need for a cool chain. *Am. J. Trop. Med. Hyg.* 74 (5): 750–754.

Kahama-Maro, J., D'Acremont, V., Mtasiwa, D. et al. (2011). Low quality of routine microscopy for malaria at different levels of the health system in Dar Es Salaam. *Malar. J.* 10: 332.

Kettelhut, M.M., Chiodini, P.L., Edwards, H., and Moody, A. (2003). External quality assessment schemes raise standards: evidence from the UKNEQAS parasitology subschemes. *J. Clin. Pathol.* 56 (12): 927–932.

Khan, M.A., Walley, J.D., Munir, M.A. et al. (2011). District level external quality assurance (EQA) of malaria microscopy in Pakistan: pilot implementation and feasibility. *Malar. J.* 10: 45.

Kilian, A.H., Metzger, W.G., Mutschelknauss, E.J. et al. (2000). Reliability of malaria microscopy in epidemiological studies: results of quality control. *Trop. Med. Int. Health* 5 (1): 3–8.

Koita, O.A., Doumbo, O.K., Ouattara, A. et al. (2012). False-negative rapid diagnostic tests for malaria and deletion of the histidine-rich repeat region of the hrp2 gene. *Am. J. Trop. Med. Hyg.* 86 (2): 194–198.

Kozycki, C.T., Umulisa, N., Rulisa, S. et al. (2017). False-negative malaria rapid diagnostic tests in Rwanda: impact of Plasmodium falciparum isolates lacking hrp2 and declining malaria transmission. *Malar. J.* 16 (1): 123.

Kumar, N., Singh, J.P., Pande, V. et al. (2012). Genetic variation in histidine rich proteins among Indian Plasmodium falciparum population: possible cause of variable sensitivity of malaria rapid diagnostic tests. *Malar. J.* 11: 298.

Kyabayinze, D.J., Asiimwe, C., Nakanjako, D. et al. (2010). Use of RDTs to improve malaria diagnosis and fever case management at primary health care facilities in Uganda. *Malar. J.* 9: 200.

Le,e, N., Baker, J., Andrews, K.T. et al. (2006). Effect of sequence variation in Plasmodium falciparum histidine-rich protein 2 on binding of specific monoclonal antibodies: implications for rapid diagnostic tests for malaria. *J. Clin. Microbiol.* 44 (2778): 2773–2778.

Lon, C.T., Alcantara, S., Luchavez, J. et al. (2005). Positive control wells: a potential answer to remote-area quality assurance of malaria rapid diagnostic tests. *Trans. R. Soc. Trop. Med. Hyg.* 99 (7): 493–498.

Lucchi, N.W., Demas, A., Narayanan, J. et al. (2010). Real-time fluorescence loop mediated isothermal amplification for the diagnosis of malaria. *PLoS One* 5 (10): e13733.

Luxemburger, C., Nosten, F., Kyle, D.E. et al. (1998). Clinical features cannot predict a diagnosis of malaria or differentiate the infecting species in children living in an area of low transmission. *Trans. R. Soc. Trop. Med. Hyg.* 92 (1): 45–49.

Mabey, D., Peeling, R.W., Ustianowski, A., and Perkins, M.D. (2004). Diagnostics for the developing world. *Nat. Rev. Microbiol.* 2 (3): 231–240.

Maguire, J.D., Lederman, E.R., Barcus, M.J. et al. (2006). Production and validation of durable, high quality standardized malaria microscopy slides for teaching, testing and quality assurance during an era of declining diagnostic proficiency. *Malar. J.* 5: 92.

Makler, M.T. and Hinrichs, D.J. (1993). Measurement of the lactate dehydrogenase activity of Plasmodium falciparum as an assessment of parasitemia. *Am. J. Trop. Med. Hyg.* 48 (2): 205–210.

Makler, M.T., Piper, R.C., and Milhous, W.K. (1998). Lactate dehydrogenase and the diagnosis of malaria. *Parasitol. Today* 14 (9): 376–377.

Maltha, J., Gamboa, D., Bendezu, J. et al. (2012). Rapid diagnostic tests for malaria diagnosis in the Peruvian Amazon: impact of pfhrp2 gene deletions and cross-reactions. *PLoS One* 7 (8): e43094.

Maltha, J., Gillet, P., Cnops, L. et al. (2011). Evaluation of the rapid diagnostic test SDFK40 (Pf-pLDH/pan-pLDH) for the diagnosis of malaria in a non-endemic setting. *Malar. J.* 10: 7.

Manjurano, A., Okell, L., Lukindo, T. et al. (2011). Association of sub-microscopic malaria parasite carriage with transmission intensity in north-eastern Tanzania. *Malar. J.* 10: 370.

Manning, L., Laman, M., Rosanas-Urgell, A. et al. (2012). Rapid antigen detection tests for malaria diagnosis in severely ill Papua New Guinean children: a comparative study using Bayesian latent class models. *PLoS One* 7 (11): e48701.

Mayxay, M., Pukrittayakamee, S., Chotivanich, K. et al. (2001). Persistence of Plasmodium falciparum HRP-2 in successfully treated acute falciparum malaria. *Trans. R. Soc. Trop. Med. Hyg.* 95 (2): 179–182.

Menegon, M., L'Episcopia, M., Nurahmed, A.M. et al. (2017). Identification of Plasmodium falciparum isolates lacking histidine-rich protein 2 and 3 in Eritrea. *Infect. Genet. Evol.* 55: 131–134.

Milne, L.M., Kyi, M.S., Chiodini, P.L., and Warhurst, D.C. (1994). Accuracy of routine laboratory diagnosis of malaria in the United Kingdom. *J. Clin. Pathol.* 47 (8): 740–742.

Mockenhaupt, F.P., Bedu-Addo, G., von Gaertner, C. et al. (2006). Detection and clinical manifestation of placental malaria in southern Ghana. *Malar. J.* 5: 119.

Mockenhaupt, F.P., Rong, B., Till, H. et al. (2000). Submicroscopic Plasmodium falciparum infections in pregnancy in Ghana. *Trop. Med. Int. Health* 5 (3): 167–173.

Mockenhaupt, F.P., Ulmen, U., von Gaertner, C. et al. (2002). Diagnosis of placental malaria. *J. Clin. Microbiol.* 40 (1): 306–308.

Moody, A. (2002). Rapid diagnostic tests for malaria parasites. *Clin. Microbiol. Rev.* 15 (1): 66–78.

Moody, A., Hunt-Cooke, A., Gabbett, E., and Chiodini, P. (2000). Performance of the OptiMAL malaria antigen capture dipstick for malaria diagnosis and treatment monitoring at the Hospital for Tropical Diseases, London. *Br. J. Haematol.* 109 (4): 891–894.

Moonen, B., Cohen, J.M., Snow, R.W. et al. (2010). Operational strategies to achieve and maintain malaria elimination. *Lancet* 376 (9752): 1592–1603.

Msellem, M.I., Martensson, A., Rotllant, G. et al. (2009). Influence of rapid malaria diagnostic tests on treatment and health outcome in fever patients, Zanzibar: a crossover validation study. *PLoS Med.* 6 (4): e1000070.

Mtove, G., Hendriksen, I.C., Amos, B. et al. (2011a). Treatment guided by rapid diagnostic tests for malaria in Tanzanian children: safety and alternative bacterial diagnoses. *Malar. J.* 10: 290.

Mtove, G., Nadjm, B., Amos, B. et al. (2011b). Use of an HRP2-based rapid diagnostic test to guide treatment of children admitted to hospital in a malaria-endemic area of northeast Tanzania. *Trop. Med. Int. Health* 16 (5): 545–550.

Mubi, M., Janson, A., Warsame, M. et al. (2011). Malaria rapid testing by community health workers is effective and safe for targeting malaria treatment: randomised crossover trial in Tanzania. *PLoS One* 6 (7): e19753.

Mueller, I., Slutsker, L., and Tanner, M. (2011). Estimating the burden of malaria: the need for improved surveillance. *PLoS Med.* 8 (12): e1001144.

Mukanga, D., Tiono, A.B., Anyorigiya, T. et al. (2012). Integrated community case management of fever in children under five using rapid diagnostic tests and respiratory rate counting: a multi-country cluster randomized trial. *Am. J. Trop. Med. Hyg.* 87 (5 Suppl): 21–29.

Murray, C.K., Gasser, R.A. Jr., Magill, A.J., and Miller, R.S. (2008). Update on rapid diagnostic testing for malaria. *Clin. Microbiol. Rev.* 21 (1): 97–110.

Mwangi, T.W., Mohammed, M., Dayo, H. et al. (2005). Clinical algorithms for malaria diagnosis lack utility among people of different age groups. *Trop. Med. Int. Health* 10 (6): 530–536.

Ndhlovu, M., Nkhama, E., Miller, J.M., and Hamer, D.H. (2015). Antibiotic prescribing practices for patients with fever in the transition from presumptive treatment of malaria to "confirm and treat" in Zambia: a cross-sectional study. *Trop. Med. Int. Health* 20 (12): 1696–1706.

Ndyomugyenyi, R., Magnussen, P., and Clarke, S. (2007). Diagnosis and treatment of malaria in peripheral health facilities in Uganda: findings from an area of low transmission in south-western Uganda. *Malar. J.* 6: 39.

Nosten, F., McGready, R., and Mutabingwa, T. (2007). Case management of malaria in pregnancy. *Lancet Infect. Dis.* 7 (2): 118–125.

Notomi, T., Okayama, H., Masubuchi, H. et al. (2000). Loop-mediated isothermal amplification of DNA. *Nucleic Acids Res.* 28 (12): E63.

Ochodo, E., Garner, P., and Sinclair, D. (2016). Achieving universal testing for malaria. *BMJ* 352: i107.

Odaga, J., Sinclair, D., Lokong, J.A. et al. (2014). Rapid diagnostic tests versus clinical diagnosis for managing people with fever in malaria endemic settings. *Cochrane Database Syst. Rev.* 4: CD008998.

Oduola, A.M., Omitowoju, G.O., Sowunmi, A. et al. (1997). Plasmodium falciparum: evaluation of lactate dehydrogenase in monitoring therapeutic responses to standard antimalarial drugs in Nigeria. *Exp. Parasitol.* 87 (3): 283–289.

Okell, L.C., Ghani, A.C., Lyons, E., and Drakeley, C.J. (2009). Submicroscopic infection in Plasmodium falciparum-endemic populations: a systematic review and meta-analysis. *J. Infect. Dis.* 200 (10): 1509–1517.

Olivar, M., Develoux, M., Chegou Abari, A., and Loutan, L. (1991). Presumptive diagnosis of malaria results in a significant risk of mistreatment of children in urban Sahel. *Trans. R. Soc. Trop. Med. Hyg.* 85 (6): 729–730.

Ouedraogo, A.L., Bousema, T., Schneider, P. et al. (2009). Substantial contribution of submicroscopical Plasmodium falciparum gametocyte carriage to the infectious reservoir in an area of seasonal transmission. *PLoS One* 4 (12): e8410.

Paris, D.H., Imwong, M., Faiz, A.M. et al. (2007). Loop-mediated isothermal PCR (LAMP) for the diagnosis of falciparum malaria. *Am. J. Trop. Med. Hyg.* 77 (5): 972–976.

Parr, J.B., Verity, R., Doctor, S.M. et al. (2017). Pfhrp2-deleted Plasmodium falciparum parasites in the Democratic Republic of the Congo: a national cross-sectional survey. *J. Infect. Dis.* 216 (1): 36–44.

Payne, D. (1988). Use and limitations of light microscopy for diagnosing malaria at the primary health care level. *Bull. World Health Organ.* 66 (5): 621–626.

Polley, S.D., Gonzalez, I.J., Mohamed, D. et al. (2013). Clinical evaluation of a loop-mediated amplification kit for diagnosis of imported malaria. *J. Infect. Dis.* 208 (4): 637–644.

Polley, S.D., Mori, Y., Watson, J. et al. (2010). Mitochondrial DNA targets increase sensitivity of malaria detection using loop-mediated isothermal amplification. *J. Clin. Microbiol.* 48 (8): 2866–2871.

Ponce, C., Kaczorowski, F., Perpoint, T. et al. (2017). Diagnostic accuracy of loop-mediated isothermal amplification (LAMP) for screening patients with imported malaria in a non-endemic setting. *Parasite* 24: 53.

Poon, L.L., Wong, B.W., Ma, E.H. et al. (2006). Sensitive and inexpensive molecular test for falciparum malaria: detecting Plasmodium falciparum DNA directly from heat-treated blood by loop-mediated isothermal amplification. *Clin. Chem.* 52 (2): 303–306.

Poschl, B., Waneesorn, J., Thekisoe, O. et al. (2010). Comparative diagnosis of malaria infections by microscopy, nested PCR, and LAMP in northern Thailand. *Am. J. Trop. Med. Hyg.* 83 (1): 56–60.

Prasad, N., Murdoch, D.R., Reyburn, H., and Crump, J.A. (2015). Etiology of severe febrile illness in low- and middle-income countries: a systematic review. *PLoS One* 10 (6): e0127962.

Proux, S., Suwanarusk, R., Barends, M. et al. (2011). Considerations on the use of nucleic acid-based amplification for malaria parasite detection. *Malar. J.* 10: 323.

Rakotonirina, H., Barnadas, C., Raherijafy, R. et al. (2008). Accuracy and reliability of malaria diagnostic techniques for guiding febrile outpatient treatment in malaria-endemic countries. *Am. J. Trop. Med. Hyg.* 78 (2): 217–221.

Research and Technical Intelligence, Malaria Action Programme, World Health Organization (1988). Malaria diagnosis: memorandum from a WHO meeting. *Bull. World Health Organ.* 66 (5): 575–594.

Reyburn, H., Mbakilwa, H., Mwangi, R. et al. (2007). Rapid diagnostic tests compared with malaria microscopy for guiding outpatient treatment of febrile illness in Tanzania: randomised trial. *BMJ* 334 (7590): 403.

Reyburn, H., Ruanda, J., Mwerinde, O., and Drakeley, C. (2006). The contribution of microscopy to targeting antimalarial treatment in a low transmission area of Tanzania. *Malar. J.* 5: 4.

Ruizendaal, E., Dierickx, S., Peeters Grietens, K. et al. (2014). Success or failure of critical steps in community case management of malaria with rapid diagnostic tests: a systematic review. *Malar. J.* 13: 229.

Sakande, J., Nikiema, A., Kabre, E. et al. (2014). Implementation of a national external quality assessment program for medical laboratories in Burkina Faso: challenges, lessons learned, and perspectives. *Am. J. Clin. Pathol.* 141 (2): 181–187.

Schneider, P., Bousema, J.T., Gouagna, L.C. et al. (2007). Submicroscopic Plasmodium falciparum gametocyte densities frequently result in mosquito infection. *Am. J. Trop. Med. Hyg.* 76 (3): 470–474.

Senn, N., Rarau, P., Manong, D. et al. (2012). Rapid diagnostic test-based management of malaria: an effectiveness study in Papua New Guinean infants with Plasmodium falciparum and Plasmodium vivax malaria. *Clin. Infect. Dis.* 54 (5): 644–651.

Singh, N. and Shukla, M.M. (2002). Short report: field evaluation of posttreatment sensitivity for monitoring parasite clearance of Plasmodium falciparum malaria by use of the determine malaria pf test in central India. *Am. J. Trop. Med. Hyg.* 66 (3): 314–316.

Sirichaisinthop, J., Buates, S., Watanabe, R. et al. (2011). Evaluation of loop-mediated isothermal amplification (LAMP) for malaria diagnosis in a field setting. *Am. J. Trop. Med. Hyg.* 85 (4): 594–596.

Skarbinski, J., Ouma, P.O., Causer, L.M. et al. (2009). Effect of malaria rapid diagnostic tests on the management of uncomplicated malaria with artemether-lumefantrine in Kenya: a cluster randomized trial. *Am. J. Trop. Med. Hyg.* 80 (6): 919–926.

Sowunmi, A. and Akindele, J.A. (1993). Presumptive diagnosis of malaria in infants in an endemic area. *Trans. R. Soc. Trop. Med. Hyg.* 87 (4): 422.

Swarthout, T.D., Counihan, H., Senga, R.K., and van den Broek, I. (2007). Paracheck-pf(R) accuracy and recently treated Plasmodium falciparum infections: is there a risk of over-diagnosis? *Malar. J.* 6 (1): 58.

Talman, A.M., Duval, L., Legrand, E. et al. (2007). Evaluation of the intra- and inter-specific genetic variability of Plasmodium lactate dehydrogenase. *Malar. J.* 6: 140.

Tamiru, A., Boulanger, L., Chang, M.A. et al. (2015). Field assessment of dried Plasmodium falciparum samples for malaria rapid diagnostic test quality control and proficiency testing in Ethiopia. *Malar. J.* 14: 11.

Thiam, S., Thior, M., Faye, B. et al. (2011). Major reduction in anti-malarial drug consumption in Senegal after nation-wide introduction of malaria rapid diagnostic tests. *PLoS One* 6 (4): e18419.

Thomson, S., Lohmann, R.C., Crawford, L. et al. (2000). External quality assessment in the examination of blood films for malarial parasites within Ontario, Canada. *Arch. Pathol. Lab. Med.* 124 (1): 57–60.

Tjitra, E., Suprianto, S., Dyer, M.E. et al. (2001). Detection of histidine rich protein 2 and panmalarial ICT Malaria Pf/Pv test antigens after chloroquine treatment of uncomplicated falciparum malaria does not reliably predict treatment outcome in eastern Indonesia. *Am. J. Trop. Med. Hyg.* 65 (5): 593–598.

UNITAID. (2016a). Creating a private sector market for quality-assured RDTs in malaria endemic countries. https://unitaid.org/project/creating-private-sector-market-quality-assured-rdts/#en (Accessed October 21, 2016).

UNITAID (2016b). *Malaria Diagnostics Landscape Update.* Geneva: World Health Organization for the UNITAID Secretariat.

Versteeg, I. and Mens, P.F. (2009). Development of a stable positive control to be used for quality assurance of rapid diagnostic tests for malaria. *Diagn. Microbiol. Infect. Dis.* 64 (3): 256–260.

Vinnemeier, C.D., Schwarz, N.G., Sarpong, N. et al. (2012). Predictive value of fever and palmar pallor for P. falciparum parasitaemia in children from an endemic area. *PLoS One* 7 (5): e36678.

Wafula, R., Sang, E., Cheruiyot, O. et al. (2014). High sensitivity and specificity of clinical microscopy in rural health facilities in western Kenya under an external quality assurance program. *Am. J. Trop. Med. Hyg.* 91 (3): 481–485.

Wanja, E., Achilla, R., Obare, P. et al. (2017). Evaluation of a laboratory quality assurance pilot programme for malaria diagnostics in low-transmission areas of Kenya, 2013. *Malar. J.* 16 (1): 221.

Warhurst, D.C. and Williams, J.E. (1996. ACP Broadsheet no 148. July 1996). Laboratory diagnosis of malaria. *J. Clin. Pathol.* 49 (7): 533–538.

Whitty, C.J., Chandler, C., Ansah, E. et al. (2008). Deployment of ACT antimalarials for treatment of malaria: challenges and opportunities. *Malar. J.* 7 (Suppl 1): S7.

WHO. (2016). Methods manual for laboratory quality control testing of malaria RDTs. http://www.who.int/malaria/publications/rdt-lab-quality-manual/en

WHO/FIND/CDC (2015). *Summary Results of WHO Product Testing of Malaria RDTs: Rounds 1–6 (2008–2015)*. Geneva, Switzerland: World Health Organization.

WHO/FIND/CDC. (2017). Malaria rapid diagnostic test performance – Results of WHO product testing of malaria RDTs: round 7 (2015–2016). http://www.who.int/malaria/publications/atoz/978924151268/en

WHO Global Malaria Program. (2015). A WHO external quality assurance scheme for malaria nucleic acid amplification testing – Meeting report. 978 92 4 150998 5. http://www.who.int/malaria/publications/atoz/9789241509985/en

WHO Global Malaria Program. (2017). False-negative RDT results and implications of new reports of P. falciparum histidine-rich protein 2/3 gene deletions. WHO reference number: WHO/HTM/GMP/2017.18. http://www.who.int/malaria/publications/atoz/information-note-hrp2-based-rdt/en

Williams, J.E., Cairns, M., Njie, F. et al. (2016a). The performance of a rapid diagnostic test in detecting malaria infection in pregnant women and the impact of missed infections. *Clin. Infect. Dis.* 62 (7): 837–844.

Williams, J., Njie, F., Cairns, M. et al. (2016b). Non-falciparum malaria infections in pregnant women in West Africa. *Malar. J.* 15: 53.

Wilson, M.L. (2012). Malaria rapid diagnostic tests. *Clin. Infect. Dis.* 54 (11): 1637–1641.

Wongsrichanalai, C., Barcus, M.J., Muth, S. et al. (2007). A review of malaria diagnostic tools: microscopy and rapid diagnostic test (RDT). *Am. J. Trop. Med. Hyg.* 77 (6 Suppl): 119–127.

World Health Organization. (2009). Bench aids for malaria microscopy. 978 92 4 154786 4. http://www.who.int/malaria/publications/atoz/9789241547864/en

World Health Organization (2010a). *Basic Malaria Microscopy – Part I: Learner's Guide*, 2e. Geneva, Switzerland: 9241547820. http://www.who.int/malaria/publications/atoz/9241547820/en.

World Health Organization. (2010b). *Basic Malaria Microscopy – Part II: Tutor's Guide*, 2e. Geneva, Switzerland: 924154791X. http://www.who.int/malaria/publications/atoz/924154791X/en.

World Health Organization (2010c). *Guidelines for the Treatment of Malaria*. Geneva: World Health Organization.

World Health Organization. (2011a). Good practices for selecting and procuring rapid diagnostic tests for malaria. 978 92 4 150112 5. http://www.who.int/malaria/publications/atoz/9789241501125/en

World Health Organization (2011b). *Universal Access to Malaria Diagnostic Testing: An Operational Manual*. Geneva: World Health Organization.

World Health Organization (2012a). *Disease Surveillance for Malaria Control: Operational Manual*. Geneva: World Health Organization 978 92 4150334 1. http://www.who.int/malaria/publications/atoz/9789241503341/en.

World Health Organization (2012b). *Disease Surveillance for Malaria Elimination: Operational Manual*. Geneva: World Health Organization 978 92 4150333 4.

World Health Organization. (2015a). Malaria Microscopy Standard Operating Procedures. http://www.wpro.who.int/mvp/lab_quality/mm_sop/en

World Health Organization (2015b). *Guidelines for the Treatment of Malaria*, 3e. Geneva: World Health Organization 978 92 4 154912 7. http://www.who.int/malaria/publications/atoz/9789241549127/en.

World Health Organization (2016). *Malaria Microscopy Quality Assurance Manual – Version 2*. Geneva: World Health Organization 978 92 4 154939 4. http://www.who.int/malaria/publications/atoz/9789241549394/en.

World Health Organization. (2017a). Lot testing: Pre and post-purchase webpage. 29 November 2017. http://www.who.int/malaria/areas/diagnosis/rapid-diagnostic-tests/evaluation-lot-testing/en/2017

World Health Organization. (2017b). Malaria microscopy webpage. 14 March 2017. http://www.who.int/malaria/areas/diagnosis/microscopy/en/2017 http://www.who.int/malaria/pubications/atoz/924154791X/en http://www.who.int/malaria/pubications/atoz/9789241547864/en

World Health Organization. (2017c). WHO-FIND malaria RDT evaluation programme webpage. 10 July 2017. http://www.who.int/malaria/areas/diagnosis/rapid-diagnostic-tests/rdt-evaluation-programme/en/20172

World Health Organization (2017d). *World Malaria Report 2017*. Geneva: World Health Organization 978 92 4 156552 3. http://www.who.int/malaria/publications/world-malaria-report-2017/report/en.

World Health Organization. (2018). Training materials on the use of malaria RDTs. http://www.who.int/malaria/areas/diagnosis/rapid-diagnostic-tests/job-aids/en

Yeboah-Antwi, K., Pilingana, P., Macleod, W.B. et al. (2010). Community case management of fever due to malaria and pneumonia in children under five in Zambia: a cluster randomized controlled trial. *PLoS Med.* 7 (9): e1000340.

Yeung, S., Patouillard, E., Allen, H., and Socheat, D. (2011). Socially-marketed rapid diagnostic tests and ACT in the private sector: ten years of experience in Cambodia. *Malar. J.* 10: 243.

Yukich, J.O., Bennett, A., Albertini, A. et al. (2012). Reductions in artemisinin-based combination therapy consumption after the nationwide scale up of routine malaria rapid diagnostic testing in Zambia. *Am. J. Trop. Med. Hyg.* 87 (3): 437–446.

Webliography

https://www.alere.com/en/home/product-details/alere-malaria-agpf.html?utm_source=offline0marketing&utm_medium=infographic&utm_campaign=world_malaria_day_2017 (accessed Dec 2017). Alere. Alere Malaria Ag P.f. webpage. 2017.

https://www.cdc.gov/dpdx/2017. Centers for Disease Control & Prevention (CDC). DPDX – Laboratory identification of parasitic diseases of public health concern. 5 December 2017.

http://www.who.int/malaria/areas/diagnosis/microscopy/en/2017 Centers for Disease Control & Prevention (CDC). Laboratory diagnosis of malaria: Preparation of blood smears.

http://www.cdc.gov/dpdx/resources/pdf/benchAids/malaria/Malaria_staining_benchaid. pdf. Centers for Disease Control & Prevention (CDC). Laboratory diagnosis of malaria: Staining for malaria parasites.

http://www.who.int/malaria/publications/atoz/9789241509985/en. WHO Global Malaria Program A WHO external quality assurance scheme for malaria nucleic acid amplification testing. Meeting report December 2015

http://www.unitaid.org/en/creating-a-private-sector-market-for-quality-assured-rdts-in-malaria-endemic-countries (accessed October 21, 2016). UNITAID. Creating a private sector market for quality-assured RDTs in malaria endemic countries. 2016.

https://unitaid.eu/project/creating-private-sector-market-quality-assured-rdts/#en2017. UNITAID. Creating a private-sector market for quality-assured rapid diagnostic tests. 31 Dec 2016 2017.

http://www.who.int/malaria/publications/atoz/9789241549394/en. WHO F. Methods manual for laboratory quality control testing of malaria RDTs. 2016.

http://www.who.int/malaria/areas/diagnosis/rapid-diagnostic-tests/evaluation-lot-testing/en/2017. WHO Global Malaria Program. A WHO external quality assurance scheme for malaria nucleic acid amplification testing – Meeting report. 2015. 978 92 4 150998 5.

http://www.who.int/malaria/publications/atoz/information-note-hrp2-based-rdt/en. WHO Global Malaria Program. False-negative RDT results and implications of new reports of P. falciparum histidine-rich protein 2/3 gene deletions. 2017. WHO reference number: WHO/HTM/GMP/2017.18.

http://www.who.int/malaria/publications/atoz/9789241549127/en. World Health Organization. Bench aids for malaria microscopy. 2009. 978 92 4 154786 4.

http://www.who.int/malaria/publications/atoz/9789241547864/en. World Health Organization. Basic malaria microscopy – Part I: Learner's guide. 2nd ed. Geneva, Switzerland; 2010a. 9241547820

http://www.who.int/malaria/publications/atoz/9241547820/en. World Health Organization. Basic malaria microscopy – Part II: Tutor's guide. 2nd ed. Geneva, Switzerland; 2010b. 924154791X.

http://www.who.int/malaria/publications/atoz/9789241501125/en World Health Organization. Good practices for selecting and procuring rapid diagnostic tests for malaria. 2011. 978 92 4 150112 5.

http://www.who.int/malaria/publications/atoz/9789241503341/en. World Health Organization. Disease Surveillance for Malaria Control: Operational Manual. Geneva, Switzerland; 2012. 978 92 4150334 1

http://www.wpro.who.int/mvp/lab_quality/mm_sop/en World Health Organization. Malaria Microscopy Standard Operating Procedures. 2015.

http://www.who.int/malaria/areas/diagnosis/rapid-diagnostic-tests/job-aids/en. World Health Organization. Malaria Microscopy Quality Assurance Manual – Version 2. Switzerland, Geneva; 2016. 978 92 4 154939 4.

http://www.who.int/malaria/publications/world-malaria-report-2017/report/en. World Health Organization. World Malaria Report 2017. Geneva, Switzerland; 2017a. 978 92 4 156552 3.

http://www.who.int/malaria/publications/atoz/924154791X/en. World Health Organization. Malaria microscopy webpage. March 2017b.

http://www.who.int/malaria/publications/atoz/978924151268/en. World Health Organization. WHO-FIND malaria RDT evaluation programme webpage. 10 July 2017c. Malaria rapid diagnostic test performance. Results of WHO product testing of malaria RDTs: round 7 (2015–2016)

http://www.who.int/malaria/publications/rdt-lab-quality-manual/en. World Health Organization. Lot testing: Pre and post-purchase webpage. 29 November 2017d.

http://www.who.int/malaria/publications/atoz/9789241501125/en. World Health Organization. Training materials on the use of malaria RDTs.

8

Rapid Diagnostic Tests for Human African Trypanosomiasis

Veerle Lejon[1], Epco Hasker[2], and Philippe Büscher[3]

[1] *Unité Mixte de Recherche IRD-CIRAD 177 INTERTRYP, Institut de Recherche pour le Développement, University of Montpellier, Montpellier, France*
[2] *Department of Public Health, Institute of Tropical Medicine, Antwerp, Belgium*
[3] *Department of Biomedical Sciences, Institute of Tropical Medicine, Antwerp, Belgium*

8.1   Introduction

Human African trypanosomiasis (HAT) is a vector-borne disease caused by two closely related protozoan parasites that are transmitted by tsetse flies (*Glossina sp.*) (Figure 8.1).

The West and Central African form of HAT (gambiense-HAT) is a chronic disease due to infection with *Trypanosoma brucei* (*T.b.*) *gambiense*. The East and South-East African form (rhodesiense-HAT) is a fulminant disease due to infection with *Trypanosoma brucei rhodesiense* (Büscher et al. 2017). Both gambiense- and rhodesiense HAT are endemic in rural sub-Saharan Africa within the limits of tsetse fly distribution. Both forms of the disease occur only in Uganda, but without overlap (Picozzi et al. 2005). Similar parasites within the *Trypanosoma* genus cause infections in animals. Some of these are transmitted mechanically by other blood-sucking flies and some sexually. Contrary to the human disease, animal trypanosomiases therefore occur beyond the tsetse fly belt in Africa (Hoare 1972). Gambiense-HAT causes the majority of HAT infections (1420 cases or 98.1% in 2017) reported in endemic countries with the Democratic Republic of the Congo, the Republic of Congo, the Central African Republic, Chad and South Sudan being the most affected (http://apps.who.int/gho/data/node.main.A1636?lang=en). The number of reported rhodesiense-HAT patients is much lower (27 cases or 1.9% of all

Revolutionizing Tropical Medicine: Point-of-Care Tests, New Imaging Technologies and Digital Health,
First Edition. Edited by Kerry Atkinson and David Mabey.
© 2019 John Wiley & Sons, Inc. Published 2019 by John Wiley & Sons, Inc.

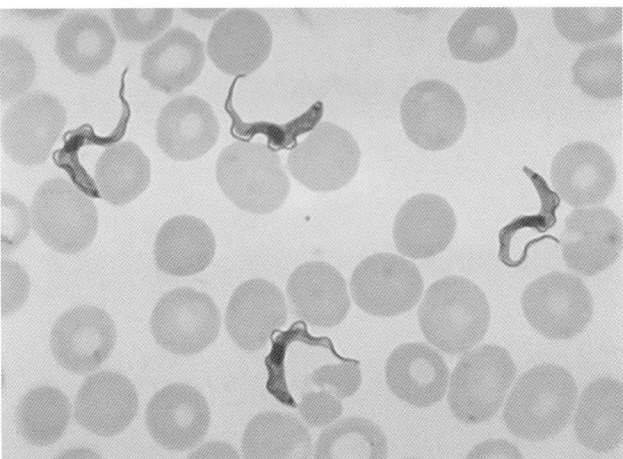

Figure 8.1 Trypanosoma parasites in a blood smear from a patient with African trypanosomiasis. Typical trypomastigote stages (the only stages found in patients), with a posterior kinetoplast, a centrally located nucleus, an undulating membrane and an anterior flagellum are visible. The trypanosomes' length ranges from 14 to 33 μm. This image, reproduced here from Wikimedia Commons, is a work of the Centers for Disease Control and Prevention, part of the United States Department of Health and Human Services, taken or made as part of an employee's official duties. As a work of the U.S. Federal Government, the image is in the public domain. The content provider is CDC/ Dr. Myron G. Schultz. https://phil.cdc.gov/phil/details.asp?pid=613.

HAT cases in 2017). Rhodesiense-HAT occasionally affects travelers who contract the disease by visiting game parks and nature reserves in Malawi, Tanzania, Uganda and Zambia (Migchelsen et al. 2011; Simarro et al. 2012).

HAT devastated entire populations in sub-Saharan Africa at the beginning of the twentieth century. HAT epidemics were brought under control in the 1960s (Simarro et al. 2008) only after the availability of the first effective drugs combined with drastic measures that included bush clearing, insecticide spraying, culling of wild animals, forced participation in screening campaigns, movement restrictions, and mass-prophylaxis. Relaxation of these efforts allowed gambiense-HAT to re-emerge in the 1990s with a peak of about 37 000 detected cases in 1998 (Franco et al. 2014).

Since 2012 gambiense-HAT has been targeted by the World Health Organization (WHO) for elimination as a public health problem by 2020 (Simarro et al. 2015). Today, already, HAT has become a rare disease and with fewer than 1500 cases detected in 2017. HAT prevalence has never been as low. This drop in prevalence can largely be attributed to the deployment of large scale control measures, and for gambiense-HAT control in particular, mass screening using innovative immunodiagnostic tests.

8.2 The Early Introduction of Immunodiagnostic Tests in the Diagnosis of HAT

The diagnosis of HAT is typically a multistep procedure. HAT is suspected based on a number of non-specific clinical signs and symptoms such as fever, an inoculation chancre at the site of the bite, general fatigue, headache, joint pain, enlarged cervical lymph nodes, behavioral changes and an abnormal sleep pattern. Clinical suspicion is followed directly by microscopic examination of lymph node aspirate and blood in order to detect the parasite (Lejon et al. 2013). Once infection is confirmed, a lumbar puncture is performed to determine the stage of the disease using white blood cell quantification and parasite detection in the cerebrospinal fluid (CSF). Stage determination is necessary to decide on the appropriate treatment which is different for first stage and second stage disease and different for gambiense-HAT and rhodesiense-HAT (Büscher et al. 2017).

This diagnostic algorithm of clinical suspicion followed by microscopic confirmation and stage determination has been used for decades with success. It was only in the 1960s, when HAT was already almost under control, that immunodiagnostic tests, mostly in the form of immunofluorescence assays (IFAs) for specific antibody detection, were introduced (Bailey et al. 1967). Such IFAs further helped to control gambiense-HAT in Gabon and Equatorial Guinea (Sarda 1980; Simarro et al. 2001). The major advantage of the IFA was its ability to identify persons without detectable parasitemia and even without clinical symptoms as "serological suspects." These suspected cases could then be followed up later for parasitological confirmation (Kegels et al. 1992; Magnus et al. 1978a). A second advantage was that an IFA could be performed on dried blood spots (DBS) collected during surveys for subsequent analysis in the laboratory (Bailey et al. 1967; Wéry et al. 1970). The primary disadvantage of the IFA was that it required a minimal laboratory infrastructure and electricity to power the fluorescence microscope. Other immunodiagnostic formats such as hemagglutination, ELISA and immunoprecipitation were never applied in routine HAT control and had similar disadvantages (WHO 1976). It is important to note that most immunodiagnostic tests for HAT were prepared using an undefined mixture of antigens and therefore were prone to non-specific reactions resulting in false positive results.

8.3 CATT/*T.b. gambiense*: A Breakthrough in the Immunodiagnosis of Gambiense-HAT

The card agglutination test for trypanosomiasis (CATT/*T.b. gambiense*) was the first rapid diagnostic test for immunodiagnosis of gambiense-HAT and was introduced in the 1970s (Magnus et al. 1978b). Its origin dates back to research on antigenic variation, a phenomenon that allows trypanosomes, as extracellular parasites, to survive the immune response of the vertebrate host. During the course of an infection, trypanosomes are completely covered by a dense membrane coat consisting of variant surface glycoproteins (VSGs). At any given time only one type of VSG is exposed on the surface of a trypanosome, thus defining its variable antigen type (VAT). The VSG coat protects the trypanosomes against effector molecules of the innate immune system such as complement, but it is also highly immunogenic. To escape the host antibody response, trypanosomes change the VAT of their surface coat on a regular basis (Horn 2014). Although trypanosomes are capable of expressing thousands of different VSGs, researchers at the Institute of Tropical Medicine (ITM) in Antwerp, Belgium, discovered that several VATs of *T.b. gambiense* such as the Lille Trypanosome Antigen Type (LiTat) 1.3, LiTat 1.5 and LiTat 1.6 are recognized by specific antibodies in the serum of almost all gambiense-HAT patients (Van Meirvenne et al. 1975, 1977; Vervoort et al. 1978). This discovery was used to develop the first serological "rapid" diagnostic test for gambiense-HAT that could be used in field conditions, the CATT/*T.b. gambiense* (Magnus et al. 1978b).

The CATT/*T.b. gambiense* reagent consists of cloned *T.b. gambiense* LiTat 1.3 trypanosomes, purified from the blood of infected rats and subsequently fixed, stained and lyophilized in volumes for 50 tests. CATT is stable at ≤ 8 °C for years and can be transported at ambient temperature without losing its activity. Once the reagent is reconstituted with buffer, however, it must be kept cool and used within a couple of days. The test is performed by mixing one drop of blood with one drop of reagent and letting it react under gentle horizontal rotation for five minutes. The reaction is based on the interaction of antibodies recognizing the surface antigens of the fixed trypanosomes. When present in a blood sample, such antibodies will form lattices with the trypanosomes, that are visible to the naked eye as blue agglutinates (Van Meirvenne 1999). CATT/*T.b. gambiense* detects both IgM and IgG isoforms, irrespective of the stage of the disease. Using CATT/*T.b. gambiense* as a screening test, laborious microscopic examinations can be limited to CATT-positive individuals only. Sensitivity of CATT/*T.b. gambiense* varies according to geographical location but it is believed to range from 78% to almost 100%. Specificity is estimated at approximately 97% (WHO 2013). As finger-prick blood samples from ten persons can be tested simultaneously, CATT/*T.b. gambiense* is particularly suited for mass-screening campaigns and a few hundred persons per day can be screened by a single mobile team (Robays et al. 2004) (Figure 8.2).

Screening based on CATT/*T.b. gambiense* has been deployed on a large scale (up to several million screenings per year) since the late 1980s, and has without doubt contributed to the rapid decline of gambiense-HAT over the last two decades. A similar test for rhodesiense-HAT does not exist, since to date a counterpart for LiTat 1.3 VSG in

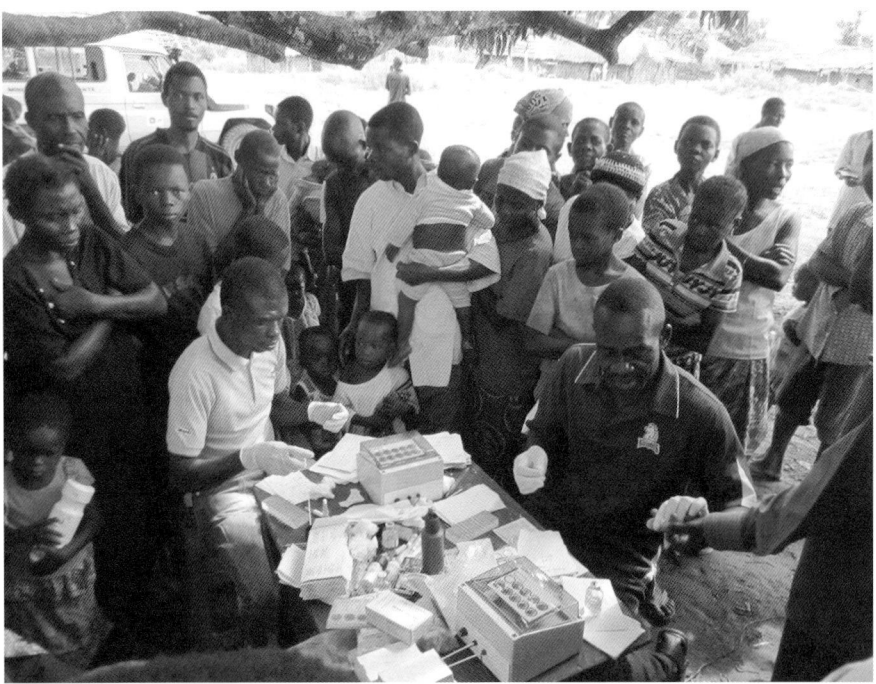

Figure 8.2 Mobile team in the Democratic Republic of the Congo, screening villagers with the CATT/*T.b. gambiense. Source:* Veerle Lejon.

T.b. gambiense has not been found in *T.b. rhodesiense* (Van Meirvenne 1999). However, in sharp contrast to gambiense-HAT, parasitemia in most cases of rhodesiense-HAT cases is sufficiently high to allow detection of parasites in a thin blood film or thick blood drop. Furthermore, the disease typically evolves so rapidly that there is little time to develop detectable antibody titers of diagnostic value.

8.4 The Changing Epidemiology of Gambiense-HAT: The Need for Improved Rapid Diagnostic Tests

Today prevalence in most HAT foci has dropped to below 0.1% and the elimination of gambiense-HAT seems within reach (Franco et al. 2017). This situation entails major challenges to the design and performance of diagnostic tests. Active screening by mobile teams becomes less cost-effective and is gradually being replaced by passive screening of clinical suspects that present to fixed health centers (Mitashi et al. 2012, 2015). These health centers are often devoid of laboratory facilities and electricity. Moreover, attendance rates, and hence the number of HAT suspects, tend to be low (Mitashi et al. 2015) (Figure 8.3).

The CATT/*T.b. gambiense* reagent, requiring refrigeration and packed in vials of 50 test doses, is not suitable in this environment. Individual rapid diagnostic tests (RDTs) are needed that meet the ASSURED criteria (**A**ffordable, **S**ensitive, **S**pecific, **U**ser-friendly, **R**apid and robust, **E**quipment-free and **D**eliverable to the end-user) (Peeling et al. 2006). Such RDTs have been developed recently and two first generation RDTs for detection of gambiense-specific antibodies are commercially available (Figure 8.4).

Figure 8.3 Laboratory table in a fixed health center in the Democratic Republic of the Congo *Source:* Philippe Büscher (2015).

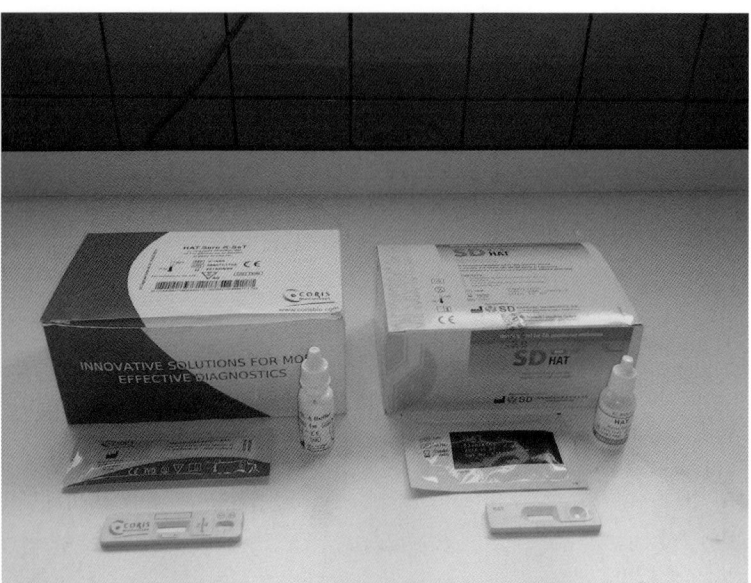

Figure 8.4 Two "first generation" rapid diagnostic tests for *Trypanosoma brucei gambiense* sleeping sickness: HAT Sero *K*-SeT and SD Bioline HAT. *Source:* Philippe Büscher (2015).

The HAT Sero *K*-SeT is produced by Coris BioConcept in Belgium (http://www .corisbio.com/Products/Human-Field/Human-African-Trypanosomiasis.php). The antigen consists of a combination of purified, native *T.b. gambiense* LiTat 1.3 and LiTat 1.5 VSG. The choice of these antigens was based on confirmation of their excellent diagnostic performance in other serological test formats such as CATT/*T.b. gambiense*, ELISA, LATEX/*T.b. gambiense* and immune trypanolysis (Van Meirvenne et al. 1995; Lejon et al. 1998; Büscher et al. 1999; Jamonneau et al. 2000). Following a successful phase I evaluation, the HAT Sero *K*-SeT underwent prospective evaluation in the Democratic Republic of the Congo and proved promising with 98.5% sensitivity and 98.6% specificity (Büscher et al. 2013a, b, 2014; Büscher 2015).

The SD Bioline HAT is produced by Standard Diagnostics in South Korea (https:// www.alere.com/en/home/product-details/sd-bioline-hat.html). This test also makes use of purified, native *T.b. gambiense* LiTat 1.3 and LiTat 1.5 VSGs that were selected for their superior diagnostic performance among a larger panel of candidate antigens (Biéler et al. 2016). In a large-scale prospective evaluation in Angola, the Democratic Republic of the Congo and the Central African Republic, the sensitivity and specificity of a prototype of the SD Bioline HAT were 89.3 and 94.6% respectively (Bisser et al. 2016). In another prospective study, conducted in the Democratic Republic of the Congo, the observed sensitivity was much lower (59.0%) while the specificity was somewhat higher (98.9%) (Lumbala et al. 2018).

Thus far both first generation RDTs have been evaluated simultaneously on a large collection of stored plasma samples from HAT patients and endemic control samples from Guinea and Côte d'Ivoire. This retrospective study confirmed the high sensitivity (>99%) of both tests while their specificity (around 88%) was lower than reported for previous prospective studies in Central Africa using fresh whole blood specimens (Jamonneau et al. 2015).

8.5 Second Generation RDTs for HAT

A major disadvantage of the first generation RDTs is the use of native antigens that are purified from trypanosomes grown in laboratory rodents, making them expensive and prone to batch-to-batch variation. Fortunately, a number of studies have shown that the native LiTat 1.3 and LiTat 1.5 VSGs can successfully be replaced in diagnostic tests by synthetic peptides or by recombinant antigens, expressed in a variety of prokaryotic and eukaryotic expression systems such as *Escherichia coli*, *Pichia pastoris*, *Spodoptera frugiperda* and *Leishmania tarentolae* (Van Nieuwenhove et al. 2011, 2012; Sullivan et al. 2013, 2014; Rogé et al. 2014; Sternberg et al. 2014; Rooney et al. 2015).

In a large-scale prospective evaluation a prototype SD Bioline HAT 2.0 containing recombinant antigens derived from the N-terminal part of LiTat 1.5 VSG and of Invariable Surface Glycoprotein 65 (ISG65) showed a similar specificity (98.1%) but a higher sensitivity (71.2%) than the SD Bioline HAT using native VSGs (98.9 and 59.0% respectively) (Lumbala et al. 2018). Recombinant LiTat 1.5 VSG and ISG65 are also used in the rHAT Sero-Strip, which is an alternative "light" version RDT in dipstick format (Gilleman et al. 2015a). Preliminary results obtained with this second generation RDT from Coris BioConcept on archived sera are promising (sensitivity 97%, specificity 95.5%) (Gilleman et al. 2015b).

8.6 Future Perspectives and Challenges

In the context of HAT elimination, the recent introduction of RDTs for gambiense-HAT has satisfied a real need. RDTs are being adopted quickly for control and surveillance of HAT. Nevertheless, several uncertainties remain around their diagnostic accuracy and optimal use. Moreover, their performance in real life situations has not been evaluated by independent research. Only one consortium of independent researchers is conducting a comparative prospective evaluation of the different commercialized and prototype RDTs for HAT and data are not yet available (http://www.ditect-hat.eu).

As indicated above, both first generation RDTs have been evaluated simultaneously on a large collection of stored plasma samples from HAT patients and endemic control samples from Guinea and Côte d'Ivoire with a specificity 88%, which was lower than reported for previous prospective studies in Central Africa using fresh whole blood specimens (Jamonneau et al. 2015). Therefore, the serial application of RDTs has been proposed to increase specificity without affecting sensitivity, as is currently done for the diagnosis of HIV infection using RDTs, preferably combining RDTs that use different antigens.

A recent prospective study using CATT/*T.b. gambiense* and the Standard Diagnostics first and prototype second generation tests (SD Bioline HAT 1.0 and 2.0) showed surprisingly low sensitivities of 63, 59 and 71% respectively although specificities of 98–99% were reported. In this study, the authors demonstrated that parallel application of the two RDTs increased the sensitivity to 90%, with a modest effect on specificity, lowering it to 97% (Lumbala et al. 2018).

Since RDTs for HAT are used in peripheral health centers where confirmation of the diagnosis by microscopy is often not available, their suboptimal specificity is a reason for concern. Indeed, the positive predictive value (PPV) of a 99% sensitive but 90% specific RDT at low-prevalence (0.1%) is only 1%. Such a low PPV is not compatible with a realistic test-and-treat scenario even in the event that new, safe, oral drugs become available (Mesu et al. 2018). It is clear that the RDTs for gambiense-HAT serodiagnosis need further development to improve their diagnostic accuracy.

Another important issue that needs to be addressed before implementing RDTs in HAT control is how to ensure high quality testing, especially in peripheral health centers that often face tremendous operational challenges. Apart from proper training of personnel and provision of logistic support, digital capturing with a camera or a test-reader of the RDT results, which must be read within five minutes after the test run, may be considered.

In conclusion, CATT/*T.b. gambiense*, the earliest rapid test developed for HAT, has contributed to the unprecedented success of HAT control obtained over the last two decades. However, ASSURED-compliant RDTs are crucial in order to achieve elimination of the disease. Expectations on simplification of the diagnostic pathway of HAT are high but, before the laborious parasitological confirmation step can be omitted, further improvement in accuracy and/or better documentation of optimal RDT application are needed.

References

Bailey, N.M., Cunningham, M.P., and Kimber, C.D. (1967). The indirect fluorescent antibody technique applied to dried blood, for use as a screening test in the diagnosis of human trypanosomiasis in Africa. *Trans. R. Soc. Trop. Med. Hyg.* 61: 696–700.

Biéler, S., Waltenberger, H., Barrett, M.P. et al. (2016). Evaluation of antigens for development of a serological test for human African trypanosomiasis. *PLoS One* 11: e0168074.

Bisser, S., Lumbala, C., Nguertoum, E. et al. (2016). Sensitivity and specificity of a prototype rapid diagnostic test for the detection of *Trypanosoma brucei gambiense* infection: a multi-centric prospective study. *PLoS Negl. Trop. Dis.* 10: e0004608.

Büscher, P. 2015. Contribution of diagnostics in sleeping sickness elimination by 2020. ISNTD *D³* conference, 20–21 May, London.

Büscher, P., Cecchi, G., Jamonneau, V., and Priotto, G. (2017). Human African trypanosomiasis. *Lancet* 390: 2397–2409. https://doi.org/10.1016/S0140-6736(17)31510-6.

Büscher, P., Gilleman, Q., and Lejon, V. (2013a). Novel rapid diagnostic test for sleeping sickness. *N. Engl. J. Med.* 368: 1069–1070.

Büscher, P., Lejon, V., Magnus, E., and Van Meirvenne, N. (1999). Improved latex agglutination test for detection of antibodies in serum and cerebrospinal fluid of *Trypanosoma brucei gambiense* infected patients. *Acta Trop.* 73: 11–20.

Büscher, P., Mertens, P., Leclipteux, T. et al. (2014). Sensitivity and specificity of HAT Sero-K-SeT, a rapid diagnostic test for serodiagnosis of sleeping sickness caused by *Trypanosoma brucei gambiense*: a case-control study. *Lancet Glob. Health* 2: e359–e363.

Büscher, P., Mumba Ngoyi, D., Pyana, P.P. et al. (2013b). New rapid tests for antibody detection serodiagnosis in *Trypanosoma brucei gambiense* sleeping sickness. *HAT Platform Newsletter* 13: 21–22.

Franco, J.R., Cecchi, G., Priotto, G. et al. (2017). Monitoring the elimination of human African trypanosomiasis: update to 2014. *PLoS Negl. Trop. Dis.* 11: e0005585.

Franco, J.R., Simarro, P.P., Diarra, A., and Jannin, J.G. (2014). Epidemiology of human African trypanosomiasis. *Clin. Epidemiol.* 6: 257–275.

Gilleman, Q., Büscher, P., Mertens, P. and Leclipteux, T. (2015a). Development of a Rapid Diagnostic Test for active case detection in sleeping sickness control: recHAT Sero-Strip. The International Society for Neglected Tropical Diseases. 74. London. Research Handbook. Drug discovery, development & diagnostics. ISNTD D^3 conference, 20–21 May, London.

Gilleman, Q., Büscher, P., Mertens, P. and Leclipteux, T. (2015b). Development of a rapid diagnostic test for active case detection in sleeping sickness control: recHAT Sero-Strip. 94. 9th European Congress on Tropical Medicine and International Health, 6–10 September 2015, Basel, Switzerland.

Hoare, C.A. (1972). *The Trypanosomes of Mammals*. Oxford: Blackwell Scientific Publications.

Horn, D. (2014). Antigenic variation in African trypanosomes. *Mol. Biochem. Parasitol.* 195: 123–129.

Jamonneau, V., Camara, O., Ilboudo, H. et al. (2015). Accuracy of individual rapid tests for serodiagnosis of *gambiense* sleeping sickness in West Africa. *PLoS Negl. Trop. Dis.* 9: e0003480.

Jamonneau, V., Truc, P., Garcia, A. et al. (2000). Preliminary evaluation of LATEX/*T. b. gambiense* and alternative versions of CATT/*T. b. gambiense* for the serodiagnosis of human African trypanosomiasis of a populationat risk in Côte d'Ivoire: considerations for mass-screening. *Acta Trop.* 76: 175–183.

Kegels, G., Criel, B., van Lerberghe, W. et al. (1992). Screening for *Trypanosoma brucei gambiense* antibodies with the indirect fluorescent antibody test (IFAT). *Ann. Soc. Belg. Med. Trop.* 72: 271–281.

Lejon, V., Büscher, P., Magnus, E. et al. (1998). A semi-quantitative ELISA for detection of *Trypanosoma brucei gambiense* specific antibodies in serum and cerebrospinal fluid of sleeping sickness patients. *Acta Trop.* 69: 151–164.

Lejon, V., Jacobs, J., and Simarro, P.P. (2013). Elimination of sleeping sickness hindered by difficult diagnosis. *Bull. World Health Organ.* 91: 718.

Lumbala, C., Biéler, S., Kayembe, S., et al. (2018). Prospective evaluation of a rapid diagnostic test for Trypanosoma brucei gambiense infection developed using recombinant antigens. *PLoS Negl. Trop. Dis.* 12: e0006386

Magnus, E., Van Meirvenne, N., Vervoort, T. et al. (1978a). Use of freeze-dried trypanosomes in the indirect fluorescent antibody test for the serodiagnosis of sleeping sickness. *Ann. Soc. Belg. Med. Trop.* 58: 103–109.

Magnus, E., Vervoort, T., and Van Meirvenne, N. (1978b). A card-agglutination test with stained trypanosomes (C.A.T.T.) for the serological diagnosis of *T.b.gambiense* trypanosomiasis. *Ann. Soc. Belg. Med. Trop.* 58: 169–176.

Mesu, V.K.B.K., Kalonji, W.M., Bardonneau, C., et al. (2018). Oral fexinidazole for late-stage African *Trypanosoma brucei gambiense* trypanosomiasis: a pivotal multicentre, randomised, non-inferiority trial. *The Lancet*, 391: 144–154

Migchelsen, S.J., Büscher, P., Hoepelman, A.I.M. et al. (2011). Human African trypanosomiasis: a review of non-endemic cases in the past 20 years. *Int. J. Infect. Dis.* 15: e517–e524.

Mitashi, P., Hasker, E., Lejon, V. et al. (2012). Human African trypanosomiasis diagnosis in first-line health services of endemic countries, a systematic review. *PLoS Negl. Trop. Dis.* 6: e1919.

Mitashi, P., Hasker, E., Mbo, F. et al. (2015). Integration of diagnosis and treatment of sleeping sickness in primary healthcare facilities in the Democratic Republic of the Congo. *Tropical Med. Int. Health* 20: 98–105.

Peeling, R.W., Holmes, K.K., Mabey, D., and Rondon, A. (2006). Rapid tests for sexually transmitted infections (STIs): the way forward. *Sex. Transm. Dis.* 82 (Suppl V): v1–v6.

Picozzi, K., Fèvre, E.M., Odiit, M. et al. (2005). Sleeping sickness in Uganda: a thin line between two fatal diseases. *Br. Med. J.* 331: 1238–1241.

Robays, J., Bilengue, M.M.C., Van der Stuyft, P., and Boelaert, M. (2004). The effectiveness of active population screening and treatment from sleeping sickness control in the Democratic Republic of Congo. *Tropical Med. Int. Health* 9: 542–550.

Rogé, S., Van Nieuwenhove, L., Meul, M. et al. (2014). Recombinant antigens expressed in *Pichia pastoris* for the diagnosis of sleeping sickness caused by *Trypanosoma brucei gambiense*. *PLoS Negl. Trop. Dis.* 8: e3006.

Rooney, B., Piening, T., Büscher, P. et al. (2015). Expression of *Trypanosoma brucei gambiense* antigens in *Leishmania tarentolae*. Potential for rapid serodiagnostic tests (RDTs). *PLoS Negl. Trop. Dis.* 9: e0004271.

Sarda, J. (1980). Interêt de l'immunofluorescence indirecte dans la trypanosomiase humaine en campagne de masse. Premiers résultats obtenus en 1979 dans le foyer de l'estuaire du Gabon. In: *Rapport Final de la 13ième Conférence Technique* (ed. OCEAC), 191–205. Yaoundé: OCEAC.

Simarro, P.P., Cecchi, G., Franco, J.R. et al. (2015). Monitoring the progress towards the elimination of *gambiense* human African trypanosomiasis. *PLoS Negl. Trop. Dis.* 9: e0003785.

Simarro, P.P., Franco, J.R., Cecchi, G. et al. (2012). Human African trypanosomiasis in non-endemic countries (2000-2010). *Journal of Travel Medicine* 19: 44–53.

Simarro, P.P., Franco, J.R., and Ndongo Asumu, P. (2001). Le foyer de trypanosomiase humaine africaine de Luba en Guinee Equatoriale a-t-il été éradique? *Med. Trop.* 61: 441–444.

Simarro, P.P., Jannin, J., and Cattand, P. (2008). Eliminating human African trypanosomiasis: where do we stand and what comes next? *PLoS Med.* 5: 174–180.

Sternberg, J.M., Gierlinski, M., Biéler, S. et al. (2014). Evaluation of the diagnostic accuracy of prototype rapid tests for human African trypanosomiasis. *PLoS Negl. Trop. Dis.* 8: e3373.

Sullivan, L., Fleming, J., Sastry, L. et al. (2014). Identification of sVSG117 as an immunodiagnostic antigen and evaluation of a dual-antigen lateral flow test for the diagnosis of human African trypanosomiasis. *PLoS Negl. Trop. Dis.* 8: e2976.

Sullivan, L., Wall, S.J., Carrington, M., and Ferguson, M.A.J. (2013). Proteomic selection of immunodiagnostic antigens for human African trypanosomiasis and generation of a protoype lateral flow immunodiagnostic device. *PLoS Negl. Trop. Dis.* 7: e2087.

Van Meirvenne, N. (1999). Biological diagnosis of human African trypanosomiasis. In: *Progress in Human African Trypanosomiasis, Sleeping Sickness* (ed. M. Dumas, B. Bouteille and A. Buguet), 235–252. Paris: Springer.

Van Meirvenne, N., Janssens, P.G., Magnus, E. et al. (1975). Antigenic variation in syringe passaged populations of *Trypanosoma (Trypanozoon) brucei*. II. Comparative studies on two antigenic-type collections. *Ann. Soc. Belg. Med. Trop.* 55: 25–30.

Van Meirvenne, N., Magnus, E., and Büscher, P. (1995). Evaluation of variant specific trypanolysis tests for serodiagnosis of human infections with *Trypanosoma brucei gambiense*. *Acta Trop.* 60: 189–199.

Van Meirvenne, N., Magnus, E., and Vervoort, T. (1977). Comparisons of variable antigenic types produced by trypanosome strains of the subgenus *Trypanozoon*. *Ann. Soc. Belg. Med. Trop.* 57: 409–423.

Van Nieuwenhove, L., Büscher, P., Balharbi, F. et al. (2012). Identification of mimotopes with diagnostic potential for *Trypanosoma brucei gambiense* variant surface glycoproteins with human antibody fractions. *PLoS Negl. Trop. Dis.* 6: e1682.

Van Nieuwenhove, L., Rogé, S., Balharbi, F. et al. (2011). Identification of peptide mimotopes of *Trypanosoma brucei gambiense* variant surface glycoproteins. *PLoS Negl. Trop. Dis.* 5: e1189.

Vervoort, T., Magnus, E. and Van Meirvenne, N. (1978). Characterization and serodiagnostic use of an ubiquitous variable antigen type of the subgenus Trypanozoon. *4th European Immunoloy Meeting, April 12–14 1978, Budapest, Hungary.*

Wéry, M., Wéry-Paskoff, S., and Van Wettere, P. (1970). The diagnosis of human African trypanosomiasis (*T. gambiense*) by the use of fluorescent antibody test. *Ann. Soc. Belg. Med. Trop.* 50: 613–634.

World Health Organization (1976). Parallel evaluation of serological tests applied in African trypanosomiasis: a WHO collaborative study. *Bull. World Health Organ.* 54: 141–147.

World Health Organization (2013). Control and surveillance of human African trypanosomiasis. WHO meeting on elimination of African trypanosomiasis (Trypanosoma brucei gambiense), Geneva, 3–5 December 2012 WHO/HTM/NTD/IDM/2013, http://apps.who.int/iris//bitstream/10665/95732/1/9789241209847_eng.pdf Accessed 11 April 2017.

Webliography

http://www.corisbio.com/Products/Human-Field/Human-African-Trypanosomiasis.php. The HAT Sero *K*-SeT produced by Coris BioConcept in Belgium.

http://www.ditect-hat.eu. An ongoing comparative prospective evaluation of the different commercialized and prototype RDTs for HAT.

http://www.standardia.com/en/home/product/rapid/infectious-disease/HAT.html). The SD Bioline HAT produced by Standard Diagnostics in South Korea.

https://www.dndi.org/diseases-projects/hat. DNDi (Drugsfor neglected diseases. About sleeping sickness.

http://apps.who.int/gho/data/node.main.A1636?lang=en. WHO. Number of new reported cases (T.b. gambiense) 2004–2015.

http://apps.who.int/iris/bitstream/10665/95732/1/9789241209847_eng.pdfLast accessed on 11-4-2017. World Health Organization 2013. Control and surveillance of human African trypanosomiasis.

9

Rapid Diagnostic Tests for Visceral Leishmaniasis

Marleen Boelaert[1, 2, 3], Suman Rijal[1, 2, 3], and François Chappuis[1, 2, 3]

[1] Institute of Tropical Medicine, Antwerp, Belgium
[2] DNDi, Geneva, Switzerland
[3] Geneva University Hospitals, Geneva, Switzerland

9.1   Introduction

Visceral leishmaniasis (VL) or kala-azar is one of the Neglected Tropical Diseases (Trouiller et al. 2002) that usually has a fatal outcome if not treated in a timely manner. The infectious agent is an obligate intracellular parasite of the genus Leishmania. The vector is a sandfly.

Four countries (India, Bangladesh, Sudan, and Brazil) report up to 90% of the world's annual cases of this vector-borne disease. In the Indian subcontinent, VL is caused by *Leishmania donovani* and in Europe, North-Africa, and Latin-America by *Leishmania infantum*. Early case detection and treatment is one of the key control strategies, as there is no vaccine to prevent the infection in humans. Early case management also a better prognosis. Unfortunately, VL is a poverty-related disease, and patients typically live in remote areas with inadequate access to health services, which leads to sometimes leads to serious delays between the onset of symptoms and the initiation of treatment (Hasker et al. 2012).

The common clinical manifestations of VL include fever, fatigue, weakness, loss of appetite and weight loss, combined with enlarged liver, spleen and lymph nodes.

Revolutionizing Tropical Medicine: Point-of-Care Tests, New Imaging Technologies and Digital Health, First Edition. Edited by Kerry Atkinson and David Mabey.
© 2019 John Wiley & Sons, Inc. Published 2019 by John Wiley & Sons, Inc.

The slowly evolving clinical picture lacks specificity as these symptoms and signs are common in other diseases that occur in VL-endemic areas (Boelaert et al. 2007). Treating patients by clinical presumption is not recommended because the medicines have potentially serious side effects. Before the advent of the Rapid Diagnostic Test (RDT) a rural health worker suspecting a case of VL had to refer the patient to a hospital for diagnostic confirmation and treatment, an effort which many patients could not afford. Another approach was to send impregnated filter paper blood samples for DAT testing in a central laboratory but this led to long delays before results were available, loss of samples and patients lost-to-follow-up. Hence the importance of the RDT for VL: better access to care, better treatment outcomes, and better control of transmission.

When discussing VL diagnostics it is always important to be aware that there is a disease spectrum from asymptomatic *L. donovani/L. infantum* infection to full-blown VL (Ho et al. 1982; Hirve et al. 2016). Detecting and treating infection in asymptomatic cases is currently not performed, as the available drugs are too toxic to justify their use in asymptomatic people, who may recover from the disease without medical intervention. We will, therefore, limit the discussion in this chapter to the diagnosis of full-blown visceral leishmaniasis[1] (kala-azar[2]).

9.2   Parasitology, a Reference Standard?

The recommended method for diagnosis of VL has been the demonstration of the amastigote stage of the *Leishmania* parasite (called LD bodies) in stained tissue smears or culture since Leishman identified them in 1901 (Figure 9.1).

Clinical guidelines for non-endemic areas continue to recommend microscopy as the method of choice to diagnose VL (Aronson et al. 2017). When flawlessly executed, the parasitological tests are very specific, but their sensitivity varies depending on the type of tissue aspirate. Examination of spleen tissue aspirates has the highest sensitivity - in the range of 95% (Zijlstra et al. 1992; Bryceson 1996; Thakur 1997). Kager et al. (1983) proposed a minimally invasive approach for splenic aspiration, practiced by Médecins Sans Frontières under field conditions in South Sudan in the 1980s. However, this diagnostic procedure has a risk of potentially fatal hemorrhage in the range of 1 per 1000 procedures (Chulay and Bryceson 1983; Boussery et al. 2001; Sundar and Rai 2002).

---

[1] Lieutenant General Sir William Boog Leishman KCB, KCMG, FRS (1865–1926) was a Scottish pathologist and British Army medical officer. He was Director-General of Army Medical Services from 1923 to 1926. He was born in Glasgow and attended Westminster School and the University of Glasgow before entering the Royal Army Medical Corps. He served in India, where he did research on enteric fever and kala-azar. He returned to the United Kingdom and was stationed at the Victoria Hospital in Netley in 1897. In 1900 he was made Assistant Professor of Pathology in the Army Medical School, and described a method of staining blood for malaria and other parasites – a modification and simplification of the existing Romanowsky method using a compound of methylene blue and eosin, which became known as Leishman's stain. In 1901, while examining pathologic specimens of a spleen from a patient who had died of kala azar (now called visceral leishmaniasis), he observed oval bodies and published his account of them in 1903. Charles Donovan of the Indian Medical Service independently found such bodies in other kala-azar patients, and they are now known as Leishman-Donovan bodies (not to be confused with Donovan bodies, which are found in granuloma inguinale caused by Klebsiella granulomatis).

[2] From late-nineteenth century Assamese: kālā "black" + āzār "disease" – because of the bronzing of the skin often associated with visceral leishmaniasis.

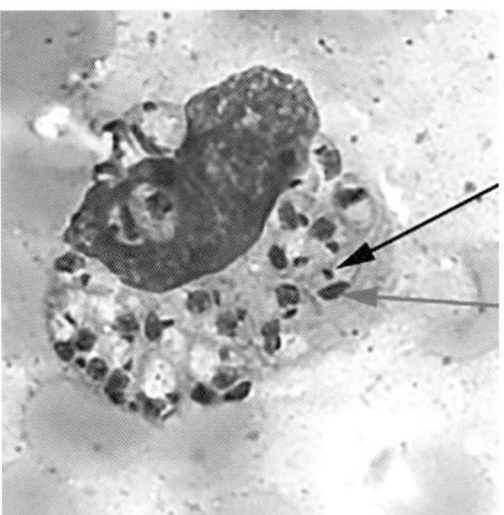

Figure 9.1 Light-microscopic examination of a stained bone marrow specimen from a patient with visceral leishmaniasis – showing a macrophage containing multiple *Leishmania* **amastigotes** (the tissue stage of the parasite). Note that each amastigote has a **nucleus** (gray arrow) and a rod-shaped **kinetoplast** (black arrow). Visualization of the kinetoplast is important for diagnostic purposes in order to be confident the patient has leishmaniasis. *Source:* This image is from the Public Health Image Library (PHIL) and the Laboratory Identification of Parasitic Diseases of Public Health Concern (DPDx), both part of the Centers for Disease Control and Prevention, USA. As work of employees of the United States Federal Government, this image is in the public domain. www.cdc.gov. (*See color plate section for the color representation of this figure.*)

Kager et al. (1983) observed one death in a series of 671 splenic aspirates in Kenya; Thakur (1997) reported three deaths in a series of 3000 in India. Thus the procedure is contraindicated in people with severe anemia and in those with a prolonged prothrombin time or significant thrombocytopenia. It has to be performed by well-trained staff in a facility where blood transfusion facilities, and ideally splenectomy, are readily available (Bryceson 1996). After a splenic aspirate under direct observation the patient must remain in the recumbent position for at least eight hours (Bryceson 1996). Additionally, splenic aspiration is not possible in individuals without a palpable spleen (Zijlstra et al. 1992) or in hyperactive children.

As an alternative, LD bodies can be searched for in aspirates of bone marrow or lymph nodes, with a sensitivity of 60–80% and 50–60% respectively. The sensitivity of reading blood smears is even lower except in HIV-coinfected patients who have higher parasitemia (Diro et al. 2017). To circumvent this low sensitivity clinicians are encouraged to repeat aspirations from a number of different tissues (Zijlstra et al. 1991). In summary, these parasitological techniques require substantial expertise from the clinician as well as the laboratory technician. Thus, they are not suitable for use in first-line health services in countries where VL is endemic.

9.3 Serological Assays

Over the past decades several antibody-detection tests have been developed. Their main advantages are their non-invasiveness and their high sensitivity for visceral disease (Kar 1995). However, their main limitation is that their specificity for active VL disease is far from optimal. Cross-reaction with other pathogens is possible; they do not discriminate between clinical and sub-clinical infection and they remain positive for months or years after cure has been achieved. The latter is a major limitation in practical terms, as it prevents the clinician from monitoring treatment. Such assessments of cure are necessary at the end of treatment and until six months after treatment, at a time when antibody levels have not yet waned. In Kenya an enzyme-linked immunosorbent assay (ELISA) was developed which had a high degree of sensitivity and specificity; however, no commercial kit became available; furthermore, ELISA techniques are not appropriate for decentralized use. The indirect immunofluorescence antibody test (IFAT) showed a very high degree of accuracy in many studies and is still the test of choice in many laboratories in the Mediterranean region, but the necessity of an immunofluorescence microscope makes it neither appropriate nor affordable for decentralized diagnosis. The first "breakthrough" field test was the Direct Agglutination Test (DAT) (El Harith et al. 1986), as it did not require sophisticated laboratory equipment to perform it. Approximately 20 years ago Sundar et al. (1998) reported high sensitivity and specificity for an immunochromatographic test in strip format based on the recombinant leishmania antigen rK39 which could detect antibody in peripheral blood.

9.4 The First Serological Test for Field Use: The Direct Agglutination Assay

In 1985 Dr. Abdallah El Harith, a Sudanese scientist, was offered the opportunity to develop a serological test for VL at the Koninklijk Instituut voor de Tropen in Amsterdam with the financial support of the European Union. El-Harith built upon a technique described by Allain and Kagan (1975), and on similar work by Magnus et al. (1978) who developed a card agglutination test for the diagnosis of African trypanosomiasis. El Harith's proof of concept paper reported very high sensitivity and specificity values (El Harith et al. 1986) that were later corroborated in phase 2 diagnostic studies on archived sera (El Safi and Evans, 1989; Hailu 1990; Singla et al. 1993; Sinha and Sehgal 1994). As the ongoing VL epidemic in Sudan (de Beer et al. 1991) created a pressing demand for better diagnostics, the DAT test was rapidly taken to the field (Seaman et al. 1992). However, contradictory reports were soon published on the specificity of DAT in the clinical setting (Zijlstra et al. 1991; El Masum and Evans 1995). A methodological problem at least partly explained the issue of low specificity in clinical suspects. Researchers who compared the DAT (or any other new VL test) to a reference standard with suboptimal sensitivity but 100% specificity (such as bone marrow or lymph node smears), always obtained an underestimation of the true specificity of the DAT (or another index test). Some true VL cases were missed by this suboptimal reference standard, but they are picked up by the DAT (Boelaert et al. 1999a). The effectiveness of the DAT proved satisfactory in a large multi-center study (Boelaert et al. 1999b), but its reproducibility was low due to its sensitivity to heat and shock (Boelaert et al. 1999c). A freeze–dried version of the test was developed to circumvent the latter problem (Meredith et al. 1995; Zijlstra et al. 1997).

9.5 The Early Development an Immunochromatographic Test Using the Recombinant Leishmania Antigen rK39

An ELISA test based on a 39 amino acid repeat recombinant leishmania antigen from *Leishmania infantum/chagasi* (rK39) showed high diagnostic accuracy (Badaro et al. 1996; Zijlstra et al. 1998). This rK39 antigen became the basis of an immunochromatographic test using a lateral flow dipstick format (Sundar et al. 1998). The initial validation study with the original rK39 ICT showed 100% sensitivity and 98% specificity in India (Sundar et al. 1998), but this particular format (Arista Biologicals, Allentown, PA, USA) never became commercially available. An rK39 ICT produced by a different company (INBIOS, Seattle, WA, USA) proved to be an excellent diagnostic guide in suspected VL cases in India (Sundar and Rai 2002) and in Bangladesh (Sarkari et al. 2002). In Nepal an early prototype of this ICT showed a specificity of only 71% in controls with clinical signs of VL (Chappuis et al. 2006). Bern et al. (2000) and Boelaert et al. (2004) obtained better specificity with another generation of the INBIOS ICT and with an ICT produced by DiaMed AG, Switzerland, respectively. In Sudan Zijlstra et al. (2001) observed a low sensitivity (67%) with a dipstick from the same manufacturer.

A meta-analysis (Chappuis et al. 2006) and a Cochrane review (Boelaert et al. 2014) concluded this RDT had excellent diagnostic accuracy.

9.6 Impact of the VL RDT

Figure 9.2 demonstrates the potential impact of such an RDT on access to care. In 2003, the Médecins Sans Frontières program in Ethiopia was able to reach three times more patients after the introduction of the RDT.

The free provision of the RDT under the VL Elimination Initiative also contributed to the rapid decline in cases in the Indian subcontinent.

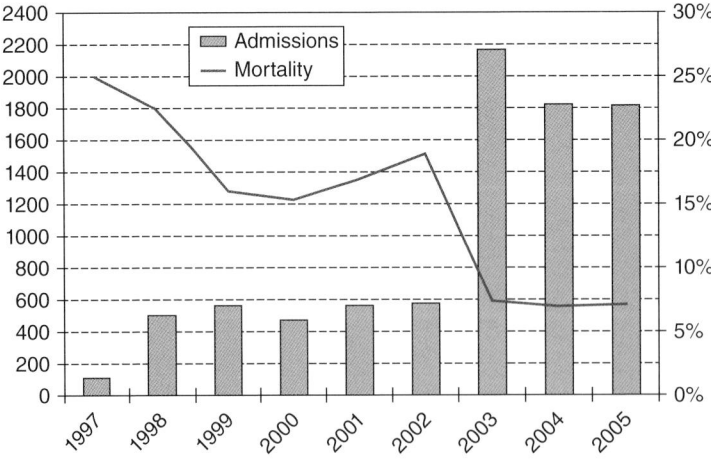

Figure 9.2 VL case numbers and case fatality in the MSF program in Ethiopia. *Source:* Courtesy K. Ritmeijer, MSF.

9.7 Challenges

There are still challenges associated with the RDT. As shown by Duthie et al. (2012), a positive rK39 will remain positive for many months after treatment. Secondly, as shown in the Cochrane review, the sensitivity of this antibody detection test is less in East African patients, most likely due to lower antibody levels (Bhattacharyya et al. 2014b). Therefore, the algorithm for diagnosing VL cases remains complex in some regions (Figure 9.3).

Various claims have been made about the superiority of RDTs based on the homologous antigen, but these seem of marginal importance. A major issue, however, is the variable performance of several brands as was illustrated during the early years of development. This is a matter that persists today (Cunningham et al. 2012).

9.8 Other Tests

Sarkari et al. (2002) described a latex antigen detection test, based on a low molecular weight, heat-stable carbohydrate present in the urine of kala-azar patients. The prototype of this test showed 100% (95% CI 98.8–100.0) specificity and a sensitivity between 64 (95% CI 42.5–82.0) to 100% (95% CI 47.8–100.0) (Attar et al. 2001). In a field trial in Sudan the test was positive in all 15 microscopy positive kala-azar cases (sensitivity 100% [95% CI 78.2–95.2]) and was negative in 41 out of 45 bone marrow and/or lymph node smear-negative clinical suspect cases of kala-azar (specificity 87.2% [95% CI 78.8–97.5]).

The sensitivity of the Kalon Biological Ltd.'s KAtex was disappointingly low in clinically suspect patients in an endemic kala-azar area in Nepal (Rijal et al. 2004). However, further work is warranted since this technique holds promise for the monitoring of therapy, which none of the serological tests so far have accomplished. Vallur et al. (2015) presented novel antibody detection tests as more specific for the acute stage of VL, as did Bhattacharyya et al. (2014a)

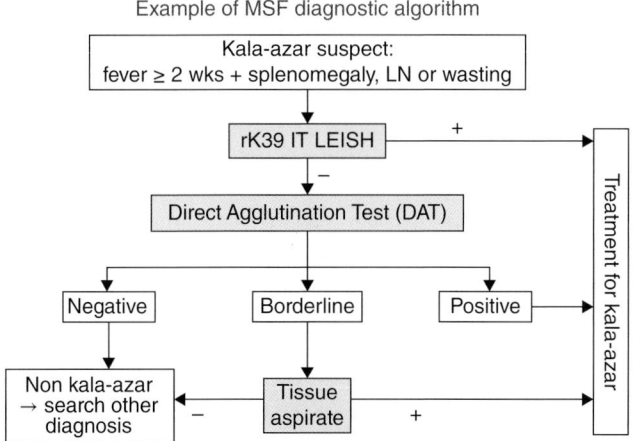

Figure 9.3 Diagnostic algorithm for visceral leishmaniasis in use in MSF programs in Sudan. Note: Visceral *leishmaniasis* is also known as *kala-azar*, black fever and Dumdum fever.

9.9   Discussion

The DAT and the rK39 RDT have been extensively evaluated in all regions affected by VL, and with consistent results. In contrast, the KATEX urinary antigen test has not been so widely evaluated.

The performance of these serological tests in HIV-coinfected patients has been extensively reviewed by Hailu and Berhe (2002) and will not be discussed here.

The DAT is an excellent test regarding diagnostic accuracy, but it has known limitations:

- It requires quite some non-reusable laboratory material
- DAT reading requires skilled, trained and supervised staff
- It takes 18 hours to read the result
- The cost is in the range of US$2–4 per test.

The issue of the low reproducibility was solved by the introduction in 1999 of a freeze-dried version. Today the DAT is mainly used in Ethiopia and Sudan and is both produced locally (for example, by The Institute of Pathobiology in Ethiopia), and is also imported. Bangladesh at one point introduced local production of the DAT, which greatly improved case management, but this production was not maintained. Control programs in India or Nepal do not use it and neither does Brazil nor Kenya. Relief organizations working under difficult conditions in Southern Sudan and Somalia use the DAT on filter paper samples which are flown out to a central laboratory. Results are communicated to field teams by radio, indicating that the DAT cannot be considered a straightforward point-of-care test.

The immunochromatographic assays are easier to use. The differences between test kits from different manufacturers, as well as different formats (generations) supplied by the same manufacturer over time, do not always make comparison easy. However, the performance characteristics of a very high sensitivity and good specificity make the rK39 dipstick an adequate test for the field diagnosis of VL in clinical suspects in the Indian subcontinent. In contrast, their sensitivity is somewhat lower in East African patients. There is little information available on the reproducibility of results. A faint reaction makes reading sometimes difficult, although Schallig et al. (2002) reported good agreement between two readers. With a cost in the range of US$1–2 per test, these RDTs are a good option for VL control programs, with these caveats: the commercial production of the dipstick has been interrupted several times (Veeken 2001; Chappuis et al. 2003), and sustainability of supply is sometimes problematic. In the Sudan the availability of the RDTs has repeatedly been hampered by the US embargo that restricts their importation. Additionally, both the DAT and the rK39 dipstick share a major limitation: as they remain positive for many months in successfully treated patients, they cannot be used to monitor cure. A latex agglutination test detecting a heat-stable leishmanial antigen from the urine of kala-azar patients presents an interesting technology in this respect. Antigen detection tests have the potential for monitoring response to treatment where the antibody-based tests are of no help. The KATEX urine test is not sensitive enough to be used in routine practice, and the further development of antigen detection tests or other markers of acute stage disease needs to be supported.

9.10 Conclusions

Effective case management and control of visceral leishmaniasis depends critically on adequate, user-friendly diagnostic tests. The introduction in 1998 of immunochromatographic tests (ICTs) based on a 39 amino-acid-repeat recombinant leishmanial antigen from *Leishmania infantum/chagasi* (rK39) was easy to use in the field, had a 15 minutes turn-around-time and was a breakthrough for the control of Visceral Leishmaniasis. Much evidence has been accumulated over the past 15 years on the diagnostic accuracy of DAT and rK39, and has consistently demonstrated their validity and thus supported their use in clinical practice.

Further research and development of point-of-care tests that allow for both diagnosis and monitoring of therapeutic progress are needed.

References

Allain, D.S. and Kagan, I.G. (1975). A direct agglutination test for leishmaniasis. *Am. J. Trop. Med. Hyg.* 24: 232–236.

Aronson, N., Herwaldt, B.L., Libman, M. et al. (2017). Diagnosis and treatment of leishmaniasis: clinical practice guidelines by the Infectious Diseases Society of America (IDSA) and the American Society of Tropical Medicine and Hygiene (ASTMH). *Am. J. Trop. Med. Hyg.* 96: 24–45.

Attar, Z.J., Chance, M.L., El Safi, S. et al. (2001). Latex agglutination test for the detection of urinary antigens in visceral leishmaniasis. *Acta Trop.* 78: 11–16.

Badaro, R., Benson, D., Eulalio, M.C. et al. (1996). rK39: a cloned antigen of Leishmania chagasi that predicts active visceral leishmaniasis. *J. Infect. Dis.* 173: 758–761.

Bern, C., Jha, S.N., Joshi, A.B. et al. (2000). Use of the recombinant rK39 dipstick test and the direct agglutination test in a setting endemic for visceral leismaniasis in Nepal. *Am. J. Trop. Med. Hyg.* 63: 153–157.

Bhattacharyya, T., Ayandeh, A., Falconar, A.K. et al. (2014a). IgG1 as a potential biomarker of post-chemotherapeutic relapse in visceral leishmaniasis, and adaptation to a rapid diagnostic test. *PLoS Negl. Trop. Dis.* 8: e3273.

Bhattacharyya, T., Bowes, D.E., El-Safi, S. et al. (2014b). Significantly lower anti-Leishmania IgG responses in Sudanese versus Indian visceral leishmaniasis. *PLoS Negl. Trop. Dis.* 8: e2675.

Boelaert, M., Bhattacharya, S., Chappuis, F. et al. (2007). Evaluation of rapid diagnostic tests; visceral leishmaniasis. *Nat. Rev. Microbiol.* 5: 30–39.

Boelaert, M., El Safi, S., Goetghebeur, E. et al. (1999a). Latent class analysis permits unbiased estimates of the validity of DAT for the diagnosis of visceral leishmaniasis. *Trop. Med. Int. Health* 4: 395–401.

Boelaert, M., El Safi, S., Jacquet, D. et al. (1999b). Operational validation of the direct agglutination test for diagnosis of visceral leishmaniasis. *Am. J. Trop. Med. Hyg.* 60: 129–134.

Boelaert, M., El Safi, S.H., Jacquet, D. et al. (1999c). Operational validation of the Direct Agglutination Test (DAT) for diagnosis of visceral leishmaniasis. *Am. J. Trop. Med. Hyg.* 60: 129–134.

Boelaert, M., Rijal, S., Regmi, S. et al. (2004). A comparative study of the effectiveness of diagnostic tests for visceral leishmaniasis. *Am. J. Trop. Med. Hyg.* 70: 72–77.

Boelaert, M., Verdonck, K., Menten, J. et al. (2014). Rapid tests for the diagnosis of visceral leishmaniasis in patients with suspected disease. *Cochrane Database Syst. Rev.* (6): CD009135.

Boussery, G., Boelaert, M., Van Peteghem, J. et al. (2001). Visceral leishmaniasis (kala-azar) outbreak in Somali refugees and Kenyan shepherds, Kenya. *Emerg. Infect. Dis.* 7: 603–604.

Bryceson, A.D. (1996). Leishmaniasis. In: *Manson's Tropical Diseases*, 20e (ed. J. Farrar, P.J. Hotez, T. Junghanss, et al.), 1213–1245. London: WB Saunders.

Chappuis, F., Rijal, S., Soto, A. et al. (2006). A meta-analysis of the diagnostic performance of the direct agglutination test and rK39 dipstick for visceral leishmaniasis. *BMJ* 333: 723.

Chappuis, F., Rijal, S., Singh, R. et al. (2003). Prospective evaluation and comparison of the direct agglutination test and an rK39-antigen-based dipstick test for the diagosis of suspected kala-azar in Nepal. *Trop. Med. Int. Health.* 8: 277–285.

Chulay, J.D. and Bryceson, A.D. (1983). Quantitation of amastigotes of Leishmania donovani in smears of splenic aspirates from patients with visceral leishmaniasis. *Am. J. Trop. Med. Hyg.* 32: 475–479.

Cunningham, J., Hasker, E., Das, P. et al. (2012). A global comparative evaluation of commercial immunochromatographic rapid diagnostic tests for visceral leishmaniasis. *Clin. Infect. Dis.* 55: 1312–1319.

De Beer, P., El Harith, A., Deng, L.L. et al. (1991). A killing disease epidemic among displaced Sudanese population identified as visceral leishmaniasis. *Am. J. Trop. Med. Hyg.* 44: 283–289.

Diro, E., Yansouni, C.P., Takele, Y. et al. (2017). Diagnosis of visceral leishmaniasis using peripheral blood microscopy in Ethiopia: a prospective Phase-III study of the diagnostic performance of different concentration techniques compared to tissue aspiration. *Am J Trop Med Hyg.* 96: 190–196.

Duthie, M.S., Raman, V.S., Piazza, F.M., and Reed, S.G. (2012). The development and clinical evaluation of second-generation leishmaniasis vaccines. *Vaccine* 30: 134–141.

El Harith, A., Kolk, A.H., Kager, P.A. et al. (1986). A simple and economical direct agglutination test for serodiagnosis and sero-epidemiological studies of visceral leishmaniasis. *Trans. R. Soc. Trop. Med. Hyg.* 80: 583–587.

El Masum, M.A. and Evans, D.A. (1995). Characterisation of Leishmania isolated from patients with kala-azar and post kala-azar dermal leishmaniasis in Bangladesh. *Trans. R. Soc. Trop. Med. Hyg.* 89: 331–332.

El Safi, S.H. and Evans, D.A. (1989). A comparison of the direct agglutination test and enzyme-linked immunosorbent assay in the sero-diagnosis of leishmaniasis in the Sudan. *Trans. R. Soc. Trop. Med. Hyg.* 83: 334–337.

Hailu, A. (1990). Pre- and post-treatment antibody levels in visceral leishmaniasis. *Trans. R. Soc. Trop. Med. Hyg.* 84: 673–675.

Hailu, A. and Berhe, N. (2002). The performance of direct agglutination tests (DAT) in the diagnosis of visceral leishmaniasis among Ethiopian patients with HIV co-infection. *Annals. Trop. Med. Parasit.* 96: 25–30.

Hasker, E., Singh, S.P., Malaviya, P. et al. (2012). Visceral leishmaniasis in rural Bihar, India. *Emerg. Infect. Dis.* 18: 1662–1664.

Hirve, S., Boelaert, M., Matlashewski, G. et al. (2016). Transmission dynamics of visceral Leishmaniasis in the Indian subcontinent - a systematic literature review. *PLoS Negl. Trop. Dis.* 10: e0004896.

Ho, M., Siongok, T.K., Lyerly, W.H., and Smith, D.H. (1982). Prevalence and disease spectrum in a new focus of visceral leishmaniasis in Kenya. *Trans. R. Soc. Trop. Med. Hyg.* 76: 741–746.

Kager, P.A., Rees, P.H., Manguyu, F.M. et al. (1983). Splenic aspiration: experience in Kenya. *Trop. Geogr. Med.* 35: 125–131.

Kar, K. (1995). Serodiagnosis of leishmaniasis. *Crit. Rev. Microbiol.* 21: 123–152.

Magnus, E., Vervoort, T., and Van Meirvenne, N. (1978). A Card-Agglutination Test with Stained Trypanosomes (CATT) for the serological diagnosis of *Trypanosoma brucei gambiense* trypanosomiasis. *Ann. Soc. Belg. Med. Trop.* 58: 169–176.

Meredith, S.E., Kroon, N.C., Sondorp, E. et al. (1995). Leish-KIT, a stable direct agglutination test based on freeze-dried antigen for serodiagnosis of visceral leishmaniasis. *J. Clin. Microbiol.* 33: 1742–1745.

Rijal, S., Boelaert, M., Regmi, S. et al. (2004). Evaluation of a urinary antigen-based latex agglutination test in the diagnosis of kala-azar in eastern Nepal. *Trop. Med. Int. Health* 9: 724–729.

Sarkari, B., Chance, M., and Hommel, M. (2002). Antigenuria in visceral leishmaniasis: detection and partial characterisation of a carbohydrate antigen. *Acta Trop.* 82: 339–348.

Seaman, J., Ashford, R.W., Schorscher, J., and Dereure, J. (1992). Visceral leishmaniasis in southern Sudan: status of healthy villagers in epidemic conditions. *Ann. Trop. Med. Parasitol.* 86: 481–486.

Singla, N., Singh, G.S., Sundar, S., and Vinayak, V.K. (1993). Evaluation of the direct agglutination test as an immunodiagnostic tool for kala-azar in India. *Trans. R. Soc. Trop. Med. Hyg.* 87: 276–278.

Sinha, R. and Sehgal, S. (1994). Comparative evaluation of serological tests in Indian kala-azar. *J. Trop. Med. Hyg.* 97: 333–340.

Sundar, S. and Rai, M. (2002). Laboratory diagnosis of visceral leishmaniasis. *Clin. Diagn. Lab. Immunol.* 9: 951–958.

Sundar, S., Reed, S.G., Singh, V.P. et al. (1998). Rapid accurate field diagnosis of Indian visceral leishmaniasis. *Lancet* 351: 563–565.

Thakur, C.P. (1997). A comparison of intercostal and abdominal routes of splenic aspiration and bone marrow aspiration in the diagnosis of visceral leishmaniasis. *Trans. R. Soc. Trop. Med. Hyg.* 91: 668–670.

Trouiller, P., Olliaro, P.L., Torreele, E. et al. (2002). Drug development for neglected diseases: a deficient market and a public-health policy failure. *Lancet* 359: 2188–2194.

Vallur, A.C., Hailu, A., Mondal, D. et al. (2015). Specific antibody responses as indicators of treatment efficacy for visceral leishmaniasis. *Eur. J. Clin. Microbiol. Infect. Dis.* 34: 679–686.

Veeken, H., Ritmeijer, K., Seaman, J., and Davidson, R. (2003). Comparison of an rK39 dipstick rapid test with direct agglutination test and splenic aspiration for teh diagnosis of kala-azar in Sudan. *Trop. Med. Int. Health* 8: 164–167.

Zijlstra, E.E., Ali, M.S., El Hassan, A.M. et al. (1991). Direct agglutination test for diagnosis and sero-epidemiological survey of kala-azar in the Sudan. *Trans. R. Soc. Trop. Med. Hyg.* 85: 474–476.

Zijlstra, E.E., Ali, M.S., El Hassan, A.M. et al. (1992). Kala-azar: a comparative study of parasitological methods and the direct agglutination test in diagnosis. *Trans. R. Soc. Trop. Med. Hyg.* 86: 505–507.

Zijlstra, E.E., Daifalla, N.S., Kager, P.A. et al. (1998). rK39 enzyme-linked immunosorbent assay for diagnosis of Leishmania donovani infection. *Clin. Diagn. Lab. Immunol.* 5: 717–720.

Zijlstra, E.E., Nur, Y., Desjeux, P. et al. (2001). Diagnosing visceral leishmaniasis with the recombinant K39 strip test: experience from the Sudan. *Trop. Med. Int. Health* 6: 108–113.

Zijlstra, E.E., Osman, O.F., Hofland, H.W. et al. (1997). The direct agglutination test for diagnosis of visceral leishmaniasis under field conditions in Sudan: comparison of aqueous and freeze-dried antigens. *Trans. R. Soc. Trop. Med. Hyg.* 91: 671–673.

10

A Rapid Diagnostic Test for Dengue

Claire Mullender[1], and James Whitehorn[2]

[1] *St George's Healthcare Trust, London, UK*
[2] *University College, London, UK*

CHAPTER MENU

10.1 Introduction

Dengue[1] is an arboviral infection of humans transmitted by *Aedes* mosquitoes (Simmons et al. 2012). The virus is a member of the flaviviridae family and has four antigenically distinct serotypes (DENV1-4) (Whitehorn and Simmons 2011). Like other flaviviruses dengue is an enveloped single-stranded positive sense RNA virus. It is globally the most important arboviral infection with approximately half of the world's population living in areas at risk of disease transmission (Simmons et al. 2012; Bhatt et al. 2013). Dengue is a major public health threat throughout the tropical world. The expansion of its geographic range is closely related to the global spread of permissive vectors (Simmons et al. 2012). The economic burden of dengue is considerable – the disease exerts a significant toll on health systems, often those in low-resource settings (Shepard et al. 2016). While some aspects of the Phase III efficacy clinical trial of the leading dengue vaccine candidate were encouraging, there remain many unanswered questions about how this vaccine will fit into immunization programs in dengue-endemic countries

---

[1] The origin of the Spanish word dengue is not certain, but it is possibly derived from dinga in the Swahili phrase Ka-dinga pepo, which describes the disease as being caused by an evil spirit.

Revolutionizing Tropical Medicine: Point-of-Care Tests, New Imaging Technologies and Digital Health, First Edition. Edited by Kerry Atkinson and David Mabey.
© 2019 John Wiley & Sons, Inc. Published 2019 by John Wiley & Sons, Inc.

(Hadinegoro et al. 2015; Simmons 2015; Katzelnick et al. 2017). Furthermore, there is no specific therapeutic agent for dengue and vector control remains the mainstay of disease prevention efforts (Whitehorn et al. 2014).

10.2 Clinical Features of Dengue

Symptomatic dengue develops after an incubation period of between four and seven days following the bite of an infected *Aedes* mosquito (Figure 10.1).

The illness is characterized by an abrupt onset of fever often accompanied by headache, myalgia and rash.

In most cases the illness is self-limiting and resolves within seven days.

Severe dengue is seen more frequently in the context of secondary infections with a different serotype or in primary infections in infants born to mothers previously infected with dengue (Simmons et al. 2012). In these situations pathogenesis is thought to be mediated by the phenomenon of antibody-dependent enhancement in which cross-reactive but non-neutralizing antibodies bind virus and have a greater potential to infect Fc receptor-bearing cells such as B lymphocytes, follicular dendritic cells, natural killer cells, macrophages, neutrophils, eosinophils, basophils, human platelets and mast cells (Whitehorn and Simmons 2011). The clinical features in young children often differ from those observed in adults with vomiting and abdominal pain observed more frequently. The critical period of illness occurs around the time of defervescence, and typically lasts for 24–48 hours. This period is characterized by a transient increase in capillary permeability and an associated rise in hematocrit. The degree of plasma

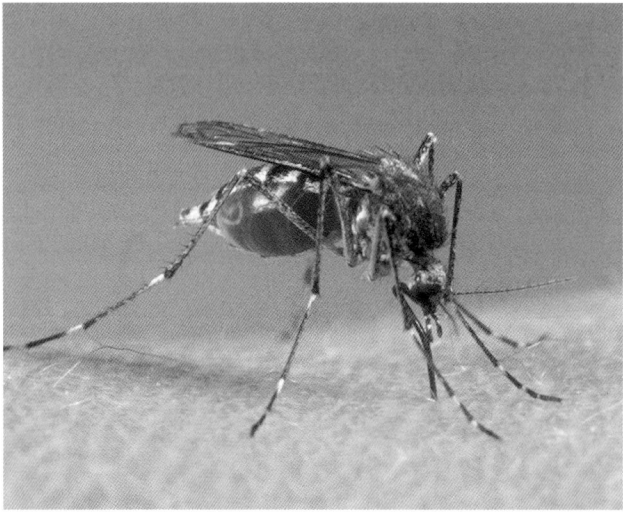

Figure 10.1 An *Aedes aegypti* mosquito biting a person. This species of mosquito transmits dengue fever, Chikungunya fever, Zika fever and yellow fever. *Source:* This image, reproduced here from Wikimedia Commons, is in the public domain because it is a work of a United States Department of Agriculture employee, taken or made as part of that person's official duties. As a work of the U.S. Federal Government. Picture from the USDA website at http://www.ars.usda.gov/is/graphics/photos/aug00/k4705-9.htm.

leak varies and is influenced by both host and viral factors. If there is critical volume loss the patient will develop clinical shock. In addition, bleeding may occur. This is usually minor mucosal bleeding, but severe hemorrhage can develop. This is observed more frequently in adult patients with dengue. After resolution of the critical period the recovery phase commences. During recovery patients can suffer from disabling fatigue. Dengue was historically classified as "dengue fever," "dengue hemorrhagic fever" or "dengue shock syndrome." There were many problems with this classification system and there were concerns that the use of it resulted in under-diagnosis of severe dengue (Deen et al. 2006; Farrar et al. 2013). The latest WHO guidelines reflect the syndromic nature of dengue and classify the disease into "dengue" and "severe dengue," and place emphasis on the warning signs that may predict progression to a more severe disease phenotype (Figure 10.2) (WHO 2009).

10.3   The Importance of Making a Rapid Diagnosis

In the early phases of dengue, it is very difficult to distinguish it from other febrile illnesses and making a clinical diagnosis is not reliable (Simmons et al. 2012). The differential diagnosis of dengue is broad and depends on the specific local epidemiology of other infectious diseases. Improving the diagnosis of dengue will greatly assist patient management by optimizing patient triage, earlier recognition of severe disease and appropriate recognition and management of other diseases such as malaria and bacterial infections. Furthermore, better diagnostics will limit inappropriate antibiotic prescribing and will assist in efforts to improve antibiotic stewardship. In addition, better

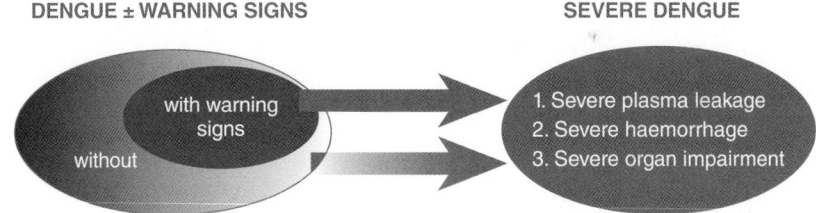

DENGUE ± WARNING SIGNS	SEVERE DENGUE
with warning signs / without	1. Severe plasma leakage 2. Severe haemorrhage 3. Severe organ impairment

CRITERIA FOR DENGUE ± WARNING SIGNS

Probable dengue
live in /travel to dengue endemic area.
Fever and 2 of the following criteria:
- Nausea, vomiting
- Rash
- Aches and pains
- Tourniquet test positive
- Leukopenia
- Any warning sign

Laboratory-confirmed dengue
(important when no sign of plasma leakage)

Warning signs*
- Abdominal pain or tenderness
- Persistent vomiting
- Clinical fluid accumulation
- Mucosal bleed
- Lethargy, restlessness
- Liver enlargement >2 cm
- Laboratory: increase in HCT concurrent with rapid decrease in platelet count

* (requiring strict observation and medical intervention)

CRITERIA FOR SEVERE DENGUE

Severe plasma leakage
leading to:
- Shock: (DSS)
- Fluid accumulation with respiratory distress

Severe bleeding
as evaluated by clinician

Severe organ involvement
- Liver: AST or ALT >= 1000
- CNS: Impaired consciousness
- Heart and other organs

Figure 10.2 The 2009 WHO recommended dengue classification. This new classification places emphasis on the fact that dengue presents as a spectrum, and on recognition of the warning signs that may indicate disease progression. *Source:* Reproduced with permission from WHO.

diagnostics will assist surveillance efforts and have an important role in research studies. Traditionally, diagnostic approaches have required considerable laboratory support which is often not available in dengue-endemic settings.

10.4 The Host Response to Infection

In order to understand the different diagnostic approaches to dengue it is necessary to understand both typical viral kinetics following infection and the host response to infection.

Following infection, dengue virus (DENV) replicates to a high titer in blood and can be detected two days prior to symptom onset and for up to five days after symptom onset (WHO 2009). In secondary (subsequent) infections the duration of viremia is shorter (Vaughn et al. 2000). Viral non-structured protein 1 (NS1) is secreted from infected cells and induces a humoral response (Libraty et al. 2002; Peeling et al. 2010). The levels of NS1 correlate approximately with the degree of viremia and have an association with the degree of clinical severity. In primary infections NS1 antigenemia persists for up to nine days (Libraty et al. 2002; WHO 2009).

The antibody response to infection differs in primary and secondary infections. In primary infections the IgM response is detected approximately five days after the onset of illness and peaks after about two weeks. In secondary infection the IgM response occurs earlier (typically two to three days after the onset of illness), but at significantly lower titers (WHO 2009; Whitehorn and Simmons 2011; Blacksell 2012). The IgG antibody response develops a few days after the onset of the IgM antibody response, is serotype-specific, and may persist for many years following a single infection (Whitehorn and Simmons 2011; Blacksell 2012). In secondary DENV infections a rapid and earlier rise in the IgG antibody response is observed, which is detectable four to five days after the onset of illness (Blacksell 2012).

10.5 Existing Diagnostic Strategies

The main laboratory methods used to diagnose dengue infection are detection of viral RNA, viral proteins (e.g. NS1) and/or detection of specific IgM/IgG antibodies. These tests are usually performed in combination, and results can be produced in 24–48 hours. Viral isolation in culture was previously considered the gold standard of testing but is seldom used in practice, as it costly and laborious. The approximate timeline of dengue infections and the appropriate diagnostic method are summarized in Figure 10.3.

Serological diagnosis is one of the most commonly used diagnostic methods. Immunoglobulin G (IgG) and Immunoglobulin M (IgM) antibody-capture enzyme-linked immunosorbant assays (ELISA) are in routine practice in many settings (Muller et al. 2017). The differences in IgM to IgG ratios allows primary infection to be distinguished from secondary infection (Tang and Ooi 2012; Changal et al. 2016). A disadvantage of serological approaches to diagnosis is the need for two samples to be taken at separate time points, which is not always practical. A major problem with serological assays is cross-reactivity with other flaviviruses such as Japanese encephalitis, yellow fever and Zika virus, which may co-circulate in dengue-endemic areas (Simmons et al. 2012; Muller et al. 2017). Using serology in combination with another testing approach

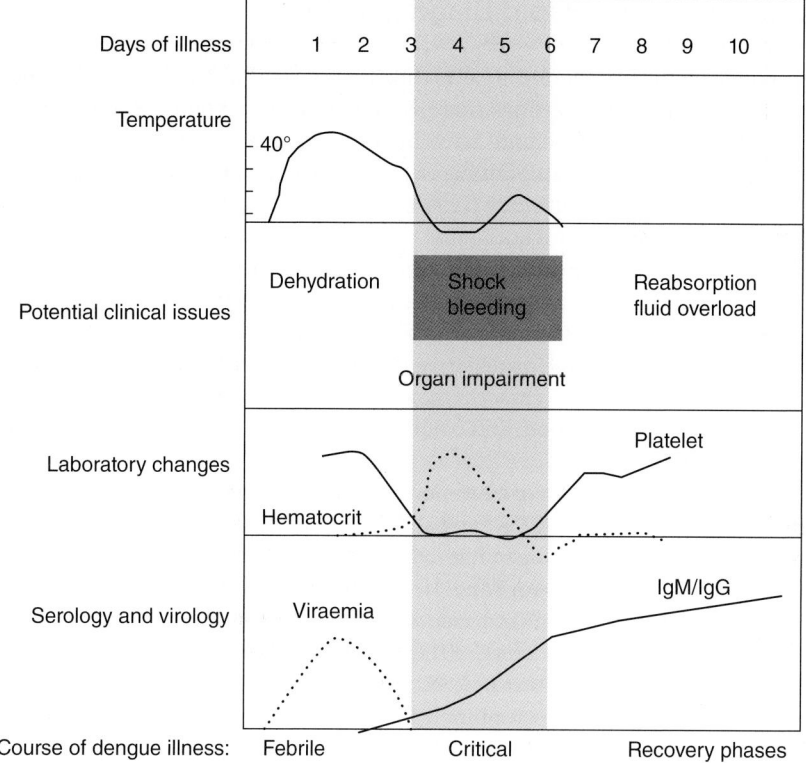

Figure 10.3 Approximate timeline of primary and secondary infections and the diagnostic options that can be used at different stages. *Source:* Reproduced with permission from WHO.

(for example, NS1 antigen detection) can improve diagnostic accuracy (WHO 2009; Muller et al. 2017).

The viral protein NS1 is a useful diagnostic target and can be used in parallel with serological assays. It is detectable before serological tests become positive and, in primary infections, can remain positive for up to nine days (Libraty et al. 2002). It is highly specific and, if a quantitative assay is used, may have an additional prognostic role (Avirutnan et al. 2006). Unfortunately, due to the production of cross-reactive antibodies, the diagnostic window is shorter in secondary infections (Changal et al. 2016; Muller et al. 2017). NS1 detection is an important component of dengue rapid diagnostic tests (RDTs).

The polymerase chain reaction (PCR) is a molecular technique used for detection of DENV. PCR is most commonly performed on blood samples, but can be performed on other samples including urine, saliva and formalin fixed tissues (Tang and Ooi 2012). It is very sensitive and specific but its use is limited to the period of time the patient is viremic (the first five days of illness). While PCR can be used to distinguish between subtypes, confirming viremia does not distinguish between primary and secondary infection – hence it is used in combination with serological tests (Peeling et al. 2010). While PCR technology is becoming increasingly available it is still largely only accessible in research laboratories or well-resourced clinical laboratories and is therefore not a commonly used diagnostic approach in most dengue-endemic areas.

10.6 Review of Existing Rapid Diagnostic Tests

The problem in many dengue-endemic areas is that laboratory diagnostics are often not readily available. In low-resource settings there is a clear need for robust RDTs. A variety of RDTs for the diagnosis of dengue have been developed and are commercially available. They are immunochromatographic assays that can be classified as those that detect a single analyte: either the antibody (IgM ± IgG) or the NS1 antigen, or those that combine tests to detect antibody and antigen.

The immunochromatography assay relies on the binding of antibody in the specimen to antigen on the assay strip. The antigen–antibody complex migrates and is captured by a second antibody leading to the appearance of a colored band that can be easily interpreted by the reader as positive or negative (Drancourt et al. 2016). They come in the form of a hand-held lateral flow cassette that requires no specialized equipment or training, uses just a few drops of blood, and can be performed in approximately 15 minutes (Blacksell 2012).

Multiple diagnostic evaluations have been performed to compare RDT performance to reference laboratory testing (RT-PCR, virus isolation and IgM/IgG ELISA testing). For single analyte RDTs results have been highly variable. The sensitivity and specificity for RDTs based on antibody detection range from 52–95% and 76–89% respectively. The sensitivity and specificity for RDTs based on NS1 detection range from 40–66% and 80–99% respectively (Blacksell et al. 2011; Hunsperger et al. 2014; Hunsperger et al. 2016).

In comparison, evaluations of the combined tests have shown more promising and consistent results with sensitivity ranging from 80 to 100% and specificity ranging from 75 to 92% (Osorio et al. 2010; Tricou et al. 2010; Wang and Sekaran 2010; Blacksell et al. 2011; Gan et al. 2014). Many RDTs using a combined diagnostic approach are available and these are being used more frequently to assist clinical practice in dengue-endemic areas. Results indicating a good degree of accuracy have been replicated in prospective studies, during suspected outbreak investigations, as well as retrospective evaluation of known positive dengue samples (Blacksell et al. 2011; Gan et al. 2014; Hunsperger et al. 2014). They have also been shown to maintain performance after long term storage at temperatures of 35 °C (Sengvilaipaseuth et al. 2017).

As mentioned previously, during primary infection NS1 is detectable from one to nine days after symptom onset, and IgM becomes detectable from 5 to 15 days after symptom onset. Combination tests therefore have an expanded window of diagnostic opportunity, which accounts for their improved performance compared to single analyte tests. The optimum time during illness to perform RDTs has yet to be established. Some studies have found a significant reduction in sensitivity from three days after the onset of symptoms (Osorio et al. 2010), while others have found an increase in sensitivity three days after symptom onset (Tricou et al. 2010).

While specificity is high, false positives due to cross- reactivity have been reported in cases of chikungunya, leptospirosis, scrub typhus, Q fever and bacteremia (Blacksell et al. 2011). This may result in a delay in appropriate treatment for serious bacterial infection. There is also a potential for a false positive IgM or IgG antibody test result to occur due to persistence of antibodies from a recent previous infection with a different dengue serotype. To date, there are no studies comparing the accuracy of RDTs across all four dengue serotypes (Blacksell et al. 2011).

Another concern with current RDTs is that, despite manufacturers' claims, there is no evidence that they are able to reliably distinguish between primary and secondary infection (Blacksell 2012). In addition, a reduction in sensitivity has been reported for the diagnosis of secondary infection (Tricou et al. 2010). This can be explained by a reduced IgM response during secondary infection. It is arguably more important to diagnose a secondary infection, which is associated with an increased risk of the severe forms of the disease.

The ideal properties of a dengue RDT are summarized in Table 10.1. Currently available RDTs are summarized in Table 10.2

Table 10.1 Features of the ideal RDT for dengue.

Ideal characteristics of a RDT for dengue virus infection
Good diagnostic performance Specific: low rate of false positives Sensitive: low rate of false negatives Performance comparable to gold standard testing (i.e. ELISA + PCR) Able to distinguish between primary and secondary infection
User-friendly Does not require skilled personnel or training Equipment-free Results easy to interpret, with minimal user bias Rapid results
Appropriate for use in dengue endemic environment Maintains good performance in hot and humid conditions No required cold storage Easily transportable Affordable

Table 10.2 Comparison of selected existing rapid diagnostic tests for dengue.

Analyte	Name of test	Company	Sensitivity (%)	Specificity (%)
NS1	Dengue NS1 Ag Strip	Biorad	58–99	94.4–100
	SD Bioline Dengue Duo Early Rapid NS1	Standard Diagnostics, Inc.	49–62	97–100
		Panbio/Alere	59–69	92–97
IgM	SD Bioline Dengue Duo Dengue Duo cassette	Standard Diagnostics, Inc.	54–80	90–100
		Panbio/Alere	65–70	80–98
NS1/IgM	SD Dengue Duo Bioline Early Rapid NS1 and Duo assay	Standard Diagnostics, Inc.	76–93	89–100
		Panbio/Alere	89–90	75
NS1/IgG/IgM	SD BIOLINE Dengue Duo	Standard Diagnostics, Inc.	80.7–83.7	89.1–97.9
	Early Rapid NS1 and Duo assay	Panbio/Alere	93	Not reported

Source: Adapted from Blacksell (2012).

10.7 Future Directions

Fortunately, most dengue infections are uncomplicated. However, it is not currently possible to accurately identify those patients in the early stages of infection who are at most risk of disease progression. Having the ability to do this would assist appropriate patient triage and allow health resources to be directed to those in most need. Currently, health facilities are often overwhelmed with dengue patients, many of whom could be managed as outpatients (WHO 2009; Simmons et al. 2012). Having the dengue equivalent of the CURB-65 score for pneumonia would be a major advance. (CURB-65 is a clinical prediction rule that has been validated for predicting mortality in community-acquired pneumonia and infection of any site. The score is an acronym for each of the risk factors measured. Confusion of new onset, blood urea nitrogen greater than $7\,\mathrm{mmol\,L^{-1}}$ ($19\,\mathrm{mg\,dL^{-1}}$), respiratory rate of 30 breaths per minute or greater; systolic blood pressure less than 90 mmHg or diastolic blood pressure 60 mmHg or less and age 65 or older.)

In combination with better RDTs such a scoring system would aid earlier identification of patients at most risk of disease progression and potentially avoid hospital admission for those at lower risk. To this end a major multicenter prospective study has been conducted which is due to report shortly (Jaenisch et al. 2016). This study aims to identify simple laboratory and clinical markers that are predictive of disease progression and can be incorporated into clinical algorithms that can be easily used in low resource settings.

Finally, various new approaches to dengue RDTs are being developed (Muller et al. 2017). These are currently in the early stages of development but may, in the long term, add to the diagnostic options for this disease. These approaches include the use of piezoelectric devices and loop-mediated isothermal amplification (LAMP) techniques (Hu et al. 2015; Pirich et al. 2017; Notomi et al. 2015). LAMP technology has the potential to be a game changer in diagnostics as it does not require a thermal cycler and therefore has lower energy requirements, is readily portable and can therefore bring the principles of molecular diagnostics to low resource settings. LAMP has been shown to have high sensitivity but, to date, has only been used in a research setting (Teoh et al. 2013).

10.8 Conclusions

An ideal RDT would be able to reliably diagnose dengue infection throughout the course of illness, and distinguish between primary and secondary infection. If these tests can also incorporate quantification of the NS1 levels, the additional data may assist in the creation and development of prognostic algorithms. The ability to diagnose dengue accurately would not only assist patient management but would also aid in surveillance and in clinical research - for example, the recruitment into prospective studies. While there has been significant improvement in available RDTs, and indeed access to more sophisticated laboratory diagnostics, there is still a need for further progress.

References

Avirutnan, P., Punyadee, N., Noisakran, S. et al. (2006). Vascular leakage in severe dengue virus infections: a potential role for the nonstructural viral protein NS1 and complement. *J. Infect. Dis.* 193: 1078–1088.

Bhatt, S., Gething, P.W., Brady, O.J. et al. (2013). The global distribution and burden of dengue. *Nature* 496: 504–507.

Blacksell, S.D. (2012). Commercial dengue rapid diagnostic tests for point-of-care application: recent evaluations and future needs? *J. Biomed. Biotechnol.* 2012: 151967.

Blacksell, S.D., Jarman, R.G., Bailey, M.S. et al. (2011). Evaluation of six commercial point-of-care tests for diagnosis of acute dengue infections: the need for combining NS1 antigen and IgM/IgG antibody detection to achieve acceptable levels of accuracy. *Clin. Vaccine Immunol.* 18: 2095–2101.

Changal, K.H., Raina, A.H., Raina, A. et al. (2016). Differentiating secondary from primary dengue using IgG to IgM ratio in early dengue: an observational hospital based clinico-serological study from North India. *BMC Infect. Dis.* 16: 715.

Deen, J.L., Harris, E., Wills, B. et al. (2006). The WHO dengue classification and case definitions: time for a reassessment. *Lancet* 368: 170–173.

Drancourt, M., Michel-Lepage, A., Boyer, S., and Raoult, D. (2016). The point-of-care laboratory in clinical microbiology. *Clin. Microbiol. Rev.* 29: 429–447.

Farrar, J.J., Hien, T.T., Horstick, O. et al. (2013). Dogma in classifying dengue disease. *Am. J. Trop. Med. Hyg.* 89: 198–201.

Gan, V.C., Tan, L.K., Lye, D.C. et al. (2014). Diagnosing dengue at the point-of-care: utility of a rapid combined diagnostic kit in Singapore. *PLoS One* 9: e90037.

Hadinegoro, S.R., Arredondo-Garcia, J.L., Capeding, M.R. et al. (2015). Efficacy and long-term safety of a dengue vaccine in regions of endemic disease. *N. Engl. J. Med.* 373: 1195–1206.

Hu, S.F., Li, M., Zhong, L.L. et al. (2015). Development of reverse-transcription loop-mediated isothermal amplification assay for rapid detection and differentiation of dengue virus serotypes 1-4. *BMC Microbiol.* 15: 265.

Hunsperger, E.A., Sharp, T.M., Lalita, P. et al. (2016). Use of a rapid test for diagnosis of dengue during suspected dengue outbreaks in resource-limited regions. *J. Clin. Microbiol.* 54: 2090–2095.

Hunsperger, E.A., Yoksan, S., Buchy, P. et al. (2014). Evaluation of commercially available diagnostic tests for the detection of dengue virus NS1 antigen and anti-dengue virus IgM antibody. *PLoS Negl. Trop. Dis.* 8: e3171.

Jaenisch, T., Tam, D.T., Kieu, N.T. et al. (2016). Clinical evaluation of dengue and identification of risk factors for severe disease: protocol for a multicentre study in 8 countries. *BMC Infect. Dis.* 16: 120.

Katzelnick, L.C., Coloma, J., and Harris, E. (2017). Dengue: knowledge gaps, unmet needs, and research priorities. *Lancet Infect. Dis.* 17: e88–e100.

Libraty, D.H., Young, P.R., Pickering, D. et al. (2002). High circulating levels of the dengue virus nonstructural protein NS1 early in dengue illness correlate with the development of dengue hemorrhagic fever. *J. Infect. Dis.* 186: 1165–1168.

Muller, D.A., Depelsenaire, A.C., and Young, P.R. (2017). Clinical and laboratory diagnosis of dengue virus infection. *J. Infect. Dis.* 215: S89–S95.

Notomi, T., Mori, Y., Tomita, N., and Kanda, H. (2015). Loop-mediated Isothermal Amplification (LAMP): principle, features, and future prospects. *J. Microbiol.* 53: 1–5.

Osorio, L., Ramirez, M., Bonelo, A. et al. (2010). Comparison of the diagnostic accuracy of commercial NS1-based diagnostic tests for early dengue infection. *Virol. J.* 7: 361.

Peeling, R.W., Artsob, H., Pelegrino, J.L. et al. (2010). Evaluation of diagnostic tests: dengue. *Nat. Rev. Microbiol.* 8: S30–S38.

Pirich, C.L., De Freitas, R.A., Torresi, R.M. et al. (2017). Piezoelectric immunochip coated with thin films of bacterial cellulose nanocrystals for dengue detection. *Biosens. Bioelectron.* 92: 47–53.

Sengvilaipaseuth, O., Phommasone, K., De Lamballerie, X. et al. (2017). Temperature of a dengue rapid diagnostic test under tropical climatic conditions: a follow up study. *PLoS One* 12: e0170359.

Shepard, D.S., Undurraga, E.A., Halasa, Y.A., and Stanaway, J.D. (2016). The global economic burden of dengue: a systematic analysis. *Lancet Infect. Dis.* 16: 935–941.

Simmons, C.P. (2015). A candidate dengue vaccine walks a tightrope. *N. Engl. J. Med.* 373: 1263–1264.

Simmons, C.P., Farrar, J.J., Nguyen, V.V., and Wills, B. (2012). Dengue. *N. Engl. J. Med.* 366: 1423–1432.

Tang, K.F. and Ooi, E.E. (2012). Diagnosis of dengue: an update. *Expert Rev. Anti-Infect. Ther.* 10: 895–907.

Teoh, B.T., Sam, S.S., Tan, K.K. et al. (2013). Detection of dengue viruses using reverse transcription-loop-mediated isothermal amplification. *BMC Infect. Dis.* 13: 387.

Tricou, V., Vu, H.T., Quynh, N.V. et al. (2010). Comparison of two dengue NS1 rapid tests for sensitivity, specificity and relationship to viraemia and antibody responses. *BMC Infect. Dis.* 10: 142.

Vaughn, D.W., Green, S., Kalayanarooj, S. et al. (2000). Dengue viremia titer, antibody response pattern, and virus serotype correlate with disease severity. *J. Infect. Dis.* 181: 2–9.

Wang, S.M. and Sekaran, S.D. (2010). Early diagnosis of dengue infection using a commercial dengue duo rapid test kit for the detection of NS1, IGM, and IGG. *Am. J. Trop. Med. Hyg.* 83: 690–695.

Whitehorn, J. and Simmons, C.P. (2011). The pathogenesis of dengue. *Vaccine* 29: 7221–7228.

Whitehorn, J., Yacoub, S., Anders, K.L. et al. (2014). Dengue therapeutics, chemoprophylaxis, and allied tools: state of the art and future directions. *PLoS Negl. Trop. Dis.* 8: e3025.

WHO (2009). *Dengue: Guidelines for Diagnosis, Treatment, Prevention and Control – New Edition*. Geneva: World Health Organisation.

11

Rapid Diagnostic Tests for Influenza

A.C. Hurt[1, 2], and I.G. Barr[1, 2, 3]

[1] *WHO Collaborating Centre for Reference and Research on Influenza, Victorian Infectious Disease Reference Laboratory (VIDRL), Doherty Institute, Melbourne, Victoria, Australia*
[2] *Department of Microbiology and Immunology, Faculty of Medicine, Dentistry, and Health Sciences, University of Melbourne, Parkville, Victoria, Australia*
[3] *Faculty of Science and Technology, Federation University, Churchill, Victoria, Australia*

CHAPTER MENU

11.1 Introduction

Rapid diagnostic tests, or point-of-care tests (often abbreviated to POCs), generally refer to simple tests that can either be carried out at the bedside or in a non-specialized local facility without the need to transport the sample to a modern laboratory for testing. Rapid influenza diagnostic tests (RIDTs) are primarily carried out in developed countries such as Japan and the USA in doctor's surgeries where they are used to confirm influenza prior to the prescription of influenza antiviral drugs such as Tamiflu®, Relenza®, or Inavir®. In addition, they are also utilized in some regional laboratories and hospital emergency wards to assist with patient triage (Biggs et al. 2010; Angoulvant et al. 2011) but unlike many other rapid diagnostic tests, RIDTs have yet to be utilized in the household setting.

The primary use of RIDTs in the has been for outbreak investigations in remote locations that have no access to laboratories and the causative agent of the disease is unknown or needs to be confirmed. With recent improvement in the performance and ease of use of RIDTS, there is potential for these tests to be more widely used in other

Revolutionizing Tropical Medicine: Point-of-Care Tests, New Imaging Technologies and Digital Health,
First Edition. Edited by Kerry Atkinson and David Mabey.

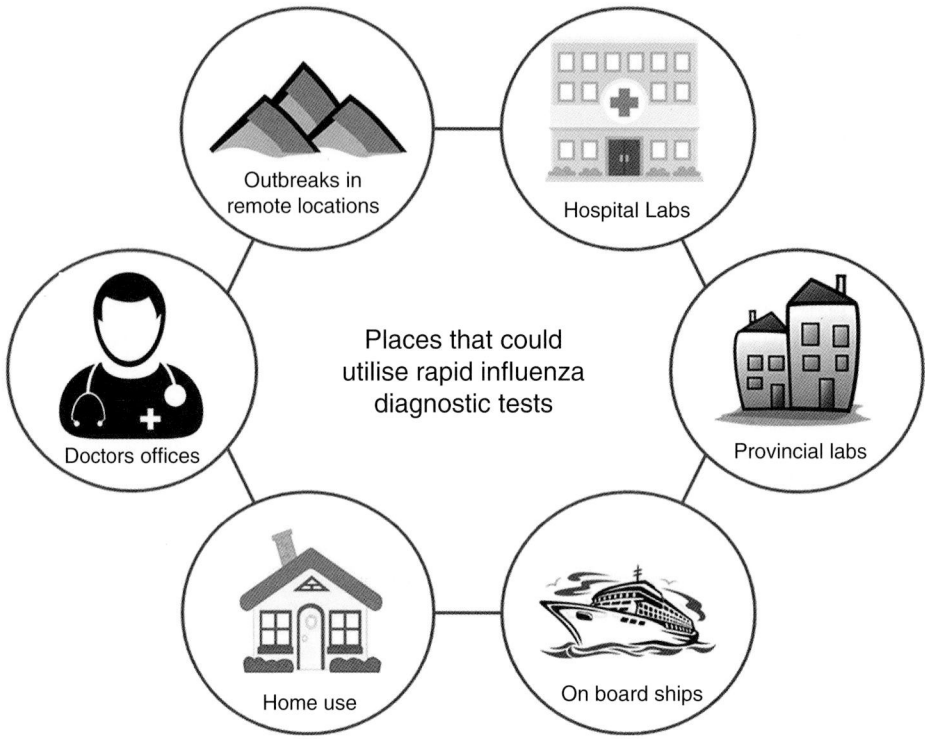

Figure 11.1 Summary of places that could utilize rapid influenza diagnostic tests.

settings such as village healthcare centers, regional laboratories and even central diagnostic laboratories for ad hoc testing (Figure 11.1).

The type of rapid test used to detect influenza will vary according to what resources are available at the test site and the purpose of the test. Many factors need to be considered in making this decision including the distance from the site where the cases or outbreaks are located, time to transport the samples to the nearest large testing facility and get results back, the availability of reliable electrical power, the potential pathogens involved, the reason for the test, the environmental conditions under which the test sample will be transported, stored and performed and the degree of technical training and competency of the available test operator(s) (Peeling and McNerney 2014).

11.2 Overview of RIDTs

The current influenza RIDTs can be broadly separated based on the detection of either influenza viral antigen or influenza nucleic acid. In their simplest form, antigen detection RIDTs (AD RIDTs) typically use chromatographic or fluorescence-based immunoassays to detect the viral nucleoprotein of either influenza A or B viruses and these tests can be further divided into those that are read by eye and those that are read by an analyzer (Table 11.1).

Table 11.1 Summary of the characteristics and performance of selected rapid influenza diagnostics tests.

Test type	Examples of tests in common use	Reader required (cost)	Complexity	Time to result (mins)	Sensitivity (vs PCR)	Specificity (vs PCR)	Storage	CLIA waived	Refs
Antigen detection									
Result read by eye	Quickvue Influenza A+B (Quidel)	No	**	10	73 A	100 A	RT	Yes	1
	BinaxNOW Influenza A & B Card (Alere)	No	*	15	46–79 A 25–81 B	99–100 A 100 B	RT	Yes	1,2,3,4,5,6
	BD Directigen EZ Flu A+B	No	*	15	72 A 48 B	100 A 100 B	RT	No	5
Antigen detection	Sofia Influenza A+B FIA (Quidel)	Yes ($$)	**	15	71–95 A 33–98 B	91–100 A 71–100 B	RT	Yes	3,6,7,8,9
Result read by analyser device	BD Veritor Flu A+B	Yes ($)	*	15	73–94 A 73–94 B	97–100 A 99–100 B	RT	Yes	2,3,7,8,9,10
	3M Rapid detection Influenza A+B	Yes (?)	*	15	70–75 A 87 B	98–100 A 99 B	RT	No	1, 4
Nucleic acid detection	Solana Influenza A+B Assay (Quidel)	Yes ($$$)	***	45	No studies	No studies	2–8	No	None
	Alere i Influenza A & B (Alere)	Yes ($$$)	***	15	78 A 75B	100 A 99 B	2–8	Yes	6

(Continued)

Table 11.1 (Continued)

Test type	Examples of tests in common use	Reader required (cost)	Complexity	Time to result (mins)	Sensitivity (vs PCR)	Specificity (vs PCR)	Storage	CLIA waived	Refs
	Cobas Liat System (Roche Diagnostics)	Yes ($$$)	*	20	99 A	100 A	2–8	Yes	11
					100 B	100 B			
	Xpert Flu (Cephied)	Yes ($$$)	*	75	97 A	100 A	RT	No	5
					100 B	100 B			
	Xpert Xpress Flu (Cephied)	Yes ($$$)	*	20–30	No studies	No studies	RT	Yes	None

Reference numbers refer to: 1. Dale et al. (2008); 2. Hassan et al. (2014); 3. Dunn et al. (2014); 4. Ginocchio et al. (2009); 5. DiMaio et al. (2012); 6. Hazelton et al. (2015); 7. Ryu et al. (2016); 8. Ryu et al. (2017); 9. Leonardi et al. (2015); 10. Ndegwa et al. (2017); 11. Binnicker et al. (2015).

In comparison, the nucleic acid detection RIDTs (NAD RIDTs), utilize either isothermal or polymerase chain reaction (PCR) technology to detect viral RNA from influenza A or B viruses, and therefore these tests incorporate a process that extracts, amplifies and detects nucleic acid from the virus, thus involving specific instrumentation.

Some rapid tests have been issued a waiver by the FDA from the requirements under the US Clinical Laboratory Improvement Amendments (CLIA) Act of 1988 (Services 2017). Without this waiver, testing facilities in the United States that perform laboratory testing on human specimens for health assessment or the diagnosis, prevention, or treatment of disease must conform to high standards of testing and compliance. CLIA-waived tests must therefore be simple and have a low risk for producing an incorrect result, but this does not mean that waived tests are completely error-proof. Errors can occur anywhere in the testing process, particularly when the manufacturers' instructions are not followed and when testing personnel are not familiar with all aspects of the test system including the sample types for which the test has been approved, compliance with the storage conditions and expiry dates of kits. A number of the RIDTs have a CLIA-waiver (Table 11.1) and these will also be highlighted in the following sections.

11.3 Antigen Detection-based RIDTs

AD RIDTs in their simplest form have been available since the early 2000s and can be performed by operators with minimum training in the field, without access to electricity or a laboratory, are read by eye, and most can be completed in approximately 15 minutes. The specificity of these tests is excellent (99–100%), indicating a high degree of confidence when a positive influenza result is generated. However, the main drawback with the AD RIDTS read by eye is their poor sensitivity, which is reported to range from between 25 and 81% when compared to reverse transcriptase polymerase chain reaction (rtPCR) (Table 11.1). In head-to-head studies using the most common AD RIDTs read by eye (QuickVue, BinaxNOW and BD Directigen) with the same specimens, the sensitivity of the three tests was reported to be very similar (Hurt et al. 2007; Cho et al. 2013) (Table 11.1, Figure 11.2).

Unfortunately, the sensitivity of the AD RIDTs read by eye to detect influenza B viruses has been consistently lower than that for influenza A viruses (Hurt et al. 2007; Cho et al. 2013; Busson et al. 2014). In addition, studies have also found variability in performance when detecting different influenza A virus subtypes. Sensitivity for the detection of H1N1pdm09 viruses, which emerged in the 2009 pandemic, was as much as 27% lower than that for the human influenza virus subtypes H3N2 and former seasonal H1N1 (Hurt et al. 2009; Chan et al. 2012). (Influenza A contains the glycoproteins hemagglutinin and neuraminidase. For this reason, they are described as H1N1, H1N2, etc. depending on the type of H or N antigens they express; pdm is an abbreviation for pandemic.)

AD RIDTs read by eye have also demonstrated lower sensitivity for the detection of novel avian H7N9 influenza viruses than for seasonal human H3N2 influenza viruses (Baas et al. 2013). Based on published cycle threshold (C_t) values for clinical specimens from influenza A (H7N9) patients, the limit of detection of the AD RIDTs read by eye was such that the majority of influenza A (H7N9) cases would not have been detected using these tests (Baas et al. 2013).

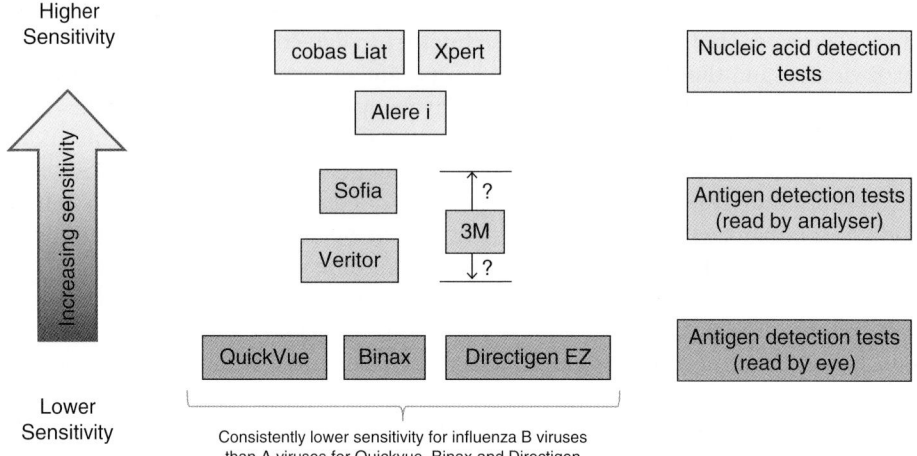

Figure 11.2 Illustration of the relative sensitivities of the different RIDTs based on studies where head-to-head comparisons have been conducted. Head-to-head sensitivity comparisons: Alere i vs Cobas (Nolte et al. 2016); Alere i vs Sofia vs Binax (Hazelton et al. 2015); Sofia vs Veritor (Leonardi et al. 2015; Ryu et al. 2016; Ryu et al. 2017); Veritor vs Binax (Hassan et al. 2014); Veritor vs Sofia vs Binax (Dunn et al. 2014); RAMP vs QuickVue versus Binax (Dale et al. 2008); Binax vs RAMP (Ginocchio et al. 2009); Xpert vs Binax vs Directigen (DiMaio et al. 2012); Binax vs Directigen vs Sofia (Lee et al. 2012); Binax vs Directigen (Busson et al. 2014); Binax vs Directigen vs QuickVue (Hurt et al. 2007). No data is available for the relative performance of the Xpert Xpress Flu test (Cepheid) or the Solana Influenza A+B assay (Quidel), both of which are NAD RIDTs.

In an effort to improve the sensitivity of the AD RIDTs read by eye, some manufacturers have developed devices that read the result of the RIDT, thereby avoiding the manual reading by eye. Simple readers, such as the battery-powered CLIA-waived BD Veritor System Reader, provide a definitive result of the test outcome on the screen, indicating that the sample has tested positive or negative for influenza A/B or has obtained a "Result Invalid" or "Control Invalid" test result, meaning that the test needs to be repeated (Becton Dickinson [BD] 2017). Other AD RIDTs, such as the Quidel Sofia system, have used a combination of immunofluorescence-based reagents and small benchtop analyzers to improve their test sensitivity and to produce definitive results (Quidel 2017). The AD RIDTs read by analyzer avoid the need to interpret the readout, eliminate person-to-person variation in reading the tests manually and increase the sensitivity due to the analyzers being able to detect results that are not be visible by eye (Table 11.1). Across a number of head-to-head studies, the AD RIDTs read by analyzer had a 23 and 20% higher mean sensitivity than the AD RIDTs read by eye for influenza A and B viruses respectively (Ginocchio et al. 2009; Lee et al. 2012; Dunn et al. 2014; Hassan et al. 2014; Hazelton et al. 2015; Ryu et al. 2016).

Four studies have also directly compared the two most widely used AD RIDTs read by analyzers - Sofia and Veritor (Dunn et al. 2014; Leonardi et al. 2015; Ryu et al. 2016; Ryu et al. 2017). The Sofia system was found to have a marginally higher sensitivity than the Veritor system for detection of influenza A viruses. The mean sensitivity of Sofia versus Veritor across four studies for Influenza A was 85.9% versus 79.6%. However, a more substantial difference for influenza B viruses was found – the mean sensitivity of Sofia versus Veritor across four studies was 91.3% versus 78.6% (Dunn et al. 2014; Leonardi et al. 2015; Ryu et al. 2016; Ryu et al. 2017) (Figure 11.2).

These studies also demonstrated improved sensitivity of the Sofia and Veritor systems for the detection of influenza B viruses compared to the AD RIDTs read by eye. The latter had sensitivities ranging from 30 to 53% (Hurt et al. 2007; Cho et al. 2013; Busson et al. 2014). There have been no published studies to date that have directly compared the 3 MAD RIDT test, which is also read by an analyzer, in order to determine its relative performance with that of the Veritor or Sofia systems (Figure 11.2). However, when it was compared to rtPCR its sensitivity ranged from 70.1% to 86.5% and it consistently performed better than AD RIDTs read by eye (Dale et al. 2008; Ginocchio et al. 2009).

11.4 Nucleic Acid Detection-based RIDTs

The newest generation of RIDTs is based on PCR amplification or isothermal amplification. These NAD RIDTs target viral nucleic acid rather than viral protein. This has produced a large increase in sensitivity of detection compared to the AD RIDTs, including those using an analyzer (Figure 11.2). The sensitivity of the Cobas Liat System is 99% for influenza A and 100% for influenza B (Binnicker et al. 2015). The Xpert Flu system achieved 97% sensitivity for influenza A and 100% for influenza B (DiMaio et al. 2012). These figures are close to those of real-time PCR assays, which are currently considered the gold standard and the main test used in large established laboratories (Table 11.1). The Alere i is another NAD RIDT, but in head-to-head comparisons it has consistently shown lower sensitivity than either the Xpert (Chapin and Flores-Cortez 2015; Jokela et al. 2015; Van Nguyen et al. 2016) or the Cobas systems (Nolte et al. 2016). However, it was superior to either the BinaxNOW or the Sofia AD RIDTs (Hazelton et al. 2015). As well as maintaining the high sensitivity of the AD RIDTs, three of the NAD RIDTs have also achieved CLIA-waived status from the FDA (Alere i Influenza A+B, the Cobas Liat and the Xpert Xpress Flu [Table 11.1]), indicating that the tests are simple to use, have a low risk for erroneous results and can be used by non-laboratory personnel.

These new NAD RIDTs allow operators at remote outbreak sites, as well as technicians in regional laboratories, to access diagnostic detection systems with comparable sensitivity to that available in large laboratories and hospitals in major cities. This dramatically reduces the need for shipment of samples to major laboratories and empowers local laboratories to have their own testing capability. However, there are still a number of issues using these NAD RIDTs in LMICs with the major disadvantage being the cost of the individual tests and the cost of the equipment to perform or read the test. Other potential issues are the need to have storage for reagents at 4 °C or less, their low throughput and the requirement for a stable electrical supply for most of these types of tests.

A positive feature of the NAD RIDTs is that manufacturers can develop assays that detect multiple viruses in the same test, something that has not been done to date for the AD RIDTs. One example is the CLIA-waived Xpert Xpress Flu/RSV system, which allows reliable diagnosis of both influenza and respiratory syncytial virus (RSV) infection in 20–30 minutes. Both the Xpert Xpress Flu/RSV and Xpert Xpress Flu tests also have a novel design feature which employs multiple gene targets for each virus. This built-in redundancy results in higher sensitivity and specificity, and reduces the impact of viral mutations that arise through replication errors or antigenic drift, which have historically been a problem with molecular tests especially for influenza. In the Xpert tests, a sample is loaded into a cartridge, which is inserted into a GeneXpert System. Different GeneXpert systems of varying capacity (one cartridge up to 80 cartridges) are available including the GeneXpert Omni, that processes a single cartridge unit and operates off battery power for up to two days, making it ideal for use in remote regions with unreliable power supply or in true field conditions.

11.5 Factors that Alter RIDTs Performance

The performance of RIDTs can differ from study to study due to a number of factors. These include the following: patient age - children shed more virus than adults or elderly people; the type of specimen analyzed (nasopharyngeal aspirate or wash, nasal wash, nasal swab, throat swab, gargle); storage conditions for the samples and test kits/reagents prior to testing and the type of laboratory assay used as the comparator (for example rtPCR positive cases versus culture-positive cases). For this reason, head-to-head studies which compare multiple RIDTs tests in the same study can be particularly informative in determining their relative performance (Figure 11.2).

Another variable that can influence the successful detection of influenza virus by an RIDT is the specific time point during viral replication when the specimen is taken. If it is too early in the course of illness (for example, during the first 12 hours after the onset of symptoms) there may not be enough viral antigen to be detected. Specimens taken 48–72 hours after symptom onset, when peak influenza virus levels are reached, are most likely to yield influenza-positive results, whereas levels of virus typically drop significantly five to seven days after symptom onset. Thus, detection of influenza at this time point will be less likely. On the other hand, the high specificity of these RIDTs gives a high level of confidence in the test if a positive result is obtained (Chartrand and Pai 2012).

11.6 The Use of RIDTs in LMICs

The detection of influenza in isolated or remote regions with no access to laboratories has led the World Health Organization (WHO) to recommend the use of rapid tests to assist in making clinical decisions, and in the early detection and control of outbreaks. Unfortunately, there have been few studies published on the use of rapid tests for the detection of seasonal influenza in LMICs, but when they have been used the results were similar to those obtained in developed countries (Ndegwa et al. 2017). An important fact to consider is the use of influenza rapid tests in regions where people may be exposed to zoonotic influenza viruses – for example, A(H5N1), A(H7N9) or A(H3N2)v carried by poultry, wild birds, swine, or other susceptible animals. The ability of RIDTs to detect these non-seasonal animal influenza A viruses appears to be poorer than for seasonal influenza A viruses (for example, A(H3N2) or A(H1N1)). Seven RIDTs (five AD and two NAD RIDTs) had a 40–60% poorer analytical sensitivity for the zoonotic avian A(H7N9) and swine A(H3N2)v viruses than for the human seasonal A(H3N2) viruses and surprisingly, the NAT RIDTs performed no better than the AD RIDTs (Chan et al. 2013). These findings correlated with other studies (Sakai-Tagawa et al. 2014; Chen et al. 2015), including one which tested six RIDTs (including five other RIDTs) against avian A(H7N9) influenza (Chen et al. 2015). These deficiencies in RIDTs could be very important as a negative test result may be interpreted as the patient being negative for all types of influenza (seasonal and zoonotic). It is possible that the patient may have been infected with a zoonotic influenza virus but that the test is not sensitive enough to detect the virus. Even though human infections with zoonotic influenza viruses are relatively rare, they can be very serious and have been fatal in 53% and 39% of human cases of A(H5N1) and A(H7N9) virus infection respectively (WHO 2017). This makes the use of RIDTs to detect these zoonotic viruses problematic. It would be of concern if false negative results from RIDTs led to delayed (or no) anti-influenza treatment for severely ill patients with zoonotic influenza infection since it may increase the likelihood of a fatal outcome.

11.7   Conclusions

With the variety of RIDTs currently available, the now have a vast array of choice for the detection of influenza viruses. Depending on the specific situation, there are options to select RIDTs that require no access to power, have excellent sensitivity and specificity and can yield a result in less than 15 minutes. However, the cost of the RIDTs remains sufficiently expensive that their widespread use in many low-income countries is unlikely. For the NAD RIDTs, the need for additional instrumentation further adds to this cost. Furthermore, many of the NAD RIDTs also need to be stored at 2–8 °C, making their use more difficult in extreme climates. If these issues can be overcome or reduced, then laboratories may be willing to invest in the use of one particular NAT RIDT platform. An example would be the Cepheid GeneXpert system with which a range of pathogens including tuberculosis, multi-resistant *Staphyloccus aureus*, enteroviral meningitis, *Clostridium difficile*, influenza A/B and RSV (and more) can be detected by simply adding the sample to an appropriate test cartridge or cartridges and then placing them on the base unit and the processing and test results are generated automatically.

While the list of RIDTs continues to grow there are still some fundamental issues with many of these tests. As indicated above, their performance with human seasonal influenza viruses ranges from acceptable to excellent, but their performance against non-human zoonotic influenza viruses is generally much poorer. While human infections with non-human or zoonotic influenza viruses are rare, they can be fatal. Thus, it is important to use alternative testing methods such as specifically designed rtPCR assays to confirm or negate infections in countries or regions where these viruses are encountered. More attention by manufacturers to overcome this problem would be welcome. Of course, no RIDT can overcome factors such as poor sampling, poor sample storage or too early or too late sample collection, so a combination of improved disease understanding and improved sample handling are needed in order to synergize with recent improvements in the RIDTs to improve healthcare overall and save lives in the due to influenza.

Acknowledgment

The Melbourne WHO Collaborating Centre for Reference and Research on Influenza is supported by the Australian Government Department of Health.

References

Angoulvant, F., Bellettre, X., Houhou, N. et al. (2011). Sensitivity and specificity of a rapid influenza diagnostic test in children and clinical utility during influenza A (H1N1) 2009 outbreak. *Emerg. Med. J.* 28 (11): 924–926.

Baas, C., Barr, I.G., Fouchier, R.A. et al. (2013). A comparison of rapid point-of-care tests for the detection of avian influenza A(H7N9) virus, 2013. *Euro Surveill* 18 (21).

BD. 2017. "BD Veritor System Instructions." Retrieved 30th June 2017, from https://www.youtube.com/watch?v=wCGUZ2Bl7pg

Biggs, C., Walsh, P., Overmyer, C.L. et al. (2010). Performance of influenza rapid antigen testing in influenza in emergency department patients. *Emerg. Med. J.* 27 (1): 5–7.

Binnicker, M.J., Espy, M.J., Irish, C.L., and Vetter, E.A. (2015). Direct detection of influenza A and B viruses in less than 20 minutes using a commercially available rapid PCR assay. *J. Clin. Microbiol.* 53 (7): 2353–2354.

Busson, L., Hallin, M., Thomas, I. et al. (2014). Evaluation of 3 rapid influenza diagnostic tests during the 2012-2013 epidemic: influences of subtype and viral load. *Diagn. Microbiol. Infect. Dis.* 80 (4): 287–291.

Chan, K.H., Chan, K.M., Ho, Y.L. et al. (2012). Quantitative analysis of four rapid antigen assays for detection of pandemic H1N1 2009 compared with seasonal H1N1 and H3N2 influenza A viruses on nasopharyngeal aspirates from patients with influenza. *J. Virol. Methods* 186 (1–2): 184–188.

Chan, K.H., To, K.K., Chan, J.F. et al. (2013). Analytical sensitivity of seven point-of-care influenza virus detection tests and two molecular tests for detection of avian origin H7N9 and swine origin H3N2 variant influenza A viruses. *J. Clin. Microbiol.* 51 (9): 3160–3161.

Chapin, K.C. and Flores-Cortez, E.J. (2015). Performance of the molecular Alere I influenza A&B test compared to that of the xpert flu A/B assay. *J. Clin. Microbiol.* 53 (2): 706–709.

Chartrand, C. and Pai, M. (2012). How accurate are rapid influenza diagnostic tests? *Expert Rev. Anti-Infect. Ther.* 10 (6): 615–617.

Chen, Y., Wang, D., Zheng, S. et al. (2015). Rapid diagnostic tests for identifying avian influenza A(H7N9) virus in clinical samples. *Emerg. Infect. Dis.* 21 (1): 87–90.

Cho, C.H., Woo, M.K., Kim, J.Y. et al. (2013). Evaluation of five rapid diagnostic kits for influenza A/B virus. *J. Virol. Methods.* 187 (1): 51–56.

Dale, S.E., Mayer, C., Mayer, M.C., and Menegus, M.A. (2008). Analytical and clinical sensitivity of the 3M rapid detection influenza A+B assay. *J. Clin. Microbiol.* 46 (11): 3804–3807.

DiMaio, M.A., Sahoo, M.K., Waggoner, J., and Pinsky, B.A. (2012). Comparison of Xpert Flu rapid nucleic acid testing with rapid antigen testing for the diagnosis of influenza A and B. *J. Virol. Methods* 186 (1–2): 137–140.

Dunn, J., Obuekwe, J., Baun, T. et al. (2014). Prompt detection of influenza A and B viruses using the BD Veritor System Flu A+B, Quidel(R) Sofia(R) influenza A+B FIA, and Alere BinaxNOW(R) influenza A&B compared to real-time reverse transcription-polymerase chain reaction (RT-PCR). *Diagn. Microbiol. Infect. Dis.* 79 (1): 10–13.

Ginocchio, C.C., Lotlikar, M., Falk, L. et al. (2009). Clinical performance of the 3M Rapid Detection Flu A+B Test compared to R-Mix culture, DFA and BinaxNOW influenza A&B test. *J. Clin. Virol.* 45 (2): 146–149.

Hassan, F., Nguyen, A., Formanek, A. et al. (2014). Comparison of the BD Veritor System for flu A+B with the Alere BinaxNOW influenza A&B card for detection of influenza A and B viruses in respiratory specimens from pediatric patients. *J. Clin. Microbiol.* 52 (3): 906–910.

Hazelton, B., Nedeljkovic, G., Ratnamohan, V.M. et al. (2015). Evaluation of the Sofia influenza A + B fluorescent immunoassay for the rapid diagnosis of influenza A and B. *J. Med. Virol.* 87 (1): 35–38.

Hurt, A.C., Alexander, R., Hibbert, J. et al. (2007). Performance of six influenza rapid tests in detecting human influenza in clinical specimens. *J. Clin. Virol.* 39 (2): 132–135.

Hurt, A.C., Baas, C., Deng, Y.M. et al. (2009). Performance of influenza rapid point-of-care tests in the detection of swine lineage A(H1N1) influenza viruses. *Influenza Other Respir. Viruses* 3 (4): 171–176.

Jokela, P., Vuorinen, T., Waris, M., and Manninen, R. (2015). Performance of the Alere i influenza A&B assay and mariPOC test for the rapid detection of influenza A and B viruses. *J. Clin. Virol.* 70: 72–76.

Lee, C.K., Cho, C.H., Woo, M.K. et al. (2012). Evaluation of Sofia fluorescent immunoassay analyzer for influenza A/B virus. *J. Clin. Virol.* 55 (3): 239–243.

Leonardi, G.P., Wilson, A.M., Mitrache, I., and Zuretti, A.R. (2015). Comparison of the Sofia and Veritor direct antigen detection assay systems for identification of influenza viruses from patient nasopharyngeal specimens. *J. Clin. Microbiol.* 53 (4): 1345–1347.

Ndegwa, L.K., Emukule, G., Uyeki, T.M. et al. (2017). Evaluation of the point-of-care Becton Dickinson Veritor rapid influenza diagnostic test in Kenya, 2013-2014. *BMC Infect. Dis.* 17 (1): 60.

Nolte, F.S., Gauld, L., and Barrett, S.B. (2016). Direct comparison of Alere i and cobas Liat influenza A and B tests for rapid detection of influenza virus infection. *J. Clin. Microbiol.* 54 (11): 2763–2766.

Peeling, R.W. and McNerney, R. (2014). Emerging technologies in point-of-care molecular diagnostics for resource-limited settings. *Expert. Rev. Mol. Diagn.* 14 (5): 525–534.

Quidel. 2017. "Sofia Test Kits." Retrieved 30 June 2017, from https://www.quidel.com/ immunoassays/sofia-tests-kits.

Ryu, S.W., Lee, J.H., Kim, J. et al. (2016). Comparison of two new generation influenza rapid diagnostic tests with instrument-based digital readout systems for influenza virus detection. *Br. J. Biomed. Sci.* 73 (3): 115–120.

Ryu, S.W., Suh, I.B., Ryu, S.M. et al. (2017). Comparison of three rapid influenza diagnostic tests with digital readout systems and one conventional rapid influenza diagnostic test. *J. Clin. Lab. Anal.* 32 (2): https://doi.org/10.1002/jcla.22234.

Sakai-Tagawa, Y., Ozawa, M., Yamada, S. et al. (2014). Detection sensitivity of influenza rapid diagnostic tests. *Microbiol. Immunol.* 58 (10): 600–606.

Services, C. f. M. M. (2017). "Clinical Laboratory Improvement Amendments (CLIA)." Retrieved 30 June 2017, from https://www.cms.gov/Regulations-and-Guidance/ Legislation/CLIA/index.html?redirect=/clia.

Van Nguyen, J.C., Camelena, F., Dahoun, M. et al. (2016). Prospective evaluation of the Alere i influenza A&B nucleic acid amplification versus xpert flu/RSV. *Diagn. Microbiol. Infect. Dis.* 85 (1): 19–22.

WHO. (2017). "Influenza at the human-animal interface, summary and assessment, 17 May 2017 to 15 June 2017." Available from http://www.who.int/influenza/human_ animal_interface/Influenza_Summary_IRA_HA_interface_06_15_2017.pdf?ua=1

12

A Rapid Diagnostic Test for Ebola Virus Disease[1]

Catherine Houlihan[1] and Colin Brown[2, 3, 4]

[1] Department of Infection and Immunity, University College London, London, UK
[2] King's Sierra Leone Partnership, King's Centre for Global Health, King's Health Partners, King's College London, London, UK
[3] National Infection Service, Public Health England, London, UK
[4] Department of Infection, Royal Free Hampstead NHS Trust, London, UK

CHAPTER MENU

12.1 Case Report

An 18 year old man attends a large hospital in North Western Sierra Leone. He has severe abdominal pain, fever and is vomiting. He reports having had contact with people in his village who were diagnosed with Ebola Virus Disease (EVD) two weeks previously. A venous blood sample is collected from him after he is admitted to an onsite Ebola Holding Unit (EHU). In the EHU he sleeps beside patients with diarrhea and vomiting who later test positive for EVD. Two days after admission his abdominal pain remains severe, he is febrile and is significantly dehydrated despite some intravenous hydration. He has not moved his bowels.

The EVD test comes back negative and he is admitted to the main hospital, still vomiting when attempting to drink water, still febrile, and still in severe pain. The surgeon is cautious about seeing him, aware that he has been admitted to the EHU and concerned that he may be incubating EVD. He is observed for a further two days and given only analgesia. There are no facilities to perform abdominal imaging.

---

1 Ebola was named after a river that flowed close to the village of Yambuka in Zaire (now the Democratic Republic of Congo) where Ebola disease was first diagnosed. In the local language, Lingala, it means "Black River."

Revolutionizing Tropical Medicine: Point-of-Care Tests, New Imaging Technologies and Digital Health,
First Edition. Edited by Kerry Atkinson and David Mabey.
© 2019 John Wiley & Sons, Inc. Published 2019 by John Wiley & Sons, Inc.

Four days after he first came to hospital he suddenly becomes hypotensive, tachycardic and drowsy. His abdomen is rigid and he is taken to theater for an emergency laparotomy where he is found to have a ruptured appendix. He survives and is discharged 14 days after his first presentation.

12.2   Introduction

EVD outbreaks were first identified in 1976 and by 2013 had cumulatively caused 2300 cases with 1546 deaths (CDC 2016). The West Africa outbreak, caused by the Zaire strain of Ebola virus (EBOV), was thought to have started in December 2013 and by its close in June 2016 had led to a staggering 28 616 cases and 11 310 deaths (WHO 2016a, b). Some of the contributing factors to the size of the outbreak were the lack of a robust surveillance system in West Africa, which would allow for identification and monitoring of suspect cases and the contacts of confirmed cases; a lack of adequate testing at a local and regional level to provide rapid diagnosis and isolation of new cases; and lack of appropriate infrastructure for the safe isolation of suspect and confirmed cases. The absence of these key systems undoubtedly contributed to onwards spread.

12.3   Diagnostic Methods to Detect Ebola Virus Disease

Diagnostic methods to detect EVD in West Africa and worldwide have until recently been limited to molecular techniques, specifically detecting viral RNA by reverse transcription polymerase chain reaction (rtPCR) in bodily fluids. Since the patient's bodily fluids can potentially transmit Ebola virus to laboratory staff and healthcare workers (HCWs), the virus has high outbreak potential, and because the disease has a high case fatality rate (between 30% and 70% in most case series) with no previous evidence of effective treatment or vaccine, EBOV is designated a Hazard Group 4 pathogen in the UK and a Biosafety level 4 pathogen in the USA (Advisory Committee on Dangerous Pathogens [ACDP] 2013; Centers for Disease Control and Prevention [CDC] 2016; Chosewood and Wilson 2009).

Samples potentially carrying viable virus require very careful sampling and handling from collection through to analysis. There are now multiple commercially available rtPCR assays which have been validated for the detection of EBOV in plasma or serum samples from patients suspected of having EVD (Shorten et al. 2016). Buccal swabs from deceased patients can also be tested on these assays and are useful for screening community deaths to more accurately assess population-level mortality from the disease, and to interrupt transmission chains. The restrictions in this diagnostic method are multiple, including the requirement for venous blood draws. HCWs collecting the venous blood sample are required to wear high-level personal protective equipment (PPE) inhibiting vision, restricting movement and leading to heat stress in tropical climates. Patients can be confused due to their illness, leading to an increased risk of needle stick injuries during venous sampling. Médecins Sans Frontières (MSF) have a protocol for collecting blood for EBOV testing which requires three staff to collect a sample from one patient. Samples must also be tested in a facility which meets the rigorous standards for processing a Biosafety level 4 pathogen: a high level of staff training, PPE, secure entry and exit to the facility, high efficiency particulate air (HEPA) filtered air, decontamination procedures on exit, and dedicated waste management. Modifications to facilities in West Africa were required, such as setting up Biosafety level 3 laboratories with small high-level isolation sections. Reagents for rtPCR testing require cold supply chain storage, posing a problem in tropical countries with an unstable electricity supply. The combined cost of setting up and maintaining laboratory facilities, sample transfer and staff training resulted in one test costing an average of US$100 (Nouvellet et al. 2015). Finally, and arguably most importantly, PCR-based testing methods take between two and six hours to process and the testing facilities are usually located in regional or national laboratories distant from the unwell EVD patient who presents to a rural healthcare facility. Samples require secure transport and a reporting infrastructure must exist to communicate results back to local health centers. During the peak of the West African epidemic less than 50% of Ebola Treatment Centers (ETCs) and no EHU had an on-site testing facility (WHO 2016a). This process is similar in high income settings: in the United Kingdom (UK) all samples are transferred to a single testing facility (Public Health England's Rare and Imported Pathogens Laboratory) for EBOV rtPCR. In the West African outbreak the combination of laboratory availability and time for PCR testing led to a prolonged turnaround time from patient testing to receiving results. In one large EHU based at a major hospital in Sierra Leone, laboratory turnaround time from sample collection to results being available was a mean of 3.0 and 2.4 days in October and November 2014, when the total case number was increasing at its most rapid rate (Brown et al. 2017). At its lowest rate toward the end of the epidemic in April 2015, the time taken to obtain results was 0.8 days. The delay, particularly early in the epidemic, clearly limited public health intervention for confirmed positive cases.

The consequence of delays in diagnosis of EVD are multiple. Patients with Ebola have non-specific symptoms which are common to multiple other pathogenic infections, many of which are life-threatening. Fever, weakness or fatigue and vomiting or nausea are among the most common presenting symptoms of EVD (Lado et al. 2015), and the case definition provided by WHO has a low sensitivity (Lado et al. 2015; Levine et al. 2015). Patients meeting a suspect case definition should be isolated pending the result of the diagnostic test in order to limit spread in the community. They are usually admitted to a dedicated EHU or a "suspect" ward in an ETC, and are at risk of nosocomial acquisition of EBOV, though evidence suggests this risk is minimal (Arkell et al. 2016; Fitzgerald et al. 2017, 2014). One third of individuals admitted to an ETC in Sierra Leone did not have EVD but likely suffered a significant delay in the diagnosis and management of their illness, or died while awaiting results (Lado et al. 2015). A further consequence of the lack of being able to provide a rapid negative test for EVD is maintaining a functioning healthcare system outside that of EVD care. Patients with EVD-like symptoms present to other specialties including obstetrics, general surgery, general medicine and pediatrics, and lead to high infection and fatality rate in HCWs (Senga et al. 2016). The ensuing lack of HCWs (due to EVD itself and absenteeism due to fear of EVD) and reduced attendance at healthcare facilities have contributed to an excess mortality from non-EVD conditions of around 24 900 per year (Phillips et al. 2016). A negative RDT, if highly sensitive, may have mitigated this and allowed healthcare provision to continue by immediate exclusion of EVD in, for example, patients with obstructed labor, acute surgical abdomen, HIV opportunistic infections and many other conditions. Only through dedicated EHUs operating a screening service for all patients at point of presentation did hospitals remain open in Sierra Leone (Brown et al. 2017).

Progress toward a rapid diagnostic test for EBOV was made with the EBOV-RPA Test, a rapid PCR which only required 15 minutes to amplify RNA. EBOV-RPA was tested in Sierra Leone in 2015 on 271 patients who were rtPCR positive and 104 who were rtPCR negative. The test was 97% sensitive (95% confidence interval [CI]: 95.5–99.3) and 97% specific (95% CI: 93.9–100) (Yang et al. 2016). The disadvantage was that full laboratory equipment and advanced training were required to perform the test. Less equipment, and the potential to run on batteries, are features of the reverse transcription-loop-mediated isothermal amplification (RT-LAMP) assay. Although an extremely high sensitivity and specificity were again demonstrated, the same high level of training and freezer storage capacity are still required, although the test was run from the back of a pick-up truck in the mobile field laboratory in West Africa (Kurosaki et al. 2016).

The Xpert Ebola assay (Cepheid) is an automated PCR assay that requires substantially less training to operate, has storage requirements of 15–25 °C and the test results are available in under 90 minutes. In this test patient samples are inoculated into a pre-filled vial which is then placed into a cartridge and loaded on to the GeneXpert platform. The kit can be used on finger prick blood samples as well as on oral swabs and on blood obtained by venepuncture. It inactivates virus load of up to 4.6×10^6 plaque forming units (PFU) of EBOV, although concerns have been raised that patient viral loads in blood are often higher (Broadhurst et al. 2016). Field evaluations in West Africa demonstrated a 100% sensitivity and 96% specificity on 211 blood samples compared to rtPCR in Sierra Leone (Semper et al. 2016). Similar results were seen in 218 blood samples in Guinea (Van den Bergh et al. 2016). Invalid results were found in approximately 10% of samples, and technicians raised concerns about viral inactivation in samples, storage requirements, and cleaning (Van den Bergh et al. 2016). Importantly, GeneXpert platforms are increasingly being used for the diagnosis of tuberculosis in settings where EVD outbreaks have occurred, allowing the use of transferable skills for this system.

Another cartridge-based system is the Idylla system developed by Biocartis, Janssen Diagnostics and the Institute for Tropical Medicine in Antwerp. This rapid real-time rtPCR has been CE marked (approved for marketing by the European Union) and the USA Food and Drug Administration (FDA). Although extensively tested for cross reactivity and stability in the laboratory, it has not been tested in West Africa and laboratory verification was not based on samples from people with EBOV (Cnops et al. 2016).

IFilmArray Biothreat-E test and FilmArray NGDS BT-E assay are closed system real-time rtPCR kits for EBOV detection which were tested during the West African outbreak. The former was field tested in 60 samples in Sierra Leone and demonstrated an 84% sensitivity and 89% specificity compared to real time rtPCR (Weller et al. 2016), although with FilmArray NGDS BT-E the authors commented that samples were heat-treated to reduce handling risk, and suggested that this may have affected the results. The same test demonstrated 100% concordance with rtPCR in 81 blood samples which were not heat treated, although only five samples were from patients with EVD and not all were from suspect patients (Leski et al. 2015). As well as the reasonably fast turn-around time and reduced skill requirement a major future advantage of this technology is the potential to test for multiple pathogens on the same sample. The Biothreat-E test has the potential to simultaneously detect 16 pathogens, allowing appropriately trained HCWs to exclude EVD as well as to identify the causative illness.

12.4 Rapid Diagnostic Tests for Ebola Virus Disease for Use in a Point-of-Care Facility

In November 2014 the WHO called on diagnostic companies and others to develop rapid diagnostic tests (RDTs) for EVD and outlined desirable criteria for this test (Table 12.1) (WHO 2014). These included that the test be suited for use in peripheral health clinics with no laboratory infrastructure in place, testing procedures that involve fewer than three steps, production of results in less than 30 minutes and without biosafety requirements beyond the wearing of personal protective equipment.

WHO further suggested that the reagents be easily stored, the tests were portable with no requirement for power, and staff training would take less than half a day. This request was accompanied by consultation with experts in diagnostics and manufacturers as well as non-government organizations (NGOs) including MSF and the Foundation for Innovative New Diagnostics (FIND) in order to most efficiently bring RDTs through development to deployment. Four months later WHO released interim technical guidance on the use of RDTs for EVD. The document mentioned two specific circumstances in which RDTs would be beneficial, whilst reaffirming that rtPCR was required for confirmation of all cases: (i) in remote settings where rtPCR was not immediately available, and (ii) in non-remote settings where it was not possible to manage the number of cases and suspects with existing facilities. The main justification in both settings was isolation of infected individuals. The document clearly stated that RDTs should not be used in settings such as airport screening, screening blood for transfusion, certifying patients "Ebola-free" before investigating or managing other conditions, and releasing patients from treatment centers. If found to be fully sensitive, de-escalation in EVD negative patients would be a key benefit.

Three RDTs for EVD were approved for emergency use by WHO in 2016 and by mid-2017 three others were in development (Table 12.2). The first test approved by WHO was ReEBOV™ Antigen Rapid Test Kit developed by Corgenix (WHO 2015). This test produces a pink to red colored line when EBOV VP40 antigen, if present in the patient's sample, binds to a polyclonal antibody against Ebola nuclear protein (NP) which is bound to a nanoparticle. The test is semi-quantitative, giving a result from low to high positivity (at five levels) depending on depth of color. 30 µL of whole blood, plasma or serum is applied to the dipstick which is then inserted into a tube containing four drops (~200 µL) of sample buffer. Results are available after 15–20 minutes and are read with the assistance of a card, requiring minimum training for use and interpretation. Cold chain supply is required and cross-reactivity with rheumatoid factor and high hematocrit have been demonstrated. However, no cross-reactivity was seen from multiple common bacterial, viral or protozoal infections or medications (Cross et al. 2016).

A laboratory in Kenema, Sierra Leone, tested 408 samples and compared the ReEBOV to rtPCR with an appropriate cut-off for positivity, demonstrating a sensitivity of 91.1% (195 of 214 samples; 95% CI: 86.5–94.6) and specificity of 90.2% (175 of 194; 95% CI: 85.1–94.0) (Boisen et al. 2015). The same test was then used in two Ebola Treatment Centers in Port Loko and Kambia (Sierra Leone) and compared ReEBOV on finger-prick samples with rtPCR on venous blood samples collected at the same time in patients presenting acutely unwell, and who met EVD criteria (Broadhurst et al. 2015). Sensitivity was 100% (95% CI: 87.7–100 on 28 samples) and specificity 92.2% (95% CI: 83.8–97.1 on 77 samples). Similar results were obtained when performed on 277 samples in the laboratory. Importantly, there was agreement between operators using the test in real time whilst wearing personal protective equipment in 100 of 105 tests.

Table 12.1 WHO summary of desirable criteria for a rapid diagnostic test for Ebola virus disease.

Feature	Desired	Acceptable
Target population	Patients presenting with fever to healthcare facilities for assessment	
Target use setting	Decentralized healthcare facilities with **no** laboratories infrastructure available	Decentralized healthcare facilities with **minimum** laboratory infrastructures available
Intended use	In Ebola outbreak setting, distinguish between symptomatic patients with acute Ebola virus infection and non-Ebola virus infection **without** the need for confirmatory testing	In Ebola outbreak setting, distinguish between symptomatic patients with acute Ebola virus infection and non-Ebola virus infection **with** the need for confirmatory testing
Sensitivity (compared to rtPCR)	>98%	>95%
Analytical specificity	>99%	>99%
Sample type	Capillary whole blood from finger stick or other less invasive sample types (saliva, buccal smear)	Whole blood from phlebotomy
Number of steps to be performed by operator	<3	<10
Biosafety	No additional biosafety in addition to PPE	No additional biosafety in addition to PPE
Time to result	<30 minutes	<3 hours
Internal control	Included	Included
Sample preparation	None or fully integrated	None or fully integrated
Operating conditions	5-50C	5-40C
Reagents reconstitution	All agents ready to use	Minimal requirements with all agents provided
Training requirements	Less than half a day for any level healthcare worker	Less than 2 days for any level of healthcare worker
Equipment	Small and portable, handheld instrument. Weight <2 kg	Small, table top device, Portable
Power requirement	None	110–220 V AC current DC power with rechargeable battery lasting up to 8 hours of testing
Maintenance	None	1 annual calibration ideally by operator

Source: Modified from WHO (2014).

A lateral flow diagnostic test for Ebola was developed using an undisclosed antigen by The UK Defense Science and Technology Laboratory (DSTL) (Walker et al. 2015). This test has the minimum skill and time requirements of other tests (Table 12.2) and was tested on 131 patients in four EHUs in Sierra Leone. Importantly, it has no cold chain requirement. Different levels of a semi-quantitative positive red band were assessed and

Table 12.2 Summary of lateral flow assays available or in development for point-of-care diagnosis of EVD.

Assay name (manufacturer)	Viral target (antigen), virus species detected	Specimen	Storage requirements	Time to results	Field testing? Sensitivity and specificity	Limit of detection
Antigen Rapid Test Kit ReEBOV (Corgenix)	EBOV (VP40)	Whole blood, serum or plasma	2–8°C with stability for six months	15–20 minutes	Multiple field tests. Sensitivity 100% (95% CI: 87.7–100), specificity 92.2% (95% CI: 83.8–97.1)	2.11E+08 RNA copies/mL
OraQuick Ebola Rapid Antigen Test Kit (OraSure Technologies, Inc.)	EBOV, Sudan, Bundibuyo (VP40)	Cadaveric oral fluid and Whole blood	2–8°C for 12 months	30 minutes	Feasibility established in Guinea. No field sensitivity data.	Not given
SD Q Line Ebola Zaire Ag (SD Biosensor Inc.)	EBOV. (GP, NP and VP40)	Cadaveric oral fluid and whole blood	1–40°C	Unclear	No field testing in West Africa	Not given
QuickNavi-Ebola (Research Center for Zoonosis Control, Japan)	EBOV, Taï forest, Bundibuyo Virus (NP)	Serum / plasma or whole blood	Does not require refrigeration	10 minutes	No field testing has taken place to date	0.03 to 0.3 $\mu g\,mL^{-1}$, depending on the isolates
NMRC EBOV LFI	EBOV; Ebola Sudan, Taï Forest, and Reston	Plasma or serum and oral samples. Not developed for whole blood	Stable at 4C° for 1 year	15 minutes	Tested in laboratory in Liberia. 290 samples: sensitivity 87.8% (75.3–94.3) and specificity 97.5% (94.7–98.9)	Limit of detection 5×10^4 to 5×10^5 plaque-forming units/mL
DSTL EVD Lateral Flow Assay	Ebola Zaire only (undisclosed antigen)	Whole blood	Does not require refrigeration. Unclear storage duration.	20 minutes	100% sensitive (95% CI: 78.2–100), and a specificity of 97% (95% CI: 91.4–99.1)	Not given
Ebola eZYSCREEN lateral flow assay (Stada Pharmm Germany and Senovas, Germay)	EBOV (GP)	Two assays: whole blood and serum/plasma	Stable for 393 days at 30°C and 120 days at 45°C	15 minutes	Sensitivity 65.3% and specificity 98.9% on whole blood; sensitivity 74.5% and specificity 100% on serum	105 plaque forming units/mL

the authors suggested that selectively using a score above 4 to determine positivity on the DSTL lateral flow assay offers 100% sensitivity (95% CI: 78.2–100), compared to rtPCT, and 97% specificity (95% CI: 91.4–99.1) (Walker et al. 2015).

The USA's Naval Medical Research Center (NMRC) used a lateral flow assay, which had been developed for environmental testing of EBOV contamination, on blood samples and oral swabs in Liberia and compared these to rtPCR results (Phan et al. 2016). This test had a good specificity of 97.5% (95% CI: 94.7–98.9) in 290 plasma samples, but a limited sensitivity of 87.8% (95% CI: 75.3–94.3), although this increased when stricter limits were placed on the comparison rtPCR cut-off. The major disadvantage is that the test has not been modified for use on whole blood and therefore did not give valid results unless plasma or serum were used. The requirement for centrifugation confines it to use only in a laboratory setting.

QuickNavi-Ebola is a lateral flow immunochromogenic assay that uses murine a monoclonal antibody against the Ebola nucleoprotein. This test gives a visible blue line when in contact with antigen from Ebola Zaire (Zaire is now the Democratic Republic of the Congo), Taï Forest virus (a close relative of EBOV) and Bundibuyo virus (another close relative of EBOV). However, it did not react with Ebola Reston (a different strain of Ebola virus that is lethal in monkeys but not in humans) or Ebola Sudan, nor with Marburg virus antigens. Published results are available from tests involving virus culture fluid and infected monkeys after day 4, but no field testing has been published to date (Yoshida et al. 2016).

OraQuick Ebola Rapid Antigen Test Kit developed by OraSure Technologies is a chromographic lateral flow assay which detects VP40 antigen on cadaveric oral fluid and in whole blood. Samples which had tested positive using rtPCR in West Africa were tested at the Centers for Disease Control (CDC) in the USA. The test had a sensitivity of 84.0% in 25 samples (95% CI: 63.92–95.46) and a specificity of 98.0% in 50 samples (95% CI: 89.4–99.9). The poor sensitivity was improved by selecting samples with higher viral loads, similar to other assays. The OraQuick was supported by WHO for the testing of symptomatically unwell patients suspected of having EBOV during the West Africa outbreak, and a large feasibility study included 1628 patients tested in Guinea found its performance to be acceptable, though no positive results were seen (Jean Louis et al. 2017).

Ebola eZYSCREEN lateral flow assay uses monoclonal antibodies against the EBOV's envelope glycoprotein and has been tested on 144 patients at an Ebola treatment center in Guinea (Gallais et al. 2017). The test had a sensitivity of 65.3% and a specificity 98.9% on whole blood, and a sensitivity 74.5% and a specificity 100% on serum, compared to rtPCR. Despite the low sensitivity, the test was developed and tested early in the epidemic and was approved by the EU after this.

Finally, SD Q Line Ebola Zaire Ag developed by SD Biosensor Inc. is another lateral flow assay. This test uses murine monoclonal antibodies to EBOV glycoprotein, nucleoprotein (NP) and VP40 which bind antigen, if present, to gold particles. Storage requirements offered the widest range of acceptable temperatures (Table 12.2), but on whole blood sensitivity was 84.9% (95% CI: 78.6–91.2) on 126 rtPCR-positive samples while specificity was 99.7% (95% CI: 99.1–100.0) on 289 samples.

12.5 Conclusions

Using traditional molecular techniques to diagnose infection with Ebola virus at the point of a patient's presentation can delay an un-infected individual's treatment and potentially put that individual at risk of acquiring EBOV. Using a highly sensitive test to rule out EVD would allow un-infected people access to treatment outside an Ebola Treatment Center (ETC) or an EHU. This would result in these healthcare settings being better staffed, since HCWs could be reassured about the risk of acquisition of EVD disease. It would also reduce the need for high bed capacity in holding units during large scale outbreaks of EVD disease. A rapid diagnostic test with high sensitivity would allow the identification and isolation of infected and infectious persons, thus preventing onward spread. Having RDTs available in remote or rural areas at risk of EVD could dramatically reduce the time from its introduction into a community to recognition of EVD and instigation of a public health response. This would limit subsequent people from becoming infected. Additionally, scale up of the response required for an EVD outbreak would be decreased. Seven lateral flow assays are in various stages of validation and several look promising. Furthermore, closed system nucleic acid amplification tests offer exciting promise with minimal reagent and power needs, and minimal operator skill requirements. FilmArray methods offer the possibility of diagnosing multiple pathogens from a single sample. Having highly sensitive and specific rapid diagnostic tests easily available for subsequent outbreaks of Ebola will prevent a repeat of the large epidemic seen in 2014–2016 in West Africa.

Bibliography

ACDP. (2013) 'The Approved List of biological agents,' *HSE Books*, Available from: www.hse.gov.uk/pubns/misc208.pdf

Arkell, P., Youkee, D., Brown, C.S., et al. (2016). 'Quantifying the risk of nosocomial infection within Ebola Holding Units: a retrospective cohort study of negative patients discharged from five Ebola Holding Units in Western Area, Sierra Leone.' *Tropical medicine & international health*. England.

Boisen, M.L., Oottamasathien, D., Jones, A.B. et al. (2015). Development of prototype filovirus recombinant antigen immunoassays. *J. Infect. Dis.* 212 (Suppl 2): S359–S367.

Broadhurst, M.J., Brooks, T.J.G., and Pollock, R. (2016). Diagnosis of Ebola virus disease: past, present, and future. *Clin. Microbiol. Rev.* 29 (4): 773–793.

Broadhurst, M.J., Kelly, J.D., Miller, A. et al. (2015). ReEBOV antigen rapid test kit for point-of-care and laboratory-based testing for Ebola virus disease: a field validation study. *Lancet* 386 (9996): 867–874. https://doi.org/10.1016/S0140-6736(15)61042-X.

Brown, C.S., Kessete, Q., Baker, P., et al. 2017. 'Bottlenecks in health systems functioning for control of Ebola Virus Disease in Connaught Hospital, Freetown, Sierra Leone,' in *European Congress of Clinical Microbiology and Infectious Diseases*. Vienna, Austria.

CDC. (2016). 'Outbreaks Chronology: Ebola Virus Disease,' Available from https://www.cdc.gov/vhf/ebola/outbreaks/history/chronology.html (accessed 30 June 2017).

Chosewood, L.C. and Wilson, D.E. (eds) (2009). Biosafety in Microbiological and Biomedical Laboratories (BMBL) 5th Edition. Available from https://www.cdc.gov/biosafety/publications/bmbl5/bmbl5_introduction.pdf

Cnops, L., Van den Eede, P., Pettitt, J. et al. (2016). Development, evaluation, and integration of a quantitative reverse-transcription polymerase chain reaction diagnostic test for Ebola virus on a molecular diagnostics platform. *J. Infect. Dis.* 214 (suppl 3): S192–S202. Available from https://academic.oup.com/jid/article-lookup/doi/10.1093/infdis/jiw150.

Cross, R.W., Boisen, M.L., Millett, M.M. et al. (2016). Analytical validation of the ReEBOV antigen rapid test for point-of-care diagnosis of Ebola virus infection. *J. Infect. Dis.* 214 (suppl 3): S210–S217. Available from https://academic.oup.com/jid/article-lookup/doi/10.1093/infdis/jiw293.

Fitzgerald, F., Wing, K., Naveed, A. et al. (2017). Risk in the "red zone": outcomes for children admitted to Ebola holding units in Sierra Leone without Ebola virus disease. *Clin. Infect. Dis.* 65 (1): 162–165.

Fitzpatrick, G., Vogt, F., Gbabai, O.B.M. et al. (2014). Describing readmission to an Ebola Case Management Centre (CMC). *Eurosurveillance* 19 (40): 1–5.

Gallais, F., Gay-Andrieu, F., Picot, V. et al. (2017). Validation sur le terrain du nouveau test de diagnostic rapide Ebola eZYSCREEN? *Bulletin de la Société de pathologie exotique* 110 (1): 38–48.

Kurosaki, Y., Magassouba, F., Oloniniyi, O.K. et al. (2016). Development and evaluation of Reverse Transcription-Loop-Mediated Isothermal Amplification (RT-LAMP) assay coupled with a portable device for rapid diagnosis of Ebola virus disease in Guinea Transcription-Loop-Mediated Isothermal Amplification (RT-LAMP) assay coupled with a portable device for rapid diagnosis of Ebola virus disease in Guinea. *PLoS Negl. Trop. Dis.* 10 (2): Available from http://www.mhlw.go.jp/stf/seisakunitsuite.

Lado, M., Walker, N.F., Baker, P. et al. (2015). Clinical features of patients isolated for suspected Ebola virus disease at Connaught hospital, Freetown, Sierra Leone: a retrospective cohort study. *Lancet Infect. Dis.* 15 (9): 1024–1033.

Leski, T.A., Ansumana, R., Taitt, C.R. et al. (2015). Use of the FilmArray system for detection of Zaire ebolavirus in a small hospital in Bo, Sierra Leone. *J. Clin. Microbiol.* 53 (7): 2368–2370.

Levine, A.C., Shetty, P.P., Burbach, R. et al. (2015). Derivation and internal validation of the Ebola prediction score for risk stratification of patients with suspected Ebola virus disease. *Ann. Emerg. Med.* 66 (3): 285–293.e1.

Jean, L.F., Huang, J.Y., Nebie, Y.K. et al. (2017). Implementation of broad screening with Ebola rapid diagnostic tests in Forécariah, Guinea. *Afr. J. Lab. Med.* 6 (1): 1–6.

Van den Bergh, V., Chaillet, R., Sow, P. et al. (2016). Feasibility of Xpert Ebola assay in Médecins Sans Frontières Ebola program, Guinea. *Emerg. Infect. Dis.* 22 (2): 210–216.

Nouvellet, P., Garske, T., Mills, H.L. et al. (2015). The role of rapid diagnostics in managing Ebola epidemics. *Nature* 528 (7580): S109–S116.

Phan, J.C., Pettitt, J., George, J.S. et al. (2016). Lateral flow immunoassays for Ebola virus disease detection in Liberia. *J. Infect. Dis.* 214 (Suppl 3): S222–S228.

Phillips, J., Steven, F.K., John, M. et al. (2016). Effects of the west africa Ebola virus disease on health-care utilization – a systematic review. *Front. Public Health* 4 (4): 222.

Semper, A.E., Broadhurst, M.J., Richards, J. et al. (2016). Performance of the GeneXpert Ebola assay for diagnosis of Ebola virus disease in Sierra Leone: a field evaluation study. *PLOS Med.* 13 (3): e1001980.

Senga, M., Pringle, K., Ramsay, A. et al. (2016). Factors underlying Ebola virus infection among health workers, Kenema, Sierra Leone, 2014–2015. *Clin. Infect. Dis.* 63 (4): 454–459.

Shorten, R.J., Brown, C.S., Jacobs, M. et al. (2016). Diagnostics in Ebola virus disease in resource-rich and resource-limited settings. *PLoS Negl. Trop. Dis.* 10 (10): 1–16.

Walker, N.F., Brown, C.S., Youkee, D. et al. (2015). Evaluation of a point-of-care blood test for identification of Ebola virus disease at Ebola holding units, western area, Sierra Leone, January to February 2015. *Eurosurveillance* 20 (12): 1–6.

Weller, S.A., Bailey, D., Matthews, S. et al. (2016). Evaluation of the biofire FilmArray BioThreat-E test (v2.5) for rapid identification of Ebola virus disease in heat-treated blood samples obtained in Sierra Leone and the United Kingdom. *J. Clin. Microbiol.* 54: 114–119.

WHO (2014). *Target Product Profile for Zaïre ebolavirus rapid, simple test to be used in the control of the Ebola outbreak in West Africa* Available from http://www.who.int/medicines/publications/target-product-profile.pdf?ua=1

WHO (2015). *Interim guidance on the use of rapid Ebola antigen detection tests* Available from http://apps.who.int/iris/bitstream/10665/160265/1/WHO_EVD_HIS_EMP_15.1_eng.pdf?ua=1

WHO (2016a). 'Ebola Situation Reports,' Available from http://apps.who.int/ebola/ebola-situation-reports (accessed 21 September 2016).

WHO (2016b). *Ebola Situation Reports: archive* Available from http://www.who.int/csr/disease/ebola/situation-reports/archive/en

Yang, M., Ke, Y., Wang, X. et al. (2016). Development and evaluation of a rapid and sensitive EBOV-RPA test for rapid diagnosis of Ebola virus disease. *Sci. Rep.* 6: 26943. https://doi.org/10.1038/srep26943.

Yoshida, R., Muramatsu, S., Akita, H. et al. (2016). Development of an immunochromatography assay (QuickNavi-Ebola) to detect multiple species of Ebola viruses. *J. Infect. Dis.* 214 (Suppl 3): S185–S191.

Webliography

www.hse.gov.uk/pubns/misc208.pdf ACDP 2013. 'The Approved List of biological agents.' *HSE Books.*

https://www.cdc.gov/vhf/ebola/outbreaks/history/chronology.htl CDC 2016 'Outbreaks Chronology: Ebola Virus Disease'.

https://www.cdc.gov/biosafety/publications/bmbl5/bmbl5_introduction.pdf Chosewood, L.C. and Wilson, D.E. (eds) 2009. Biosafety in Microbiological and Biomedical Laboratories.https://www.cdc.gov/labs/pdf/CDC-BiosafetyMicrobiologicalBiomedical Laboratories-2009-P.pdf

http://www.who.int/medicines/publications/target-product-profile.pdf?ua=1 WHO 2014. Target Product Profile for Zaïre ebolavirus rapid, simple test to be used in the control of the Ebola outbreak in West Africa.

http://apps.who.int/iris/bitstream/10665/160265/1/WHO_EVD_HIS_EMP_15.1_eng. pdf?ua=1 WHO 2015. Interim guidance on the use of rapid Ebola antigen detection tests.

http://apps.who.int/ebola/ebola-situation-reports Accessed 21 September 2016. WHO 2016a. 'Ebola Situation Reports'.

http://www.who.int/csr/disease/ebola/situation-reports/archive/en WHO 2016b. Ebola Situation Reports: archive.

13

Rapid Diagnostic Tests for Yaws

Michael Marks

Clinical Research Department, London School of Hygiene and Tropical Medicine, London, UK

13.1   Introduction

Yaws[1], caused by *Treponema pallidum subspecies* (ssp.) *pertenue,* is one of the four treponemal diseases affecting humans (Marks et al. 2014c; Mitjà et al. 2013). The disease is closely related to syphilis, which is caused by the almost identical *T. pallidum* ssp. *pallidum.* In contrast to syphilis, yaws is not sexually transmitted, but is believed to be spread by skin to skin contact between individuals with active disease. Again, unlike syphilis, mother-to-child transmission of yaws does not occur.

Yaws was the first disease to be targeted for eradication by the World Health Organization (WHO). A worldwide mass screening and treatment program was led by WHO and UNICEF. The campaign examined more than 300 million individuals of whom 50 million were treated. Although yaws was not eradicated, by the end of the

---

[1] The name is believed to be of Carib origin, "yaya" meaning sore. Caribs are an indigenous people of the Lesser Antilles in the Caribbean.

Revolutionizing Tropical Medicine: Point-of-Care Tests, New Imaging Technologies and Digital Health,
First Edition. Edited by Kerry Atkinson and David Mabey.
© 2019 John Wiley & Sons, Inc. Published 2019 by John Wiley & Sons, Inc.

major campaign in 1964, the burden of yaws had been significantly reduced to approximately 2.5 million cases (Asiedu et al. 2008). Following this initial success control efforts and surveillance dropped down the public health agenda internationally and domestically in many countries. In the 1970s and 1980s there was a resurgence of cases in some countries in West and Central Africa (Hervé et al. 1992; Meheus and Antal 1992). This led to a renewal of control efforts, which again reduced the burden of the disease but did not eradicate it. Following the failure of these previous eradication efforts a variety of factors have contributed to the disease re-emerging as an important public health problem in West Africa, South East Asia and the Pacific over the last several years (Asiedu et al. 2008).

More recently a single oral dose of azithromycin has been shown to be highly effective in treating early disease (Mitjà et al. 2012). Access to an easy to deliver oral treatment has prompted renewed calls for yaws eradication and in 2012 the WHO launched a new strategy based on mass drug administration (MDA) with azithromycin. A major barrier to scaling up interventions has been the lack of adequate epidemiological data on the current distribution and number of cases of yaws worldwide (Mitjà et al. 2015). The need for improved data to guide yaws eradication activities has also prompted renewed interest in the development and validation of diagnostic tests for yaws that can be used both in routine clinical care but also in programmatic activities such as mapping and post-MDA surveillance.

13.2 Epidemiology

Yaws is found in warm and humid environments (Hackett 1953). The disease primarily affects rural communities with low standards of hygiene, with the incidence declining as social and economic status rise. Yaws is currently known to be endemic in at least 12 countries worldwide. The majority of cases are reported from West and Central Africa and the Pacific. At least 76 countries previously reported yaws but do not currently have adequate data on the presence or absence of the disease. Two countries have reported interrupting transmission of which only India has been certified by WHO as having officially interrupted transmission (Figure 13.1).

The disease most commonly affects young children with the majority of cases reported in individuals aged between 2 and 15 years old. The disease is spread by direct skin to skin, non-sexual, contact, often after a cut or abrasion in the lower legs (Perine et al. 1984). Treponemal infections closely related to yaws and syphilis have been identified in primates, but there is no evidence to suggest zoonotic transmission between humans and non-human primates occurs (Harper et al. 2012; Klegarth et al. 2017). Unlike syphilis there is no evidence that mother-to-child transmission of yaws occurs.

The early lesions of yaws are the most infectious and it is believed that individuals are infectious for a period up to 12–18 months following primary infection (Perine et al. 1984) although disease relapse can extend this period. By comparison the destructive lesions of late yaws are not infectious. The disease is highly focal and pockets of transmission are well recognized. Studies in both the Solomon Islands and Papua New Guinea have demonstrated that village level endemicity is a major risk factor for infection and re-infection following treatment (Marks et al. 2015c; Mitjà et al. 2011a).

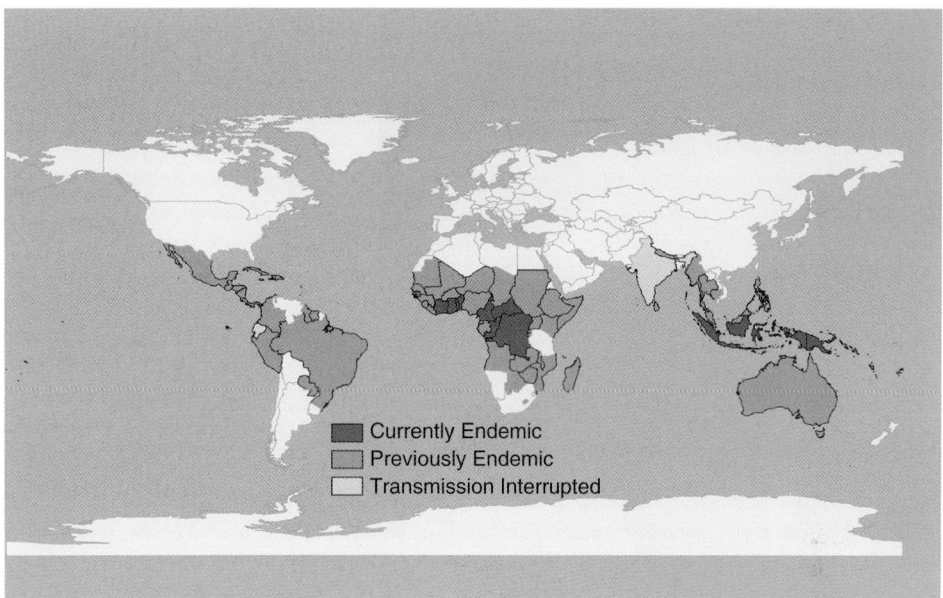

Figure 13.1 Current distribution of yaws worldwide. *Source:* Reproduced with permission from Marks et al. (2015b). (*See color plate section for the color representation of this figure.*)

13.3 Clinical Features

As with other treponemal diseases, the clinical features of yaws may be conveniently divided into primary, secondary and tertiary disease (Marks et al. 2015d; Mitjà et al. 2013; Perine et al. 1984). Although this classification is clinically useful in clinical practice patients may present with a mixture of clinical signs.

13.3.1 Primary Yaws

The initial lesion of primary yaws is a papule appearing at the site of inoculation after about 21 days (range 9–90 days) (Perine et al. 1984). This initial lesion may evolve either into an exudative papilloma, 2–5 cm in size, or degenerate to form a single, crusted non-tender ulcer (Figure 13.2).

The lower limbs are the commonest site for primary yaws lesions but other parts of the body may all be affected. In the absence of treatment lesions heal spontaneously over three to six months leaving a pigmented scar (Sehgal 1990).

13.3.2 Secondary Yaws

After a period of one to two months, hematogenous and lymphatic spread of treponemes may result in progression to secondary yaws, which predominantly affects the skin and bones (Mitjà et al. 2011b, 2011c), often with general malaise and lymphadenopathy. A wide range of skin manifestations has been described in secondary yaws including disseminated papillomatous or ulcerative lesions, scaly macular lesions or hyperkeratotic lesions on the palms and soles (Mitjà et al. 2011c; Perine et al. 1984).

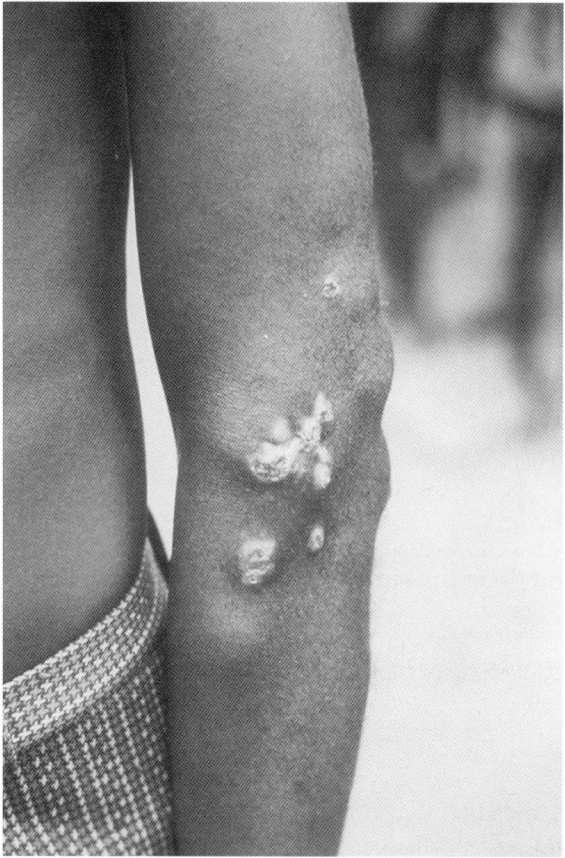

Figure 13.2 Juxta-articular nodules on the elbow resulting from a *Treponema pallidum* ssp. *pertenue* infection. *Source:* This image, reproduced here from Wikimedia Commons, is in the public domain and was obtained from the CDC Public Health Image Library. Image credit: CDC/Dr. Peter Perine (PHIL #3842), 1979. (http://phil.cdc.gov/phil/home.asp Public Health Image Library). Image credit: CDC/Dr. Peter Perine (PHIL #3842).

Involvement of the bones is also a major feature of secondary yaws manifesting classically as osteoperiostitis, involving the fingers (resulting in dactylitis) or long bones (forearm, fibula, tibia) which results in bony swelling and pain (Mitjà et al. 2011b). In most patients multiple bones are affected. In a study from Papua New Guinea (Mitjà et al. 2011c) 75% of children with secondary yaws had joint pain. Following treatment of primary or secondary yaws, skin lesions usually resolve within two to four weeks and bone pain may begin to resolve in as little as 48 hours (Marks et al. 2014c; Mitjà et al. 2013).

As in all treponemal infections, untreated patients may develop latent infection, with positive serology but no clinical signs. Latent cases can relapse, usually in the first five years (rarely up to 10 years) after infection (Perine et al. 1984). Relapsing lesions tend to occur around the axillae, anus and mouth. Diagnosis in these individuals relies entirely on serological testing (see below).

13.3.3   Tertiary Yaws

The destructive lesions of tertiary yaws were previously reported to occur in up to 10% of untreated patients but are now rarely seen. As in other stages of the disease, the skin is most commonly affected. Nodular lesions may occur near joints and ulcerate, causing tissue necrosis (Hackett 1953). Destructive lesions of the face were one of the most marked manifestations of late stage yaws. *Gangosa*, a destructive osteitis of the palate and nasopharynx results in mutilating facial ulceration. *Goundou*, which was rarely reported even when yaws was hyperendemic, is characterized by exostoses of the maxillary bones (Mitjà et al. 2013).

13.4   Diagnostic Quandaries

The differential diagnosis of tropical ulcerative lesions consistent clinically with yaws is broad (Lupi et al. 2006). The primary lesions of yaws may be mistaken for cutaneous leishmaniasis, tropical ulcer caused by fusobacteria and Treponema vincentii, or pyoderma. A particularly important differential diagnosis is infection with *Haemophilus ducreyi*, the causative organism of the sexually transmitted disease chancroid, which is now recognized to be an extremely common cause of non-genital skin lesions in a number of countries where yaws is endemic (Ghinai et al. 2015; Marks et al. 2014a; Mitjà et al. 2014). Clinical differentiation of these diseases is challenging even for experts and highlights the need for access to appropriate diagnostic tests both to guide clinical management and improve the quality of yaws surveillance and reporting data.

13.5   Diagnostic Tests for Yaws

T. pallidum is not viable ex vivo, which has limited the value of direct diagnostic methods used for other bacterial infections. The classic diagnostic test was dark-field microscopy to allow visualization of spirochetes (Perine et al. 1984). This technique has a number of limitations; it can only be applied to active infectious lesions, cannot be used to diagnose latent infection, and requires a high degree of technical skill and access to equipment that are not available in most lower income settings.

13.5.1   Serology

As with venereal syphilis, serological assays remain the mainstay of laboratory diagnosis (Larsen et al. 1995). Serological diagnosis of yaws requires detection of two distinct sets of antibodies: one against a treponemal antigen and one against a non-treponemal antigen. Treponemal antibody tests include ELISAs as well as the more traditional *T. pallidum* particle agglutination (TPPA) and hemagglutination (TPHA) assays. These tests are highly specific for infection with *T. pallidum* but once positive remain so for life regardless of whether the individual has received curative treatment or not.

The so called "non-treponemal" antibody tests include the venereal disease research laboratory (VDRL) and rapid plasma reagin (RPR) tests. The tests use an antigen of cardiolipin, lecithin and cholesterol. Patient-derived antibodies produced against lipid in the cell surface of *T. pallidum* react with antigen to cause visible flocculation. The VDRL is read microscopically whereas the RPR can be read with the naked eye. Titers of non-treponemal assays rise initially during infection and then fall following success-ful treatment. A fourfold fall in non-treponemal antibody titer by six months is consid-ered evidence of cure and titers may become zero, especially after treatment of early infection (Romanowski et al. 1991). RPR titers are generally higher in primary than sec-ondary yaws (Perine et al. 1984).

Titers may be slower to fall in those individuals with a lower baseline RPR/VDRL titer and some individuals may remain serofast despite curative treatment. (Serofast is defined as a <4-fold decline in non-treponemal antibody titers at 6–12 months or as persistently low titers after treatment.) Currently no diagnostic tests can accurately dif-ferentiate the serofast state from ongoing infection in individuals with persistently ele-vated non-treponemal antibody titers.

A further challenge is that the pathogenic treponemes are serologically indistinguish-able from each other. A positive diagnostic test may therefore represent infection with yaws, syphilis or one of the other endemic treponematoses. Clinical and epidemiologi-cal factors may assist the clinician in making a more accurate diagnosis but if there is any doubt the individual should be treated as if they have syphilis.

13.6   Rapid Diagnostic Tests for Yaws

Whilst current serological assays do not require sophisticated equipment, they do require access to laboratory facilities, which are rarely available to the remote commu-nities where yaws is endemic. As yaws and syphilis are serologically indistinguishable there has been interest in using tests originally developed for syphilis as a test for yaws. A large number of rapid diagnostic tests (RDTs) have been developed and evaluated for syphilis many of which meet the ASSURED criteria for diagnostic tests (Marks and Mabey 2017). (The World Health Organization Sexually Transmitted Diseases Diagnostics Initiative [SDI] developed the ASSURED criteria as a benchmark to decide if tests address disease control needs: Affordable, Sensitive, Specific, User-friendly, Rapid and robust, Equipment-free and Deliverable to end-users.)

The majority of tests developed for syphilis are point-of-care tests which can be per-formed on a finger-prick capillary blood sample. Most rapid syphilis tests are designed to detect treponemal antibodies only and therefore a positive test indicates exposure to one of the human treponematoses. These tests cannot differentiate infection with yaws from infection with syphilis and like other treponemal antibody tests they will remain positive for life even after successful treatment. Although no rapid syphilis test has been formally evaluated for use in diagnosing yaws it is anticipated that the overall performance characteristics would be similar to their performance when diagnosing syphilis.

13.6.1   The DPP Syphilis Screen and Confirm Kit

Currently a single commercially available RDT has been formally evaluated for yaws. Unlike most other syphilis RDTs, the DPP (dual path platform) Syphilis Screen and Confirm Kit (Chembio Diagnostic System Inc., New York, USA) is able to detect both treponemal and non-treponemal antibodies. This allows the kit to differentiate previous exposure from current infection. The test has been formally evaluated in several countries (Ayove et al. 2014; Marks et al. 2014b) and a meta-analysis of the test performance has also been conducted (Marks et al. 2016).

Overall the sensitivity of the DPP RDT is 83.1% for the treponemal component and 77.7% for the non-treponemal component with a specificity of 96.4 and 94.5% for the two components respectively. The performance of the test is significantly affected by disease titer (Table 13.1) with a sensitivity of greater than 95% in patients with an RPR titer of ≥1 : 16. The majority of cases of active infectious yaws have titers of ≥1 : 16 and the performance of the test is therefore good in this group of patients. Individuals with latent infection and those who are serofast following successful treatment frequently have lower titers and might be missed using the RDT.

13.6.2 Luminex

Assays have also been developed which allow detection of treponemal and non-treponemal antibodies from a capillary blood sample collected on to filter paper (a dried blood spot [DBS]). These assays have reasonable sensitivity and specificity and may be multiplexed with multiple other diagnostic tests from a single DBS (Cooley et al. 2016). However, they are not suitable for routine clinical care and their use is likely to be limited to large scale surveillance activities.

13.7 Molecular Assays

In high-income settings nucleic acid amplification tests (NAATs) have largely replaced dark field microscopy for the detection of *T. pallidum* from infectious lesions of treponemal diseases. There is a potential role for these tests in the management of patients with suspected yaws as clinical differentiation of yaws from other common skin diseases is challenging. Whilst a point-of-care serological assay may help, atypical skin lesions in the presence of reactive serology may represent either infectious yaws or latent yaws with an alternative cause for the current skin lesion. As such molecular assays might help identify patients who definitively do or do not have yaws and guide appropriate treatment and onward referral as needed.

T. pallidum ssp. *pallidum* and *T. pallidum* ssp. *pertenue* have extremely high DNA sequence homology (Cejková et al. 2012) making differentiation based on standard targets of molecular assays difficult. Despite this limitation a number of molecular assays have been developed for the diagnosis of yaws, which target variable areas of the *T. pallidum* genome. The currently validated assays include both standard and real-time multiplex PCR assays (Chi et al. 2015; Mitjà et al. 2014). These assays can be used to detect the DNA of *T. pallidum* ssp. *pertenue* from the infectious lesions of primary and secondary yaws. These tests have no role in diagnosing individuals with tertiary yaws. Attempts have been made to utilize molecular assays in individuals with reactive serology but no clinical signs of disease to differentiate latent infection from individuals with a serofast state following treatment but currently no assay is capable of this (Marks et al. 2015a). As well as primary diagnosis, molecular assays have also been developed that detect the point mutations in the 23s rRNA gene of *T. pallidum* which are associated with azithromycin resistance (Chen et al. 2013). Unfortunately access to these diagnostic tests remains extremely limited even in high income settings and they are frequently

Table 13.1 Performance of the DPP Screen and Confirm Test for the serological diagnosis of yaws.

Disease	Number	Reference Test Positive	DPP Positive	Sensitivity (95% CI)	Reference Test Negative	DPP Negative	Specificity (95% CI)
Overall							
Treponemal Test	1611	716	596	83.1% (80.3–85.9%)	895	863	96.4% (95.0–97.5%)
Non-Treponemal Test	1611	596	463	77.7% (74.1–81.0%)	1015	964	94.5% (93.4–96.2%)
RPR <1:16							
Treponemal Test	1316	426	313	73.5% (69.0–77.6%)	890	859	96.5% (95.1–97.6%)
Non-Treponemal Test	1316	301	178	59.1% (53.3–64.7%)	1015	964	95.0% (93.4–96.2%)
RPR ≥1:16							
Treponemal Test	295	290	283	97.6% (95.1–99.0%)	5	4	80.0% (28.4–99.5%)
Non-Treponemal Test	295	295	285	96.6% (93.8–98.4%)			

Source: Adapted with permission from Marks et al. (2016).

available only at reference laboratories. Outside the setting of research studies there is extremely limited access to these tests in low income settings where yaws is endemic.

Loop mediated isothermal amplification is an alternative NAAT which allows detection of DNA in the field without the requirement for sophisticated laboratory equipment. LAMP assays for the detection of *T. pallidum* and *H. ducreyi* (a common differential diagnosis, see above) are in development but no assay has yet been adequately validated or made commercially available.

References

Asiedu, K., Amouzou, B., Dhariwal, A. et al. (2008). Yaws eradication: past efforts and future perspectives. *Bull. World Health Organ.* 86: 499–499A.

Ayove, T., Houniei, W., Wangnapi, R. et al. (2014). Sensitivity and specificity of a rapid point-of-care test for active yaws: a comparative study. *Lancet Glob. Health* 2: e415–e421. https://doi.org/10.1016/S2214-109X(14)70231-1.

Cejková, D., Zobaníková, M., Chen, L. et al. (2012). Whole genome sequences of three *Treponema pallidum* ssp. *pertenue* strains: yaws and syphilis treponemes differ in less than 0.2% of the genome sequence. *PLoS Negl. Trop. Dis.* 6: e1471. https://doi.org/10.1371/journal.pntd.0001471.

Chen, C.-Y., Chi, K.-H., Pillay, A. et al. (2013). Detection of the A2058G and A2059G 23S rRNA gene point mutations associated with azithromycin resistance in *Treponema pallidum* by use of a TaqMan real-time multiplex PCR assay. *J. Clin. Microbiol.* 51: 908–913. https://doi.org/10.1128/JCM.02770-12.

Chi, K.-H., Danavall, D., Taleo, F. et al. (2015). Molecular differentiation of *Treponema pallidum* subspecies in skin ulceration clinically suspected as yaws in Vanuatu using real-time multiplex PCR and serological methods. *Am. J. Trop. Med. Hyg.* 92: 134–138. https://doi.org/10.4269/ajtmh.14-0459.

Cooley, G.M., Mitja, O., Goodhew, B. et al. (2016). Evaluation of multiplex-based antibody testing for use in large-scale surveillance for yaws: a comparative study. *J. Clin. Microbiol.* 54: 1321–1325. https://doi.org/10.1128/JCM.02572-15.

Ghinai, R., El-Duah, P., Chi, K.-H. et al. (2015). A cross-sectional study of "yaws" in districts of Ghana which have previously undertaken azithromycin mass drug administration for trachoma control. *PLoS Negl. Trop. Dis.* 9: e0003496. https://doi.org/10.1371/journal.pntd.0003496.

Hackett, C.J. (1953). Extent and nature of the yaws problem in Africa. *Bull. World Health Organ.* 8: 127–182.

Harper, K.N., Fyumagwa, R.D., Hoare, R. et al. (2012). *Treponema pallidum* infection in the wild baboons of East Africa: distribution and genetic characterization of the strains responsible. *PLoS One* 7: e50882. https://doi.org/10.1371/journal.pone.0050882.

Hervé, V., Kassa Kelembho, E., Normand, P. et al. (1992). Resurgence of yaws in Central African Republic. Role of the pygmy population as a reservoir of the virus. *Bull. Soc. Pathol. Exot.* 85: 342–346.

Klegarth, A.R., Ezeonwu, C.A., Rompis, A. et al. (2017). Survey of Treponemal infections in free-ranging and captive macaques, 1999-2012. *Emerg. Infect. Dis.* 23: 816–819. https://doi.org/10.3201/eid2305.161838.

Larsen, S.A., Steiner, B.M., and Rudolph, A.H. (1995). Laboratory diagnosis and interpretation of tests for syphilis. *Clin. Microbiol. Rev.* 8: 1–21.

Lupi, O., Madkan, V., and Tyring, S.K. (2006). Tropical dermatology: bacterial tropical diseases. *J. Am. Acad. Dermatol.* 54: 559. 578-580. doi:https://doi.org/10.1016/j.jaad.2005.03.066.

Marks, M., Chi, K.-H., Vahi, V. et al. (2014a). *Haemophilus ducreyi* associated with skin ulcers among children, Solomon Islands. *Emerg. Infect. Dis.* 20: 1705–1707. https://doi.org/10.3201/eid2010.140573.

Marks, M., Goncalves, A., Vahi, V. et al. (2014b). Evaluation of a rapid diagnostic test for yaws infection in a community surveillance setting. *PLoS Negl. Trop. Dis.* 8: e3156. https://doi.org/10.1371/journal.pntd.0003156.

Marks, M., Solomon, A.W., and Mabey, D.C. (2014c). Endemic treponemal diseases. *Trans. R. Soc. Trop. Med. Hyg.* 108: 601–607. https://doi.org/10.1093/trstmh/tru128.

Marks, M., Katz, S., Chi, K.-H. et al. (2015a). Failure of PCR to detect *Treponema pallidum* ssp. *pertenue* DNA in blood in latent yaws. Failure of PCR to detect *Treponema pallidum* ssp. *pertenue* DNA in blood in latent yaws. *PLoS Negl. Trop. Dis.* 9: e0003905. https://doi.org/10.1371/journal.pntd.0003905.

Marks, M., Mitjà, O., Solomon, A.W. et al. (2015b). Yaws. *Br. Med. Bull.* 113: 91–100. https://doi.org/10.1093/bmb/ldu037.

Marks, M., Vahi, V., Sokana, O. et al. (2015c). Mapping the epidemiology of yaws in the Solomon Islands: a cluster randomized survey. *Am. J. Trop. Med. Hyg.* 92: 129–133. https://doi.org/10.4269/ajtmh.14-0438.

Marks, M., Lebari, D., Solomon, A.W., and Higgins, S.P. (2015d). Yaws. *Int. J. STD AIDS* 26: 696–703. https://doi.org/10.1177/0956462414549036.

Marks, M., Yin, Y.-P., Chen, X.-S. et al. (2016). Metaanalysis of the performance of a combined treponemal and nontreponemal rapid diagnostic test for syphilis and yaws. *Clin. Infect. Dis.* 63: 627–633. https://doi.org/10.1093/cid/ciw348.

Marks, M. and Mabey, D.C. (2017). The introduction of syphilis point of care tests in resource limited settings. *Expert. Rev. Mol. Diagn.* 17: 321–325. https://doi.org/10.1080/14737159.2017.1303379.

Meheus, A. and Antal, G.M. (1992). The endemic treponematoses: not yet eradicated. *World Health Stat. Q. Rapp. Trimest. Stat. Sanit. Mond.* 45: 228–237.

Mitjà, O., Hays, R., Ipai, A. et al. (2011a). Outcome predictors in treatment of yaws. *Emerg. Infect. Dis.* 17: 1083–1085. https://doi.org/10.3201/eid/1706.101575.

Mitjà, O., Hays, R., Ipai, A. et al. (2011b). Osteoperiostitis in early yaws: case series and literature review. *Clin. Infect. Dis.* 52: 771–774. https://doi.org/10.1093/cid/ciq246.

Mitjà, O., Hays, R., Lelngei, F. et al. (2011c). Challenges in recognition and diagnosis of yaws in children in Papua New Guinea. *Am. J. Trop. Med. Hyg.* 85: 113–116. https://doi.org/10.4269/ajtmh.2011.11-0062.

Mitjà, O., Hays, R., Ipai, A. et al. (2012). Single-dose azithromycin versus benzathine benzylpenicillin for treatment of yaws in children in Papua New Guinea: an open-label, non-inferiority, randomised trial. *Lancet* 379: 342–347. https://doi.org/10.1016/S0140-6736(11)61624-3.

Mitjà, O., Asiedu, K., and Mabey, D. (2013). Yaws. *Lancet* 381: 763–773. https://doi.org/10.1016/S0140-6736(12)62130-8.

Mitjà, O., Lukehart, S.A., Pokowas, G. et al. (2014). *Haemophilus ducreyi* as a cause of skin ulcers in children from a yaws-endemic area of Papua New Guinea: a prospective cohort study. *Lancet Glob. Health* 2: e235–e241. https://doi.org/10.1016/S2214-109X(14)70019-1.

Mitjà, O., Marks, M., Konan, D.J.P. et al. (2015). Global epidemiology of yaws: a systematic review. *Lancet Glob. Health* 3: e324–e331. https://doi.org/10.1016/S2214-109X(15)00011-X.

Perine, P.L., Hopkins, D.R., Niemel, P.L.A. et al. (1984). *Handbook of Endemic Treponematoses : Yaws, Endemic Syphilis and Pinta*. Geneva, Switzerland: World Health Organization.

Romanowski, B., Sutherland, R., Fick, G.H. et al. (1991). Serologic response to treatment of infectious syphilis. *Ann. Intern. Med.* 114: 1005–1009. https://doi.org/10.7326/0003-4819-114-12-1005.

Sehgal, V.N. (1990). Leg ulcers caused by yaws and endemic syphilis. *Clin. Dermatol.* 8: 166–174.

14

Rapid Diagnostic Tests for the Detection of Sickling Hemoglobin

Amina Nardo-Marino[1] and Tom N. Williams[2,3]

[1] Department of Haematology, Herlev and Gentofte Hospital, University of Copenhagen, Herlev, Denmark
[2] The KEMRI/Wellcome Trust Research Programme, Kilifi, Kenya
[3] Imperial College London, St Mary's Hospital, London, UK

CHAPTER MENU

14.1   Sickle Cell Disease

Sickle hemoglobin (hemoglobin S, HbS) is a structural variant of normal adult hemoglobin (HbA). HbS is the result of a point mutation in the hemoglobin subunit beta (HBB) gene located on chromosome 11 that results in the substitution of valine for glutamic acid at position 6 of the β-globin subunit (βS) of the hemoglobin molecule. Sickle cell disease (SCD) refers to any condition in which HbS is the predominant hemoglobin within the red blood cells (RBCs).

HbS polymerizes when oxygen tensions are low, leading to deformation of the affected RBCs, so-called "sickling." The rigid sickled RBCs are less deformable in the microcirculation, causing microvascular occlusion and hemolysis. SCD is characterized by recurrent episodes of acute illness, chronic anemia and organ damage, as well as a significant reduction in life expectancy. The most common and severe form of SCD results from homozygosity for the sickle mutation (HbSS) and is typically referred to as sickle cell anemia (SCA). Carriers of only one sickle allele (HbAS) are often referred to as having sickle cell trait (SCT) and typically have no symptoms. They do, however, have

Revolutionizing Tropical Medicine: Point-of-Care Tests, New Imaging Technologies and Digital Health,
First Edition. Edited by Kerry Atkinson and David Mabey.

increased resistance to *Plasmodium falciparum* infections, explaining why the geographic distribution of HbS coincides with that of malaria (Piel et al. 2010).

Other forms of SCD occur when HbS is inherited heterozygously alongside another hemoglobin mutation. Examples of common compound heterozygous forms of SCD include: HbS/β^0-thalassaemia, HbS/β^+-thalassaemia, and HbSC-disease (Piel et al. 2017). (Mutated alleles in people with thalassemia are called β^+ when partial Hb function is conserved or β^0, when no functioning Hb is produced.)

Another compound form is with the gene for HbC (producing HbSC disease). HbC is due to a different variation in the β-chain gene. Since HbC does not polymerize as readily as HbS, there is less sickling and fewer acute vaso-occlusive events occur in people with HbSC disease. However, persons with HbSC disease have more significant retinopathy, ischemic necrosis of bone, and priapism than those with homozygous HbSS disease.

In countries with the highest predicted rates of disease, there are currently no global data outlining the burden of SCD worldwide. It has recently been estimated, however, that over 300 000 children are born with SCA every year, of whom at least 280 000 are born in resource-limited settings. In contrast, less than 1% of all children with SCD are born in Europe and North America (Piel et al. 2013).

In developed countries newborn screening and systematic clinical follow up with prevention of sepsis and organ damage have led to an increased life expectancy for patients with SCD. Nevertheless, the majority of children with SCD are still born in resource-limited settings and, due to poor diagnostic facilities and low priority in health policy plans, most of these children continue to die undiagnosed from potentially preventable complications within the first five years of life (Grosse et al. 2011).

14.2 Diagnosing Sickle Cell Disease

A number of methods for diagnosing SCD are available, the most common being hemoglobin electrophoresis, high performance liquid chromatography (HPLC), isoelectric focusing and molecular approaches such as polymerase chain reaction (PCR) (Ryan et al. 2010). All these methods require comprehensive laboratory facilities as well as trained staff, well-maintained equipment, and electricity. Commercial costs typically run between $5–10 per test (Williams 2015). SCD testing is therefore often limited, with simple blood films and so called "sickling tests" and sickle solubility methods still being used as primary diagnostic modalities in many resource-limited settings.

14.2.1 The Blood Film

The blood film from a patient with SCD is often normal at birth and in the early neonatal period, as the percentage of hemoglobin F (HbF) is still high and HbS levels are low. Abnormalities are usually detectable from approximately six months of age. The blood film from an adult with SCD will show a variable number of crescent or sickle-shaped cells, representing irreversibly sickled cells that have not regained their normal shape on exposure to atmospheric oxygen. The number of sickled cells is variable, ranging from only occasional cells to 30–40% of all RBCs. Other poikilocytes, such as elongated cells pointed at one or both ends, may also be seen, in addition to features

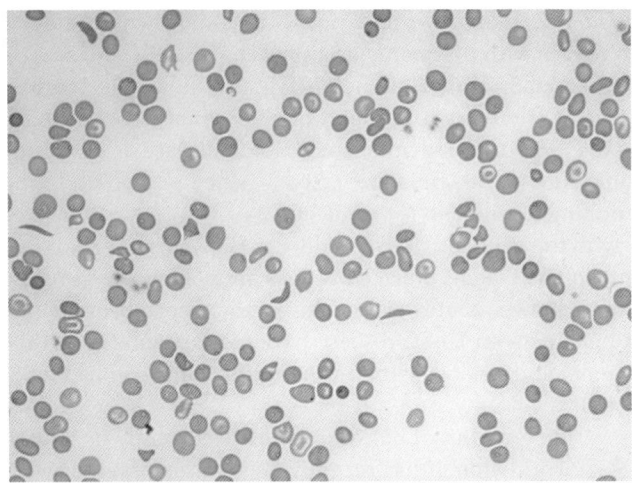

Figure 14.1 Blood film from a patient with SCA showing characteristically sickled RBCs. *Source:* Photo by Peter Nørgaard.

of hyposplenism - specifically Howell-Jolly[1,2] bodies and target cells (Bain 2006). Figure 14.1 shows a blood film from a patient with SCD.

14.2.2 The Sodium Metabisulfite Sickling Test

The "sickling test" or "sickle cell test" is a simple and rapid method of producing sickling of RBCs in wet cover slip preparations of blood from patients with suspected SCD. In order to perform the test, a microscope, microscopic slides and coverslips, and a chemical reducing agent are required (Hicks et al. 1973).

The test is based on the principle that the degree of sickling of abnormal RBCs correlates with the concentration of deoxygenated HbS. By adding a chemical reducing agent to the patient's blood, the production of reduced hemoglobin in the RBCs increases. The reducing agent most commonly used is sodium metabisulfite ($Na_2O_5S_2$), which can be prepared locally at a low cost. Sodium metabisulfite must be freshly made and used in a concentration that does not exceed 2% (Daland and Castle 1948).

The test is performed by adding a drop of 2% sodium metabisulfite to a small drop of blood on a glass microscope slide. Capillary blood, taken directly from the patient's ear or fingertip, or venous blood in any anticoagulant can be used. After mixing, a cover slip is dropped on the preparation. Sickling typically occurs within 15 minutes at room temperature. The result is assessed by direct light microscopic observation. In positive samples the typical sickle-shaped RBCs will appear (Greenberg et al. 1972). Figure 14.2 illustrates a positive sodium metabisulfite sickling test. The sodium metabisulfite sickling test identifies the presence of HbS but does not distinguish between SCT and SCD. The test cannot be used in the neonatal period, as low levels of HbS may lead to false negative results (Clark 1972). Furthermore, the test may be unreliable if blood samples are stored too long, exposed to heat or bacterial contamination, or allowed to dry during testing (Schneider et al. 1967).

---

[1] William Howell (1860–1945), American physiologist and physician.
[2] Justin Marie Jolly (1870–1953), French hematologist and histopathogist.

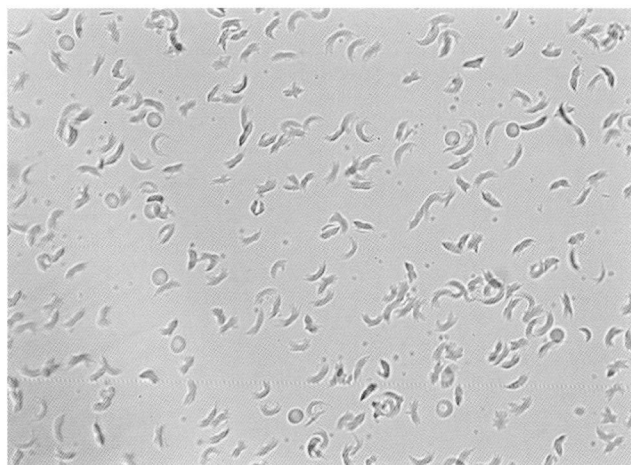

Figure 14.2 Positive sodium metabisulfite sickling test under direct light microscopy. *Source:* Photo by Peter Nørgaard.

14.2.3 The Sickle Solubility Test (SICKLEDEX)

Sickle solubility testing methods are simple screening tests for detecting HbS when present in sufficient concentrations. Several commercial kits using sodium hydrosulfite ($Na_2O_4S_2$) are available. Among these is the widely used SICKLEDEX (Streck, Inc., Omaha, NE, USA), as first proposed by Diggs and colleagues in 1968 (Diggs et al. 1968).

Sickle solubility kits are relatively quick and easy to use. The reagents used are a phosphate buffer solution and reagent powder, containing saponin and sodium hydrosulfite. Buffer and reagent powder are mixed in a test tube and the test is completed by adding fresh or anticoagulated whole blood to the solution. The basic principle of the test is to lyse the RBCs using saponin, thereby releasing hemoglobin into the test tube solution. Hemoglobin is then reduced, using sodium hydrosulfite. Reduced HbS precipitates and becomes insoluble when exposed to the concentrated phosphate buffer, whereas HbA (normal Hb) will remain in solution. Interpretation is based on a visual inspection of the test tube; a cloudy, turbid suspension indicates the presence of HbS, whereas a negative test is revealed by a clear, transparent pink solution. A positive result should occur within 15 minutes (Canning and Huntsman 1970).

False negative results of sickle solubility tests have been reported in patients with severe anemia, patients with an HbS fraction under 10%, and in patients with high HbF levels. False positive results can occur in patients suffering from conditions associated with increased serum viscosity (for example, polycythemia and multiple myeloma) and in some variant hemoglobins such as hemoglobin I, hemoglobin Bart's and hemoglobin Jamaica-Plain) (Hicks et al. 1973).

SICKLEDEX identifies HbS with both high sensitivity and specificity. The test is limited, however, by the fact that a positive result simply indicates the presence of HbS and does not distinguish between heterozygous and homozygous states. Furthermore, the test cannot identify other hemoglobin variants. SICKLEDEX, as well as the sodium metabisulfite sickling test, should therefore only be used to screen for HbS. All positive or equivocal results require further evaluation by an alternative technique, such as hemoglobin electrophoresis or HPLC. Table 14.1 summarizes the principles of SICKLEDEX and the sodium metabisulfite sickling test.

Table 14.1 Available point-of-care diagnostic methods for identifying HbS.

Methods	Principle	Sensitivity for HbS	Specificity for HbS	Cost ($)	References
Sickling test *Slide test*	Qualitative test revealing sickled RBCs on a microscope slide after reducing HbS using sodium metabisulfite. A light microscope is required to read the result.	97.3%	99.6%	0.33	Daland and Castle (1948) Schneider et al. (1967) Hicks et al. (1973)
SICKLEDEX *Solubility test*	Qualitative test in which RBCs are first lysed using saponin. HbS is then reduced using sodium hydrosulfite, causing it to become insoluble when exposed to a concentrated phosphate buffer. A cloudy, turbid suspension indicates the presence of HbS.	99%	99.9%	0.36	Canning and Huntsman (1970) Hicks et al. 1973 Tubman et al. (2015)

14.2.4   Novel Point-of-Care Tests

In recent years a number of new point-of-care (POC) diagnostic tests for SCD have been developed. Among these are an optimized solubility testing method using a paper based assay (Yang et al. 2013; Piety et al. 2017) and a cell density-based aqueous multiphase system (Kumar et al. 2014). The most promising test, however, is a lateral flow immunoassay device marketed as Sickle SCAN® (BioMedomics, Inc., Durham, NC, USA).

The Sickle SCAN device is a rapid, qualitative lateral flow immunoassay kit that can be used to diagnose the most common forms of SCD (HbSS and HbSC), as well as SCT (HbAS), using a capillary blood sample. The test uses a sandwich format chromatographic immunoassay approach for the qualitative measurement of HbA, HbS and HbC. Alpha-globin is used as a positive control. Results are available in less than five minutes and can be detected at the bedside by visual inspection (Kanter et al. 2015). The test requires only 5 μL of blood taken by fingerprick, heelstick, or venipuncture. The blood sample is added to a buffer-loaded tube using a provided capillary sampler. In the tube the RBCs are lysed and hemoglobin released into the solution. A few drops of the hemolysed solution are then added to the sample inlet of the Sickle SCAN cartridge. From here it flows through the test cartridge in order to interact with antibody-conjugated colorimetric detector nanoparticles and travel to the capture zones, identified by lines on the device. Four detection lines, representing HbA, HbS, HbC and a control line, are possible. Blood samples containing two hemoglobin variants will have both hemoglobin variants detected (e.g. HbSC) (Kanter et al. 2015).

On evaluation, Sickle SCAN has performed very well. In one laboratory-based study of 71 patient samples, the test identified the presence of HbA, HbSS, and HbC with an overall diagnostic accuracy of 99% when compared to genotypes identified by gold standard diagnostic testing (hemoglobin electrophoresis or HPLC) (Kanter et al. 2015).

In a more recent study 139 blood samples were tested with Sickle SCAN and compared to results using capillary zone electrophoresis (McGann et al. 2016). The test was found to be simple and rapid with high sensitivity and specificity for the detection of HbA, HbS, and HbC, accurately detecting HbS and HbC at concentrations as low as 1–2%. The test demonstrated 98.4% sensitivity and 98.6% specificity for the diagnosis of HbSS disease and 100% sensitivity and specificity for the diagnosis of HbSC disease. Furthermore, most variant hemoglobins, including high concentrations of HbF, were found not to interfere with the test's ability to correctly detect HbS or HbC, suggesting that it may also be suitable for neonatal screening. Further studies are needed to confirm the accuracy of Sickle SCAN under real-life conditions in low resource communities.

14.3   Conclusions

SCD is a neglected, chronic disease of growing global health importance. Early diagnosis of SCD is necessary in order to allow for parental education and the initiation of preventive treatments. Despite their strong limitations as diagnostic tools, the sodium metabisulfite sickling test and SICKLEDEX are still widely used to screen for SCD in many resource-limited settings. This is in large part due to a lack of alternative high quality, and yet affordable, diagnostic techniques. A cheap and reliable POC diagnostic test, such as Sickle SCAN, could facilitate screening for SCD in lower income countries, with a combination approach potentially being most feasible in settings where resources are constrained. By screening for the presence of HbS with SICKLEDEX and further evaluating all positive results with Sickle SCAN, a rapid and accurate diagnosis could be obtained at an affordable price.

Bibliography

Bain, B.B. (2006). *Hemoglobinopathy Diagnosis*, 2e. Hoboken, New Jersey: Blackwell Publishing.

Canning, D.M. and Huntsman, R.G. (1970). An assessment of Sickledex as an alternative to the sickling test. *J. Clin. Pathol.* 23 (8): 736–737.

Clark, K.G. (1972). An improved solubility test for hemoglobin S. *J. Clin. Pathol.* 25 (8): 730–731.

Daland, G.A. and Castle, W.B. (1948). A simple and rapid method for demonstrating sickling of the red blood cells; the use of reducing agents. *J. Lab. Clin. Med.* 33 (9): 1082–1088.

Diggs, L.W., Schorr, J.B., and Ascari, W.Q. (1968). A new diagnostic test for hemoglobin S. Presented as an exhibit before the 22nd joint annual meeting of the American Society of Clinical Pathologists and the College of American Pathologists, Miami Beach, Fl.

Greenberg, M.S., Harvey, H.A., and Morgan, C. (1972). A simple and inexpensive screening test for sickle hemoglobin. *N. Engl. J. Med.* 286 (21): 1143–1144.

Grosse, S.D., Odame, I., Atrash, H.K. et al. (2011). Sickle cell disease in Africa: a neglected cause of early childhood mortality. *Am. J. Prev. Med.* 41 (6 Suppl 4): S398–S405.

Hicks, E.J., Griep, J.A., and Nordschow, C.D. (1973). Comparison of results for three methods of hemoglobin S identification. *Clin. Chem.* 19 (5): 533–535.

Kanter, J., Telen, M.J., Hoppe, C. et al. (2015). Validation of a novel point of care testing device for sickle cell disease. *BMC Med.* 13: 225.

Kumar, A.A., Patton, M.R., Hennek, J.W. et al. (2014). Density-based separation in multiphase systems provides a simple method to identify sickle cell disease. *Proc. Natl. Acad. Sci. U.S.A.* 111 (41): 14864–14869.

McGann, P.T., Schaefer, B.A., Paniagua, M. et al. (2016). Characteristics of a rapid, point-of-care lateral flow immunoassay for the diagnosis of sickle cell disease. *Am. J. Hematol.* 91 (2): 205–210.

Piel, F.B., Hay, S.I., Gupta, S. et al. (2013). Global burden of sickle cell anemia in children under five, 2010-2050: modelling based on demographics, excess mortality, and interventions. *PLoS Med.* 10 (7): e1001484.

Piel, F.B., Patil, A.P., Howes, R.E. et al. (2010). Global distribution of the sickle cell gene and geographical confirmation of the malaria hypothesis. *Nat. Commun.* 1: 104.

Piel, F.B., Steinberg, M.H., and Rees, D.C. (2017). Sickle cell disease. *N. Engl. J. Med.* 376 (16): 1561–1573.

Piety, N.Z., George, A., Serrano, S. et al. (2017). A paper-based test for screening Newborns for sickle cell disease. *Sci. Rep.* 7: 45488.

Ryan, K., Bain, B.J., Worthington, D. et al. (2010). Significant hemoglobinopathies: guidelines for screening and diagnosis. *Br. J. Haematol.* 149 (1): 35–49.

Schneider, R.G., Alperin, J.B., and Lehmann, H. (1967). Sickling tests. Pitfalls in performance and interpretation. *JAMA* 202 (5): 419–421.

Williams, T.N. (2015). An accurate and affordable test for the rapid diagnosis of sickle cell disease could revolutionize the outlook for affected children born in resource-limited settings. *BMC Med.* 13: 238.

Yang, X., Kanter, J., Piety, N.Z. et al. (2013). A simple, rapid, low-cost diagnostic test for sickle cell disease. *Lab Chip* 13 (8): 1464–1467.

15

Progress Toward the Development of Rapid Diagnostic Tests for Lymphatic Filariasis and Onchocerciasis

Roger B. Peck, Dunia Faulx, and Tala de los Santos

PATH (www.path.org)

15.1   Introduction

Lymphatic filariasis (LF), also known as elephantiasis, and onchocerciasis, commonly known as river blindness, are profoundly disfiguring neglected tropical diseases primarily affecting communities in low-resource settings in Africa and Southeast Asia. Both diseases are caused by parasitic worms that use insect vectors as the intermediate host. A clinical overview of these two diseases is shown in Table 15.1.

During a blood meal by the insect vector, parasite larvae are transmitted to humans, the primary host. In the human the larvae mature into adult female and male filariae. The adults breed, producing microfilariae that migrate to tissues, the lymphatic system, and/or the blood system. Microfilariae are then taken up by the insect vector during a blood meal where they develop into larvae capable of being transmitted to a human, thus continuing the transmission cycle.

Both diseases can be controlled and ultimately eliminated through vector control and mass drug administration (MDA). Used in combination, these population-based treatment and surveillance approaches are a feasible strategy when humans are the sole reservoir for a disease and preventing them from becoming infective is key to breaking the cycle of transmission. Control and elimination of LF and onchocerciasis requires engagement over several years with communities at risk and other stakeholders,

Revolutionizing Tropical Medicine: Point-of-Care Tests, New Imaging Technologies and Digital Health,
First Edition. Edited by Kerry Atkinson and David Mabey.

Table 15.1 Clinical overview of lymphatic filariasis and onchocerciasis.

	Lymphatic filariasis	Onchocerciasis
Number infected and at risk	An estimated 120 million people infected. Almost 25 million with hydrocele. Almost 15 million with lymphedema	Approximately 215 million people live in communities at risk
Parasitic worm species	*Wuchereria bancrofti* *Brugia malayi* *Brugia timor*	*Onchocerca volvulus*
Vector	Mosquitoes, primarily genus *Aedes, Anopheles,* and *Culex*	Blackflies, genus *Simulium*
Symptoms	Asymptomatic, Fever Lymphatic damage Kidney damage Hydrocele Chylocele Lymphedema	Asymptomatic Rash Itching Skin nodules Eye lesions leading to blindness
Location of adult filariae infection	Lymphatic system	Subcutaneous nodules
Location of microfilariae infection	Lymphatic system Blood	Skin Soft tissues Lymphatic system blood Urine Sputum
Treatment	Annual mass drug administration with albendazole with ivermectin or diethylcarbamazine citrate; at least 5 rounds are recommended	Annual or semiannual mass drug administration with ivermectin; 10 to 15 rounds of treatment are recommended
Regions affected	Africa Southeast Asia	Africa Latin America Yemen

including national control and elimination programs, the World Health Organization and pharmaceutical companies.

Exciting progress and new partnerships have been made in the effort to rid the world of these debilitating diseases (WHO 2012a). The World Health Organization has developed guidelines for disease control and elimination (WHO 2011, 2016). Pharmaceutical companies have committed to making drugs available to communities at risk of LF and onchocerciasis for as long as it takes to eliminate these diseases as public health problems. Ministries of health, non-government organizations, implementing partners and funders are engaged in coordinated activities toward overcoming the global impact and eventual elimination of LF and onchocerciasis (WHO 2012b).

Diagnostic tools are critical for determining how advanced a country program is in its disease control and elimination efforts and where additional support, either in the form of treatment or technical assistance, would be beneficial. Approaches to diagnostic testing include direct filariae or microfilariae identification, detection of parasite-specific antigens, detection of parasite nucleic acid or indirect detection of parasite exposure by measuring the human antibody response. Diagnostic platforms

Table 15.2 Characteristics of diagnostic test platforms for lymphatic filariasis and onchocerciasis.

Platform	Benefits	Challenges
Microscopy	• Historical reference • Provides direct evidence of infection • Can be conducted at or near the point-of-care • Finger stick for lymphatic filariasis	• Complex logistics • Equipment • Training • Good specificity, poor sensitivity • Timing of specimen collection must be linked to mass drug administration • Challenging sample collection (night blood and skin snips)
Blood filtration (lymphatic filariasis only)	• Provides direct evidence of infection • More sensitive than finger stick microscopy	• Venous blood • Large sample volume • Logistics
Polymerase chain reaction (PCR)	• Highly sensitive, even when microfilariae are not visible • Finger stick; dried blood spot specimen	• Complex • Laboratory-based • Logistics
Enzyme-linked immunosorbent assay (ELISA)	• Finger stick; dried blood spot specimen • Indicative of prior exposure • Timing of specimen collection not linked to mass drug administration	• Equipment • Training • Logistics • Antigen detection cross-reactive with other species • Antibody detection not necessarily indicative of active infection
Rapid strip test	• Easy to use • Well suited for testing at or near the point-of-care	• Potentially lower performance • Antigen detection cross-reactive with other species • Antibody detection not necessarily indicative of active infection

for LF and onchocerciasis include light microscopy, polymerase chain reaction (PCR), enzyme-linked immunosorbent assay (ELISA), and rapid strip tests. Each platform has benefits and challenges (Table 15.2), but rapid strip tests are by far the best suited for point-of-care use in remote settings.

The stages in a control and elimination program are shown in Table 15.3.

Table 15.3 Stages in a control and elimination program.

1. Mapping

   Determining where to treat the level of endemicity; primarily assessing current infection.

2. Monitoring

   Monitoring the impact of mass drug administration

3. Transmission assessment/stopping decision

   Determining when transmission is no longer sustainable and mass drug administration can be stopped

4. Post-treatment surveillance

   Monitoring for disease recrudescence

15.2   The Development of Rapid Diagnostic Tests

Rapid test development, optimization and implementation is dependent on the disease and the end use. Understanding that a test will be used in a community that is actively engaged in a surveillance activity provides context for the development. The test must be robust enough to withstand heat and humidity and ideally it should have a rapid turnaround time to provide a meaningful feedback loop to the community participants.

15.3   Rapid Diagnostic Tests for Lymphatic Filariasis

The relatively quick transition from identifying microfilariae with microscopy to using a rapid test was largely driven by the logistical difficulty in taking night blood samples. There was a definite lack of enthusiasm among communities and surveillance workers for nighttime sampling. Furthermore, the microscopy method did a fairly poor job of detecting very low microfilaria density in the blood. Early versions of rapid tests detected circulating filarial antigen, which were originally detected using laboratory-based methods such as immunoradiometric assay (IMRA) or ELISA. These early tests transitioned to rapid card tests in the mid-1990s (Weil et al. 1997). The transition from nocturnal blood smears to antigen-detection rapid tests significantly reduced the logistical complexities required to run a surveillance campaign for LF. Rapid tests are more easily field-deployable, can be used at any time, and require minimal equipment.

15.3.1   Antigen Detection

Two rapid diagnostic tests are currently available for the detection of LF antigens: the BinaxNOW® Filariasis ICT (immunochromatographic card test) and the Alere™ Filariasis Test Strip (FST). The BinaxNOW Filariasis ICT is a rapid diagnostic test that detects LF using blood, serum, or plasma. It tests for circulating filarial antigens and can provide results in 10 minutes (Weil et al. 2013). This test was the official test used by the World Health Organization's Global Program to Eliminate Lymphatic Filariasis for mapping, determining when to stop MDA, and for post-treatment surveillance. Although the test is effective, if the results are not read within a relatively short time frame, false positives are common. The Alere FST (Alere 2017) is similar to the BinaxNOW Filariasis ICT in that it detects the circulating filarial antigens, but the test is a strip as opposed to a card. In a study comparing the BinaxNOW card test to the Alere test strip in the laboratory and in the field, it was found that the test strip was more sensitive, particularly in the field. The test strip also uses less blood than the card test and has a longer shelf life at ambient temperatures (Weil et al. 2013). Figure 15.1 shows the Alere FST being used in the field.

   Despite technological advances, there are no antigen-detection tests for the *Brugia* species of LF. Both antigen detection tests described above have a high false positive rate in individuals infected with the "eye worm" *Loa loa* (Bakajika et al. 2014). This limits the use of these tests in central Africa where *Loa loa* is endemic. Another limitation of antigen detection is that it is able to detect infections only after the development of adult parasites, which may not happen until a year and a half after an individual becomes infected.

Figure 15.1 The Alere Filariasis Test Strip (FST) being used in the field.

15.3.2 Antibody Detection

Antibody-detection diagnostic tests for LF detect antibodies to the parasite. These tests do not distinguish between current and previous infection and therefore are indicative only of exposure (Haarbrink et al. 1995). The detection of antibodies is relevant for disease mapping in untreated areas. Epidemiological surveillance is relevant once transmission is assumed to have been interrupted and seroprevalence is obtained by age groups. If young people test positive for exposure to LF, it indicates that either transmission has not been interrupted or that infected people have migrated into the community. Antibody tests can thus be used to monitor for recrudescence, enabling an early programmed response to the recrudescence. Detecting antibodies to third-stage larvae would allow for early detection of LF in sentinel populations. This would improve surveillance after MDA has stopped and programs are monitoring for potential recrudescence (Adjobimey and Hoerauf 2010).

Elevated levels of immunoglobulin G4 (IgG4), an antibody that is particularly relevant in filarial diseases, can be indicative of asymptomatic infection (Adjobimey and Hoerauf 2010). There are several options for antibody detection: antibodies to *Brugia malayi* 14 (Bm14) (Weil et al. 2011) *B. malayi* 33 (Bm33) (Krushna et al. 2009) and *Wuchereria bancrofti* SXP-1 (WbSXP-1) (Pandiaraja et al. 2010) are a few that have been detected using different tools including ELISA. The Filaria Detect™ IgG4 ELISA system, developed by InBios International, is a laboratory-based test currently in development that detects Wb123 IgG4 antibodies to LF (InBios 2017). Determining appropriate biomarkers to be used in diagnostic assays requires testing samples from diverse geographic location in various stages of control and elimination in order to validate that the biomarker is appropriate for use. Bm14 and WbSXP-1 were chosen for this test because they tend to elicit strong immune reactions in individuals (Lammie et al. 2004).

The absence of an antigen-detection test for *Brugia* infections is less than ideal. Countries endemic for *Brugia* encouraged the development of a rapid diagnostic test in the mid-2000s, which led to the use of antibody detection earlier than in countries where only *W. bancrofti* occurred (Weil and Ramzy 2007). The *Brugia* rapid antibody-detection test, manufactured by Reszon Diagnostics International, was developed to detect anti-filarial IgG4 antibodies to a recombinant *B. malayi* antigen (BmR1). The test was evaluated in areas where *Brugia timori* occurred and was found to be accurate for both *Brugia* species. Requiring only a small amount of blood, the test provides results in 25 minutes and is used in areas where *Brugia* is found, such as in Southeast Asia.

Recently PATH, a global health nonprofit, developed a rapid test for LF that was commercialized by Standard Diagnostics, a subsidiary of Alere. The SD BIOLINE Lymphatic Filariasis IgG4 rapid test is a qualitative test that detects antibodies to Wb123 antigen. The test requires 30 minutes to run, and the results can be read for up to 24 hours, which is conducive for the logistics of field evaluation.

Rapid test development, optimization and implementation is dependent on the disease and the end use. Understanding that a test will be used outside of a laboratory in a community that is actively engaged in a surveillance activity provides a context for its development. The test must be robust enough to withstand heat and humidity and ideally it should have a rapid turnaround time to provide a meaningful feedback loop to the community participants. During design the test was validated against a set of ideal and real-world benchmarks to truly understand how it might perform during implementation. By challenging the boundaries of test performance, the developers are better able to guide country programs on use of the test and how to interpret results. These factors were taken into consideration during the development and optimization of the SD BIOLINE Lymphatic Filariasis IgG4 rapid diagnostic test, and the result is a robust, easy-to-use test with a turnaround time of 30 minutes. Diagnostic tests for lymphatic filariasis (LF) are shown in Table 15.4.

15.4 Rapid Diagnostic Tests for Onchocerciasis

Significant barriers exist to the control and elimination of onchocerciasis, including its large geographic area, the difficulty and variety of its terrain (especially in Africa), human and fly migration patterns, vector efficiency, and co-endemicity with *Loa loa* infections (Dadzie 2002; Traore et al. 2012). Recent results from Latin America, Senegal, Mali and Uganda, however, suggest that elimination of onchocerciasis in Africa using ivermectin is feasible (Diawara et al. 2009; Lakwo et al. 2013).

Onchocerciasis control programs rely on population-level surveillance to determine treatment steps, with entire communities surveyed for disease in order to understand prevalence rates. There is no reliable gold standard for diagnosing onchocerciasis (Udall 2007).

15.4.1 Antigen Detection

The adult worms take roughly a year and a half to mature to the point of creating microfilariae, so there is a substantial time lapse from when an individual is infected to when he or she tests positive by skin snip microscopy (Toé et al. 2000; WHO 1976). Treatment with ivermectin has an influence on the result of skin snip microscopy: in communities being treated with ivermectin, skin snips often do not have any microfilariae (Taylor et al. 1989). Furthermore, skin snip microscopy is not sensitive enough to detect light or early infections (Remme et al. 1986). Skin snips or skin scrapings subjected to PCR testing have a higher specificity and sensitivity (Toé et al. 1998; Zimmerman et al. 1994), but this method is expensive and has to be performed in a moderate- to high-complexity laboratory. In addition, skin snips are painful and invasive, and communities

Table 15.4 Overview of tests for lymphatic filariasis.

Diagnostic tool	Use	Description
Ultrasonography	Identification of adult worms primarily in the scrotum of asymptomatic males	Ultrasonography is a field-implemented test that is rarely used; it may be beneficial, however, in clinical care for some patients
Night blood smears	Detection of microfilaria in the night blood for disease mapping and monitoring of program progress	Field-based parasite detection by light microscopy from specimens collected at night. Night blood smears are common diagnostic procedures for the surveillance of LF. Although this test has been used across the world, the logistical complexity of using it in villages at night has made it difficult to implement.
Polymerase chain reaction (PCR) for *Wuchereria bancrofti*	Detection of DNA in the night blood to increase sensitivity	A laboratory-based test that can be used to identify *W. bancrofti* DNA in the blood. May be used to confirm positive test results or to identify positive individuals who did not test positive for night blood smears. Used primarily for research purposes. Detects DNA indicative of current infections
BinaxNOW Filariasis ICT	Disease mapping, monitoring program progress, determining when to stop MDA, and for surveillance after MDA has stopped. Detects antigens indicative of current infection	The BinaxNOW Filariasis card test is a field-deployable rapid test that detects antigens of, or current infection with, *W. bancrofti*. The diagnostic test must be read within a short time frame or the risk of false positives goes up.
Alere Filariasis Test Strip	Disease mapping, monitoring program progress, determining when to stop MDA, and for surveillance after MDA has stopped. Detects antigens indicative of current infection	A field-deployable rapid test that detects antigens of, or current infection with, *W. bancrofti*. This diagnostic is more robust than its predecessor, the BinaxNOW card test, but has the risk of false positives due to other filarial infections
Antibody-detection Brugia Rapid	Disease mapping, monitoring program progress, determining when to stop MDA, and for surveillance after MDA has stopped	Due to the lack of an antigen-detection test for Brugia timori or Brugia malayi, this field-based antibody detection test has been used as a field-friendly alternative to night blood smears
InBios Filaria Detect IgG4 ELISA	Diagnosing exposure to Wb123 through an IgG4 response.	A laboratory-based test that requires a well-equipped laboratory, the Wb123 ELISA is used to map areas of low-endemicity and for post-treatment surveillance after MDA has stopped
SD BIOLINE Lymphatic Filariasis IgG4 Rapid Test	Diagnosing exposure to Wb123 through an IgG4 response	The Wb123 rapid test is a field-deployable antibody detection tool. Recent to the market, this test is available to evaluate LF programs in areas where *W.bancrofti* is the primary infective agent

Abbreviations used: MDA, mass drug administration; ELISA, enzyme-linked immunosorbent assay; LF, lymphatic filariasis.

are becoming increasingly opposed to participating in surveys that employ this method (Boatin et al. 1998).

15.4.2 Antibody Detection

Over the last 30 years several antigens have been evaluated as targets for antibody-based onchocerciasis tests in various formats (Nde et al. 2002; Weil et al. 2000). The Ov16 antigen was the most promising and is currently the most advanced (Burbelo et al. 2009; Weil et al. 2000). As noted in results first published in the early 1990s, antibodies to the Ov16 antigen appear rapidly after infection, with Ov16 being highly expressed by the third and fourth larval stages, making Ov16 a potential indicator of early infection (Lobos et al. 1991; Rodríguez-Peréz et al. 2010). Immune response in humans to filarial infections is characterized by, among other things, a high level of parasite-specific immunoglobulin G (IgG) (Gbakima et al. 1996). This response is often dominated by IgG4, which has been validated as a biomarker for exposure to onchocerciasis, and is believed to be the most specific biomarker.

An ELISA test was developed to detect Ov16 IgG4. This laboratory-based test requires significant infrastructure and a highly skilled laboratory technician, so deploying it is difficult in low-resource settings. To perform Ov16 ELISA testing dried blood spots must be sent to central testing facilities, causing issues with logistics and shipping costs. The results are received weeks or months later, which prevents a timely feedback loop with the community. Despite these challenges, the Ov16 ELISA has been successfully used by the Onchocerciasis Elimination Program for the Americas (Gonzalez et al. 2009; Rodríguez-Pérez et al. 2003). The Carter Center has also used the Ov16 ELISA in the field in Guatemala, Mexico, Ecuador, and Colombia with good results (Lindblade et al. 2007; Lobos et al. 1991; Rodríguez-Pérez et al. 2010).

Although the laboratory-based ELISA format has proven successful in the Americas, a rapid test would be ideal for several reasons. Rapid tests require no laboratory infrastructure or instrumentation. They need relatively minimal training and technical skills and they can be read within 30 minutes to an hour, giving surveillance teams and communities nearly instant results. Positive tests that need to be followed up in order to monitor for recrudescence can be handled promptly.

A rapid card test for Ov16 was developed but never commercialized by its original manufacturer AmRad (later purchased by Binax). In 2013 PATH was awarded a grant to develop and commercialize a new test. To do so, the Ov16 ELISA test was first optimized in the laboratory and the technology was then transferred to the rapid strip test platform. The Ov16 rapid test, commercialized by Standard Diagnostics (a subsidiary of Alere) as the SD BIOLINE Onchocerciasis IgG4, has undergone extensive testing in the laboratory and in the field, with several thousands of tests used in the West African nation of Togo. The premise of the test is simple: using a finger prick, a small drop of blood is placed on to a cassette, followed by a few drops of buffer. If an individual is positive for Ov16 antibodies, a test line and a control line appear. If an individual is negative, then only a control line appears. The test can be read within 20 minutes, and can be read reliably for up to 24 hours.

The steps in performing the SD BIOLINE Onchocerciasis IgG4 rapid test are shown in Figure 15.2.

The benefits of the SD BIOLINE Onchocerciasis IgG4 are numerous. The test is small and easily portable, making it field-deployable with relatively minimal required

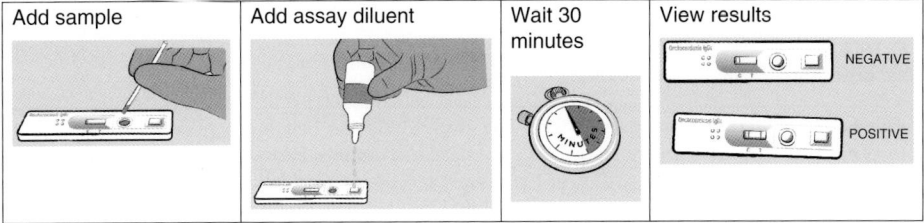

Add sample	Add assay diluent	Wait 30 minutes	View results
			NEGATIVE
			POSITIVE

Figure 15.2 The steps in performing the SD BIOLINE Onchocerciasis IgG4 rapid test.

ancillary equipment. The performance of the Ov16 rapid test in the field is acceptable: in one study the test was 96.8% sensitive and 99.1% specific (Faulx et al. 2014). Community members are more accepting of the rapid test compared to to skin snip microscopy (Dieye et al. 2017). During a clinical evaluation of test prototypes in Togo, one of the authors witnessed a woman accepting the finger prick for the rapid test without a problem; however, once the woman realized that a skin snip was next, she left the trial immediately. In addition to being less invasive than skin snip microscopy, the quick results feedback loop promotes a feeling of trust among participants and encourages others to be tested.

The test is currently being used globally by early adopters as country programs move into elimination phases of onchocerciasis. The current World Health Organization guidelines for stopping MDA and verifying elimination of onchocerciasis include further evaluations of the Ov16 rapid test as a possible ELISA replacement (WHO 2016).

Diagnostic tools for onchocerciasis are summarized in Table 15.5.

Table 15.5 Diagnostic tools for onchocerciasis.

Diagnostic tool	Use	Description
Node palpation	Rapid mapping of onchocerciasis	Node palpation was commonly used in the field for mapping areas endemic for onchocerciasis, but it is not relevant for routine surveillance or program monitoring.
Onchocercomectomy	Identification of infection with *Onchocerca volvulus*	Onchocercomectomy is invasive but has the added benefit of removing the adult worms from the body, thus reducing their ability to give birth to microfilariae
Skin snip microscopy	Diagnoses active infection by identifying microfilaria in the skin	Field-based skin snip microcoscopy is a highly specific diagnostic that identifies active infection, but poor sensitivity and low community acceptance are barriers
Polymerase chain reaction (PCR) on skin snips	Increases the sensitivity of skin snips	Improves the sensitivity of skin snips; PCR is a high-complexity, laboratory-based test that increases the logistical difficulty of field surveillance work. It requires a well-equipped laboratory.
Luciferase immunoprecipitation systems (LIPS)	Identifies multiple antibodies for diagnosis of exposure to four *Onchocerca volvulus* antigens	LIPS and the rapid LIPS (QLIPS) are laboratory-based assays that are able to detect antibodies to multiple antigens. Although highly sensitive and specific, the test is relatively new and requires a well-equipped laboratory

(Continued)

Table 15.5 (Continued)

Diagnostic tool	Use	Description
Antibody-detection enzyme-linked immunosorbent assay (ELISA)	Diagnoses exposure to Ov16 through an IgG4 response	The Ov16 ELISA has been used throughout the Americas successfully. However, it is a test that requires a well-equipped laboratory, which reduces its usability in other geographic locations, such as sub-Saharan Africa.
SD BIOLINE Onchocerciasis IgG4 Rapid Test	Diagnoses exposure to Ov16 through an IgG4 response	The Ov16 rapid test is a field-deployable antibody detection test. Recent to the market, this test has the potential to be used for mapping and verifying elimination of onchocerciasis across the world

15.5 Next tests and Steps

While the Ov16 rapid test is used by onchocerciasis programs across the world, the next-generation test for LF incorporates an IgG4 antibody specific for LF, Wb123. LF and onchocerciasis often overlap geographically. Additionally, the two diseases share ivermectin as a medication for MDA. If an area is no longer believed to have LF, the final transmission assessment for LF cannot occur until the region has stopped receiving ivermectin for onchocerciasis. An integrated test that detects IgG4 antibodies to both Ov16 and Wb123 may drive the integration of the two programs on a national level, a strategy recognized as a best option for many neglected tropical diseases.

Bibliography

Adjobimey, T. and Hoerauf, A. (2010). Induction of immunoglobulin G4 in human filariasis: an indicator of immunoregulation. *Ann. of Trop. Med. & Para.* 104 (6): 455–464.

Alere. BinaxNOW Filariasis. 2017 [cited 2017 March 18]; Available from: http://www.alere.com/ww/en/product-details/binaxnow-filariasis.html

Bakajika, D.K., Nigo, M.M., Lotsima, J.P. et al. (2014). Filarial Antigenemia and Loa loa night blood Microfilaremia in an area without Bancroftian Filariasis in the Democratic Republic of Congo. *Am. J. of Trop. Med. & Hyg.* 91 (6): 1142–1148.

Boatin, B.A., Toé, L., Alley, A.S. et al. (1998). Diagnostics in onchocerciasis: future challenges. *An. of Trop. Med. & Para.* 92 (suppl. 1): S41–S45.

Burbelo, P.D., Leahy, H.P., Iadarola, M.J., and Nutman, T.B. (2009). A four-antigen mixture for rapid assessment of Onchocerca volvulus infection. *PLOS Neg. Trop. Dis.* 3 (5): 1–10.

Dadzie, Y. 2002. Final report of the conference on the eradicabilty of onchocerciasis. The Carter Center: Atlanta, GA.

Diawara, L., Traoré, M.O., Badji, A. et al. (2009). Feasibility of onchocerciasis elimination with Ivermectin treatment in endemic foci in Africa: first evidence from studies in Mali and Senegal. *PLOS Neg. Trop. Dis.* 3 (7): e497. https://doi.org/10.1371/journal.pntd.0000497.

Dieye, Y., Storey, H.L., Barrett, K.L. et al. (2017). Feasibility of utilizing the SD BIOLINE onchocerciasis IgG4 rapid test in onchocerciasis surveillance in Senegal. *PLOS Negl. Trop. Dis.* 11 (10): e0005884.

Faulx, D.G.A., Valdez M., Peck R., et al. (2014). Field evaluation of Standard Diagnostics' onchocerciasis IgG4 rapid diagnostic test prototypes, in American Society of Tropical Medicine and Hygiene Annual Conference. New Orleans, LA, USA.

Gbakima, A.N., Nutman, T.B., Bradley, J.E. et al. (1996). Immunoglobulin G subclass responses of children during infection with Onchocerca volvulus. *Clinical and Vaccine Immunology* 3 (1): 98.

Gonzalez, R.J., Cruz-Ortiz, N., Rizzo, N. et al. (2009). Successful interruption of transmission of Onchocerca volvulus in the Escuintla-Guatemala focus, Guatemala. *PLOS Neg. Trop. Dis.* 3 (3): e404.

Haarbrink, M., Abadi, K., van Beers, S. et al. (1995). IgG4 antibody assay in the detection of filariasis. *Lancet* 346 (8978): 853–854.

InBios. Filaria DetectIgG4 ELISA Kit. (2017). Available from http://www.inbios.com/filaria-detect-igg4-elisa-usa

Krushna, N.S., Shiny, C., Dharanya, S. et al. (2009). Immunolocalization and serum antibody responses to Brugia malayi pepsin inhibitor homolog (Bm-33). *Microb. & Imm.* 53 (3): 173–183.

Lakwo, T.L., Garms, R., Rubaale, T. et al. (2013). The disappearance of onchocerciasis from the Itwara focus, western Uganda after elimination of the vector Simulium neavei and 19 years of annual ivermectin treatments. *Acta Tropica* 126 (3): 218–221.

Lammie, P.J., Weil, G., Noordin, R. et al. (2004). Recombinant antigen-based antibody assays for the diagnosis and surveillance of lymphatic filariasis – a multicenter trial. *Filaria J.* 3 (1): 9.

Lindblade, B.K.A., Zea-Flores, G., Rizzo, N. et al. (2007). Elimination of Onchocercia volvulus transmission in the Santa Rosa focus of Guatemala. *Am. J. Trop. Med. & Hyg.* 77 (2): 334–341.

Lobos, E., Weiss, N., Karam, M. et al. (1991). An immunogenic Onchocerca volvulus antigen: a specific and early marker of infection. *Science* 251 (5001): 1603–1605.

Nde, P.N., Pogonka, T., Bradley, J.E. et al. (2002). Sensitive and specific serodiagnosis of onchocerciasis with recombinant hybrid proteins. *Am. J. Trop. Med. & Hyg.* 66 (5): 566–571.

Pandiaraja, P., Arunkumar, C., Hoti, S.L. et al. (2010). Evaluation of synthetic peptides of WbSXP-1 for the diagnosis of human lymphatic filariasis. *Diag. Micro. & Inf. Dis.* 68 (4): 410–415.

Remme, J., Ba, O., Dadzie, K.Y., and Karam, M. (1986). A force-of-infection model for onchocerciasis and its applications in the epidemiological evaluation of the onchocerciasis control programme in the Volta River basin area. *Bull. WHO* 64 (5): 667–681.

Rodríguez-Pérez, M.A.U., Thomas, R., Domínguez-Vázquez, A. et al. (2010). Interruption of transmission of Onchocerca volvulus in the Oaxaca focus, Mexico. *Am. J. Trop. Med. & Hyg.* 83 (1): 21–27.

Rodríguez-Pérez, M.A., Domínguez-Vázquez, A., Méndez-Galván, J. et al. (2003). Antibody detection tests for Onchocerca volvulus: comparison of the sensitivityof a cocktail of recombinant antigens used in the indirect enzyme-linked immunosorbent assay with a rapid-format antibody card test. *Trans. R. Soc. Trop. Med. & Hyg.* 97 (5): 539–541.

Steel, C., Golden, A., Kubofcik, J. et al. (2013). Rapid Wuchereria bancrofti-specific antigen Wb123-based IgG4 immunoassays as tools for surveillance following mass drug administration programs on lymphatic Filariasis. *Clin. & Vacc. Imm.* 20 (9): 1155–1161.

Supali, T., Rahman, N., Djuardi, Y. et al. (2004). Detection of filaria-specific IgG4 antibodies using Brugia rapid test in individuals from an area highly endemic for Brugia timori. *Acta Trop.* 90 (3): 255–261.

Taylor, H.R., Munoz, B., Keyvan-Larijani, E., and Greene, B.M. (1989). Reliability of detection of microfilariae in skin snips in the diagnosis of onchocerciasis. *Am. J. Trop. Med. & Hyg.* 41 (4): 467–471.

Toé, L., Boatin, B.A., Adjami, A. et al. (1998). Detection of Onchocerca volvulus infection by O-150 polymerase chain reaction analysis of skin scratches. *J. Inf. Dis.* 178 (1): 282–285.

Toé, L., Adjami, A., Boatin, B.A. et al. (2000). Topical application of diethylcarbamazine to detect onchocerciasis recrudescence in West Africa. *Trans. R. Soc. Trop. Med. & Hyg.* 94 (5): 519–525.

Traore, M.O., Sarr, M.D., Badji, A. et al. (2012). Proof-of-principle of onchocerciasis elimination with Ivermectin treatment in endemic foci in Africa: final results of a study in Mali and Senegal. *PLOS Neg. Trop. Dis.* 6 (9): e1825.

Udall, D.N. (2007). Recent updates on onchocerciasis: diagnosis and treatment. *Clin. Inf. Dis.* 44 (1): 53–60.

Weil, G.J. and Ramzy, R.M. (2007). Diagnostic tools for filariasis elimination programs. *Trs. in Paras.* 23 (2): 78–82.

Weil, G.J., Lammie, P.J., and Weiss, N. (1997). The ICT Filariasis test: a rapid-format antigen test for diagnosis of bancroftian filariasis. *Paras. Today* 13 (10): 401–404.

Weil, G.J., Steel, C., Liftis, F. et al. (2000). A rapid-format antibody card test for diagnosis of onchocerciasis. *J. Inf. Dis.* 182 (6): 1796–1799.

Weil, G.J., Curtis, K.C., Fischer, P.U. et al. (2011). A multicenter evaluation of a new antibody test kit for lymphatic filariasis employing recombinant Brugia malayi antigen Bm-14. *Acta Tropica* 120: S19–S22.

Weil, G.J., Curtis, K.C., Fakoli, L. et al. (2013). Laboratory and field evaluation of a new rapid test for detecting Wuchereria bancrofti antigen in human blood. *Am. J. Trop. Med. & Hyg.* 89 (1): 11–15.

WHO (1976). Epidemiology of Onchocerciasis: Report of a WHO Expert Committee.

WHO (2011). Monitoring and epidemiological assessment of mass drug administration in the global programme to eliminate lymphatic filariasis: a manual for national elimination programmes.

WHO (2012a). London declaration on neglected tropical diseases, http://www.who.int/neglected_diseases/London_Declaration_NTDs.pdf

WHO (2012b). Accelerating Work to Overcome the Global Impact of Neglected Tropical Diseases. A Roadmap for Implementation.

WHO (2016). Guidelines for Stopping Mass Drug Administration and Verifying Elimination of Human Onchocerciasis.

Zimmerman, P.A., Guderian, R.H., Arujo, E. et al. (1994). Polymerase chain reaction-based diagnosis of Onchocerca volvulus infection: improved detection of patients with onchocerciasis. *J. Inf. Dis.* 169 (3): 686–689.

Webliography

http://www.alere.com/ww/en/product-details/binaxnow-filariasis.html Alere. BinaxNOW Filariasis. 2017 (cited March 18th 2017).

http://www.inbios.com/filaria-detect-igg4-elisa-usa nBios. Filaria DetectIgG4 ELISA Kit. 11 December 2017.

https://en.wikipedia.org/wiki/London_Declaration_on_Neglected_Tropical_DiseasesCrompton DWT (editor) 2012. Accelerating Work to Overcome the Global Impact of Neglected Tropical Diseases – A Roadmap for Implementation (PDF). Geneva: WHO Press, World Health Organization.

Part III

Other Tests that Can Be Performed Rapidly at the Primary-Point-of-Care

16

Point-of-Care Testing for Blood Counts, HbA1c, Renal Function, Electrolytes, Acid–Base Balance and Hepatitis

Mark Shephard, Lara Motta, Brooke Spaeth, Heather Halls, and Lauren Duckworth

Flinders University, Adelaide, South Australia, Australia

CHAPTER MENU

Revolutionizing Tropical Medicine: Point-of-Care Tests, New Imaging Technologies and Digital Health,
First Edition. Edited by Kerry Atkinson and David Mabey.
© 2019 John Wiley & Sons, Inc. Published 2019 by John Wiley & Sons, Inc.

16.1   Introduction

This chapter describes five different point-of-care (POC) tests or test profiles that are useful for the detection and monitoring of selected chronic, acute and infectious disease markers and that can be readily performed in the primary care setting. For each test or test profile, there are now a wide range of small, portable POC testing devices or test systems that are available on the global market. For a POC testing device to be most useful in the LMICs, it should ideally have some, if not all, of the following specifications:

- be lightweight (<10 kg), portable and have a small footprint;
- require just a few microliters of sample (blood or fluid) to perform the test;
- have a turnaround time for result of <10 minutes;
- have the capacity to use battery (as well as, or instead of, AC power);
- use reagents that can be stored at room temperature rather than requiring refrigeration;
- deliver results of sound analytical quality (that is, with accuracy and precision equivalent to that expected of an accredited pathology laboratory).

The capacity to perform POC testing can provide clinical, operational, and cultural benefits for patients in LMICs. Clinically, a rapid POC test result can be used to serially monitor the efficacy of treatment (for chronic conditions), safely, and effectively triage the patient for appropriate medical care (acute markers) or reduce the time to initiate treatment (for infectious markers). Operationally, POC provides improved access to pathology testing in countries or locations where laboratories may be long distances from the point of primary patient care delivery. Culturally, POC testing may decrease patient loss to follow-up, with patients often failing to revisit the clinic for discussion of their pathology results once they are returned from the laboratory (which may take several days/weeks).

Each of the five tests or profiles described in this chapter will include discussion of the clinical use of the test or test profile; device options for POC testing, and where possible, an example of a working POC testing model with capacity for translation in the LMICs or a case study illustrating the value of POC testing.

16.2   Point-of-Care Testing for Blood Counts

16.2.1   Clinical Use of Blood Counts

A full blood count (FBC), also known as a full blood evaluation (FBE) or complete blood count (CBC), is one of the most frequently requested pathology tests, reflecting the wide range of common and less common clinical conditions that may be associated with abnormalities in FBC results. The FBC is a panel of tests that, in simple terms, counts the number of red blood cells, white blood cells, and platelets in a patient sample. For red blood cells, the FBC also measures the concentration of hemoglobin, as well as a number of red blood cell indices. For white blood cells (WBCs) (leukocytes), the FBC also provides a count of the different subtypes (or populations) of WBCs present, usually as either a three-part differential count reporting the numbers of lymphocytes, monocytes and granulocytes (neutrophils, eosinophils and basophils) or five-part differential reporting counts of lymphocytes, neutrophils, monocytes, eosinophils and basophils.

Red blood cell (RBC) counts and hemoglobin may be raised in conditions such as polycythemia, severe dehydration, surgery and burns, and may be lowered, for example, in anemia and hemorrhage. Abnormalities in white cell counts may indicate bacterial, viral or parasitic infections, allergic reactions, inflammatory responses or leukemia. Platelet counts are useful in the diagnosis and monitoring of disease states that may affect the process of blood clotting. A high platelet count (thrombocytosis) may be found in some cancers, hemorrhage, iron deficiency or hemolytic anemias and rheumatoid arthritis, while platelets may be lowered (thrombocytopenia), for example, in malaria, viral infections, bone marrow disorders, sepsis or autoimmune diseases. The unit of measurement for red and white cell and platelet counts is 10^9 cells/L or 10^6 cells/μL.

For FBC counts, significant increases in result variability are observed for samples measured greater than 72 hours after blood collection (ICSH 2014). For this reason, a FBC by POC testing has significant benefits over laboratory testing, as laboratories in LMICs are generally located long distances from the primary site of patient care. For example, a common use of POC FBC testing in the LMICs is for the detection and management of dengue fever, which requires serial measurement of hematocrit and leukocytes. The World Health Organization recommends that these tests should be easily accessible and results should be available within two hours in severe cases of dengue infection (WHO 2009).

16.2.2 POC Testing Device Options for Hemoglobin Estimation

POC testing for hemoglobin is widely used in the LMICs to screen for anemia in at-risk populations. One of the most commonly used POC testing devices is the HemoCue Hemoglobin (Hb) meter, which include the Hb 201+ and Hb 301+ models (Briggs et al. 2012). The HemoCue Hb analyzers provide a hemoglobin concentration result in ≤10 seconds using capillary or venous whole blood. The devices weigh approximately 500 g and work from battery or AC power, making them highly portable. The most recently released Hb 301+ device has an operating temperature range of 15–40 °C with consumables (microcuvettes) that are able to be stored for both short (six weeks at −18 to 50 °C) and long periods (10–40 °C) if unopened, making the test system suitable for environmental extremes. The HemoCue devices are also simple to operate, even for non-laboratory trained staff, making them highly suitable for use in LMICs (Briggs et al. 2012). Several recent peer-reviewed studies have confirmed the reliability and analytical quality of the HemoCue Hb analyzers (Kok et al. 2015; Rappaport et al. 2016).

16.2.3 POC Testing Device Options for Full Blood Count Estimation

Despite being one of the most frequently requested pathology tests, the development of reliable and accurate FBC POC testing devices has only occurred in relatively recent times, particularly for devices that include a platelet count. There are a number of larger "bench-top" hematology analyzers available on the market that can provide a FBC including platelet count; however, these devices should be considered satellite laboratory analyzers rather than POC analyzers due to their size and weight (generally ≥20 kg). As previously mentioned, for LMICs the most suitable POC testing analyzers are small and portable or hand-held weighing ≤10 kg, preferably with the option of battery or AC power usage. A summary of current FBC POC testing devices that weigh ≤10 kg is given in Table 16.1.

While several POC testing options provide a total and a three or five-part differential white cell count, very few have been evaluated in LMICs or low resource settings. For example, the Chempaq XBC provides Hb, total WBC, five-differential and platelet count in less than three minutes; however, it relies on AC power and evaluations of the device have only been conducted in highly resourced hospital or laboratory settings (Rao et al. 2008, 2011). The only POC testing device measuring a total WBC and five-part differential count evaluated outside of the tertiary hospital setting is the HemoCue

Table 16.1 POC testing devices weighing less than 10 kg for FBC estimation.

Device name	POC tests	Method	Description/ weight/ power source	Time to results	Sample size/ type
Chempaq XBC	Hb, WBC, NEU, LYM, MON, BAS, EOS, Platelet count	Impedance and photometry measurement	Small bench-top device, 1.9 kg, AC only	<3 mins	20 μL capillary or whole blood
HemoCue WBC DIFF*	WBC, NEU, LYM, MON, BAS, EOS	Cell recognition technology	Small bench-top device, battery operated, 1.3 kg, battery or AC	<5 mins	10 μL capillary or whole blood
Norma Icon 3	WBC, NEU, LYM, GRA, RBC, HCT, MCV, PCT	Volumetric impedance and micro-fluidics	Medium bench-top device, 10 kg, AC only	45–60 tests per hour	9.6 μL whole blood
Norma Icon 5	WBC, NEU, LYM, MON, BAS, EOS, RBC, HCT, MCV, PCT	Laser based flow cytometry and volumetric impedance	Medium bench-top device, 9.4 kg, AC only	60 tests per hour	30 μL whole blood
QBC STAR	Hb, MCHC, PCT, WBC, GRA, LYM, MON	Various methods depending on test type	Large bench-top device, 8.6 kg, AC only	4.5 mins	70 μL capillary or venous whole blood

Abbreviations used: Hb, hemoglobin; WBC, total white blood cells; GRA, granulocytes; NEU, neutrophils; LYM, lymphocytes; MON, monocytes; BAS, basophils; EOS, eosinophils; HCT, hematocrit; MCV, mean corpuscular volume; MCHC, mean corpuscular hemoglobin concentration; PCT, platelet count.

WBC DIFF (Spaeth et al. 2015). The HemoCue WBC DIFF is also the only POC testing device measuring these parameters that can work on batteries as well as AC power. However, a limitation of the device is that it cannot measure RBC/hemoglobin or platelets on the same platform.

16.2.4 A Case Study Illustrating the Value of POC Testing for White Blood Cell Counts

The following patient case describes how the HemoCue WBC DIFF POC testing device assisted in the diagnosis of septicemia in a young child in a remote Australian community.

16.2.5 Presentation/History

A nine year old girl with recent knee trauma (after she dived into water and hit her right knee on a rock) presented to a remote health service with her right knee swollen and hot. The Rural Medical Practitioner's preliminary diagnosis was septic arthritis and he noted that the patient's brother had had a recent history of acute rheumatic fever. The patient was able to walk and there was no abrasion over the knee. There were no skin lesions, except for an old dry scabies wound on her right ankle, and no history of recent tonsillitis.

16.2.6 Examinations

Initial investigations included the following: temperature 38.7 °C; heart rate 116 beats min^{-1}; respiratory rate 20 breaths per/min; blood pressure 117/95 mmHg; oxygen saturation 100%; blood glucose level 7.2 mmol L^{-1}; Hb 137 g L^{-1}; Targeted Real-Time Early Warning (TREW) Score for septic shock 0 (low risk).

A total and five-part differential white cell count test was performed using the HemoCue WBC DIFF and showed an elevated total white cell count of 17.2×10^9 cells/L (reference interval $4–10 \times 10^9$ cells/L) and neutrophils 13.6×10^9 cells/L (reference interval $2.5–7.5 \times 10^9$ cells/L).

16.2.7 Treatment/Follow-up

The on-call pediatrician advised the immediately evacuation of the patient to the closest hospital (~500 km away, 90 minutes flight) without antibiotics unless signs of sepsis (such as tachycardia) developed. However, the air evacuation service advised that they would be unable to land due to bad weather. Several hours later (evening) the patient was described as stable (at home) and, due to the delay in evacuation, the pediatrician advised starting antibiotics (flucloxacillin 50 mg kg^{-1}). The patient was evacuated several hours later.

16.2.8 Outcome

In hospital the child was diagnosed as having severe septic arthritis and osteomyelitis involving her right knee which required a long hospitalization. The hospital laboratory results the same day showed WCC 17.7×10^9 cells/L and neutrophils 14.3×10^9 cells/L.

16.2.9 Summary

The treating medical practitioner at the remote health center confirmed that the WBC DIFF POC test results had assisted with the decision to evacuate the patient as the total WCC and neutrophils were elevated.

16.3 Point-of-Care Testing for HbA1c

16.3.1 Clinical Use of Hemoglobin A1c

Hemoglobin A1c (HbA1c) is defined as glucose that is attached to the N-terminal valine of the beta chain of hemoglobin in the red blood cell. For more than three decades, HbA1c has been accepted by the medical and scientific fraternity as the "gold standard" pathology test for monitoring glycemic control in patients with established diabetes mellitus. The circulating HbA1c level reflects a "time-weighted" estimate of a patient's average glycemic control over a three to four month window (Goodall et al. 2007). An HbA1c of <7% ($53\,mmol\,mol^{-1}$) is considered the optimal target for clinical management of patients with diabetes.

More recently, the clinical use of HbA1c as a marker for diagnosing diabetes has taken center stage in the global literature, with many countries now adopting the recommendation that an HbA1c of 6.5% ($48\,mmol\,mol^{-1}$) can be used as a cut-off for diagnosing diabetes (WHO 2011; John and the UK Department of Health Advisory Committee on Diabetes 2012; ADA 2015); the rationale for using the 6.5% cut-off is that, beyond this level, the prevalence of moderate retinopathy in diabetes patents begins to rise exponentially (Colagiuri et al. 2011). The use of HbA1c POC testing for the monitoring of patients with diabetes is well established, but its use in diagnosing diabetes remains contentious (Shephard et al. 2016c). In past years, some professional organizations have placed a "blanket ban" on using POC testing devices for diagnosis but, in more recent times, a more rational approach has been taken with individual devices being assessed separately for their "fitness for purpose" in diagnosing diabetes (Shephard et al. 2016c).

While the debate continues to rage about the ability of HbA1c POC testing devices to measure a glucose level of 6.5% accurately, in practice our experience suggests that most people with newly diagnosed diabetes in LMICs are more likely have HbA1c levels of the order of 8–10% when first tested. Thus the value of opportunistic HbA1c POC testing in the primary care setting should not be underestimated. However, the key to maximizing the benefits of POC testing is to ensure that the result is acted upon clinically in a timely fashion and that medications are available in LMICs to support those identified in most need of diabetes care.

16.3.2 POC Testing Device Options for Measuring HbA1c

There are at least 15 POC testing devices for HbA1c on the global market, with the number likely to increase beyond 20 in the coming five years. Ideal specifications for a POC testing device in the LMICs were outlined in the Introduction to this chapter. In addition to those listed, HbA1c testing devices should also

- have a wide measuring range (of the order of 5% ($31\,mmol\,mol^{-1}$) to 15% ($140\,mmol\,mol^{-1}$);
- employ a method that is not subject to interference from hemoglobinopathies.

There is no single device which fits all these specifications "perfectly," but certainly the Siemens DCA Vantage and the Alere Afinion HbA1c POC analyzers have been shown to consistently exhibit sound analytical performance in international studies as well as meeting most of the above criteria (Lenters-Westra and Slingerland 2010, 2014; Shephard et al. 2016c). In Australia, the Siemans DCA test system has consistently met the desired analytical goal for imprecision of <3% in a large POC testing program called Quality Assurance for Aboriginal and Torres Strait Islander Medical Services (QAAMS) (www.qaams.org.au) conducted in indigenous medical services located mainly in rural and remote locations of Australia (please see section 16.3.3.) (Shephard et al. 2016a). The Roche cobas b 101, which uses "room temperature" reagents, also performed well analytically in the 2014 evaluation by Lenters-Westra and Slingerland. The SD A1cCare, which uses battery power and has "room temperature" reagents, shows promise for use in LMICs.

16.3.3 An Example of a POC Testing Model with Application in the LMICs

The QAAMS Program is a national Australian POC testing program that has enabled blood HbA1c and the urinary albumin: creatinine ratio (ACR) to be performed by trained indigenous health workers and nurses at more than 200 sites in over 150 indigenous medical services across Australia (Shephard et al. 2016a). This program has been in operation since 1999. QAAMS is underpinned by four core elements: continuous education, training and competency assessment for indigenous POC device operators; monthly surveillance of analytical quality of POC testing (through sites conducting both quality control [QC] and external proficiency testing [PT]); intensive technical and scientific support for services by a team of scientists; and ongoing research to evaluate the effectiveness of the POC testing program. This research has shown that the program has remained

- analytically sound, with no statistically significant difference being observed between the quality of HbA1c testing in QAAMS facilities compared to that achieved by Australian laboratories using the same PT testing material over the past 15 years (Shephard et al. 2016a);
- clinically effective, as evidenced by statistically significant improvements in glycemic control (falls in HbA1c) within and across groups of diabetes patients at different services following the introduction of POC testing (Shephard et al. 2015);
- operationally effective, with improvements in timeliness of diabetes care after POC testing compared to laboratory testing, both in terms of turnaround times for reporting HbA1c results to patients and turnaround time for patient follow-up and consultation (Spaeth et al. 2015); and
- culturally effective, through qualitative surveys among doctors, device operators, and indigenous patients with diabetes which showed statistically significant improvements in satisfaction with POC testing after its introduction (Shephard 2006a; Shephard et al. 2016b).

The growth and sustainability of the program has been due to continuous funding and support for the program provided by the Australian Government for the past 18 years and the provision of Medicare rebates (Government reimbursement) for participating services to cover the cost of HbA1c (and urine ACR) testing cartridges.

Table 16.2 Improvements in glycemic control in diabetes patients from different countries following introduction of POC testing in the ACE Program.

Country	HbA1c (%) baseline POC test	HbA1c (%) most recent POC test	Average months since POC introduced	Number of patients
South Africa	9.7% ± 2.4	8.4%[a] ± 2.4	15	131
Papua New Guinea	9.2% ± 2.1	8.6%[a] ± 2.4	16	154
Canada	8.4% ± 2.2	8.1%[a] ± 2.0	12	577

a) Results statistically significant ($p < 0.05$).

This Australian-based model, which operates mainly in the challenging clinical environment of rural and remote Australia, was translated internationally in 2011, initially to 35 indigenous communities in seven countries (Canada, South Africa, Papua New Guinea, East Timor, Thailand, Solomon Islands and Western Samoa). The "ACE" (Analytical and Clinical Excellence) Program was founded on similar core principles as QAAMS (Motta and Shephard 2015). The program continues to operate successfully in Canada (north west Ontario), South Africa (through the HOPE clinic in Johannesburg) and Papua New Guinea (Morobe Province), where POC testing has resulted in improvements in glycemic control among indigenous clients with diabetes accessing the service (Table 16.2) (Motta et al. 2017a, 2017b).

In Papua New Guinea the local provincial Morobe government has strongly supported the initiative, while in Canada, the North West Ontario Local Health Integration Network and St. Joseph's Care Group have provided much needed support to maintain the program. In East Timor, HbA1c POC testing has been used for the diagnosis, as well as the management, of indigenous patients with diabetes (Dawkins et al. 2015). At the National Eye Centre in Dili, Timor-Leste (East Timor), 283 adult patients being assessed for cataract surgery had POC testing for HbA1c performed as part of their clinical assessment. 43 of 283 patients tested (15%) were found to have diabetes. Of these, 27 (63%) were newly diagnosed with diabetes based on their POC testing results, with the mean HbA1c of this newly-diagnosed group being 9.7% ($83 \, \text{mmol} \, \text{mol}^{-1}$).

16.4   Point-of-Care Testing for Renal Function

16.4.1   Clinical Summary

Chronic kidney disease (CKD) is a growing global public health problem, with an estimated prevalence of 15% in the United States and 10% in Australia (AIHW 2016; CDC 2017). However, rates of CKD are known to vary by geographical location. Populations living in remote and low socioeconomic areas have over two times the mortality rate of those living in metropolitan areas (AIHW 2013). CKD is often referred to as the "silent disease," as symptoms may not appear until more than half the kidney function is lost, meaning many individuals are unaware they have the condition until it is well progressed. Risk factors for the development of CKD include diabetes, high blood pressure, obesity, cardiac disease and family history. There are five categorized stages of CKD, with the most severe, stage 5, indicating end-stage kidney disease (Kidney Health Australia 2017). Patients are diagnosed and categorized by CKD stage through the use of two kidney function biomarkers: estimated glomerular filtration rate (eGFR) calculated from blood or serum creatinine, and a urine test for albumin (the main protein in the urine), which is usually measured as the urine ACR. Both creatinine and urine ACR can be measured using POC testing devices.

16.4.2 POC Testing Device Options for Creatinine

Currently there are at least 14 different POC testing devices on the global market that measure creatinine in whole blood, serum or plasma. Usually, devices can be divided into "blood gas" analyzers (that, in addition to blood gasses, measure creatinine as one of their metabolites) and "non-blood gas" analyzers (which also measure a range of tests including creatinine but not blood gasses). In terms of size these devices range from benchtop devices to handheld devices. The environmental conditions in the LMICs, including poor access to power, space, refrigeration or even a basic healthcare clinic, generally restrict the use of most benchtop devices which require AC power. This narrows the selection of creatinine POC testing devices down to three: the Abbott i-STAT®, the Alere epoc® and the Nova StatSensor® Creatinine. All three devices are handheld, battery operated and have a similar method principle which involves a cascade of enzymatic reactions followed by electrochemical detection.

The Abbott i-STAT is a "blood gas" analyzer that measures creatinine on the CHEM8+ cartridge profile using 90 μL of whole blood. The device has a measuring range of $18-1768\,\mu mol\,L^{-1}$. The i-STAT will generate a creatinine result in two minutes; however, it does not provide an automatic calculation of the eGFR. i-STAT consumables generally require refrigeration (but they can be kept at room temperature for up to 14 days), and the device has an operating temperature range of between 16 and 30°C. This may limit its use in settings where high or low temperatures are experienced.

The Alere epoc is also a "blood gas" analyzer that measures creatinine on the "BGEM" test card using 92 μL of whole blood. The device has a measuring range for creatinine of $27-1326\,\mu mol\,L^{-1}$. The epoc measures the creatinine in only 30 seconds and also provides an automatic calculation of eGFR. The epoc reagents do not require refrigeration but the analyzer does also have a "limited" operating temperature for LMICs of between 15 and 30°C.

The Nova StatSensor is a "non-blood gas" analyzer as it only measures creatinine. The device uses a dry strip which requires just 1 μL of whole blood. The StatSensor has a measuring range of $27-1056\,\mu mol\,L^{-1}$. The device takes 30 seconds to provide both a creatinine and calculated eGFR result. The device has a slightly wider operating temperature range of 15°C and 40°C but its reagents require refrigeration (Table 16.3).

In terms of analytical quality, the goals for all creatinine POC testing devices are based on biological variation and are assessed in terms of total error (TE), which combines both accuracy and precision. For creatinine, the minimum TE goal is 11.4%, the desirable TE is 7.6% and the optimal TE is 3.8% (Myers et al. 2006). Studies have shown that both the i-STAT and epoc met analytical goals for creatinine (Shephard 2011; Tirimacco et al. 2015). However, the Nova StatSensor did not perform as well analytically as the other devices, tending particularly to underestimate creatinine results at the high end (Shephard 2011; Kosack et al. 2015).

Table 16.3 Devices for the measurement of blood creatinine.

Device	Range of creatinine measurable ($\mu mol\,L^{-1}$)	Amount of blood required (μL)	Analysis time	eGFR calculation	Storage conditions	Operating temperature range
Abbott i-STAT	18–1768	90	2 minutes	No	Refrigeration or room temperature for up to 14 days	16–30 °C
Alere epoc	27–1326	92	30 seconds	Yes	Refrigeration not required	15–30 °C
Nova StatSensor	27–1056	1	30 seconds	Yes	Reagents require refrigeration	15–40 °C

16.4.3 POC Testing Device Options for Urine Albumin

POC testing devices that measure urine albumin or ACR can be split into two broad categories, qualitative (or semi-quantitative) and quantitative. Urine dipsticks, such as the Roche Combur-Test and the Siemens Multistix, provide a qualitative measurement of albumin along with other metabolites in the urine. Although they are simple to use in all types of environmental settings, urine pads are subject to a high level of operator variability. This is due to the pads on the dipstick having to be read visually by the operator (although reading devices are available that can remove the subjectivity of the visual read). Dipsticks therefore only provide a crude estimate of protein in urine.

As for quantitative devices, these can be split into those that measure urine albumin and those that measure urine ACR. The two main POC testing devices that measure ACR are the Siemens DCA Vantage and the Alere Afinion. Both devices use ACR cartridges that measure urine albumin by immunoassay and urine creatinine colormetrically; they are ~60× more sensitive at detecting urine albumin than dipsticks. The DCA Vantage has a sample volume of $40\,\mu L$, a measuring range of 5–$300\,mg\,L^{-1}$ and 1–$44\,mmol\,mol^{-1}$ of urine albumin and urine creatinine, respectively and produces results in seven minutes. The DCA Vantage is a benchtop device that requires AC power, and its reagents require refrigeration, but can be stored at room temperature for up to three months. The Afinion has a sample volume of $35\,\mu L$, a measuring range of 5–$200\,mg\,L^{-1}$ and 2–$30\,mmol\,mol^{-1}$ for albumin and creatinine, respectively and produces results in five minutes. The Afinion also requires AC power, and its reagents require refrigeration, but can be kept at room temperature for three days.

In terms of analytical quality, the goals for imprecision that are used in the QAAMS program (mentioned earlier in this chapter) are <10% for albumin, <6% for creatinine and <12% for ACR (Shephard 2006a). Both the DCA Vantage and Alere Afinion have been shown to meet these analytical goals (Omoruyi et al. 2012).

16.4.4 A Case Study Illustrating the Value of POC Testing for Renal Function

The following case involved a 51 year old male presenting to a primary healthcare clinic in remote Australia.

16.4.4.1 Presentation

The patient presented with a two days history of diarrhea and vomiting associated with muscle cramps and general weakness. He had no abdominal pain and was afebrile.

16.4.4.2 POC Testing Performed

A CHEM8+ test on the i-STAT was performed to check the patient's hydration status and the results were as follows

- Sodium 134 mmol L^{-1} (reference interval 135–145 mmol L^{-1})
- Potassium 4.1 mmol L^{-1} (3.8–5.0 mmol L^{-1})
- Urea 16.1 mmol L^{-1} (3–8 mmol L^{-1})
- Creatinine 578 µmol L^{-1} (60–110 mmol L^{-1})

16.4.4.3 Diagnosis

The patient was diagnosed as being clinically dehydrated due to acute gastroenteritis with secondary acute renal impairment. The patient was commenced on intravenous normal saline for rehydration and was evacuated to the nearest hospital for assessment of renal impairment.

16.4.4.4 Treatment/Follow-up

In hospital the patient was continued on fluid replacement. Two days later pathology results showed an improvement in renal function with serum creatinine dropping to 184 µmol L^{-1} and then 84 µmol L^{-1} two days later.

16.4.4.5 Outcome

Following treatment the patient was discharged home with regular check-ups scheduled.

16.4.4.6 Summary

This case demonstrates how access to POC testing allowed the identification of acute renal failure without the presentation of symptoms. Without access to POC testing, diagnosis of acute renal failure may have been missed, which could have led to further renal complications.

16.5 Point-of-Care Testing for Electrolytes and Acid–Base Balance

16.5.1 Clinical Use of Tests for Electrolytes and Acid–Base Balance

The measurement of an electrolyte profile with at least one indicator of acid–base balance or status is one of the most commonly performed and useful pathology tests, whether in a tertiary referral hospital or in a primary care setting in the LMICs. The reasons for performing these tests may be very different in these two settings, markedly influencing the choice of device.

The most frequently measured electrolytes in blood are sodium (Na^+), potassium (K^+), chloride (Cl^-), and bicarbonate (HCO_3^-). This panel will be referred to as the standard electrolyte profile (ELP). These tests are often measured along with glucose, urea and creatinine, (ionized) calcium and hemoglobin on POC testing devices. Increases in other physiologically important anions such as lactate and ketoacids (acetoacetate and β-hydroxybutyrate) can also be measured by POC devices that have the ability to calculate the anion gap. (The anion gap = (Na+ + K+) – (Cl– + HCO_3^-). The anion gap can be normal, high, or low. A high anion gap indicates metabolic acidosis – increased acidity of the blood due to metabolic processes.)

The homeostatic mechanisms of the body act to maintain the body's pH within a narrow range. The determination of acid–base status requires the measurement of not only the pH of the fluid of interest (usually arterial or venous whole blood), but also the components of the major buffering system of the body, in order to determine the underlying cause of any abnormality. This therefore requires the measurement of carbon dioxide (as pCO_2) in addition to pH and HCO_3^-.

Bicarbonate can be measured either as part of the electrolyte profile (as mentioned previously) or as part of a blood gas profile (BGP), which usually includes the measurement of pH, oxygen (pO_2), pCO_2, HCO_3^-, and the calculation of base excess (BE). Although POC testing devices can measure a BGP in either venous or arterial whole blood samples, a venous sample is sufficient in most primary care situations, provided the appropriate reference range is used. pO_2 is significantly reduced in a venous, compared to an arterial sample. However, monitoring a patient's degree of oxygenation is frequently performed by pulse oximetry. Therefore in this chapter determination of an acid–base profile (ABP) is defined as a BGP minus pO_2.

Common indications for the measurement of a standard electrolyte profile likely to be found in the LMICs are shown in Table 16.4. In each of these conditions an ABP is also indicated due to potential acidosis or alkalosis from a metabolic cause (Klutts and Scott 2008; Brandis 2015).

Respiratory conditions, resulting in potential respiratory acidosis or alkalosis and where the measurement of an ABP is indicated, are likely to be seen in a low resource setting such as chronic obstructive pulmonary disease and asthma, pulmonary and other infections, central nervous system and chest trauma, drugs, snake venom, pulmonary embolism and upper airways obstruction.

Presentations of diarrhea with evidence of dehydration in remote areas where health service delivery is challenged are very common. In 2016 Shephard et al. (2016d) estimated that 4.7% of all residents of remote communities serviced by government-funded remote health services presented at least once with this condition. A high prevalence of abnormalities in the electrolyte profile and blood gas profile has also been found in re-emerging diseases such as Ebola virus disease (EVD) and dengue fever. In a recent study of recovering EVD patients in Guinea, the prevalence of hypokalemia was 33%, hyponatremia 77.6%, with decreased total CO_2 and a high anion gap, possibly explained by lactic acidosis which is also very frequent (Van Griensven et al. 2016).

Table 16.4 Indications for determination of ELP and for ABP when due to a metabolic cause.

Generalized symptoms or causes	Detailed cause/condition
Vomiting and/or diarrhea caused by an infectious condition	Gastrointestinal infections: *Escherichia coli*, Salmonella, Giardia, Cholera Other infections: Dengue Fever, Tuberculosis, Malaria, Ebola
Vomiting and/or diarrhea caused by a non-infectious gastrointestinal disorder.	Appendicitis, obstruction, liver or kidney disease, cholecystitis
Other causes of dehydration	Lack of water or exposure to temperature extremes
Accidents and shock	General accidents and injuries, disasters resulting in blood loss trauma and rhabdomyolysis
Toxins and reactions to therapeutic drugs	Alcohol, drugs, snake bites, diuretics, laxatives, anti-psychotics
Chest pain and shortness of breath	May be secondary to or accompanied by an electrolyte abnormality
Hyper- or hypo- glycaemia	Diabetes type 1 and 2 including ketoacidosis
Acute and Chronic Renal Failure	Acute kidney injury, chronic glomerulonephritis
Malnutrition	Starvation, prolonged gastrointestinal condition, cancer
Other less common symptoms and conditions in this setting	Addison's Disease, Cushing's syndrome

16.5.2 POC Testing Device Options for Electrolyte Profile and Acid–Base Balance

POC testing options are detailed in Table 16.5 and may be considered under the broad classifications of large bench top devices (>20 kg), medium-sized bench top devices (<20 kg) and hand-held devices.

The first POC testing devices to measure either electrolytes or blood gasses were relatively large devices weighing >20 kg and were located in intensive care units or emergency departments of large hospitals. These devices are expensive and not appropriate for the needs of most LMICs. However, device manufacturers have since developed medium-sized benchtop POC testing devices which measure blood gas and electrolyte profiles using cutting edge ion sensitive electrode technology.

Medium-sized analyzers are simple to use and maintain, with all consumables for a batch of tests provided in one reagent pack. Such devices (like the Instrumentation Laboratories GEM Premier 4000 which uses reagents that do not require refrigeration) may be of benefit in a regional center in LMIC or a field hospital which has a relatively high workload. In situations where a wider range of tests is necessary, for example where liver abnormalities are prevalent or where lipid profiles are required, alternative benchtop POC testing devices (such as the Abaxis Piccolo) are available which can provide these POC tests in addition to an ELP. In many settings, however, a hand-held device which is able to perform both ELP and ABP to the required analytical specifications is optimal in terms of portability, ease of use, robustness and cost-effectiveness. Such hand-held devices include the Abbott i-STAT and the Alere

Table 16.5 POC testing device options for electrolytes and acid–base balance.

Device	Bench top (BT) or hand-held (HH)/ (Weight in kg)	Tests performed in addition to ELP and ABP[a]	Comments
Siemens RapidPoint 500	BT/16.5 kg	Hb, Bilirubin	Cartridge life 28 days Replacement measurement and AQC cartridges require refrigeration (2-10 °C)
Instrumentation Laboratories GEM Premier™ 4000	BT/19 kg	Hct, Hb, Bilirubin	Cartridge life 30 days. Allows for various test configurations. Smallest cartridge packs test 75 samples. Replacement cartridges are stored at room temperature (15–25 °C)
Radiometer ABL90 FLEX	BT/11 kg	Hb, Bilirubin, Hct calculated	Sensor Pack and Solution Pack life in device 30 days Replacement sensor pack stored at 2–10 °C, Solution Pack 2–25 °C
Abaxis Piccolo	BT/5 kg	Bilirubin, creatinine, urea Does not perform ABP (apart from TCO_2)	Will perform over 10 test profiles including lipids and liver function tests
Samsung LABGEO^{PT10}	BT/2 kg	HbA1c, Creatinine Does not perform ABP (apart from TCO_2)	Similar cartridge profiles to Piccolo but includes HbA1c
Abbott i-STAT	HH/0.7 kg	ELP (Chem8+ cartridge) profile, also includes Creatinine, Hct and Hb (calculated from Hct). ABP (CG4+ cartridge) profile, also includes lactate. INR, Troponin I, CK-MB, BNP, HCG are available on separate cartridges.	A range of cartridges is available dependent on country Storage for 2 weeks only at RT
Alere epoc	HH/0.7 kg	Creatinine, Hct with calculation of Hb	Tests all on one BGEM test card, storage at RT

Legend: ELP: Electrolyte profile, ABP: Acid–Base Profile, BT: Bench top, HH: Hand held, Hb: Hemoglobin, Hct: Hematocrit, INR: International normalized ratio, TCO_2: Total carbon dioxide, CK-MB: Creatine kinase MB isoenzyme, BNP: B-type natriuretic peptide, HCG: human chorionic gonadotropin.
a) All devices will also measure glucose, lactate and calcium (ionized or total).

epoc. Both devices will also give a measure of the anion gap (and therefore determine the possibility of ketoacidosis), base excess and lactate concentrations.

All devices in Table 16.4 utilize a blood sample of 65–100 µL. A capillary sample should not be used for these tests profiles due to the risk of hemolysis or dilution with interstitial fluids. The choice of device will depend on the exact clinical setting (regional/remote/disaster) and whether additional POC tests are required. The predicted test throughput and the cost of the individual tests must also be taken into account. For example with hand-held devices, while the epoc has the advantage of room temperature storage for its test cards, it cannot measure cardiac troponin or INR.

16.5.2.1 Recent evidence-base for the effectiveness of POC testing for electrolytes and acid–base balance (with relevance to the LMICs)

The Abbott i-STAT, using the Chem8+ cartridge, was used to test all patients diagnosed with EVD who were admitted to the Médecins san Frontières treatment unit in Guinea in 2015 (Van Griensven et al. 2016). A high prevalence of test abnormalities was found and, as well as guiding treatment, risk criteria were developed to determine which patients required a higher level of care. The use of the Piccolo to measure a panel of tests including a full ELP and creatinine was found to decrease the turnaround time for results and for the time for clinical decision making in teaching hospitals in Korea (Lee et al. 2011). The Piccolo has also been used successfully in Antarctica by the Polar Medicine Unit of the Australian Antarctic Division (Watzl and Ayton 2016).

The Flinders University International Centre for Point-of-Care Testing, in partnership with the Northern Territory (NT) Department of Health, manages an i-STAT POC testing program in 72 remote health facilities across the NT of Australia (Spaeth et al. 2017). The ELP is measured using the Chem8+ cartridge and the BGP using the CG4+ cartridge, with testing for INR and troponin I also being conducted in the program using separate cartridge types. All testing is carried out under a training, competency, and quality management framework (with training available in flexible formats including e-learning). Electrolytes comprise 27% and blood gasses 10% of the total number of tests performed in the program (Spaeth et al. 2017). The clinical and economic effectiveness of the program has been demonstrated with a reduction in costly medical retrievals to a major hospital as a result of the availability of these on-site POC tests. In a study of six remote NT health centers from July to December 2015, tests for ELP and ABP prevented 10 of 28 evacuations for patients who were assessed because they missed dialysis and 2 of 25 evacuations for patients who presented with acute diarrhea. These saved medical evacuations equated to cost savings of AU\$7734 per "missed dialysis" patient and AU\$757 per patient with acute diarrhea (Shephard et al. 2016d). When extrapolated to the whole of the NT, the per annum savings were AU\$6.21 m for patients who missed dialysis and AU\$1.51 m for those with acute diarrhea (Shephard et al. 2016d).

16.6 Point-of-Care Testing for Hepatitis

16.6.1 Clinical Use of Hepatitis Testing

Hepatitis can be caused by infection, alcohol or drug use. Viral hepatitis, caused by hepatitis A, B, C, D, or E viruses, is at the top of the list of ten leading causes of mortality globally, claiming more than one million lives a year. The five hepatitis viruses differ greatly in their modes of transmission, the populations they affect and their health outcomes. Viral hepatitis B (HBV) and hepatitis C (HCV) are blood-borne infections with significant transmission occurring in early life (from mother to child at birth or during the first five years of life), through unsafe injections and medical procedures, or less commonly through sexual contact. Worldwide over 350 million people are living with chronic HBV and HCV infection (WHO 2016), meaning that the combined global burden of hepatitis greatly exceeds the burden contributed by HIV/AIDS (33 million) and other sexually transmitted and blood-borne infections (137 million) (Pai et al. 2016).

The burden of hepatitis remains disproportionately high in low- and middle-income countries. HBV prevalence is highest in sub-Saharan Africa and East Asia, with 5–10% of the adult population infected. High rates (2–5% of the adult population) are also found in South America, Eastern and Central Europe, the Middle East and India (WHO 2016), the Pacific Islands, indigenous communities in Australia and Inuit communities in Canada (Shivkumar et al. 2012). Approximately 95% of adults infected with HBV clear the infection spontaneously; the remaining 5% develop chronic HBV infection (Pai et al. 2016). Effective treatment is available for people with chronic HBV, but most people require lifelong treatment. Immunization is the most effective strategy for HBV prevention.

HCV is found worldwide, with Central and East Asia, North Africa, and the Middle East being the most affected regions with over 3% of the general population infected. HCV in these areas is most often caused by unsafe medical procedures. HCV epidemics relating to unsafe injecting drug use occur worldwide, with approximately 67% of all injecting drug users having been infected with HCV. There is currently no vaccine to prevent HCV. However, effective (and expensive) treatments with cure rates over 90% are available. Despite this, the prevalence of HCV is increasing worldwide (WHO 2016).

16.6.2   Pathology Tests for Hepatitis

Unfortunately, despite effective treatments being available, few have access to them, with less than 1% of people with chronic hepatitis infection receiving antiviral treatment. This is largely due to the high costs, especially for chronic HBV infection where the treatment is lifelong. In terms of hepatitis testing, simple and effective testing strategies are lacking, with over 95% of people with chronic hepatitis undiagnosed (WHO 2016). In 2016, the World Health Organization launched a global strategy to eliminate viral hepatitis as a major public health threat by 2030. This aims to halt transmission and to provide everyone living with viral hepatitis with access to safe, affordable and effective care and treatment (WHO 2016).

While serological POC tests for hepatitis screening are available, POC tests for virological analyses are lacking. These tests are a mandatory requirement for treatment and without them, it is virtually impossible distinguish between acute and chronic hepatitis infection. Thus, the management of hepatitis infection still relies heavily on laboratory support (Johannessen 2015).

16.6.3   POC Testing for Hepatitis B

In order to diagnose HBV, serological tests are used to screen for detection of HBV markers, including hepatitis B surface antigen (HBsAg), hepatitis B surface antibody (anti-HBs), hepatitis B core antibody (anti-HBc), and hepatitis B envelope antigen (HBeAg). HBsAg is the first serological marker to appear in acute HBV and persistence of it for more than six months characterizes chronic HBV infection. The presence of anti-HBs indicates HBV immunity due to either past resolved HBV infection, or vaccination. The presence of anti-HBc may indicate HBV immunity due to past resolved

HBV infection, low-level chronic HBV infection, or resolving acute infection. The presence of HBeAg indicates high levels of viral replication and high infectivity. The diagnostic algorithm for HBV is detailed and expensive, and requires laboratory confirmatory testing and an HBV DNA viral load test before treatment initiation and monitoring. Whilst there are many serological POC tests available, there are currently no options for virological HBV testing outside the laboratory (Pai et al. 2016).

Regardless, inexpensive rapid POC tests are an excellent tool for first line HBV screening in low-resource settings, and for hard-to-reach populations such as injecting drug users (Shivkumar et al. 2012). There are over 20 POC testing options now available for detecting HBsAg by rapid immunoassay. A systematic review pooling data from over 15 different HBsAg POC tests showed a sensitivity of 94.7% (95% CI: 93.7–95.6) and specificity of 99.4% (95% CI: 99.2–99.6) (Shivkumar et al. 2012). The Alere Determine test is currently the most popular HBsAg test worldwide, detecting HBsAg by lateral flow in 50 µL of whole blood, serum or plasma collected from a finger-stick or venipuncture, with results available in 15 minutes. The test is easy to use, requiring minimal training and no equipment, has an 18-month shelf life and can be stored at room temperature (2–30 °C), making it suitable for low-resource settings. The Determine test has been extensively evaluated, with studies demonstrating a sensitivity between 92.7 and 99.6% and specificity between 96.0 and 100%. Importantly, it has been validated in many different countries including Asia and Africa, and no significant differences in sensitivity or specificity were seen when its use in LMICs was compared to its use in developed countries (Pai et al. 2016).

There are also a number of POC tests for anti-HBs and HBeAg. However, their use is less common and fewer evaluations have been conducted. The pooled sensitivity and specificity of four POC tests for anti-HBs were 92.7% (95% CI: 89.7–95.0) and 87.4% (95% CI: 83.5–90.7), respectively (Shivkumar et al. 2012). The only HBeAg test to be included in the systematic review was the Binax NOW which simultaneously detects both HBsAg and HBeAg with a pooled sensitivity of 97.0% (95% CI: 95.6–98.0) and a specificity of 99.7% (95% CI: 99.5–99.9).

16.6.4 POC Testing for Hepatitis C

While only 5% of people infected with HBV go on to develop chronic HBV, up to 85% of people infected with HCV proceed to chronic HCV. However, unlike chronic HBV which can remain latent, requiring lifelong treatment, there is no reservoir for HCV, meaning that it is fully curable with new antiviral drugs.

There are over 25 rapid immunoassay tests for detecting anti-HCV antibody available on the market. These serological tests are easy to use, require only a small amount of whole blood, serum, plasma or oral fluid and are relatively inexpensive (ranging from $0.50–$2.00 for blood-based tests and $10 for oral fluid tests). They are recommended by the WHO for use in settings with limited access to laboratories, hard-to-reach and rural populations. A systematic review pooling data from 32 studies on 25 rapid HCV tests showed a sensitivity of 99% (95% CI: 98–100) and a specificity of 100% (95% CI: 100–100) (Easterbrook 2016). There are currently only two rapid anti-HCV antibody tests prequalified by WHO including the SD Bioline HCV and the OraQuick HCV Rapid Antibody Test.

The SD Bioline HCV test is an immunochromatographic rapid assay for the qualitative detection of HCV antibodies in 10 µL of whole blood, serum or plasma, with results available in 5 minutes. The test is easy to use, requires minimal training and no equipment, has a long 24-month shelf life, can be stored at room temperature (1–30 °C), and is stable in humid conditions. It is intended for use in high prevalence populations or in people who have a history of HCV risk exposure including pregnant women. Its use has not been established in infants or children. Pooled data from three studies demonstrate a sensitivity of 93.5% (95% CI: 73.2–98.7) and a specificity of 99.5% (95% CI: 97.7–99.9) (Khuroo et al. 2015).

The OraQuick HCV test is also an immunochromatographic rapid assay for the qualitative detection of anti-HCV antibodies. This test is more expensive than the SD Bioline, but enables testing to be performed on one drop of oral fluid, whole blood, serum or plasma. The addition of oral fluid allows non-invasive testing to be performed which may help improve access to, and acceptability of, HCV screening. The test is easy to use, requires minimal training and no equipment, has an 18-month shelf life, can be stored at room temperature (2–30 °C) and results are available in 20 minutes. Pooled data from 12 studies demonstrated a sensitivity of 99.5% (95% CI: 98.9–99.8) and a specificity of 99.8% (95% CI: 99.6–99.9) (Khuroo et al. 2015). Of note, this test was the only rapid HCV test in eight different studies to identify acute HCV at the same time as the reference laboratory immunoassay. The other rapid HCV tests lagged behind by a few days to a few weeks (Khuroo et al. 2015).

However, the detection of anti-HCV antibody is only the first step in the HCV diagnostic pathway. This test does not distinguish between acute and chronic HCV infection, and thus cannot determine which patients require treatment (Pai et al. 2016). Following initial screening, all patients who test positive for anti-HCV antibody, should undergo confirmatory nucleic acid testing for HCV RNA (either quantitative or qualitative) or HCV core antigen (HCVcAg) testing to confirm the presence of HCV infection. An HCV viral load test (a quantitative HCV RNA test) is also recommended to establish a baseline prior to beginning antiviral therapy, with subsequent HCV viral load tests used to monitor response to treatment. Until very recently these HCV viral load tests could only be performed by a laboratory with turnaround times of up to one week.

In April 2017 the WHO added the Xpert HCV Viral Load performed on the Cepheid GeneXpert to their list of prequalified *in vitro* diagnostics to help in the management of HCV-infected patients undergoing antiviral therapy at the point-of-care. This assay performs RNA extraction, reverse transcription and real-time PCR targeting the 5′ untranslated region of the HCV genome, with all components of the test contained in a single-use cartridge. The test can be performed on 1 mL of serum or plasma with results in 105 minutes. A test cartridge is available for venipuncture or finger-stick capillary whole blood but this cartridge is currently marked for research purposes only. The test is easy to use (one minute hands-on time), and the cartridges have a 12-month shelf life when stored at room temperature (2–28 °C). When compared to the reference laboratory, the sensitivity and specificity of the Xpert HCV Viral Load was good, with a sensitivity 100% (95% CI: 92.0–100) and a specificity 99.1% (95% CI: 94.9–100) for venipuncture plasma and 95.5% (95% CI: 84.5–99.4) and 98.1% (95% CI: 93.4–99.8) for finger-stick capillary whole blood (Grebely et al. 2017). In addition to HCV Viral Load, the GeneXpert platform is able to perform a growing list of tests for different conditions including healthcare-associated infections such as *Clostridium difficile* and multi-resistant *Staphyloccus aureus* (MRSA), selected oncology and genetic tests, sexually transmitted infections such as HIV, chlamydia, gonorrhea, and trichomoniasis and other critical infectious diseases including influenza and Ebola virus disease. While the GeneXpert platform requires connection to power and a computer, has a relatively long turnaround time to result compared to most POC tests and requires more training, its use at the point-of-care (particularly for chlamydia and gonorrhea) has been validated in international settings including rural/remote and indigenous communities (Shephard et al. 2016a).

Overall POC testing for hepatitis has shown excellent performance compared to reference laboratory testing, providing access to quick and simple screening for hepatitis at the point-of-care, and the ability to diagnose and initiate treatment for HCV during a single visit to the clinic.

16.7 Conclusions

This chapter has articulated the benefits of utilizing POC testing in the LMICs for a range of chronic, acute, and infectious clinical conditions. However the enthusiasm for and opportunities provided by POC testing in LMICs need to be weighed against a range of potential barriers and limitations to its implementation.

There needs to be mobilization of political will to support POC testing at the primary care level (where it is most needed). This support relates particularly to the costs of purchasing devices, reagents and quality testing materials for POC testing networks through either direct funding or governmental reimbursement schemes.

The sustainability of POC testing services are linked to the maintenance of training and quality frameworks (as well as costs associated with POC testing as mentioned above). With regard to training, there needs to be innovative and creative thinking to deliver flexible modes of training delivery and a range of different types of training resources to support a network, with particular consideration given to the skills and knowledge of the POC device operator group and awareness of cultural issues that may impact on how POC training is conducted.

There are many logistic barriers to the safe, efficient and timely delivery of devices and consumables in LMICs, especially where large diagnostic companies do not have a major presence and where maintenance of cold chain delivery is critical.

A POC testing device for use in an LMIC should ideally undergo an initial assessment and evaluation at the clinical site in which it is to be used and by the device operators who will be performing patient testing routinely. This ideally requires collaboration with either the local laboratory or a specialist POC testing provider.

Consideration also needs to be given to the safe disposal of all waste associated with POC testing, particularly where reagents may contain potentially toxic or harmful chemicals.

Finally, for POC testing to be clinically effective, it should be embedded within clinical pathways that support decision making on the relevant clinical condition, notably in relation to when, why and how frequently POC testing should be used within the clinical algorithm for detection, management and treatment. POC testing performed in isolation or without a clinical framework may lead to inappropriate levels of testing and waste of reagents.

These barriers can and need to be overcome for POC testing to reach its full potential in the LMICs. Fortunately, there are now many long-standing evidence-based POC testing models that set an international precedent for translation into the LMICs.

Bibliography

American Diabetes Association (2015). Standards of medical care in diabetes – 2015. *Diabetes Care* 38 (Supp 1): S1–S93.

Australian Institute of Health and Welfare (AIHW). (2013). Chronic kidney disease: regional variation in Australia. Cat. No. PHE 172. Canberra: AIHW.

Australian Institute of Health and Welfare (AIHW). (2016). Chronic Kidney Disease. Viewed 6/6/17 from www.aihw.gov.au/chronic-kidney-disease

Brandis, K. (2015). Acid-Base physiology. www.anaesthesiamcq.com

Briggs, C., Kimber, S., and Green, L. (2012). Where are we at with point-of-care testing in haematology. *Brit. J. of Haema.* 158: 679–690.

CDC. (2017). National Chronic Kidney Disease Fact Sheet, 2017. Atlanta, GA: US Department of Health and Human Services, Centers for Disease Control and Prevention.

Colagiuri, S., Lee, C., Wong, T. et al. (2011). Glycemic thresholds for diabetes-specific retinopathy: implications for diagnostic criteria for diabetes. *Diabetes Care* 34 (1): 145–150.

Dawkins, R.C.H., Oliver, G.F., Sharma, M. et al. (2015). An estimation of the prevalence of diabetes mellitus and diabetic retinopathy in adults in Timor-Leste. *BMC Research Notes* 8: 249.

Easterbrook, Phillipa on behalf of the WHO Guidelines Development Group (2016). Who to test and how to test for chronic hepatitis C infection: 2016 WHO testing guidance for low- and middle-income countries. *J. Hepat.* 65 (1): S46–S66.

Goodall, I., Colman, P.G., Schneider, H.G. et al. (2007). Desirable performance standards for HbA1c analysis - precision, accuracy and standardisation. *Clin. Chem. & Lab. Med.* 45 (8): 1083–1097.

Grebely, J., Lamoury, F.M.J., Hajarizadeh, B. et al. (2017). Evaluation of the Xpert HCV Viral Load point-of-care assay from venepuncture-collected and finger-stick capillary whole-blood samples: a cohort study. *Lancet Gastro. & Hepat.* 2 (7): 514–520.

International Council for Standardization in Haematology (ICSH) (2014). Stability of complete blood count parameters with storage: toward defined specifications for different diagnostic applications. *Int. J. Lab. Hemat.* 36 (2): 111–113.

Johannessen, A. (2015). Where are we with point-of-care testing. *J. V. Hep.* 22: 362–365.

John, W.G. and the UK Department of Health Advisory Committee on Diabetes (2012). Expert position statement: use of HbA1c in the diagnosis of diabetes mellitus in the UK. The implementation of World Health Organization guidance 2011. *Dia. Med.* 29 (11): 1350–1357.

Khuroo, M.S., Khuroo, N.S., and Khuroo, M.S. (2015). Diagnostic accuracy of point-of-care tests for hepatitis C virus infection: a systematic review and meta-analysis. *PLoS One* 10 (3): e0121450.

Kidney Health Australia. 2017. Stages of chronic kidney disease. Viewed 6/6/17 from http://kidney.org.au/your-kidneys/detect/kidney-disease/stages-of-chronic-kidney-disease-787

Klutts, J.S. and Scott, M.G. (2008). Physiology and disorders of water, electrolyte, and acid-base metabolism. In: *Tietz Textbook of Clinical Chemistry*, 6e (ed. C.A. Burtis, E.A. Ashwood and D.E. Burns), 655–674. WB Saunders: Philadelphia.

Kok, J., Ng, J., Li, S.C. et al. (2015). Evaluation of point-of-care testing in critically unwell patients: comparison with clinical laboratory analysers and applicability to patients with Ebolavirus infection. *Pathology* 47 (5): 405–409.

Kosack, C.S., de Kieviet, W., Bayrak, K. et al. (2015). Evaluation of the Nova StatSensor® XpressTM creatinine point-of-care handheld analyzer. *PLoS One* 10 (4): e0122433. https://doi.org/10.1371/journal.pone.0122433.

Lee, E., Shin, S.D., Song, K. et al. (2011). A point-of-care chemistry test for reduction of turnaround and clinical decision time. *Am. J. of Emergency Medicine* 29 (5): 489–495.

Lenters-Westra, E. and Slingerland, R.J. (2010). Six of eight haemoglobin A1c point-of-care instruments do not meet the general accepted analytical performance criteria. *Clin. Chem.* 56 (1): 44–52.

Lenters-Westra, E. and Slingerland, R.J. (2014). Three of 7 hemoglobin A1c point-of-care instruments do not meet generally accepted analytical performance criteria. *Clin. Chem.* 60 (8): 1062–1072.

Motta, L. and Shephard, M. (2015). The international ACE Program – Point-of-care testing for diabetes management. *P. of C.* 14: 76–80.

Motta, L., Shephard, M., Albert, L., Timothy, C. and Harou, S. (2017a). Point-of-care testing for diabetes management in Papua New Guinea. Presented as a poster at the 14th National Rural Health Conference. Cairns, Australia, 26–29 April 2017.

Motta, L., Shephard, M., Brink, J. et al. (2017b). Point-of-care testing improves diabetes management in a primary care clinic in South Africa. *Prim. Care Diab.* 11 (3): 248–253.

Myers, G., Miller, W.G., Coresh, J. et al. (2006). Recommendations for improving serum creatinine measurement: a report from the Laboratory Working Group of the National Kidney Disease Education Program. *Clin. Chem.* 52 (1): 5–18.

Omoruyi, F., Mustafa, G., Okorodudu, A., and Petersen, J. (2012). Evaluation of the performance of urine albumin, creatinine and albumin-creatinine ratio assay on two POCT analyzers relative to a central laboratory method. *Clinica Chimica Acta.* 413 (5–6): 625–629.

Pant Pai, N., Baum, P., and Luo, R. (2016). Point-of-care testing for hepatitis. In: *A Practical Guide to Global Point-of-Care Testing* (ed. M. Shephard), 245–255. Melbourne: CSIRO Publishing.

Rao, L.V., Ekberg, B.A., Connor, D. et al. (2008). Evaluation of a new point of care automated complete blood count (CBC) analyser in various clinical settings. *Clinica Chimica Acta* 389 (1–2): 120–125.

Rao, L.V., Moiles, D., and Snydr, M. (2011). Finger-stick complete blood counts: comparison between venous and capillary blood. *P. of C.* 10 (3): 120–122.

Rappaport, A.I., Karakochuk, C.D., Whitfield, K.C. et al. (2016). A method comparison study between two hemoglobinometer models to measure hemoglobin concentrations and estimate anemia prevalence among women in Preah Vihear, Cambodia. *Int. J. Lab. Hemat.* 39: 95–100.

Shephard, M. (2006a). Analytical goals for point-of-care testing used for diabetes management in Australian health care settings outside the laboratory. *P. of C.* 5 (4): 177–185.

Shephard, M. (2006b). Clinical and cultural effectiveness of the 'QAAMS' point-of-care testing model for diabetes management in Australian Aboriginal medical services. *Clin. Biochem. Rev.* 27: 161–170.

Shephard, M. (2011). Point-of-care testing and creatinine measurement. *Clin Biochem. Rev.* 32: 109–114.

Shephard, M., Causer, L., and Guy, R. (2016a). Point-of-care testing in rural, remote and indigenous settings. In: *A Practical Guide to Global Point-of-Care Testing* (ed. M. Shephard), 343–354. Melbourne: CSIRO Publishing.

Shephard, M., O'Brien, C., Burgoyne, A. et al. (2016b). A review of the cultural safety of a national indigenous point-of-care testing program for diabetes management. *Aus. J. of Prim. Health* 22 (4): 368–374.

Shephard, M., Shaw, J., and Zimmet, P. (2016c). Point-of-care testing for diabetes: haemoglobin A1c. In: *A Practical Guide to Global Point-of-Care Testing* (ed. M. Shephard), 119–132. Melbourne: CSIRO Publishing.

Shephard, M., Spaeth, B., Kaambwa, B., et al. (2016d). Point-of-Care Testing for Better Management of Acutely Ill Remote Patients. Emergency Medicine Foundation Grant, Final Report, October 2016.

Shephard, M., Spaeth, B., Motta, L., and Shephard, A. (2015). Point-of-care testing in Australia: practical advantages and benefits of community resiliency for improving outcomes. In: *Global Point-of-Care: Strategies for Disasters, Complex Emergencies, and Public Health Resilience* (ed. G. Kost), 527–535. Washington, DC: AACC Press.

Shivkumar, S., Peeling, R., Jafari, Y. et al. (2012). Rapid point-of-care first-line screening tests for hepatitis B infection: a meta-analysis of diagnostic accuracy (1980-2010). *Am. J. Gastro.* 107: 1306–1313.

Spaeth, B., Shephard, M., Auld, M., and Omond, R. (2017). Immediate pathology results now available for all remote Northern Territorians. *Aus. J. of Rural Health*.

Spaeth, B., Shephard, M., McCormack, B., and Sinclair, G. (2015). Evaluation of hemocue white blood cell differential counter at a remote health centre in Australia's Northern Territory. *Pathology* 47 (1): 91–95.

Tirimacco, R., Simpson, P.A., Siew, L. et al. (2015). Evaluation of creatinine and chloride on the epoc blood gas and electrolyte point of care analyser. *Clin. Biochem. Rev.* 36: S30.

Van Griensven, J., Bah, E., Haba, N. et al. (2016). Electrolyte and metabolic disturbances in Ebola patients during a clinical trial, Guinea, 2016. *Eme. Inf. Dis.* 22 (12): 2120–2127.

Watzl, R. and Ayton, J. (2016). Point-of-care testing and extreme environments – the Australian Antarctic Division. In: *A Practical Guide to Global Point-of-Care Testing* (ed. M. Shephard), 406–419. Melbourne: CSIRO Publishing.

World Health Organisation (2009). *Dengue: Guidelines for Diagnosis, Treatment, Prevention and Control*. Geneva, Switzerland: WHO Press, World Health Organisation.

World Health Organisation (2011). Use of glycated haemoglobin (HbA1c) in the diagnosis of diabetes mellitus. *Dia. Res. & Clin. Prac.* 93: 299–309.

World Health Organization (2016). *Global Health Sector Strategy on Viral Hepatitis 2016–2021: Towards Ending Viral Hepatitis*. Geneva: WHO Press.

Webliography

www.aihw.gov.au/chronic-kidney-disease Australian Institute of Health and Welfare (AIHW). 2016. Chronic Kidney Disease.

Brandis, K. 2015. Acid-Base physiology. www.anaesthesiamcq.com

http://kidney.org.au/your-kidneys/detect/kidney-disease/stages-of-chronic-kidney-disease-787 Kidney Health Australia. 2017. Stages of chronic kidney disease.

Kosack, C.S., de Kieviet, W., Bayrak, K. et al. (2015). Evaluation of the Nova StatSensor® XpressTM Creatinine Point-Of-Care Handheld Analyzer. *PLoS ONE* 10 (4): e0122433. https://doi.org/10.1371/journal.pone.0122433.

www.qaams.org.au - Quality Assurance for Aboriginal and Torres Strait Islander Medical Services.

http://www.who.int/tdr/publications/documents/dengue-diagnosis.pdf?ua=1 - World Health Organisation. 2009. Dengue: Guidelines for Diagnosis, treatment, prevention and control. Geneva, Switzerland: WHO Press, World Health Organisation.

17

Microscopy Skills: Cell Counts, Gram Stains, Ziehl-Neelsen Staining (ZN) and Blood Films

Michael Harrison

Sullivan Nicolaides Pathology, Brisbane, Queensland, Australia

---

CHAPTER MENU

17.1   Introduction

Microscopy is one of the longest used forms of point-of-care testing (POCT) and in this role has truly stood the test of time. The combination of an adequate microscope, a few basic reagents and an astute and competent observer can produce a surprising amount of invaluable high-quality laboratory test information to guide the management of patients with serious disease at the point of their care. If done poorly, however, both the rapid result availability and the immediate therapeutic implications can mean poor outcomes for some patients. Some of these POCT microscopy tests can be used to screen samples and enable early intervention based on an interim result and with the definitive result becoming available from a more distant laboratory later on. An example is a direct Ziehl-Neelsen[1] (ZN) smear examination of sputum followed by culture for tuberculosis (TB). This may be a reasonable compromise.

---

1 First described by two German doctors: the bacteriologist Franz Ziehl (1859–1926) and the pathologist Friedrich Neelsen (1854–1898).

Revolutionizing Tropical Medicine: Point-of-Care Tests, New Imaging Technologies and Digital Health, First Edition. Edited by Kerry Atkinson and David Mabey.
© 2019 John Wiley & Sons, Inc. Published 2019 by John Wiley & Sons, Inc.

17.2 Microscopy

Microscopy appears to be deceptively simple and indeed the process of making a wet mount or a Gram stain or a stained blood film is relatively straightforward, but the interpretation of what is seen down the microscope is often far from easy. Well trained and experienced laboratory technicians who also receive regular feedback on their microscopy evaluations will be the first to acknowledge that some of the things they look at are very hard to interpret and often require second and even more opinions. Many microscopy findings will need a further and definitive test to be performed resolve the issue.

To be competent in basic microscopy the microscopist must be carefully trained against predetermined learning objectives and be properly assessed at the end of this training. They will then need regular feedback on their performance either by review of the results of the examinations they have made or by further testing on the patient sample being performed in an assisting laboratory. Regular quality assurance (QA) exercises which are relevant to the specimens examined will also need to be performed.

Unfortunately, these steps are not always followed. An example is a study of point-of-care error in which the authors (Meier and Jones 2005) reported on practitioner-performed microscopy in POCT laboratories in a resource-rich country – the USA. They noted significant levels of operator incompetence with findings that included 25% of certificate holders not having documented practitioners' ability to carry out the microscopic examinations indicated, 28% not having written operating procedures to which they could adhere to, and 36% not performing the required maintenance of the microscopes and other equipment that they used. Thus, this was not simply a matter of whether there were financial resources available to perform these vital functions but rather there was a lack of appreciation of the critical nature of proper training, feedback, careful following of protocols and adequate maintenance of equipment to ensure the delivery of an accurate result using basic microscopy.

A number of studies and reviews have been performed on the use of microscopy in specific areas of diagnostic testing in low- and middle-income countries (LMICs) including detection of urinary tract infections, detection of various fecal parasites, detection of TB and detection of malaria.

There is a resource produced by the WHO for this purpose – the Manual of Basic Techniques for a Health Laboratory, 2nd edition, 2003, which covers most aspects of setting up of a peripheral health laboratory, the use of a microscope and basic microscopic tests for stool parasites and parasites of skin and blood as well as Gram stains, examination of sputum and throat swabs, genital specimens aspirates and exudates and urine. Direct TB microscopy is also covered as is basic hematology and immunoserology. Although it is a simple "how to" manual, it should not be assumed that an untrained person could follow its instructions and competently perform microscopy.

Another useful resource is Essentials of Clinical Laboratory Management in developing regions from the The International Federation of Clinical Chemistry and Laboratory Medicine (IFCC). This practical document focuses on improving quality and functionality of already established clinical laboratories. In particular it discusses how the range of tests performed in the laboratory should be selected, how total quality management is applied and how the technical requirements of The International Organization for Standardization (ISO) 15189 are met.

These publications are valuable resources but the actual training is very dependent on a supportive environment with experienced and competent trainers following a documented training program and of course lots of "hands and eyes on" experience. Microscopy is an experientially developed skill and although it is widely and successfully taught and learnt by many, some people are simply unable to master it. Competency assessment during and on-completion of training is the most important way to ensure the quality of the results of microscopy examination. Competency assessment must also be performed periodically after the initial training.

Before embarking on setting up a peripheral laboratory to perform microscopic tests and training staff to do these tests it is important that the range of tests to be performed has already been carefully and realistically considered. Mabey et al. (2004) examined the way that the need for diagnostic tests can be quantified and also where microscopy fits into this scheme (Figure 17.1).

As can be seen in this figure the highest needs are for TB case management and malaria case management, both of which are dependent on accurate microscopy and other testing, something that can be performed in a basic laboratory with adequate equipment and well-trained staff.

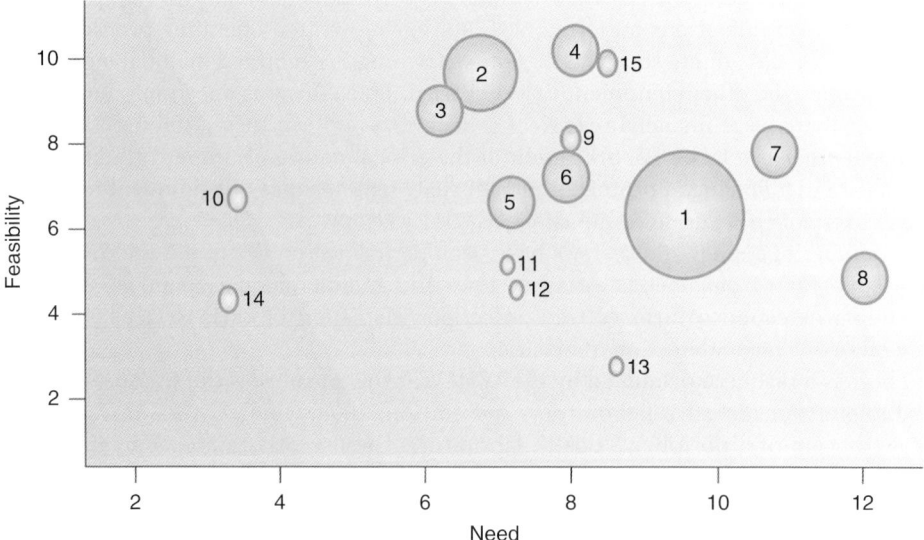

Figure 17.1 Scheme to assess priorities for diagnostics development for selected diseases in the developing world. The horizontal axis shows the need for the test: the vertical axis shows the feasibility of developing such a test; and the size of the circle indicates the relative burden of disease in disability adjusted life-years (DALYs) that could be prevented if it were widely used. Both scales are arbitrary. 1, tuberculosis case-management; 2, malaria case-management; 3, malaria test-of-cure; 4, syphilis case-management; 5, visceral leishmaniasis test-of-cure; 6, schistosomiasis case-management; 7, visceral leishmaniasis case-management; 8, African trypanosomiasis case-management; 9, lymphatic filariasis case-management; 10, dengue case-management; 11, leprosy case-management; 12, lymphatic filariasis test-of-cure; 13, leprosy – detection of latent disease; 14, Chagas disease (South America trypanosomiasis); 15, Onchocerciasis case-management. *Source:* Mabey et al. (2004).

17.3   Microscopy in a POC Testing Laboratory

The microscope used in a POCT laboratory must be suitable for the intended task. A modern binocular microscope with its own illumination source will significantly improve the quality of microscopic interpretation and allows for longer periods of careful microscopy to be undertaken with better results and less fatigue. However, even a monocular microscope using a mirror and an external light source can be adequate for the task if the amount of microscope work required is fairly small. At a minimum the lower lenses (known as objectives) should include a 10X, a 40X, and a 100X (oil immersion) lenses. It is critical that the lenses are kept clean – in particular the oil needs to be wiped off the oil immersion lens at least daily using soft paper tissue and other lenses may need to be cleaned if they become contaminated with microscope oil using a cleaning solution (80% alcohol, 10% ether and 10% acetone). Failure to keep oil off the other microscope lenses is a major cause of poor optical clarity leading to difficulty in making the correct diagnosis. Microscopes need to be well maintained and kept dust-free by being covered when not in use.

Each person who uses a microscope should be familiar with its basic operation. There are many excellent resources on the internet which demonstrate the proper use of a microscope and this should be the first component of a training program. Important issues to cover include using lower power (10X) and coarse focus to bring the slide into focus and then fine focus for any higher magnification objectives. Other factors needing attention are the adjustment of the binocular head to fit the operator's inter-pupillary distance, and the function and set-up of the condenser. Microscope lamps should be operated at the lowest effective power setting and turned off when the microscope is not in use, thus preserving their life. Microscope light bulbs are a critical spare part which must be carried at all times and all operators should know the correct procedure to change a blown light bulb. Keeping the microscope clean and limiting its use to people who are trained in its operation are other important principles to follow. Standard microscopes use bright field illumination but if the microscope is used regularly for urine microscopy then a phase contrast microscope will greatly aid in accurate identification of cell types and detection of renal casts. Lowering the condenser in a bright field microscope can mimic this effect; however, the condenser does need to be returned to its normal position when examining Gram[2] stains and blood films.

Essential reagents include glass slides, glass cover slips, Gram stain reagents and a blood film stain (either Diff-Quick or Giemsa). Either a reusable glass counting chamber with a matched coverslip (kept in a jar of alcohol between uses) or disposable counting chambers will allow for the examination of urine samples and other body fluids. Every slide/counting chamber MUST be labeled with the patient's name, date of test and other unique identifiers (if available) at the time of testing. All stain reagents need to be replaced at their expiry date and control slides should be available to test their performance at least daily or if testing is less frequently, performed before each use. Day-to-day management of the quality of reagents is critical in the performance of these important tests.

Typical instructions for the performance of various stains and examinations are shown below.

---

2 Gram staining was invented by the Danish bacteriologist Hans Christian Gram (1853–1938).

17.4 Gram Staining

1. Label the slide/smear with patient's surname, first name, laboratory number and date.
2. Air dry and heat fix slide (exception is body fluids where methanol fixation is the preferred method).
3. Flood with crystal violet for 30 seconds.
4. Rinse with tap water.
5. Flood with Lugol's iodine[3] for 30 seconds.
6. Tip off iodine and, while holding slide, add Gram decolorizer quickly (acetone: ethanol 1 : 1) until no more purple streams out.
7. Rinse immediately with tap water.
8. Flood with dilute carbol fuchsin for one minute.
9. Rinse with tap water and allow to dry.
10. A quality control slide – for example a mixture of *Staphylococcus* and *E.coli* prepared by the parent laboratory – should be stained and examined either daily or with each use of the stains (when Gram staining is performed less frequently than daily).
11. Scan slide at 10X magnification to find the most suitable area to examine. Leukocytes can be identified at 40X; however, oil immersion (100X) must be used to examine for bacteria.

Below are suggested reporting parameters for the microscopic findings of the Gram stain and various wet preparations (Table 17.1, and 17.2).

Note that for the Gram stain of sputum epithelial cell numbers should be reported using a 10X objective. All other findings should be reported in a semi-quantitative way using a 100X objective.

Table 17.1 Suggested reporting parameters for the microscopic findings of the Gram stain and various wet preparations.

Report	Gram stain	Wet preparation (feces)	Gram stain (sputum only)	Wet preparation
	Leukocytes, epithelial cells (not sputum), bacteria, yeasts	Leukocytes, erythrocytes, crystals	Epithelial cells	Casts
	100X objective oil immersion	40X objective	10X objective	10X objective
Occasional	<1 per field	<5 per field	<5 per field	<1 per field
+	1–5 per field	5–30 per field	5–10 per field	1–5 per field
++	5–30 per field	30–150 per field	10–25 per field	5–10 per field
+++	30 per field	>150 per field	>25 per field	>10 per field

Note that for the Gram stain of sputum epithelial cell numbers should be reported using a 10X objective. All other findings should be reported in a semi-quantitative way using a 100X objective.

---

3 Lugols iodine was first made in 1829 by the French physician Jean Lugol (1786–1851).

Table 17.2 Suggested reporting parameters for the microscopic findings of protozoa and helminths.

Report	Wet preparation - protozoa	Wet preparation - helminths
	40X objective	10X objective
Occasional	1–5 per coverslip	1–5 per coverslip
+	1 per 5 to 10 fields	1 per 5 to 10 fields
++	1 per 2 to 3 fields - 1 to 2 per field	1–2 per field
+++	>3 per field	>3 per field

The typical useful application of the Gram stain (without an accompanying culture) is quite limited. Most samples contain a range of bacteria and microscopy cannot distinguish pathogens from non-pathogens. Only if the material is from "sterile sites" will Gram stain microscopy be informative – for example, CSF, joint aspirates and gastric aspirates from neonates. Gram stain deposit can mimic Gram positive cocci and easily lead to false positive results being reported. Alternatively, Gram negative bacteria – for example *Haemophilus influenzae* and *Neisseria meningitidis* – may be extremely hard to see, thus resulting in an incorrect false negative "no bacteria seen" result being reported.

17.5   Ziehl-Neelsen Stain (ZN) for *Mycobacterium tuberculosis*

Method

1. Prepare and heat fix slide.
2. Either heat loosely capped bottle of carbol fuchsin (0.5%) in microwave to approximately 70 °C using Table 17.5 as a guide (place bottle in plastic container while heating) or gently heat reagent on slide with a flame, e.g. Bunsen burner, alcohol-soaked swab.
3. Flood with heated carbol fuchsin for five minutes (take care, bottle is hot).
4. Rinse with tap water.
5. Decolorize with 5% acid alcohol until stain no longer streams from smear (approximately three minutes).
6. Rinse with tap water.
7. Flood with methylene blue for one minute.
8. Rinse with tap water and allow to dry.
9. Examine 300 oil immersion fields for acid fast bacilli.

Further details are shown in Table 17.3 and Table 17.4.

It should be noted that direct smears of sputum have a significantly lower sensitivity (35–70%) compared to culture. The sensitivity can be improved by digesting the sputum with bleach and either centrifuging the specimen or simply standing it overnight and then making a smear of the deposit. Bleach has the advantage of reducing the risk of transmission of TB to laboratory staff (Foulds and O'Brien 1998).

Table 17.3 Methods for ZN staining for *Mycobacterium tuberculosis*.

Number of slides	Quantity of stain (mL)	Container/capacity	Heating time (seconds)
Circa 28	125	Pyrex bottle/250 mL	50
Circa 14	50	Wheaton bottle/125 mL	28
4	25	Wheaton bottle/125 mL	23

Table 17.4 Interpretation of ZN staining for *Mycobacterium tuberculosis*.

Number of AFBs seen per X100 oil immersion fields (HPFs)	Report
Nil in at least 100 HPFs	No AFBs seen
1–9 per 100 HPFs	Occasional (exact number per ×100 HPFs)
10–99 per HPFs	1+
1–10 per HPF in at least 50 HPFs	2+
More than 10/HPF in at least 20 HPFs	3+

Run positive and negative control slides with each batch of stains. For example use *Mycobacterium gordonae* as a positive control and *E.coli* as a negative control.

It is important to clean the microscope lens thoroughly after examining the positive control and other positive smears. Failure to do this may lead to transfer of positive material via the immersion oil on to the next slide.

17.6   Blood Film Preparation, Staining and Reporting

Method
- Mix ethylenediaminetetraacetic acid (EDTA) tube thoroughly by inverting tube 10 times until all red blood cells are completely resuspended.

17.6.1   Manual Preparation of a Blood Film

- Place a small drop of well-mixed EDTA anticoagulated blood in the center of a glass slide about 1 cm from the frosted end.
- Spread the drop of blood immediately as delay results in partial drying of the blood.
- Move the spreader back at an angle of about 45° until it makes contact with the drop of blood.
- When the drop of the blood has spread out along the line of contact between the spreader and the slide, push the spreader forward with an even steady motion.
- Regulate the thickness of the film by the angle of the spreader (the greater the angle, the thicker the film) and by the speed of spreading (the faster the film is spread, the thicker the film).
- Quickly air dry the slide. Slow drying may cause distortion of the erythrocytes.

- Label the film with the patient's last name, first name, laboratory number and date. In the event that a blood film is being made from a non-EDTA tube (i.e. citrate tube, whole blood or an apheresis sample) this needs to be indicated on the blood film label details.
- The spreader must be thoroughly cleaned and dried in between making blood films.

17.6.2   Staining of the Blood Film

Use Wright's Giemsa stain[4].

The protocol below is the recommended procedure and may be used for either manual or automated staining (for example, Merck Midas, Poly Stainer). Manual procedures may need adjustment of volumes if staining containers used vary in size. However, the ratio of solutions should remain constant (refer to local procedures manual for site-specific modifications). The details are shown in Table 17.5.

Notes:

- Change methanol daily to minimize water absorption.
- Change diluted Wright's Giemsa stain at least daily.
- Cover stain baths to reduce evaporation.
- Use freshly made buffer.
- Some local variations may be necessary to obtain optimal staining.
- If staining of films is sub-optimal, replace contents of all baths.
- Freshly made stain must sit for a minimum of 15 minutes prior to use.

17.6.3   Alternative Diff-Quick Method

- Air dry slide.
- Dip in fixative solution for five seconds (five dips for one second). Allow excess to drain.

Table 17.5 Recommended procedure for manual or automated staining using Wright's Giemsa stain.

Station number	Contents	Duration of staining
1	Absolute methanol	30 seconds
2	Wright's Giemsa stain	2 minutes
3	1 part Wright's Giemsa stain to 2 parts diluted[a] phosphate buffer pH 6.8 (for example, 100 mL Wright's Giemsa to 200 mL buffer)	5 minutes
4	Tap water rinse	Approximately 15 seconds
5	Diluted phosphate buffer pH 6.8[a]	Approximately 30 seconds
6	Dry	2 minutes or air dry

a)   Buffer to be diluted prior to use as per manufacturer's guidelines.

---

4 James Homer Wright (1869–1928) was an American pathologist. Gustav Giemsa (1867–1948) was a German chemist and bacteriologist.

- Dip slide in Solution I for five seconds (five dips for one second). Allow excess to drain.
- Dip slide into Solution II for five seconds (five dips for one second). Allow excess to drain.
- Rinse with distilled water and allow to dry.
- Cover using mounting media and glass cover slip.
- Macroscopically check the slide for quality: compare platelet count, total leukocyte count and hemoglobin values with film findings for plausibility. The minimum required for this would be a hemoglobin or hematocrit value; however, a total white cell count would be highly desirable. These can be obtained on basic POCT analyzers.

The rule of thumb is that there will be between one or two white blood cells per 1000 red blood cells and one or two platelets per 20 red blood cells. Platelet clumps can make the apparent number of platelets lower than expected.

Malarial parasite detection requires significant training and, in most cases, diligent examination of the thin area of a blood film for up to 30 minutes is required before the diagnosis can be reliably excluded. Even then significant numbers of false negative (low parasite count) cases will occur. Positive cases should be referred to a reference laboratory for confirmation of species identification although rapid antigen tests may also be helpful when microscopic speciation is uncertain. External quality assurance programs are essential to determine the quality of testing and the microscopist's competence. This process has been well documented (Ashraf et al. 2012) and includes the following steps:

- Routine testing of microscopist competency through use of reference slide sets or slide bank,
- remedial training of microscopists who fail routine evaluation,
- international accreditation of competency of microscopists at national reference level,
- cross-checking of slides submitted to reference or higher level microscopists in numbers that are sustainable and statistically relevant, and
- procurement and provision of quality microscopy supplies.

Results must be documented at the place of testing and should include the patient's full name, date of birth and date and time of testing.

Regular quality control needs to be performed. This will be determined by the type of testing done.

Any abnormalities detected should be referred to an associated laboratory for confirmation.

Staining quality of blood films is affected by a multitude of factors and Table 17.6 sets out how these faults can be determined.

17.7 Conclusions

The application of microscopy as a point-of-care test has been described in three different circumstances – the limited application of Gram stain examination to material removed from usually sterile sites, examination of ZN-stained sputum smears and malaria microscopy of stained thin smear blood films.

Table 17.6 Factors involved in faulty staining of blood films.

Problem	Possible causes and solutions
RBC artifact	• Change the methanol, it may be contaminated with water • Ensure spreader is completely dry before films are made
Stain precipitates	• Buffer concentration may be too high; try a fresh batch • Solvent content may be too low; try adding a small amount of methanol to the stain in use • Inadequate washing • Stain solution left uncovered too long; if old stain, replace stain • Stain solution not filtered; ensure solution was filtered
Pale staining	• Error in dilution/concentration too high • Old staining solution • Dye content too low: poor batch of stain (incorrect preparation of stock – manufacturing issue) • Impure dyes, especially Azure A and/or C (manufacturing issues)
Leukocyte nuclei light blue or neutrophils agranular	• pH too low • Staining time too short • Dye content too low- poor batch of stain (impure dyes/manufacturing/transport issues) • Insufficient azure B
Too blue	• Incorrect preparation of stock; eosin concentration too low (manufacturing issue) • Stock exposure to bright day light • Film too thick • Inadequate time in buffer solution
Too pink	• Incorrect proportion of azure B: eosin Y • Impure dyes • Buffer pH too low • Excess washing in buffer solution
Neutrophils appear toxic	• pH too high • Staining time too long • Azure B concentration too high or poor batch of stain
RBC and eosinophils appear blue	• pH too high • Buffer concentration too high • Stain time too long
Blue background	• Inadequate fixation or prolonged storage before fixation • Blood collected into heparin as anticoagulant

The critical success factors are

- Good equipment and reagents and basic QC/QA which is always performed as per written protocols
- Training of staff by experienced and expert trainers
- Competency assessment of microscopist skills on completion of training and regularly thereafter
- Retraining if any issues are identified.

If basic microscopy in these circumstances is to be successfully used then these requirements must be met.

If these requirements cannot be met then POCT should be restricted to non-microscopy applications – for example modern rapid diagnostic tests using antibodies and color indicating technologies.

Bibliography

Ashraf, S., Kao, A., Hugo, C. et al. (2012). Developing standards for malaria microscopy: external competency assess for malaria microscopists in the Asia-Pacific. *Malaria Journal* 11: 352.

Foulds, J. and O'Brien, R. (1998). New tools for the diagnosis of tuberculosis: the perspective of developing countries. *Int. J. Tuberc. Lung Dis.* 2 (10): 778–783.

de Kievert, W., Frank, E. and Stekel, H. n.d. Essentials of Clinical Laboratory Management in developing regions. http://www.ifcc.org/media/185572/2008%20-%20C-CLM%20Monograph.pdf

Mabey, D., Peeling, R., Ustianowski, A., and Perkins, M. (2004). Diagnostics for the developing world. *Nat. Rev.* 2: 231–240.

Meier, F.A. and Jones, B.A. (2005). Point-of-care testing error: sources and amplifiers, taxonomy, prevention strategies, and detection monitors. *Arch. Pathol. Lab. Med.* 129: 1262–1267.

WHO. (2003). Manual of basic techniques for a Health Laboratory, 2nd Edition. http://www.who.int/iris/handle/10665/42295

Webliography

http://www.ifcc.org/media/185572/2008%20-%20C-CLM%20Monograph.pdf n.d. de Kievert, W., Frank, E. and Stekel, H. Essentials of Clinical Laboratory Management in developing regions.

http://www.who.int/iris/handle/10665/42295 WHO. 2003. Manual of basic techniques for a Health Laboratory, 2nd Edition.

18

India Ink Stain and Cryptococcal Antigen Test for Cryptococcal Infection

Hannah K. Mitchell[1], Joseph N. Jarvis[1,2,3,4], and Mark W. Tenforde[5,6]

[1] Botswana-UPenn Partnership, Gaborone, Botswana
[2] Department of Clinical Research, Faculty of Infectious and Tropical Diseases, London School of Hygiene and Tropical Medicine, London, UK
[3] University of Botswana, Gaborone, Botswana
[4] Division of Infectious Diseases, Perelman School of Medicine, University of Pennsylvania, Philadelphia, PA, USA
[5] Division of Allergy and Infectious Diseases, University of Washington School of Medicine, Seattle, WA, USA
[6] Department of Epidemiology, University of Washington School of Public Health, Seattle, WA, USA

18.1   Introduction

Cryptococcus neoformans is an encapsulated yeast found worldwide in organic matter including soil and rotting wood. It is a major cause of morbidity and mortality in immunocompromised hosts, primarily causing a severe and often fatal meningoencephalitis. A second species, *Cryptococcus gattii*, has a more limited geographic distribution, and in addition to causing disease in immunocompromised hosts has been associated with disease outbreaks in immunocompetent individuals (Williamson et al. 2016). Asymptomatic pulmonary infection occurs after inhalation of the fungus and most people who are exposed will clear *Cryptococcus* or contain it within pulmonary granulomas without developing clinical disease (Baker 1976; Goldman et al. 2000). In immunosuppressed individuals, particularly those with defects in T cell-mediated immunity such as advanced HIV/AIDS infection, the disease frequently disseminates and progresses to a

Revolutionizing Tropical Medicine: Point-of-Care Tests, New Imaging Technologies and Digital Health,
First Edition. Edited by Kerry Atkinson and David Mabey.
© 2019 John Wiley & Sons, Inc. Published 2019 by John Wiley & Sons, Inc.

life-threatening invasive fungal infection (Tenforde et al. 2017). Cryptococcal infection has been reported in almost all body sites. Cryptococcal meningitis, caused by dissemination of yeast to the central nervous system, is the most severe manifestation of disease.

In resource-limited settings with high HIV prevalence, the vast majority of cases of cryptococcal meningitis are seen in individuals with advanced HIV/AIDS (typically with CD4$^+$ T cell counts <100 cells/mm^3). HIV-associated cryptococcal meningitis is a major public health concern, with an estimated 223 100 incident cases in 2014, primarily in sub-Saharan Africa and Southeast Asia (Rajasingham et al. 2017). Up to 15–20% of deaths in HIV-infected adults in sub-Saharan Africa are caused by cryptococcal meningitis (Lawn et al. 2008; Rajasingham et al. 2017) and cryptococcal meningitis is the most common microbiologically-confirmed cause of adult meningitis in this setting (Jarvis et al. 2010a; Wall et al. 2014).

Symptoms of headache, neck stiffness, confusion, visual disturbance, photophobia, or nausea and vomiting in an individual with known or suspected advanced HIV, typically developing over several days to weeks, should prompt laboratory investigation for cryptococcal meningitis. Once diagnosed, guidelines recommend a 14 days course of intravenous (IV) amphotericin B with oral flucytosine, which is associated with a clear survival benefit (WHO 2011; Day et al. 2013; Infectious Diseases Society of America [IDSA] 2013). This is followed by a continuation phase of oral fluconazole until immune restoration on antiretroviral therapy (ART). After commencing treatment for cryptococcal meningitis ART initiation should be delayed for five weeks (Boulware et al. 2014b).

Due to prohibitive drug costs, difficulty with prolonged IV therapy and drug monitoring, as well as the lack of flucytosine in most resource-poor countries, many settings instead use amphotericin B with high-dose fluconazole or high-dose fluconazole alone for treatment. Serial lumbar punctures (LPs) to lower raised intracranial pressure (ICP) and close laboratory monitoring are also important in management and may reduce mortality (Meda et al. 2014; Rolfes et al. 2014). Three month mortality is estimated at up to 55% in Southeast Asia and 70% in sub-Saharan Africa (Park et al. 2009). One major factor driving this devastatingly high mortality in resource-limited settings is lack of timely and accurate diagnosis (Tenforde et al. 2016). Over the past decade, major progress has been made in the development and implementation of cheap and highly accurate point-of-care testing for cryptococcal disease that can be utilized readily in resource poor settings. Screening high risk HIV-infected individuals for early cryptococcal infection with the cryptococcal antigen (CrAg) lateral flow assay and targeted pre-emptive fluconazole may also prevent meningitis and lower mortality.

18.2 Diagnosis of Cryptococcal Meningitis

Prompt diagnosis and initiation of antifungal therapy is essential for reducing morbidity and mortality from cryptococcal meningitis. A variety of methods exist for the diagnosis of cryptococcal infection including;

- Fungal culture
- India ink staining

- CrAg testing (detection of cryptococcal capsular antigen) by;

   i) Enyzme-linked immunosorbent (ELISA) assay
   ii) Latex agglutination (LA) assay
   iii) Lateral flow assay (LFA).

Current World Health Organization (WHO) guidelines recommend LFA or LA CrAg testing for the rapid diagnosis of cryptococcal meningitis (WHO 2011). In settings where LP can be readily performed and is not clinically contraindicated, LP with cerebrospinal fluid (CSF) testing is recommended. LP has the important additional benefit of allowing measurement of ICP by manometry and drainage of CSF in order to decrease raised pressure, and should always be encouraged. In settings where LP cannot be performed easily or is contraindicated, a presumptive diagnosis of meningitis is made with the appropriate clinical context and a positive blood LFA or LA test. India ink staining of CSF is recommended for rapid diagnosis of cryptococcal meningitis only when these tests are unavailable.

Fungal culture of CSF is still important in the diagnosis of cryptococcal meningitis, but should be performed together with a rapid test because culture takes several days to yield a positive result. Following treatment, CrAg assays remain positive for weeks to months because they detect capsular polysaccharide antigen from either living or dead *Cryptococci*. Therefore, culture is required to investigate relapsed or refractory cryptococcal meningitis and to confirm fungal clearance in CSF. Rapid CrAg assays and India ink stain will be considered for the remainder of this Chapter.

18.3   Cryptococcal Antigen Testing (CrAg)

A number of commercially available tests have been developed for the detection of cryptococcal capsular polysaccharide antigen (glucuronoxylomannan) by LA, ELISA, and LFA.

18.3.1   Latex Agglutination (LA) and Enzyme-Linked ImmunoSorbent Assays (ELISA)

LA and ELISA tests have been commercially available for decades and can deliver same day results. Both assays are generally based on polyclonal antibodies and have a sensitivity of 93–100% and a specificity of 93–98% for the diagnosis of cryptococcal meningitis (Tanner et al. 1994; Jarvis et al. 2011; Boulware et al. 2014a; Kabanda et al. 2014). They can be used on CSF, serum, and plasma. CSF CrAg titers derived from LA testing of serial sample dilutions are closely correlated with viable organism load derived from quantitative cryptococcal cultures at disease presentation, and high titres are strongly associated with increased risk of mortality (Jarvis et al. 2014). During effective antifungal treatment cryptococcal colony forming unit (CFU) counts decline rapidly, but CrAg titres decline much more slowly, often staying positive for several years, and do not correlate with the rate of decline in CFUs (Brouwer et al. 2005; Jarvis et al. 2011). Thus, serial monitoring of CrAg titers during treatment cannot be used reliably to determine treatment responses and is not recommended.

However, these assays also have major limitations that restrict their use in resource-limited settings. The latex agglutination assay requires laboratory infrastructure and technical expertise. CSF samples are initially centrifuged and pre-heated at 100°C before performance of the assay and serum samples require pre-treatment with pronase. (Pronase is a commercially available mixture of proteases isolated from the extracellular fluid of *Streptomyces griseus*.) Test reagents require refrigeration, necessitating a stable cold supply chain. ELISA testing is also impractical and rarely performed in resource-limited settings as a spectrophotometer is required to deliver a result and the test again requires a stable cold chain. Both tests are relatively expensive, particularly if performed in small volumes, further limiting their use in most resource-limited settings.

18.3.2 Lateral Flow Assay

The CrAg lateral flow assay (IMMY, Norman, OK, USA) was developed in 2009 and approved for use by the United States Food and Drug Administration in July 2011. The LFA is similar in appearance to a home urine pregnancy test.

The test strip contains gold-conjugated monoclonal antibodies that cross-react with *Cryptococcus* polysaccharide capsular glucoroxylomannan (GXM) antigen. One drop (40 μL) of CSF is added to one drop of manufacturer's diluent and, in the presence of cryptococcal antigen, two red lines appear on the strip within 10 minutes. One control line appears in a negative test. Through serial dilutions, semi-quantitative results can also be obtained.

The LFA test has several notable advantages over the LA, ELISA and the India ink stain and is the best test for accurate and rapid diagnosis of cryptoccoccal disease in resource-limited settings. Test kits can be stored at room temperature and have a shelf life of two years. The assay does not require a stable cold supply chain, sample boiling, or a microscope, centrifuge, or other expensive centralized laboratory equipment. Furthermore, it is simple to perform and interpret, requiring minimal operator training. The manufacturer sells the test kits in resource-limited settings for approximately US$2 per kit. In South Africa the estimated total cost including consumables and laboratory costs is US$4.28 per blood LFA test (Cassim et al. 2017), which is significantly lower than the costs of the LA, ELISA or CrAg assays (Meya et al. 2010).

The LFA detects all *Cryptococcus* serotypes (A-D) and is highly sensitive across all subtypes of *C. neoformans* and *C. gattii* (Percival et al. 2011). The test has been validated for use on CSF and blood. A pooled analysis of early validation studies showed a median sensitivity of 100% and specificity of 97.7% in CSF, and a sensitivity of 100% and specificity of 99.5% in blood samples, with higher sensitivity than either LA or ELISA (Vijayan et al. 2013). The slightly lower observed specificity is almost certainly artefactual, and due to higher sensitivity of the LFA test than the comparator test of culture or LA methods (Boulware et al. 2014a). As with the LA test, LFA titres at the time of antifungal treatment initiation correlate closely with quantitative cryptococcal culture results and are predictive of mortality (Kabanda et al. 2014).

More recently the test was evaluated in a Ugandan study on capillary whole blood obtained by finger prick. This study showed 100% agreement between capillary whole blood and reference serum/plasma samples (Williams et al. 2015), suggesting that blood-based diagnosis can be performed with rapid point-of-care finger testing using a simple lancet without a need for phlebotomy supplies for venipuncture.

Several non-invasive sampling sources have been evaluated for LFA testing, including saliva and urine. Unfortunately, testing on saliva has very poor sensitivity (Kwizera et al.

2014). In addition to limited sensitivity, LFA testing of urine samples was found to have poor specificity, which could result in a significant number of false positive tests and unwarranted treatment (Magambo et al. 2014; Longley et al. 2016).

18.4 India Ink Stain

India ink is a dark black liquid made from soot. It was first used around 3000 BCE in China. Its principle uses over past millennia have been for non-medical purposes, such as for art, calligraphy, printing, and tattooing.

India ink can be used as a stain for *Cryptococcus*, forming an outline around the thick polysaccharide capsule of the yeast. Following placement of one drop of India ink on a slide one loop of CSF is added and mixed with the ink, a cover slip is placed and the slide examined using a light microscope (WHO 2009). A counting chamber can also be used. If *Cryptococcus* is present, white "halo" shaped circles can be appreciated on the slide under the microscope, as the polysaccharide capsule does not take up the black dye (Figure 18.1).

The yeast can also be seen on Gram stain, but India ink stain is preferable as it allows clear visualization of the capsule, distinguishing it from other yeasts.

A comparison between the CrAg test and the India ink test are shown in Table 18.1.

In the context of suspected cryptococcal meningitis, sensitivity of India ink is poor. Published sensitivity estimates range from 30 to 86% (Dominic et al. 2009; Jarvis et al. 2010a; Boulware et al. 2014a). Differences in observed sensitivity are likely due to operator training and experience in preparing slides and detecting yeast by microscope.

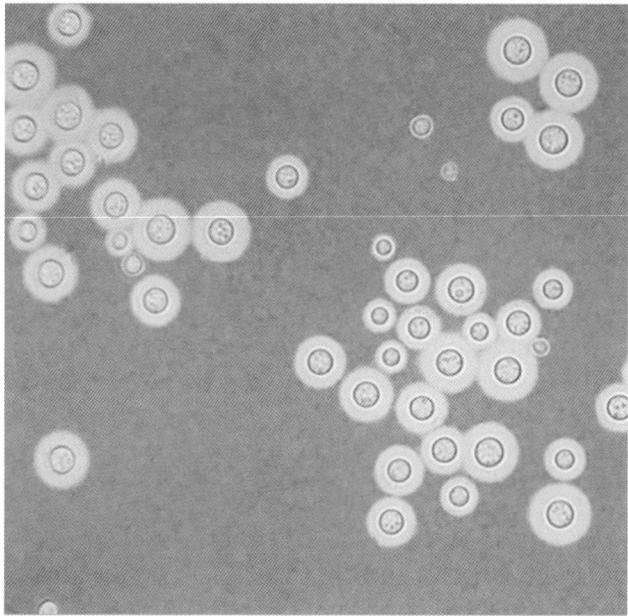

Figure 18.1 Cryptococci stained with India ink. *Source:* Centers for Disease Control and Prevention (2013). This image is by an employee of the USA Federal Government in the course of their work and as such the image is in the public domain.

Table 18.1 A comparison of CrAg testing and India ink testing.

Parameter	CrAg test	India ink test
Performance on CSF	100% sensitivity, 97.7% specificity	CSF: 30–86% sensitivity, close to 100% specificity
Performance on blood	100% sensitivity, 99.5% specificity	
Cost	US$2 for test kit; US$3–4 with laboratory and distribution costs	<US$1 for reagent
Time to result	10 minutes	Minutes
Equipment	Test kit, pipette, microtubules	Light microscope, slides, ink, pipette
Specimen sources	CSF, finger prick capillary blood, venipunture blood (serum or plasma)	CSF

Patients presenting with cryptococcal meningitis in resource-limited settings frequently have advanced immunosuppression with high CSF fungal burdens, and sensitivity may be higher in this context but still unacceptably low for use as a reliable diagnostic test.

Advantages of the India ink stain include low cost of reagent (the cost of India ink per sample is a fraction of US$1) and no need for a cold supply chain since India ink can be stored at room temperature. However, poor sensitivity potentially leading to treatment delay, need for operator expertise, and requirement of a light microscope make India ink an inferior rapid diagnostic test compared to the CrAg LFA. Additionally, India ink stain cannot be performed on blood, so diagnosis cannot be confirmed using this method when lumbar puncture cannot be performed or is contraindicated.

Difficulty in obtaining India ink has been reported in resource-limited settings. An alternative to India ink has been proposed in the form of mascara, which was found to be at least equivalent to India ink in a small study but is likely of limited clinical value (Ibembe and Wiggin 2015).

18.5 CrAg Testing for the Prevention of Cryptococcal Meningitis

Several cohort studies of ART-naive, HIV-infected patients have demonstrated that asymptomatic cryptococcal infection, detected using blood CrAg testing, strongly predicts the development of cryptococcal meningitis following ART initiation (French et al. 2002; Jarvis et al. 2009).

The median observed period between positive CrAg blood testing and clinically apparent meningitis in these studies was three to five weeks, providing a window of opportunity to preemptively treat the subclinical infection before the development of life-threatening meningoencephalitis. Using pooled data from published studies detection of CrAg was found in 7.2% of patients with advanced HIV infection in Africa and a similar prevalence was found in Southeast Asia (Meya et al. 2015).

In 2011 the WHO conditionally recommended screening serum or plasma for CrAg in ART-naive, HIV-infected adults with a CD4 count of <100 cells/mm^3 before initiation of ART in regions with CrAg prevalence >3% in order to enable detection of early cryptococcal infection and treatment before disease progression (WHO 2011).

In areas with high CrAg prevalence samples with $CD4^+ < 100$ cells/mm^3 processed by a laboratory should be automatically screened for cryptococcal antigen. Far higher CrAg screening rates are seen with routine "reflex" screening of all samples with $CD4^+ < 100$ cells/mm^3 compared to provider-initiated screening when clinicians have to individually request CrAg tests (Vallabhaneni et al. 2016).

Patients with a positive CrAg test should be screened for symptoms and signs of cryptococcal meningitis (headache, confusion, neck stiffness), which if present would require further evaluation with a lumbar puncture and appropriate treatment for cryptococcal meningitis if diagnosed (Jarvis et al. 2012). Figure 18.2 is an algorithm for cryptococcal screening and implementation of treatment (CDC 2013).

In asymptomatic patients who are CrAg-positive, a delay in ART initiation for two weeks is recommended along with a tapering course of fluconazole to prevent progression of infection to meningitis:

- 800 mg fluconazole once daily for two weeks; then
- 400 mg fluconazole once daily for eight weeks; then
- 200 mg fluconazole once daily until $CD4^+$ count is >200 cells/mm^3 for six months.

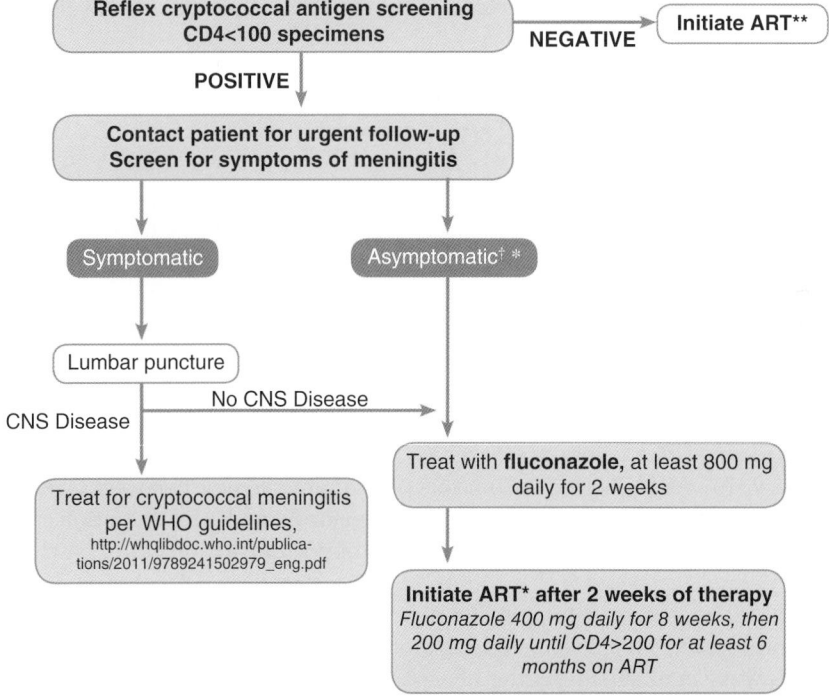

† If resources are available, a lumbar puncture should also be offered to asymptomatic patients with appropriate counseling.
*Populations who require special attention include: patients on tuberculosis medications or nevirapine, patients with a previous history of cryptococcal meningitis, pregnant women or breastfeeding mothers, patients with liver disease, and children.
**Initiate ART if not already started

Figure 18.2 Infographic for cryptococcal screening implementation (CDC). Abbreviation used: ART, antiretroviral therapy. *Source:* This image is by an employee of the USA Federal Government in the course of their work and as such the image is in the public domain.

CrAg screening and targeted pre-emptive fluconazole therapy for CrAg-positive patients has been shown to reduce the number of cases of incident cryptococcal meningitis (Longley et al. 2016). A large multicenter randomized study in Tanzania and Zambia also showed that a group receiving community support including LFA CrAg screening prior to ART initiation had 28% lower mortality than a standard care comparator group (Mfinanga et al. 2015). This suggested a mortality benefit from CrAg screening. Screening has also been shown to be highly cost-effective in diverse settings in sub-Saharan Africa and Southeast Asia (Meya et al. 2010; Jarvis et al. 2013; Rugemalila et al. 2013; Smith et al. 2013; Ramachandran et al. 2017).

Areas of uncertainty remain for CrAg screening for prevention of HIV-associated cryptococcal meningitis. First, an ideal antifungal treatment regimen is not definitively established. Previous dose escalation studies demonstrated a greater rate of fungal killing with higher dose of fluconazole (1200 mg daily compared to 800 mg daily) with no increase in observed drug toxicity (Longley et al. 2008). Either higher doses of fluconazole or high-dose fluconazole with oral flucytosine may be superior. Second, lack of meningitis symptoms at the time of positive CrAg screening does not definitively rule out meningitis detectable by CSF testing (Longley et al. 2016). Third, the efficacy of CrAg screening for the prevention of cryptococcal meningitis in HIV-infected patients previously started on ART is not established. Lastly, the optimal period of delay before initiation of ART is unclear.

18.6 Logistical Challenges of CrAg Screening

Logistical challenges must be overcome to effectively scale-up CrAg screening for the prevention of HIV-associated cryptococcal meningitis. The best approach to screening will depend on the unique context and delivery models of local health systems.

Centralized laboratory-based CrAg screening may lead to delays in diagnosis of early infection and treatment. Laboratories must first run samples, then notify clinicians of positive test results, then patients must be contacted to arrange follow-up and additional evaluation and treatment. Loss to follow-up has been identified as a major obstacle in the scale-up of CrAg screening, with a study in South Africa showing that 48% of patients who had a positive laboratory-based CrAg test did not return for follow-up (Govender et al. 2015). As high-dose fluconazole is recommended for two weeks before starting ART, additional delays between starting fluconazole and ART initiation may also lead to a prolonged period of high risk for other opportunistic infections, such as tuberculosis.

Decentralized LFA CrAg screening at the point-of-care probably has the greatest potential for preventing HIV-associated cryptococcal meningitis in patients with advanced HIV (Wake et al. 2015). As the LFA performs well on finger prick capillary blood and results are available in less than 10 minutes (Boulware et al. 2014a), LFA screening could be implemented in a clinic or other community settings for HIV-infected patients with a documented or suspected low CD4 count, newly diagnosed HIV, or based on WHO clinical staging. Rapid point-of-care CD4 count testing using commercial assays could also guide reflex CrAg screening. Point-of-care LFA testing facilitates same-day symptoms screening and initiation of pre-emptive fluconazole in asymptomatic patients or timely evaluation by lumbar puncture in symptomatic patients.

Effective rollout of CrAg screening programs will require major shifts in clinical care models, as HIV testing and treatment services are frequently disconnected and HIV-related laboratory testing is often performed at stand-alone reference laboratories. Furthermore, in symptomatic patients who require an LP, local capacity is necessary to perform appropriate evaluation.

18.7 Non-Meningeal Cryptococcal Disease

Pulmonary cryptococcal disease is common in patients with advanced HIV infection. An estimated 10–55% of patients with cryptococcal meningitis also have pulmonary involvement (Jarvis and Harrison 2008). An autopsy series of 8421 South African miners with a high prevalence of HIV found that 7% had cryptococcal pneumonia, of whom only 1.2% were diagnosed prior to death (Wong et al. 2007). Co-infection was also common, observed in 16.5% of miners with cryptococcal lung infection and primarily due to *Pneumocystis jirovecii* and pulmonary TB.

Diagnosis of non-meningeal cryptococcal disease is challenging in resource-limited settings. Cryptococcal pneumonia should be considered in immunosuppressed patients presenting with respiratory symptoms and abnormal pulmonary imaging or smear-negative tuberculosis that is not responding to treatment (Jarvis et al. 2010b). Definitive diagnosis of pulmonary cryptococcosis requires isolation of *Cryptococcus* from bronchoalveolar lavage (BAL) culture or histology showing encapsulated yeast forms (Jarvis and Harrison 2008). Limited data suggest that detection of CrAg in BAL using the LA test is also potentially of clinical utility, with an LA CRAG titer of >1:8 being 100% sensitive and 98% specific for cryptococcal pneumonia (Baughman et al. 1992). However, in resource-limited settings these investigations are often not possible. In such circumstances, patients with suspected pulmonary cryptococcosis may have serum CrAg checked (either by LA or LFA) and treatment initiated if positive. Cases of smear-negative suspected TB should also undergo CrAg testing of blood to evaluate the possibility of pulmonary cryptococcosis (Jarvis et al. 2011).

All immunocompromised individuals diagnosed with pulmonary cryptococcosis require a lumbar puncture to exclude dissemination with cryptococcal meningitis (Perfect et al. 2010). Mild-to-moderate pulmonary disease can be managed with 6–12 months of oral fluconazole, whereas severe pulmonary disease or concomitant meningitis necessitates IV amphotericin-based induction therapy.

18.8 Conclusions

The diagnosis of cryptococcal meningitis has been transformed in recent years with the development of the cryptococcal antigen lateral flow assay, an affordable test that delivers results within 10 minutes, has a sensitivity equivalent to or better than culture (Boulware et al. 2014a), and is ideally suited for use in resource-poor settings with limited infrastructure or trained laboratory personnel. Use of the LFA test has the ability to save many lives through prompt diagnosis of cryptococcal meningitis.

The rapid LFA shows additional promise for screening early infection in patients with advanced HIV enabling pre-emptive treatment to prevent the development of meningitis. A health system focus should be on disseminating tests more widely, training staff on their use and guidelines and incorporating testing into national HIV programs so that early asymptomatic cryptococcal infection can be detected and managed before development of overt clinical disease.

References

Baker, R.D. (1976). The primary pulmonary lymph node complex of cryptococcosis. *American Journal of Clinical Pathology* 65 (1): 83–92.

Baughman, R.P., Rhodes, J.C., Dohn, M.N. et al. (1992). Detection of cryptococcal antigen in bronchoal veolar lavage fluid: a prospective study of diagnostic utility. *The American Review of Respiratory Disease* 145: 1226–1229.

Boulware, D.R., Meya, D.B., Muzoora, C. et al. (2014a). Timing of antiretroviral therapy after diagnosis of cryptococcal meningitis. *New England Journal of Medicine* 370 (26): 2487–2498.

Boulware, D.R., Rolfes, M.A., Rajasingham, R. et al. (2014b). Multisite validation of cryptococcal antigen lateral flow assay and quantification by laser thermal contrast. *Emerging Infectious Diseases* 20 (1): 45–53.

Brouwer, A.E., Teparrukkul, P., Pinpraphaporn, S. et al. (2005). Baseline correlation and comparative kinetics of cerebrospinal fluid colony-forming unit counts and antigen titers in cryptococcal meningitis. *The Journal of Infectious Diseases* 192 (4): 681–684.

Cassim, N., Schnippel, K., Coetzee, L.M. et al. (2017). Establishing a cost-per-result of laboratory-based, reflex cryptococcal antigenaemia screening (CrAg) in HIV+ patients with CD4 counts less than 100 cells/μl using a Lateral Flow Assay (LFA) at a typical busy CD4 laboratory in South Africa. *PLoS One.* 12 (2): e0171675.

Centers for Disease Control and Prevention (2013). Cryptococcal Screening Program Training Manual for Healthcare Providers. US: CDC.

Day, J.N., Chau, T.T.H., Wolbers, M. et al. (2013). Combination antifungal therapy for cryptococcal meningitis. *New England Journal of Medicine.* 368 (14): 1291–1302.

Dominic, R.S., Prashanth, H., Shenoy, S. et al. (2009). Diagnostic value of latex agglutination in cryptococcal meningitis. *Journal of Laboratory Physicians.* 1 (2): 67–68.

French, N., Gray, K., Watera, C. et al. (2002). Cryptococcal infection in a cohort of HIV-1-infected Ugandan adults. *AIDS.* 16 (7): 1031–1038.

Goldman, D.L., Lee, S.C., Mednick, A.J. et al. (2000). Persistent Cryptococcus neoformans pulmonary infection in the rat is associated with intracellular parasitism, decreased inducible nitric oxide synthase expression, and altered antibody responsiveness to cryptococcal polysaccharide. *Infection and Immunity.* 68 (2): 832–838.

Govender, N., Roy, M., Mendes, J. et al. (2015). Evaluation of screening and treatment of cryptococcal antigenaemia among HIV-infected persons in Soweto, South Africa. *HIV Medicine.* 16 (8): 468–476.

Ibembe, I.N. and Wiggin, T.R. (2015). An alternative to India ink stain. *Tropical Doctor.* 45 (3): 206–207.

Infectious Diseases Society of America (2013). *Guidelines for Prevention and Treatment of Opportunistic Infections in HIV-Infected Adults and Adolescents.* United States: IDSA.

Jarvis, J.N., Bicanic, T., Loyse, A. et al. (2014). Determinants of mortality in a combined cohort of 501 patients with HIV-associated cryptococcal meningitis: implications for improving outcomes. *Clinical Infectious Diseases.* 58 (5): 736–745.

Jarvis, J.N., Govender, N., Chiller, T. et al. (2012). Cryptococcal antigen screening and preemptive therapy in patients initiating antiretroviral therapy in resource-limited settings: a proposed algorithm for clinical implementation. *Journal of the International Association of Physicians in AIDS Care.* 11 (6): 374–379.

Jarvis, J. and Harrison, T. (2008). Pulmonary cryptococcosis. *Seminars in Respiratory and Critical Care Medicine* 29 (2): 141–150.

Jarvis, J.N., Harrison, T.S., Lawn, S.D. et al. (2013). Cost effectiveness of cryptococcal antigen screening as a strategy to prevent HIV-associated cryptococcal meningitis in South Africa. *PLoS One* 8 (7): e69288.

Jarvis, J.N., Lawn, S.D., Vogt, M. et al. (2009). Screening for cryptococcal antigenemia in patients accessing an antiretroviral treatment program in South Africa. *Clinical Infectious Diseases.* 48 (7): 856–862.

Jarvis, J.N., Meintjes, G., Williams, A. et al. (2010a). Adult meningitis in a setting of high HIV and TB prevalence: findings from 4961 suspected cases. *BMC Infectious Diseases.* 10: 67.

Jarvis, J.N., Percival, A., Bauman, S. et al. (2011). Evaluation of a novel point-of-care cryptococcal antigen test on serum, plasma, and urine from patients with HIV-associated cryptococcal meningitis. *Clinical Infectious Diseases* 53 (10): 1019–1023.

Jarvis, J.N., Wainwright, H., Harrison, T.S. et al. (2010b). Pulmonary cryptococcosis misdiagnosed as smear-negative pulmonary tuberculosis with fatal consequences. *International Journal of Infectious Diseases.* 14 (Suppl 3): e310–e312.

Kabanda, T., Siedner, M.J., Klausner, J.D. et al. (2014). Point-of-care diagnosis and prognostication of cryptococcal meningitis with the cryptococcal antigen lateral flow assay on cerebrospinal fluid. *Clinical Infectious Diseases.* 58 (1): 113–116.

Kwizera, R., Nguna, J., Kiragga, A. et al. (2014). Performance of cryptococcal antigen lateral flow assay using saliva in Ugandans with CD4. *PLoS One.* 9 (7): e103156.

Lawn, S.D., Harries, A.D., Anglaret, X. et al. (2008). Early mortality among adults accessing antiretroviral treatment programmes in sub-Saharan Africa. *AIDS.* 22 (15): 1897–1908.

Longley, N., Jarvis, J.N., Meintjes, G. et al. (2016). Cryptococcal antigen screening in patients initiating ART in South Africa: a prospective cohort study. *Clinical Infectious Diseases* 62 (5): 581–587.

Longley, N., Muzoora, C., Taseera, K. et al. (2008). Dose response effect of high-dose fluconazole for HIV-associated cryptococcal meningitis in southwestern Uganda. *Clinical Infectious Diseases.* 47 (12): 1556–1561.

Magambo, K.A., Kalluvya, S.E., Kapoor, S.W. et al. (2014). Utility of urine and serum lateral flow assays to determine the prevalence and predictors of cryptococcal antigenemia in HIV-positive outpatients beginning antiretroviral therapy in Mwanza, Tanzania. *Journal of the International AIDS Society* 17: 19040.

Meda, J., Kalluvya, S., Downs, J.A. et al. (2014). Cryptococcal meningitis management in Tanzania with strict schedule of serial lumber punctures using intravenous tubing sets. *Journal of Acquired Immune Deficiency Syndromes.* 66 (2): e31–e36.

Meya, D.B., Manabe, Y.C., Castelnuovo, B. et al. (2010). Cost-effectiveness of serum cryptococcal antigen screening to prevent deaths among HIV-infected persons with a CD4$^+$ cell count ≤100 cells/μL who start HIV therapy in resource-limited settings. *Clinical Infectious Diseases.* 51 (4): 448–455.

Meya, D., Rajasingham, R., Nalintya, E. et al. (2015). Preventing cryptococcosis-shifting the paradigm in the era of highly active antiretroviral therapy. *Current Tropical Medicine Reports.* 2 (2): 81–89.

Mfinanga, S., Chanda, D., Kivuyo, S.L. et al. (2015). Cryptococcal meningitis screening and community-based early adherence support in people with advanced HIV infection starting antiretroviral therapy in Tanzania and Zambia: an open-label, randomised controlled trial. *The Lancet.* 385 (9983): 2173–2182.

Park, B.J., Wannemuehler, K.A., Marston, B.J. et al. (2009). Estimation of the current global burden of cryptococcal meningitis among persons living with HIV/AIDS. *AIDS* 23 (4): 525–530.

Percival, A., Thorkildson, P., and Kozel, T.R. (2011). Monoclonal antibodies specific for immunorecessive epitopes of glucuronoxylomannan, the major capsular polysaccharide of Cryptococcus neoformans, reduce serotype bias in an immunoassay for cryptococcal antigen. *Clinical and Vaccine Immunology.* 18 (8): 1292–1296.

Perfect, J.R., Dismukes, W.E., Dromer, F. et al. (2010). Clinical practice guidelines for the management of cryptococcal disease: 2010 update by the Infectious Diseases Society of America. *Clinical Infectious Diseases.* 50 (3): 291–322.

Rajasingham, R., Smith, R.M., Park, B.J. et al. (2017). Global burden of disease of HIV-associated cryptococcal meningitis: an updated analysis. *The Lancet Infectious Diseases.* 17 (8): 873–881.

Ramachandran, A., Manabe, Y., Rajasingham, R. et al. (2017). Cost-effectiveness of CRAG-LFA screening for cryptococcal meningitis among people living with HIV in Uganda. *BMC Infectious Diseases* 17 (1): 225.

Rolfes, M.A., Hullsiek, K.H., Rhein, J. et al. (2014). The effect of therapeutic lumbar punctures on acute mortality from cryptococcal meningitis. *Clinical Infectious Diseases.* 59 (11): 1607–1614.

Rugemalila, J., Maro, V.P., Kapanda, G. et al. (2013). Cryptococcal antigen prevalence in HIV-infected Tanzanians: a cross-sectional study and evaluation of a point-of-care lateral flow assay. *Tropical Medicine & International Health* 18 (9): 1075–1079.

Smith, R.M., Nguyen, T.A., Ha, H.T.T. et al. (2013). Prevalence of cryptococcal antigenemia and cost-effectiveness of a cryptococcal antigen screening program – Vietnam. *PLoS One.* 8 (4): e62213.

Tanner, D.C., Weinstein, M.P., Fedorciw, B. et al. (1994). Comparison of commercial kits for detection of cryptococcal antigen. *Journal of Clinical Microbiology.* 32 (7): 1680–1684.

Tenforde, M.W., Scriven, J.E., Harrison, T.S. et al. (2017). Immune correlates of HIV-associated cryptococcal meningitis. *PLoS Pathogens.* 13 (3): e1006207.

Tenforde, M.W., Wake, R., Leeme, T. et al. (2016). HIV-associated cryptococcal meningitis: bridging the gap between developed and resource-limited settings. *Current Clinical Microbiology Reports* 3 (2): 92–102.

Vallabhaneni, S., Longley, N., Smith, M. et al. (2016). Implementation and operational research: evaluation of a public-sector, provider-initiated cryptococcal antigen screening and treatment program, Western Cape, South Africa. *Journal of Acquired Immune Deficiency Syndromes* 72 (2): e37–e42.

Vijayan, T., Chiller, T., and Klausner, J.D. (2013). Sensitivity and specificity of a new cryptococcal antigen lateral flow assay in serum and cerebrospinal fluid. *MLO: Medical Laboratory Observer* 45 (3): 16–20.

Wake, R., Glencross, D.K., Sriruttan, C. et al. (2015). Cryptococcal antigen screening in HIV-infected adults – let's get straight to the point-of-care. *AIDS* 30 (3): 1.

Wall, E.C., Everett, D.B., Mukaka, M. et al. (2014). Bacterial meningitis in Malawian adults, adolescents, and children during the era of antiretroviral scale-up and *Haemophilus influenzae* type b vaccination, 2000–2012. *Clinical Infectious Diseases* 58 (10): e137–e145.

World Health Organisation (2009). Laboratory manual for diagnosis of fungal opportunistic infections in HIV/AIDS patients. India: WHO.

World Health Organisation (2011). Rapid advice: Diagnosis, prevention and management of cryptococcal disease in HIV-infected adults, adolescents and children. Switzerland: WHO.

Williams, D.A., Kiiza, T., Kwizera, R. et al. (2015). Evaluation of fingerstick cryptococcal antigen lateral flow assay in HIV-infected persons: a diagnostic accuracy study. *Clinical Infectious Diseases* 61 (3): 464–467.

Williamson, P.R., Jarvis, J.N., Panackal, A.A. et al. (2016). Cryptococcal meningitis: epidemiology, immunology, diagnosis and therapy. *Nature Reviews Neurology* 13 (1): 13–24.

Wong, M.L., Back, P., Candy, G. et al. (2007). Cryptococcal pneumonia in African miners at autopsy. *The International Journal of Tuberculosis and Lung Disease: The Official Journal of the International Union Against Tuberculosis and Lung Disease* 11 (5): 528–533.

19

Mid Upper Arm Circumference Tapes for Assessment of Severe Acute Malnutrition

Jane Crawley[1], Martha Mwangome[2], James Berkley[1,2,3], and André Briend[4,5]

[1] Centre for Tropical Medicine and Global Health, University of Oxford, Oxford, UK
[2] KEMRI/Wellcome Trust Research Programme, Kilifi, Kenya
[3] The Childhood Acute Illness and Nutrition Network (CHAIN), Nairobi, Kenya
[4] University of Tampere School of Medicine and Tampere University Hospital, Tampere, Finland
[5] Department of Nutrition, Exercise and Sports, Faculty of Science, University of Copenhagen, Copenhagen, Denmark

---

CHAPTER MENU

19.1 Introduction

Malnutrition is often used to specifically refer to undernutrition where an individual is not getting enough calories, protein or micronutrients. If undernutrition occurs during pregnancy, or before two years of age, it may result in permanent problems with physical and mental development. Extreme undernourishment, known as starvation, may have symptoms that include a short height, thin body, very poor energy levels, and swollen legs and abdomen; the latter two are due to edema and ascites respectively, as a result of a low serum albumin level. The symptoms of micronutrient deficiencies depend on the micronutrient that is lacking.

Undernourishment is most often due to not enough high-quality food being available to eat. This is often related to high food prices and poverty. A lack of breastfeeding may contribute, as may a number of infectious diseases such as gastroenteritis, pneumonia, malaria, and measles, which increase nutrient requirements.

Revolutionizing Tropical Medicine: Point-of-Care Tests, New Imaging Technologies and Digital Health,
First Edition. Edited by Kerry Atkinson and David Mabey.
© 2019 John Wiley & Sons, Inc. Published 2019 by John Wiley & Sons, Inc.

There are two main types of undernutrition: protein-energy malnutrition and dietary deficiencies. Protein-energy malnutrition has two severe forms: marasmus (a lack of protein and calories) and kwashiorkor[1] (a lack of protein). Common micronutrient deficiencies include a lack of iron, iodine, vitamin A and vitamin B.

Wasting, or acute malnutrition, which results from recent weight loss or failure to gain weight adequately, is potentially treatable. Worldwide in 2016 there were 52 million children under the age of five who were wasted, and 17 million who were severely wasted (UNICEF 2017). Since malnutrition contributes to nearly half of all deaths among children below five years (Black et al. 2013), early detection and treatment is of paramount importance. Malnutrition predominantly occurs in children under five years old, and assessment and treatment programs are typically targeted at children aged six months to five years. A child with severe acute malnutrition (SAM) and wasting is shown in Figure 19.1a. A child with edematous SAM (kwashiorkor) and her normally nourished younger sibling below her is shown in Figure 19.1b.

Malnutrition may also occur in young infants, especially if there is early introduction of complementary foods or suboptimal breastfeeding. Infants were traditionally excluded from nutrition surveys and treatment programs, largely because of a paucity of disease prevalence data and the assumption that they were protected from malnutrition by breastfeeding. With the introduction of the WHO Growth Standards (WHO 2006), which are internationally applicable and reflect infant and child growth under optimal environmental conditions, an additional three million infants under six months were recognized as being severely wasted (Kerac et al. 2011). Malnutrition in this age group is associated with increased mortality (Mwangome et al. 2012a; Berkley et al. 2016; Mwangome et al. 2017).

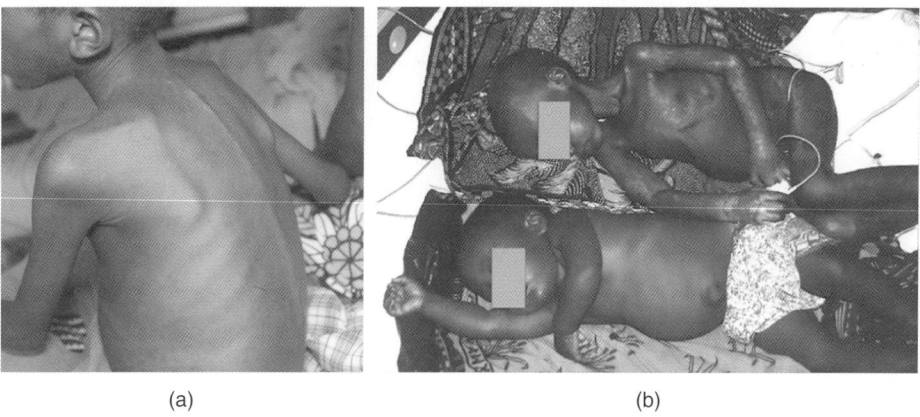

(a)   (b)

Figure 19.1 (a) A child with severe acute malnutrition and wasting. Photograph Dr. Jane Crawley. (b) A child with edematous severe acute malnutrition (kwashiorkor) and her normally nourished younger sibling (below her). *Source:* Photograph Dr. Jane Crawley.

---

1 The name Kwashiorkor is derived from the Ga language of coastal Ghana, translated as "the sickness the baby gets when the new baby comes" or "the disease of the deposed child," reflecting the development of the condition in an older child who has been weaned from the breast when a younger sibling is born.

In many parts of the world school-age children, adolescents and adults are at risk of acute malnutrition arising from food insecurity, conflict and natural disasters (FAO, IFAD, UNICEF, WFP and WHO 2017), while malnutrition and food insecurity are associated with increased mortality and poor clinical outcome among adults living with HIV/AIDS (Benzekri et al. 2015).

19.2 Mid Upper Arm Circumference (MUAC)

Mid upper arm circumference (MUAC) tapes are simple, inexpensive tools that can be used in both community- and facility-based settings to rapidly identify individuals of all ages who are acutely malnourished and at increased risk of death.

19.3 Comparison of MUAC with other Anthropometric Indices

Traditionally, nutritional assessment has utilized weight for height z-scores (WHZ) or body mass index (BMI, weight in kg/height in m^2) to classify individuals as wasted or underweight respectively (WHO 1995, 2000). However, both of these indices involve measurement of length or height, which, in young children, can be difficult and time-consuming, requiring two trained people if accurate measurements are to be obtained. The subsequent conversion of weight and height measurement to WHZ, or calculation of BMI, introduces further potential sources of error (Mwangome and Berkley 2014). Another practical constraint is that height/length measuring boards are bulky and not easily portable. In contrast, measurement of mid upper arm circumference (MUAC) can be rapidly and accurately performed by one trained person. The simple, inexpensive MUAC tapes can be used in both community- and facility-based settings by minimally trained health workers (Velzeboer et al. 1983; Mwangome et al. 2012b), and recent work from rural Niger suggests that mothers can be successfully taught to use the tapes to measure MUAC on their own children, thus increasing the feasibility of regular malnutrition screening, and, consequently, the likelihood of its early detection (Blackwell et al. 2015). MUAC has the added advantage of being less affected by acute dehydration (Mwangome et al. 2011) and body shape (Myatt et al. 2009) than weight-based indices.

19.4 MUAC: A Brief Historical Perspective

MUAC measurement was introduced in the 1960s as a tool for the rapid assessment of nutritional status of young children, the underlying assumption being that MUAC was closely related to muscle mass, and varied little with age (Jelliffe 1966). MUAC fell from favor when the assumption of age independence was shown to be incorrect, and WHZ became the gold standard for assessment of acute malnutrition (WHO 2000). MUAC subsequently regained popularity as a diagnostic tool when its value as a predictor of mortality (discussed in more detail below) was recognized.

The mid 1990s saw the development of ready-to-use therapeutic feeds (RUTF), an alternative to therapeutic milks (Formula [F] 75 and F100) that can be consumed safely at home or in the community for the treatment of acute malnutrition. RUTF enabled uncomplicated cases of SAM to be treated at home, instead of all cases being treated in facilities. This paved the way for the treatment of uncomplicated cases of SAM by community management of acute malnutrition (CMAM) programs, leaving a small minority of patients with severe infections or metabolic derangements to be treated in hospital. CMAM was highly effective when piloted in Ethiopia during the 2000 humanitarian emergency and elsewhere (Collins and Sadler 2002; Manary et al. 2004), and was subsequently endorsed by United Nations agencies (WHO 2007). It is now considered the standard of care for managing acute malnutrition in emergency and development contexts. This has led to a paradigm shift, where the emphasis has shifted from using WHZ to detect the "most malnourished" children, as defined by statistical deviation from a standard, to a risk-based strategy of using MUAC to detect those malnourished children who are in most urgent need of treatment to prevent death (Briend et al. 2012, 2017). As a simple, low-cost means of rapidly screening and monitoring at-risk populations, MUAC is an ideal tool for use by nutrition programs in community and hospital settings, and in humanitarian emergencies (Goossens et al. 2012). Its simplicity and low cost facilitates high program coverage.

19.5   Technique for Measuring MUAC

As suggested in the original WHO monograph (Jelliffe 1966), MUAC has traditionally been measured on the left upper arm. The midpoint of the upper arm has been located by measurement, and defined as the point half way between the acromion process ("tip of the shoulder") and the olecranon ("tip of the elbow") (Figure 19.2).

The underlying assumption was that subcutaneous fat distribution around the upper arm is not uniform, and that the size of the triceps and biceps in the left arm, the non-dominant arm in the majority of people, is less influenced by physical activity. Although these considerations are less relevant in children, the recommendations have continued to be reiterated in the majority of training manuals. In a recent study, however, mothers in rural Niger were trained to screen their own children for malnutrition, using either arm, and locating the midpoint of the upper arm by visual inspection rather than by measurement (Blackwell et al. 2015). Mothers were able to correctly classify their child's nutritional status with high sensitivity and specificity for both SAM and moderate acute malnutrition (MAM). Concordance between their assessments and those of community health workers (CHWs) given conventional training in MUAC measurement was high. The findings were confirmed when this approach was examined under programmatic conditions in a large pragmatic intervention trial (Alé et al. 2016), and overall costs were reduced. Consequently, many nutrition programs have now abandoned the use of measurement in favor of visual inspection to locate the midpoint of the upper arm.

The choice of MUAC tape is important, as tapes should, ideally, be made of thin, flexible, non-stretchable material. Color-coded insertion tapes (Zerfas 1975), incorporating a window through which the arm circumference can easily be read between arrows, are

1. Find midpoint of upper arm

Step 1a

Step 1b

Step 1c

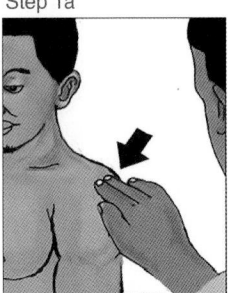

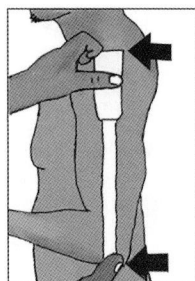

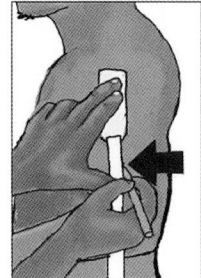

Always use left arm. Bend arm to a 90 degree angle. Find arm endpoints at the tip of the shoulder and tip of the elbow.

Use thumbs to place tape at endpoints.

Make a mark on the arm's midpoint.

2. Measure circumference

Step 2a

Step 2b

Step 2c

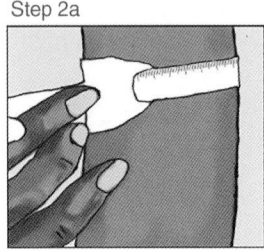

Too loose

Too tight

Straighten the arm. Wrap the tape around the midpoint and thread it through the window.

Adjust the tension of the tape so that it is not too tight or too loose.

Record the measurement in mm where the arrows point inward.

Figure 19.2 Technique for measuring MUAC on adolescents and adults. *Source:* The Figure is from a Malawi Ministry of Health poster, reproduced by courtesy of the FANTA project: https://www.fantaproject.org

generally preferred. UNICEF tapes cost $0.06, and are color-coded to indicate nutritional status (Figure 19.3).

19.6 MUAC, Mortality Risk, and Definitions of Severe Acute Malnutrition

Definitions of SAM in different age groups are based on the anthropometric measurements that best predict mortality (Table 19.1).

The value of MUAC as a predictor of death is now increasingly recognized, and MUAC cut-offs are now included in several age-specific definitions of SAM and are used to guide admission to, and discharge from, hospitals and therapeutic feeding programs (Goossens et al. 2012).

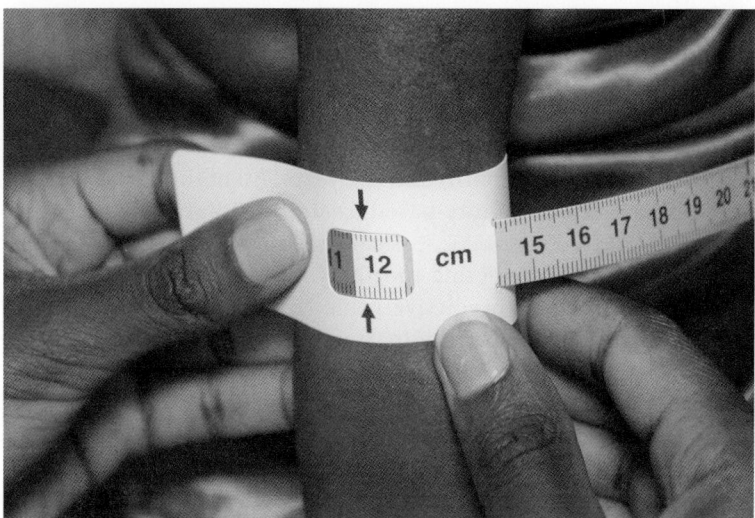

Figure 19.3 Measurement of mid-upper arm circumference. *Source:* figure 19.1, Blackwell et al. (2015). Licensed under Creative Commons, CC-BY-4.0. Note: For measurements on young children (6–60 months), many nutrition programs now suggest that the midpoint of the upper arm is located by visual inspection instead of measurement.

Table 19.1 MUAC and WHO definitions of severe acute malnutrition (SAM) for different age groups.

Age group	WHO definition of SAM	MUAC cut-off for SAM	References
<6 months	WHZ less than –3 and/or bilateral pitting edema	Less than 110 mm[a]	Mwangome et al. (2013, 2017)
6–60 months	WHZ less than –3 and/or MUAC less than 115 mm and/or edema	Less than 115 mm	WHO (2009a)
5–19 years	BMI for age z-score less than –3	MUAC for age z-score less than –3[a]	Mramba et al. (2017)
20–60 years	BMI less than 16 kg m^{-2}	MUAC <160 mm or MUAC ≤185 mm plus one or more of the following: bilateral pitting edema, inability to stand or sunken eyes	Collins et al. (2000) WHO (2009b)

a) Data published since latest WHO guidelines WHO (2013); MUAC cut-off not endorsed by WHO.

19.6.1 Young Children (6–60 Months)

WHO defines SAM in a young child between 6 and 60 months as WHZ less than –3 and/or MUAC less than 115 mm, with or without edema (Table 19.1). The use of this definition to guide admission to therapeutic feeding programs is, however, unnecessarily cumbersome, since it involves measuring two different anthropometric indices. A number of community studies, conducted in Bangladesh, Senegal, Uganda, Malawi and the Democratic Republic of Congo in the 1980s and 1990s (summarized in Briend et al. 2017) examined the relationship between anthropometry and mortality in children between the ages of six months and five years. All of these studies took place before the development of community-based treatment programs, and reflect outcome in the absence of treatment. MUAC was consistently superior to WHZ and weight for age z-scores (WAZ) in predicting mortality, and correcting MUAC for age or height for did not improve its predictive value (Briend and Zimicki 1986). Following the release of the new WHO Growth Standards in 2006 (WHO 2006), reanalysis of data from Senegal and Congo confirmed the superiority of MUAC over other anthropometric indices as a predictor of death (Briend et al. 2012; Schwinger et al. 2016). In hospital studies, MUAC has been comparable or superior to other nutritional indices in predicting inpatient death (Briend et al. 1986; Berkley et al. 2005; Sachdeva et al. 2016). The combination of WHZ and MUAC as admission criteria does not improve the predictive value for mortality over that of MUAC or WHZ assessed individually (Briend et al. 2012; Sachdeva et al. 2016). Use of MUAC alone as an admission criterion would save time and be easier logistically than MUAC plus WHZ.

The superior performance of MUAC at identifying high-risk children could be due to its preferential selection of younger and/or more stunted children, or to the close relationship of MUAC with muscle and fat mass (Briend et al. 2015). These two explanations are not mutually exclusive, as young and stunted children tend to have a low muscle mass in relation to body weight, which makes them more vulnerable to malnutrition. Muscle constitutes only 23% of the body weight of newborn infants compared to 53% in adults, and this proportion increases slowly with age (FAO-WHO-UNU Expert Consultation 1985). Consequently, in children, variations in muscle mass will have a relatively small impact on WHZ compared to MUAC, while WHZ and other weight-based indices will be more affected by changes in hydration status. This is particularly true for malnourished children, who frequently have an excess of body water (Waterlow, 1992).

19.6.2 Infants (Below Six Months)

In the current WHO guidelines (WHO 2013) the definition of SAM in infants below six months is confined to WHZ less than −3 and/or the presence of bilateral pitting edema, and does not include MUAC (Table 19.1). Recent evidence suggests, however, that MUAC <110 mm is a useful predictor of mortality in this younger age group (Mwangome et al. 2012a, 2017). Among 2876 Gambian infants below 6 months of age, a single MUAC measurement taken between 6 and 14 weeks at the time of routine immunization was better than WHZ at predicting mortality during the first year of life. MUAC <110 mm identified 14% of the study population whose risk of death was nearly 10 times higher than that of well-nourished infants. Risk of death was increased four-fold in infants with MUAC <115 mm, but this cut-off identified 27% of the study population and therefore lacked specificity (Mwangome et al. 2012a). This suggests that MUAC <110 mm may be a useful screening tool to rapidly identify SAM in infants aged between two and six months.

19.6.3   Older Children and Adolescents (5–19 Years)

WHO recommends using BMI to assess nutritional status in older children, adolescents and adults, and defines SAM as BMI <16 kg m^{-2} (WHO 1995). The 2007 WHO Growth References for 5–19 year olds did not include MUAC, and until recently there was no internationally recognized MUAC reference for this age group. However, researchers have now constructed MUAC for age z-score curves for 5–19 year olds (Mramba et al. 2017). These converge with the 2006 WHO Growth Standards, and were validated using two prospective cohorts, 685 HIV-positive children in Uganda and Zimbabwe, and 1174 children discharged from a rural district hospital in Kenya. MUAC for age was comparable to BMI in predicting mortality, and a MUAC z-score of –3 classified a greater number of children as being severely malnourished than did BMI, reflecting its higher sensitivity (Mramba et al. 2017). MUAC is a potentially useful tool for diagnosing malnutrition in school-age children and adolescents, a MUAC for age z-score of less than –3 indicating SAM and the need for treatment.

19.6.4   Adults (20–60 Years)

There are no published community studies examining the relationship between different anthropometric indices and mortality of malnourished adults in the absence of nutritional support. However, a number of studies on adults in inpatient facilities (Powell-Tuck and Hennessy 2003; Gustafson et al. 2007; Irena et al. 2013) suggest that MUAC is a better prognostic predictor for death than BMI. In non-malnourished adults, arm muscle area is better than BMI at identifying patients at risk of dying (Miller et al. 2002; Soler-Cataluña et al. 2005). A simple clinical model was developed for assessing the nutritional status of adults aged 20–60 years in emergency settings (Collins et al. 2000), and was subsequently incorporated into the Integrated Management of Adult and Adolescent (IMAI) acute care guidelines (WHO 2009b). BMI was considered too complicated for use in emergency settings, and the criteria proposed for adult admission to therapeutic feeding centers comprised MUAC <160 mm, or MUAC ≤185 mm plus one of bilateral pitting edema, inability to stand or sunken eyes. However, in common with other proposed MUAC cut-offs for identifying undernutrition in adult populations (Tang et al. 2016, 2017), this model has never been validated.

19.7   Conclusions: Use of MUAC in Different Settings

MUAC tapes are simple, portable and inexpensive tools that can be used in a wide variety of community settings and inpatient facilities to diagnose and monitor children and adults of all ages with SAM, who are at increased risk of dying. Teaching mothers in impoverished rural settings to use MUAC tapes to screen their own children for malnutrition in place of CHWs could, potentially, increase the frequency of screening and facilitate earlier detection. In emergency settings, MUAC can be used to identify individuals requiring CMAM or inpatient care, to monitor response to treatment and to guide discharge. In hospitals MUAC measurements may prompt the need for further investigation and identify those at increased risk of post-discharge mortality. Finally, MUAC measurements provide a useful, standardized assessment of nutritional status for clinical research.

References

Alé, F.G., Phelan, K.P., Issa, H. et al. (2016). Mothers screening for malnutrition by mid-upper arm circumference is non-inferior to community health workers: results from a large-scale pragmatic trial in rural Niger. *Arch. Public Health* 74: 38.

Benzekri, N.A., Sambou, J., Diaw, B. et al. (2015). High prevalence of severe food insecurity and malnutrition among HIV-infected adults in Senegal, West Africa. *PLoS One* 10 (11): e0141819. https://doi.org/10.1371/journal.pone.0141819.

Berkley, J., Mwangi, I., Griffiths, K. et al. (2005). Assessment of severe malnutrition among hospitalized children in rural Kenya: comparison of weight for height and mid upper arm circumference. *JAMA* 294 (5): 591–597.

Berkley, J.A., Ngari, M., Thitiri, J. et al. (2016). Daily co-trimoxazole prophylaxis to prevent mortality in children with complicated severe acute malnutrition: a multicentre, double-blind, randomised placebo-controlled trial. *Lancet Glob. Health* 4: e464–e473.

Black, R., Victora, C., Walker, S.P. et al. (2013). Maternal and child undernutrition and overweight in low-income and middle-income countries. *Lancet* 382: 427–459.

Blackwell, N., Myatt, M., Allafort-Duverger, T. et al. (2015). Mothers Understand And Can do it (MUAC): a comparison of mothers and community health workers determining mid-upper arm circumference in 103 children aged from 6 months to 5 years. *Arch. Public Health* 73: 26. https://doi.org/10.1186/s13690-015-0074-z.

Briend, A. and Zimicki, S. (1986). Validation of arm circumference as an indicator of risk of death in one to four year old children. *Nutr. Res.* 6: 249–261.

Briend, A., Dykewicz, C., Graven, K. et al. (1986). Usefulness of nutritional indices and classifications in predicting death of malnourished children. *Br. Med. J.* 293: 373–375.

Briend, A., Maire, B., Fontaine, O. et al. (2012). Mid-upper arm circumference and weight-for-height to identify high-risk malnourished under-five children. *Matern. Child. Nutr.* 8: 130–133.

Briend, A., Khara, T., and Dolan, C. (2015). Wasting and stunting - similarities and differences: policy and programmatic implications. *Food Nutr. Bull.* 36 (1 Suppl): S15–S23.

Briend, A., Mwangome, M., and Berkley, J. (2017). Using mid-upper arm circumference to detect high-risk malnourished patients in need of treatment. In: *Handbook of Famine, Starvation, and Nutrient Deprivation* (ed. V.R. Preedy and V.B. Patel), 1–17. Springer International Publishing doi: 10.1007/978-3-319-40007-5_11-1.

Collins, S., Duffield, A. and Myatt, M. (2000). Assessment of nutritional status in emergency-affected populations. UN ACC/Sub-Committee on Nutrition. https://www.unscn.org/web/archives_resources/files/AdultsSup.pdf (accessed 10 December 2017).

Collins, S. and Sadler, K. (2002). Outpatient care for severely malnourished children in emergency relief programmes: a retrospective cohort study. *Lancet* 360: 1824–1830.

FAO-WHO-UNU Expert Consultation (1985). Energy and Protein Requirements. Geneva. World Health Organization.

FAO, IFAD, UNICEF, WFP and WHO (2017). *The State of Food Security and Nutrition in the World 2017*, Building resilience for peace and food security. Rome: FAO http://www.fao.org/3/a-I7695e.pdf (accessed 16 December 2017).

Goossens, S., Bekele, Y., Yun, O. et al. (2012). Mid-upper arm circumference based nutrition programming: evidence for a new approach in regions with high burden of acute malnutrition. *PLoS One* 7 (11): e49320. https://doi.org/10.1371/journal.pone.0049320.

Gustafson, P., Gomes, V.F., Vieira, C.S. et al. (2007). Clinical predictors for death in HIV-positive and HIV-negative tuberculosis patients in Guinea-Bissau. *Infection* 35: 69–80.

Irena, A.H., Ross, D.A., Salama, P. et al. (2013). Anthropometric predictors of mortality in undernourished adults in the Ajiep Feeding Programme in southern Sudan. *Am. J. Clin. Nutr.* 98: 335–339.

Jelliffe, D.B. (1966). The assessment of nutritional status of the community. WHO Monograph Series No 53. Geneva.

Kerac, M., Blencowe, H., Grijalva-Eternod, C. et al. (2011). Prevalence of wasting among under 6-month-old infants in developing countries and implications of new case definitions using WHO growth standards: a secondary data analysis. *Arch. Dis. Child.* 96: 1008–1013.

Manary, M.J., Ndekha, M.J., Ashorn, P. et al. (2004). Home-based therapy for severe malnutrition with ready to use food. *Arch. Dis. Child.* 89: 557–561. https://doi.org/10.1136/adc.2003.034306.

Miller, M.D., Crotty, M., Giles, L.C. et al. (2002). Corrected arm muscle area: an independent predictor of long-term mortality in community-dwelling older adults? *J. Am. Geriatr. Soc.* 50: 1272–1277.

Mramba, L., Ngaria, M., Mwangome, M. et al. (2017). A growth reference for mid upper arm circumference for age among school age children and adolescents, and validation for mortality: growth curve construction and longitudinal cohort study. *BMJ* 358: j3423. https://doi.org/10.1136/bmj.j3423.

Mwangome, M.K., Fegan, G., Prentice, A.M. et al. (2011). Are diagnostic criteria for acute malnutrition affected by hydration status in hospitalized children? A repeated measures study. *Nutr. J.* 10: 92. https://doi.org/10.1186/1475-2891-10-92.

Mwangome, M., Martha, K., Fegan, G. et al. (2012a). Mid upper arm circumference at age of routine infant vaccination to identify infants at elevated risk of death: a retrospective cohort study in the Gambia. *Bull. World Health Organ.* 90: 887–894.

Mwangome, M., Fegan, G., Mbunya, R. et al. (2012b). Reliability and accuracy of anthropometry performed by community health workers among infants under 6 months in rural Kenya. *Tropical Med. Int. Health* 5: 622–629.

Mwangome, M.K. and Berkley, J.A. (2014). The reliability of weight-for-length/height Z scores in children. *Matern. Child. Nutr.* 10: 474–480.

Mwangome, M., Ngari, M., Fegan, G. et al. (2017). Diagnostic criteria for severe acute malnutririon among infants aged under 6 months. *Am. J. Clin. Nutr.* 105: 1415–1423.

Myatt, M., Duffield, A., Seal, A. et al. (2009). The effect of body shape on weight-for-height and mid-upper arm circumference based case definitions of acute malnutrition in Ethiopian children. *Ann. Hum. Biol.* 36 (1): 5–20. https://doi.org/10.1080/03014460802471205.

Powell-Tuck, J. and Hennessy, E.M. (2003). A comparison of mid upper arm circumference, body mass index and weight loss as indices of undernutrition in acutely hospitalized patients. *Clin. Nutr.* 22: 307–312.

Sachdeva, S., Dewan, P., Shah, D. et al. (2016). Mid-upper arm circumference versus weight-for-height z-score for predicting mortality in hospitalized children under 5 years of age. *Public Health Nutr.* 19: 2513–2520.

Schwinger, C., Fadnes, L.T., and Van den Broeck, J. (2016). Using growth velocity to predict child mortality. *Am. J. Clin. Nutr.* 103: 801–807.

Soler-Cataluña, J.J., Sánchez-Sánchez, L., Martínez-García, M.A. et al. (2005). Mid-arm muscle area is a better predictor of mortality than body mass index in COPD. *Chest* 128: 2108–2115.

Tang, A.M., Chung, M., Dong, K., et al. (2016). Determining a global mid-upper arm circumference cut-off to assess malnutrition in pregnant women. Washington, DC: FHI 360/Food and Nutrition Technical Assistance III Project (FANTA). https://www.fantaproject.org/research/muac-adolescents-adults (accessed 16 December 2017).

Tang, A.M., Chung, M., Dong, K., et al. (2017). Determining a global mid-upper arm circumference cutoff to assess underweight in adults (men and non-pregnant women). Washington, DC: FHI 360/FANTA.

UNICEF/ WHO/ World Bank Group. (2017). Levels and trends in child malnutrition: 2017 edition. http://www.who.int/nutgrowthdb/jme_brochoure2017.pdf?ua=1 (accessed 10 December 2017).

Velzeboer, M.I., Selwyn, B.J., Sargent, F. et al. (1983). The use of arm circumference in simplified screening for acute malnutrition by minimally trained health workers. *J. Trop. Paed.* 29 (3): 159–166.

Waterlow, J.C. (1992). *Protein Energy Malnutrition*. London: Edward Arnold.

WHO (1995). Physical Status: The Use and Interpretation of Anthropometry. Technical Report Series 854. Geneva: World Health Organization.http://www.who.int/childgrowth/publications/physical_status/en (accessed 2 December 2017).

WHO (2000). The management of nutrition in major emergencies. World Health Organization. http://www.who.int/nutrition/publications/emergencies/9241545208/en (accessed 2 December 2017).

WHO (2006). Child Growth Standards: methods and development. World Health Organization. http://www.who.int/childgrowth/publications/technical_report_pub/en (accessed 13 December 2017).

WHO (2007). Community-based management of severe acute malnutrition: joint statement by the World Health Organization, the World Food Programme, the United Nations System Standing Committee on Nutrition and the United Nations Children's Fund. World Health Organization. https://www.unicef.org/publications/files/Community_Based_Management_of_Sever_Acute__Malnutirtion.pdf (accessed 2 December 2017).

WHO (2009a). Child growth standards and the identification of severe acute malnutrition in infants and children: joint statement by the World Health Organization and the United Nations Children's Fund. World Health Organization. http://www.who.int/nutrition/publications/severemalnutrition/9789241598163/en (accessed 6 December 2017).

WHO (2009b). Integrated management of adolescent and adult illness: interim guidelines for first-level facility health workers at health centre and district outpatient clinic: acute care. World Health Organization. http://www.who.int/hiv/pub/imai/primary_acute/en/index.html (accessed 13 December 2017).

WHO (2013). Guideline: Updates on the management of severe acute malnutrition in infants and children. Geneva: World Health Organization. http://www.who.int/nutrition/publications/guidelines/updates_management_SAM_infantandchildren/en (accessed 3 December 2017).

Zerfas, A.J. (1975). The insertion tape: a new circumference tape for use in nutritional assessment. *Am. J. Clin. Nutr.* 28: 782–787.

Webliography

http://www.fao.org/3/a-I7695e.pdf (accessed 16 December 2017) FAO, IFAD, UNICEF, WFP and WHO. The state of food security and nutrition in the world 2017. Building resilience for peace and food security. Rome, FAO 2017.

http://www.who.int/childgrowth/publications/physical_status/en (accessed 2 December 2017). WHO. Physical Status: The Use and Interpretation of Anthropometry. Technical Report Series 854. Geneva: World Health Organization: 1995.

http://www.who.int/childgrowth/publications/technical_report_pub/en (accessed 13 December 2017). WHO Child Growth Standards: methods and development. World Health Organization 2006.

http://www.who.int/hiv/pub/imai/primary_acute/en/index.html (accessed 13 December 2017). WHO. Integrated management of adolescent and adult illness: interim guidelines for first-level facility health workers at health centre and district outpatient clinic: acute care. World Health Organization 2009b.

http://www.who.int/nutgrowthdb/jme_brochoure2017.pdf?ua=1 (accessed 10 December 2017) UNICEF/ WHO/ World Bank Group. Levels and trends in child malnutrition: 2017 edition.

http://www.who.int/nutrition/publications/emergencies/9241545208/en (accessed 2 December 2017). WHO. The management of nutrition in major emergencies. World Health Organization 2000.

http://www.who.int/nutrition/publications/guidelines/updates_management_SAM_infantandchildren/en (accessed 3 December 2017). WHO. Guideline: Updates on the management of severe acute malnutrition in infants and children. Geneva: World Health Organization 2013.

http://www.who.int/nutrition/publications/severemalnutrition/9789241598163/en (accessed 6 December 2017). WHO. Child growth standards and the identification of severe acute malnutrition in infants and children: joint statement by the World Health Organization and the United Nations Children's Fund. World Health Organization 2009a.

https://www.fantaproject.org/research/muac-adolescents-adults (accessed 16 December 2017). Tang, A.M., Chung, M., Dong, K., et al. Determining a global mid-upper arm circumference cut-off to assess malnutrition in pregnant women. Washington, DC: FHI 360/Food and Nutrition Technical Assistance III Project (FANTA) 2016.

https://www.unicef.org/publications/files/Community_Based_Management_of_Sever_Acute__Malnutirtion.pdf (accessed 2 December 2017) WHO. Community-based management of severe acute malnutrition: joint statement by the World Health Organization, the World Food Programme, the United Nations System Standing Committee on Nutrition and the United Nations Children's Fund. World Health Organization 2007.

https://www.unscn.org/web/archives_resources/files/AdultsSup.pdf (accessed 10 December 2017) Collins S, Duffield A and Myatt M. Assessment of nutritional status in emergency-affected populations. UN ACC/Sub-Committee on Nutrition 2000.

20

Spirometry for Chronic Obstructive Pulmonary Disease Due to Inhalation of Smoke from Indoor Fires Used for Cooking and Heating

Janet G. Shaw[1,2], Annalicia Vaughan[1,2], Emma Smith[1,2], Cai Fong[2], Svetlana Stevanovic[3], and Ian A. Yang[1,2]

[1] Department of Thoracic Medicine, The Prince Charles Hospital, Brisbane, Queensland, Australia
[2] Thoracic Research Centre, Faculty of Medicine, The University of Queensland, Brisbane, Queensland, Australia
[3] International Laboratory for Air Quality and Health, Queensland University of Technology, Brisbane, Queensland, Australia

20.1 Introduction

Almost one third of the world's population uses biomass fuel for cooking and heating, contributing significantly to exposure from indoor air pollution. Common biomass fuels

Revolutionizing Tropical Medicine: Point-of-Care Tests, New Imaging Technologies and Digital Health,
First Edition. Edited by Kerry Atkinson and David Mabey.
© 2019 John Wiley & Sons, Inc. Published 2019 by John Wiley & Sons, Inc.

include wood, dung and crop residues. Burning of biomass fuels in poorly ventilated indoor household areas leads to the emission of toxic particulate matter and volatile gasses which, when inhaled, result in inflammation and oxidative stress in the lungs. This toxicity can, like that of cigarette smoking, lead to the development of chronic obstructive pulmonary disease (COPD), which is characterized by exertional breathlessness, cough and sputum production, with the presence of chronic airflow obstruction on spirometry.

Early detection of COPD, whether from biomass smoke exposure or from cigarette smoking, remains an important goal in at-risk, symptomatic individuals, in order to institute preventive measures to avoid exposure, as well as to treat symptoms and to reduce risk of exacerbations. In low- and middle-income countries (LMICs) the challenge is to implement spirometry programs for case detection of biomass-related COPD, using either handheld or desktop spirometers or simple FEV1/FEV6 devices to detect airflow obstruction. Clinical guidelines for the detection and management of COPD recognize inhalation of biomass smoke, an indoor air pollutant, as a cause of COPD. However, much more needs to be done to establish cost-effective approaches for detection of biomass smoke-related COPD in countries where biomass is commonly used as a fuel, as well as in implementing public health measures to reduce the impact of combustion of biomass fuel.

20.2 Indoor Air Pollution from Burning Biomass

20.2.1 Burning of Biomass Fuels

Approximately 40% of the world's population, or three billion people and up to 90% of rural households in LMICs, rely on biomass burning for the production of energy used for cooking, heating and lighting (Figure 20.1, adapted from (Martin et al. 2013)). This number is expected to rise further by 2030 (WHO 2002). Usage of traditional biomass fuels, such as wood, charcoal, animal dung, and crop residues, is very high in Asia, Latin America and sub-Saharan Africa. In many of these countries, over 75% of energy consumed is generated from traditional biomass fuels (WHO 2002).

Biomass is defined as plant- or animal-based material, consisting of carbohydrate components such as cellulose, hemicelluloses, proteins, simple sugars, starches, lignin, lipid, hydrocarbon components, water, ash, and other related chemical compounds (Simoneit 2002). For plant-based biomass, the chemical composition depends on the species, place of growth, climate and type of plant tissue. The major components of biomass all play an important role in producing outdoor air pollutants during biomass burning.

20.2.2 Indoor Air Pollution and Biomass Burning

Biomass burning is also a major contributor of trace gasses and aerosols indoors (Andreae 2001). As a result of incomplete combustion, apart from CO_2, biomass burning produces a range of trace gasses including carbon monoxide, methane (CH4), nitrous oxides (NOx), sulfur oxides (SOx), coarse, fine, and ultrafine particles, transition metals, polycyclic aromatic hydrocarbons, volatile organic compounds (particularly non-methane organic compounds) and bioaerosols (Smith et al. 2000; Morawska and Zhang 2002).

The composition of biomass burning emissions will depend on the chemical composition of the combustion precursor, burning conditions (flaming versus smoldering) and subsequent atmospheric transformation through photochemical aging and dilution after these are released from the source (Reid et al. 2005; Weimer et al. 2008). The non-methane organic compounds include non-methane hydrocarbons and oxygenated volatile organic compounds. Oxygenated organic compounds carry a higher toxic potential than hydrocarbon-like compounds (Stevanovic et al. 2013). Furthermore, the oxidative potential of biomass burning is different for different stages of combustion, and the heating profile is most toxic during a cold start (Miljevic et al. 2010). Carbonaceous aerosols include organic carbon and elemental carbon (Formenti et al. 2003; Bond et al. 2004; Reid et al. 2005; Sandradewi et al. 2008). Composition of biomass burning particulate matter is reported to be approximately 50–60% organic carbon and approximately 5–10% elemental carbon (Reid et al. 2005).

The choice of a particular fuel is driven by several factors including the price and availability of the fuel, technical characteristics of stoves, as well as cooking practices and cultural preferences (Masera et al. 2000). Use of biomass fuels indoors is associated with the high levels of indoor air pollution that is ultimately linked to a range of adverse health effects that include respiratory and cardiovascular diseases, tuberculosis, stroke, cancer, low birthweight and mortality and morbidity in adults and children. Poor air quality in these cases is closely linked to poor ventilation or complete lack of ventilation,

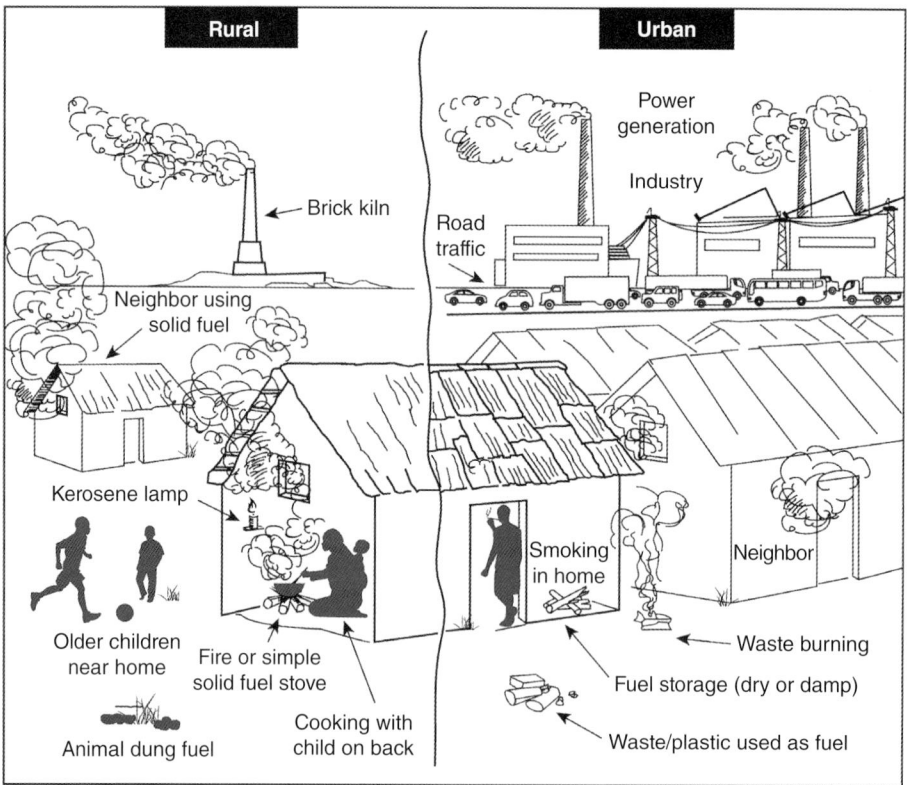

Figure 20.1 Examples of indoor and outdoor air pollution in various settings. A range of indoor and outdoor air pollutants need to be measured and controlled. In the case of biomass smoke, this is a prominent and important source of indoor air pollutants in the developing world, which leads to the onset of COPD. *Source:* Figure image from Martin et al. (2013), with adaptation of the Figure legend.

and the design of stoves that normally do not contain fume hoods. In many LMICs, indoor concentrations of particulate matter 10 μm or less in diameter (PM10) can often exceed 3000 μg m^{-3}, which is 60 times higher than the safe threshold set by the World Health Organization. Health effects of exposure to biomass burning will depend on age and gender of the individual who is exposed, and the duration and level of biomass smoke exposure.

Current research in the field is focused on the quantification and characterization of secondary pollution. Primary organic aerosols directly emitted from biomass burning react with non-methane organic compounds, ozone and hydroxyl radical and the resulting particles have changed size, mass, composition and radiative properties (Chen et al. 2017).

20.3 Mechanisms of Lung Damage from Exposure to Biomass Smoke

20.3.1 Cellular and Molecular Mechanisms Linking to Biomass Smoke and the Pathogenesis of COPD

The airway epithelium of the lungs is the first barrier of defense against the adverse effects of inhaled biomass smoke (Tam et al. 2011). The airway epithelium actively works to remove biomass smoke components, specifically the particulate matter (PM), from the airways using two main mechanisms – mucociliary clearance and the initiation of innate immune signaling (Tam et al. 2011; Silva et al. 2015). Dysregulation of these processes can arise from prolonged oxidative stress and inflammation in the bronchial epithelium due to chronic exposure to biomass smoke, which has been linked to an increased risk of COPD (Silva et al. 2015).

Biomass smoke can cause inflammation and oxidative stress within the airway epithelium via three main cellular/molecular mechanisms (Silva et al. 2015), as outlined in Figure 20.2.

Firstly, inflammatory pathways are initiated by the activation of the toll-like cell surface receptors TLR4 (Shoenfelt et al. 2009) and transient receptor potential A1 (TRPA1) (Shapiro et al. 2013). The biomass smoke PM binds to these receptors, initiating innate immune signaling pathways by activating nuclear factor kappa-light-chain-enhancer of activated B cells (NF-κB) and Activator protein 1 (AP-1) signaling pathways (von Knethen et al. 1999; Silva et al. 2015). The activation of these pathways causes the release and synthesis of pro-inflammatory cytokines and chemokines, including tumor necrosis factor-alpha (TNFα) and interleukin-1 (IL-1) (Lawrence 2009). These pro-inflammatory proteins recruit cells of the innate immune system including macrophages, neutrophils, and eosinophils to the airways (Barnes 2011). Inflammatory cell activity produces reactive oxygen species (ROS), and if prolonged, leads to oxidative stress within the tissue (Rahman and Adcock 2006; Silva et al. 2015).

Secondly, biomass smoke contains both fine PM (<2.5 μm) and ultrafine PM (<0.1 μm) (Saarikoski et al. 2007). This PM can diffuse across the mucosal layer, enter the airway epithelium and aggregate in the bronchial epithelial cells (Rothen-Rutishauser et al. 2007). These particles have endogenous ROS (Verma et al. 2012), which are released from the particle when the PM enters the cell. The PM also has organic compounds (OCs) on the particle surface that dissociate from the particle once it enters the cell (Saarikoski et al.

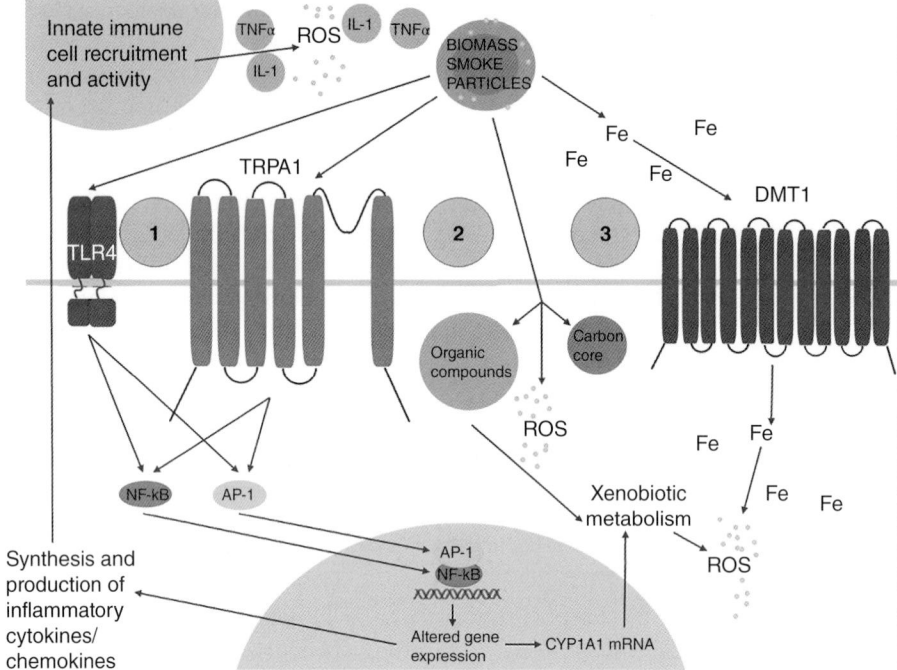

Figure 20.2 Key cellular and molecular mechanisms of bronchial epithelial cell toxicity caused by biomass smoke exposure. Biomass smoke can cause inflammation and oxidative stress within the airway epithelium via three main cellular/molecular mechanisms (number in the diagram): (1) Inflammatory pathways are initiated by the activation of cell surface receptors; (2) Biomass smoke particulate matter enters bronchial epithelial cells, leading to release of reactive oxygen species and activation of xenobiotic metabolism pathways; (3) Biomass smoke alters iron homeostasis within cells. See text for further details. Abbreviations used: TNFα, tumor necrosis factor alpha; ROS, reactive oxygen species; IL, interleukin; TLR, toll-like receptor; TRP, transient receptor potential; Fe, iron; DMT, divalent metal transporter; Nf-κB, nuclear factor kappa-light-chain-enhancer of activated B cells; AP, activator protein; CYP1A1, cytochrome P450, family 1, subfamily A, polypeptide 1. (*See color plate section for the color representation of this figure.*)

2007). These OCs bind and activate the aryl hydrocarbon receptor, which facilitates xenobiotic metabolism through the expression and activation of cytochrome P450 1A1 (CYP1A1) (Bittner et al. 2011; Lal et al. 2011; Totlandsdal et al. 2012). ROS is a common by-product of CYP1A1 activity and xenobiotic metabolism (Nebert et al. 2000).

Thirdly, biomass smoke has been shown to alter iron homeostasis within the cell, leading to the activation of iron transporters such as divalent metal transporter-1 (DMT1), leading to additional ROS production (Fariss et al. 2013; Silva et al. 2015). These inflammatory processes and the interaction of the PM components within the cell result in excessive ROS production, and if prolonged, lead to oxidative stress, cell death and injury to the airway epithelium, predisposing to COPD.

20.3.2 Epidemiological Link Between Biomass Smoke Exposure and Airflow Obstruction

Inflammation and oxidative stress have been identified as key cellular and molecular mechanisms in the pathogenesis of COPD (Caram et al. 2016). These mechanisms occur in response to environmental toxins including cigarette smoke and diesel emissions (Ristovski et al. 2012; Caram et al. 2016). Cigarette smoke exposure is considered to be the main risk factor for COPD. However, depending on the setting, up to 17–39% of COPD patients may be never-smokers (Silva et al. 2015). In China and other LMICs, there is an equal number of deaths attributed to biomass smoke exposure as there is for cigarette smoking (Salvi and Bames 2010). A meta-analysis of epidemiology studies has investigated the association between biomass smoke exposure and COPD (Hu et al. 2010). This analysis showed that people who had long-term exposure to biomass smoke had a 2.5 times greater risk of COPD than people without exposure to biomass smoke (Hu et al. 2010). Undoubtedly women and children in LMICs have a high risk for biomass smoke exposure (Hu et al. 2010), as they may spend up to eight hours a day in a poorly-ventilated kitchen with an open stove while biomass is burning. However, this analysis found that males had a higher risk of COPD, likely due to the additive effects of cigarette smoking. In many LMICs, smoking is much more prevalent in men, when compared to women (Hu et al. 2010). It is likely that the exposure to biomass smoke interacts with cigarette smoking in males in the pathogenesis of COPD (Hu et al. 2010), whereas in women and children, exposure to biomass smoke is the main risk factor leading to COPD.

Recent studies of the mechanisms and epidemiological associations of biomass smoke with COPD are summarized in Table 20.1.

20.4 Biomass Smoke-Related Chronic Obstructive Pulmonary Disease (COPD)

20.4.1 Patterns of Biomass Smoke-Related COPD

Tobacco smoke exposure has long been the primary cause of COPD, but recent evidence from several studies suggests that, especially in lower socioeconomic regions such as sub-Saharan Africa, Southeast Asia and South America, biomass smoke exposure may be a key contributor to the prevalence of COPD and is strongly correlated with disease (Shahab et al. 2006; Lim et al. 2012; Denguezli et al. 2016; Kalagouda Mahishale et al. 2016; Walia et al. 2016; Lopez-Campos et al. 2017; Sinha et al. 2017). The major source of biomass in LMICs is particulate matter derived from the burning of biofuels and wood, and PM generated from cooking with solid fuels is believed to be the largest contributor to household air pollution (HAP) contributing to one in three premature deaths attributable to COPD (Assad et al. 2015; Ni et al. 2015; Balcan et al. 2016; Kalagouda Mahishale et al. 2016). By increasing exposure to PM because of social and cultural practices, HAP may explain the increased rates of COPD in non-smoking women in parts of the Middle East, Africa, and Asia (Po et al. 2011; Kodgule and Salvi 2012; Kalagouda Mahishale et al. 2016). Increasing the complexity of the pathogenesis of COPD is the fact that PM derived from different biomass sources have different molecular compositions, which may contribute to varied disease presentation (Sussan et al. 2014; Assad et al. 2015).

20.4.2 Clinical Presentation of COPD

The clinical presentation of COPD is typically manifested by the classical symptoms of cough, dyspnea, and sputum production. Diagnosis can also be made using spirometry,

Table 20.1 Examples of studies of biomass smoke exposure.

Author	Year	Location	Cohort	Exposure	Findings	Citation
Heinzerling	2016	Guatamala	557 children, mean age 5.4 years; Gender: 245 males (48.7%)	Open wood-fire stove	Spirometry showed ↑ PEF ↓ FEV1 ↓ Lung function	Heinzerling et al. (2016)
Muala	2015	Sweden	14 healthy adults 21–35 years old; 8 males (57%), never smokers	Wood-fire smoke in exposure chamber	Bronchial wash fluid, broncho-alveolar lavage fluid and bronchial biopsies showed ↓ macrophage, neutrophil and lymphocyte numbers ↓ sICAM-1, MPO and MMP-9 ↓ CD3⁺, CD8⁺ and submucosal mast cells	Muala et al. (2015)
Guamieri	2014	Guatamala	45 women, mean age: 26 years	Open wood-fire stove	Spirometry and induced sputum showed that ↑ exhaled CO was associated with ↑ mRNA expression of IL-8, TNFα and MMP-9, ↓ intervention reduced MMP-9 expression	Guarnieri et al. (2014)
Kurmi	2013	Nepal	1392 adults, mean age 35 years. Gender: 656 males (47%)	Open wood-fire stove	Spirometry showed ↓ FEV1 ↓ FVC ↓ forced expiratory flow rate	Kurmi et al. (2013)
Sehlstedt	2010	Sweden	16 healthy adults, mean age 24 years; Gender: 10 males (53%)	Wood-fire smoke in controlled exposure chamber	↑ glutathione concentration in BAL; ↑ mucosal symptoms; No acute inflammatory responses	Sehlstedt et al. (2010)
Ekici	2005	Turkey	397 women	Wood, grass, crops and animal dung	Symptoms and spirometry showed ↑ chronic airways disease; 23% of chronic airways disease attributed to biomass smoke exposure	Ekici et al. (2005)
Perez-Padilla	1996	Mexico	127 women, age over 40 years	Open wood stove	Symptoms and signs showed ↑ risk of chronic bronchitis alone and chronic bronchitis and chronic airway obstruction together	Perez-Padilla et al. (1996)

Abbreviations used: PEF, peak expiratory flow; sICAM, soluble intercellular adhesion molecule; MPO, myeloperoxidase; MMP, matrix metalloproteinase; CD, cluster of differentiation; CO, carbon monoxide; mRNA, messenger ribonucleic acid; IL, interleukin; TNF, tumor necrosis factor; FEV1, forced expiratory volume in one second; FVC, forced vital capacity; BAL, bronchoalveolar fluid.

where the diagnosis is defined by a decrease in the post-bronchodilator forced expiratory volume in one second (FEV1)/forced vital capacity (FVC) ratio. For the diagnosis of COPD, the Global Initiative for chronic Obstructive Lung Disease (GOLD) guidelines have suggested that a threshold of 0.7 or below be used to diagnose airflow obstruction. It should be noted that this standard is arbitrary and has not been clinically validated (Rabe et al. 2007; Nazir and Erbland 2009). However, this fixed cut-off does not consider age-related changes in FEV1/FVC ratios due to loss of lung elasticity, and hence an FEV1/FVC ratio of less than the lower limit at the fifth percentile of the normal (lower limit of normal, LLN) has been suggested to diagnose COPD.

This ability to diagnose at risk subjects utilizing spirometry may be especially important as prevalence studies, such as Burden of Lung Disease (BOLD), The Latin American Project for the Investigation of Obstructive Lung Disease (PLATINO) and Prevalence Study of COPD in Colombia (PREPOCOL), indicate that in both LMICs and developed regions, COPD may be underdiagnosed by physicians, especially as smoking is generally considered the leading risk factor (Salvi and Bames 2010; Lim et al. 2012, Bernd et al. 2015; Kalagouda Mahishale et al. 2016; Sinha et al. 2017). In fact, across several of the BOLD prevalence studies, non-smoking COPD as defined by GOLD standards were often misdiagnosed as asthma or undiagnosed due to erroneous patient self-reporting of symptoms or cultural or social barriers to access to spirometry (Shahab et al. 2006; Bernd et al. 2015; Balcan et al. 2016). As indicated by the PLATINO studies, under-diagnosis of COPD is consistent in both smokers and never-smokers, indicating a need for more effective diagnostic techniques (Bernd et al. 2015). Particularly in lower income countries, where access to spirometry is often restricted, implementation of the routine use of spirometry in the diagnosis of COPD may lead to a more accurate diagnosis of disease and allow earlier intervention, thus improving quality of life (QOL) by more effective management (Bernd et al. 2015; Kalagouda Mahishale et al. 2016).

20.4.3 Complications of Biomass Smoke Exposure

Exposure to biomass PM can result in a wide array of complications for patients. Biomass PM has been linked with an increased incidence of COPD phenotypes including chronic bronchitis and emphysema, as well as an increased risk of other pulmonary diseases such as lung cancer (Assad et al. 2015). Notably, biomass exposure has also been suggested to be linked to a novel type of COPD, bronchial anthracofibrosis (BAF) (Kim et al. 2009; Gupta and Shah 2011; Assad et al. 2015). BAF has been described in women from rural areas in Southeast Asia and the Middle East, where the most common symptom was an obstruction-based defect in ventilation as described by GOLD standards (Kim et al. 2009; Sigari and Mohammadi 2009; Gupta and Shah 2011; Assad et al. 2015). The burden of disease in non-smoking COPD has been suggested to be disproportionately higher than in smoking-related COPD: a Mexican study suggested that non-smoking COPD patients reported worse respiratory symptoms and decreased QOL relative to levels of emphysema when compared to smoking-related COPD (Camp et al. 2014). This is particularly of note as exposure to biomass PM is far more prevalent than exposure to tobacco PM, with approximately three billion people exposed to biomass PM compared one billion exposed to tobacco PM (Salvi and Bames 2010). Exposure to biomass smoke in early childhood may also cause a decrease in lung function and growth, resulting in significantly decreased QOL in adulthood (Assad et al. 2015). Air pollution has also been linked to acute exacerbations in COPD due to particulates, sulfur dioxide, nitrogen dioxide and ozone (Ni et al. 2015). Chronic exposure to biomass may exacerbate underlying hypoxia and pulmonary hypertension; however, studies in these areas are inconclusive and still need further verification (Assad et al. 2015).

20.4.4 Treatment of Biomass-Related COPD

Management of biomass-associated COPD clearly requires a multifactorial approach. Reduction in the use of biomass- producing fuels is an obvious candidate for reducing the incidence of COPD, although this may not be feasible in LMICs where the economic burden of clean fuel is too high (Assad et al. 2015). Several alternative interventions that improve ventilation in the kitchen environment and utilize fuels that are cleaner have been trialed with varying success to improve household air quality (Albalak et al. 1999: Chapman et al. 2005; Ward et al. 2008; Romieu et al. 2009; Smith et al. 2011; Alexander et al. 2014; Zhou et al. 2014). Examples include introduction of windows, chimneys and fans. A prospective Chinese study found that over nine years subjects who introduced clean fuels or improved ventilation had a decreased incidence of COPD, as well as a decrease in the rate of decline of FEV1 (Zhou et al. 2014). Individual level interventions predominantly involve preventative actions to reduce the risk of exposure of biomass PM. Additionally, some active treatments have been trialed (Laumbach et al. 2015). For example, antioxidants such as quercetin, a plant polyphenol, may help reduce the oxidative burden of COPD, whilst the dietary intake of short chain fatty acids (SCFA) may suppress lung inflammation and attenuate emphysema and COPD (Miles et al. 2014, Ghorbani et al. 2015; Tomoda et al. 2015; Jin et al. 2016). Greater access to spirometry by healthcare professionals in LMICs will enable earlier and more accurate diagnosis of COPD, which in turn will facilitate earlier intervention and allow better disease management by primary care providers (Bernd et al. 2015).

20.5 Detecting Airflow Obstruction in Biomass Smoke-Related COPD

20.5.1 Spirometry and Other Measurement Devices to Detect Airflow Obstruction

Spirometry testing is recognized as the gold standard for diagnosing and grading the severity of airflow obstruction (Levy et al. 2009). The test involves the measurement of the forced expiratory volume in one second (FEV1), the forced vital capacity (FVC) and the ratio between these measurements (FEV1/FVC). A post-bronchodilator FEV1/FVC ratio below the lower limit of normal predicted for an individual suggests the presence of airflow limitation typical of COPD.

Increasingly, the forced expiratory volume at *six* seconds of exhalation (FEV6) is emerging as a comparable substitute for the FVC in screening for COPD (Ferguson et al. 2000; Jing et al. 2009; Kaufmann et al. 2009; Bhatt et al. 2014). FEV1/ FEV6 ratios <0.70–<0.80 (Jing et al. 2009; Frith et al. 2011; Sichletidis et al. 2011; van den Bemt et al. 2014) have been used to detect airflow obstruction in primary care settings.

A range of portable and inexpensive machines that can measure a small selection of respiratory function parameters are available (Table 20.2 and Figure 20.3). Handheld spirometers can measure FEV1, FEV6 and the FEV1/FEV6 ratio and can provide a practical, low-cost alternative to more specialized equipment that may not be readily available or accessible in all countries.

Table 20.2 Comparison of alternative devices to laboratory-based lung function testing.

Feature	Office (desktop) spirometers	Handheld spirometers	Peak flow devices
Cost[a]	Require larger financial investment	Inexpensive (20–70 USD)	Inexpensive (<22 USD)
Portability	Low	Easy	Easy
Power source	Require stable source of energy	Batteries	None
Visual information	Yes (flow-volume loops, volume-time graphs)	Rare although some can connect to computers or PDAs to generate graphs	None
Quality checks/ feedback	Instant feedback	Rare	None
Ability of machine to show data trend for individual patient	Yes	No	No
Calibration	Available for most if required	No	No
Training required	Yes	Yes	Yes
Measurements	FEV1, FVC, FEV1/ FVC, PEF	FEV1, FEV6, FEV1/ FEV6, PEF	PEF
Predicted values and lower limit of normal	Well validated and available on screen; can be calculated for precise age and height of patient	Well validated but not displayed on device; nomograms and predicted values charts will need to be available in fixed increments (e.g. 160 cm, 165 cm, etc.) for interpretation	Not well validated in adults
Rate of false positives	Low	Low	High
Clinical utility of measurements	Preferred option when laboratory-based equipment not available	Comparable to office-based spirometers	Poor index of early distal airways obstruction; does not discriminate between asthma and COPD

a) Adapted from an online presentation from the Federation of International Respiratory Societies. Available at: http://www.who.int/gard/news_events/5-5___Firs%20Task%20Force%20on%20Simple %20Pulmonary%20Function%20Tests-Dr.pdf

Abbreviations used: PDA, personal digital assistant; FEV1, forced expiratory volume in one second; FEV6, forced expiratory volume in six seconds; FVC, forced vital capacity; PEF, peak expiratory flow.

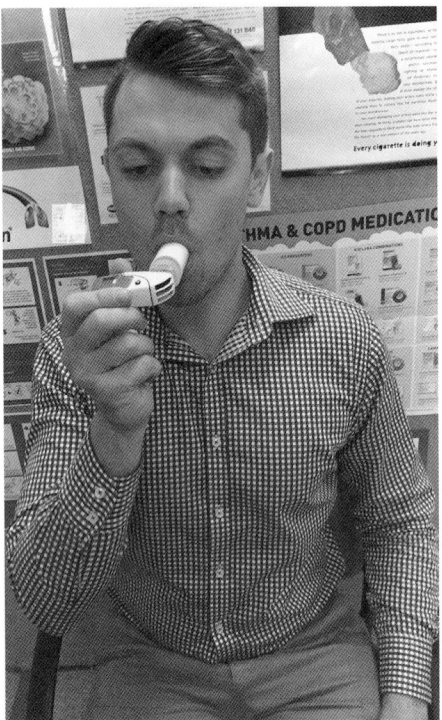

Figure 20.3 Use of a handheld FEV1/FEV6 device to detect airflow obstruction. Handheld spirometers or FEV1/FEV6 devices (shown) are feasible for use in early detection programs, as they are portable, relatively inexpensive, require only batteries for power, and need minimal training for healthcare workers in the field.

20.5.2 Facilities and Methods for Spirometry Testing

The facility requirements for handheld spirometers are minimal and can easily be established in primary care in a range of urban and rural settings. The patient requires a chair so that the test can be performed in an upright position, and where possible, some privacy should be provided while performing the test.

The spirometer should be cleaned at least once monthly, and disinfected as per the manufacturer's instructions between patients. A check for particulate matter build-up should be performed regularly. A new one-way disposable mouthpiece should be used for each patient to reduce the risk of cross-infection (Miller et al. 2005).

20.5.3 Training of Health Professionals to Undertake Spirometry

Any healthcare worker can be trained to perform spirometry. The test does not need to be performed by a physician, nurse or technologist. Minimal training is required to operate handheld spirometers. However, the use of an instruction sheet and completion of an online training program is recommended (see the Lung Foundation Australia position paper on targeted COPD case finding in community settings. Available at: http://lungfoundation.com.au/wp-content/uploads/2014/02/Position-Paper-1.pdf).

While Internet access in LMICs is rapidly becoming more available, it can be constrained by a number of factors including connection problems, low bandwidth, slow downloads, and high cost of access. Regardless, the majority of healthcare workers have at least limited access to the Internet, enabling the possibility of distance-based and self-paced learning (Ajuwon 2015).

Most handheld spirometers do not require calibration. However, as part of a quality control program, a designated healthcare worker with no known lung disease and the ability to obtain reproducible results should be available to regularly monitor the performance of the machine using himself/herself as the test subject. Use of the same spirometer for longitudinal testing of a patient is recommended for comparable results as variation in output does occur between spirometers. Alternative energy sources for medical devices, such as solar power generated devices (rather than batteries and electrical power), will make these devices more accessible to remote and poorly resourced areas. This has been recommended and supported by the World Health Organization.

20.5.4 Other Early Detection Methods for COPD

While spirometry performed on a handheld device is a relatively quick test to perform, it can be used in conjunction with short (less than one page) screening questionnaires to identify patients with a high pre-test probability of COPD. The combination of questionnaire and respiratory test has been shown to have a higher accuracy of diagnosing COPD than either tool alone (Sichletidis et al. 2011; Nelson et al. 2012; Martinez et al. 2017).

20.6 Lessons Learnt from Clinical Guidelines for the Detection of Cigarette Smoking-Related COPD

20.6.1 Clinical Guidelines for the Diagnosis of COPD

International clinical guidelines such as the Global Initiative for Chronic Obstructive Lung Disease (GOLD) Report emphasize the importance of history taking and spirometry for making the diagnosis of COPD (Global Initiative for Chronic Obstructive Lung Disease 2017). In at-risk individuals who have been exposed to inhaled toxins (cigarette smoke, biomass smoke, occupational dusts and gasses, and other outdoor air pollutants), symptoms consistent with COPD are exertional dyspnoea, cough, sputum production, wheeze and repeated chest infections and exacerbations. Physical examination is abnormal in more severe COPD, with chest hyperinflation, reduced intensity of breath sounds on auscultation, and sometimes wheeze. COPD from any cause is diagnosed when spirometry demonstrates a post-bronchodilator FEV1/FVC ratio of <0.70 in an individual with persistent respiratory symptoms (Global Initiative for Chronic Obstructive Lung Disease 2017).

The severity of COPD is classified based on symptoms and frequency of exacerbations (Global Initiative for Chronic Obstructive Lung Disease 2017). A similar emphasis on symptoms and spirometry are also promoted by national clinical guidelines, e.g. COPD guidelines in Australia (http://copdx.org.au). However, there are many challenges in the dissemination of, and adherence to, clinical guidelines for COPD, including limited clinician awareness and uptake of spirometry, scarce resources and the question of applicability of guidelines to individual patients (Overington et al. 2014). These barriers need to be overcome to achieve the successful translation of the guidelines to clinical and public health practice.

Biomass smoke-related COPD is well recognized in the international clinical guidelines for COPD. The GOLD Report highlights use of biomass fuel for indoor cooking as a major risk factor for COPD, especially in housing that is not optimally ventilated (Global Initiative for Chronic Obstructive Lung Disease 2017). This international guideline elaborates further on the large at-risk population globally who use biomass as energy for household activities, thus emphasizing the public health significance of biomass-related COPD and the need for its early detection.

20.6.2 Early Detection Programs for COPD in LMICs: Checklists for Risk Factors and Symptoms

Screening of asymptomatic at-risk individuals is not recommended for COPD (U.S. Preventive Services Task Force 2016), whereas case finding or early case detection of patients with symptoms and risk factors is advised (Global Initiative for Chronic Obstructive Lung Disease 2017). However, to date this has been difficult to implement. The GOLD Report summarized key indicators for considering a diagnosis of COPD, and these are useful as a checklist for early detection programs for COPD (Table 20.3). These symptoms are highly relevant to biomass-related COPD, as demonstrated by observational studies (Regalado et al. 2006; Moreira et al. 2013; Elhefny and Elessawy 2015).

Other resources have incorporated symptoms and risk factors in interactive online checklists that can be used by people in the community, including checklists adapted for indigenous populations (for example, Lung Foundation Australia: http://lungfoundation.com.au/lung-health-checklist). People who meet the criteria in the checklists would then ideally be invited to undertake lung function testing in order to

Table 20.3 Suggested checklist of risk factors and symptoms consistent with COPD: for implementation of early detection programs for biomass-related COPD[a].

Risk factors
• Age over 40 years
and
• Exposure to smoke from biomass fuels used for home cooking and heating
• Cigarette smoking
• Exposure to occupational dusts and fumes
• Family history
Symptoms
Any of
• Breathlessness – exertional, progressive, persistent, but not variable
• Chronic cough
• Chronic sputum production
• Repeated exacerbations or respiratory infections

a) Adapted from the GOLD Report (Global Initiative for Chronic Obstructive Lung Disease 2017) with modification to emphasize biomass smoke exposure as a risk factor and to implement prevention in LMICs.

detect any persistent airflow limitation. This could confirm the diagnosis of COPD and thus lead to chronic disease management and risk factor minimization.

20.6.3 Implementing Programs for the Early Detection of Biomass-Related COPD

Establishing and maintaining early detection programs for COPD remains a global challenge. In LMICs resource and access issues limit the translation of the large spirometry programs recommended in developed countries. Considerations in LMICs need to include

- Identifying at-risk and symptomatic individuals, particularly in rural and remote areas
- Selection of appropriate health professionals to train to perform spirometry
- Determination of the type and cost of equipment to detect airflow obstruction
- Provision of access to treatment and preventive strategies once biomass-related (or smoking-related) COPD is identified.

20.7 Conclusions

Biomass smoke-related COPD is an important disease globally, due to the frequent use of biomass fuels for household cooking and heating in the LMICs. This results in indoor air pollution. Biomass smoke contains coarse, fine and ultrafine particles, which lead to airway inflammation and oxidative stress when inhaled over a long period of time. This airway epithelial injury predisposes to the development of COPD. People with COPD have chronic exertional dyspnea, cough, and sputum production as well as frequent acute exacerbations. The presence of symptoms is required in order to diagnose COPD. Spirometry is the gold standard test required to demonstrate chronic airflow obstruction.

Spirometry for detecting COPD due to inhalation of biomass-related smoke from indoor fires used for cooking and heating is therefore clearly an important initiative in the LMICs. A range of spirometers are available for use in detection programs. The use of handheld spirometers is feasible in low resource countries, in terms of cost, power required and portability. Efficient training programs need to be established, in order to

Table 20.4 Recommendations for research, clinical practice and public health.

Increase understanding of the pathogenesis of COPD from biomass smoke exposure
Develop cost-effective alternatives to burning of biomass fuels for cooking and heating in the LMICs
Improve ventilation of indoor areas to reduce exposure to biomass smoke
Develop early detection programs consisting of symptom questionnaires and spirometry at the primary-point-of-care to increase the prevention and treatment of biomass smoke-induced COPD
Train healthcare workers to effectively use handheld spirometers or peak flow meters at the primary-point-of-care
Implement clinical practice guidelines and public health policies to promote the detection and treatment of biomass smoke-induced COPD in LMICs

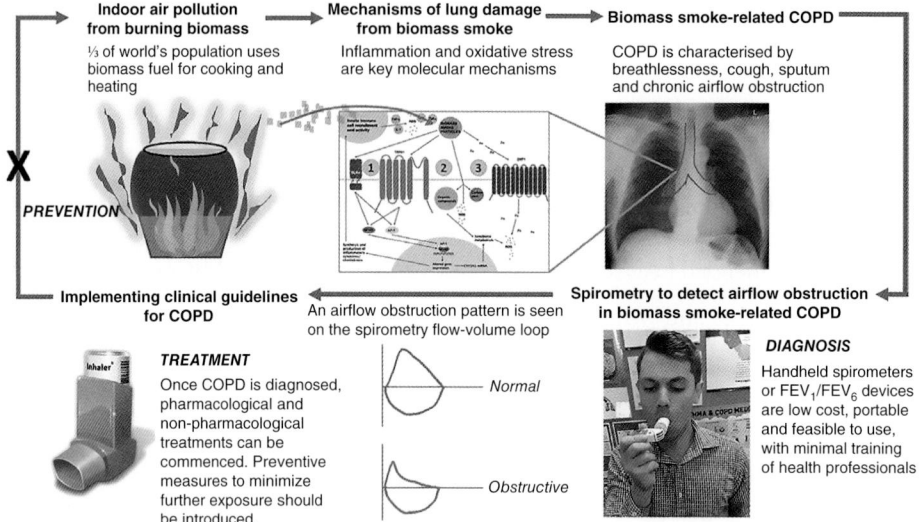

Figure 20.4 Overall approach to understanding, diagnosing and treating biomass smoke-related COPD. Biomass smoke is a common and important source of air pollution in the LMICs. Inhaled biomass smoke results in inflammation and oxidative stress in the airways of the lungs, leading to chronic obstructive pulmonary disease (COPD) in susceptible individuals. COPD causes breathlessness, cough, sputum and repeated infective exacerbations, and is detected by the presence of chronic airflow obstruction on spirometry. Early detection of biomass smoke-related COPD is feasible with handheld spirometers or FEV1/FEV6 devices, leading to implementation of treatment for COPD, and preventive and public health measures to reduce the air pollution source.

adequately train healthcare workers to perform spirometry and interpret results. Clinical guidelines for the detection of biomass smoke-induced COPD should be implemented, in order to effectively prevent or detect this condition.

Key priorities for translational research, diagnosis and public health are listed in Table 20.4 and the suggested overall approach is shown in Figure 20.4.

Acknowledgments

We sincerely thank the patients and staff involved in our research programs. Ian Yang is supported by a National Health and Medical Research Committee (NHMRC) Career Development Fellowship 1026215, NHMRC project grant APP1121740, The Prince Charles Hospital Foundation and The University of Queensland. Annalicia Vaughan is supported by a PhD Scholarship from The Prince Charles Hospital Foundation.

Bibliography

Ajuwon, G.A. (2015). Internet Accessibility and Use of Online Health Information Resources by Doctors in Training Healthcare Institutions in Nigeria. Library Philosophy and Practice (e-journal) Paper 1258: http://digitalcommons.unl.edu/libphilprac/1258.
Albalak, R., Frisancho, A., and Keeler, G. (1999). Domestic biomass fuel combustion and chronic bronchitis in two rural Bolivian villages. *Thorax* 54 (11): 1004–1008.

Alexander, D., Linnes, J.C., Bolton, S., and Larson, T. (2014). Ventilated cookstoves associated with improvements in respiratory health-related quality of life in rural Bolivia. *J. Public Health (Oxf.)* 36 (3): 460–466.

Andreae, M.O. and Merlet, P. (2001). Emission of trace gases and aerosols from biomass burning. *Glob. Biogeochem. Cycles* 15 (4): 955–966.

Assad, N.A., Balmes, J., Mehta, S. et al. (2015). Chronic obstructive pulmonary disease secondary to household air pollution. *Semin. Respir. Crit. Care Med.* 36 (3): 408–421.

Balcan, B., Akan, S., Ugurlu, A.O. et al. (2016). Effects of biomass smoke on pulmonary functions: a case control study. *Int. J. Chron. Obstruct. Pulmon. Dis.* 11: 1615–1622.

Barnes, P. (2011). Inflammation in COPD. *Clin. Respir. J.* 5 (s1): 1–2.

Bernd, L., Joan, B.S., Michael, S. et al. (2015). Determinants of underdiagnosis of COPD in national and international surveys. *Chest* 148 (4): 971–985.

Bhatt, S.P., Kim, Y.I., Wells, J.M. et al. (2014). FEV(1)/FEV(6) to diagnose airflow obstruction. Comparisons with computed tomography and morbidity indices. *Ann. Am. Thorac. Soc.* 11 (3): 335–341.

Bittner, M., Macikova, P., Giesy, J.P., and Hilscherova, K. (2011). Enhancement of AhR-mediated activity of selected pollutants and their mixtures after interaction with dissolved organic matter. *Environ. Int.* 37 (5): 960–964.

Bond, T.C., Streets, D.G., Yarber, K.F. et al. (2004). A technology-based global inventory of black and organic carbon emissions from combustion. *J. Geophys. Res.-Atmos.* 109 (D14).

Camp, P.G., Ramirez-Venegas, A., Sansores, R.H. et al. (2014). COPD phenotypes in biomass smoke- versus tobacco smoke-exposed Mexican women. *Eur. Respir. J.* 43 (3): 725–734.

Caram, L., Ferrari, R., Nogueira, D. et al. (2016). Identification of oxidative stress markers and their association with inflammation in smokers and early chronic obstructive pulmonary disease (COPD). *Eur. Respir. J.* 48 (suppl 60): https://doi.org/10.1183/13993003.

Chapman, R.S., He, X., Blair, A.E., and Lan, Q. (2005). Improvement in household stoves and risk of chronic obstructive pulmonary disease in Xuanwei, China: retrospective cohort study. *BMJ* 331 (7524): 1050.

Chen, J., Li, C., Ristovski, Z. et al. (2017). A review of biomass burning: emissions and impacts on air quality, health and climate in China. *Sci. Total Environ.* 579: 1000–1034.

Denguezli, M., Daldoul, H., Harrabi, I. et al. (2016). COPD in nonsmokers: reports from the Tunisian population-based burden of obstructive lung disease study. *PLoS One* 11 (3): e0151981.

Ekici, A., Ekici, M., Kurtipek, E. et al. (2005). Obstructive airway diseases in women exposed to biomass smoke. *Environ. Res.* 99 (1): 93–98.

Elhefny, R.A. and Elessawy, A.F. (2015). Assessment of chronic obstructive pulmonary disease in rural women. *Egypt. J. Chest Dis. Tuberc.* 64 (2): 343–346.

Fariss, M.W., Gilmour, M.I., Reilly, C.A. et al. (2013). Emerging mechanistic targets in lung injury induced by combustion-generated particles. *Toxicol. Sci.* 132 (2): 253–267.

Ferguson, G.T., Enright, P.L., Buist, A.S., and Higgins, M.W. (2000). Office spirometry for lung health assessment in adults: a consensus statement from the National Lung Health Education Program. *Respir. Care* 45 (5): 513–530.

Formenti, P., Elbert, W., Maenhaut, W. et al. (2003). Inorganic and carbonaceous aerosols during the Southern African Regional Science Initiative (SAFARI 2000) experiment:

chemical characteristics, physical properties, and emission data for smoke from African biomass burning. *J. Geophys. Res.-Atmos.* 108 (D13): https://doi.org/10.1029/2002JD002408.

Frith, P., Crockett, A., Beilby, J. et al. (2011). Simplified COPD screening: validation of the PiKo-6(R) in primary care. *Prim. Care Respir. J.* 20 (2): 190–198.

Ghorbani, P., Santhakumar, P., Hu, Q. et al. (2015). Short-chain fatty acids affect cystic fibrosis airway inflammation and bacterial growth. *Eur. Respir. J.* 46 (4): 1033–1045.

Global Initiative for Chronic Obstructive Lung Disease. (2017). "Global strategy for the diagnosis, management, and prevention of chronic obstructive pulmonary disease". Retrieved 6 July 2017, from http://goldcopd.org.

Guarnieri, M.J., Diaz, J.V., Basu, C. et al. (2014). Effects of woodsmoke exposure on airway inflammation in rural Guatemalan women. *PLoS One* 9 (3): e88455.

Gupta, A. and Shah, A. (2011). Bronchial anthracofibrosis: an emerging pulmonary disease due to biomass fuel exposure. *Int. J. Tuberc. Lung Dis.* 15 (5): 602–612.

Heinzerling, A.P., Guarnieri, M.J., Mann, J.K. et al. (2016). Lung function in woodsmoke-exposed Guatemalan children following a chimney stove intervention. *Thorax* 71 (5): 421–428.

Hu, G., Zhou, Y., Tian, J. et al. (2010). Risk of COPD from exposure to biomass smoke: a Metaanalysis. *Chest* 138 (1): 20–31.

Jin, X., Su, R., Li, R. et al. (2016). Amelioration of particulate matter-induced oxidative damage by vitamin c and quercetin in human bronchial epithelial cells. *Chemosphere* 144: 459–466.

Jing, J.Y., Huang, T.C., Cui, W. et al. (2009). Should FEV1/FEV6 replace FEV1/FVC ratio to detect airway obstruction? A metaanalysis. *Chest* 135 (4): 991–998.

Kalagouda Mahishale, V., Angadi, N., Metgudmath, V. et al. (2016). The prevalence of chronic obstructive pulmonary disease and the determinants of underdiagnosis in women exposed to biomass fuel in India- a cross section study. *Chonnam Med. J.* 52 (2): 117–122.

Kaufmann, M., Hartl, S., Geyer, K. et al. (2009). Measuring FEV(6) for detecting early airway obstruction in the primary care setting. Quality and utility of the new PiKo-6 device. *Respiration* 78 (2): 161–167.

Kim, Y.J., Jung, C.Y., Shin, H.W., and Lee, B.K. (2009). Biomass smoke induced bronchial anthracofibrosis: presenting features and clinical course. *Respir. Med.* 103 (5): 757–765.

Kodgule, R. and Salvi, S. (2012). Exposure to biomass smoke as a cause for airway disease in women and children. *Curr. Opin. Allergy Clin. Immunol.* 12 (1): 82–90.

Kurmi, O.P., Devereux, G.S., Smith, W.C.S. et al. (2013). Reduced lung function due to biomass smoke exposure in young adults in rural Nepal. *Eur. Respir. J.* 41 (1): 25–30.

Lal, K., Mani, U., Pandey, R. et al. (2011). Multiple approaches to evaluate the toxicity of the biomass fuel cow dung (kanda) smoke. *Ecotoxicol. Environ. Saf.* 74 (7): 2126–2132.

Laumbach, R., Meng, Q., and Kipen, H. (2015). What can individuals do to reduce personal health risks from air pollution? *J. Thorac. Dis.* 7 (1): 96–107.

Lawrence, T. (2009). The nuclear factor NF-kappaB pathway in inflammation. *Cold Spring Harb. Perspect. Biol.* 1 (6): a001651.

Levy, M.L., Quanjer, P.H., General Practice Airways et al. (2009). Diagnostic spirometry in primary care: proposed standards for general practice compliant with American Thoracic Society and European Respiratory Society recommendations: a General Practice Airways Group (GPIAG)1 document, in association with the Association for Respiratory Technology & Physiology (ARTP)2 and Education for Health3 1 www.gpiag. org 2 www.artp.org 3 www.educationforhealth.org.uk. *Prim. Care Respir. J.* 18 (3): 130–147.

Lim, S.S., Vos, T., Flaxman, A.D. et al. (2012). A comparative risk assessment of burden of disease and injury attributable to 67 risk factors and risk factor clusters in 21 regions, 1990–2010: a systematic analysis for the global burden of disease study 2010. *Lancet* 380 (9859): 2224–2260.

Lopez-Campos, J.L., Fernandez-Villar, A., Calero-Acuna, C. et al. (2017). Occupational and biomass exposure in chronic obstructive pulmonary disease: results of a cross-sectional analysis of the on-Sint study. *Arch. Bronconeumol.* 53 (1): 7–12.

Martin, W.J. II, Glass, R.I., Araj, H. et al. (2013). Household air pollution in low- and middle-income countries: health risks and research priorities. *PLoS Med.* 10 (6): e1001455.

Martinez, F.J., Mannino, D., Leidy, N.K. et al. (2017). A new approach for identifying patients with undiagnosed chronic obstructive pulmonary disease. *Am. J. Respir. Crit. Care Med.* 195 (6): 748–756.

Masera, O.R., Saatkamp, B.D., and Kammen, D.M. (2000). From linear fuel switching to multiple cooking strategies: a critique and alternative to the energy ladder model. *World Dev.* 28 (12): 2083–2103.

Miles, S.L., McFarland, M., and Niles, R.M. (2014). Molecular and physiological actions of quercetin: need for clinical trials to assess its benefits in human disease. *Nutr. Rev.* 72 (11): 720–734.

Miljevic, B., Heringa, M.F., Keller, A. et al. (2010). Oxidative potential of logwood and pellet burning particles assessed by a novel profluorescent nitroxide probe. *Environ. Sci. Technol.* 44 (17): 6601–6607.

Miller, M.R., Crapo, R., Hankinson, J. et al. (2005). General considerations for lung function testing. *Eur. Respir. J.* 26 (1): 153–161.

Morawska, L. and Zhang, J. (2002). Combustion sources of particles. 1. Health relevance and source signatures. *Chemosphere* 49 (9): 1045–1058.

Moreira, M.A.C., Barbosa, M.A., Jardim, J.R. et al. (2013). Chronic obstructive pulmonary disease in women exposed to wood stove smoke. *Rev. Assoc. Med. Bras.* 59: 607–613.

Muala, A., Rankin, G., Sehlstedt, M. et al. (2015). Acute exposure to wood smoke from incomplete combustion--indications of cytotoxicity. *Part. Fibre Toxicol.* 12: 33.

Nazir, S.A. and Erbland, M.L. (2009). Chronic obstructive pulmonary disease. *Drugs Aging* 26 (10): 813–831.

Nebert, D.W., Roe, A.L., Dieter, M.Z. et al. (2000). Role of the aromatic hydrocarbon receptor and [Ah] gene battery in the oxidative stress response, cell cycle control, and apoptosis. *Biochem. Pharmacol.* 59 (1): 65–85.

Nelson, S.B., LaVange, L.M., Nie, Y. et al. (2012). Questionnaires and pocket spirometers provide an alternative approach for COPD screening in the general population. *Chest* 142 (2): 358–366.

Ni, L., Chuang, C.-C., and Zuo, L. (2015). Fine particulate matter in acute exacerbation of COPD. *Front. Physiol.* 6: 294.

Overington, J.D., Huang, Y.C., Abramson, M.C. et al. (2014). Implementing clinical guidelines for chronic obstructive pulmonary disease: barriers and solutions. *J. Thorac. Dis.* 6 (11): 1586–1596.

Perez-Padilla, R., Regalado, J., Vedal, S. et al. (1996). Exposure to biomass smoke and chronic airway disease in Mexican women. A case-control study. *Am. J. Respir. Crit. Care Med.* 154 (3 Pt 1): 701–706.

Po, J.Y.T., Fitzgerald, J.M., and Carlsten, C. (2011). Respiratory disease associated with solid biomass fuel exposure in rural women and children: systematic review and meta-analysis. *Thorax* 66 (3): 232.

Rabe, K.F., Hurd, S., Anzueto, A. et al. (2007). Global strategy for the diagnosis, management, and prevention of chronic obstructive pulmonary disease: GOLD executive summary. *Am. J. Respir. Crit. Care Med.* 176 (6): 532–555.

Rahman, I. and Adcock, I. (2006). Oxidative stress and redox regulation of lung inflammation in COPD. *Eur. Respir. J.* 28 (1): 219–242.

Regalado, J., Pérez-Padilla, R., Sansores, R. et al. (2006). The effect of biomass burning on respiratory symptoms and lung function in rural Mexican women. *Am. J. Respir. Crit. Care Med.* 174 (8): 901–905.

Reid, J.S., Koppmann, R., Eck, T.F., and Eleuterio, D.P. (2005). A review of biomass burning emissions part II: intensive physical properties of biomass burning particles. *Atmos. Chem. Phys.* 5 (3): 799–825.

Ristovski, Z.D., Miljevic, B., Surawski, N.C. et al. (2012). Respiratory health effects of diesel particulate matter. *Respirology* 17 (2): 201–212.

Romieu, I., Riojas-Rodriguez, H., Marron-Mares, A.T. et al. (2009). Improved biomass stove intervention in rural Mexico: impact on the respiratory health of women. *Am. J. Respir. Crit. Care Med.* 180 (7): 649–656.

Rothen-Rutishauser, B., Muhlfeld, C., Blank, F. et al. (2007). Translocation of particles and inflammatory responses after exposure to fine particles and nanoparticles in an epithelial airway model. *Part. Fibre Toxicol.* 4: 9.

Saarikoski, S., Sillanpää, M., Sofiev, M. et al. (2007). Chemical composition of aerosols during a major biomass burning episode over northern Europe in spring 2006: experimental and modelling assessments. *Atmos. Environ.* 41 (17): 3577–3589.

Salvi, S. and Barnes, P.J. (2010). Is exposure to biomass smoke the biggest risk factor for COPD globally? *Chest* 138 (1): 3–6.

Sandradewi, J., Prévôt, A.S.H., Weingartner, E. et al. (2008). A study of wood burning and traffic aerosols in an Alpine valley using a multi-wavelength Aethalometer. *Atmos. Environ.* 42 (1): 101–112.

Sehlstedt, M., Dove, R., Boman, C. et al. (2010). Antioxidant airway responses following experimental exposure to wood smoke in man. *Part. Fibre Toxicol.* 7: 21.

Shahab, L., Jarvis, M.J., Britton, J., and West, R. (2006). Prevalence, diagnosis and relation to tobacco dependence of chronic obstructive pulmonary disease in a nationally representative population sample. *Thorax* 61 (12): 1043–1047.

Shapiro, D., Deering-Rice, C.E., Romero, E.G. et al. (2013). Activation of transient receptor potential ankyrin-1 (TRPA1) in lung cells by wood smoke particulate material. *Chem. Res. Toxicol.* 26 (5): 750–758.

Shoenfelt, J., Mitkus, R.J., Zeisler, R. et al. (2009). Involvement of TLR2 and TLR4 in inflammatory immune responses induced by fine and coarse ambient air particulate matter. *J. Leukoc. Biol.* 86 (2): 303–312.

Sichletidis, L., Spyratos, D., Papaioannou, M. et al. (2011). A combination of the IPAG questionnaire and PiKo-6(R) flow meter is a valuable screening tool for COPD in the primary care setting. *Prim. Care Respir. J.* 20 (2): 184–189.

Sigari, N. and Mohammadi, S. (2009). Anthracosis and anthracofibrosis. *Saudi Med. J.* 30 (8): 1063–1066.

Silva, R., Oyarzún, M., and Olloquequi, J. (2015). Pathogenic mechanisms in chronic obstructive pulmonary disease due to biomass smoke exposure. *Arch. Bronconeumol.* 51 (6): 285–292.

Simoneit, B.R.T. (2002). Biomass burning – a review of organic tracers for smoke from incomplete combustion. *Appl. Geochem.* 17 (3): 129–162.

Sinha, B., Singla, R., and Chowdhury, R. (2017). An epidemiological profile of chronic obstructive pulmonary disease: a community-based study in Delhi. *J. Postgrad. Med.* 63 (1): 29–35.

Smith, K.R., McCracken, J.P., Weber, M.W. et al. (2011). Effect of reduction in household air pollution on childhood pneumonia in Guatemala (RESPIRE): a randomised controlled trial. *Lancet* 378 (9804): 1717–1726.

Smith, K., Samet, J., Romieu, I., and Bruce, N. (2000). Indoor air pollution in developing countries and acute lower respiratory infections in children. *Thorax* 55 (6): 518–532.

Stevanovic, S., Miljevic, B., Surawski, N.C. et al. (2013). Influence of oxygenated organic aerosols (OOAs) on the oxidative potential of diesel and biodiesel particulate matter. *Environ. Sci. Technol.* 47 (14): 7655–7662.

Sussan, T.E., Ingole, V.E., Kim, J.-H. et al. (2014). Source of biomass cooking fuel determines pulmonary response to household air pollution. *Am. J. Respir. Cell Mol. Biol.* 50 (3): 538–548.

Tam, A., Wadsworth, S., Dorscheid, D.R. et al. (2011). The airway epithelium: more than just a structural barrier. *Ther. Adv. Respir. Dis.* 5 (4): 255–273.

Tomoda, K., Kubo, K., Dairiki, K. et al. (2015). Whey peptide-based enteral diet attenuated elastase-induced emphysema with increase in short chain fatty acids in mice. *BMC Pulm. Med.* 15: 64.

Totlandsdal, A.I., Herseth, J.I., Bolling, A.K. et al. (2012). Differential effects of the particle core and organic extract of diesel exhaust particles. *Toxicol. Lett.* 208 (3): 262–268.

U. S. Preventive Services Task Force, Siu, A.L., Bibbins-Domingo, K. et al. (2016). Screening for chronic obstructive pulmonary disease: US preventive services task force recommendation statement. *JAMA* 315 (13): 1372–1377.

van den Bemt, L., Wouters, B.C., Grootens, J. et al. (2014). Diagnostic accuracy of pre-bronchodilator FEV1/FEV6 from microspirometry to detect airflow obstruction in primary care: a randomised cross-sectional study. *NPJ Prim. Care Respir. Med.* 24: 14033.

Verma, V., Rico-Martinez, R., Kotra, N. et al. (2012). Contribution of water-soluble and insoluble components and their hydrophobic/hydrophilic subfractions to the reactive oxygen species-generating potential of fine ambient aerosols. *Environ. Sci. Technol.* 46 (20): 11384–11392.

von Knethen, A., Callsen, D., and Brüne, B. (1999). NF-κB and AP-1 activation by nitric oxide attenuated apoptotic cell death in RAW 264.7 macrophages. *Mol. Biol. Cell* 10 (2): 361–372.

Walia, G.K., Vellakkal, R., and Gupta, V. (2016). Chronic obstructive pulmonary disease and its non-smoking risk factors in India. *COPD* 13 (2): 251–261.

Ward, T., Palmer, C., Bergauff, M. et al. (2008). Results of a residential indoor PM2.5 sampling program before and after a woodstove changeout. *Indoor Air* 18 (5): 408–415.

Weimer, S., Alfarra, M.R., Schreiber, D. et al. (2008). Organic aerosol mass spectral signatures from wood-burning emissions: influence of burning conditions and wood type. *J. Geophys. Res.-Atmos.* 113 (D10): https://doi.org/10.1029/2007JD009309.

WHO (2002). *The World Health Report, 2002: Reducing Risks, Promoting Health Life.* Geneva, Switzerland: WHO.

Zhou, Y., Zou, Y., Li, X. et al. (2014). Lung function and incidence of chronic obstructive pulmonary disease after improved cooking fuels and kitchen ventilation: a 9-year prospective cohort study. *PLoS Med.* 11 (3): e1001621.

Webliography

http://copdx.org.au - COPD guidelines in Australia
http://lungfoundation.com.au/lung-health-checklist
http://lungfoundation.com.au/wp-content/uploads/2014/02/Position-Paper-1.pdf -
    training in spirometry use

21

Point-of-Care Pulse Oximetry for Children in Low-Resource Settings

Carina King[1], Hamish Graham[2], and Eric D. McCollum[3,4]

[1] *Institute for Global Health, University College London, London, UK*
[2] *Centre for International Child Health, University of Melbourne, Murdoch Research Children's Institute, The Royal Children's Hospital, Melbourne, Victoria, Australia*
[3] *Department of Pediatrics, Eudowood Division of Pediatric Respiratory Sciences, Johns Hopkins School of Medicine, Baltimore, MD, USA*
[4] *Department of International Health, Johns Hopkins Bloomberg School of Public Health, Baltimore, MD, USA*

CHAPTER MENU

21.1 Introduction

Oxygen is fundamental to the healthy functioning of the human body, and when the levels of oxygen within the blood (oxygen saturation) fall below the normal range (approximately 95–100%) there can be serious morbidity and mortality consequences. The ability to measure oxygen saturation, with a non-invasive, low-cost and reusable device therefore has the ability to improve the diagnosis, treatment and management of children at risk of hypoxemia. This device is called a pulse oximeter. It has already transformed clinical care in high-resource settings, but its real life-saving potential will only be realized when it is scaled up globally (Box 21.1 and Figure 21.1).

Revolutionizing Tropical Medicine: Point-of-Care Tests, New Imaging Technologies and Digital Health,
First Edition. Edited by Kerry Atkinson and David Mabey.
© 2019 John Wiley & Sons, Inc. Published 2019 by John Wiley & Sons, Inc.

Box 21.1 Key Definitions

Oxygen saturation: measure of the relative proportion of oxygen-saturated hemoglobin (i.e. oxyhemoglobin) relative to the total hemoglobin in blood. This can be measured peripherally and non-invasively using a pulse oximeter device.

Peripheral oxygen saturation (SpO₂): SpO_2 = oxygenated hemoglobin / (oxygenated hemoglobin + deoxygenated hemoglobin)

Pulse oximeter: a device that uses spectrophotometry to non-invasively estimate the hemoglobin percentage saturated with oxygen in the peripheral blood.

Hypoxemia: a low peripheral oxygen saturation, defined as less than 90% according to the World Health Organization.

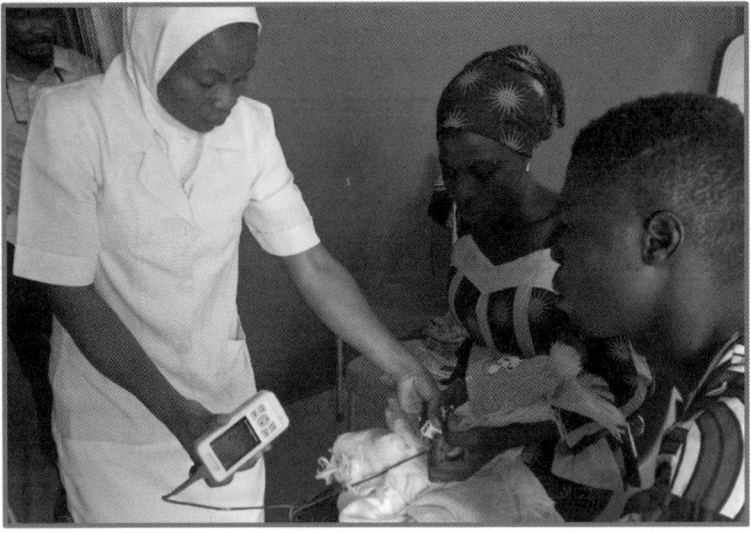

Figure 21.1 Nigerian nurse measuring the oxygen saturation of a child with respiratory distress.

21.2 Hypoxemia

Hypoxemia, a peripheral arterial oxyhemoglobin saturation (SpO_2) of less than 90%, is an indicator of severe disease, most commonly involving the pulmonary system, and is associated with increased mortality in children. It is included in the 2014 World Health Organization (WHO) Integrated Management of Childhood Illnesses (IMCI) guideline booklet as a sign for referral (WHO 2014) and as a danger sign in the 2013 WHO Pocketbook of Hospital Care for Children indicating that more intensive treatment is required (WHO 2013). It is associated with increased mortality from pneumonia, as well as other illnesses like malaria, sepsis and anemia (Orimadegun et al. 2014; Hooli et al. 2016). In 2016 pneumonia was estimated to cause almost half a million deaths in children under five years of age, more than human immunodeficiency virus, tuberculosis, and diarrhea combined (Liu et al. 2016). As hypoxemia and malnutrition are often cited as the key risks for pneumonia-related mortality, effectively identifying hypoxemic children is a key policy approach to reducing this burden.

From a review conducted in 2009, the median prevalence of hypoxemia for children hospitalized with WHO-defined pneumonia was 13%. At the time this corresponded to an estimated 1.5–2.7 million episodes of hypoxemic pneumonia annually (Subhi et al. 2009a). However, due to the lack of routinely collected data on peripheral oxygen saturation (SpO_2) from low-income countries and limited use of pulse oximeters within the full range of healthcare settings, this estimate was based mostly on clinical trials and observational research data from hospital pediatric wards or hospital clinics. Additionally, this review found a wide range of hypoxemia prevalence reported between studies.

Studies from sub-Saharan Africa, including Kenya, the Gambia and Malawi, report prevalence from as low as 5.7 up to 59% in hospital settings. Evidence from South Asia has shown consistently higher hypoxemic presentations at hospital, ranging from 17–53%. In spite of these differences in prevalence seen across low-income settings, the risk of inpatient mortality among hypoxemic children is consistently high, with increased risks of death of 4.9 (95% CI: 1.0, 17.8), 4.5 (95% CI: 3.8, 5.3) and 4.6 (95% CI: 2.2, 9.6), reported from sub-Saharan African settings (Onyango et al. 1993; Usen et al. 1999; Mwaniki et al. 2009). In Uganda a study of pneumonia and asthma admissions found the hazard of inpatient death was 10.7 (95% CI: 1.4–81.1) higher in hypoxemic children (Nantanda et al. 2014). Evidence from Malawi also indicated that moderate hypoxemia, defined as a SpO_2 90–92%, is also predictive of mortality among children hospitalized with pneumonia (Hooli et al. 2016) (Box 21.2 Figure 21.2).

Hypoxemia is notoriously difficult to predict clinically. Systematic reviews on hypoxemia consistently conclude that no single clinical sign or symptom is both sensitive and specific enough to accurately predict hypoxemia, and there is significant variability between studies regarding which signs are best (Usen and Weber 2001; Rojas-Reyes et al. 2014). Respiratory and general danger signs (such as cyanosis, nasal flaring and lethargy) typically have high specificity, while signs of respiratory distress (including fast breathing and chest indrawing) are reasonably sensitive. Various combinations of signs and symptoms have been proposed, but while these can provide better predictive value than individual signs, they also have limitations (Usen and Weber 2001; Rojas-Reyes et al. 2014).

Firstly, combinations of clinical signs vary in their ability to predict hypoxemia across different age groups (Onyango et al. 1993; Mwaniki et al. 2009). Since some signs are more relevant for younger children than older children (for example, grunting, inability to feed), combinations that are highly predictive of hypoxemia in very young children are often less predictive in older children (and vice versa) (Onyango et al. 1993; Mwaniki et al. 2009).

Box 21.2 Pulse Oximetry at Altitude

Although it is well recognized that the peripheral oxygen saturation is lower at higher altitudes compared to lower altitudes, there is little outcome evidence on what hypoxemia thresholds are clinically important when at higher altitude. There is also ongoing debate about what thresholds are most relevant at lower altitudes. It is therefore less clear at what oxygen saturation values supplemental oxygen treatment should be administered when at altitude. Complex compensatory mechanisms can also play an important role in the body's adaptation to altitude (Subhi et al. 2009a). This is an area of active research.

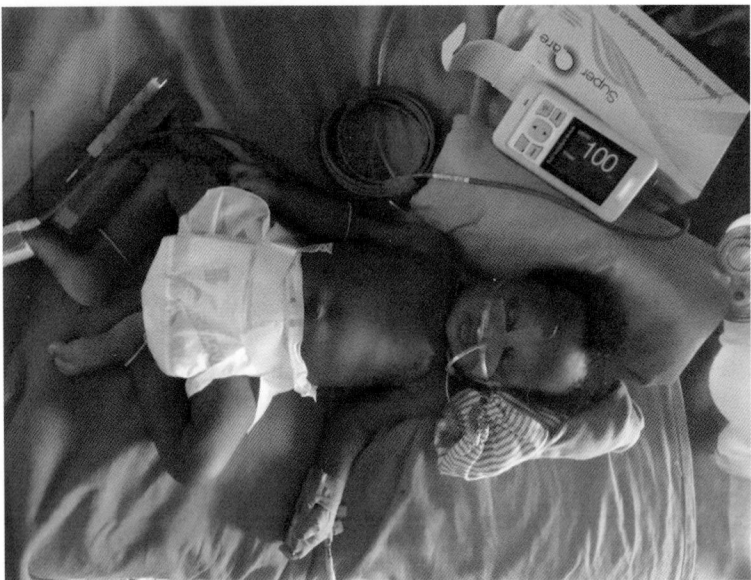

Figure 21.2 A pulse oximeter being used to check the oxygen saturation of a hospitalized child in Nigeria.

Secondly, all combinations of clinical signs to predict hypoxemia require a trade-off between sensitivity and specificity. The most common syndromic approaches to identifying hypoxemic children tend to favor sensitivity over specificity. WHO guidelines recommend using the following signs to identify children with hypoxemia: fast breathing (>70 breaths per minute), cyanosis, severe chest indrawing, and inability to feed. This combination typically provides moderately high sensitivity (~60–80%) but low specificity (<50%) (Rojas-Reyes et al. 2014), making it difficult to prioritize the right children when oxygen is scarce (which is a common scenario in low-income settings). Furthermore, combinations that use respiratory rate or signs of respiratory distress typically lose sensitivity in older populations due to differences in what is "normal."

Thirdly, clinical signs are largely subjective and there is often poor inter-rater reliability. This was well-demonstrated by Wang et al. (1992) who reported poor correlation between observers in rating clinical signs of hospitalized children with respiratory illness, while there was good agreement in SpO_2 measurements (by pulse oximetry) (Wang et al. 1992). This lack of agreement between providers in classifying clinical signs of respiratory distress and general danger signs is likely more pronounced in settings where healthcare providers have more limited training and less ongoing supervision.

Given the importance of hypoxemia as a sign of severe illness, healthcare workers need a more reliable test to non-invasively measure blood oxygen levels at the bedside. Pulse oximetry serves as this standard.

21.3 Pulse Oximetry

A pulse oximeter is a non-invasive device that can be used to estimate a SpO_2 to determine if a patient is hypoxemic (Duke et al. 2009). The devices most commonly display a

SpO$_2$, pulse rate, and a signal quality metric to assist the user in determining whether the SpO$_2$ reading is likely to be valid. SpO$_2$ oximeters incorporate sophisticated microprocessors, rapid cycling light-emitting diodes, and detectors that interpret the ratios of infrared and red light wavelengths left unabsorbed by oxygenated and deoxygenated hemoglobin during pulsatile and nonpulsatile blood flow through body tissues (usually a finger or toe) (Fouzas et al. 2011). The calculation used to estimate the SpO$_2$ is often referred to as the "r-curve" with most pulse oximeter manufacturers having developed specific calibrated r-curves such that different devices may provide modestly different SpO$_2$ estimates on the same child. The oximeter may also be augmented with proprietary-specific software that accounts for states of low perfusion, such as shock or dehydration, or high movement states such as a moving child.

The pulse oximeter sensor, or probe, contains the light-emitting diodes and detector and comes as either a single use or reusable instruments (Duke et al. 2009; Fouzas et al. 2011) (Figure 21.3). It is usually designed as a clip that closes across the anterior and posterior aspects of a digit Figure 21.4) or a wrap that encircles a digit (Figure 21.5), and is sized to fit either adults, children or neonates according to weight-based sensitivities (Figures 21.3–21.5) (WHO 2016).

Therefore, depending on the age and size of the patient there can be different combinations of oximeters and probes used and this can add potential complexity to obtaining the SpO$_2$.

While oximetry technology development started in the 1930s, the first clinically-relevant pulse oximeter was developed by Takuo Aoyagi, a Japanese engineer, in the 1970s. The significant role of pulse oximeters in medical monitoring and practice were quickly recognized, and two leading USA anesthetists, Barker and Tremper, noted in 1987 that "pulse oximetry may soon be a standard of practice for routine monitoring in any clinical setting in which the patient is at risk of hypoxemia" (Wukitsch 1987).

Pulse oximeters have several advantages over other methods for measuring oxygenation (for example, transcutaneous monitoring or blood gas measurements). These are due to them being reliable, almost instantaneous in giving a result, truly non-invasive, relatively easy to use, and adaptable to different patient populations (Bowes et al. 1989). By the early 1990s pulse oximetry had become an established standard of care – first in anesthesia and then throughout critical care, including emergency medicine, neonatal and other care areas.

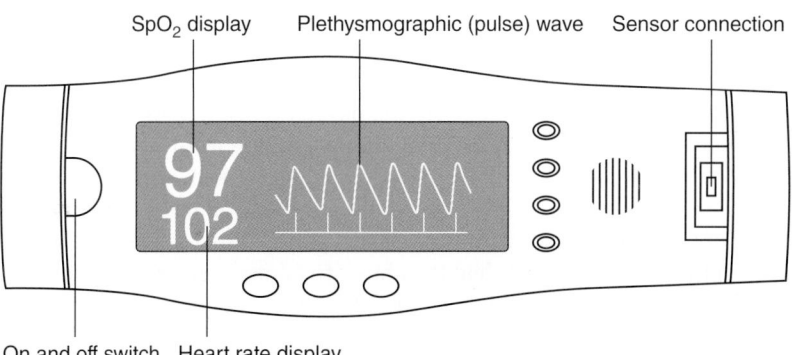

Figure 21.3 The pulse oximeter sensor, or probe, contains the light-emitting diodes and detector.

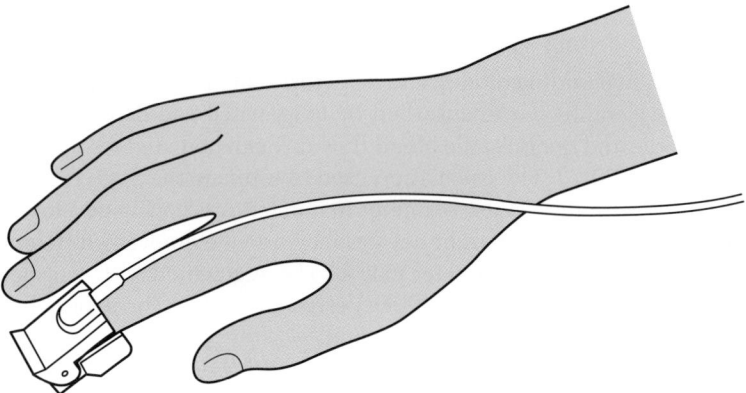

Figure 21.4 The pulse oximeter is usually designed as a clip that closes across the anterior and posterior aspects of a digit.

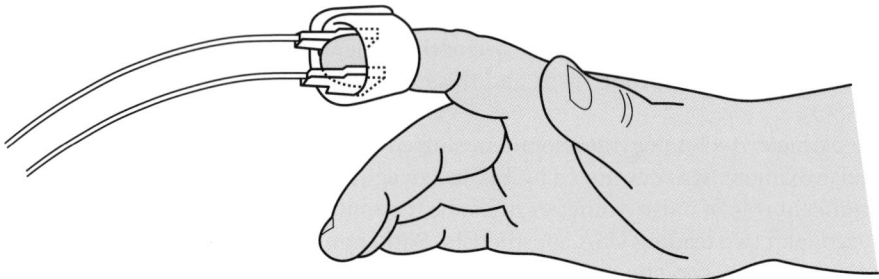

Figure 21.5 The pulse oximeter can also be designed as a wrap that encircles a digit.

21.4 Current Situation in Low-Resource Settings

Despite the importance of SpO_2 in determining severity of illness and treatment decisions, and global recognition of pulse oximetry as a basic standard of hospital care for children, pulse oximeters are poorly available in hospitals globally, especially in low-resource settings where the greatest burden of hypoxemia exists. Recent data from hospital surveys in Zambia, Malawi, Uganda and Nigeria show that 90% of hospitals have no access to pulse oximeters for children, and 80% of children requiring oxygen do not receive it – even if oxygen is available on site. No hospitals had pulse oximetry protocols or training. Specifically, a cross-sectional study at a rural hospital in Zambia found that none of the children with hypoxemia were receiving oxygen treatment as pulse oximeters were not routinely used (Foran et al. 2010). A similar finding from Malawi in 2013 showed that 78.5% of hypoxemic children were not receiving oxygen, with only one of the five included hospitals recording SpO_2 on admission (McCollum et al. 2013). In Uganda, of 11 district hospitals surveyed, only one had a pulse oximeter available on the pediatric ward, even though all hospitals had the ability to administer oxygen (Nabwire et al. 2017). In Nigeria only 1 of 12 hospitals recorded SpO_2 for children, and only 20% of hypoxemic children received oxygen (Graham et al. 2018).

Implementation of pulse oximetry in primary care outpatient settings has been even slower, and pulse oximetry is conspicuously absent from WHO and UNICEF primary

Figure 21.6 A Malawian community health worker measuring the oxygen saturation of a child in an informal rural community clinic.

care guidelines (for example, the WHO integrated community case management [ICCM] guidelines). However, recent studies have confirmed both the utility and feasibility of pulse oximetry at peripheral healthcare facilities and with community healthcare workers in both Malawi and Tanzania (McCollum et al. 2016; Keitel et al. 2017). For example, in Malawi outpatient healthcare providers successfully measured the SpO_2 on more than 94% of more than 14 000 children and it was found that providers were more than twofold more likely to correctly refer a child to hospital when their SpO_2 was found to be low (McCollum et al. 2016). Perhaps most importantly, this study also found that more than two-thirds of hypoxemic children would not have been referred to hospital if providers had applied the 2014 WHO IMCI guidelines in the absence of pulse oximetry-measured SpO_2 (Figure 21.6 and 21.7).

21.5 Current Challenges and Future Opportunities

21.5.1 The Device

Despite growing evidence of cost-effectiveness (Burn et al. 2014; Floyd et al. 2015), cost remains a commonly cited barrier to pulse oximetry implementation and use in LMICs. Hospitals and governments are faced with the choice of cheap oximeters of questionable quality (for example, adult fingertip oximeters are available online for under $10 USD) or more expensive higher quality oximeters from companies based in high-income countries that have specialty sensors for neonates and children. The less-expensive oximeters often do not meet industry-wide device standards, break easily, are not designed specifically for children, and are rarely covered by warranty. On the other hand, the higher quality oximeters are typically designed for use in high-resource settings, with expensive "single-use" disposable probes or high-end specialty probes for smaller children and costly repairs (making sustainability challenging even in high-income countries [Russell and Graydeal 1995]).

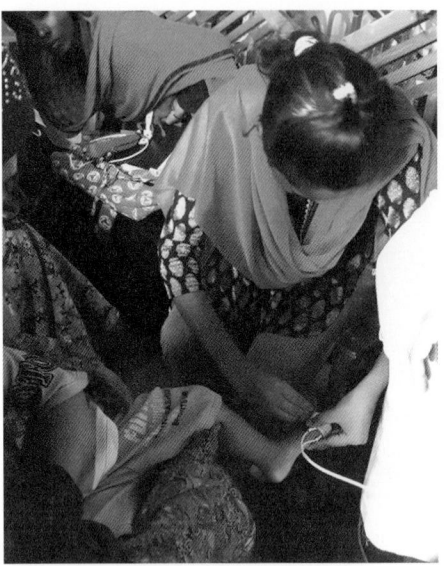

Figure 21.7 A Bangladeshi community health worker measuring the oxygen saturation of a child in a rural household.

To scale-up pulse oximetry globally, pulse oximeters are needed that are robust and durable in settings with poor and intermittent power, that are affordable to purchase and repair, and are designed to fit all sizes of children without having to use multiple devices. We also need pulse oximeters that are built for use in LMICs. Specifically, oximeters that can be quickly and reliably used for spot-checks on multiple children, rather than built for ongoing continuous monitoring of a single child, with minimal extraneous information and low cleaning and maintenance requirements.

We have seen some progress toward affordable and appropriate pulse oximetry technology for low-resource settings – thanks largely to the work of the Lifebox Foundation. Since 2011 the Lifebox Foundation has been providing quality, affordable pulse oximeters to resource-limited hospitals, as part of their mandate to make surgery safer (Enright et al. 2016; www.lifebox.com). By providing simple, robust pulse oximeters at a price of no more than $250 USD, with reusable probes and a good warranty, Lifebox has set a new standard for pulse oximetry implementation in low-resource settings. Other work has focused on improving the technology, including better motion tolerance, better performance during poor perfusion, better probe design, and integration of clinical decision-making aids (Salyer 2003; Nitzan et al. 2014). These are welcome advances but have mostly benefited hospital-based critical care settings. The future must also include better oximeters for spot-checking the SpO$_2$ of children in hospital wards and in facility-based outpatient and community environments in low-resource settings.

To this end there are several ongoing projects aiming at addressing this key pediatric pulse oximeter gap. Masimo Corporation (www2.masimo.com) is a medical technology company that specializes in non-invasive monitoring devices, including pulse oximeters. In collaboration with the Bill and Melinda Gates Foundation, Masimo is developing a pediatric pulse oximeter device for low-resource settings. The device is expected to have a lower price point, be sufficiently robust for outpatient use, and to be available in 2018. Lifebox Foundation, in collaboration with Acare Technology Co. Ltd., a

Taiwanese-based medical technology company (http://www.acaretech.com/index.html), and with the Bill and Melinda Gates Foundation, has developed a low-cost, reusable pediatric pulse oximeter probe for neonates, infants and children. The probe was developed using a human centered design approach in Malawi, Bangladesh, and the United Kingdom (King et al. 2018), and is also expected to be available in 2018. Lastly, Philips Healthcare (https://www.usa.philips.com/healthcare) is developing a pediatric pulse oximeter monitoring device targeting low-resource settings and this device is also expected to be introduced in the near future. While these exciting innovations are likely to spark further interest in revisiting the feasibility of implementing pulse oximetry more widely in low-resource settings, including outside of hospitals, having suitable and robust child-specific devices are only one of the key aspects to successfully introducing and optimizing the true potential of oximetry in low-resource settings. Thoughtfully addressing the users, facilities, and broader health systems in low-resource settings will be just as critical.

21.5.2 Users

With limited availability of pulse oximeters globally, it is no surprise that healthcare workers have low awareness and knowledge about pulse oximeters (Ginsburg et al. 2012). Feasibility studies have shown that nurses, non-physician clinicians, and community health workers are able to use oximetry successfully following basic training (Emdin et al. 2015; McCollum et al. 2016). However, experience from Papua New Guinea (Duke et al. 2008; Matai et al. 2008), Laos (Gray et al. 2017), and Nigeria (Graham et al. 2017) suggests that training alone may not be sufficient to trigger or sustain actual change in clinical pulse oximetry practice.

Emerging data suggest that the implementation of pulse oximetry requires consideration of broader individual and institutional factors (Graham et al. 2017). For example, initial reactions to the introduction of pulse oximetry in hospitals may be resentment, viewing it as additional workload for already overburdened staff. Staff are unlikely to fully accept pulse oximetry as part of routine care until they recognize the benefits – and this may take time. The benefits of oximeters are not only in detecting hypoxemia and directing oxygen care. They are also a helpful tool more broadly – monitoring patients, detecting deterioration and communication with families. Much more work is needed in this area.

There are also opportunities to make oximetry reading more intuitive and fail-safe for users to understand and interpret. For example, most devices require some understanding of the underlying science behind pulse oximetry in order to judge whether the SpO_2 signal is of adequate quality and therefore valid. This can be done in several ways, including by judging the plethysmography pattern, which is commonly presented as a waveform or bouncing bar on most devices. These patterns are generated from the arterial pulsewave and can be distorted, for example, by movement artifact, low perfusion or vasoconstriction. Further, judging whether the measurement is valid can also require taking into account whether the SpO_2 reading is likely to be biologically plausible for that child or not. Making this determination can be done by matching the child's clinical status to the SpO_2 value, where a severely low SpO_2 may represent an incorrect reading for a child who is otherwise clinically well. An abnormally low pulse rate, when considering the child's age, may also suggest an invalid SpO_2 measurement.

Requiring the user to rely on their judgment introduces unnecessary human error into the SpO_2 interpretation. This is inherently more problematic in low-resource settings with weak health systems as many healthcare workers do not get the training, or on-going mentorship, needed to understand how oximeters work, posing a limitation for implementation (Sinex 1999; Toffaletti and Zijlstra 2007; Fouzas et al. 2011; Chan et al. 2013). The ability to integrate oximetry into clinical decision making, in order to support informed treatment and referral, can be nuanced and complex, especially in situations where pulse oximetry is less reliable or accurate such as children in shock who have poor peripheral perfusion. This contradiction in the opportunity and limitations of an apparently simple-to-use technology poses important questions for implementation, and also brings forward additional opportunities to further innovate and simplify, and perhaps automate, the interpretation of the SpO_2 value.

An overview of the positions of the stakeholders in the use of pulse oximetry is shown in Figure 21.8.

21.5.3 Implementers/Facilities

Most hospitals or frontline facilities have no guidelines on the use of pulse oximetry, or training to support its adoption (Ginsburg et al. 2012), despite its inclusion in guidelines such as the WHO Pocketbook of Hospital Care for Children (WHO 2013), WHO IMCI guidelines booklet (WHO 2014), WHO oxygen therapy for children (WHO 2016), and

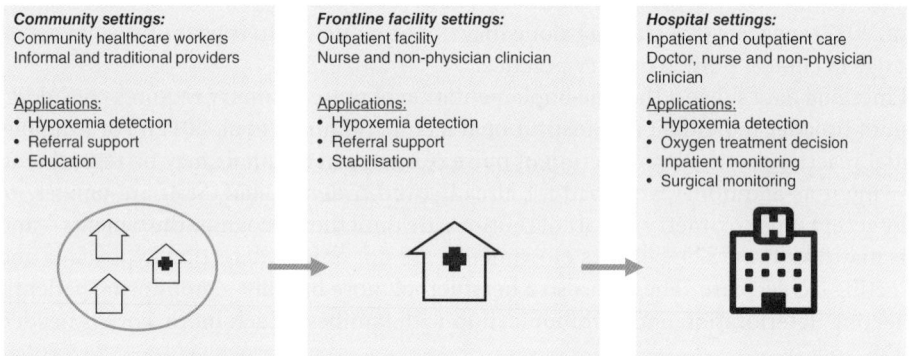

Community settings:
Community healthcare workers
Informal and traditional providers

Applications:
• Hypoxemia detection
• Referral support
• Education

Frontline facility settings:
Outpatient facility
Nurse and non-physician clinician

Applications:
• Hypoxemia detection
• Referral support
• Stabilisation

Hospital settings:
Inpatient and outpatient care
Doctor, nurse and non-physician clinician

Applications:
• Hypoxemia detection
• Oxygen treatment decision
• Inpatient monitoring
• Surgical monitoring

The child

It was late afternoon by the time the family made it to hospital. The child had been unwell for two days. At first, it was just fevers. They gave her vegetable broth and let her rest. But today her breathing had got worse, so her parents hitched a ride into town to seek help.

The community health worker

"When the mother brought the child I could see her breathing was not normal. When we checked her oxygen level [using pulse oximetry] we found it was low, and we knew the patient should go to the hospital. We didn't used to do pulse oximetry. We didn't even know what it was. Now we do it on all the patients, and it helps us know who is really sick, who has low oxygen levels, who needs extra care."

The doctor

"Her parents said she had a fever, so I thought it was probably just malaria. But I did notice that she was breathing a bit fast. Maybe she was sicker than I thought. I wondered what her oxygen level was? Luckily, we all do pulse oximetry now - it is one of our vital signs, along with heart rate, breathing rate and temperature."

The nurse

"When we checked her oxygen saturation it was only 82%. That is very low, so we gave her oxygen immediately. She was with us for 2 days before she was well enough to stop oxygen, and then go home. Pulse oximetry makes us confident in monitoring patients like this. Before we had pulse oximetry we wouldn't know - we were working in the dark."

The mother

"When the doctor first said she needed oxygen I said no. Other mothers had told me that children who got oxygen died, and I was scared. But then the nurse tested the oxygen level [using pulse oximetry]. She showed me that my baby's oxygen level was low – and then when they gave her oxygen it came up. She explained that they would keep checking the oxygen level to make sure baby was getting better. Now I make sure that every nurse comes and does the check."

Figure 21.8 An overview of the positions of the stakeholders in the use of pulse oximetry.

the WHO Surgical Safety Checklist, which contains pulse oximetry as one of the 19 checks for improving safe surgery (WHO 2009). There is a substantial gap in the lack of clear guidance on how to use pulse oximetry, particularly in primary care and for newborns, who are at risk of potentially permanent organ damage from excessive oxygen administration. Therefore, better training materials and clinical support aids are needed to facilitate effective implementation, especially as we know that sustained good oximetry practice is difficult to achieve, even in very well-resourced settings (Goldsmith and Greenspan 2007).

Recent qualitative work from Malawi and Bangladesh has found interesting opportunities for implementation, which were different between healthcare settings (King et al. 2018). Healthcare providers placed a value on oximeters, but this was more apparent among those working in communities and outpatient primary care, rather than those in hospital settings where there is a higher level of training and more readily available access to diagnostic support. In higher levels of care, healthcare workers were more confident in their clinical judgment for diagnosing severe illness – despite the widespread evidence that hypoxemia does not necessarily correspond to visible signs and symptoms. This poses important considerations for scaling-up, as training needs to balance how oximetry sits into the wider pediatric clinical assessment.

Another interesting opportunity in implementation is the trust which could be developed between healthcare providers and caregivers of children, through the use a pulse oximeter. As an objective device, healthcare providers reported that they could educate caregivers about the severity of disease in their children, and that this in turn improved acceptance of referrals to hospital and administration of oxygen therapy. In community settings, where healthcare providers have little support or diagnostic ability, this is an important aspect of implementation, with communities being included and engaged in scale-up.

Key challenges were highlighted by healthcare providers from low-income settings. First, the charging systems of available devices were considered unsuitable for low-income settings, which can have limited and intermittent power supplies. The second was the issue of devices not yet being designed as spot-check devices, but rather as continuous monitoring devices used for spot-check scenarios. There was a desire for devices to be able to store measurements, and to stop at a "final" result. Finally, there was a desire for low-cost devices to be quicker at producing accurate measurements in smaller children, and those who are moving. Currently many pulse oximeters have sophisticated motion tolerance software to improve their functioning in mobile children, but these devices are currently prohibitively expensive for low-resource settings. This is especially important in outpatient settings, where children may be more mobile than in critical care or surgical settings, which devices have traditionally been developed for.

Effective implementation of pulse oximetry will require a much better understanding of how oximetry is adopted by healthcare workers in different geographical and sociocultural environments – and more testing of what works in these various contexts. Work is ongoing in these key areas and more will likely be needed.

21.5.4 Policy/Health Systems

In general, there are few examples of low- and middle-income countries making a widespread and sustained commitment to pulse oximetry implementation at the policy and health system levels. More recently, however, there are notable exceptions of political leadership now emerging, with the Ethiopian and Nigerian Ministries of Health launching initiatives to ensure oximetry and oxygen are available across their healthcare systems, in 2016 and 2017, respectively (Ginsburg et al. 2016a; Nigeria Federal Ministry of Health 2017).

Somewhat surprising is the current lack of robust evidence that implementation of pulse oximeters leads to a reduction in mortality for children in low-income settings

Table 21.1 The barriers, current challenges and emerging solutions and opportunities for the use of pulse oximetry in low- and middle-income countries.

Barrier	Current challenges	Emerging solutions and opportunities
Device	Oximeters not designed for use in low-resource settings (for example, need to be robust, designed for spot-check readings, improved battery life, quality with small children and movement artifact, probes should fit all sizes of children)	The Gates Foundation funded efforts to improve oximeter quality and appropriateness (Lifebox, Masimo).
	Cost versus quality	Lifebox Foundation provides low-cost, high-quality oximeters to hospitals.
Users	Low knowledge / skills in using pulse oximetry	Successful experiences in various projects show that healthcare workers can effectively use oximeters with simple training, but that institutional adoption into routine care requires more work).
	Resistance to adding extra tasks to already busy workload	Recent findings show that, with experience, healthcare workers can appreciate oximetry as a "help," not an additional burdensome task. They recognize the benefits of oximeters in detecting hypoxemia and directing oxygen care, as well as broader benefits such as monitoring patients, detecting deterioration and communication with families.
Implementers, managers	High cost	High-quality devices becoming cheaper, <USD$250.
	Procurement challenges - cost versus quality	Need independent register of devices showing which ones meet essential quality standards.
	Lack of guidelines and training material on pulse oximetry	Need for quality guidelines and training material, endorsed by WHO/UNICEF
	Lack of skills in promoting adoption of new practice	Need for guidance material to help hospitals effectively introduce oximetry into routine care.

Table 21.1 (Continued)

Barrier	Current challenges	Emerging solutions and opportunities
Policy / Systems / Advocacy	Poor access to oximeters	Global coalitions increasing focus on oxygen and pulse oximetry, and child pneumonia.
	Pulse oximeters not on essential medications lists	WHO essential medications list recently added broader listing for oxygen. Need to advocate for similar change for essential diagnostic equipment such as pulse oximeters.
	Lack of oxygen strategy documents	Some countries (Ethiopia and Nigeria) have adopted national strategies (with substantial donor support).
	Market failures	Need independent, trusted authority to provide information and guidance on pulse oximeters.

(Enoch et al. 2016). Addressing this evidence gap will enable better cost-effectiveness calculations and encourage more widespread integration into national policies and health plans. These issues are summarized in Table 21.1.

21.6   Conclusions

Pulse oximeters are essential medical devices that have important applications at every level of global health systems. The prospect of universal access, improved devices and better clinical use of oximetry is exciting. Future pulse oximeters will be more intuitive, more accurate, provide more information and be integrated into data capture and clinical decision-making tools. Oximeters will non-invasively measure a wide range of biological markers in addition to oxygen saturation and pulse rate, including total hemoglobin level, respiratory rate and possibly temperature, all from a single probe – multiwavelength oximeters (Barker and Badal 2008; Shamir et al. 2012; Rice et al. 2013). Support of clinical decision making by oximeters is something which is particularly relevant in community and primary care settings where healthcare providers work more remotely with less support and needs to be integrated into mobile electronic data capture tools (Dunsmuir et al. 2012; Ginsburg et al. 2016b). Lessons learnt from the current implementation and scale-up of oximeters, as well as incentives to manufacture low-cost but high-quality devices suitable for low-income settings, will be invaluable for this next generation of technology.

Acknowledgments

We thank the caregivers, children, healthcare providers, the Ministries of Health and other participants involved in the studies referenced in this Chapter.

Bibliography

Barker, S.J. and Badal, J.J. (2008). The measurement of dyshemoglobins and total hemoglobin by pulse oximetry. *Curr. Opin. Anaesthesiol.* 21 (6): 805–810.

Bowes, W.A. III, Corke, B.C., and Hulka, J. (1989). Pulse oximetry: a review of the theory, accuracy, and clinical applications. *Obstet. Gynecol.* 74 (3 Pt 2): 541–546.

Burn, S.L., Chilton, P.J., Gawande, A.A., and Lilford, R.J. (2014). Peri-operative pulse oximetry in low-income countries: a cost-effectiveness analysis. *Bull. World Health Organ.* 92 (12): 858–867.

Chan, E.D. and Chan, M.M. (2013). Pulse oximetry: understanding its basic principles facilitates appreciation of its limitations. *Respir. Med.* 107 (6): 789–799.

Chhibber, A.V., Hill, P.C., Jafali, J. et al. (2015). Child mortality after discharge from a health facility following suspected pneumonia, meningitis or septicaemia in rural Gambia: a cohort study. *PLoS One* 10 (9): e0137095.

Duke, T., Subhi, R., Peel, D., and Frey, B. (2009). Pulse oximetry: technology to reduce child mortality in developing countries. *Ann. Trop. Paediatr.* 29 (3): 165–175.

Duke, T., Wandi, F., Jonathan, M. et al. (2008). Improved oxygen systems for childhood pneumonia: a multihospital effectiveness study in Papua New Guinea. *Lancet* 372 (9646): 1328–1333.

Dunsmuir, D., Petersen, C., Karlen, W., et al. (2012). The phone oximeter for mobile spot-check. Retrieved March 28, 2018, from http://www.phoneoximeter.org

Emdin, C.A., Mir, S., Sultana, S. et al. (2015). Utility and feasibility of integrating pulse oximetry into the routine assessment of young infants at primary care clinics in Karachi, Pakistan: a cross-sectional study. *BMC Pediatr.* 15: 141.

Enoch, A.J., English, M., and Shepperd, S. (2016). Does pulse oximeter use impact health outcomes? A systematic review. *Arch. Dis. Child.* 101 (8): 694–700.

Enright, A., Merry, A., Walker, I., and Wilson, I. (2016). Lifebox: a global patient safety initiative. *A A Case Rep.* 6 (12): 366–369.

Floyd, J., Wu, L., Hay Burgess, D. et al. (2015). Evaluating the impact of pulse oximetry on childhood pneumonia mortality in resource-poor settings. *Nature* 528 (7580): S53–S59.

Foran, M., Ahn, R., Novik, J. et al. (2010). Prevalence of undiagnosed hypoxemia in adults and children in an under-resourced district hospital in Zambia. *Int. J. Emerg. Med.* 3 (4): 351–356.

Fouzas, S., Priftis, K.N., and Anthracopoulos, M.B. (2011). Pulse oximetry in pediatric practice. *Pediatrics* 128 (4): 740–752.

Ginsburg, A.S., Gerth-Guyette, E., Mollis, B. et al. (2014). Oxygen and pulse oximetry in childhood pneumonia: surveys of clinicians and student clinicians in Cambodia. *Tropical Med. Int. Health* 19 (5): 537–544.

Ginsburg, A.S., Izadnegahdar, R., and Klugman, K.P. (2016a). World pneumonia day 2016: pulse oximetry and oxygen. *Lancet Glob. Health* 4 (12): e893–e894.

Ginsburg, A.S., Tawiah Agyemang, C., Ambler, G. et al. (2016b). mPneumonia, an innovation for diagnosing and treating childhood pneumonia in low-resource settings: a feasibility, usability and acceptability study in Ghana. *PLoS One* 11 (10): e0165201.

Ginsburg, A.S., Van Cleve, W.C., Thompson, M.I.W., and English, M. (2012). Oxygen and pulse oximetry in childhood pneumonia: a survey of healthcare providers in resource-limited settings. *J. Trop. Pediatr.* 58: 389–393.

Goldsmith, J.P. and Greenspan, J.S. (2007). Neonatal intensive care unit oxygen management: a team effort. *Pediatrics* 119 (6): 1195–1196.

Graham, H.R., Ayobami, A.I., Bakare, A.A. et al. (2017). Improving oxygen therapy for children and neonates in secondary hospitals in Nigeria: study protocol for a stepped-wedge cluster randomised trial. *Trials* 18: 502. https://doi.org/10.1186/s13063-017-2241-8.

Graham, H.R., Bakar, A.A., Gray, A. et al. (2018). Adoption of paediatric and neonatal pulse oximetry by 12 hospitals in Nigeria: a mixed-methods realist evaluation. *BMJ Glob. Health* 3: e000812.

Gray, A.Z., Morpeth, M., Duke, T. et al. (2017). Improved oxygen systems in district hospitals in Lao PDR: a prospective field trial of the impact on outcomes for childhood pneumonia and equipment sustainability. *BMJ Pes Open 2017,* 1: e000083. https://doi.org/10.1136/bmjpo-2017-000083.

Hooli, S., Colbourn, T., Lufesi, N. et al. (2016). Predicting hospitalised paediatric pneumonia mortality risk: an external validation of RISC and mRISC, and local tool development (RISC-Malawi) from Malawi. *PLoS One* 11 (12): e0168126.

Keitel, K., Kagoro, F., Samaka, J. et al. (2017). A novel electronic algorithm using host biomarker point-of-care tests for the management of febrile illnesses in Tanzanian children (e-POCT): a randomized, controlled non-inferiority trial. *PLoS Med.* 14 (10): e1002411.

King, C., Boyd, N., Walker, I. et al. (2018). Opportunities and barriers in paediatric pulse oximetry for pneumonia in low-resource clinical settings: a qualitative evaluation from Malawi and Bangladesh. *BMJ Open.* 8: e019177.

Liu, L., Oza, S., Hogan, D. et al. (2016). Global, regional, and national causes of under-5 mortality in 2000–15: an updated systematic analysis with implications for the sustainable development goals. *Lancet.* 388 (10063): 3027–3035.

Matai, S., Peel, D., Wandi, F. et al. (2008). Implementing an oxygen programme in hospitals in Papua New Guinea. *Ann. Trop. Paediatr.* 28: 71–78.

McCollum, E.D., Bjornstad, E., Preidis, G.E. et al. (2013). Multicenter study of hypoxemia prevalence and quality of oxygen treatment for hospitalized Malawian children. *Trans. R. Soc. Trop. Med. Hyg.* 107 (5): 285–292.

McCollum, E.D., King, C., Deula, R. et al. (2016). Pulse oximetry for children with pneumonia treated as outpatients in rural Malawi. *Bull. World Health Organ.* 94 (12): 893–902.

Mwaniki, M.K., Nokes, D.J., Ignas, J. et al. (2009). Emergency triage assessment for hypoxaemia in neonates and young children in a Kenyan hospital: an observational study. *Bull. World Health Organ.* 87 (4): 263–270.

Nabwire, J., Namasopo, S., and Hawkes, M. (2017). Oxygen availability and nursing capacity for oxygen therapy in Ugandan Paediatric wards. *J. Trop. Pediatr.* 64 (2): 97–103. https://doi.org/10.1093/tropej/fmx033.

Nantanda, R., Ostergaard, M.S., Ndeezi, G., and Tumwine, J.K. (2014). Clinical outcomes of children with acute asthma and pneumonia in Mulago hospital, Uganda: a prospective study. *BMC Pediatr.* 14: 285.

Nigeria Federal Ministry of Health (2017). National strategy for the scale-up of medical oxygen in health facilities: 2017–2022.

Nitzan, M., Romem, A., and Koppel, R. (2014). Pulse oximetry: fundamentals and technology update. *Med. Devices (Auckl.)* 7: 231–239.

Onyango, F.E., Steinhoff, M.C., Wafula, E.M. et al. (1993). Hypoxaemia in young Kenyan children with acute lower respiratory infection. *BMJ* 306: 612–615.

Orimadegun, A., Ogunbosi, B., and Orimadegun, B. (2014). Hypoxemia predicts death from severe falciparum malaria among children under 5 years of age in Nigeria: the need for pulse oximetry in case management. *Afr. Health Sci.* 14 (2): 397–407.

Rice, M.J., Gravenstein, N., and Morey, T.E. (2013). Noninvasive hemoglobin monitoring: how accurate is enough? *Anesth. Analg.* 117 (4): 902–907.

Rojas-Reyes, M.X., Granados Rugeles, C., and Charry-Anzola, L.P. (2014). Oxygen therapy for lower respiratory tract infections in children between 3 months and 15 years of age. *Cochrane Database Syst. Rev.* 12: CD005975.

Russell, G.B. and Graybeal, J.M. (1995). Accuracy of laminated disposable pulse-oximeter sensors. *Respir. Care* 40 (7): 728–733.

Salyer, J.W. (2003). Neonatal and pediatric pulse oximetry. *Respir. Care.* 48 (4): 386–396.

Shamir, M.Y., Avramovich, A., and Smaka, T. (2012). The current status of continuous noninvasive measurement of total, carboxy, and methemoglobin concentration. *Anesth. Analg.* 114 (5): 972–978.

Sinex, J.E. (1999). Pulse oximetry: principles and limitations. *Am. J. Emerg. Med.* 17 (1): 59–67.

Subhi, R., Adamson, M., Campbell, H. et al. (2009a). The prevalence of hypoxaemia among ill children in developing countries: a systematic review. *Lancet Infect. Dis.* 9: 219–227.

Subhi, R., Smith, K., and Duke, T. (2009b). When should oxygen be given to children at high altitude? A systematic review to define altitude-specific hypoxaemia. *Arch. Dis. Child.* 94 (1): 6–10.

Toffaletti, J. and Zijlstra, W.G. (2007). Misconceptions in reporting oxygen saturation. *Anesth. Analg.* 105 (6 Suppl): S5–S9.

Usen, S. and Weber, M. (2001). Clinical signs of hypoxaemia in children with acute lower respiratory infection: indicators of oxygen therapy. *Int. J. Tuberc. Lung. Dis.* 5 (6): 505–510.

Usen, S., Weber, M., Mulholland, K. et al. (1999). Clinical predictors of hypoxaemia in Gambian children with acute lower respiratory tract infection: prospective cohort study. *BMJ* 318 (7176): 86–91.

Wang, E.E., Milner, R.A., Navas, L., and Maj, H. (1992). Observer agreement for respiratory signs and oximetry in infants hospitalized with lower respiratory infections. *Am. Rev. Respir. Dis.* 145 (1): 106–109.

World Health Organization (2009). Implementation Manual: WHO Surgical Safety Checklist 2009. Safe Surgery Saves Lives. Retrieved 28/03/2018, from http://www.who.int/patientsafety/safesurgery/checklist_implementation/en.

World Health Organization (2013). *Pocket Book of Hospital Care for Children: Guidelines for the Management of Common Childhood Illnesses.* Geneva: World Health Organization.

World Health Organization (2014). *Integrated Management of Childhood Illness: Chart Booklet.* Geneva: World Health Organization.

World Health Organization (2016). *Oxygen Therapy for Children.* Geneva: World Health Organization.

Wukitsch, M.W. (1987). Pulse oximetry: historical review and Ohmeda functional analysis. *Int. J. Clin. Monit. Comput.* 4 (3): 161–166.

Webliography

http://www.phoneoximeter.org. Dunsmuir, D., Petersen, C., Karlen, W., et al. 2012. The phone oximeter for mobile spot-check. Retrieved March 28, 2018.

http://www.who.int/patientsafety/safesurgery/checklist_implementation/en. World Health Organisation. 2009. Implementation Manual: WHO Surgical Safety Checklist 2009. Safe Surgery Saves Lives. Retrieved 28/03/2018.

22

The Use of Near-Infrared Spectroscopy to Monitor Tissue Oxygenation, Metabolism and Injury in Low Resource Settings

Gemma Bale, and Ilias Tachtsidis

Department of Medical Physics and Biomedical Engineering, University College, London, UK

CHAPTER MENU

22.1   Introduction

In a low resource setting there are many barriers for providing the standard of healthcare that is normal in developed countries. Modern healthcare largely depends on reliable and accurate medical devices that can diagnose disease, monitor treatment and provide accurate prognostic information. Often medical devices are associated with high costs for purchasing and maintenance that often includes costs for a large number of consumables. In addition, a particular infrastructure is required and highly trained staff are needed to operate these devices. To be successful and useful in low- and middle-income countries (LMICs), a medical device needs to be

- inexpensive
- portable
- safe
- non-invasive

Revolutionizing Tropical Medicine: Point-of-Care Tests, New Imaging Technologies and Digital Health, First Edition. Edited by Kerry Atkinson and David Mabey.

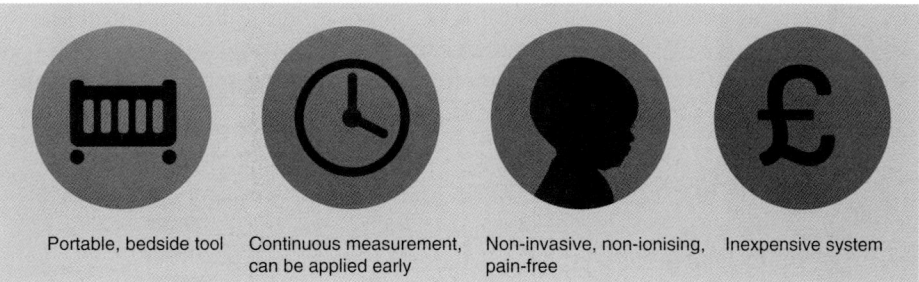

| Portable, bedside tool | Continuous measurement, can be applied early | Non-invasive, non-ionising, pain-free | Inexpensive system |

Figure 22.1 Schematic of the benefits of NIRS.

- easy to use and interpret
- battery-powered.

Near-infrared spectroscopy (NIRS) is a medical device that fulfills all of these requirements (Figure 22.1) and thus has great potential to make an impact on healthcare in LMICs. Its most important use to date has been in neurological monitoring. Its success in this field is due to its low cost, ease of use and ability to be used in a low resource environment.

Other neuromonitoring techniques include electroencephalography (EEG), magnetoencephalography (MEG), positron emission tomography (PET), functional magnetic resonance imaging (fMRI), and functional near-infrared spectroscopy (fNIRS). The first two techniques monitor brain function directly, by measurements of electrical brain activity, while the other modalities monitor brain function indirectly, by parameters that are related to neuronal activity, such as glucose uptake (PET) or surrogates of cerebral blood flow (fMRI, fNIRS). MEG, PET, and fMRI all require expensive equipment that requires a lot of space as well as highly skilled technicians. EEG is a portable and inexpensive technique, but it requires electrical shielding and highly trained technicians to operate it and clinicians to interpret its measurements.

A schematic illustration of an NIRS system and the path of light traveling through the brain is shown in Figure 22.2.

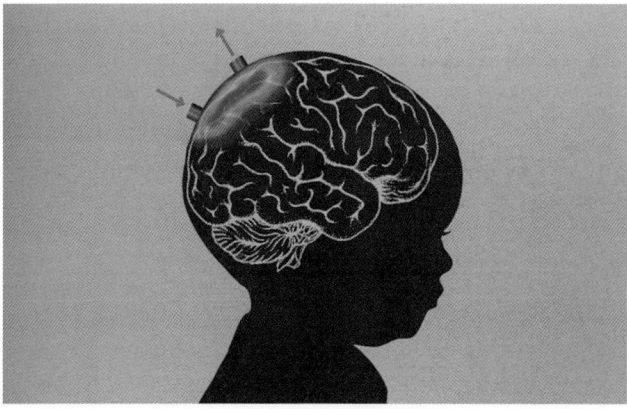

Figure 22.2 Schematic illustration of NIRS system and the path of light traveling through the brain.

This Chapter describes the technology of NIRS, its use in healthcare and research, current applications in low resource settings and diseases prevalent in the LMICs, as well as looking to the future of NIRS technology in low resource settings.

22.2 Near-Infrared Spectroscopy

22.2.1 Scientific Background

NIRS is a spectroscopic method that uses the near-infrared region of the electromagnetic spectrum. NIRS is based on molecular overtone and combination vibrations. The molecular overtone and combination bands seen in the near-IR region are typically very broad, leading to complex spectra. (Spectroscopy is the study of the interaction between matter and electromagnetic radiation. The electromagnetic spectrum is the range of frequencies [the spectrum] of electromagnetic radiation and their respective wavelengths and photon energies).

NIRS can be used as a non-invasive optical neuromonitoring technique that provides continuous measurements of brain oxygenation and hemodynamics (Scholkmann et al. 2014). It uses the near-infrared region of the electromagnetic spectrum (from 600 to 1000 nm). There are many different techniques including continuous wave (CW), broadband CW, spatially resolved spectroscopy (SRS), frequency domain (FD) and time resolved (TR).

All NIRS techniques depend on two key principles.

1) Biological tissue is relatively transparent in the near infrared (NIR) region; therefore, if NIR light is shone on to the head, some of it will travel into the brain and be reflected back out (Figure 22.3a, b).
2) Back-scattered light from the tissue contains information about light absorbers within the tissue. The two main absorbers in the NIR region are oxygenated- and deoxygenated-hemoglobin (HbO2 and HHb respectively). When hemoglobin is bound to oxygen it has a different absorption spectrum (Figure 22.3a, b) so it can be spectrally distinguished from deoxygenated hemoglobin.

NIRS instruments record the back-scattered light and use this to determine concentrations of absorbers within the tissue. CW NIRS technology is based on a light intensity measurement. Near-infrared light is directed into the tissue and the intensity of the reflected light is measured. CW NIRS uses an algorithm called the modified Beer–Lambert law (MBLL) which can estimate the relative concentration changes of HbO_2 and HHb inside the tissue from the attenuation change:

$$\Delta A = \sum \Delta c_n \cdot \alpha_n \cdot d.$$

The MBLL describes the change in attenuation of the light coming out of the head (ΔA) as a product of the change in concentration of the absorbers (Δc), their absorption spectra (α) and the distance that the light has traveled (d). Thus, by detecting the change in light attenuation in at least two wavelengths, it is possible to calculate the change in concentration of absorbers, typically HbO_2 and HHb. In this way the cerebral hemodynamics can be monitored. An example of concentration change data is shown in Figure 22.4.

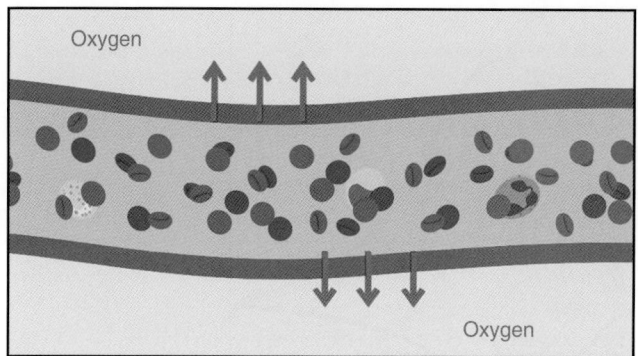

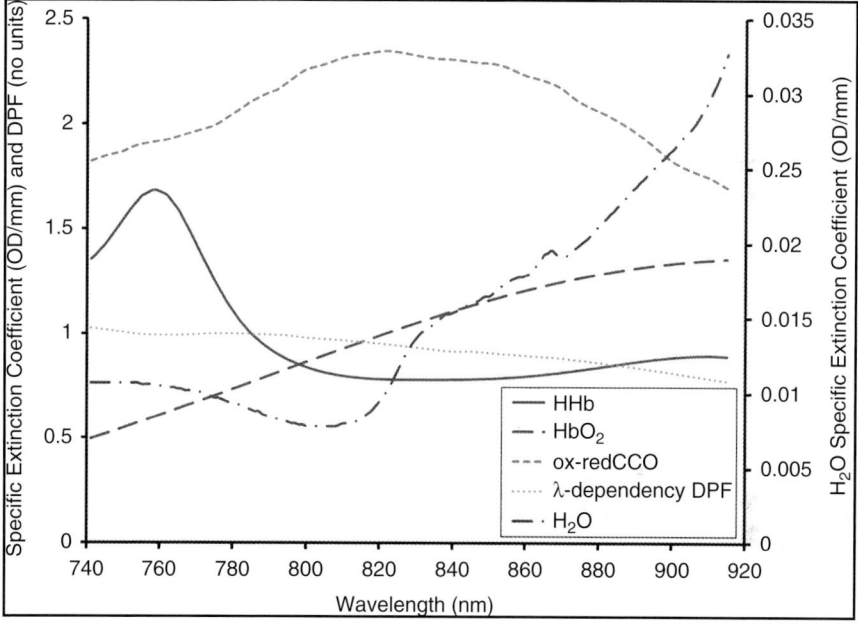

Figure 22.3 (a) Image showing hemoglobin color change with change in oxygenation. (b) Chromophore extinction spectra in the NIR. *Source:* Picture used with permission from Bale et al. (2016). (*See color plate section for the color representation of this figure.*)

It is also possible to monitor changes in metabolism additional to hemoglobin changes using CW broadband NIRS (Bale et al. 2016). Cytochrome-c-oxidase (CCO), an enzyme in the mitochondrial membrane, is another absorber in the NIR spectrum. CCO is the terminal electron acceptor in the electron transport chain and passes electrons to oxygen. As it changes its redox state, its absorption spectrum changes (Redox, short for reduction–oxidation reaction, is a chemical reaction in which the oxidation states of atoms are changed. Any such reaction involves both a reduction process and a complementary oxidation process, two key concepts involved with electron transfer processes.)

It is thus possible to monitor the oxidation of CCO and, therefore, to monitor oxidative metabolism. However, the concentration of hemoglobin is 10 times greater than CCO *in vivo*, so it is difficult to extract this metabolic signal with only a few wavelengths.

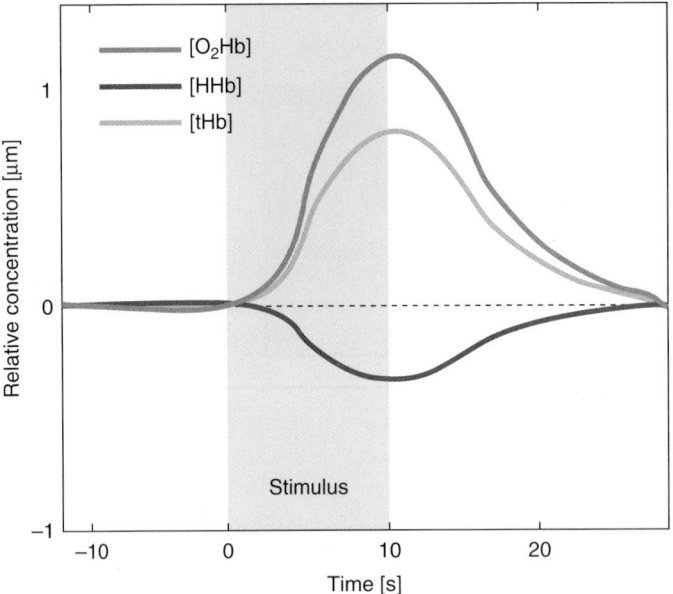

Figure 22.4 Illustration of the hemodynamic response function due to neuronal stimulation. *Source:* Picture used with permission from Scholkmann et al. (2014).

Because of this, many wavelengths must be used to observe the small attenuation changes due to CCO: this is called broadband NIRS.

The disadvantage of CW NIRS systems is that they cannot fully determine the optical properties of the tissue since only changes in absorption are measured. Therefore, they cannot resolve absolute concentrations but only changes in concentration. A further extension of CW NIRS, cerebral oximetry, uses SRS to give a measurement of absolute tissue saturation. The hemoglobin saturation measured by this method is called the tissue oxygenation index (TOI) or tissue oxygen saturation (StO_2):

$$StO_2 = \frac{HbO_2}{HbO_2 + HHb} \times 100\%$$

This is the oxygenated percentage of the tissue hemoglobin interrogated by the NIR light. SRS uses a multi-distance approach to obtain an absolute measurement of tissue oxygenation. An example is shown in Figure 22.5 comparing the image in a child with cerebral malaria to that of a child with uncomplicated malaria.

Frequency domain and time resolved NIRS obtain more information from the tissue than CW techniques to resolve absolute concentrations. FD NIRS modulates the emitted light intensity and then measures the intensity of the detected light as well as the phase shift, which corresponds to the time of flight. TR NIRS emits an extremely short pulse of light into the tissue and measures the arrival times of the photons that emerge from the tissue. These technologies yield more information than CW NIRS, but the technology is more complex and expensive.

A recent development in NIRS technology is a new method to measure blood flow, called diffuse correlation spectroscopy (DCS). DCS is the measurement of the diffusing temporal field autocorrelation function to obtain information about tissue dynamics

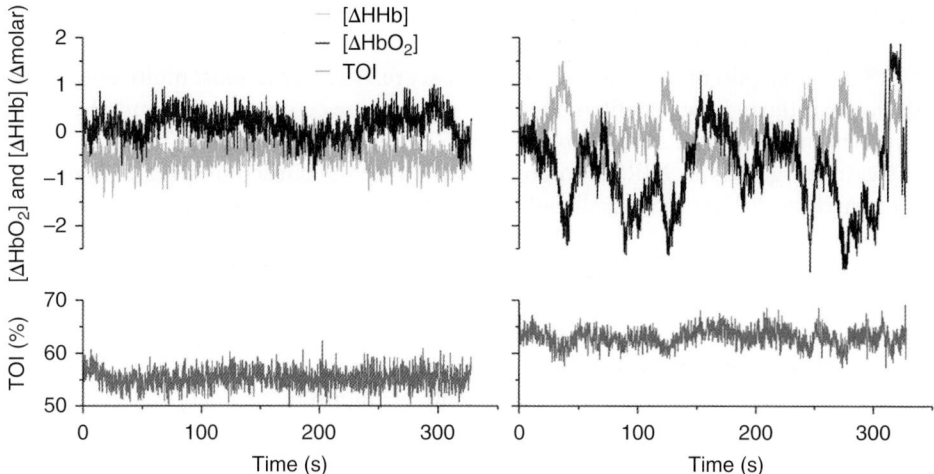

Figure 22.5 NIRS-measured concentration changes and tissue oxygen saturation (TOI) in patients with cerebral malaria (left) and uncomplicated malaria (right). There is a lack of spontaneous oscillations in the cerebral malaria pattern that is present in uncomplicated malaria pattern. *Source:* Picture used with permission from Kolyva et al. (2013).

(Buckley et al. 2014). DCS detects blood flow by quantifying temporal fluctuations of light fields emerging from the tissue surface. Typically, these light fields are generated by illuminating the brain surface with NIR laser light. Some of the NIR light propagates through the scalp and skull and into the brain where it is scattered by moving red blood cells in tissue vasculature before emerging from the tissue surface. This "dynamic" scattering from moving cells causes the detected intensity to temporally fluctuate, and the time scale of these fluctuations is quantified by the intensity temporal autocorrelation function of the collected light, thus allowing a recovery of blood flow.

22.2.2 Equipment

A typical CW NIRS system consists of a light source with two or more wavelengths (for example, laser diodes or light-emitting diodes [LEDs]). The light is directed into the head using optical fibers which are held onto the head. The light that is reflected back from the head is picked up by a detector (for example, a photodiode). All of these components are relatively inexpensive, small, and can be battery-powered so NIRS is therefore very suitable for use in low-resource settings.

   NIRS systems can monitor a few centimeters into human tissue. Therefore, it is possible to monitor the cerebral cortex or any other tissue of interest that is close to the surface, such as skeletal muscle. A single source-detector pair or "channel" can monitor the local region between and beneath them. To build up an image of the whole brain or tissue, multiple channels are used to give whole head coverage. NIRS systems can be sufficiently flexible to allow the user to choose the area or areas to monitor.

22.3 Clinical Applications

In more economically developed countries NIRS has been used both in healthcare and in research. The main use of NIRS in healthcare is as a cerebral oxygenation monitor.

NIRS is used routinely during cardiac surgery to monitor the adequacy of cerebral perfusion (Zheng et al. 2013).

NIRS has great potential in neonatal intensive care. An international multicenter clinical trial investigated the benefit and harm of monitoring cerebral oxygenation by NIRS in preterm infants in the first 72 hours of life (Plomgaard et al. 2015). The trial found that treatment guided by NIRS oximetry could reduce the burden of cerebral hypoxia.

A research study has used broadband NIRS to monitor metabolism (using CCO measurement) in neonatal brain injury (Mitra et al. 2017). The authors found that blood pressure passivity of NIRS-measured metabolism was strongly correlated to the outcome of the injury. Additionally, NIRS-measured CCO has been shown to correlate strongly with electrical activity during seizures, showing potential to be a new seizure monitoring device (Mitra et al. 2016) (more details can be found at www.metabolight.org). The addition of metabolism to NIRS measurements could be important in its clinical utility.

22.4   Research Applications

NIRS is particularly useful in neuroscientific research as a functional brain monitor since it provides similar information to the blood oxygen level-dependent (BOLD) signal from functional magnetic resonance spectroscopy (fMRI). Functional NIRS (fNIRS) is able to monitor brain activity indirectly via the concentration changes in HbO_2 and HHb. When the subject is presented with a stimulus, there is local neural activation which requires energy. Hence there is an increase in local cerebral blood flow to supply oxygen to the region. This tight coupling between neural activity and cerebral blood flow is known as neurovascular coupling, and this concept enables indirect monitoring of brain activation by measurement of local hemodynamics. The increase in local cerebral blood flow leads to an increase in HbO_2 and decrease in HHb, which is known as the hemodynamic response function (illustrated in Figure 22.4).

One of the key advantages of NIRS is its ability to recover an analogue to the BOLD signal but in a more natural environment, and this is a great benefit in neuroscience. It is also relatively inexpensive and does not need a technician for its operation. Furthermore, it has a higher temporal resolution than fMRI, although it does have a lower spatial resolution.

fNIRS has been able to give insights into areas of research that would not be possible with other methods, such as neonates and infants who cannot be easily studied using fMRI. A notable example is the use of NIRS to identify signs of autism in infants at four to six months of age (Lloyd-Fox et al. 2013). Currently there are no strong predictors of autism in early infancy and diagnosis is not reliable until around three years of age. However, NIRS was able to demonstrate a reduced neural sensitivity to social stimuli in young infants at risk for autism via a reduced hemodynamic response function. This shows the utility of having a neuromonitoring technique that is flexible and can adapt to situations in unchartered areas of research.

22.5   The Use of NIRS in Low Resource Settings

We carried out a literature search on the use of NIRS research on diseases that are prevalent in the LMICs. Diseases searched included malaria, HIV/AIDS, tuberculosis, respiratory tract infections, diarrheal diseases, meningitis, syphilis, hepatitis, measles, whooping cough, tetanus, yaws, yellow fever, dengue fever, bladder disease and hydrocephalus. The findings from the search are presented in this section.

22.5.1 Malaria

A major pathophysiological mechanism of severe malaria is the obstruction of blood flow leading to tissue hypoxia, which is exacerbated by impaired microvascular function (Yeo et al. 2013). A study in Papua, Indonesia used NIRS to assess microvascular function in skeletal muscle in patients with moderate to severe *falciparum* malaria (Yeo et al. 2013). Impaired microvascular function was found to be associated with increased mortality among individuals with severe malaria, while oxygen consumption was increased. A further study used the same technique to investigate microvascular function in children with *falciparum* malaria and found similar results (Yeo et al. 2014).

Cerebral malaria is the most severe and life-threatening neurological complication of *Plasmodium falciparum* malaria (Idro et al. 2010). As early as 1997 NIRS has been used to investigate changes in regional cerebral perfusion and oxygenation due to cerebral malaria (Kampfl et al. 1997). Cerebral oxygenation, as measured by NIRS, was found to be lower on the affected cortex compared to the contralateral one. A study of NIRS-measured HbO_2 oscillations in patients with *falciparum* malaria was carried out in Rourkela, India. The results showed that patients with cerebral malaria had a significantly reduced power spectrum density at certain frequencies compared to cases with non-cerebral severe malaria or uncomplicated malaria (Kolyva et al. 2013) (Figure 22.5). This suggested that cerebral malaria can disrupt autonomic function. Additionally, this study found that the power spectrum density was increased to normal levels once patients recovered.

These investigations show that NIRS can be a useful tool in the understanding of microvascular function and oxygenation in malaria, and, further, can be used to stratify patients with different severity of disease (Kolyva et al. 2013; Yeo et al. 2013, 2014).

22.5.2 Meningitis

Meningitis is prevalent in the LMICs, particularly across the "meningitis belt" that spans sub-Saharan Africa. The disease can result in neurological morbidity in infected children and adults. Increased intracranial pressure or vascular thrombosis can lead to decreased cerebral oxygenation and a subsequent decreased StO_2 level. A NIRS study of StO_2 in children with meningitis found that children with persistently decreased StO_2 developed cerebral infarction (Watzman et al. 1999).

Other NIRS research in meningitis has investigated cerebral metabolism (by CCO) in animal models and perhaps can be used as a proof of principle for studies in humans (Tureen et al. 1996; Park et al. 1999). One study observed a significant reduction in the oxidation state of CCO in rabbits 12–18 hours after infection with bacterial meningitis (Tureen et al. 1996). This was accompanied by a trend of decreasing oxygenation which could be the cause of the reduced metabolism. Another explanation is that meningitis may cause uncoupling of oxidative phosphorylation in mitochondrial enzymes. However, another NIRS study of bacterial meningitis in newborn piglets found an increase in the oxidation state of CCO four hours after infection (Park et al. 1999). This discrepancy is likely due to the difference in the timings of the measurement. At four hours after infection there is likely to be a compensatory neuroprotective mechanism against a lack of oxygen, which cannot be maintained beyond 12 hours. This animal work shows that NIRS-measured metabolism could have a role in monitoring the progression of meningitis.

22.5.3 Dengue Fever

The manifestation of dengue virus ranges from a self-limiting fever to a potentially life-threatening plasma leakage syndrome, known as dengue hemorrhagic fever (DHF) (Soller et al. 2014). If detected early, plasma leakage can be managed with careful use of intravenous fluids. However, current detection methods, including x-rays and laboratory tests, are often not available in resource-poor settings and preventative fluid administration is often too late. A symptom of DHF is decreased peripheral perfusion and hence NIRS has been explored as a technique to identify plasma leakage (Soller et al. 2014) (Figure 22.6).

A study in Bangkok, Thailand found that low muscle StO_2 (less than 48%) was able to identify DHF in children with suspected dengue virus infection. NIRS was used to monitor muscle saturation continuously and could inform in real-time the changes in saturation that related to plasma leakage.

22.5.4 Bladder Disease

Bladder outlet obstruction is common in the LMICs. It is also commonly diagnosed late, which can cause further complications and even premature death. Stothers et al. (2010) found that NIRS measurements of bladder oxygenation correlated with the presence of bladder outlet obstruction and thus created a diagnostic algorithm for NIRS (Stothers et al. 2010). A new NIRS-based detection method for bladder obstruction was developed for use in rural Uganda (Stothers et al. 2016). The team worked with local staff in a Ugandan medical clinical to modify the device and orient them with the device and

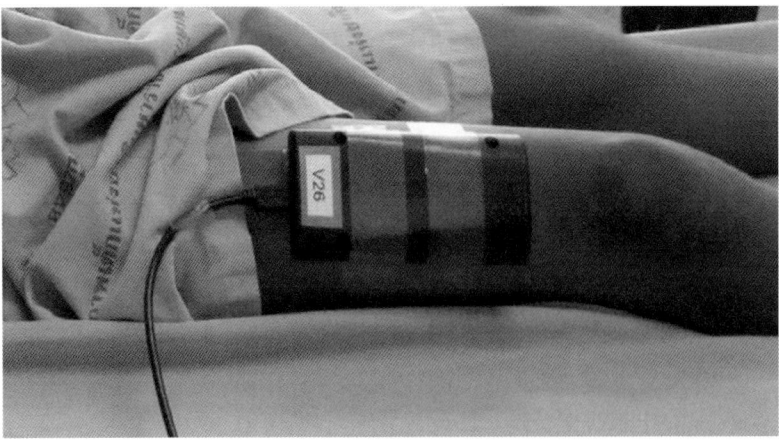

Figure 22.6 Photograph of NIRS sensor applied to a child's thigh to monitor peripheral perfusion to identify plasma leakage in dengue haemorrhagic fever in Bangkok, Thailand. *Source:* Picture used with permission from Soller et al. (2014).

method for screening. The NIRS system and protocol was successfully implemented and local staffs were trained quickly in the procedure, exemplifying the simplicity of NIRS technology. This proof of principle study showed the feasibility of the measurement, although further work needs to be done to verify its clinical utility in this disease.

22.5.5   Hydrocephalus

Each year in sub-Saharan Africa over 200 000 infants are affected by post-infectious hydrocephalus. The majority are untreated, leading to severe brain damage and death. For those treated with a surgical procedure, uncertainty remains over which children will respond well. A research team recently studied infants in Mbale, Uganda before and after surgery to determine if NIRS was able to predict treatment and neurological outcomes (Lin et al. 2017). The team performed FD NIRS combined with DCS to measure blood flow in 35 infants with hydrocephalus, in order to estimate the cerebral metabolic rate of oxygen consumption ($CMRO_2$) (Figure 22.7).

FD NIRS-DCS presurgical measurements demonstrated high accuracy in detecting brain structural damage from hydrocephalus. Differences in tissue scattering after surgery could predict treatment failure within six months. Furthermore, brain regions with higher $CMRO_2$ tended to recover better than regions with low $CMRO_2$, which was in agreement with CT brain scans taken at six months after surgery. NIRS-DCS measurements have facilitated immediate improvement in the care of hydrocephalus in infants in both low- and middle-income countries and developed countries (Lin et al. 2017).

22.5.6   Neurodevelopment

fNIRS has recently become a very valuable technique for monitoring the development of the infant brain, due to its ease of use in infants, good spatial resolution (compared to

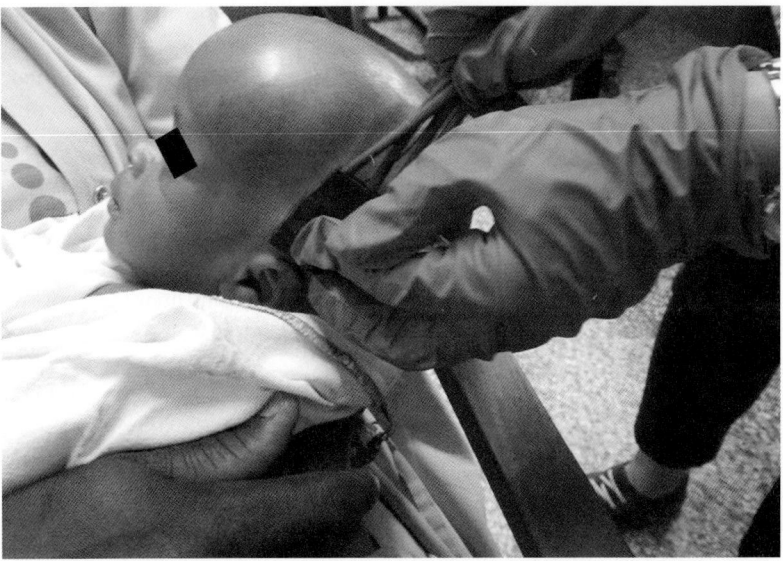

Figure 22.7 Infant with hydrocephalus during FD NIRS-DCS scan in Mbale, Uganda. *Source:* Picture used with permission from Maria-Angela Franceschini (personal communication).

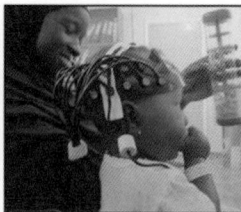

Figure 22.8 NIRS headgear on a newborn, 6-month-old infant and 13 month old infant and a 2 year old toddler (left to right) in the Gambia. *Source:* Picture used with permission from Lloyd-Fox et al. (2017).

EEG) and suitability for use in remote settings (Lloyd-Fox et al. 2014). Additionally, fNIRS enables researchers to study neurodevelopment outside the laboratory in previously inaccessible settings in under-studied populations (Jasinska and Sosthene 2018) (Figure 22.8).

Cognitive impairment associated with childhood malnutrition and stunting is generally considered irreversible. Research into neurodevelopment during malnutrition has been explored using fNIRS (Lloyd-Fox et al. 2014; Roberts et al. 2017). Using the ultra-portability of NIRS equipment, researchers have been able to study populations in rural Gambia (Lloyd-Fox et al. 2014, 2017) and Guinea-Bissau (Roberts et al. 2017). One team measured functional brain activity in response to visual-social and vocal stimuli in infants aged four to eight months in rural Gambia compared to an age-matched UK cohort (Lloyd-Fox et al. 2014). The aim of The Bright Project (http://www.globalfnirs. org/the-bright-project) is to develop biomarkers of cognitive development throughout childhood as a standard against which to assess future interventions. Another group has used fNIRS to assess cognitive function after nutritional intervention in undernourished and stunted children in rural Guinea-Bissau (Roberts et al. 2017). This study demonstrated that there was an association between improved working memory and supplement consumption. These research efforts have established the utility of NIRS to monitor cerebral function during interventions in low resource settings.

New fNIRS studies have been set up across the world to monitor neurodevelopment in adverse conditions. The INDIA (Infant Neurodevelopment and Dyadic Interaction Assessment) project examines how early influences affect brain development and growth in Shivgarh, India (John Spencer, personal communication). (A dyad is a group of two people, the smallest possible social group). The researchers are enrolling six- and nine-month old infants from low- and high-socioeconomic status families from the same micro cultural context. They are assessing the development of visual working memory using fNIRS over the first years of life (Figure 22.9). The ultimate goal of the project is to develop tools to move from early assessment of adversity on the emergence of the functional brain network to early intervention.

Finally, an investigation of child language, reading and cognitive development using fNIRS in the Côte d'Ivoire has been initiated recently (Jasinska and Sosthene 2018). In order to help future research efforts in low-resource settings, Jasinska and Sosthene (2018) have published their protocols for transporting and setting up a mobile NRIS laboratory, with guidelines for establishing a successful study. They emphasized that including local researchers will increase the chance of success because of their knowledge of local social and cultural values and systems. Their contribution is very important in designing culturally-appropriate protocols. Further, they recommend developing

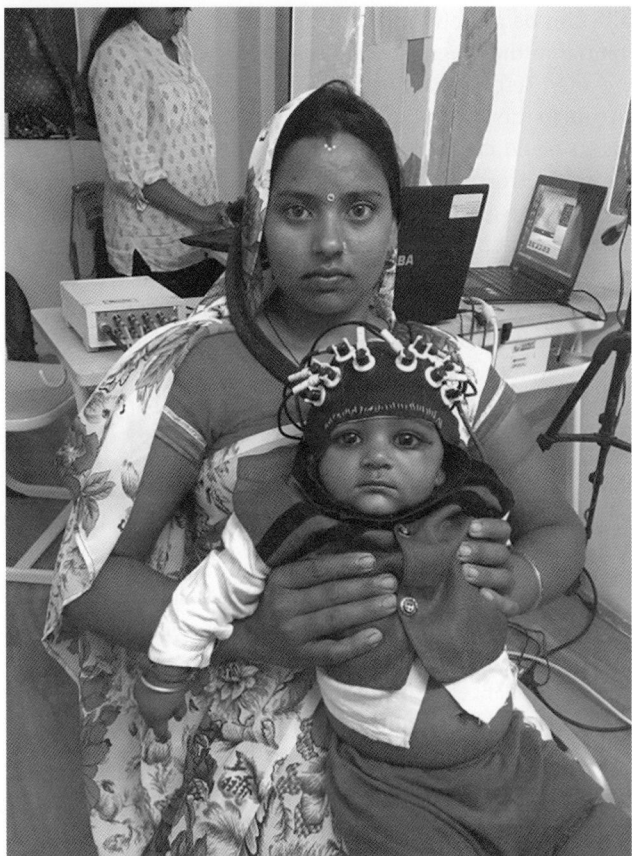

Figure 22.9 Photograph of a nine month old infant during an fNIRS study as part of the INDIA project in Shivgarh, India. *Source:* Picture used with permission from John Spencer (personal communication).

suitable informed consent protocols and failure to adopt an appropriate approach can adversely affect the research. Finally, the authors endorsed sharing research findings with community members and government partners, as partnerships built on continued dialog will aid in the eventual translation of research findings into policy.

22.6 Conclusions

In this chapter we have demonstrated that

- NIRS is a non-invasive, portable, easy to use and inexpensive technique to monitor tissue (most commonly cerebral) hemodynamics, oxygenation and metabolism in real time (Figure 22.1).
- NIRS studies have been performed successfully in low resource settings (Côte d'Ivoire, Gambia, Guinea-Bissau, India, Indonesia, Thailand and Uganda).
- NIRS can provide important information in diseases that are a burden in LMICs including malaria, meningitis, dengue fever, bladder disease, hydrocephalus, and malnutrition.

The evidence shows that NIRS can be a useful monitor of health in low resource settings and holds significant potential in the diagnosis of disease as well as in monitoring treatment. The most immediate and useful impact of NIRS could be in malaria. Studies have found that NIRS is a useful tool in the understanding of cerebral microvascular function and oxygenation in this disease, and can be used to stratify patients with different severity of disease. This could improve treatment monitoring and identify those patients in need of more intensive treatment. The global hunt for brain injury biomarkers is, and will be, moving to LMICs, so having an appropriate neuromonitoring modality that is suitable for low resource settings will become increasingly important. The low-cost, ease of use and portability of NIRS makes it an ideal neuromonitoring tool for LMICs. Indeed, NIRS was recently used to perform the first functional brain study in a remote rural location, and the authors of that study stated that "fundamentally this work was only achievable due to the choice of neuroimaging method: fNIRS" (Lloyd-Fox et al. 2014). The future of neuromonitoring in global healthcare looks brighter with NIRS technology.

Bibliography

Bale, G., Elwell, C.E., and Tachtsidis, I. (2016). From Jöbsis to the present day: a review of clinical near-infrared spectroscopy measurements of cerebral cytochrome-c-oxidase. *J. Biomed. Opt.* 21 (9): 91307.

Buckley, E.M., Parthasarathy, A.B., Grant, P.E. et al. (2014). Diffuse correlation spectroscopy for measurement of cerebral blood flow: future prospects. *Neurophotonics* 1 (1): 11009.

Idro, R., Marsh, K., John, C.C., and Newton, C.R. (2010). Cerebral malaria; mechanisms of brain injury and strategies for improved neuro-cognitive outcome. *Pediatr. Res.* 68 (4): 267–274.

Jasinska, K. and Sosthene, G. (2018). Neuroimaging field methods using functional near infrared spectroscopy (NIRS) neuroimaging to study global child development: rural Sub-Saharan Africa. *J. Vis. Exp.* https://doi.org/10.3791/57165.

Kampfl, A., Pfausler, B., Haring, H.P. et al. (1997). Impaired microcirculation and tissue oxygenation in human cerebral malaria: a single photon emission computed tomography and near-infrared spectroscopy study. *Am. J. Trop. Med. Hyg.* 56 (6): 585–587.

Kolyva, C., Kingston, H., Tachtsidis, I. et al. (2013). Oscillations in cerebral hemodynamics in patients with falciparum malaria. *Adv. Exp. Med. Biol* 765: 101–108.

Lin, P.-Y., Sutin, J., Farzam, P. et al. (2017). Noninvasive optical method can predict hydrocephalus treatment and brain outcomes-initial experiences with post-infectious hydrocephalus infants in Uganda. *J. Cereb. Blood Flow Metab.* 37: 242–243.

Lloyd-Fox, S., Begus, K., Halliday, D. et al. (2017). Cortical specialisation to social stimuli from the first days to the second year of life: a rural Gambian cohort. *Developmental Cognitive Neuroscience* 25: 92–104.

Lloyd-Fox, S., Papademetriou, M., Darboe, M.K. et al. (2014). Functional near infrared spectroscopy (fNIRS) to assess cognitive function in infants in rural Africa. *Sci. Rep.* 4: 4740.

Lloyd-Fox, S., Blasi, A., Elwell, C.E. et al. (2013). Reduced neural sensitivity to social stimuli in infants at risk for autism. *Proc. R. Soc. B Biol. Sci.* 280 (1758): 20123026–20123026.

Mitra, S., Bale, G., Mathieson, S. et al. (2016). Changes in cerebral oxidative metabolism during neonatal seizures following hypoxic–ischemic brain injury. *Front. Pediatr.* 4: 1–7.

Mitra, S., Bale, G., Highton, D. et al. (2017). Pressure passivity of cerebral mitochondrial metabolism is associated with poor outcome following perinatal hypoxic ischemic brain injury. *J. Cereb. Blood Flow Metab.* https://doi.org/10.1177/0271678X17733639.

Park, W.S., Chang, Y.S., Ko, S.Y. et al. (1999). Effects of microbial invasion on cerebral hemodynamics and oxygenation monitored by near infrared spectroscopy in experimental Escherichia coli meningitis in the newborn piglet. *Neurol. Res.* 21 (4): 391–398.

Plomgaard, A.M., van Oeveren, W., Petersen, T.H. et al. (2015). The SafeBoosC II randomised trial: treatment guided by near-infrared spectroscopy reduces cerebral hypoxia without changing early biomarkers of brain injury. *Pediatr. Res.* 79 (4): 1–8.

Roberts, S.B., Franceschini, M.A., Krauss, A. et al. (2017). A pilot randomized controlled trial of a new supplementary food designed to enhance cognitive performance during prevention and treatment of malnutrition in childhood. *Curr. Dev. Nutr.* 1 (11): e000885.

Scholkmann, F., Kleiser, S., Metz, A.J. et al. (2014). A review on continuous wave functional near-infrared spectroscopy and imaging instrumentation and methodology. *NeuroImage* 85 ((Pt 1): 6–27.

Soller, B., Srikiatkachorn, A., Zou, F. et al. (2014). Preliminary evaluation of near infrared spectroscopy as a method to detect plasma leakage in children with dengue hemorrhagic fever. *BMC Infect. Dis.* 14 (1): 396.

Stothers, L., Macnab, A., Mutabazi, S. et al. (2016). Near-infrared spectroscopic screening for bladder disease in Africa: training rural clinic staff to collect data of diagnostic quality. *J. Spectro.* 2016: 1–7.

Stothers, L., Guevara, R., and Macnab, A. (2010). Classification of male lower urinary tract symptoms using mathematical modelling and a regression tree algorithm of noninvasive near-infrared spectroscopy parameters. *Eur. Urol.* 57 (2): 327–333.

Tureen, J., Liu, Q., and Chow, L. (1996). Near-infrared spectroscopy in experimental pneumococcal meningitis in the rabbit: cerebral hemodynamics and metabolism. *Pediatr. Res.* 40 (5): 759–763.

Watzman, H., Costarino, A.T., Priestley, M. et al. (1999). Cerebral oxygen saturation in children with meningitis. *Crit. Care Med.* 27 (1): 79.

Yeo, T.W., Lampah, D.A., Kenangalem, E. et al. (2014). Decreased endothelial nitric oxide bioavailability, impaired microvascular function, and increased tissue oxygen consumption in children with falciparum malaria. *J. Infect. Dis.* 210 (10): 1627–1632.

Yeo, T.W., Lampah, D.A., Kenangalem, E. et al. (2013). Impaired skeletal muscle microvascular function and increased skeletal muscle oxygen consumption in severe falciparum malaria. *J. Infect. Dis.* 207 (3): 528–536.

Zheng, F., Sheinberg, R., Yee, M.S. et al. (2013). Cerebral near-infrared spectroscopy monitoring and neurologic outcomes in adult cardiac surgery patients: a systematic review. *Anesth. Analg.* 116: 663–676.

Webliography

Bale, G., Elwell, C.E., and Tachtsidis, I. (2016). From Jöbsis to the present day: a review of clinical near-infrared spectroscopy measurements of cerebral cytochrome-c-oxidase. *Journal of Biomedical Optics* 21 (9): 91307. http://biomedicaloptics.spiedigitallibrary.org/article.aspx?doi=10.1117/1.JBO.21.9.091307.

Buckley, E.M., Parthasarathy, A.B., Grant, P.E. et al. (2014). Diffuse correlation spectroscopy for measurement of cerebral blood flow: future prospects. *Neurophotonics* 1 (1): 11009. http://neurophotonics.spiedigitallibrary.org/article.aspx?articleid=1884410.

http://www.ajtmh.org/content/journals/10.4269/ajtmh.1997.56.585 [Accessed December 20, 2017].Kampfl, A., Pfausler, B., Haring, H.-P. et al. (1997). Impaired Microcirculation and Tissue Oxygenation in Human Cerebral Malaria: A Single Photon Emission Computed Tomography and Near-Infrared Spectroscopy Study. *Am. J. of Trop. Med. and Hyg.* 56 (6): 585–587.

Kolyva, C., Palm, F., Bruley, D.F. et al. (2013). Oscillations in Cerebral Hemodynamics in Patients with Falciparum Malaria. *Adv. Exp. Med. Biol.* 756: 101–108. http://link. springer.com/10.1007/978-1-4614-4989-8.

http://www.ncbi.nlm.nih.gov/pubmed/28017265 [Accessed January 3, 2018]Lloyd-Fox, S., Begus, K., Halliday, D. et al. (2017). Cortical specialisation to social stimuli from the first days to the second year of life: A rural Gambian cohort. *Dev. Cog. Neuro* 25: 92–104.

https://www.ncbi.nlm.nih.gov/pmc/articles/PMC5381189/pdf/srep04740.pdf [Accessed December 21, 2017]Lloyd-Fox, S., Papademetriou, M., Darboe, M.K. et al. (2014). Functional near infrared spectroscopy (fNIRS) to assess cognitive function in infants in rural Africa. *Scientific Reports* 4: 4740.

Lloyd-Fox, S., Blasi, A., Elwell, C.E. et al. (2013). Reduced neural sensitivity to social stimuli in infants at risk for autism. *Proceedings of the Royal Society B: Biological Sciences* 280 (1758): 20123026–20123026. http://rspb.royalsocietypublishing.org/cgi/doi/10.1098/ rspb.2012.3026.

Mitra, S., Bale, G., Mathieson, S. et al. (2016). Changes in Cerebral Oxidative Metabolism during Neonatal Seizures Following Hypoxic–Ischemic Brain Injury. *Frontiers in Pediatrics* 4 (August): 1–7. https://www.ncbi.nlm.nih.gov/pmc/articles/PMC4978952/.

http://www.tandfonline.com/doi/full/10.1080/01616412.1999.11740949 [Accessed December 20, 2017]Park, W.S., Chang, Y.S., Ko, S.Y. et al. (1999). Effects of microbial invasion on cerebral hemodynamics and oxygenation monitored by near infrared spectroscopy in experimental Escherichia coli meningitis in the newborn piglet. *Neurological Research* 21 (4): 391–398.

Plomgaard, A.M., van Oeveren, W., Petersen, T.H. et al. (2015). The SafeBoosC II randomised trial: treatment guided by near-infrared spectroscopy reduces cerebral hypoxia without changing early biomarkers of brain injury. *Pediatric Research* (January): 1–8. http://www.nature.com/doifinder/10.1038/pr.2015.266.

http://cdn.nutrition.org/content/asnaoa/1/11/e000885.full.pdf [Accessed December 21, 2017]Roberts, S.B., Franceschini, M.A., Krauss, A. et al. (2017). Pilot Randomized Controlled Trial of a New Supplementary Food Designed to Enhance Cognitive Performance during Prevention and Treatment of Malnutrition in Childhood. *Current Developments in Nutrition.* 1 (11): e000885.

http://www.ncbi.nlm.nih.gov/pubmed/23684868 [Accessed May 27, 2014]Scholkmann, F., Kleiser, S., Metz, A.J. et al. (2014). A review on continuous wave functional near-infrared spectroscopy and imaging instrumentation and methodology. *NeuroImage* 85 (Pt 1): 6–27.

http://bmcinfectdis.biomedcentral.com/articles/10.1186/1471-2334-14-396 [Accessed December 20, 2017]Soller, B., Srikiatkachorn, A., Zou, F. et al. (2014). Preliminary evaluation of near infrared spectroscopy as a method to detect plasma leakage in children with dengue hemorrhagic fever. *BMC Inf. Dis.* 14 (1): 396.

http://www.hindawi.com/journals/jspec/2016/1241862/ [Accessed December 20, 2017]Stothers, L., Macnab, A., Mutabazi, S. et al. (2016). Near-Infrared Spectroscopic Screening for Bladder Disease in Africa: Training Rural Clinic Staff to Collect Data of Diagnostic Quality. *J. of Spect* 1–7.

http://www.nature.com/doifinder/10.1203/00006450-199611000-00016 [Accessed December 20, 2017]Tureen, J., Liu, Q., and Chow, L. (1996). Near-Infrared Spectroscopy in Experimental Pneumococcal Meningitis in the Rabbit: Cerebral Hemodynamics and Metabolism. *Pediatric Research* 40 (5): 759–763.

https://academic.oup.com/jid/article-lookup/doi/10.1093/infdis/jiu308 [Accessed December 20, 2017]Yeo, T.W., Lampah, D.A., Kenangalem, E. et al. (2014). Decreased Endothelial Nitric Oxide Bioavailability, Impaired Microvascular Function, and Increased Tissue Oxygen Consumption in Children with Falciparum Malaria. *J. of Inf. Dis.* 210 (10): 1627–1632.

Zheng, F., Sheinberg, R., Yee, M.S. et al. (2013). Cerebral near-infrared spectroscopy monitoring and neurologic outcomes in adult cardiac surgery patients: a systematic review. *Anesth. Analg.* 116: 663–676. http://www.pubmedcentral.nih.gov/articlerender.fcgi?artid=3863709&tool=pmcentrez&rendertype=abstract.

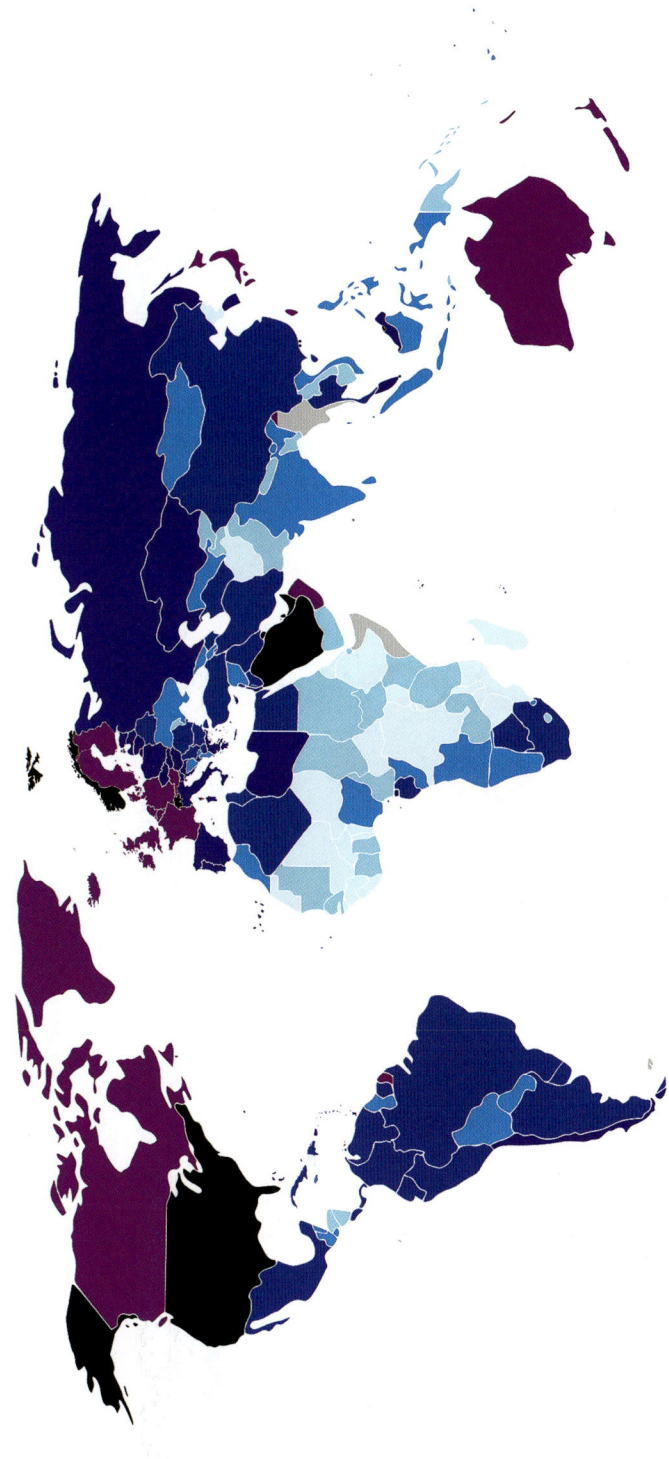

Figure 1.1 World map of GDP per Capita by Country. Gross domestic product (GDP) is converted to international dollars using purchasing power parity rates (PPP). World Bank, International Comparison Program database (2011–2014). GDP per capita - PPP in international US $ 50000 or more 35000–50000 20000–35000 10000–20000 5000–10000 2000–5000 less than 2000 no data available. *Source:* This image, reproduced here from Wikimedia Commons, is in the public domain because its creator, Rfassbind, has placed it there. https://commons.wikimedia.org/wiki/File:World_map_GDP_per_capita.svg.

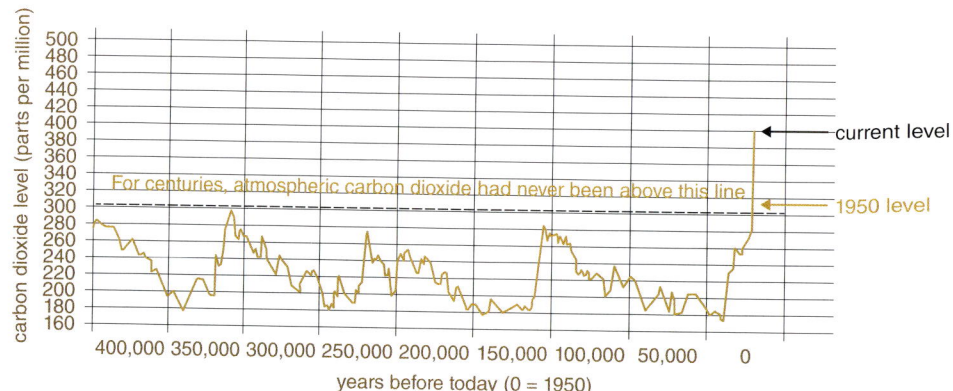

Figure 38.3 The increase in atmospheric CO_2 since the Industrial Revolution. *Source:* Credit: Vostok ice core data/J.R. Petit et al.; NOAA Mauna Loa CO_2 record. This image was created by a USA Government employee in the course of their job with the National Oceanic and Atmospheric Administration (NOAA), USA, and, as such, it is the public domain. The Industrial Revolution was the transition to new manufacturing processes in the period from about 1760 to sometime between 1820 and 1840.

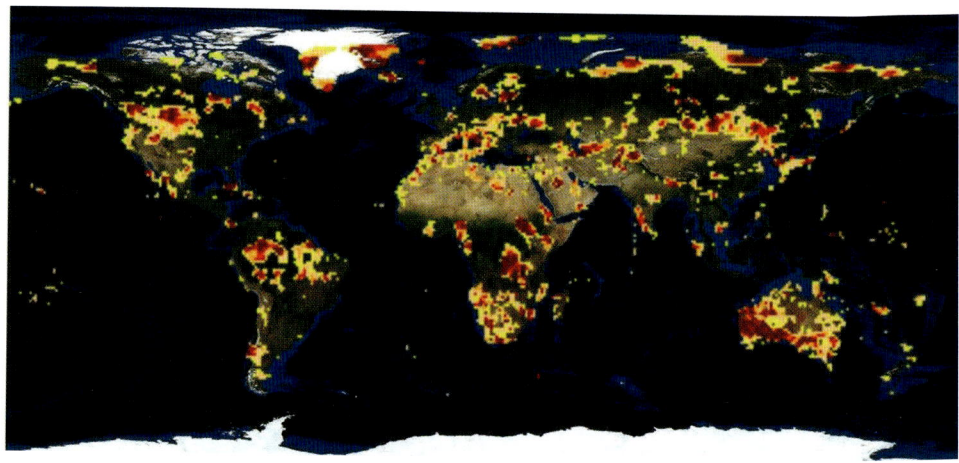

Figure 38.4 Global drought map. Global Drought Information System. 24 month standardized precipitation index (SPI). High SPI value (closer to 3): heavy precipitation (rainfall) event over time period specified. Medium SPI value (approximately = 0): normal precipitation event over time period specified. Low SPI value (closer to − 3): low precipitation event over time period specified (drought?) https://www.drought.gov/gdm/current-conditions.

Part IV

Cheap Imaging Technologies

23

The Use of Point-of-Care Ultrasound in the Resource-Limited Setting

Tom Heller[1], Michaëla A.M. Huson[2], Sabine Bélard[3,4], Dan Kaminstein[5], and Elizabeth Joekes[6]

[1] Lighthouse Clinic, Kamuzu Central Hospital, Lilongwe, Malawi
[2] Center of Tropical Medicine and Travel Medicine, Division of Infectious Diseases, Academic Medical Center, University of Amsterdam, Amsterdam, The Netherlands
[3] Department of Pediatric Pneumology and Immunology, Charité-Universitätsmedizin Berlin, Berlin, Germany
[4] Berlin Institute of Health, Berlin, Germany
[5] Department of Emergency Medicine and Hospitalist Service, Medical College of Georgia at Augusta University, Augusta, GA, USA
[6] Department of Radiology, Royal Liverpool University Hospitals, Liverpool, UK

Revolutionizing Tropical Medicine: Point-of-Care Tests, New Imaging Technologies and Digital Health,
First Edition. Edited by Kerry Atkinson and David Mabey.
© 2019 John Wiley & Sons, Inc. Published 2019 by John Wiley & Sons, Inc.

23.1   Introduction to Point-of-Care Ultrasound (POCUS)

Point-of-care ultrasound (POCUS) applications have allowed the use of ultrasound well beyond diagnostic imaging departments since the 1990s. Ultrasound is a rapid and inexpensive technology which can be easily repeated and involves no radiation exposure to the patient. Many specialties including pediatrics, rheumatology, cardiology, nephrology and anesthesiology are now using ultrasound as a bedside diagnostic tool and to assist procedures relevant to their specialty (Moore and Copel 2011). In recent years Emergency Medicine is the specialty that has most advanced the use of clinical ultrasound at the patient bedside. Ultrasound equipment is used in the emergency room (ER) to aid diagnostic decisions and facilitate therapeutic interventions. The advantage of POCUS is that clinicians can use real-time images rather than images recorded and reported by a sonographer or a radiologist, thus allowing immediate correlation of findings with signs and symptoms.

While the environment of an emergency department in a high resource setting may look and feel very different from a small clinic in rural Africa there are several similarities. Clinicians in both settings must make decisions with limited information and decide in a timely manner on treatment options. Tools such as ultrasound that can support these decisions are invaluable.

The fundamental difference between POCUS and conventional comprehensive ultrasound exams is that the clinician seeks to answer simple but clinically relevant and usually binary questions, for example, "Is a pericardial effusion present? Yes or no?" Findings that are included in POCUS protocols usually need to fulfill three criteria: they must be prevalent in the patient population, must entail decisions on relevant treatment and management, and they must be easy enough to be recognized by medical staff without lengthy imaging training. POCUS findings that are commonly accepted as time-sensitive and key to immediate management decisions, as well as easy to perform by less experienced staff, can be grouped as level 1 or primary applications. More complex techniques and applications can be grouped as level 2 or secondary applications (Table 23.1). Clearly there is direct applicability for many of the applications in resource-limited settings (and the remainder of this chapter will focus on these), while others are less relevant, due to lower prevalence or lack of treatment options.

Table 23.1 POCUS applications.

System	Primary level application	Secondary level application
Abdominal	FAST ascites Hydronephrosis Abdominal aortic aneurysm	FASH Focal liver lesions Bladder changes in schistosomiasis
Pelvic	Early intrauterine pregnancy Basic third trimester ultrasound	Adnexal masses
Cardiothoracic	FAST1-Pericardial effusion FAST1-Pleural effusion Pneumothorax	Hypotension-RUSH Cardiomegaly Lung consolidations Intrathoracic masses Pulmonary edema
Soft tissue and vascular	Deep vein thrombosis	Abscess localization Soft tissue masses

Abbreviations: FAST, focused assessment with sonography for trauma; FASH, focused assessment with sonography for HIV associated tuberculosis.

23.2 Physics and Technical Aspects of Ultrasound

23.2.1 Physics

A basic understanding of the physics of ultrasound is necessary to understand how the image is generated. The core of the ultrasound transducer is a piezoelectric crystal, which is capable of transforming electrical energy into sound (vibration) energy and vice versa. The crystal generates short ultrasound pulses (duration 10^{-6} seconds of time or 1 microsecond or 1 millionth of a second), which are sent into the body. Each tissue has a certain ability to promote sound waves; at interfaces between different tissues part of the acoustic energy is reflected. The crystal receives the reflected waves between pulses (duration 10^{-3} seconds of time or one millisecond) and reverts them back into short currents. The localization of each image point is calculated from the time elapsed between sending and receiving the signal and the brightness of each point is calculated from the intensity of the received signal. The energy used for diagnostic ultrasound is very low and has no relevant side effects when used in POCUS indications.

The probe frequencies used for diagnostic purposes range from 2.5 MHz (echocardiography) to 16 MHz (for superficial small body parts). The higher the frequency the better the resolution, but the more limited the depth penetration of the beam. For visualization of deeper anatomical structures (for example the aorta), low frequency probes have to be used; for superficial structures (for example the jugular vein), a high frequency transducer is preferred.

The shape of the transducer influences the shape of the image on the screen. Abdominal ultrasound is usually performed using a curvilinear probe; for small parts linear transducers are preferred. For cardiac scans a sector probe with a small footprint is preferred to allow for the geometry of the ribs. If possible, at least two transducers are desirable in order to scan thoracic, abdominal and superficial structures included in POCUS protocols (Figure 23.1).

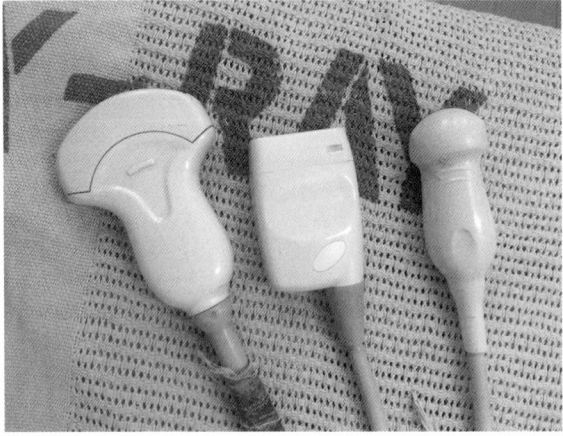

Figure 23.1 Three different ultrasound probes. From left to right: Curvilinear "abdominal" probe, linear superficial probe, cardiac probe with small footprint.

The commonly encountered black and white screen image is known as the B (brightness) mode. It is possible to visualize the Doppler signal of blood flow in color (known as "color flow mapping"), but this is beyond the scope of this chapter.

To describe and interpret ultrasound images, the brightness of the anatomical or pathological structures is characterized in comparison to surrounding tissues (for example, normal liver tissue). If the structure generates more echo-signal (i.e., it is brighter) it is called hyperechoic or echogenic (Figure 23.2a). If it is darker it is called hypoechoic or echo poor (Figure 23.2b) and if it is completely black (ie, no echo signal is returned) it is described as anechoic (Figure 23.2c). If the structure is almost similar to the surrounding tissue it is called isoechoic (Figure 23.2d) and will be difficult to distinguish from the surrounding tissues.

Gas/air and bones have completely different sound propagation properties compared to other tissues. Consequently, the image that results from the beam interacting with air or bone interfaces shows distinct types of artifacts (so called "dirty" and "clean" shadowing). The presence of shadowing yields important diagnostic information, but also obscures any structures behind the gas or bone.

23.2.2 Ultrasound Image Orientation

Understanding the orientation of an ultrasound image and how it reflects the anatomy within the body is one of the most difficult concepts for an ultrasound novice; unfortu-

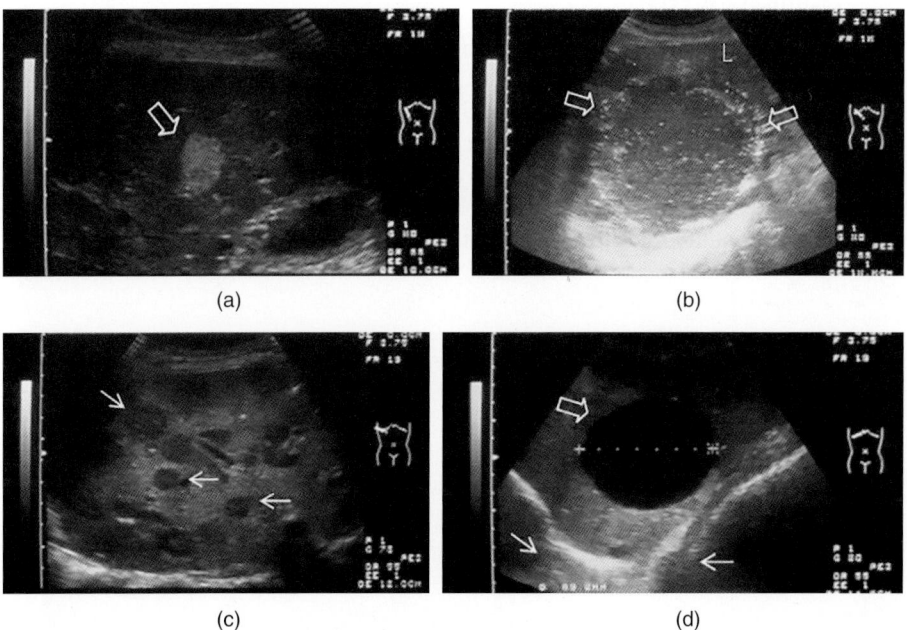

(a)   (b)

(c)   (d)

Figure 23.2 Examples of echogenicity of lesions in comparison to the surrounding tissue. (a) A well-defined echogenic lesion (open arrow) in the liver = hyperechoic; (b) A large lesion (open arrow) resembling echogenicity of the surrounding liver (L) = isoechoic; (c) Multiple lesions (arrows) with lower echogenicity = hypoechoic; (d) A large well-defined lesion (open arrow) without internal echoes and a dorsal acoustic enhancement (between arrows) = anechoic.

nately it is also one of the most important. Structures are commonly examined in two orthogonal planes, transverse (TS) and longitudinal (LS).

Always ensure that the transducer marker corresponds to the left side of the image and remember that in any scan plane, the left side of the image should correspond to the side of the probe with the marker. If a longitudinal image is required, place the transducer marker toward the patient's head (Figure 23.3a). This will project cranial structures on the left side of the screen and caudal structures to the right (Figure 23.3b). If a transverse image is required, rotate the transducer anti-clockwise from the longitudinal position (Figure 23.3a).

On a correctly oriented image the cranial parts of the liver will be projected on the left side of the screen while the caudal edge of the liver will be visible on the right side of the screen (Figure 23.3b).

In the midline, this will result in the transducer marker pointing to the right and right sided structures projecting on the left side of the screen. However, in the flank (for example when scanning the right kidney) this will result in the marker pointing to the back of the patient and posterior structures will be projected on the left side of the screen.

All structures located in the "near-field," i.e. close to the transducer, will appear in the upper parts of the screen, while deeper structures in the "far field" project at the bottom of the screen.

23.2.3   Technical Aspects

In recent years the costs of ultrasound equipment have dropped substantially and portable, robust, affordable black-and-white scanners have become available. High-end ultrasound machines provide numerous additional options, such as Doppler (flow) and color technologies, image post-processing software solutions and other support functions for the user. Although some of these are helpful in certain conditions, they are not essential in most POCUS applications. All ultrasound machines have more or less the same basic functionality and it is important to localize a few relevant buttons to optimize and adapt the image. Make sure you find at least the following on your machine.

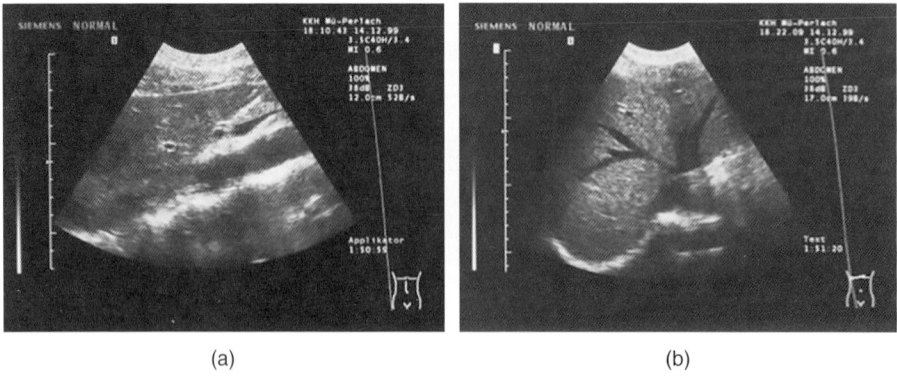

(a)   (b)

Figure 23.3 (3a): On a correctly oriented image the cranial parts of the liver will be projected on the left side of the screen, the caudal edge of the liver is visible on the right side of the screen. (3b): On a correctly oriented image the right lobe of the liver will be projected on the left side of the screen, the smaller left lobe extends to the right side of the screen.

23.2.3.1 Freeze
This freezes the image frame so that measurements can be made, structures can be examined in more detail or printouts can be made.

23.2.3.2 Gain
Increasing the gain increases the amplification of the returning echoes. The image becomes brighter. However, artifacts might also become more pronounced. Most machines offer two or more gain level control options: either near gain and far gain (which control brightness in the near and far field of the image) or multi-level time gain compensation (TGC), which allows adjustment of gain at multiple depth levels in the image (for example, to optimize visibility of pelvic organs, when scanning through the bladder).

23.2.3.3 Depth
This changes the maximum scan depth displayed on the screen. Thus, smaller structures in the near field can be "enlarged" by reducing the scan depth. To visualize deeper structures, scan depth needs to be increased.

23.2.3.4 Focus
By electronic manipulation of the ultrasound beam, the area of optimal resolution of the image can be changed. The focus should be set at or just below the level of the structure examined.

23.2.3.5 Measurements
It is important to measure the actual size of structures (at least approximately) to avoid misinterpretation. As the image size can be changed (see depth) small structures may appear large on the screen and vice versa. Measurements are usually done using the set button and the trackball.

23.3 Most Relevant POCUS Applications in the Resource-Limited Setting

23.3.1 Patients after Trauma with Suspected Internal Bleeding: The FAST Scan

23.3.1.1 The Clinical Problem
Each year intentional and unintentional injuries account for nearly 1 in 10 deaths worldwide (WHO: Injuries, and Violence Prevention Dept 2002). The mortality burden from all types of trauma in high-income countries is only one tenth of that in low/middle-income countries. Among young adult men, road traffic injuries are the leading cause of trauma-related death worldwide, most of them located in low- and middle-income countries (LMICs) (WHO: Injuries, and Violence Prevention Dept 2002). Interpersonal violence, another source of trauma, is a leading cause of death and disability in both male and female residents of low/middle-income countries (Hofman et al. 2005). All trauma victims, irrespective of cause may suffer from internal bleeding, which can lead to hemorrhagic shock and death if not diagnosed and treated in time. Intra-abdominal, but also intra-thoracic, bleeding is of concern as body cavities allow large volume

hemorrhage. Bleeding should be suspected in any trauma victim with tachycardia and hypotension as signs of possible hypovolemic shock. A standard assessment in all trauma patients, irrespective of the clinical presentation, can detect internal bleeding before patients become symptomatic.

Different techniques besides sonography to assess intra-abdominal bleeding all have advantages and disadvantages. Diagnostic peritoneal lavage is fast, has a high specificity if blood is aspirated and does not need dedicated equipment (except a syringe and needle), but it has the potential for iatrogenic injuries and a low sensitivity. Computed tomography has a higher sensitivity and allows evaluation of the intra-abdominal organs and the retroperitoneum to detect the source of bleeding, but it is expensive, time consuming and in most resource-limited settings simply not available. Ultrasonography compares favorably to the above-mentioned diagnostic modalities, as it is an accurate, non-invasive, repeatable technique that can be performed at the bedside. A POCUS protocol is therefore frequently used in trauma patients and is well established.

The intent of Focused Assessment with Sonography for Trauma ("FAST") is to detect free fluid in the body cavities using standardized views as a rapid, "minimal" ultrasound examination. The goal is the detection of free intra-abdominal fluid, pericardial fluid or pleural fluid. Detection of any free fluid in a trauma victim suggests internal bleeding until a different explanation for the fluid is found. As FAST is not time-consuming and has no side effects for the patient it is basically indicated in every form of trauma (acute blunt or penetrating torso trauma, trauma in pregnancy, pediatric trauma). FAST is also valuable in the pre-hospital setting for an adequate time-sensitive referral of the patient if bleeding is confirmed.

The value of the FAST scan is well established and extensively studied. As early as the 1980s, surgeons in Germany developed bedside ultrasound for evaluation of trauma patients and reported sensitivity and specificity as 84–100 and 88–100%, respectively (Halbfass et al. 1981). American groups in the 1990s confirmed the accuracy to be high (Ma et al. 1995). Bedside ultrasound for detection of pericardial fluid also has high accuracy in trauma patients. This information can significantly shorten time to surgical intervention (Plummer et al. 1992). Studies comparing FAST with chest X-rays (CXRs) for the detection of pleural fluid showed comparable accuracy of 99% (Ma and Mateer 1997). The average time to perform a complete FAST examination is 2.1–4.0 minutes – so it is fast in the true sense of the word (Ma et al. 1995). It is important to remember that smaller amounts of free fluid may be missed especially by less experienced examiners. The FAST exam should therefore be liberally repeated whenever clinical signs suggest this is necessary as blood may have accumulated over time.

Technique and Typical Ultrasound Findings Technique   Four probe positions are used the FAST protocol. The examiner assesses all of them for the presence of anechoic fluid.

23.3.1.2   Cardiac View

The view is obtained in a subxiphoid position with the convex transducer in a transverse orientation, tipped cranially, aiming toward the patient's left shoulder, using the liver as an acoustic window. The patient should try to relax the abdominal muscles (arms placed beside the body); pressure is applied with the transducer parallel to the abdominal wall, while aiming the beam behind the ribs. Asking the patient to take a deep breath can help displace the heart caudally and improve visualization.

23.3.1.3 Right Flank View

The patient is asked to put the arms behind the head for better access to the flanks; additionally, this position widens the space between individual ribs. The transducer is positioned in the longitudinal plane, posterior to the right mid-axillary line at the caudal part of the thorax to detect pleural fluid above the right hemidiaphragm. Scan plane can be adjusted so that the long axis of the probe is parallel to the ribs, allowing an intercostal view. It is important to place the transducer as far back as possible (close to the examination couch), as fluid collects in the dependent parts due to gravity. Fluid in the abdomen following gravity can be detected in Morrison's pouch[1] (the hepato-renal recess) as this is dependent in a supine patient.

23.3.1.4 Left Flank View

This position is analogous to the previous one. The transducer is now positioned in the longitudinal plane, posterior to the left mid-axillary line at the caudal part of the thorax. Pleural fluid can be detected above the left hemidiaphragm, and abdominal fluid in the dependent spleno-renal recess.

23.3.1.5 Pelvic View

Finally, the probe is placed on the lower abdomen touching the upper rim of the symphysis pubis. The pelvic region should be scanned in the transverse axis. For visualization of structures deeper in the pelvis, the transducer needs to be tipped so that the ultrasound beam is directed caudally into the small pelvis. If the patient was walking or sitting upright prior to the examination the pouch of Douglas[2] in females (recto-uterine pouch) and the recto-vesical pouch in males are of particular importance as free fluid will have predominantly collected there.

A normal subxiphoid view of the heart is shown in Figure 23.4. A normal right flank view is shown in Figure 23.5a and a normal left flank view is shown in Figure 23.5b. A normal pelvic view is shown in Figure 23.6.

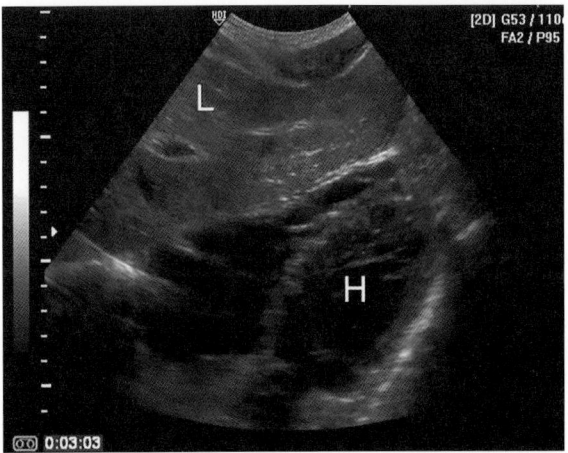

Figure 23.4 Normal sub-xiphoid view of the heart (H). The liver (L) is used as an acoustic window in the near field.

---

1 James Morrison, a British surgeon, 1853–1939.
2 James Douglas, a Scottish physician and anatomist, 1675–1742.

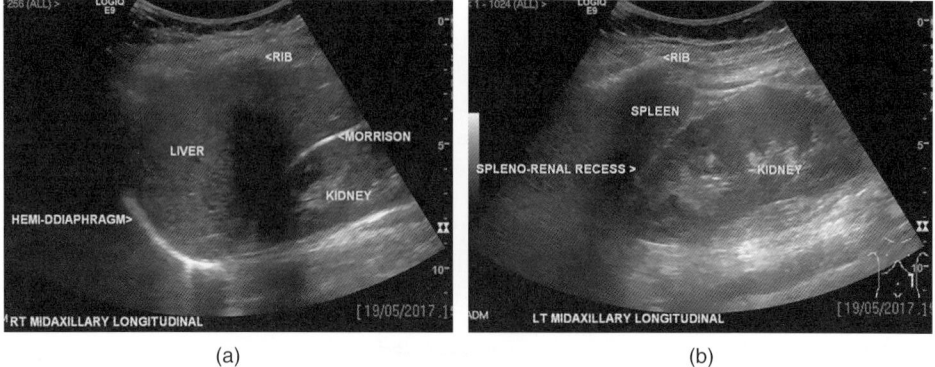

(a)   (b)

Figure 23.5 (a) Normal flank view on the right used to detect right-sided pleural fluid above the right hemidiaphragm, as well as intra-abdominal fluid in Morrison's pouch between liver and kidney. (b) Normal flank view on the left used to detect left-sided pleural fluid as well as intra-abdominal fluid in spleno-renal recess between spleen and kidney.

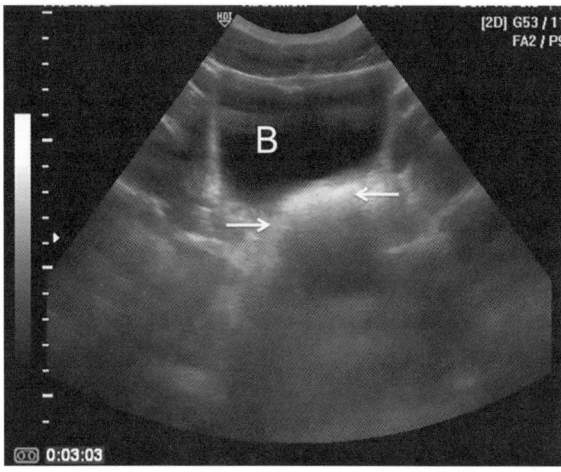

Figure 23.6 Normal pelvic view. The bladder (B) is seen as an anechoic fluid filled structure. The area behind the bladder (arrow) or the uterus in female patients should be screened for pelvic fluid which is not present in this patient.

23.3.1.6 Pathological Findings: Pericardial Fluid

Pericardial fluid shows as an anechoic, black rim around the heart separating the visceral and parietal pericardium (Figure 23.7). Small amounts of fluid are seen mainly next to the right atrium and ventricle; the anechoic rim might surround the entire heart as the amount of fluid increases. Signs of tamponade, such as diastolic collapse of the right atrium and ventricle, may be seen due to increased pericardial pressure (and see Section 23.3.6).

23.3.1.7 Pleural Fluid

Anechoic fluid may be visible in the costo-phrenic space (Figure 23.8). It may be completely echo-free, as in transudates or simple para-pneumonic effusions, but may also contain internal echoes caused by fibrin strands or areas of echogenic clotted blood.

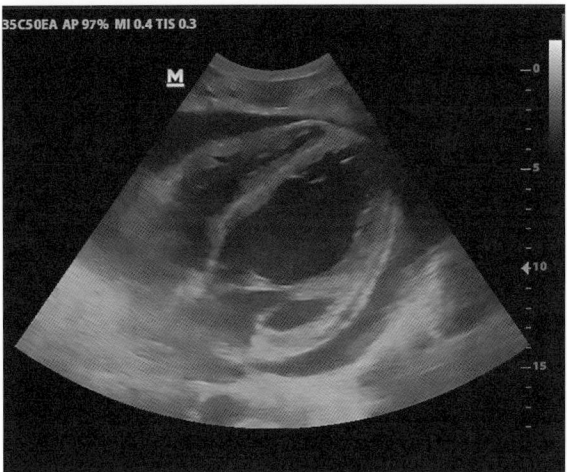

Figure 23.7 Pericardial effusion. Sub-xiphoid transverse view shows the effusion as an anechoic, black rim around the heart separating the visceral and parietal pericardium.

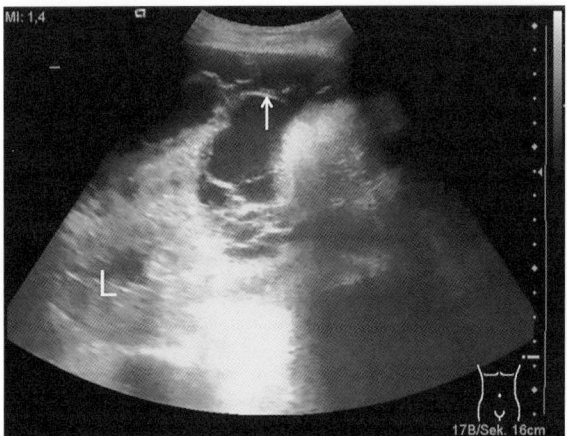

Figure 23.8 Pleural effusion. Left mid axillary view shows anechoic effusion with multiple fibrin strands (arrow). The lung (L) is collapsed and visible as a hyperechoic triangular shape in the effusion.

23.3.1.8 Abdominal Fluid

Anechoic spaces between the liver and the kidney (Figure 23.9) or between spleen and kidney are signs of free abdominal fluid. In the pelvic view, fluid might be seen between the bladder and rectum (male patient) or in Douglas' pouch, behind or surrounding the uterus (female patient) (Figure 23.10). Differentiating between bladder and free fluid can be difficult for smaller amounts of fluid. Emptying the bladder may be helpful to confirm the presence of free fluid. Again, echogenic material like streaks or strands can be seen floating in the fluid, representing fibrin and/or clot.

23.3.1.9 Diagnostic and Therapeutic Decisions

The diagnostic and therapeutic decisions based on the result of the FAST exam will largely depend on the clinical context and the setting and availability of resources.

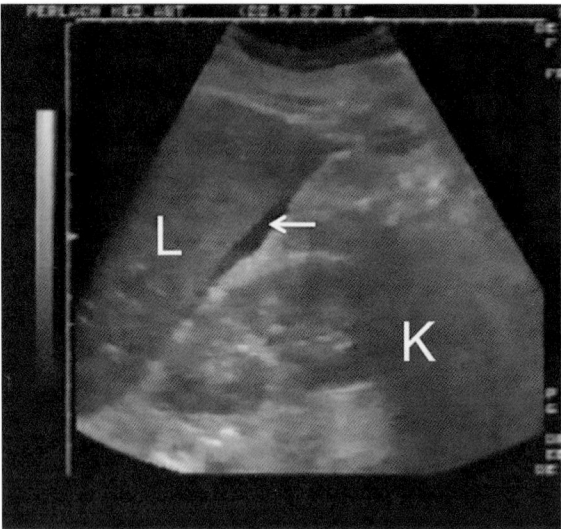

Figure 23.9 Free abdominal fluid. Right longitudinal mid-axillary view shows an anechoic fluid rim (arrow) between liver (L) and kidney (K) (Morrison's pouch).

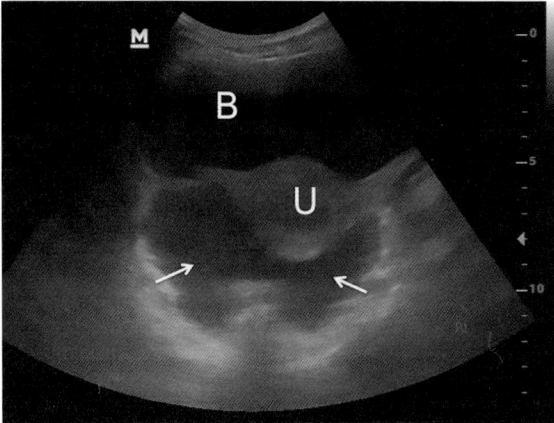

Figure 23.10 Free pelvic fluid. Transverse suprapubic view shows Douglas' pouch (arrows) behind the uterus (U) and the bladder (B).

When free fluid is detected, the patient may require surgery and blood transfusion and should generally be treated in a place that is equipped for such cases. Circulatory monitoring should be intensified and stable vascular access should be in place. In the case of pericardial effusion, cardiac tamponade may be developing, which will require further consideration and evaluation (see Section 23.3.6).

Other indications for scanning to detect free fluid due to causes other than trauma in resource-limited settings include a search for ascites or peritonitis possibly due to ruptured viscera.

Several organs in the abdomen can contain fluid, which might be mistaken for free fluid. In the upper abdomen the gallbladder, inferior vena cava and fluid in the stomach and, rarely, the colon can be seen. In the pelvis, fluid in the bladder, large ovarian cysts or the seminal vesicles can pose diagnostic challenges. In premenopausal women small amounts of free fluid in the pouch of Douglas may be normal. Hysterectomies will alter pelvic anatomy. Finally, as mentioned above, not all "free fluid" is blood. Once fluid is identified, the finding has to be correlated with the clinical situation. Additionally, it is easy to aspirate a small amount of the fluid to verify the diagnosis. Ultrasound alone cannot distinguish between blood and other types of free fluid.

23.3.1.10 Pearls and Pitfalls
- Do not perform FAST when immediate surgical treatment is indicated.
- Do not over-rely on a negative FAST exam - use clinical judgment and serial examinations.
- Know your limitations – obesity and bowel gas reduce penetration of the ultrasound beam.
- Be aware of other causes of fluid such as ascites in hepatic cirrhosis.
- Do not hesitate to use a needle to clarify the issue by paracentesis.
- Be cautious in pregnant women as intrauterine fluid can cause confusion.
- For pleural fluid, make sure that the transducer is posterior and cranial enough and that you can identify the hemi-diaphragm below the fluid.
- Remember that FAST is only designed to detect free fluid, not solid or hollow viscous injuries. Splenic laceration, bowel wall contusion, pancreatic trauma, renal pedicle injury will not be visible.

23.3.2 Patients with Weight Loss or Night Sweats: The FASH Scan

23.3.2.1 The Clinical Question
Weight loss is a frequently encountered clinical problem in resource-limited settings, in particular in populations with a high prevalence of HIV and tuberculosis. A variety of conditions can cause weight loss and wasting, ranging from nutritional (for example, malnutrition, chronic diarrhea), metabolic (for example, diabetes mellitus) and infectious diseases (for example, HIV, TB, parasitic infections), and to malignancies.

When loss of weight is accompanied by fever, night sweats and possibly also cough, TB is highly likely, particularly in HIV co-infected patients who have a substantially higher risk of TB. For this reason, commonly used TB screening algorithms include this triad of symptoms to identify TB suspects (Cuevas 2011). Obviously, other HIV-related pathologies can cause similar symptoms: malignancies such as HIV-associated non-Hodgkin lymphomas[3] and disseminated Kaposi's sarcoma[4] or other opportunistic infections.

Identifying a TB suspect will prompt a search for further evidence to support the diagnosis of TB by searching for a positive acid-fast bacilli (AFB) stain from sputum, by GeneXpert technology (see Chapter 5) or by a positive urinary LAM test (see Chapter 5). If these tests are negative or inconclusive, CXR is recommended to identify radiological patterns suggestive of TB including infiltrates, cavities or a miliary pattern.

---

3 Thomas Hodgkin, a British physician, 1798–1886.
4 Moritz Kaposi, a Hungarian dermatologist who worked in Vienna, 1837–1902.

In recent years POCUS has proved to be a useful tool in evaluating HIV patients suspected of co-infection with extra-pulmonary TB (EPTB) (Heller et al. 2012). Unilateral pleural (Luzze et al. 2001) or pericardial (Reuter et al. 2007) effusions are suggestive of pleural or pericardial EPTB; although less specific, ascitic fluid may be a sign of peritoneal TB. Detection of abdominal lymphadenopathy and splenic microabscesses are typical findings in abdominal TB (Heller et al. 2010a). Since all of these findings are easily identified by ultrasound, a logical consequence has been the development of a focused POCUS protocol for EPTB (FASH, Focused Assessment with Sonography for HIV-associated TB). The FASH exam combines the ultrasound views of FAST (Section 23.3.1) with additional views for detection of upper abdominal lymphadenopathy and splenic microabscesses (Heller et al. 2012). FASH should be considered for all patients with a high clinical probability of disseminated and EPTB (WHO 2006).

23.3.2.2 Technique and Typical Ultrasound Findings
Pericardial and Pleural Effusions The technique for detecting pericardial and pleural effusions has been described above in Section 23.3.1.

23.3.2.3 Abdominal Lymph Nodes
For the detection of enlarged upper retroperitoneal and periportal abdominal lymph nodes, the transducer is tilted to a transverse position at the level of the epigastric view of the heart. In this position the upper abdominal and periportal area can be visualized by angling the ultrasound beam first up through the left lobe of the liver toward the diaphragm, then down along the aorta. The liver serves as a good ultrasound "window" to detect periportal nodes in the liver hilum and around the coeliac axis. To minimize the distance between the probe and the retroperitoneal areas, the abdominal wall should be relaxed with the arms positioned next to the abdomen.

23.3.2.4 Spleen
The spleen is best examined in a supine or right lateral decubitus position, with the left arm placed over the head. A lateral and posterior probe position is used, angling the beam anteriorly to identify the spleen. If the probe is placed anteriorly, the spleen is usually obscured by gas in the stomach or colon. The transducer is turned parallel to the ribs. Asking the patient to breathe in is often counterproductive: the spleen usually disappears behind the expanding lung. The spleen should be scanned using an abdominal probe. A linear transducer can be helpful for better resolution of parenchymal changes and detection of microabscesses.

23.3.2.5 Pathological Findings
Pericardial Effusion Pericardial effusions can be recognized as an anechoic, black rim around the heart as described in the section on FAST. In patients with tuberculous pericarditis echogenic fibrin streaks may be seen floating in the anechoic effusion. These may be attached to the pericardium and move as the heart moves. Occasionally, the whole effusion appears echogenic due to a high particle content in purulent exudative effusions. Once an effusion has been identified, it is important to determine whether it is hemodynamically relevant (see Section 23.3.6).

Pleural Effusion Anechoic fluid may be visible in the costophrenic spaces, as described in Section 23.2. As in tuberculous pericarditis, it may contain internal echoes such as strands, septae or turbid "smoke," due to fibrinous structures or cells within the effusion.

Ascites   Ascites can be seen in the most dependent pockets of the abdominal cavity: Morrison's pouch, the spleno-renal pouch, and the pouch of Douglas. Since ascites due to tuberculous peritonitis has a high protein content (and a predominance of lymphocytes) fibrin strands, septations and webs are commonly visible.

23.3.2.6   Enlarged Lymph Nodes
Lymph nodes larger than 1.5–2 cm are considered pathological in an HIV-infected adult. Normal nodes are oval with a thin cortex and often have a hyperechoic hilum. Pathological, enlarged nodes in TB are often hypoechoic and rounded (Figure 23.11). They can be differentiated from adjacent tubular vessels by this rounded appearance, as the transducer moves across the node and along the vessels. This gives the lymph nodes a "blinking" appearance. The size of the node is measured at its maximal diameter in the short axis.

23.3.2.7   Splenic Lesions
In patients with disseminated TB, multiple splenic microabscesses are observed in the form of small hypoechoic lesions of a few millimeters in size (Figure 23.12). The microabscesses are distributed throughout the spleen and represent miliary seeding.

23.3.2.8   Diagnostic and Therapeutic Implications
In an HIV-infected patient with a pericardial effusion, empirical anti-TB treatment is warranted, particularly in the presence of severe immunosuppression. Additionally, corticosteroid treatment should be given, especially in cases of large and hemodynamically relevant pericardial effusions (Heller et al. 2010b). It should be noted that malignant effusions might also partially respond to steroid treatment, but patients will tend to relapse after the steroids are tapered off. A pleural effusion in HIV-infected patients is also highly suggestive of TB, especially when unilateral; anti-TB treatment should be initiated. Ascites without other supporting findings of disseminated TB, such as enlarged lymph nodes or splenic microabscesses, may be caused by various conditions and needs to be interpreted in the light of other clinical and laboratory findings. If no

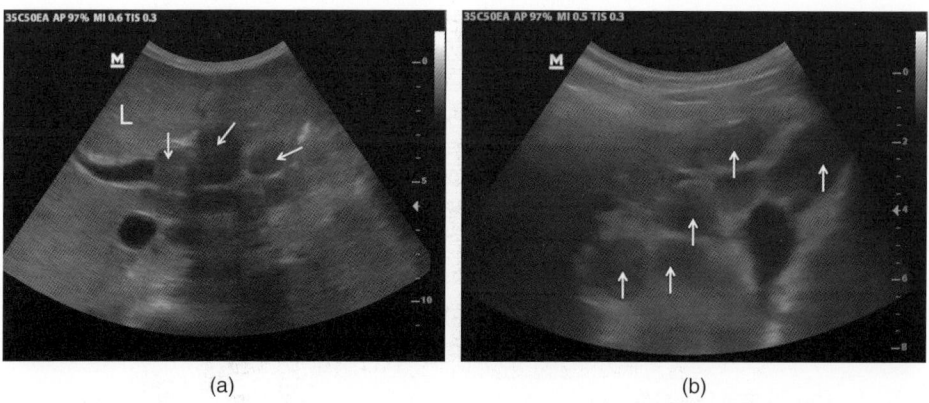

(a)   (b)

Figure 23.11 (a) hypoechoic nodular lymph nodes (arrows) on a transverse view through the liver (L) hilum. (b) multiple hypoechoic nodular lymph nodes (arrows) on a transverse view of the mid-abdomen.

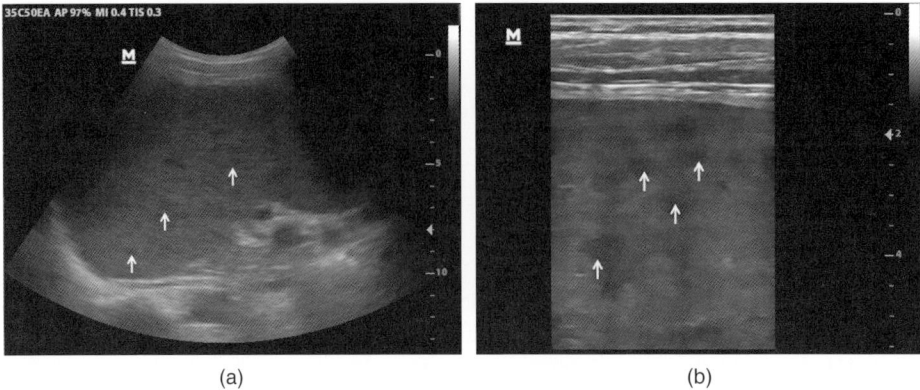

(a)   (b)

Figure 23.12 Small hypoechoic lesions representing microabscesses in the spleen of patient with disseminated TB. (a) lesions (arrows) can be seen with a standard 3–5 MHz convex probe when they are large enough, (b) the high-frequency linear probe provides better resolution and is superior to detect small lesions.

other explanation is found and the clinical picture makes TB likely, empiric treatment may be started in individual cases.

TB treatment without further laboratory confirmation is indicated in patients with enlarged abdominal lymph nodes and/or splenic microabscesses (Heller et al. 2012), especially when the local TB prevalence is high and further diagnostic work-up would delay or prevent treatment, increasing morbidity or mortality.

23.3.2.9 Pearls and Pitfalls

- Usually there are multiple enlarged lymph nodes, rather than a single one. They may also be matted together.
- In a significant proportion of patients with positive FASH findings abnormalities are also visible on the CXR (for example, an enlarged cardiac shadow due to pericardial effusion, pleural effusions, miliary TB in disseminated disease or classic features of pulmonary TB). However, approximately 25% of HIV patients with positive FASH findings have a normal CXR (Heller et al. 2013).
- Sonographic findings might initially persist or even increase during treatment as a result of the immune reconstitution inflammatory syndrome (IRIS). This can be very pronounced when the patient starts antiretroviral treatment (ART) at the same time and it therefore does not necessarily indicate an incorrect diagnosis.
- In approximately 75% of patients, sonographic findings resolve by three months. In patients with persistent findings after this period, non-compliance, an incorrect drug regimen, drug resistance, immune reconstitution inflammatory syndrome or alternative diagnoses need to be considered (Heller et al. 2014).

23.3.3 Patients with an Enlarged Cardiac Shadow on Chest X-Ray and Signs of Heart Failure: The Cardiac Scan

23.3.3.1 The Clinical Question

Ultrasound is an excellent tool for evaluating the etiology of dyspnea and the source of cardiomegaly seen on a CXR. Congestive heart failure often leads to peripheral edema, but there are many alternative causes for leg edema, including hepatic cirrhosis, malignancy, filariasis, kidney disease and venous thrombosis. In a tropical setting common cardiac causes of dyspnea and edema are pericardial effusion with tamponade (most commonly due to TB), dilated cardiomyopathy (for example, peripartum or HIV-cardiomyopathy) or resulting from infections such as trypanosomiasis, post-rheumatic heart disease and cor pulmonale (Heller et al. 2016).

Technique and Typical Ultrasound Findings Sonographic examination of the heart is generally done using intercostal and parasternal approaches or "windows" using a phased array transducer. A limited overview is also possible from the abdomen using a 3.5 MHz convex transducer. This view is obtained in a subxiphoid position with the convex transducer in a transverse orientation, tipped cranially, aiming toward the patients' left shoulder, using the liver as an acoustic window. In the subxiphoid view the right heart is closest to the surface and the left heart is visible in the far field (Figure 23.13a).

Intercostal ultrasound of the heart with a phased array probe generally uses four views, also called windows; the subxiphoid view (as described above), the left parasternal long axis and short axis, and the apical four-chamber view (Figures 23.13b–d).

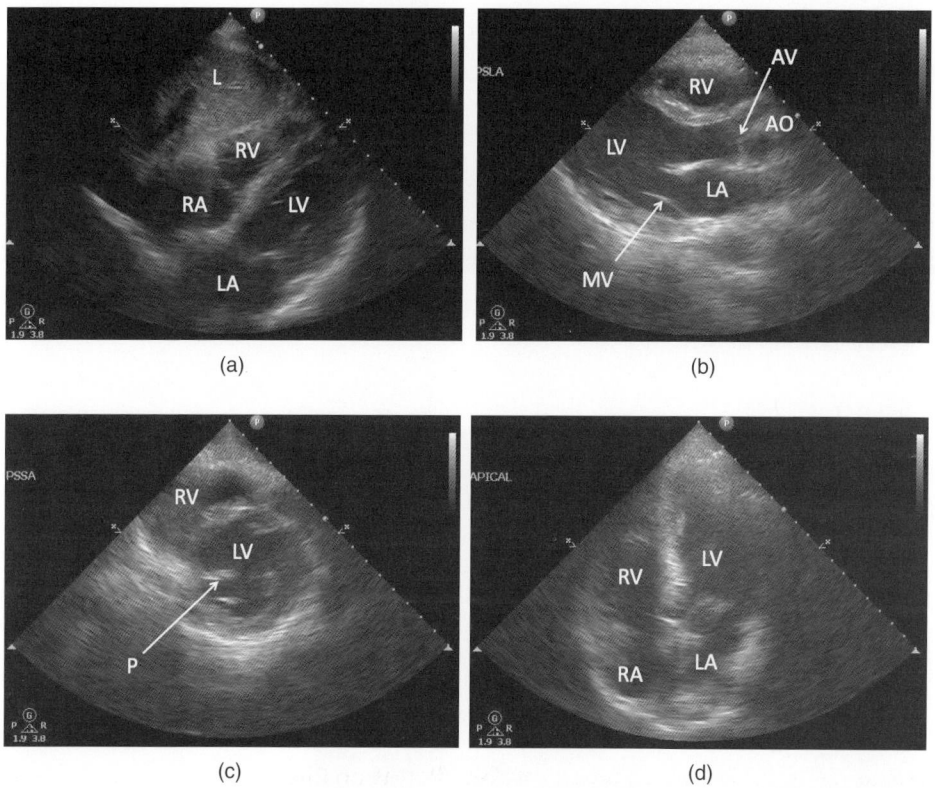

(a)   (b)

(c)   (d)

Figure 23.13 Standard views in cardiac ultrasound: (a) sub-xiphoid view of a normal heart. Note that the right side of the heart is in the near-field and the left side in the far-field of the image, (b) parasternal long axis (PLAX), (c) parasternal short axis (SAX), (d) apical four chamber view. (L: liver, RV: right ventricle, RA: right atrium, LA: left atrium, LV: left ventricle, MV: mitral valve, AV: aorta valve, AO: aorta outflow tract, P: papillary muscles of mitral valve).

The parasternal long axis view is found by placing the probe in the third or fourth intercostal space, directly to the left of the sternum, turning the probe to align between the right shoulder and the left hip. The angle of the probe is then adjusted to visualize the mitral valve, aortic valve and, if possible, the apex of the left ventricle in one view. The parasternal long axis is well suited for differentiating pericardial and pleural effusions. Large pleural effusions can appear to surround the heart, but will taper to the descending aorta, whereas pericardial effusions will continue anterior to the descending aorta (Figure 23.14). Furthermore, several measurements can be taken (Figure 23.15, Table 23.2) and this view can be used to assess left ventricular contractility.

From the parasternal long axis, with the mitral valve in the center of the image, the probe is rotated 90° in a clockwise fashion to obtain the parasternal short axis view,

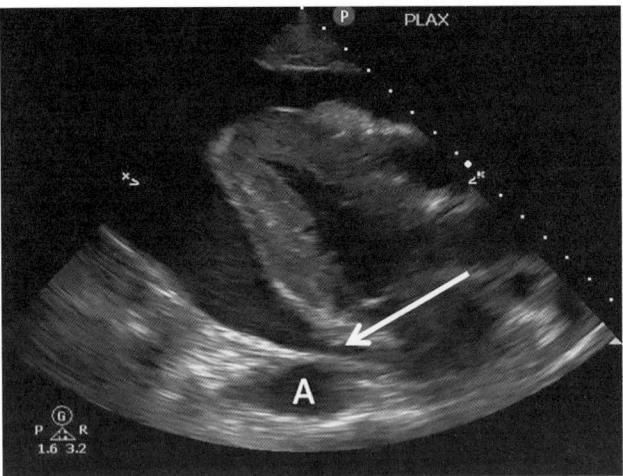

Figure 23.14 Pericardial effusion: Note that the effusion (arrow) continues anterior to the aorta (A), which would not be the case in a pleural effusion.

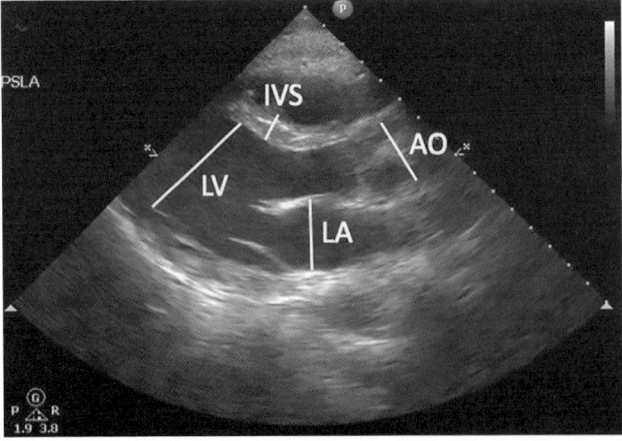

Figure 23.15 Measurement levels in the parasternal long axis view: LV: left ventricle, LA: left atrium, AO: aorta outflow tract, IVS: inter ventricular septum.

Table 23.2 Normal values for measurements of the heart.

Left ventricle			
Internal diameter	End-diastolic		3.5–5.5 cm
Wall thickness	End-diastolic	Septum	0.6–1.2 cm
Left atrium			
Diameter			2.0–4.0 cm
Aortic root			
Diameter			2.0–4.0 cm

visualizing the left chamber and mitral valve in a circumferential plane, with the right chambers adjacent. This view can provide additional information about left ventricular contractility and right ventricular size and pressure. The apical four-chamber view is obtained at the apex of the heart, which is usually located below the nipple in the fifth intercostal space, but can be displaced in patients with cardiomegaly. This view can also be found by sliding down toward the apex from the parasternal short axis and then tilting the probe toward the patients' right shoulder. This view is the most optimal for assessing the size and dimensions of the atria and ventricles. Normally the size of the left ventricle is approximately double that of the right ventricle ("2:1") and the atria are smaller than the ventricles. Finally, ultrasound of the heart should include a view of the inferior vena cava (IVC) to assess the intravascular volume status. Several protocols for point-of-care cardiac ultrasound have been developed, including the SIMPLE protocol (Mok 2016) (Table 23.3) and FATE (Focused Assessment of Transthoracic Echo) protocol (Jensen et al. 2004). The latter also offers a free download card with the different views and pathologies (http://www.fate-protocol.com).

23.3.3.2 Pathological Findings

POCUS of the heart aims to answer three important questions. Is a pericardial effusion present?" Is LV contractility impaired? What is the relative size of the atria and ventricles?

The evaluation of pericardial effusion and tamponade are described in more detail in Section 23.3.6.

Table 23.3 The SIMPLE approach for evaluation of key elements during focused cardiac ultrasound sound in shock patients.

S	Chamber *s*ize and *s*hape, particularly LV and RV *s*ize
I	*I*VC size and collapsibility *I*VS movement *I*ntimal flaps inside the aorta, suggestive of aortic dissection
M	*M*ass in the heart chambers (commonly intramural clots and atrial myxoma) *M*yocardium (motion and thickness)
P	*P*ericardial effusion *P*leural effusion
L	*L*eft ventricular systolic function
E	Abdominal aorta in the *e*pigastrium

Abbreviations: IVC, inferior vena cava; IVS, interventricular septum.

To assess the left ventricular (LV) function, a formal ejection fraction can be calculated by several methods but this is beyond the scope of POCUS. A classification of contractility as hyperdynamic, normal, moderately impaired and severely impaired, can be made by assessing whether contraction is symmetrical toward the center, whether the myocardium thickens as it contracts, and whether the mitral valve opens normally during diastole. Measurement of the parasternal long and short axis is most suitable for the purpose of "guesstimating" LV function. Reduced LV function can be observed in various cardiomyopathies and in ischemic heart disease. A hyperdynamic heart is suggestive of hypovolemic or distributive shock and is characterized by small chambers and hyperkinetic contractions that may obliterate the ventricle in systole.

Finally, the size and dimensions of the atria and ventricles in relation to each other can provide clues to underlying cardiac pathology.

Examples of cardiac pathologies evaluable by ultrasound testing are given below.

A global enlargement of both atria and ventricles is found, for example, in generalized cardiomyopathies such as postpartum or HIV-related cardiomyopathy (Figure 23.16).

Pronounced dilatation of the right ventricle in comparison to the left suggests cor pulmonale (Figure 23.17). A right ventricle/left ventricle (RV/LV) ratio of greater than 0.7 indicates a dilated right ventricle. In addition, movement of the septum away from the right ventricle during diastole indicates increased right ventricular pressure, resulting in a D-shaped left ventricle in the parasternal short axis.

Mitral stenosis or regurgitation, as observed commonly in rheumatic heart disease, results in a dilated left atrium (Figure 23.18) with a normal sized left ventricle. The right heart may also be enlarged due to the secondary increased pulmonary pressure. Often a thickened mitral valve can be identified.

A thickened ventricular septum (Figure 23.19) can be observed in patients with long-standing chronic hypertension for example, due to kidney disease or in hypertrophic cardiomyopathies.

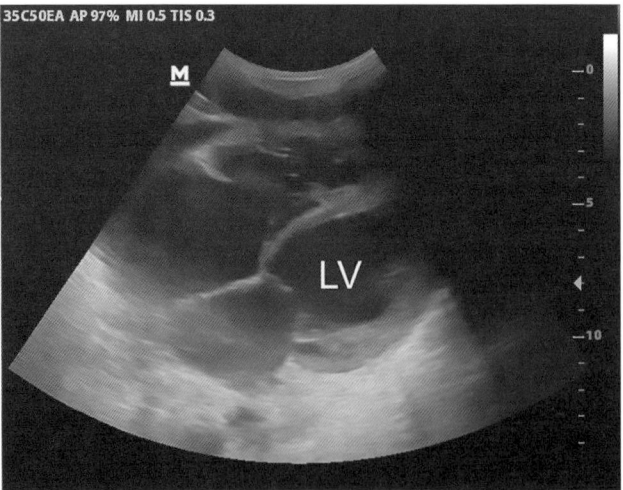

Figure 23.16 Subxiphoid view: dilated cardiomyopathy characterized by global enlargement of atria and ventricles with reduced left ventricular (LV) function (visible as reduced contractility during scanning).

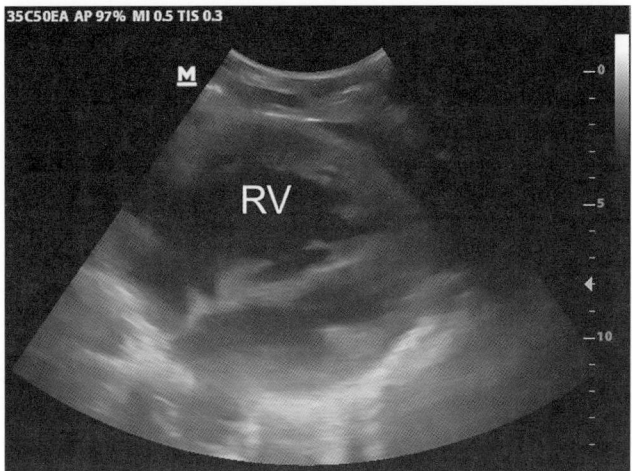

Figure 23.17 Subxiphoid view: cor pulmonale with pronounced dilatation of the right ventricle (RV) in comparison to the left.

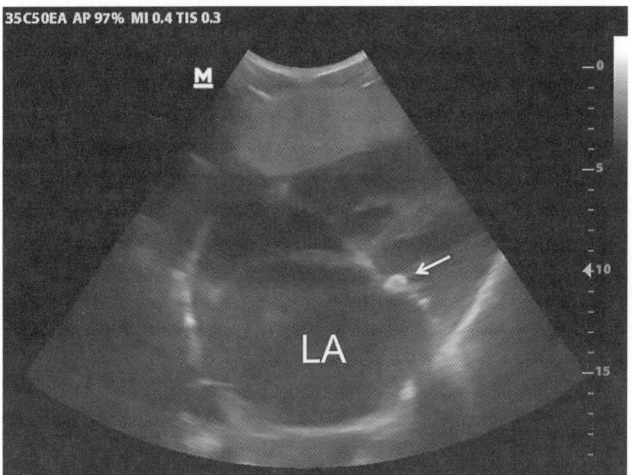

Figure 23.18 Subxiphoid view: post-rheumatic mitral valve stenosis characterized by pronounced dilatation of the left atrium (LA). The thickened mitral valve (arrow) can be seen "doming" toward the left ventricle.

Endomyocardial fibrosis, a condition primarily observed in African children, causes progressive fibrosis of the ventricular endocardium - either right, left or biventricular. This results in mitral and/or tricuspid regurgitation and pronounced atrial dilatation (Grimaldi et al. 2016).

23.3.3.3 Diagnostic and Therapeutic Implications

The presence of pericardial effusion may prompt empiric treatment for tuberculosis in an endemic setting, especially in HIV patients with severe immune suppression (Section 23.3.2) (Heller et al. 2010b). In patients with evidence of tamponade emergency pericardiocentesis can be lifesaving.

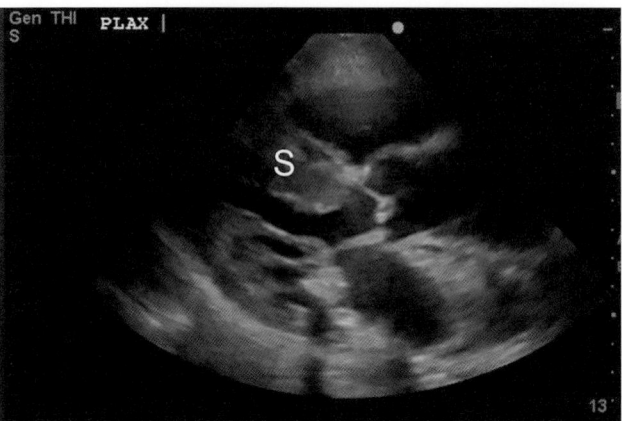

Figure 23.19 PLAX view: LV hypertrophy characterized by pronounced thickening of the septum (S) and the free wall of the ventricle.

Patients with impaired contractility should be started on long-term treatment for congestive heart failure, including a diuretic, an angiotensin-converting enzyme (ACE) inhibitor and if possible a low-dose beta blocker (Ponikowski et al. 2016). Spironolactone can be added in cases resistant to initial therapy. Nitrates and antiplatelet therapy may be added for patients with suspected ischemic heart disease (Ponikowski et al. 2016). In patients with acute heart failure, aggressive intravenous diuretic therapy is warranted, as well as oxygen administration or ventilatory support if available. An acute etiology, such as severe anemia, a hypertensive crisis or an arrhythmia should always be sought for in severe heart failure. If features of cor pulmonale are identified it can aid the diagnosis of pulmonary embolism and prompt anticoagulant treatment (see Section 23.3.5). If valvular pathologies or endomyocardial fibrosis are suspected, referral to cardiac specialist care may be indicated.

23.3.3.4 Pearls and Pitfalls
- When using the subxiphoid view, increase scan depth to cover the entire outline of the heart.
- When the heart does not come into view in the subxiphoid position, flatten the probe against the abdominal wall and let the patient bend the knees to relax the abdominal wall. If the view is obstructed by the stomach, slide to the right to use the liver as an acoustic window.
- For the parasternal and apical four-chamber views, let the patient lie on his/her left side in order to bring the heart closer to the anterior chest wall.

23.3.4 Patients with Acute Shortness of Breath: The Chest Scan

23.3.4.1 The Clinical Question
The classic imaging modality for examining pathological processes in the thorax is the chest X-ray (CXR). However, in some settings CXR is not immediately available, or if available, the findings may be ambiguous. Ultrasound of the chest can be helpful in the detection and characterization of processes located in the thoracic cavity, especially those close to, or in, the chest wall. For some indications, ultrasound may even be

superior to CXR. An example is the diagnosis of pneumothorax in trauma patients (Zhang et al. 2006). Other presentations in which ultrasound may be helpful include acute respiratory distress, dullness on percussion, reduced breath sounds and ambiguous opacities on CXR to differentiate suspected pleural effusions from large areas of consolidation or an elevated hemidiaphragm.

23.3.4.2 Technique and Typical Ultrasound Findings

The patient is scanned in a sitting or supine position using an intercostal approach. Anterior, lateral and posterior scanning in the longitudinal plane needs to be performed for the detection of effusions. It is important to be sure that the hemidiaphragm is identified above the liver and spleen to be sure that any fluid seen lies in the hemithorax and not in the subphrenic spaces. In a supine patient fluid accumulates posteriorly and may not be visible in the lateral views unless it is a large effusion. As the pleural space is located only a few centimeters below the skin, a linear high-frequency transducer may be suitable in slim patients, but lower frequency cardiac or abdominal probes may be necessary in larger patients or to assess deeper structures. In a patient with normal anatomy, the subcutaneous fat will be hypoechoic. The ribs show as hyperechoic (ossified) structures with a clean acoustic shadow. Between the ribs, intercostal muscle is visible as a linear striation. Air in normal lung parenchyma obscures deeper views by causing massive artifacts (reverberation echoes parallel to the pleura called A-lines). The pleural line, representing the visceral and parietal pleura, can be identified on intercostal views, just behind the posterior border of the ribs (Figure 23.20). Once this view is obtained the transducer is held still to observe lung sliding at the pleural interface during the respiratory cycle.

23.3.4.3 Pathological Findings

As a broad classification, pathological findings can be grouped by their air/fluid ratio. Uncomplicated effusions contain fluid only; lung consolidation contains more fluid

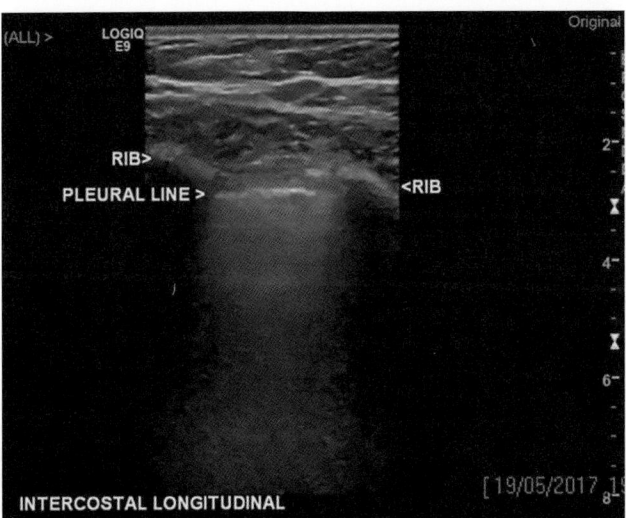

Figure 23.20 Normal view of the pleural line between two ribs; the pleural line can be seen as a thin, horizontal echogenic line, immediately behind the ribs. Rib shadows are noted on either side of the pleural line.

than air; interstitial abnormalities such as interstitial edema contain mainly air with interstitial fluid; normal lung parenchyma contains air with normal interstitium and a pneumothorax consists soley of air. The air/fluid ratio and the distribution of fluid and air determine the ultrasound phenomena and the artifacts that are detected. Examples are described below.

Transudates and uncomplicated para-pneumonic effusions appear as anechoic homogenous fluid. Empyemas, hemothorax and complex parapneumonic effusions often contain dense strands and appear multi-loculated. On occasion, they can be difficult to distinguish from consolidated lung. Ultrasound can be used to guide aspiration of small effusions.

Consolidated lung presents as hypoechoic, a "tissue-like" replacement of normal lung, without signs of lung collapse (Figure 23.21).

The consolidation may be small, focal and subpleural or it may be large and surrounded by para-pneumonic fluid. In pneumonia pathologists have long used the term "hepatization" to describe the gross appearance of the lung as it becomes solid with a consistency similar to that of the liver. This is a helpful reminder, since on ultrasound examination a consolidated lung also resembles liver parenchyma. Within the consolidation the larger bronchi remain filled with air and show as intensely echogenic bands, equivalent to the air bronchogram seen on CXR. Consolidation may be caused by infection (i.e. pneumonia) or, less often, malignancy. Lung collapse (atelectasis) can sometimes be difficult to differentiate from consolidation, as the two often occur together in pneumonia. However, unlike consolidation, healthy collapsed lung (for example, in obstruction of central airways or secondary to pleural fluid) is hyperechoic, with very little volume of the collapsed part. In partial collapse, the tip of the lung is triangular in shape and borders normal appearing parenchyma. The lung tip may re-expand on inspiration and can often be seen to "float" in the fluid.

Interstitial syndromes are sonographically characterized by so-called "B-lines" (Figure 23.22). B-lines occur when sound waves encounter a mixture of air and water (as

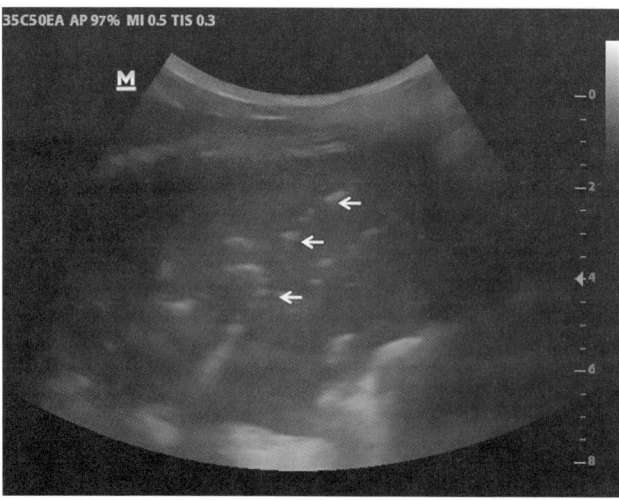

Figure 23.21 Consolidated lung with a hypoechoic appearance similar to liver parenchyma. Hyperechoic flecks of air (arrows) within the bronchi are present, equivalent to an air-bronchogram on CXR.

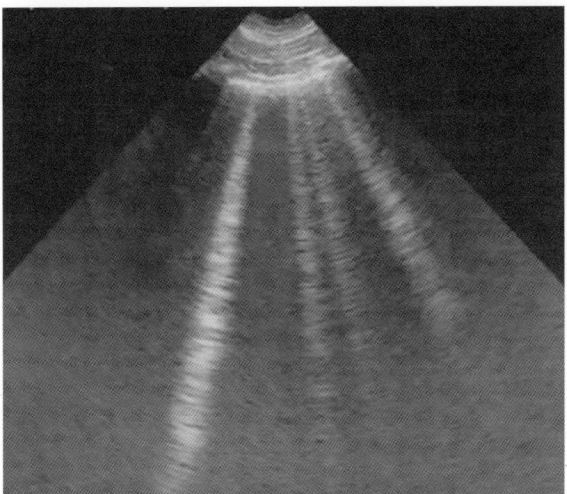

Figure 23.22 B-lines are comet-tail artifacts arising from the pleural line and extending into the far field of the image.

in interstitial edema, where the increased fluid in the interstitium is surrounded by aerated alveoli). This causes an artifact, which appears as a laser-like vertical hyperechoic line, arising from the pleura and extending to the bottom of the screen without fading. The lines move synchronously with lung sliding. As B-lines are frequently encountered in normal subjects only the presence of more than three B-lines in one intercostal space in the longitudinal plane is considered pathologic. In addition to interstitial edema, they can also be seen in interstitial pneumonias, partial atelectasis, or with lung contusions. Generalized B-lines are present in pulmonary edema, ARDS or diffuse pneumonitis. Differentiating cardiogenic from non-cardiogenic pulmonary edema or from pneumonitis is not possible.

Finally, ultrasound is a very useful tool for diagnosing pneumothorax (Lichtenstein 2015; Zhang et al. 2006). In the healthy lung pleural sliding is visible during the respiratory cycle. These movements will be absent if the lung has collapsed due to pneumothorax and only static A-lines will be visible. If the patient is examined in a supine position air in the pleural space rises to the anterior chest wall. Therefore, the investigation of the parasternal and mid-clavicular regions is particularly important to exclude a significant pneumothorax. If normal lung sliding is observed pneumothorax can be excluded at that interspace. However, if no lung sliding is observed this may be due to pneumothorax, but absent ventilation, pleural adhesions, bullae or pleurodesis may also cause loss of sliding. If the edge of the collapsed lung can be seen moving into and out of the scan beam, this is an additional sign of pneumothorax. The use of M-mode in this situation is extremely valuable, but its description is outside the scope of this chapter. (An M-mode diagnostic ultrasound presentation shows the temporal changes in echoes in which the depth of echo-producing interfaces is displayed along one axis and time is displayed along the second axis, recording motion of the interfaces toward and away from the transducer.)

23.3.4.4 Diagnostic and Therapeutic Implications

Chest ultrasound can be very useful for guiding drainage of fluid collections such as effusions or empyemas, which can provide immediate relief to the patient and aid in furthering the diagnostic process. A unilateral pleural effusion in a TB-endemic area is highly suggestive of TB, especially in HIV patients, and this finding can prompt initiation of anti-TB treatment (see Section 23.3.2). Observation of lung consolidation in a patient with signs of pneumonia will prompt antibiotic treatment. Interstitial syndromes need to be interpreted in light of the clinical picture. If generalized and suggestive of pulmonary edema, diuretic treatment should be initiated. Identification of a pneumothorax, especially in trauma patients, may warrant tube drainage.

23.3.4.5 Pearls and Pitfalls
- If no A-lines or B-lines are visualized initially, try sliding or angling the probe.
- Always confirm the position of the hemidiaphragm to ensure that any fluid seen lies in the chest.
- When searching for pneumonia or empyemas remember to scan the entire hemithorax, as they may be focal and in a non-dependent location.

23.3.5 Patients with Unilateral Leg Swelling: The Compression Deep Vein Thrombosis Scan

23.3.5.1 The Clinical Question
Bedside assessment of deep venous thrombosis (DVT) is one of the most useful POCUS applications as it is easy to perform, and has immediate therapeutic consequences (Heller et al. 2016). A DVT exam is indicated in patients with unilateral leg swelling or with signs of pulmonary embolism such as dyspnea or chest pain. Limited data is available on the prevalence of thromboembolic disease in resource-limited settings, but the few studies available suggest that it is an important problem, and that prophylaxis is underused. Patients with a number of diseases including cancer, stroke, nephrotic syndrome as well as infectious diseases such as HIV and TB, have an increased risk of thromboembolic disease (Heller et al. 2016).

23.3.5.2 Technique and Typical Ultrasound Findings
A simplified protocol to screen for DVT has been developed by emergency physicians and includes assessment of the common femoral vein in the groin and of the popliteal vein at the knee (Blaivas et al. 2000). A high frequency linear array probe is most suitable. The veins are scanned in a transverse plane and will appear as anechoic tubular structures (Figure 23.23).

Occasionally slow flow can create an impression of "smoke" in the lumen. This should not be mistaken for a clot. Thrombosis is confirmed when the vein contains hyperechoic clot and is not compressible with the probe, while applying enough pressure to deform the adjacent arterial wall (Figure 23.24).

In acute thrombosis the vein is generally distended, whereas chronic occlusion leads to narrowing of the occluded vein.

23.3.5.3 Diagnostic and Therapeutic Implications
The presence of DVT warrants immediate therapy with anticoagulants to prevent further thromboembolic complications. Heparin injection and if possible oral anticoagulation with warfarin should be initiated, although measurement of the international normalized ratio (INR) poses a problem in many settings. A negative scan result in a patient with leg swelling, without an obvious alternative cause, does

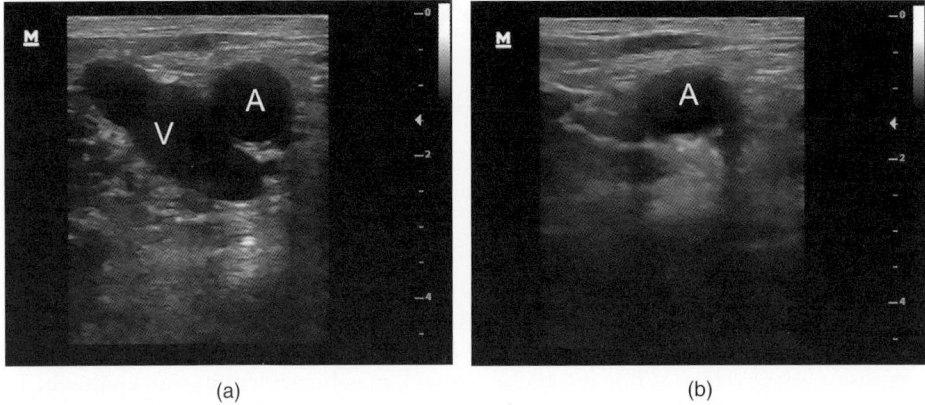

Figure 23.23 Transverse view of the left groin. Normal femoral vein (V) and artery (A): (a) Without compression: anechoic tubular structures representing the veins and the artery are visible, (b) once pressure is applied, a normal femoral vein collapses completely.

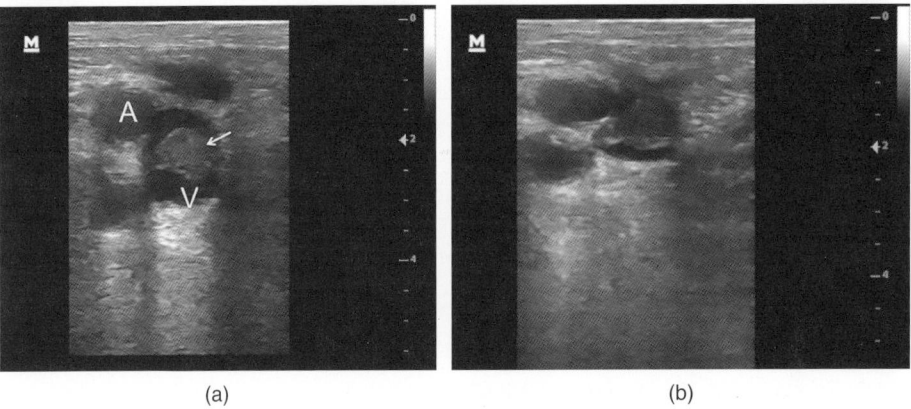

Figure 23.24 Transverse view of the right groin. DVT in the femoral vein: (a) Without compression: hyperechoic thrombus (arrow) is present in the vein (V), while the artery (A) shows normal anechoic fluid; (b) once pressure is applied, the vein does not collapse due to the intraluminal thrombus.

not exclude DVT and repeating the exam in five to seven days may detect a DVT which had originated from the deep calf veins and subsequently propagated to the larger veins (Blaivas et al. 2000).

23.3.5.4 Pearls and Pitfalls

- Let a supine patient rotate the leg externally to better visualize the common femoral vein.
- If possible, scan the entire popliteal vein in a prone position, with the ankle on a support to slightly flex the knee. This will make compression easier.
- Make sure that the veins fully collapse with the walls falling together in order to exclude DVT. If the vein compresses partially, repeat the maneuver and if it persists, consider DVT as a diagnosis.
- Lymph nodes in the groin and Baker's cyst in the popliteal fossa can mimic vessels in the transverse plane, but they can easily be identified by rotating the probe to a longitudinal view.

23.3.6 Patients with Hypotension of Unknown Cause: The RUSH Scan

23.3.6.1 The Clinical Question

Hypotension is a common finding in critically ill patients, and rapid determination of its cause is vital to ensure adequate resuscitation. Hypotension or shock can be broadly categorized into five types: (i) hypovolemic, as caused by hemorrhagic (for example, trauma or ectopic pregnancy) or non-hemorrhagic fluid loss (for example, profuse diarrhea); (ii) obstructive, for example due to pericardial effusion with tamponade, pulmonary embolism, or tension pneumothorax; (iii) cardiogenic, which may be cardiomyopathic (for example, myocardial infarction or postpartum cardiomyopathy); (iv) mechanical (for example, mitral regurgitation in rheumatic heart disease), or arrhythmic (for example, ventricular tachycardia); and (v) distributive, for example due to sepsis or anaphylaxis. While the cause of shock may be evident in some patients, in others with shock of unknown cause ultrasound examination can help guide the diagnosis, fluid management and other life-saving interventions such as pericardiocentesis (Seif et al. 2012; Shokoohi et al. 2015).

Technique and Typical Ultrasound Findings   The RUSH (Rapid Ultrasound in SHock) protocol provides a multisystem approach to the patient with shock of unknown cause, combining views from cardiac, chest, abdominal and DVT ultrasound.

23.3.6.2 Cardiac Status ("Pump")

The RUSH protocol incorporates a goal-directed echocardiogram to assess the presence of pericardial fluid and cardiac tamponade, left ventricular contractility, and right ventricle size (as described in Section 23.3.3). For the purpose of the RUSH examination, the subxiphoid view often provides sufficient information, although the apical four-chamber view is best suited to assess the relative dimensions of the left and right ventricles.

23.3.6.3 Intravascular Volume Status ("Tank")

Assessment of the intravascular volume status includes measurement of the IVC and detection of any free fluid, pulmonary edema or pneumothorax, as the latter will compromise the tank by reducing venous return to the heart. The IVC can be found by placing the probe in a subxiphoid view of the heart, turning it 90° clockwise to a longitudinal axis and sliding slightly to the right of the midline. The IVC can be differentiated from the aorta by its thinner walls and continuity with the right atrium. In transverse section it lies to the right of the aorta. The IVC diameter is measured 2 cm from the entry into the right atrium and normally has a diameter of ~1.5 cm. It partially collapses on normal or forceful inspiration or sniffing. Assessment of free fluid in the thorax and abdomen is described in more detail in Section 23.3.1. Ultrasound of the lung parenchyma is included to look for signs of pulmonary interstitial edema ("leakiness") and pneumothorax (Section 23.3.4).

23.3.6.4 Examination of the Vessels ("Pipes")

Examination of the vessels includes screening for abdominal aortic aneurysms (AAAs) and DVTs. The abdominal aorta should be examined along its entire course to rule out an aneurysm, paying special attention to the area below the renal arteries where most AAAs are located. The transducer is positioned perpendicular to the skin in the epigastrium and then slowly moved caudally to assess the aorta in the short axis. The diameter of the abdominal aorta should not exceed 3 cm. Finally, the patient is assessed for DVT (Section 23.3.5).

23.3.6.5 Pathological Findings

Cardiac Status ("Pump") Pericardial effusion usually shows as a black, anechoic rim around the heart. If this is present the heart should be evaluated for signs of tamponade. The first step of this evaluation involves looking at the IVC. If the IVC is dilated without respiratory collapse, obstructive shock and tamponade may be present. The next step is to focus on the movement of the right atrium and ventricle. Both can show collapse during diastole and impaired filling in patients with tamponade (Figure 23.25).

It has to be remembered that the size of the pericardial effusion cannot determine tamponade. As collagen fibers of the pericardium are tight and need time to adjust, rapid development of 50–100 mL of fluid can cause tamponade, whereas during slow accumulation several hundred mL can be tolerated. Further assessment of the "Pump" follows the information described in Section 23.3.3.

23.3.6.6 Intravascular Volume Status ("Tank")

One gauge of the "Tank" is the IVC as an indicator of the venous preload (Figure 23.26).

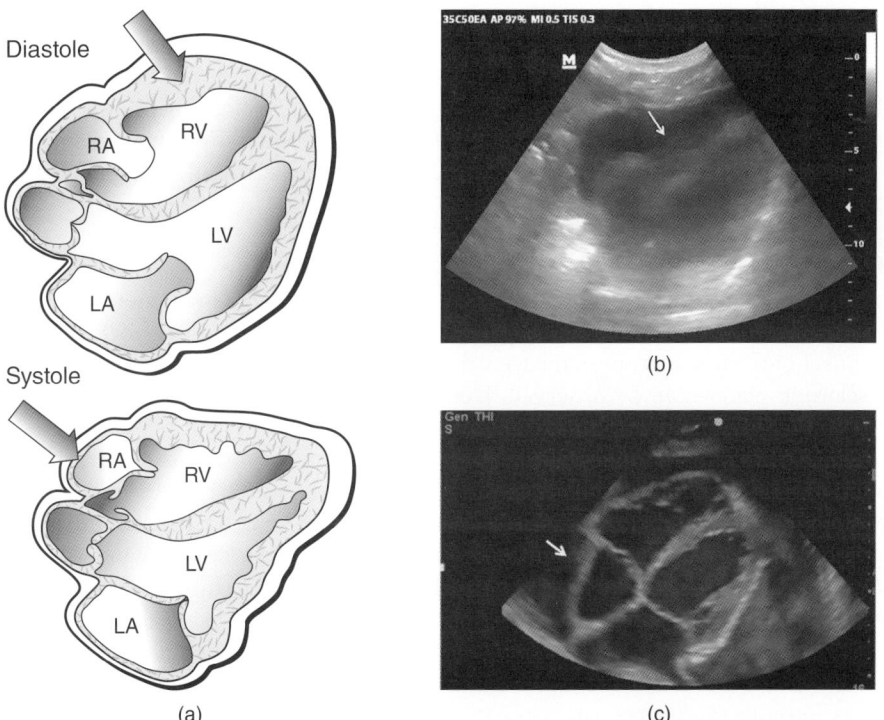

Figure 23.25 Cardiac tamponade: (a) Tamponade is characterized by impaired movement of the right ventricle and/or the right atrium due to increased pressure in the pericardial sac, (b) Sub-xiphoid view of impaired movement with compression of the right ventricle (arrow), and (c) the right atrium (arrow).

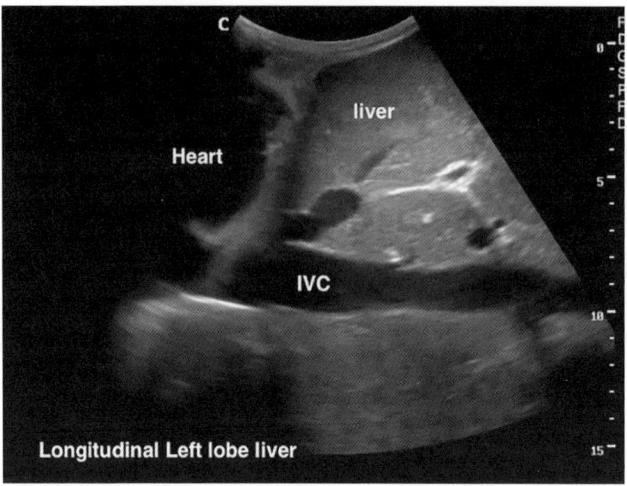

Figure 23.26 Inferior vena cava on a longitudinal view. Normally the size is between 1.5–2.5 cm and a respiratory variation in size can be seen during scanning.

Assessment of the IVC includes size and diameter variation during the respiratory cycle. A small IVC (<1.5 cm) with possible collapse during inspiration, especially in combination with a hyperdynamic left ventricle, suggests hypovolemia. A dilated IVC (>2.5 cm) with absent size variation during respiration suggest hypervolemia. Free fluid in the thorax, such as hemothorax after trauma or surgery, or in the abdomen, such as that due to ectopic pregnancy or abdominal sepsis, may suggest bleeding or infection as a cause of hypovolemia. Increased numbers of B-lines on lung ultrasound can suggest pulmonary edema and hypervolemia (Section 23.3.4). Absence of lung sliding may indicate pneumothorax.

23.3.6.7 Examination of the Vessels ("Pipes")
An aortic aneurysm shows up as dilatation of the aorta above 3 cm in diameter, and the risk of rupture increases substantially when it is larger than 5 cm. The aortic lumen can be anechoic, but sometimes an eccentric hyperechoic thrombus (Figure 23.27) or atherosclerotic plaque can be observed. Deep venous thrombosis can be assessed as described above in Section 23.3.5.

23.3.6.8 Diagnostic and Therapeutic Implications
The RUSH protocol aims to facilitate emergency management of a patient with shock. In a patient with tamponade emergency pericardiocentesis can be lifesaving. Ideally, this should be performed by an experienced physician under electrocardiographic and echocardiographic control with a dedicated pericardiocentesis set using a subxiphoidal approach. However, in resource-limited settings an ultrasound-guided intercostal approach with a simple cannula can be used (Heller et al. 2010b). Notably, the aim is to relieve intrapericardial pressure and improve cardiac output, for which draining 50–100 mL of pericardial fluid can be sufficient (Heller et al. 2010b).

Other therapeutic decisions will depend on the findings. Patients with a small IVC and a hyperdynamic heart will profit from volume substitution, either as crystalloid fluid or blood depending on the underlying cause. Patients with a dilated IVC and

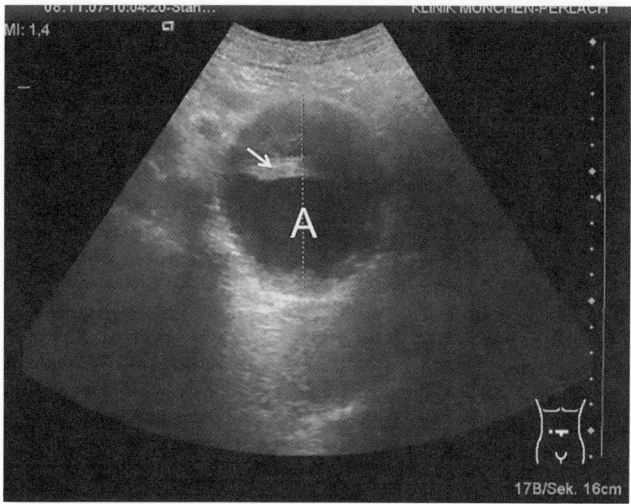

Figure 23.27 Transverse view of the abdominal aorta (A) showing an increased diameter (7 cm) and echogenic thrombotic material (arrow) especially toward the anterior wall.

possibly signs of pulmonary edema or reduced LV function can be treated with diuretics and inotropes (if available). The patient with an enlarged right ventricle and DVT will benefit from immediate initiation of anticoagulant therapy. Patients with suspected hemothorax, intra-abdominal hemorrhage or signs of an aortic aneurysm will need urgent referral for surgical or gynecological review. Acute respiratory distress and absence of lung sliding indicates pneumothorax and needs to be treated with emergency thoracostomy.

23.3.6.9 Pearls and Pitfalls
- Measure the IVC in two planes (longitudinal and transverse), as a single longitudinal measurement may be off-center, incorrectly underestimating the diameter.
- If bowel gas is obstructing the view of the abdominal aorta, apply pressure, let the patient bend his/her knees, jiggle the probe or try imaging from a different angle.
- Combine images from several different organ systems to develop a "big picture" of the primary pathology contributing to the patient's shock state.

23.3.7 Patient with Flank Pain: The Renal Scan

23.3.7.1 The Clinical Question
Renal and urinary tract disease is common in adults and children throughout all continents (Burdmann and Jha 2017). Protocols for renal and bladder point-of-care ultrasound examination have been developed for patients with flank pain or with abdominal pain and urinary symptoms to look for hydronephrosis, distended bladder and stone disease (van Hoving 2014; International Federation for Emergency Medicine 2014). The etiologic spectrum leading to urinary tract obstruction is broad. Common causes of urinary tract obstruction in resource-limited settings include external compression of the ureters from lymphadenopathy due to malignancy, tuberculosis or other masses, as well as obstruction of the ureters from stone disease and chronic infections such as

urinary schistosomiasis. Stone disease or urinary tract calcification may be a consequence of TB, arise from metabolic conditions or from side effects of drugs such as the protease inhibitor atazanavir (Rockwood et al. 2011). Bladder POCUS commonly assesses bladder volume for evaluation of urinary retention.

In a tropical endemic setting ultrasound can be used to visualize bladder and ureter pathology that arises from infection by *Schistosomia haematobium*, and which is a potential cause of hydronephrosis. More than 20 years ago the WHO published a detailed guideline on the evaluation of schistosomiasis (WHO, Special Programme for Research & Training in Tropical Diseases 1996). Simplifications to these guidelines have been suggested in order to facilitate ultrasound examination by less skilled examiners (Akpata et al. 2015).

23.3.7.2 Technique and Typical Ultrasound Findings

The Kidneys   The kidneys are retroperitoneal organs which lie below the liver on the right side and below the spleen on the left side. The basic architecture of the kidney comprises the cortex, the pyramids, and the renal sinus/pelvis. Renal ultrasound is usually performed with the patient in the supine position using a curvilinear probe. Both kidneys are evaluated in longitudinal and transverse views. The right kidney is assessed by placing the probe in the mid-axillary line in the lower intercostal space using the liver as an acoustic window. The left kidney is assessed by placing the probe in the posterior axillary line in the lower intercostal space. To obtain the longitudinal view, visualizing the longest craniocaudal diameter of the kidney, the probe may need to be slightly rotated. To obtain the transverse view the probe is then turned counterclockwise 90°; the probe is then moved and/or angled up and down to visualize the renal hilum and to assess the renal pelvis for dilatation. In a normal image the renal cortex and pyramids will appear hypoechoic; the renal sinus area will appear echogenic. Calyces of the collecting system surround the pyramids, but are not visible in a normal kidney. The proximal ureters are not visualized by ultrasound unless they are distended.

23.3.7.3 The Bladder

The bladder is assessed in transverse and longitudinal views. For the transverse view the probe is placed horizontally above the pubic symphysis. Transverse sweeps through the bladder are performed to assess its shape and wall thickness, as well as the distal ureters. For the longitudinal view the probe is turned to the midline. A normal fully distended bladder has a regular, rectangular shape with a wall not thicker than 5 mm. Normal distal ureters are not visible. Bladder volume can be measured by using the transverse and longitudinal images to obtain height, width and depth. Once the three dimensions are measured the formula $H \times W \times D \times 0.7$ is used to calculate volume. If the bladder is not fully distended, its wall cannot be assessed reliably.

23.3.7.4 Pathological Findings

Hydronephrosis   In hydronephrosis a dilated renal pelvis presents as an increased anechoic space at the center of the kidney (Figure 23.28) which is continuous with dilated calyces.

Categorization into different degrees of hydronephrosis, depending on the extent of dilation, is commonly practiced. In kidneys with multiple parapelvic cysts, care should be taken not to confuse the cysts with a hydronephrosis.

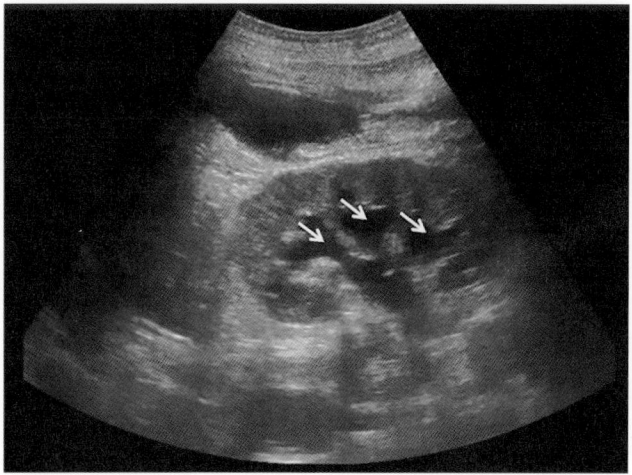

Figure 23.28 Hydronephrosis: dilatation of the urinary tract can be seen as an anechoic clover-leaf structure in the kidney (arrows).

23.3.7.5 Stone Disease/Calcifications

Calcifications and stones appear as hyperechoic structures with a posterior acoustic shadow. Stones at the pelvi-ureteric junction will often show a dilated renal pelvis with a collapsed proximal ureter. Stones within the ureter are rarely seen but the resulting hydronephrosis can be easily detected. Stones at the ureterovesical junction can be seen as echogenic structures behind the bladder wall at the orifice of the ureters.

23.3.7.6 Bladder

The most common pathology found in the bladder is bladder distention. This is characterized by an enlarged bladder which may reach up to the mid-abdomen. It can be secondary to bladder outlet obstruction or to neurological disease.

23.3.7.7 Schistosomiasis

Schistosomiasis-related urinary pathologies include a rounded or irregular shape of the bladder, bladder wall thickening with diffuse or focal thickening of >5 mm, bladder wall calcifications, and masses or pseudopolyps protruding into the bladder lumen (Figure 23.29). The distal ureters are considered pathological when thickened or dilated.

23.3.7.8 Diagnostic and Therapeutic Implications

Visualization of urinary tract obstruction should prompt further investigations on the underlying etiology, which may guide therapeutic management. A distended bladder is easily identified and can be treated by insertion of a Foley's catheter.[5]

Management of stone disease often relies on analgesia and hydration. Sonographic features suggestive of schistosomiasis are not pathognomonic, but in a high endemic area should prompt anti-helminthic treatment, which prevents further deterioration and may lead to improvement of sonographic findings.

---

5 Frederic Foley was an American urologist (1891–1966).

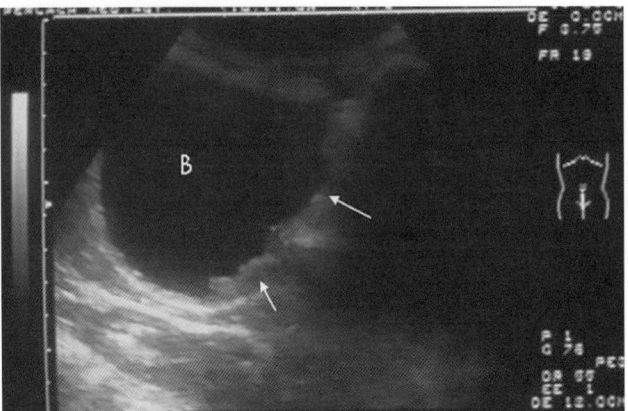

Figure 23.29 Schistosomiasis: irregular thickening (arrows) of the bladder (B) wall due to polypoid protrusions.

23.3.7.9   Pearls and Pitfalls

- Stones in the ureters can be easily overlooked as a cause of obstruction.
- Hydronephrosis may be absent in dehydration or early obstruction.
- Not all hydronephrosis is due to obstruction. In children especially, reflux is a common cause.
- Back pressure from a distended bladder may lead to physiological dilation of the renal collecting system. If hydronephrosis is identified in the presence of a full bladder, a repeat scan 20 minutes after voiding is required to confirm persistent hydronephrosis.
- Right sided dilation is commonly seen in pregnancy and may or may not be symptomatic.

23.3.8   Female Patient of Childbearing Age with Pelvic Pain or Vaginal Bleeding: The Ectopic Pregnancy (EP) Scan

23.3.8.1   The Clinical Question

In women of childbearing age who present with pelvic pain and/or vaginal bleeding, with or without hypotension, ectopic pregnancy must always be excluded. Ectopic pregnancy often presents as an emergency and is a potentially life-threatening event. Pelvic inflammatory disease associated with sexually transmitted diseases is the most important risk factor for ectopic pregnancy in LMICs (Goyaux et al. 2003). In African resource-limited countries, a majority of hospital-based studies have reported ectopic pregnancy case fatality rates of around 1–3%, 10 times higher than that reported in industrialized countries. Late diagnosis, leading in almost all cases to major complications, and delayed emergency surgical treatments are key elements accounting for such high fatality rates.

POCUS as a diagnostic test for ectopic pregnancy provides excellent sensitivity and negative predictive value; visualization of an intrauterine pregnancy (IUP) is generally sufficient to rule out ectopic pregnancy (Stein et al. 2010). Therefore, first trimester POCUS protocols for women of childbearing age with signs of ectopic pregnancy are established in emergency medicine curricula in many countries (IFEM, http://www .ifem.cc/wp-content/uploads/2016/03/IFEM-Point-of-Care-Ultrasound-Curriculum-Guidelines-2014-2.pdf). The aim of first trimester POCUS is to definitively rule in IUP and thereby rule out ectopic pregnancy.

23.3.8.2 Technique and Typical Ultrasound Findings

Transabdominal scans are performed with a 3.5–5 MHz curvilinear probe. If a transabdominal scan does not provide all the necessary information, it should be complemented by a transvaginal scan if available. Ideally the patient should have a full bladder. The probe is located superior to the pubic symphysis and longitudinal and transverse views need to be obtained.

The longitudinal view visualizes the bladder, the uterus in the craniocaudal axis, and the vaginal stripe. In most patients the uterus will be anteflected. The transverse view visualizes the bladder, the uterus in an axial view, and lateral to the uterus the ovaries. In order to assess the whole uterus in the axial view the beam needs to angle craniocaudally. The ovaries can be difficult to identify and the fallopian tubes are rarely seen, unless pathological.

The following three criteria establish the diagnosis of a definitive IUP if visualized inside the uterus:

1) Yolk sac: the yolk sac is of 5–6 mm in diameter located within a bigger sac called the gestational sac. Both the gestational sac and the yolk sac are anechoic, round structures with an echogenic rim. The yolk sac becomes visible six to seven weeks after the last menstrual period (LMP) (Table 23.4) and regresses and disappears by 12 weeks of gestation. To ensure that the gestational sac with the yolk sac are located within the uterus, the myometrial mantle should be clearly visualized.
2) Fetal pole: the fetal pole becomes visible by seven weeks as an echogenic structure between the yolk sac and gestational sac.
3) Intrauterine heartbeat: by the end of the seventh gestational week the embryo has a size of 5–10 mm and cardiac motion should become visible.

23.3.8.3 Pathological Findings

Pathological findings on transabdominal sonography are the failure to identify the above-mentioned criteria of intrauterine pregnancy in combination with a positive βHCG-pregnancy test. Failure to identify a definitive IUP (Figure 23.30) by transabdominal ultrasound in patients with a positive βHCG test is either due to a very early normal IUP, an abnormally developing IUP, an active or complete miscarriage, or an ectopic pregnancy.

Table 23.4 Correlation of transabdominal sonography, βHCG level and approximate gestational age.

Embryonic parameter	Gestational age (weeks)	βHCG level	Time first detectable by ultrasound examination (weeks of gestation)
Gestational sac	5.5–6	1700–6000	4.5
Yolk sac	6–6.5	8000–15 000	6–7
Fetal pole	7	13 000–15 000	7
Cardiac activity	7	16 000–25 000	7
Fetal parts	>8	29 000–39 000	5.5

Source: Adapted from Noble and Nelson (2011).

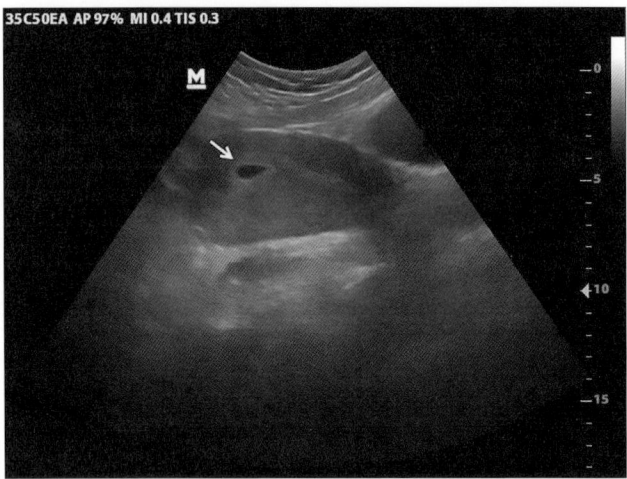

Figure 23.30 Uterus with a small amount of intrauterine fluid (arrow). No fetal pole or yolk sack was identified, which in presence of a positive ßHCG -pregnancy test suggests extra-uterine pregnancy.

In all cases the patient should be evaluated further. Free fluid may be present in ectopic pregnancy in the pouch of Douglas (Figure 23.10), especially as a sign of rupture. If no IUP is seen a FAST exam should be performed.

23.3.8.4 Diagnostic and Therapeutic Implications
If findings are unclear or in the case of pathologic findings, a formal obstetric consultation and a transvaginal scan (if available) is warranted. In the case of unstable patients, emergency management is necessary, including timely surgical intervention.

23.3.8.5 Pearls and Pitfalls
- The goal of a first trimester POCUS in women of childbearing age with pelvic complaints is to confirm an IUP and to EXCLUDE an ectopic pregnancy; if you cannot confirm an IUP, be highly suspicious of an ectopic pregnancy.
- Pelvic structures are often not midline, scan from right to left to find the best plane.
- Identifying the gestational sac alone is not confirmation of a normal IUP gestation; it can also be present in an ectopic pregnancy. A yolk sac or the fetus must be seen within the gestational sac to diagnose IUP.
- Beware of the limitations of transabdominal ultrasound in early pregnancy. Transabdominal ultrasound is less sensitive than transvaginal ultrasound.
- Patients with hypotension have an increased suspicion for ruptured ectopic pregnancy.
- Look for free fluid not only in the pouch of Douglas, and also perform a FAST examination.

23.3.9 Patient with Soft Tissue Pain or Swelling: The Soft Tissue Scan

23.3.9.1 The Clinical Question
The skin, subcutaneous tissues and many parts of the musculoskeletal system are relatively superficial structures and ideal targets for ultrasound examination

(Chen et al. 2016). Patients with soft tissue infection may have fever, chills and leukocytosis in addition to redness, local heat, and swelling at the infected sites. Cellulitis is the most common type of soft tissue infection and is confined within the subcutaneous compartment (Chen et al. 2016). An abscess is a collection of pus within tissue and develops as the primary site of infection or may arise as a complication from cellulitis; a painful red mass is usually present; it may be tender, fluctuating and is often warm. The differentiation between an abscess and cellulitis is important as management differs (incision and drainage versus antibiotics); bedside ultrasound is an option to differentiate these two conditions in the absence of other imaging modalities. POCUS has been shown to alter patient management in up to half of patients with suspected abscesses (Tayal et al. 2006).

23.3.9.2 Technique and Typical Ultrasound Findings

For sonographic assessment of superficial structures, a high frequency linear transducer is used. For assessment of deeper structures, a curvilinear transducer with lower frequency is required. The location of interest should always be scanned in two planes and should be compared to the healthy side if possible. The epidermis and dermis appear as thin hyperechoic layers. The subcutaneous fat layer usually appears hypoechoic and nodular with hyperechoic linear septa. Striated muscles are hypoechoic with hyperechoic striation, best seen in the long axis of the muscles (Figure 23.31).

23.3.9.3 Pathological Findings

In cellulitis the dermis appears thick and bright, has blurred tissue margins and may show a characteristic cobblestone pattern; a hyperechoic, hyperemic pattern of the inflamed subcutaneous fat intersected by anechoic fluid tracking along the connective tissue is seen (Figure 23.32).

Abscesses are usually anechoic or hypoechoic focal areas with posterior acoustic enhancement (Figure 23.33).

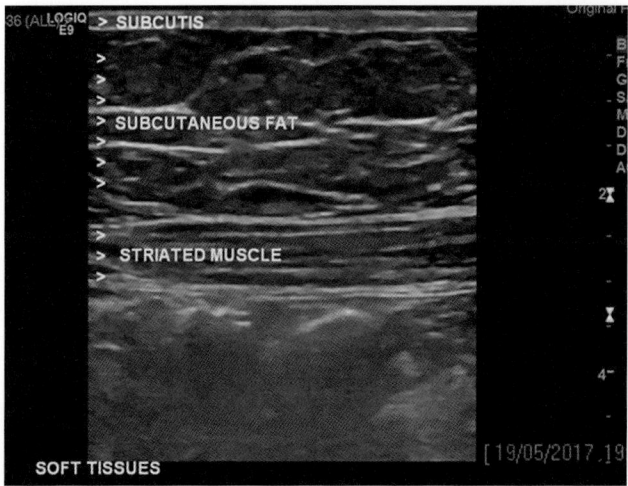

Figure 23.31 Soft tissue ultrasound showing the echogenic cutis in the near field, the hypoechoic subcutaneous fat and the parallel fibers of the muscle.

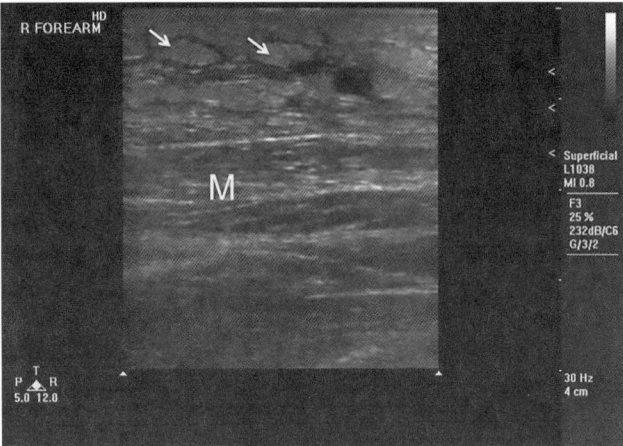

Figure 23.32 Cellulitis with cobblestone pattern. In the subcutaneous layers echogenic "islands" (arrows) surrounded by hypoechoic edema are seen (M = muscle).

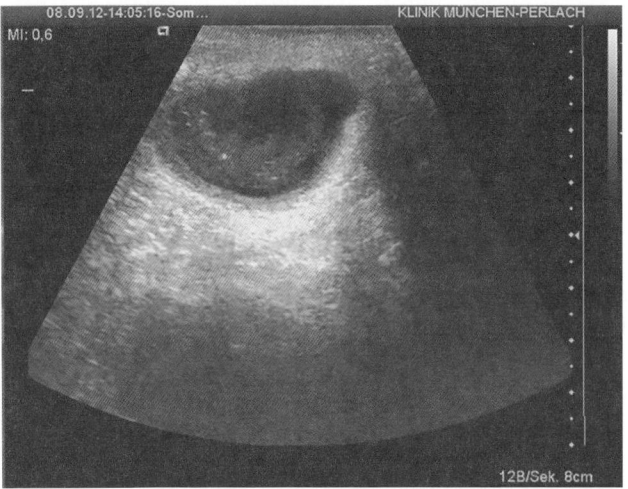

Figure 23.33 Subcutaneous abscess: hypoechoic well-defined focal area with acoustic enhancement. Ultrasound-guided aspiration for microbiological diagnosis is easily possible.

They may also contain mixed internal echogenicity, including hyperechoic foci with posterior shadowing, representing gas inclusions (Figure 23.34).

The margins have a rather irregular shape and in complex abscesses different layers may be distinguishable. The "squish sign" is a movement of echogenic particles in response to compression and can be used to differentiate a liquefied abscess from a soft tissue mass. Differential diagnoses of cellulitis should include deep vein thrombosis (see Section 23.3.5) and edema which may also show a cobblestone pattern. Differential diagnoses of abscesses in the groin should include hernias and enlarged lymph nodes. The former may sometimes show peristalsis in herniated bowel loops. Reactive lymph nodes typically show a bright hilum, discrete borders and no posterior

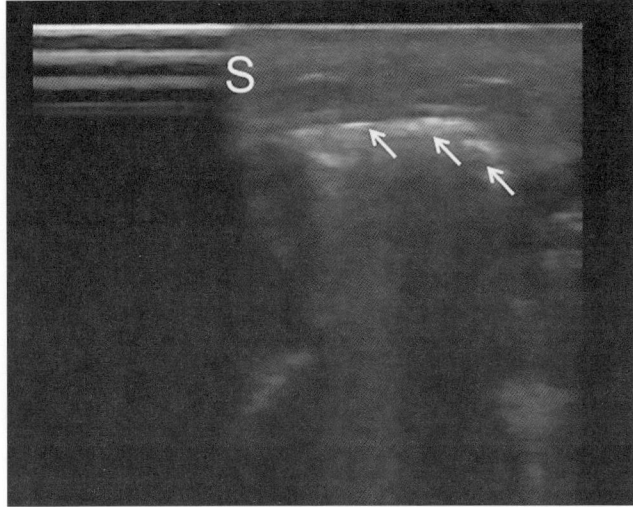

Figure 23.34 Subcutaneous infection with gas formation: in the subcutaneous layers (S) echogenic gas (arrows) in the tissue is visible. The shadow caused by gas is typically white and often referred to as "dirty" shadow.

acoustic enhancement on ultrasound. Necrotic malignant or tuberculous nodes, however, have a very similar appearance to abscesses (Figure 23.35).

23.3.9.4 Diagnostic and Therapeutic Implications

Uncomplicated cellulitis is treated with antibiotics. Uncomplicated and accessible abscesses are usually managed by blind incision and drainage (Marx 2014). POCUS can be used to diagnose occult abscesses as well as to decide on the safest route for incision and drainage.

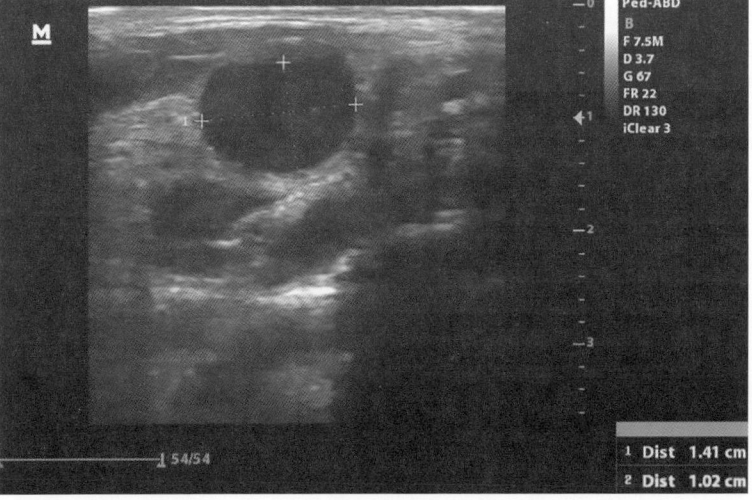

Figure 23.35 Lymph node: hypoechoic round lymph node in the region of the neck.

23.3.9.5   Pearls and Pitfall
- In case of doubt compare the area of interest to the contralateral normal area.

23.4   Considerations for Teaching and Implementation

Ultrasound examination is a highly operator-dependent tool and in the absence of significant competence its performance can put patients at significant risk of harm. For a basic range of ultrasound skills, the WHO recommends that physicians should undergo training for three–six months, including 300–500 ultrasound examinations. However, such training programs are too time- consuming to allow wide application by clinicians in low resource settings. Training should be targeted to healthcare professionals working on the front line, which in many cases will be a general physician in a district hospital or clinic, with little or no experience of imaging technology, but also little time to be away from his or her clinical duties. Focused POCUS applications do not require as much training and thus allow a degree of training flexibility. Physicians can be trained in short modules, which are easier to combine with day-to-day duties (Maru et al. 2010).

A variety of associations including the American College of Emergency Physicians (ACEP 2008) and the European Federation of Societies for Ultrasound in Medicine and Biology (EFSUMB 2009) endorse this approach of focused courses and a section of the International Federation for Emergency Medicine (IFEM) has recently developed a useful consensus document drawing on regional and national guidelines on standards and training requirements, to ensure an appropriate level of training (IFEM 2014).

Based on these recommendations, further training schedules should be developed for indications outside emergency medicine to accelerate the spread of the technology. Defining POCUS curricula appropriate to specific low resource settings will be important for targeted allocation of training resources. A suggested method to identify the most useful topics in a local curriculum is to assign weights to three components - prevalence of disease, diagnostic impact of the scan and technical difficulty - for each of the POCUS applications (Table 23.5). Although this multifactorial weighting model has not yet been validated and the input of the weights is subjective in itself, it is easy to apply.

As face-to-face POCUS training modules are by definition short and the trainer is usually only available for a short period of time, methods to ensure continuous support and quality assurance need to be considered when designing or implementing training. With increasing availability of internet access, the use of telemedicine solutions can be an important component to support trainees in areas where direct supervision is absent

Table 23.5 Weighting of prevalence, diagnostic impact and difficulty of POCUS applications.

Weighting	Disease prevalence	Diagnostic impact of ultrasound examination	Degree of difficulty of ultrasound examination
1	Rare	Minor or no treatment change	Advanced
2	Relatively common	Treatment change	Moderate
3	Very common	Urgent treatment change (possibly life-threatening)	Easy

(Janssen et al. 2013). Less secure, but widely available in the resource-limited setting, is the transmission of ultrasound clips via open access data-sharing platforms in order to obtain second opinions. While technically easy and fast, patient confidentiality is a potential issue and more secure platforms should be used whenever possible.

A number of introductory texts covering the use of ultrasound in the resource-limited setting and in infectious diseases exist (Partners in Health 2011; WHO 2011). They are aimed at physicians with limited ultrasound experience working in resource-limited settings and are available free or at low-cost. The further development of standardized training guidelines, for indications other than emergency medicine, based on previous experience and on training material available is a key priority in the field. This will allow much needed wider dissemination of POCUS, while maintaining operator competency.

23.5 Conclusions

POCUS is intended to rapidly assess patients when access to other diagnostic imaging modalities is limited. What makes POCUS unique is that the clinician is acquiring and interpreting the images and then directly applying them to diagnosis and treatment of the patient at the bedside. It is this integrated application combining a clinician's knowledge, experience and direct interaction with the patient that provides POCUS with its true power.

Bibliography

ACEP (2008). Abstracts of the American College of Emergency Physicians (ACEP) Research Forum Scientific Assembly. August 27–28. Chicago, Illinois. *Ann. Emerg. Med.* 52 (4 Suppl): S41–S172. https://doi.org/10.1016/j.annemergmed.2008.06.066.

Akpata, R., Neumayr, A., Holtfreter, M.C. et al. (2015). Erratum to: the WHO ultrasonography protocol for assessing morbidity due to Schistosoma haematobium. Acceptance and evolution over 14 years. Systematic review. *Parasitol. Res.* 114 (5): 2045–2046.

Blaivas, M., Lambert, M.J., Harwood, R.A. et al. (2000). Lower-extremity Doppler for deep venous thrombosis--can emergency physicians be accurate and fast? *Acad. Emerg. Med.* 7 (2): 120–126.

Burdmann, E.A. and Jha, V. (2017). Acute kidney injury due to tropical infectious diseases and animal venoms: a tale of 2 continents. *Kidney Int.* 91 (5): 1033–1046.

Chen, K.C., Lin, A.C., Chong, C.F., and Wang, T.L. (2016). An overview of point-of-care ultrasound for soft tissue and musculoskeletal applications in the emergency department. *J. Intensive Care* 4: 55.

Cuevas, L.E. (2011). The urgent need for new diagnostics for symptomatic tuberculosis in children. *Indian J. Pediatr.* 78 (4): 449–455.

EFSUMB (2009). European Federation of Societies for Ultrasound in Medicine and Biology 2009. Building a European ultrasound community.1 Appendix 13: Intensive care ultrasound minimum training requirements for the practice of medical ultrasound in Europe. http://www.efsumb.org/guidelines/2009-07-06apx13.pdf

Goyaux, N., Leke, R., Keita, N., and Thonneau, P. (2003). Ectopic pregnancy in African developing countries. *Acta Obstet. Gynecol. Scand.* 82 (4): 305–312.

Grimaldi, A., Mocumbi, A.O., Freers, J. et al. (2016). Tropical endomyocardial fibrosis: natural history, challenges, and perspectives. *Circulation* 133 (24): 2503–2515.

Halbfass, H.J., Wimmer, B., Hauenstein, K., and Zavisic, D. (1981). Ultrasonic diagnosis of blunt abdominal injuries. *Fortschr. Med.* 99 (41): 1681–1685.

Heller, T., Goblirsch, S., Wallrauch, C. et al. (2010a). Abdominal tuberculosis: sonographic diagnosis and treatment response in HIV-positive adults in rural South Africa. *Int. J. Infect. Dis.* 14 (Suppl 3): e108–e112.

Heller, T., Lessells, R.J., Wallrauch, C., and Brunetti, E. (2010b). Tuberculosis pericarditis with cardiac tamponade: management in the resource-limited setting. *Am. J. Trop. Med. Hyg.* 83 (6): 1311–1314.

Heller, T., Wallrauch, C., Goblirsch, S., and Brunetti, E. (2012). Focused assessment with sonography for HIV-associated tuberculosis (FASH): a short protocol and a pictorial review. *Crit. Ultrasound J.* 4 (1): 21.

Heller, T., Goblirsch, S., Bahlas, S. et al. (2013). Diagnostic value of FASH ultrasound and chest X-ray in HIV-co-infected patients with abdominal tuberculosis. *Int. J. Tuberc. Lung Dis.* 17 (3): 342–344.

Heller, T., Wallrauch, C., Brunetti, E., and Giordani, M.T. (2014). Changes of FASH ultrasound findings in TB-HIV patients during anti-tuberculosis treatment. *Int. J. Tuberc. Lung Dis.* 18 (7): 837–839.

Heller, T., Mtemang'ombe, E.A., Huson, M.A. et al. (2016). Ultrasound for patients in a high HIV/TB prevalence setting – a needs assessment and review of focused applications for sub-Saharan Africa. *Int. J. Infect. Dis.* 56: 229–236.

Hofman, K., Primack, A., Keusch, G., and Hrynkow, S. (2005). Addressing the growing burden of trauma and injury in low- and middle-income countries. *Am. J. Public Health* 95 (1): 13–17.

van Hoving, D.J., Lamprecht, H.H., Stander, M. et al. (2014). Adequacy of the emergency point-of-care ultrasound core curriculum for the local burden of disease in South Africa. *Emerg. Med. J.* 30 (4): 312–315.

International Federation for Emergency Medicine (2014). Point-of-Care Ultrasound Curriculum Guidelines, *Volucella* 2017.

Janssen, S., Grobusch, M.P., and Heller, T. (2013). 'Remote FASH' tele-sonography – a novel tool to assist diagnosing HIV-associated extrapulmonary tuberculosis in remote areas. *Acta Trop.* 127 (1): 53–55.

Jensen, M.B., Sloth, E., Larsen, K.M., and Schmidt, M.B. (2004). Transthoracic echocardiography for cardiopulmonary monitoring in intensive care. *Eur. J. Anaesthesiol.* 21 (9): 700–707.

Lichtenstein, D.A. (2015). BLUE-protocol and FALLS-protocol: two applications of lung ultrasound in the critically ill. *Chest* 147 (6): 1659–1670.

Luzze, H., Elliott, A.M., Joloba, M.L. et al. (2001). Evaluation of suspected tuberculous pleurisy: clinical and diagnostic findings in HIV-1-positive and HIV-negative adults in Uganda. *Int. J. Tuberc. Lung. Dis.* 5 (8): 746–753.

Ma, O.J. and Mateer, J.R. (1997). Trauma ultrasound examination versus chest radiography in the detection of hemothorax. *Ann. Emerg. Med.* 29 (3): 312–315.

Ma, O.J., Mateer, J.R., Ogata, M. et al. (1995). Prospective analysis of a rapid trauma ultrasound examination performed by emergency physicians. *J. Trauma* 38 (6): 879–885.

Maru, D.S., Schwarz, R., Jason, A. et al. (2010). Turning a blind eye: the mobilization of radiology services in resource-poor regions. *Glob. Health* 6: 18.

Marx, J.A. (2014). Skin and soft tissue infections. In: *Rosen's Emergency Medicine: Concepts and Clinical Practice*, 8e (ed. R. Walls and R. Hockberger), 1851–1863. Elsevier/Saunders.

Mok, K.L. (2016). Make it SIMPLE: enhanced shock management by focused cardiac ultrasound. *J. Intensive Care* 4: 51.

Moore, C.L. and Copel, J.A. (2011). Point-of-care ultrasonography. *N. Engl. J. Med.* 364 (8): 749–757.

Noble, V.E. and Nelson, B.P. (2011). *Manual of Emergency and Critical Care Ultrasound.* Cambridge: Cambridge University Press.

Partners in Health (2011). Manual of Ultrasound for Resource-Limited Settings.

Plummer, D., Brunette, D., Asinger, R., and Ruiz, E. (1992). Emergency department echocardiography improves outcome in penetrating cardiac injury. *Ann. Emerg. Med.* 21 (6): 709–712.

Ponikowski, P., Voorts, A.A., Anker, S.D. et al. (2016). 2016 ESC guidelines for the diagnosis and treatment of acute and chronic heart failure. *Rev. Esp. Cardiol.* 69 (12): 1167.

Reuter, H., Burgess, L.J., Louw, V.J., and Doubell, A.F. (2007). The management of tuberculous pericardial effusion: experience in 233 consecutive patients. *Cardiovasc. J. S. Afr.* 18 (1): 20–25.

Rockwood, N., Mandalia, S., Bower, M. et al. (2011). Ritonavir-boosted atazanavir exposure is associated with an increased rate of renal stones compared with efavirenz, ritonavir-boosted lopinavir and ritonavirboosted darunavir. *AIDS* 25: 1671–1673.

Seif, D., Perera, P., Mailhot, T. et al. (2012). Bedside ultrasound in resuscitation and the rapid ultrasound in shock protocol. *Crit. Care Res. Pract.* 2012: 503254.

Shokoohi, H., Boniface, K.S., Pourmand, A. et al. (2015). Bedside ultrasound reduces diagnostic uncertainty and guides resuscitation in patients with undifferentiated hypotension. *Crit. Care Med.* 43 (12): 2562–2569.

Stein, J.C., Wang, R., Adler, N. et al. (2010). Emergency physician ultrasonography for evaluating patients at risk for ectopic pregnancy: a meta-analysis. *Ann. Emerg. Med.* 56 (6): 674–683.

Tayal, V.S., Hasan, N., Norton, H.J., and Tomaszewski, C.A. (2006). The effect of soft-tissue ultrasound on the management of cellulitis in the emergency department. *Acad. Emerg. Med.* 13 (4): 384–388.

World Health Organization (2006). Improving the Diagnosis and Treatment of Smear-Negative Pulmonary and Extrapulmonary Tuberculosis among Adults and Adolescents: Recommendations for HIV-Prevalent and Resource-Constrained Settings.

World Health Organization (2011). WHO manual of diagnostic ultrasound.

World Health Organization (1996), Special Programme for Research & Training in Tropical Diseases. Ultrasound in schistosomiasis: a practical guide to the standard use of ultrasonography for assessment of schistosomiasis-related morbidity.

World Health Organization (2002). Injuries, and Violence Prevention Department. The injury chart book: A graphical overview of the global burden of injuries.

Zhang, M., Liu, Z.-H., Yang, J.-X. et al. (2006). Rapid detection of pneumothorax by ultrasonography in patients with multiple trauma. *Crit. Care* 10 (4): R112.

Webliography

http://www.ifem.cc/wp-content/uploads/2016/03/IFEM-Point-of-Care-Ultrasound-Curriculum-Guidelines-2014-2.pdf

24

The Use of Obstetric Ultrasound in Low Resource Settings

Helen Allott

The Liverpool School of Tropical Medicine, Liverpool, UK

24.1   Introduction

Given the extent to which ultrasound has become integrated into everyday obstetric practice in high-income settings, it seems a reasonable assumption that the introduction of obstetric ultrasound to resource-limited settings is likely to be of substantial benefit. However, the evidence of benefit from such an intervention merits rigorous examination, particularly where difficult choices regarding the deployment of scarce financial and human resources must be made. An example of a normal ultrasound image of a 16 week old human fetus is shown in Figure 24.1.

24.2   Pregnancy-Related Problems for Which Portable Ultrasound may be Useful

Many pregnancy-related problems and abnormalities might be detected by the provision of routine ultrasound services (Table 24.1). For example, some women are unsure of their last menstrual period dates, or may not have had any menses in the period prior to conception due to breastfeeding or the use of depot contraceptive preparations or other factors. In this circumstance, an ultrasound scan prior to 20 weeks of gestation is

Revolutionizing Tropical Medicine: Point-of-Care Tests, New Imaging Technologies and Digital Health,
First Edition. Edited by Kerry Atkinson and David Mabey.
© 2019 John Wiley & Sons, Inc. Published 2019 by John Wiley & Sons, Inc.

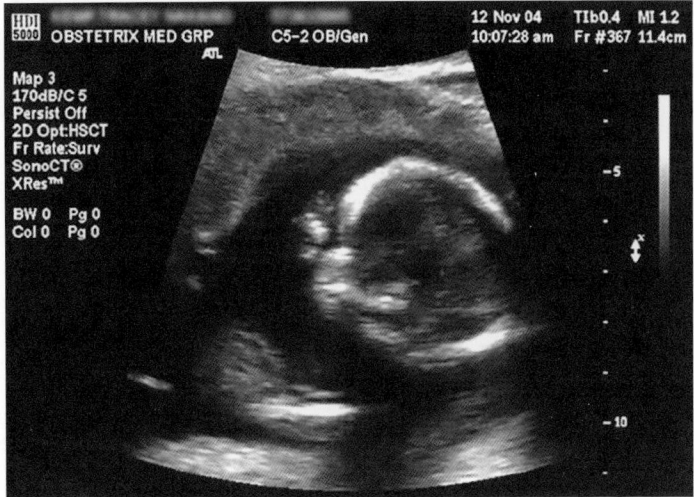

Figure 24.1 Obstetric ultrasound scan of a human fetus at 16 weeks of gestation. The bright white circle center-right is the head, which faces to the left. Features include the forehead at 10 o'clock and the right hand covering the eyes. *Source:* This image, reproduced here from Wikimedia Commons, is in the public domain, because its creator, Jeremy Kemp, has placed it there. Wikipedia (http://www.wikipedia.org).

Table 24.1 Obstetric and gynecological conditions diagnosable by ultrasound scan.

1) Gestational age
2) Placenta previa
3) Fetal malposition
4) Multiple gestation
5) Ectopic pregnancy
6) Retained products of conception
7) Fetal anomalies
8) Fetal growth restriction
9) Poly- and oligo-hydramnios
10) Intrauterine fetal death
11) Fibroids and ovarian cysts

likely able to predict the expected date of delivery with an accuracy to within 1 week. This would reduce both the risks of failure to intervene in cases of post-maturity as well as premature intervention resulting in iatrogenic preterm delivery.

24.3 Problems with the Use of Ultrasound Scanning in Limited Resource Settings

Ultrasound scans are investigations, not treatments. Investigations are only of value if they are appropriate, accurate, relevant and result in the formulation and enactment of clinical decisions. Such action requires the presence of a fully functional and affordable health system with intact and adequate referral pathways. Investigations can only be of benefit if they lead to decisions relating to patient care in a system in which they are followed up and acted upon. Furthermore, ultrasound scanning is an investigation that requires real-time interpretation by the operator, as opposed to a machine-generated blood test result. Such interpretation depends upon the skill and training of the operator and also, to an extent, the quality of the image obtainable with the available technology.

Considering this pathway from investigation to action, and even if either routine or targeted ultrasound scanning was to be made universally available in low resource settings, there are many constraints that may render it of limited benefit. Indeed, introducing such a costly intervention may well be detrimental to the health system as a whole, consuming both finances and personnel that could otherwise have been deployed elsewhere to much greater benefit. It is therefore essential that any new technology should be introduced only after rigorous studies have been conducted to ascertain that the intervention is of verifiable benefit before funds are diverted into up-scaling.

There have been many undoubtedly well-meant attempts to introduce obstetric ultrasound scanning in low resource settings, often reporting various measures of success. Yet it is only recently that an adequately powered cluster-randomized trial investigating whether routine antenatal ultrasound can improve pregnancy outcomes in low-income country settings has been undertaken (McClure et al. 2017). Prior to this "first look" trial, no previous studies had been sufficiently powered to detect differences in maternal and newborn mortality. The trial by McClure and colleagues enrolled almost 57 000 women between the intervention and control arms in 58 clusters in five countries: Pakistan, Kenya, Zambia, the Democratic Republic of Congo and Guatemala. During the 18 month period of the study, 78% of women delivering in the intervention clusters received at least one scan and 60% received two. The study found that the introduction of routine antenatal clinic ultrasound scanning did not increase antenatal clinic use or hospital births nor did it improve the composite outcome nor any component of that outcome including stillbirth, neonatal mortality, near miss or maternal mortality compared to control sites. This was despite the fact that 71% of women who were referred for an ultrasound-detected condition attended that referral. Conditions noted on ultrasound scanning included twins, placenta previa, oligo- and polyhydramnios and abnormal fetal lie. It might reasonably have been expected that such findings would lead to improved outcomes, and the lack of improvement raises many questions that have yet to be answered. For example, was it the case that the intervention failed to demonstrate benefit because of health system deficiencies resulting in complications receiving less than adequate treatment?

Findings reported in the Cochrane reviews of routine ultrasound for fetal assessment in both early and late pregnancy in high-income settings also failed to report any reduction in adverse outcomes for babies, despite an improvement in the detection of both multiple pregnancies and major fetal abnormalities and a reduction in induction for "post term" pregnancy. Routine late pregnancy scans were also not associated with improvements in overall perinatal mortality (Bricker et al. 2015; Whitworth et al. 2015). It is likely that functional health systems were involved in the studies included in these reviews, suggesting that health system factors cannot be entirely responsible for the lack of improved outcomes.

Nevertheless, valuable lessons were learned in the course of the multi-country trial by McClure et al. A case study from the Democratic Republic of Congo, one of the five countries in the trial, illustrated clearly the challenges to implementation of the intervention (Swanson et al. 2017). The equipment used (GE Healthcare Logic-e ultrasound unit), including transducer, printer and accessories was valued at US$22 000 at the time of the study. Because of this, special security measures were required. The machines could not be left in health centers. They had to be locked up each night at a staff member's house and transported to health centers daily by motorcycle. Provision of electricity was another issue. Solar panels had to be bolted on 10-ft poles next to the health centers because the thatched roofs of the centers were not suitable for rooftop installation. Batteries and inverters had to be secured in locked boxes in the centers, which were also protected by security personnel. Provision had to be made for servicing and repair, and extra machines provided to cover the extended time period required to send a machine back to Kinshasa for repairs. The process was protracted due to the complexities of importing replacement parts. For a regular service to be maintained, significant planning for maintenance and repair needs to be factored in. Adequate time is also required for the procurement of disposables including gel, wipes, and thermal printer paper. Due to the staffing constraints in many of the health centers it was necessary to train and appoint staff from elsewhere to undertake the scans for the study. This would not be sustainable without permanent extra funding. Other challenges included issues relating to the referral of patients to functional comprehensive emergency obstetric care units when abnormalities were suspected by the sonographers. Bearing in mind both geographical and financial obstacles, close co-ordination with these units is essential in order to reduce barriers to patient attendance, as well as to ensure quality assurance of the ultrasound diagnostic images. Issues that further complicate the situation included transfer of staff from referral units during the course of the study. While solutions to all these obstacles were implemented during the course of the trial, in the real world of scale-up and permanent service provision, implementation without the extra funds will be challenging.

It has been suggested by several authors that the introduction of ultrasound services in low resource settings may have collateral benefits including an increase in attendance for antenatal consultations and facility deliveries conducted by skilled birth attendants, resulting in a potential reduction in maternal and neonatal mortality and morbidity. A study in Uganda reported a substantial increase in clinic visits and attended deliveries following the introduction of a low-cost ultrasound program (Ross et al. 2013). However, this result should be interpreted with caution since the intervention was introduced in a private clinic where attendees are pre-selected because of their ability to pay for services. While the evidence from the study undoubtedly demonstrated increased attendance, such results may not be generalizable in other settings. As detailed above, the cluster-randomized trial by McClure et al. (2017) failed to demonstrate any increase in antenatal clinic use or hospital births as a result of antenatal ultrasound scanning.

24.4 Provision of Trained Sonographers

The issue of the provision of appropriate training of sonographers is of paramount importance. In a review of training opportunities, LaGrone et al. (2012) found that general and obstetric physicians, as well as non-medical personnel, conducted many scans with little or no formal training in ultrasonography. Even when training had been undertaken, it often failed to meet the WHO criteria regarding the number of supervised scans that should be undertaken prior to certification. Follow-up supervision of trainees was also very variable. Despite this, some training programs did result in good outcomes with regard to diagnostic accuracy and knowledge retention. For example, Greenwold et al. (2014) reported success in implementing a basic obstetric ultrasound training program in rural Mozambique, involving an 8 weeks training course supported by training videos and followed up by 10 months of remotely supported supervision. A week of formal lectures was followed by seven weeks of practical hands-on training. Results from the supervised scans demonstrated that the trained staff performed basic scans with a high degree of accuracy.

24.5   The Perspective of the Pregnant Woman to Antenatal Ultrasound Scanning

Despite the lack of evidence of improved outcomes, routine ultrasound has been much advocated in both early and late pregnancy, and indeed scans have become routine practice in many high-income settings, where an ultrasound scan is usually a much-anticipated event, regarded by the woman and her family as an opportunity to "see" the baby for the first time. For some this anticipation will be tinged with anxiety on the basis of either experiences of abnormalities detected in past pregnancies or concerns regarding symptoms in the index pregnancy. This is no less the case in low resource settings. For example, women in rural Kenya described both anxiety and relief regarding their experience of antenatal ultrasound scans (Olouch et al. 2015). Some women were scared that the scan might use harmful rays that could affect their babies and others were simply anxious because they did not know what to expect. However, women were pleased to have viability confirmed and to be reassured that all was well. On the other hand, news of the detection of an abnormality such as placenta previa was an understandable cause for worry and concern. Similar findings were found in studies from Botswana (Tautz et al. 2000) and Thailand (Rijken et al. 2012). In Botswana there was minimal communication or explanation provided to the patients by the sonographer, and patients were sometimes frightened when the light in the scanning room was switched off for the procedure. In Thailand women were generally happy with the scanning process, despite sonographers conveying minimal information regarding the scan findings. Women were less anxious about the process due to their trust in the healthcare providers, although sometimes embarrassed about undergoing exposure of their abdomen, a practice not generally accepted in the local culture. A common theme in studies related to women's views of scanning is that of poor communications to the women by the sonographers, which suggests that training in communication skills is required as an essential component of sonographer education.

24.6   Abuse of Ultrasound Scanning in Pregnancy

Without appropriate guidance women may sometimes have a tendency to over-estimate the predictive power of ultrasound, leading to erroneous perceptions of the ability of scans to produce accurate information. This imbalance of knowledge between recipient and provider can lead to exploitation by some providers. An article by Ugwu et al. (2014) described Nigerian women's perception of the accuracy of ultrasound dating in late pregnancy, where many women booked late in pregnancy and often self-referred for private scans prior to attending an antenatal clinic. These investigators found that a majority of women were unaware of the limitations of ultrasound with respect to prediction of the expected date of delivery in late pregnancy. This led to some women refusing induction of labor for post-maturity if the expected date of delivery that had been derived from a scan in late pregnancy was not in agreement with expected date of delivery derived from menstrual dates. There was concern that such refusal could result in adverse consequences from unduly prolonged pregnancy, including intrauterine fetal death, fetal macrosomia, increased Cesarean section and instrumental deliveries, birth asphyxia and perinatal death. (Fetal macrosomia is a term used to describe a newborn who is significantly larger than average. A baby diagnosed with fetal macrosomia has a birth weight of more than 8 lbs, 13 oz [4 kg], regardless of his or her gestational age.) Ugwu and colleagues highlighted the need for legislation to prevent uncontrolled proliferation of ultrasound services by poorly trained personnel with consequent adverse effects. There are anecdotal reports of private ultrasound practitioners awarding financial commissions to junior obstetricians referring patients from public hospitals for scans, without first having obtained the patient's history or undertaken any clinical examination. In one such case, a woman "rescued" from such a referral for abdominal pain was found to be in advanced labor (Ameh 2017, personal communication).

Other abuses include the use of ultrasound scanning for gender determination as a precursor to gender-specific termination of pregnancy. Obstetric ultrasound in India has been subject to regulation consequent to the 2003 Pre-conception and Pre-natal Diagnostic Techniques (Prohibition of Sex Selection) Act that made it illegal to determine the gender of the unborn child. Under this act it is mandatory that all ultrasound facilities must be registered and since 2004 manufacturers and distributors of ultrasound machinery have explicit responsibility to ensure its proper use. The place of use of such machinery must be registered and qualified users must be named. The buyers of the ultrasound equipment must have a valid certificate and sign an affidavit stating they will not disclose the gender of a fetus. Despite this apparently robust legislation, there have been multiple instances where the law has been weakly upheld and gender selection has been practiced. For example, a population-based study in Delhi found that 2.3% of ultrasound scans were performed to determine fetal gender, despite this being illegal in India (Chaturvedi et al. 2007). Of those seeking gender determination, 30% claimed to be unaware that scans for this purpose were illegal. Half of those found to be expecting a female fetus went on to obtain an induced abortion.

24.7 Advances in Ultrasound Technology (and See Chapter 23)

Ultrasound technology has advanced rapidly in recent years. In particular, the use of portable ultrasound has evolved from laptop-like devices to the point where wireless or wired scan probes can be connected to either android or iPhones uploaded with appropriate applications. Examples of devices employing mobile phone linkages include the Philips Lumify, the Clarius C3 convex scanner or the Sonostar Uprobe-2.

At the same time the quality of imaging has increased considerably. Several companies now market such devices and software applications, and the cost of subscriptions is well within the reach of many would-be sonographers in low-resource settings. In common with many technological advances, such availability has the potential for both benefit and danger, depending upon the context in which it is utilized. The ability to perform obstetric ultrasound scans at the bedside with a device transported in the pocket of the sonographer undoubtedly has a multiplicity of potential clinical benefits in aiding rapid diagnosis in challenging clinical and logistical circumstances. Such benefits were supported by a study where the use of a pocket–sized ultrasound machine as a tool to assess early pregnancy complicated by pain or bleeding, pregnancy advanced beyond 14 weeks, and gynecological pathology was assessed (Sayasneh et al. 2012). They found that the use of a hand-held device resulted in measurements of gestation sac diameter, crown-rump length and femur length were in close agreement with those measured using a high specification conventional ultrasound machine.

However, this relatively cheap and accessible technology also opens the door to abuse by unregistered operators performing scans without adequate regulation or training. The motivation here is primarily for financial gain, but using poor quality scans for inappropriate indications, often at the expense of patients' limited finances and clinical outcomes.

24.8 Targeted Ultrasound Scanning

Although routine antenatal sonography has not been demonstrated to be of proven advantage in limited resource settings (McClure et al. 2017) this finding is in contrast to the benefits to be gained from *targeted* ultrasound scanning, undertaken for individual patients presenting with a clinical problem, such as bleeding in pregnancy or cessation of fetal movements. Contingent upon an intact referral chain, ultrasound diagnosis of placenta previa should allow for referral to a comprehensive facility with recourse to blood transfusion and where a safe Cesarean section could then be undertaken. Likewise, removing uncertainty as to fetal viability could contribute to important decisions regarding mode of delivery and lead to the avoidance of women being subjected to unnecessary Cesarean section. Physicians in Tanzania recognized that limitations in the availability of scans posed restrictions with regard to their ability to manage these and other complications (Åhman et al. 2016). Portable or hand-held scanning devices capable of use for point-of-care scanning have particular benefits, and as technology improves, the use of such devices has increasing potential for bedside scanning in emergency situations (Wanyonyi et al. 2017). Clinical indications are not the only prompt for targeted ultrasound scanning and indeed the need for ultrasound can sometimes only be determined retrospectively after a scan has been performed, which presents its own residual conundrum.

24.9 Conclusions

In common with many emerging technologies, the use of ultrasound scanning for obstetric diagnosis has a worldwide reach. The use and abuse of ultrasound in pregnancy has already proliferated in resource-limited settings and will continue to do so, with widening availability fueled by both technological improvements and decreasing costs. In order to maximize benefit for patients and stem the potential for abuse, the obstetric ultrasound community must work to define globally recognized standards for training, regulation and implementation in accordance with practice that is evidence-based.

References

Åhman, A., Kidanto, H., Ngarina, M. et al. (2016). "Essential but not always available when needed" – an interview study of physician's experiences and views regarding use of obstetric ultrasound in Tanzania. *Global Health Action* 9: 31062.

Bricker, L., Medley, N., and Pratt, J. (2015). Routine ultrasound in late pregnancy (after 24 weeks' gestation). *Cochrane Database of Systematic Reviews* 29 (6): CD001451.

Chaturvedi, S., Chhabra, P., Bharadwaj, S. et al. (2007). Fetal sex-determination in Delhi: a population-based investigation. *Tropical Doctor* 37: 98–100.

Greenwold, N., Wallace, S., Prost, A., and Jauniaux, E. (2014). Implementing an obstetric training program in rural Africa. *International Journal of Gynecology and Obstetrics* 124: 274–277.

LaGrone, L., Sadasivam, V., Kushner, A., and Groen, R. (2012). A review of training opportunities for ultrasonography in low and middle income countries. *Tropical Medicine and International Health* 17 (7): 808–819.

McClure, E., Goldenberg, R., Swanson, D. et al. (2017). Routine antenatal ultrasound in low/middle income countries: a cluster randomized trial. *American Journal of Obstetrics and Gynecology* 216 (1 Supp): S3.

Olouch, D., Mwangome, N., Kemp, B. et al. (2015). "You cannot know if it's a baby or not a baby": uptake, provision and perceptions of antenatal care and routine antenatal ultrasound scanning in rural Kenya. *BMC Pregnancy and Childbirth* 15: 127.

Rijken, M., Gilder, M., Thwin, M. et al. (2012). Refugee and migrant women's views of antenatal ultrasound on the Thai-Burmese border: a mixed methods study. *PLoS One* 7 (4): e34018.

Ross, A., DeStigter, K., Rielly, M. et al. (2013). A low-cost ultrasound program leads to increased antenatal clinic visits and attended deliveries at a health care clinic in rural Uganda. *PLoS One* 8 (10): e78450.

Sayasneh, A., Preisler, J., Smith, A. et al. (2012). Do pocket-sized ultrasound machines have the potential to be used as a tool to triage patients in obstetrics and gynecology? *Ultrasound in Obstetrics and Gynecology* 40: 145–150.

Swanson, D., Lokangaka, A., Bauserman, M. et al. (2017). Challenges of implementing antenatal ultrasound screening in a rural study site: a case study from the Democratic Republic of Congo. *Global Health Science and Practice* 5 (2): 315–324.

Tautz, S., Jahn, A., Molokomme, I., and Görgen, R. (2000). Between fear and relief: how rural pregnant women experience foetal ultrasound in a Botswana district hospital. *Social Science and Medicine* 50: 689–701.

Ugwu, E., Odoh, G., Dim, C. et al. (2014). Women's perception of accuracy of ultrasound dating in late pregnancy: a challenge to prevention of prolonged pregnancy in a resource poor Nigerian setting. *International Journal of Women's Health* 6: 195–200.

Wanyonyi, S., Mariara, C., Vinayak, S., and Stones, W. (2017). Opportunities and challenges in realizing universal access to obstetric ultrasound in Sub-Saharan Africa. *Ultrasound International Open* 3: E52–E59.

Whitworth, M., Bricker, L., and Mullan, C. (2015). Ultrasound for fetal assessment in early pregnancy. *Cochrane Database of Systematic Reviews* 14 (7): CD007958.

25

Examining the Optic Fundus and Assessing Visual Acuity and Visual Fields Using Mobile Technology

Nigel M. Bolster[1], and Andrew Bastawrous[2]

[1] *Biomedical Engineering, University of Strathclyde, Glasgow, UK*
[2] *Clinical Research Department, Faculty of Infectious & Tropical Diseases, International Centre for Eye Health (ICEH), London School of Hygiene and Tropical Medicine, London, UK*

25.1   Introduction: The Ascent of Mobile Technology

The proliferation of portable, consumer-level wireless technology has been one of the major technology events of this century. The expansion in the use of mobile technology has been made possible by the availability of affordable mobile devices, including mobile phones, smartphones, tablet computers and personal digital assistants (PDAs), coupled with the rapid and extensive expansion of mobile network coverage (GSMA Intelligence 2014). In conjunction with this phenomenon a vast ecosystem of mobile application-based services (mServices) has arisen granting mobile device users access to services in areas as diverse as banking to livestock management (Hughes and Lonie 2007; Hwang and Yoe 2011).

By far the most common mobile device is the mobile phone (Pew Research Center 2015). These portable handsets initially combined voice calls with short messaging and multimedia services (SMS and MMS), simple applications and basic Internet browsing. The widespread adoption of the smartphone, following the release of the iPhone (Apple Inc., Cupertino, CA, USA) in 2007, has greatly enhanced the features available to mobile phone users. In addition to the simple communication abilities of their predecessors, today's smartphone handsets incorporate more powerful computer processing, allowing more sophisticated application (app) installations and upgrades, as well as touch-screen

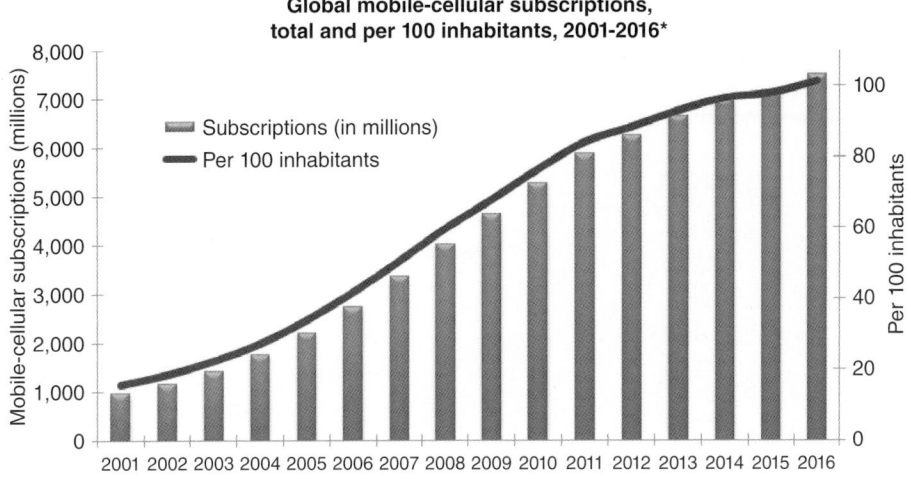

Figure 25.1 Global mobile phone subscriptions 2001–2016.

interfaces, global positioning systems (GPS), high-resolution cameras and a number of other sensors including accelerometers and fingerprint scanners.

Having started the twenty-first century on rough parity with landline fixed telephones, the number of mobile phone subscriptions globally has exploded, as shown in Figure 25.1, and there are now almost as many mobile phone subscriptions in the world as there are people (International Telecommunication Union 2016).

Notably, this phenomenon is a universal one, with it being estimated that even those nations classified as LMICs by the United Nations Statistics Division had over 94 subscriptions per 100 inhabitants by the end of 2016 (International Telecommunication Union 2016).

Similarly, with their powerful computer processing ability, access to mobile broadband, a host of sensors, low cost and lightweight form, tablet personal computers (PCs) offer affordable access to more powerful mobile computing. (A tablet computer, commonly shortened to tablet, is a mobile computer with a touchscreen display, which is usually in color, processing circuitry, and a rechargeable battery in a single thin, flat package.) Indeed, it is estimated that global sales of tablet computers in 2015 exceeded that of laptop and desktop computers combined (Palma 2015). Furthermore, a large proportion of these sales have been in low- and middle-income countries (LMICs). Indeed, as can be seen in Figure 25.2, it has been predicted that by 2018 nearly 20% of the world population will use tablet PCs regularly with four of the top five countries by tablet ownership being classified as middle-income countries (eMarketer 2015).

The spread of such mobile devices has been facilitated by the rapid expansion of mobile networks, which can be built more quickly and at a much lower cost than wired fixed telephone networks. It is this fact that has allowed LMICs to "leapfrog" the older technologies that preceded the introduction of mobile technology in high-income countries (HICs) and thus rapidly bridged the global inequality in access to portable communications technology.

Indeed, access to broadband Internet is steadily rising in LMICs on the back of mobile technology with the International Telecommunication Union (ITU) estimating that in

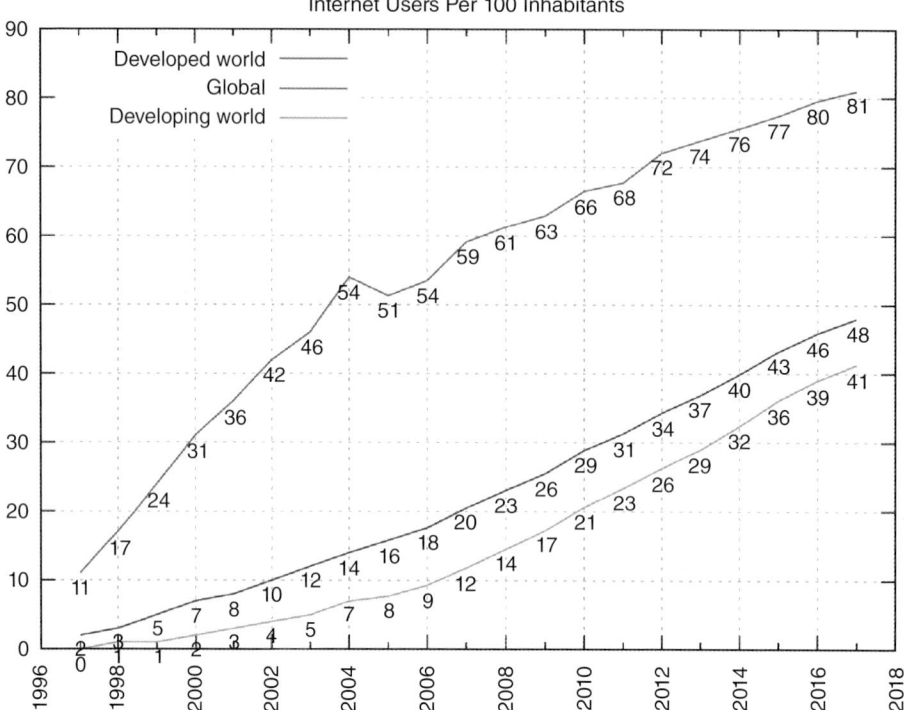

Figure 25.2 Internet users per 100 inhabitants. *Source:* International Telecommunication Union. By Jeff Ogden (W163) and Jim Scarborough (Ke4roh) – Own work, CC BY-SA 3.0, https://commons. wikimedia.org/w/index.php?curid=18972898

2016 there were over 40 mobile broadband subscriptions per 100 inhabitants in nations classified as LMICs (Figure 25.3) (International Telecommunication Union 2015). This was also estimated to constitute over 80% of all broadband subscriptions in such countries (International Telecommunication Union 2015).

The sudden arrival of so many new Internet users onto the market has posed a significant opportunity for innovators throughout the world. Today a vast number of both SMS and application-based services (mServices) are available to the mobile device user. To give a sense of the scale of this industry, the two main application marketplaces for mobile devices, the Apple "App Store" and Google "Play Store," have upwards of two million apps each (Statista 2016). This does not include SMS-based services or applications for operating systems other than iOS and Android. (iOS is an operating system used for mobile devices manufactured by Apple Inc. Both the iPod touch and the iPad are iOS devices. Android is an open-source operating system used for smartphones and tablet computers. Google's apps are all available on iOS, whereas Apple does not use Android at all.)

Whilst in HICs many of the services offered represent a convenient means of accessing existing services, in LMICs mServices are opening access to particular services to millions of users for the very first time. For example, in East Africa the mFinance service "M-PESA" (SafariCom, Nairobi, Kenya) has seen phenomenal growth, with over six million subscribing customers in Kenya alone during its first two years of operation (Mas and Morawczynski 2009). It is now seen as an integral part of the national

The developed/developing country classifications are based on the UN M49, see: http://www.itu.int/en/ITU-D/Statistics/Pages/
definitions/regions.aspx.html
Note: *Estimate
Source: ITU World Telecommunication/ICT indicators database

Figure 25.3 Active mobile broadband subscriptions per 100 people, 2007–2016.

economy (Jacob 2016). Such has been the success of the service that it has fostered its own ecosystem of innovations including, for example, "BusinessOS" (Kopo Kopo, Morningside Office Park, Ngong Rd., Nairobi, Kenya), which provides small and medium enterprises (SMEs) with automated bookkeeping, financial analytics, fore-casting and targeted marketing capabilities. These mServices have no direct compari-son in HICs and consequently it has been suggested that, increasingly, innovation and scaling of mServices will occur in LMICs, where the demand and potential for rapid growth is greatest.

It is into this fertile but rapidly changing environment that mHealth, the provision of healthcare via mobile technology, has come. The adoption of mHealth in eye care has shown particular promise in recent years. The drivers for the specialty's suitability for adoption are twofold. Firstly, as with the example of mFinance mentioned above, eye services are sorely lacking in many LMICs. Indeed, 90% of the world's 285 million visu-ally impaired people live in LMICs whilst HICs have a far larger number of ophthal-mologists per capita than LMICs (Resnikoff et al. 2012). Furthermore, 80% of this blindness is preventable with many of the major causes of blindness requiring simple treatment or surgery. As such, screening tools that are inexpensive, can be used by low-skilled healthcare workers and can transmit results from remote locations would have a transformative impact on the delivery of eye care in LMICs. Finally, many of the diag-nostic devices used in eye care, such as the ophthalmoscope and Snellen[1] visual acuity chart, have changed little in close to a hundred years. The lack of complexity in these designs therefore makes them ideal for integration with mobile technology. In this Chapter, we present a review of recent research efforts to develop such mobile screen-ing tools. Specifically, we look at three eye tests that have shown particular promise, namely visual acuity, visual field testing and fundoscopy.

---

[1] Herman Snellen was a Dutch ophthalmologist, 1834–1908.

25.2 Visual Acuity

Visual acuity is an objective, quantitative measure of visual function. There are a number of standardized charts for measuring visual acuity based on distance vision including the Snellen, Landholt's C, Tumbling E and logMAR ETDRS vision charts (Ferris et al. 1982; International Standards Organisation 2009b; Snellen 1873). In each case, visual acuity is tested in each eye individually at six meters. As shown in Table 25.1, The World Health Organization (WHO) classifies visual function into either no visual impairment, mild visual impairment, moderate visual impairment, severe visual impairment or blindness. Blindness is also further sub-categorized into those who can count fingers at 1 m (approximately 1/60), those who can detect hand movements, those who can only perceive light and those with no perception of light. Low vision refers to either moderate or severe visual impairment and visual impairment refers to either low vision or blindness (WHO 2016).

The potential benefits of hosting a visual acuity test on a smartphone or tablet PC are portability, reduction of cost compared to standard back-illuminated charts and the ability to reconfigure the test for different sequences or test distances in real time. However, before such benefits can be realized it is first necessary to consider whether the display screens of mobile devices are an appropriate technology for displaying a visual acuity chart.

There are two display technologies which are commonly used in mobile device displays. Thin film transistor liquid crystal displays (TFT LCD), a display technology that is particularly common in low cost smartphones, consist of a twisted nematic liquid crystal sandwiched between two crossed linear polarizing filters. Under normal conditions these liquid crystals are twisted and cause the polarization of incident light to change by 90° and therefore pass through the second polarizer unimpeded. However, if electrically charged the liquid crystal begins to straighten and when fully charged will allow light to pass through without any shift in the polarization angle. Thus, the second polarizer blocks the perpendicularly polarized light originating from the LED backlight causing the display to appear black.

Table 25.1 Categorization of visual impairment according to Section H54 of the ICD-10, of WHO disability standards.

		Presenting distance vision		
WHO category		Worse than	Equal to or better than	Overall category
0	Mild or no visual impairment		6/18	
1	Moderate visual impairment	6/18	6/60	
2	Severe visual impairment	6/60	3/60	Low vision
3	Blindness	3/60	1/60	Visual Impairment
4	Blindness	No light perception	Light perception	
5	Blindness		No light perception	

Source: World Health Organisation (2016).

In TFT LCD displays each picture element (pixel) consists of a layer of insulating liquid crystal placed between two layers of the transparent conductor indium tin oxide (ITO). Each pixel is controlled by a transistor switch contained within the pixel. The pixels are arranged in an active matrix where each pixel can be individually activated by addressing its row and delivering a charge down the appropriate column. The charge delivered is sufficient to maintain the pixel's state until the next cycle. Furthermore, by controlling the voltage supplied to the crystal it is common to be able to achieve a gray-scale of 256 brightness levels per pixel. Therefore, by adding a red, green or blue color filter to each pixel a total of 16.78 million colors can be produced.

Whilst LCDs can be lit by ambient light by placing a mirror behind them, those designed for mobile devices are almost exclusively lit by a white backlight. No polarizing filter is 100% efficient and thus some light is transmitted through each pixel even when fully discharged, with the intensity of light transmitted being proportional to the backlight's intensity. Another drawback of LCDs is their limited viewing angle. This results from the increased distance that light rays at wider angles travel through the liquid crystal causing it to arrive at the second polarizer with a different polarization compared to those light rays perpendicular to the polarizer. Thus, each pixel has a different illuminance when viewed from different angles.

Active matrix organic light emitting diode (AMOLED) displays similarly use a matrix of thin-film transistors to activate individual pixels. Light is generated in each pixel by a layer of a semiconducting emissive electroluminescent organic compound situated between two electrodes. Usually the anode is made of ITO allowing light to be transmitted. Typically, two transistors are used per a pixel, one to charge a storage capacitor and another to provide the voltage required to deliver a constant current to the pixel. As AMOLED displays turn off pixels completely they tend to have a better contrast ratio than LCDs. However, unlike LCDs they are prone to "screen burn-in" where a permanent artifact is left when a particular image is overused.

As both contrast and screen luminance are known to vary according to display technology and since both are known to affect acuity measurements (Johnson and Casson 1995; Sheedy et al. 1984), it is important to assess the performance of mobile devices with respect to these parameters. Such an assessment was conducted by Livingstone and colleagues who measured performance with respect to both of the parameters mentioned above on the iPad 3, iPad 4 and iPad Air 2 (Apple Inc.) tablet PCs. The authors benchmarked their findings against the Early Treatment Diabetic Retinopathy Study (ETDRS) specifications, which are based on the International Council of Ophthalmology (ICO) recommendations, and the British specifications for visual acuity charts, BS 4274–1:2003 (International Standards Organisation 2003). The former specifies that acuity charts have a luminance of not less than $80\,cd/m^2$ with the black optotypes having a luminance of less than 15% of the surrounding field whilst the latter specifies that the Weber contrast $[(L_{bckgrd} - L_{letter})/L_{bckgrd}]$ should be 90% or greater with any variance in luminance across the chart being 20% or less. (An optotype is a series of figures or letters of different sizes used in testing the acuity of vision – the Snellen chart is an example. The SI unit for luminance is candela per square meter "$[cd/m^2]$.") A luminance meter was used to measure the luminance at various points on each tablet's screen and at various screen brightness levels whilst displaying a black and white checkerboard pattern whose color inverted at regular intervals. A similar method was used to ascertain the luminance and contrast of three EDTRS charts. It was found that all of the tablets met the contrast requirements mentioned above and met the luminance requirements providing that the brightness level was set to 50% or greater. Luminance uniformity was found to be either close to, or exceeded, the requirements of BS 4274–1:2003. By comparison, the EDTRS charts were found to have lower contrast and poorer uniformity with two of the three charts falling below that which is deemed acceptable. Finally, turning on room lights was found to have little effect on the luminance and contrast of the tablets and marginally improved the luminance uniformity.

Aslam and colleagues also sought to determine the suitability of iPad 3 (Apple inc.) tablet PCs for visual assessment by investigating the consistency of a screen's performance with respect to different areas of the screen and at different viewing angles (Aslam et al. 2013). It was found that, although absolute illuminance was up to 23% less in the corners of the tablet compared to the center for the same pixel intensity, contrast between background and foreground was remarkably stable, varying by under 1.2%. The authors found similar results with respect to viewing angle.

There are more than 100 vision test apps on the Google Play Store alone with a similar number of such apps on the Apple App Store (Brady et al. 2015; Livingstone et al. 2016). Few of these, however, provide information pertaining to clinical validation (Brady et al. 2015).

Perera and colleagues found that of the eleven visual acuity apps identified in the Australian Apple App Store, none were capable of consistently predicting visual acuity within one line of that measured by a six meter Snellen visual acuity chart (Perera et al. 2015). Furthermore, the authors found that only three of the applications had optotypes within 10% of the required dimensions and even the acuity recorded by the most accurate of these was significantly different to those regarded by a 6 m Snellen visual acuity chart and was especially underestimated at lower acuities.

A similar study by Tofigh and colleagues examined a near visual acuity app, EyeHandBook, on an iPhone 5 running iOS7 (Apple inc.). Their comparison of the corrected monocular near acuity recorded for one hundred subjects compared with that recorded by a Latham and Phillips Ophthalmic (LPO) Rosenbaum card found that the application consistently overestimated the near acuity by a statistically significant average of 0.11 logMAR ($p < 0.0001$), with those with lower vision having a particularly overestimated acuity.

In contrast to the above studies, Gounder and colleagues found good agreement with a Snellen light box chart in their investigation of the EyeSnellen application (Version 1.6; Steve Colley, Perth, Australia). Using a wall-mounted first-generation iPad Mini with an iPhone 5 s acting as a controller (Apple Inc.), the authors reported a mean difference of 0.001 logMAR between the two tests (95% CI: −0.169–0.171). Whilst the accuracy reported represented an important improvement compared to previously studied applications, the need to mount the tablet as shown in Figure 25.4 and to conduct the test at six meters reduced the potential for the test to be used in scenarios where the use of traditional Snellen charts may be difficult, such as conducting vision tests in patients' homes.

Given the large number of deficient applications that are available for public download on the most popular application stores, it is important that clinicians assess the available results of peer-reviewed clinical validation before using such applications in clinical practice. Furthermore, it can be strongly argued that such software should be classified as a medical device and therefore should require regulatory approval from the relevant authorities in the regions where it is distributed (Medicines and Healthcare Products Agency 2014).

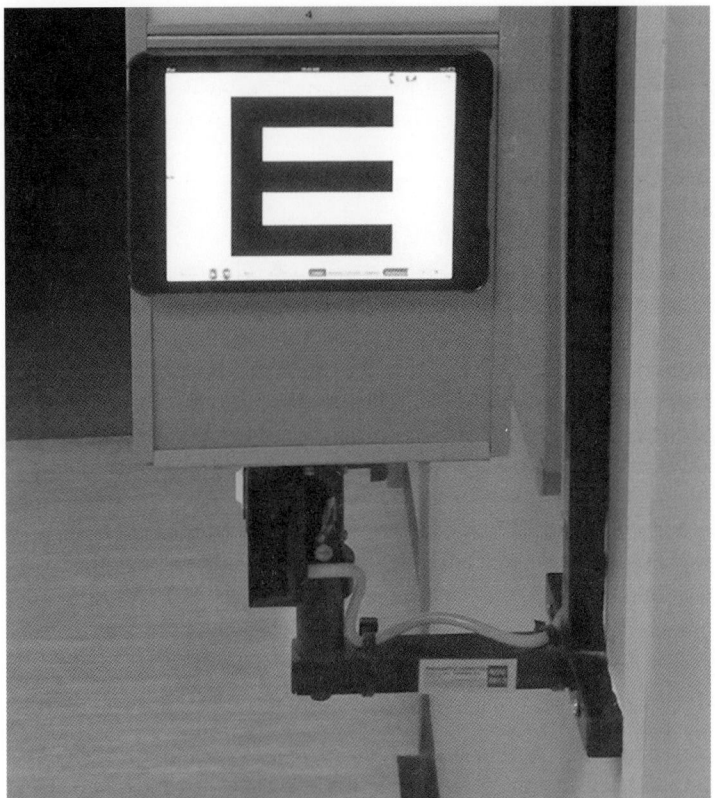

Figure 25.4 The EyeSnellen application running on a first-generation iPad mini (Apple Inc.) as mounted by Gounder and colleagues in their hospital-based validation study.

To the authors' knowledge, only one application, the Peek Acuity Pro (Peek Vision Ltd., London, United Kingdom; Figure 25.4) has been granted such regulatory approval for measurement of visual acuity, having obtained approval as a Class 1 medical device from the European Union.

This application also differs from the applications described above in its use of a staircase algorithm to determine the size of the optotype displayed based on the patient's performance within the test as opposed to manual selection by the operator. This allows visual acuity to be measured without the operator needing to determine which size of optotype the patient needs to be asked to read at every stage of the test. Indeed, the operator does not even need to look at the smartphone's display during the test which has the additional advantage of removing the possibility for operator bias.

The Peek Acuity device has been validated for measurement of monocular visual acuity by comparing its test–retest variability (TRV) and measurement time against the Snellen chart and the ETDRS-based tumbling E logMAR chart (Bastawrous et al. 2015). The study was nested within a follow-up of a community eye disease cohort study in central Kenya with comparisons being made both in patients' homes and in a temporary eye clinic (Bastawrous et al. 2014). In total, 272 participants underwent all three tests in the clinic with 233 (86%) of these having undergone tests with Peek Acuity and a port-

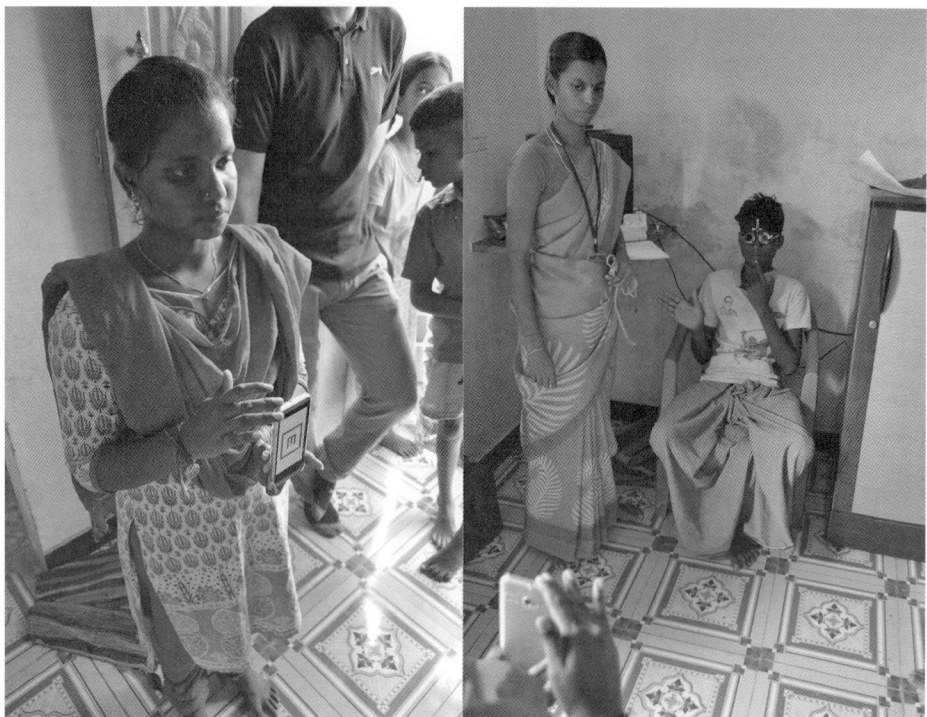

Figure 25.5 The Peek Acuity Pro device in use in the field to measure visual acuity.

able Snellen chart in their home the previous day. In each case the tests were conducted by non-healthcare workers who had been trained to use the test.

Bland-Altman and Pearson correlation analysis of pairwise comparisons of the various tests showed that Peek Acuity was comparable in accuracy to the reference standard ETDRS chart (0.055 difference of average with 95% CI of 0.023–0.088; 95% limits of agreement being −0.438–0.549; κ = 0.917 with 95% CI of 0.893–0.935). Furthermore, the smartphone-based test was found to be no slower than the Snellen test with the mean testing time for both eyes being measured as 77 seconds (95% CI: 71–84 seconds) compared to 82 seconds (95% CI: 73–91 seconds) respectively. Finally, Peek Acuity was shown to be repeatable and consistent when tested in the community and in the clinic (−0.054 difference of average with 95% CI of −0.083 to −0.025; 95% limits of agreement being −0.498–0.390; κ = 0.933 with 95% CI of 0.914–0.948).

Despite the agreement measured between Peek Acuity and the reference standard, a significant impact on the effectiveness of eye care will only be achieved when the test can be integrated into the full eye care system. To that end, a comprehensive application for school screening, Peek School Screening (Peek Vision Ltd.; application workflow pictured in Figure 25.6), has been built around the Peek Acuity test and is currently undergoing evaluation in Kenya, Botswana and India (Morjaria et al. 2017). This involves a mechanism for capturing patient details within the app for students failing the test and forwarding this to a central server. From here, a bulk SMS or voice messaging server, depending on local literacy levels, sends referral details to the child's parents and a contact at the school. Following this, further messages are sent to remind both of

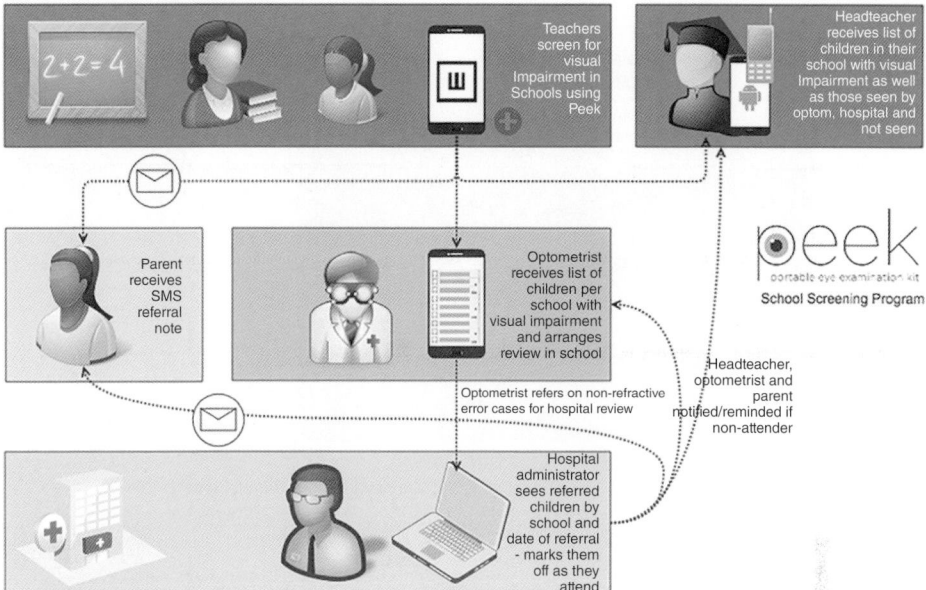

Figure 25.6 Schematic depicting the use of the Peek School Screening application workflow.

the child's contacts of an upcoming clinic where they can receive follow-up examinations and treatment. Once the child arrives at a triage clinic the details recorded during screening can be recalled using a data capture app and further actions, such as a subjective refraction test or dispensing of simple medications, can be recorded. Again, these details are uploaded to the central server allowing spectacles to be ordered or reminders of a referral to a tertiary eye center to be issued if required. In addition to vision testing, data capture and referral system, School Screening also incorporates PeekSim (Peek Vision Ltd.; print out shown in Figure 25.7), an app that generates images simulating the visual blur measured by Peek Acuity, together with information on how to book a clinic visit.

In a trial of the system in Kitale, Kenya teachers were taught to screen visual acuity reliably with Peek School Screening when compared to ophthalmic nurses conducting screening with standard vision charts and using current referral practices (unpublished data). Furthermore, it was found that a combination of the text messaging reminders and printed Polaroid pictures of PeekSim's output improved uptake of referrals to eye care providers by two and a half times compared to the control arm.

It is also well known that children's adherence to spectacle wear is major issue in screening programs (Congdon et al. 2011; Wedner et al. 2008). A study in Hyderabad, India is currently evaluating the adherence to spectacle wear at three months when a health education package delivered using Peek Acuity is used alongside spectacle dispensing (Morjaria et al. 2017). The package is given to children aged 11–15 years, their parents and teachers and consists of PeekSim images with images of relevance to Indian children of this age group. These include images of role models and classroom settings to demonstrate the severity of the child's impairment and the improvement that can be achieved through spectacle wear.

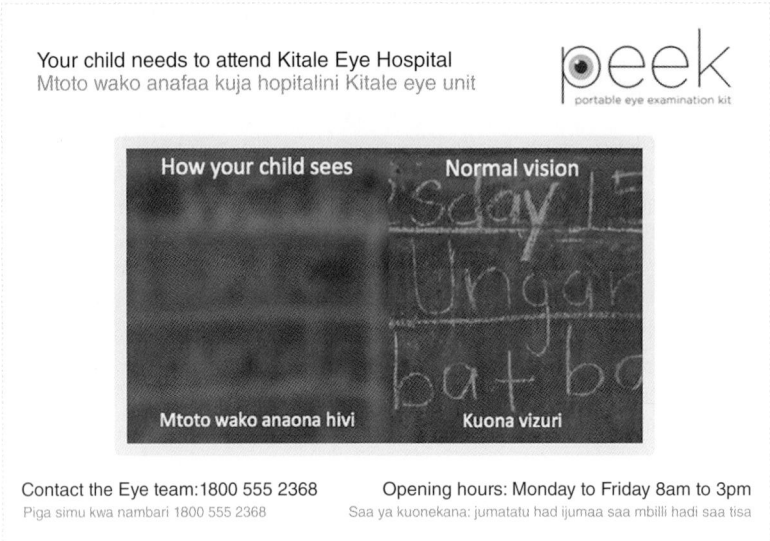

Figure 25.7 A printout of PeekSim, an app that generates images simulating the visual blur measured by Peek Acuity, together with information on how to book a clinic visit. (*See color plate section for the color representation of this figure.*)

25.3 Visual Fields

Many eye diseases and conditions result in specific blind spots, referred to as scotomas, or a loss of peripheral vision in advance of central vision loss. Visual acuity is therefore not the most appropriate measure of such vision loss as it primarily tests central vision. The two most common tests of visual fields are the Goldmann visual fields test and the Humphrey visual field (HVF) analyzer. In each test the patient places their head on a chin rest inside a uniformly illuminated hemispherical dome, has one eye covered and is asked to fixate on a central point. The patient is then asked to respond when they perceive a visual stimulus by pressing a handheld button.

In the Goldmann test the operator activates lights from the periphery inwards along a particular axis until the patient acknowledges that they can see the stimuli. By repeating this across a number of different angles a map of the extent of the patient's peripheral vision can be produced. The Humphrey visual field analyzer instead automatically activates a series of visual stimuli throughout the entire visual field and thus is able to produce a full map of the patient's visual field. In both tests, the contrast of the stimuli compared to the background is varied to give a measure of the retinal sensitivity throughout the visual field.

Both tests are reliant on equipment that is expensive and bulky. For this reason, assessing visual fields outside of eye clinics in well-resourced settings is rare. An inexpensive, mobile test that can be operated reliably by medical personnel with minimal optometry training is therefore highly sought after. Furthermore, the tests' requirements for the patient to fixate centrally and respond by pressing a button are problematic for non-compliant patients or those who have motor difficulties such as someone who has had a stroke. Thus, a visual field analyzer that is effective within such patient populations is desirable.

Tahir and colleagues tested the effect of viewing angle and screen reflections on the gamma function and screen uniformity of three tablet computer models (Tahir et al. 2014). They found that a wide range of contrasts could be generated, although these were limited in number at the lowest contrasts. Particularly relevant to visual field testing was their finding that, although there is considerable non-uniformity in absolute luminance, there was minimal variance in contrast when the edges of the displays were viewed at an angle. However, the introduction of a light source above the tablets while the tablets were tilted at 45° did have a significant impact on the measured contrast. This highlights the importance of controlling ambient lighting and tablet position during testing.

Visual Fields Easy (GLANCE Optical, Melbourne, Australia), an application for the iPad (Apple Inc.), uses a variable fixation point to test both the central and peripheral visual fields without the need for bulky or expensive equipment. The central 30° of vision is tested using a $64\,mm^2$ (size V) fixation target on a $10\,cd/m^2$ background. The test is capable of presenting a 16 decibel (dB) suprathreshold target for rapid screening or of testing 96 locations for more comprehensive testing of the visual field. The patient responds to these stimuli by touching the screen. Following the testing of the central field, the fixation target is moved to each corner of the display in turn in order to test each quadrant of the peripheral visual field, as shown in Figure 25.8. Thus, a full field of 34° horizontal and 25° vertical can be tested.

A comparison of the results recorded with Visual Fields Easy to those recorded with a Humphrey Field Analyzer for 200 healthy, 200 glaucomatous and 50 diabetic eyes was reported by Johnson et al. (2014). A correlation of missed targets of 0.75 ($p < 0.0001$) and 0.60 ($p < 0.0001$) was reported for the mean and pattern standard deviation respectively, as measured by the Humphrey Visual Field analyzer.

A subsequent version of this technology, Melbourne Rapid Fields (MRF; GLANCE Optical), is a perimeter test based on an iPad (Apple Inc.) (Kong et al. 2016). The test also employs variable fixation, setting the fixation target to the center of the tablet's display and then each one of the corners in sequence. In so doing the test can be used to

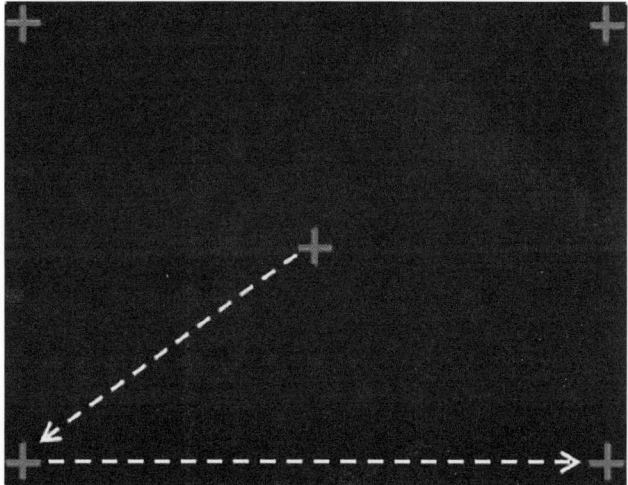

Figure 25.8 Visual fields tests based on tablet PCs move the fixation target in order to increase the tested visual field beyond the angle subtended by the display. *Source:* Vingrys et al. (2016).

test 66 locations through 30°, despite the iPad's display only subtending $17.4° \times 12.9°$ at this distance. Stimuli were round and displayed for 300 milliseconds (ms).

As the test uses a radial grid, additional test points are used to fill any gaps between peripheral points if the patient returns abnormal results. Additionally the test provides threshold estimates by using seven contrast steps between 0 and 30 dB (0, 6, 12, 17, 22, 26, 30 dB) according to the Bayes methods (Vingrys et al. 2016). Thresholding begins at a single initial luminance with the subsequent luminance being determined by a modified Zippy Estimation using the Sequential Testing (ZEST) procedure (King-Smith et al. 1994).

Fixation is tested at the beginning of the test using a blind spot monitor and throughout the central fixation test by presenting a stimulus in the location of the blind spot. As this technique cannot be implemented on the peripheral tests, a voice instruction is instead played at regular intervals during these tests. False positive results are also detected by including periods in which no stimulus is presented. Stimulus size is increased with eccentricity so that a constant threshold is expected across the field (Sloan 1961).

The patient registers by pressing the screen with a finger or by pressing the spacebar on a Bluetooth keyboard. The test takes four to six minutes in total depending on the field loss experienced by the patient. As with the previous app, calibration using a radiometer is required. Kong and colleagues compared outcomes obtained from 90 patients undergoing glaucoma investigation (age range 18–91 years, mean 69.5, standard deviation = 12.5) using the MRF hosted on an iPad 3 (Apple Inc.) to those attained using a HFA using the Swedish Interactive Thresholding Algorithm (SITA). All patients had prior experience of the HFA test and were excluded if they were not fluent in English. One eye having an acuity of 6/12 or better was selected at random for each patient and gave a reliable HFA-SITA result within three months of MRF testing. Testing was conducted by a clinician with the tablet and keyboard placed on a typing stand and the patients head unrestrained after being positioned at 33 cm.

The authors found the mean defect recorded by MRF to have an overall strong correlation with those returned by HFA (ICC = 0.93); however, they also found that MRF returned a less negative mean defect than HFA, with a bias of 1.4 dB. Good overall agreement was found between pattern deviation indices (ICC = 0.86). However, it was noted that this was weaker in eyes with more severe defects (ICC = 0.53) although the sample size for this subgroup was relatively small (n = 18). Test–retest reliability was found to be comparable to HFA (ICC = 0.93; 0.89 for mean defect and pattern deviation respectively with a bias of 0.1–0.5 dB in each case).

The authors additionally found that although MRF was quicker (5.7 ± 0.1 compared to $6.3 \pm 0.$one minutes) fixation losses occurred more often ($36 \pm 4\%$ compared to $6 \pm 4\%$), leading the authors to suggest that the blind spot monitor was corrupted by free space viewing. It was therefore suggested that future development include eye-tracking technology. "Eyecatcher" is an alternative tablet-based visual fields app that also makes use of a variable fixation target to map the visual fields (Smith et al. 2016). However, instead of using static fixation and a screen or button press to register that the patient has observed a stimulus, the patient is instructed to look at a stimulus when they see it. The test then uses an infrared eye tracker (Tobii Eye Tracker, Stockholm, Sweden) to detect if the patient has moved their eyes from the original fixation point to the stimulus. This stimulus then becomes the fixation point for the next test and so on until a full map of the visual field is built. Smith and colleagues tested the app for the 20° of central vision on a group of 12 glaucomatous and 6 healthy participants and compared the results

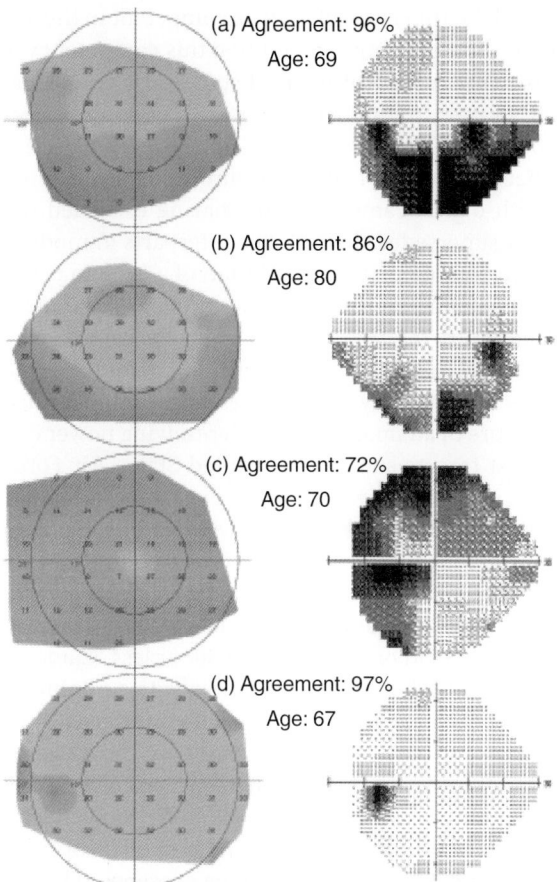

Figure 25.9 Visual fields results acquired using "Eyecatcher" for three glaucoma patients (a–c) and one healthy patient (d). Green and red regions in the "Eyecatcher" output (left column) represent areas of the visual field where the stimuli were seen and not seen respectively. Superimposed numbers represent the Humphrey sensitivity values in decibels. Right column shows the corresponding Humphrey Visual Fields results in greyscale. (*See color plate section for the color representation of this figure.*)

against standard automated perimetry testing, as shown in Figure 25.9. A median suprathreshold defect concordance between Eyecatcher and standard automated perimetry was reported to be 83% (95%CI: 62–90%) and 96% (95% CI: 92–97%) for diseased and healthy eyes respectively. Whilst these only represent preliminary results, the prospect of an accurate visual fields test that is low-cost, mobile and only requires eye movements is a promising development with potential applications in remote locations, low-income settings and within low-mobility patient groups.

25.4 Smartphone Ophthalmoscopy

Ophthalmoscopy has long been established as an important procedure in the detection of eye disorders such as glaucoma, diabetic retinopathy and macular degeneration as well as other systemic diseases including hypertension, malaria and some neurological conditions. For this reason, every medical doctor is expected to be competent in the use of a direct ophthalmoscope. However, it should be noted that often this competency is lacking, with it being reported that as many as 47% of medical students in their final years stating that they were "not at all" or only "a little" confident in performing direct ophthalmoscopy (Gupta and Lam 2006).

Images of the ocular fundus have been photographed using film cameras since the nineteenth century (Gerloff 1891). With digital camera sensors having exceeded the resolution of 35 mm film these have how superseded their film counterparts in modern retinal cameras. However, it is important to note that the resolution of a camera is not solely determined by the number of pixels found on the sensor despite many manufacturers and even some peer-reviewed literature quoting this as a proxy for camera performance. Factors affecting image quality include aberrations with the camera's optics, distortion and field curvature. Ultimately the quality is fundamentally limited by diffraction. An accurate and objective measure of an imaging system's resolving power can be obtained using specially designed test targets such as the United States Air Force (USAF) 1951 resolution test chart. The resolving power is found by finding the minimum resolvable separation of a set of three parallel black lines, with a width equal to their separation, on a white background (Armed Forces Supply Support Center 1959). The current international standards covering image quality for fundus cameras specify that a minimum resolving power of 80 lp/mm (line pairs per millimeter) in the center of the image to 40 lp/mm at the periphery for a field of view (FoV) of 30° or less. Cameras with a wider FoV are not required to have such a high resolving power, with a minimum of 60 lp/mm at the center to 25 lp/mm at the periphery being specified (International Standards Organization 2009a).

Mobile phones started to include digital cameras in 2000. There are profound engineering challenges associated with designing a camera that is viable for smartphone integration and capable of general photography. Plastic lenses and apertures of fixed sizes are generally used to keep costs low and allow for a wide range of scenes. The full camera system also needs to be thin, limiting the space available between optical components. As with any digital camera, image encoding and compression can also degrade the quality. In addition, the associated algorithms need to be able to run within the computing and storage constraints of the host mobile device.

The introduction of digital imaging technologies allows electronic storage and transmission of fundus photographs. As such, it is now standard practice in many countries for retinal images to be captured by a retinal photographer and sent electronically for grading by a trained retinal grader at a separate grading center. Furthermore, the fact that digital images are machine-readable opens up the possibility of automated screening algorithms with algorithms for diabetic retinopathy having already been demonstrated (Fleming et al. 2010; Gulshan et al. 2016).

However, standard fundus cameras are expensive, require technical skills, are bulky and are difficult to transport. Modern smartphones by contrast are inexpensive, mobile and almost universally contain integrated digital cameras for operation by those with little skill in photography. The prospect of a smartphone-based ophthalmoscope is therefore a very attractive one.

The simplest way to use a smartphone in fundus photography is simply to insert its camera into an indirect biomicroscope in place of the operator's eye. (In ophthalmology a biomicroscope is an instrument consisting of a microscope combined with a light source that can be narrowed into a slit [slit lamp].) A wide selection of attachments are available which allow the phone to be held steady at the slit lamp's eyepiece. These range from commercial products costing up to hundreds of U.S. dollars to do-it-yourself (DIY) kits costing less than $10 USD (Chan et al. 2014). Such adapters allow images of both the anterior and posterior segments to be quickly captured, stored and shared with colleagues without needing to send the patient for another examination on a separate instrument. This is particularly beneficial in clinical environments that have tight time or resource constraints and where separate devices for retinal image documentation are not available. Whilst the ability to add an inexpensive digital photography capability to a slit lamp has a number of advantages the full potential of smartphone ophthalmoscopy can only truly be realized if it is able to operate in the absence of bulky and expensive equipment designed for hospital use.

Blackenberg and Scheffer made an early attempt to rely instead upon less expensive and portable equipment by attaching a digital camera to a "Panoptic" monocular indirect ophthalmoscope (Welch-Allyn Inc., Skaneateles Falls, NY, USA) (Blackenberg, Worst and Scheffer 2011). Following this work, Welch Allyn released the "iExaminer" attachment that allows an iPhone 4 or 4s to be replace the eyepiece and captures images using the iExaminer app. In this way quality fundus images can be captured, emailed, printed and stored without the need for large hospital-based equipment. The iExaminer was also the first smartphone ophthalmoscope adapter to achieve regulatory approval having obtained USA FDA approval. Nevertheless, the full system, excluding the phone, costs the end user in the region of $800 USD (www.medicaldevicedepot.com) which, although cheaper than standard fundus photography equipment, is in the price range of premium ophthalmoscopes. Furthermore, updates to the iExaminer's compatibility tends to lag significantly behind new iPhone releases and at time of writing was only compatible with iPhone 6 and 6s series, more than nine months after the release of the iPhone 7 (www.welchallyn.com, accessed 1 July 2017). This highlights one of the major difficulties encountered by smartphone ophthalmoscope designers, namely the speed at which the mobile phone market develops.

Lord and colleagues demonstrated a simpler system for capturing fundal images when they used a 20 diopter lens (Volk Optical Inc., Mentor, OH, USA), iPhone (Apple inc.) and pen torch alone to acquire fundal images. Their method involved positioning the lens close to the eye, as per standard indirect ophthalmoscopy, and holding the phone and torch at an appropriate distance from the lens in order to provide imaging and illumination respectively. The need for complex imaging optics, ordinarily worn on the operators head, is removed with the phone's autofocus bringing the image of the fundus into focus on the imaging sensor. However, the technique requires a high level of skill and capturing high quality images was found to be difficult (Bastawrous 2012). As with standard indirect ophthalmoscopy, producing an image is sensitive to the distance between the cornea and the 20 diopter (D) lens, which varies according to the patient's refractive error, and the distance between the lens and the phone's optics. The need to hold all three items at once also contributes to the method's difficulty. Fortunately, integrated light-emitting diode (LED) flashes soon became standard with smartphones. (Flashing LEDs are used as attention-seeking indicators that do not require external electronics.) By capturing a video stream with the flash enabled in the phone's stock camera application, the need for a pen torch to be held alongside was therefore eliminated (Bastawrous 2012). The ease-of-use and image quality of the technique can be further improved by using any one of a number of specialist photography apps allowing control over settings such as focus, exposure, white balance and flash light intensity. For example, Haddock and colleagues found that they were able to produce consistently higher quality images by using the camera app "Filmic Pro" (Cinegenix LLC, Seattle, WA, USA) (Haddock et al. 2013). As well as allowing independent control over the settings mentioned above the app also allowed the extraction of individual frames after recording a period of video. Thus, even if only a fleeting glimpse of a feature of interest is achieved a still image can still be acquired without the operator needing to react quickly during the examination itself.

Indirect smartphone ophthalmoscopy's potential for use in challenging patients was demonstrated by Lin and colleagues when they applied the technique to screening neonatal patients for the retinopathy of prematurity (Lin et al. 2014). As screening for this condition requires a wide field view of the retina, and thus the combination of a smartphone with a 30D lens, it was not sufficient to gain a wide enough field of view. However, with scleral indentation a full view of the retina could be achieved allowing photographic records to be made. Thus, the technique shows promise as it does not rely on written notes alone or require an expensive wide-field retinal camera.

Nevertheless, the precise alignment needed between eye, lens and phone is a limiting factor in making indirect smartphone fundus photography feasible for the non-specialist. To improve the usability of the method, Myung and colleagues developed a lightweight arm that holds the lens at the correct distance from the camera optics (Myung et al. 2014). The authors produced both 3 dimensional printed and machined arms, which held a 20D lens at an adjustable fixed distance from an iPhone 5 running Filmic Pro. Using the native LED flash, the authors were able to produce a selection of good quality images from both normal and abnormal dilated eyes.

Clinical validation of Myung and colleagues' prototype, which they named "EyeGo," was undertaken on 144 patients presenting at an eye hospital in Hyderabad, India. Each of the patients included in the study had their pupils dilated to pupil diameters of 6 mm or more. Either an ophthalmic technician (with two years optometric training), a medical student, an optometrist or an ophthalmologist imaged the study participants using EyeGo. Images were then graded by a single grader on a desktop display who was able to adjust the color and brightness of images as deemed necessary.

The image quality of 86.7% of the videos acquired with EyeGo were of sufficient quality to exclude the presence of optic disc edema, optic disc pallor, retinal vascular occlusion, intraocular hemorrhages, and grade III/IV hypertensive retinopathy. By comparison, the images taken with a desktop fundus camera were of sufficient quality to exclude the presence of the same features in 97.7% of eyes. Following the EyeGo's clinical evaluation, an FDA-approved commercial version named "Paxos Scope" (Digisight, San Francisco, CA, USA) was released together with image capture, cloud storage and encryption software and available, as of writing, for $79 USD per month.

One drawback of smartphone ophthalmoscopy compared to standard retinal photographs is the reduced field-of-view. Maamari and colleagues demonstrated how a software method called image mosaicking could increase the field presented by a single image (Maamari et al. 2014). Images captured using the "Ocular Cellscope," which consists of a system of independent LEDs, lenses and polarizers, were inputted into the image mosaicking software installed on a desktop computer. Although the software was originally developed for use with desktop retinal cameras, the authors reported mosaics that both encompassed a large field and were of good clarity. Having demonstrated the viability of the technique with smartphone ophthalmoscopes, the next step is to develop such software capable of being deployed on the phone itself and even building a mosaic in real-time during the examination itself.

Whilst smartphone ophthalmoscopes using the principal of indirect ophthalmoscopy have shown significant promise, they are generally either difficult for the non-eye professional to use or significantly more expensive. Smartphone ophthalmoscopes using the principles of direct ophthalmoscopy offer provide both ease-of-use and affordability.

D-Eye (D-Eye Srl, Padova Italy; Figure 25.10) re-directs the flash of a smartphone to a co-axial alignment with the camera optics by inserting a beam splitter in the optical path (Russo and Civili 2015). As the position and distance of both the camera and LED flash on the back face of the phone vary according to make and model, it is necessary for the position of the optics to be adjusted. D-Eye overcomes this problem by screwing a "bumper" to the perimeter of the phone and attaching a unit containing the optics using magnets. By using a different bumper for differing phone models co-axial alignment of the camera and LED flash's optical path can be achieved. The drawback of this design is that a different bumper design needs to be manufactured for every handset for which compatibility is desired. As a result, the number of compatible smartphones is limited, with only Apple handsets being compatible at time of writing. Russo and colleagues compared D-eye mounted on an iPhone 5 (Apple Inc.) to biomicroscopy for the assessment of 240 eyes in 120 out-patients attending a diabetic eye center in Spedali Civili di Brescia, Italy (Russo et al. 2015). A sensitivity of 90% (95% CI: 82–94%) and specificity of 96% (95% CI: 90–98%) was reported for a retinal specialist's detection of diabetic retinopathy. For the assignment of a particular grade of diabetic retinopathy by a retinal specialist a simple κ of 0.78 (CI: 95%: 0.71–0.84, p < 0.001) was reported with 3.75% of eyes being ungradable with D-Eye compared to 1.7% with biomicroscopy. With regard to the detection of significant cystoid macular edema, sensitivity and specificity were 81% (95% CI: 54–94%) and 98% (95% CI: 95–99%) respectively with a simple κ of 0.79 (95% CI: 0.65–0.93).

As the images were acquired and graded by a retinal specialist it is difficult to assess how D-Eye performs in the hands of general practitioners, nurses or non-clinical staff compared to biomicroscopy. Indeed, it would also be useful to understand how use by the non-clinician compares to standard retinal photography by a clinical photographer, which is the most common mode of retinal imaging in diabetic retinopathy screening programmes (Peto and Tadros 2012).

Giardini and colleagues took an alternative approach with their direct ophthalmoscope smartphone adapter (Giardini et al. 2014). Rather than using a beam splitter their design instead used a dove prism bringing the optical path of the LED flash into close,

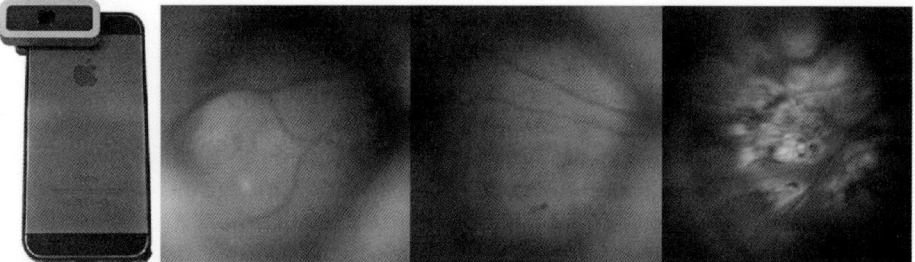

Figure 25.10 Example of mydriatic retinal images taken with the D-Eye on an iPhone 6 (far left) by a retinal specialist. Mild nonproliferative diabetic retinopathy (second from left), moderate nonproliferative diabetic retinopathy (second from right), and panretinal photocoagulation scars on a retina with proliferative diabetic retinopathy (far right) are shown. (*See color plate section for the color representation of this figure.*)

rather than exact, alignment with that of the camera. Bastawrous and colleagues demonstrated the effectiveness of this design with 3D printed adapters including a reverse dove prism mounted on a Galaxy S3 (Samsung Group, Seoul, South Korea). Their study involved a comparison of optic nerve images acquired by a lay examiner using the smartphone ophthalmoscope and a standard desktop retinal camera (CentreVue+ Digital Retinal System, Haag-Streit, Köniz, Switzerland) operated by an ophthalmic technician. The validation study was nested within a population-based cohort study of eye disease in Nakuru, Kenya with vertical cup-to-disc ratios being graded remotely at Moorfields Eye Hospital Reading Centre (London, UK). A weighted κ of 0.71 was reported, and Bland–Altman analysis showed there was an average difference of −0.2 with 95% limits of agreement between −0.21 and 0.16. Thus, it was concluded that there was good agreement between the lay-operated smartphone ophthalmoscope and the reference standard.

Peek Retina (Peek Ltd.; Figure 25.11) uses a principle similar to that employed by Giardini and colleagues. Instead of using a dove prism to re-direct the flash, Peek Retina uses a ring of LEDs that are placed close to the camera optics. In doing so, near-alignment can be achieved, allowing the fundus to be imaging through dilated pupils. Additionally, as the adapter uses a configuration of six light sources, rather than a single light source, the shadow seen during use of Giardini and colleague's adapter is eliminated. The unit also uses an attachment mechanism for the adapter's position on the phone to be adjusted, allowing its use with a wide variety of handsets (Table 25.2).

25.5 Discussion

The efficacy of a number of mobile device-based eye tests have been demonstrated and been made available to medical practitioners for purchase. Overall, these show significant promise for increasing access to eye services and reducing the cost to patients and health services. These costs include the direct costs of the equipment itself, which is much cheaper than desktop hospital equipment. Also, the portability of the discussed

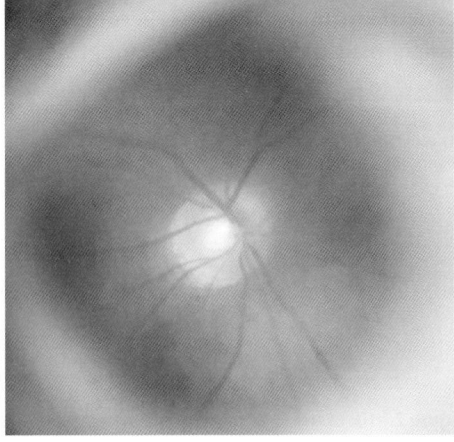

Figure 25.11 Example image of a healthy fundus (right) acquired using Peek Retina (left) under dilation. Images re-produced by permission of Peek Vision Ltd.

Table 25.2 A selection of popular, regulatory approved adapters for smartphone ophthalmoscopy.

Adapter	Manufacturer	Ophthalmoscopy technique	Compatible handsets	Regulatory approval	Price
Panoptic with iExaminer	Welch Allyn	Monocular indirect	iPhone 4, 4 s, 6, 6 s, 6+, 6 s+	FDA/CE	Approx. 795 USD
Paxos Scope	Digiscope Technologies Inc.	Indirect	iPhone 6, 6 s, 7, 7 + iPod 6	FDA	79 USD / month
D-Eye	D-Eye Srl	Direct	iPhone 5, 5 s, SE, 6, 6 s, 6+, 6 s +, 7	FDA/CE	395 EUR
Peek Retina	Peek Vision Ltd.	Direct	All Apple and most Android handsets	CE/FDA pending	180 GBP

Abbreviations used: FDA, Food and Drug Administration (USA); CE, Conformité Européene (European Conformity) – i.e. approved by the European Union; USD, US dollar; GBP, Great Britain Pound; EUR, Euro.

technology reduces the indirect costs often imposed on patients, such as travel and time off work, by allowing screening to occur in even the most remote communities rather than in centralized clinics.

However, to be truly effective in tackling the burden of avoidable blindness such tests will need to be fully integrated into the healthcare systems. Furthermore, they need to be implemented in a way that causes minimal disruption to care delivery, whilst promoting the improvement of clinical pathways and practice. Therefore, the provision of mere tools will not be enough to change the delivery of care but software, training and management systems that support the entire care flow will be required.

To that end it is commendable that iExaminer, Paxos Scope and D-Eye all have associated software that allows storage and sharing of fundus images. In particular, it is encouraging to see that the Paxos Scope application has achieved Health Insurance Portability and Accountability Act (HIPAA) compliance, although at $79 USD per month the cost of its subscription is beyond that which healthcare providers in low-income settings can afford.

Another requirement of the smartphone ophthalmoscopy solutions that are discussed in this chapter that has prevented wide-spread adoption is the need for dilation. So far, clinical evaluation of each of the solutions discussed involved dilation which in many time-constrained environments, such as emergency medicine and general practice, is not feasible. A non-mydriatic solution is therefore highly sought after and would also open up new applications, such as home screening, which has the potential to even further broaden access.

Finally, the authors believe the work of Peek Vision Ltd. in incorporating their visual acuity app into a much broader system capable of delivering screening at regional and national scale is the next important step in mHealth's development. Such a solution will consist of both technology tools, such as patient management and text message reminder systems, and training and management materials. Hence, it will provide all of the assets required to positively change behaviors throughout the clinical workflow rather than merely providing the potential for change. The authors call on other innovators of mHealth solutions to take similar steps to realize the promise that mHealth could be one of the most impactful healthcare technologies of our time.

25.6 Conclusions

Global connectivity has been transformed by the rapid proliferation of mobile devices and the cellular networks that they connect to. The impact of this technology revolution has been most profound in LMICs where the availability of low-cost mobile phones, tablet PCs and smartphones has brought billions of people access to telephony, the internet and a range of other services for the first time. Eye care has the potential to reap huge rewards from mobile device technology. Hardware adapters and software applications are beginning to emerge that offer the possibility for low-skilled workers to conduct imaging and screening tasks that could previously only be conducted by trained eye workers. Several mobile device-based visual acuity, visual fields and fundus imaging technologies have been clinically validated, approved by regulatory authorities and are now available commercially. However, for the full potential for work force task-shifting at the primary point-of-care to be realized, these technologies will need to be integrated into systems that span the full care flow and provide the training and management features that are required for eye health screening at scale.

25.6.1 Competing Interests

The authors are both paid employees of Peek Vision Ltd. of which Andrew Bastawrous is also a company director. Nigel Bolster is a named co-inventor on a patent application relating to smartphone fundoscopy with publication number WO2016151288 A1. Andrew Bastawrous is a named co-inventor on a patent application relating to smartphone fundoscopy with publication number WO 2014181096 A1. No writing assistance was utilized in the production of this material.

Bibliography

Armed Forces Supply Support Center (1959). Military Standard Photograph Lenses. MIL-STD-150A.

Aslam, T.M., Murray, I., Lai, M.Y. et al. (2013). An assessment of a modern touch-screen tablet computer with reference to core physical characteristics necessary for clinical vision testing. *Journal of The Royal Society Interface* 10: 20130239.

Bastawrous, A. (2012). Smartphone Fundoscopy. *Ophthalmology.* 19 (2): 432–433.

Bastawrous, A., Mathenge, W., Peto, T. et al. (2014). The Nakuru eye disease cohort study: methodology and rationale. *BMC Ophthalmology* 14: 60.

Bastawrous, A., Rono, H.K., Livingstone, I.T. et al. (2015). Development and validation of a smartphone-based visual acuity test (peek acuity) for clinical practice and community-based fieldwork. *JAMA Ophthalmology* 133: 930–937.

Blackenberg, M., Worst, C. and Scheffer, C. (2011). Development of a Mobile Phone Based Ophthalmoscope for Telemedicine. *33rd Annual International Conference of the IEEE EMBS: 2011*; Boston, Massachusetts USA.

Brady, C.J., Eghrari, A.O., and Labrique, A.B. (2015). Smartphone-based visual acuity measurement for screening and clinical assessment. *JAMA* 314: 2682–2683.

Chan, J.B., HO, H.C., Ngah, N.F., and Hussein, E. (2014). DIY-smartphone slit-lamp adaptor. *Journal of Mobile Technology in Medicine* 3: 16–22.

Congdon, N., Li, L., Zhang, M. et al. (2011). Randomized, controlled trial of an educational intervention to promote spectacle use in rural China: the see well to learn well study. *Ophthalmology* 118: 2343–2350.

Emarketer. (2015). Tablet Users to Surpass 1 Billion Worldwide in 2015 [Online]. Available: http://www.emarketer.com/Article/Tablet-Users-Surpass-1-Billion-Worldwide-2015/1011806 [Accessed 16th September 2016].

Ferris, F.L., Kassoff, A., Bresnick, G.H., and Bailey, I. (1982). New visual acuity charts for clinical research. *American Journal of Ophthalmology* 94: 91–96.

Fleming, A.D., Goatman, K.A., Philip, S. et al. (2010). Automated grading for diabetic retinopathy: a large-scale audit using arbitration by clinical experts. *British Journal of Ophthalmology* 94 (12): 1606–1610.

Gerloff, O. (1891). Über die Photographie des Augenhintergrundes. *Zeitschrift für Psychologie und Physiologie der Sinnesorgane* 29: 397–403.

Giardini, M.E., Livingstone, I. A., Jordan, S., et al. (2014). A smartphone based ophthalmoscope. 36th Annual International Conference of the IEEE Engineering in Medicine and Biology Society (EMBC), 2014. 2177–2180.

Gsma Intelligence. (2014). Mobile broadband reach expanding globally.

Gulshan, V., Peng, L., Coram, M. et al. (2016). Development and validation of a deep learning algorithm for detection of diabetic retinopathy in retinal fundus photographs. *JAMA* 316: 2402–2410.

Gupta, R. and Lam, W.-C. (2006). Medical students' self-confidence in performing direct ophthalmoscopy in clinical training. *Canadian Journal of Ophthalmology/Journal Canadien d'Ophtalmologie* 41: 169–174.

Haddock, L.J., Kim, D.Y., and Mukai, S. (2013). Simple, inexpensive technique for high-quality smartphone fundus photography in human and animal eyes. *Journal of Ophthalmology* 2013: 518479.

Hughes, N. and Lonie, S. (2007). M-PESA: mobile money for the "unbanked" turning cellphones into 24-hour tellers in Kenya. *Innovations* 2: 63–81.

Hwang, J.-H. and Yoe, H. (2011). Design and implementation of ubiquitous pig farm management system using iOS based smart phone. International Conference on Future Generation Information Technology, 2011. Springer, 147–155.

International Standards Organisation 2003. Test charts for clinical determination of distance visual acuity – Specification. *Visual acuity test types*. Geneva, Switzerland.

International Standards Organisation (2009a). *Ophthalmic Instruments: Fundus Cameras*. Geneva, Switzerland.

International Standards Organisation (2009b). Ophthalmic optics — Visual acuity testing — Standard optotype and its presentation (ISO 8596:2009). Geneva, Switzerland.

International Telecommunication Union (2016). Global ICT Developments. In: STATISTICS.

International Telecommunication Unions (2015). Key ICT indicators for developed and developing countries and the world (totals and penetration rates). ICT Statistics.

Jacob, F. (2016). The role of M-Pesa in Kenya's economic and political development. In: *Kenya after 50: Reconfiguring Education, Gender, and Policy* (ed. M.M. Koster, M.M. Kithinji and J.P. Rotich), 89–100. New York: Palgrave Macmillan US.

Johnson, C.A. and Casson, E.J. (1995). Effects of luminance, contrast, and blur on visual acuity. *Optometry and Vision Science* 72: 864–869.

Johnson, C., Thapa, S., and Robin, A. (2014). Visual field screening to detect glaucoma and diabetic retinopathy in Nepal using an iPad application program. *American Academy of Optometry* 140073.

King-Smith, P.E., Grigsby, S.S., Vingrys, A.J. et al. (1994). Efficient and unbiased modifications of the QUEST threshold method: theory, simulations, experimental evaluation and practical implementation. *Vision Research* 34: 885–912.

Kong, Y.X.G., He, M., Crowston, J.G., and Vingrys, A.J. (2016). A comparison of Perimetric results from a tablet perimeter and Humphrey field Analyzer in glaucoma patients. *Translational Vision Science and Technology* 5: 2.

Lin, S.-J., Yang, C.-M., Yeh, P.-T., and Ho, T.-C. (2014). Smartphone fundoscopy for retinopathy of prematurity. *Taiwan Journal of Opthalmology* 4 (2): 82–85.

Livingstone, I.A.T., Tarbert, C.M., Giardini, M.E. et al. (2016). Photometric compliance of tablet screens and retro-illuminated acuity charts as visual acuity measurement devices. *PLoS ONE* 11: e0150676.

Maamari, R.N., Keenan, J.D., Fletcher, D.A., and Margolis, T.P. (2014). A mobile phone-based retinal camera for portable wide field imaging. *British Journal of Ophthalmology* 98: 438–441.

Mas, I. and Morawczynski, O. (2009). Designing mobile money services lessons from M-PESA. *Innovations* 4: 77–91.

Medicines And Healthcare Products Regulatory Agency (2014). Medical devices: software applications (apps). London.

Morjaria, P., Bastawrous, A., Murthy, G.V.S. et al. (2017). Effectiveness of a novel mobile health education intervention (peek) on spectacle wear among children in India: study protocol for a randomized controlled trial. *Trials* 18: 168.

Myung, D., Jais, A., He, L. et al. (2014). 3D printed smartphone indirect lens adapter for rapid, high quality retinal imaging. *Journal of Mobile Technology in Medicine* 3: 9–15.

Palma, M.J. (2015). Worldwide Tablet Semiconductor 2015–2019 Forecast. International Data Corporation.

Perera, C., Chakrabarti, R., Islam, F.M.A., and Crowston, J. (2015). The eye phone study: reliability and accuracy of assessing Snellen visual acuity using smartphone technology. *Eye* 29: 888–894.

Peto, T. and Tadros, C. (2012). Screening for diabetic retinopathy and diabetic macular edema in the United Kingdom. *Current Diabetes Reports* 12: 338–345.

Pew Research Center (2015). *Device Ownership Over Time* [Online]. Available: http://www.pewinternet.org/data-trend/mobile/device-ownership [Accessed 16th September 2016].

Resnikoff, S., Felch, W., Gauthier, T.-M., and Spivey, B. (2012). The number of ophthalmologists in practice and training worldwide: a growing gap despite more than 200 000 practitioners. *British Journal of Ophthalmology* 96: 783–787.

Russo, A. and Civili, P.S. (2015). A Novel Device to Exploit the Smartphone Camera for Fundus Photography. http://dx.doi.org/10.1155/2015/823139.

Russo, A., Morescalchi, F., Costagliola, C. et al. (2015). Comparison of smartphone ophthalmoscopy with slit-lamp biomicroscopy for grading diabetic retinopathy. *American Journal of Ophthalmology* 159: 360–364. e1.

Sheedy, J.E., Bailey, I.L., and Raasch, T.W. (1984). Visual acuity and chart luminance. *Optometry and Vision Science* 61: 595–600.

Sloan, L. (1961). Area and luminance of test object as variables in examination of the visual field by projection perimetry. *Vision Research* 1: 121–IN2.

Smith, N., Bi, W., and Crabb, D. (2016). Novel perimetry using eye tracking on a tablet computer–a feasibility study. *Investigative Ophthalmology & Visual Science* 57.

Snellen, H. (1873). *Probebuchstaben zur bestimmung der sehschärfe*. Berlin: H. Peters.

Statista. (2016). *Number of apps available in leading app stores as of June 2016* [Online]. Available: https://www.statista.com/statistics/276623/number-of-apps-available-in-leading-app-stores [Accessed 12 November 2016].

Tahir, H.J., Murray, I.J., Parry, N.R.A., and Aslam, T.M. (2014). Optimisation and assessment of three modern touch screen tablet computers for clinical vision testing. *PLoS One* 9: e95074.

Vingrys, A.J., Healey, J.K., Liew, S. et al. (2016). Validation of a tablet as a tangent perimeter. *Translational Vision Science and Technology* 5: 3.

Wedner, S., Masanja, H., Bowman, R. et al. (2008). Two strategies for correcting refractive errors in school students in Tanzania: randomised comparison, with implications for screening programmes. *British Journal of Ophthalmology* 92: 19–24.

World Health Organisation (2016). H54 Visual impairment including blindness (binocular or monocular). *International Statistical Classification of Disease Related Health Problems 10th Revision (ICD-10).* Geneva.

Webliography

www.medicaldevicedepot.com – a medical supply store in the USA.

www.welchallyn.com, accessed:1 July 2017 – an American manufacturer of medical diagnosis devices, patient monitoring systems and miniature precision lamps.

Part V

Telemedicine

26

Telemedicine for Clinical Management of Adults in Remote and Rural Areas

Farhad Fatehi, Monica Taylor, Liam J. Caffery, and Anthony C. Smith

Centre for Online Health, Faculty of Medicine, The University of Queensland, Brisbane, Queensland, Australia

Revolutionizing Tropical Medicine: Point-of-Care Tests, New Imaging Technologies and Digital Health,
First Edition. Edited by Kerry Atkinson and David Mabey.
© 2019 John Wiley & Sons, Inc. Published 2019 by John Wiley & Sons, Inc.

26.1   Introduction

Universal health coverage, which was committed to by all World Health Organisation (WHO) member countries in 2005, aimed to provide access for all people to health services that they needed, with financial risk protection. Despite some progress made, healthcare systems in many nations fall far short of universal coverage, especially in low- and middle-income countries (LMICs). Moreover, it is evident that the residents of remote and rural areas have inferior health status compared to their counterparts living in metropolitan areas. Mortality and complications of chronic diseases increase with remoteness. Several factors play a role in degrading the health indices of these regions, most notably geographical and socio-economic disparities. Inequality in access to healthcare is one of the disparities observed among developed countries and LMICs. Three distinct dimensions are defined for access to healthcare: physical accessibility, financial affordability, and acceptability.

Geographical separation, high travel cost and distorted resource allocation limit physical accessibility of people to quality healthcare, especially in remote and rural areas. Disparity in the availability of health workers among different countries has been highlighted in the WHO World Health Report 2006. Countries with the greatest burden of disease and thus the highest relative need, have the lowest numbers of health workers. Uneven distribution of health workers can be seen within regions and countries as well. In Vietnam, for example, 37 out of 61 provinces have less than one health service provider per 1000 individuals, while this ratio in some other provinces is four per 1000 individuals. Similar situations can be observed in other countries as well. This inequality can be attributed to insufficient training of physicians and health workers, maldistribution of healthcare resources, and, for many LMICs, brain drain migration.

Affordability of needed healthcare is another limiting factor for many individuals and families. It is estimated that more than one billion people, mostly living in low- and middle-income countries, cannot afford their required healthcare services. The United States and Canada, with 14% of the world's population, spend around 50% of the global healthcare expenditure, whereas 11% of the world's population living in sub-Saharan Africa spend only 1% of the global healthcare dollars. LMICs have the highest rates of out-of-pocket payments for healthcare, facing many people with financially catastrophic charges.

Acceptability of a health service can be suboptimal due to long queues, perceived ineffective care and disrespectful treatment. Social, cultural and religious issues may prevent patients from seeking proper healthcare, even if those services are accessible and affordable. For instance, in some communities women are culturally or religiously obliged to receive healthcare from a female provider. In addition to these issues, privacy and confidentiality are also important aspects of an acceptable health service, in particular for people living with stigmatized diseases such as HIV/AIDS or mental health disorders. In less urbanized areas and small communities where people are more likely to know each other, disease stigmata are thought to be a strong barrier to seeking healthcare.

There is no doubt that the traditional models of healthcare cannot easily overcome the above mentioned obstacles and meet the increasing demand of health services in most countries, thus innovative and more efficient models are required for providing equitable access to healthcare services for underserved populations. One of the most promising solutions to improve access to healthcare and clinical services is the use of information and communications technology (ICT) to deliver health services, and is known as telemedicine. Telemedicine can address several barriers of access to healthcare by bridging the geographical gap between healthcare providers and consumers, decreasing the cost of services particularly from a patient's perspective, and increasing acceptability of service in a number of conditions and situations. According to WHO, it is now clear that the aims of universal health coverage are not achievable without the help of ICT (WHO 2010).

26.2 Definitions

Telemedicine is the provision of medical services using ICT where the provider (usually a doctor) and the recipient (could be a patient or another healthcare provider) are geographically separated. Another similar term is telehealth, which is the delivery of healthcare or health-related services at a distance by means of ICT. Although there is no clear distinction between the terms telehealth, telemedicine, and Electronic Health, it is widely considered that Electronic Health (eHealth) is defined as the use of ICT for the support of health and health-related fields in a cost-effective and secure way. The application of eHealth goes beyond the delivery of care and extends to education, research and surveillance. Thus eHealth embraces several inter-related domains such as the exchange of health-related data in an electronic format (e.g. electronic health records – EHRs), eLearning in health sciences, telehealth, and so on (Fatehi and Wooten 2012; WHO 2016a). Mobile Health (mHealth) is a subset of eHealth in which mobile phones or other portable electronic devices are used to support medical and public health practice. More recently, the emergence of digital and wireless technology and the Internet of Things (IoT) has led to the introduction of several new terms such as digital health and connected health, adding to the ambiguity of the terminology in this field (WHO 2016b). In an effort to not overcomplicate the matter, for the purposes of this chapter we will use the term "telemedicine" to refer to the use of ICT for the provision of any health-related service.

Appropriate use of technology is an important success factor for implementing telemedicine services. Various forms of telecommunication systems have been used for connecting healthcare providers and consumers. For several decades, telephone was the main telecommunication means of health and medical information exchange. Just three years after the invention of telephone, The Lancet published the first case of remote diagnosis by a doctor in 1879. In addition to voice calls, telephone lines were also used for electronic communication of health and medical data. One of the earliest experiments of telemedicine is the transmission of electrocardiography (EKG) information over telephone lines by Einthoven in 1906 This is regarded as the start of modern telecardiology. Years later, after the introduction of television in the United States, the Nebraska Psychiatric Institute for the first time started using television systems for distance education in psychiatry in 1955, and in 1964 they used a two-way closed-circuit television established by microwave link for medical purposes. Microwave and satellite telecommunication made it possible to establish two-way interactive videoconferencing between two or more health facilities hundreds of kilometers apart. In 1966 the US National Aeronautics and Space Administration (NASA) launched the first approved satellite for telemedical purposes (ATS-1 Satellite). Satellite-based telecommunication technology overcame the limitations of microwave links and facilitated international and intercontinental telemedicine activities. However, the high cost of these technologies restricted their use to very specific situations and settings. Later on, the development of the internet transformed communication in the health sector, as it did in other sectors, and a range of telecommunication solutions such as Dial-up, Broadband, and ADSL were introduced. In recent years, superfast Internet through fiber optic cables became available mainly to the residents of developed countries, whereas LMICs showed much more growth in mobile (cellular) Internet connectivity compared to developed countries.

The availability of high-bandwidth connectivity, along with the increased processing power of computers and mobile phones and development of audio/video compression algorithms, made videoconferencing a favorite means of telecommunication for telemedicine (Fatehi et al. 2015a).

26.3 Types of Service

Telemedicine services are divided into three main categories based on the synchronicity of the interaction between the provider and the recipient:

1) Store-and-forward (asynchronous) solutions in which the information from one party is acquired, stored, and communicated to the other party to review and to respond to. A good example of this type of service is teleradiology. It is best implemented in facilities that lack radiologists. In such situations a radiographer takes the radiographic images. The images are electronically stored and forwarded, along with other required patient data, to a radiologist to read and provide comments back to the requesting physician. Teleradiology, telepathology, and teledermatology (see Chapters 26–29) are among the most successfully implemented store-and-forward telemedicine services worldwide.

2) Real-time (synchronous) solutions in which the provider and recipient interact with each other in real-time such as videoconferencing. A typical example of this service is a video consultation between a general practitioner (GP) in a remote area and a specialist physician in a tertiary hospital. In this model the GP can present the patients to the consultant physician (with or without them present at the video consultation session) and seek advice on the management of complex patients. With the decreasing cost of videoconferencing equipment and the increasing availability of broadband connections and alternative solutions such as *Skype*, this type of telemedicine is gaining more popularity.

3) Hybrid solutions use both store-and-forward and real-time solutions for the delivery of care. An example of this service is when the patient's data are sent by fax to a specialist before a video consultation.

26.4 Purposes of Telemedicine

The main purpose of telemedicine is the provision of clinical services, including activities for screening, prevention, diagnosis, treatment, follow-up and rehabilitation. Also, in many instances, telemedicine has been used for seeking second opinions or specialist recommendations for managing complex cases. Efficacy of the care provided via telemedicine has been demonstrated in both acute care, chronic care, and emergency departments. However, people with chronic diseases would benefit more from telemedicine due to the need for frequent contact with healthcare providers for a long duration of time. In addition to clinical services, telemedicine has also successfully been used for medical training and continuing medical education.

26.5   Telemedicine for Improving Access to Care

The original impetus of telemedicine was providing access to specialized care that was not available otherwise. Uneven distribution of doctors and specialist health workers is a common issue in most countries. Such staff are costly to train and difficult to retain in remote and rural areas. Relative lower demand of clinical care due to a lower population density in small towns and isolated small communities cannot justify capital investment and employment of expensive healthcare professionals similar to that available in large cities. Lack of support and feeling of isolation are among the concerns of the healthcare providers who reside in, and cater for, less urbanized areas. These issues make training, recruiting and maintaining healthcare professionals a common challenge of health systems in LMICs especially. Thus, the required services are provided via outreach programs, which are often costly and not timely, or the patient travels to a place where the service is available, adding a financial burden and inconvenience of travel for patients and their families. Telemedicine has shown promise for improving access to specialized care for people living in areas that lack specialist clinicians and several studies show both providers and recipients are generally satisfied by these "virtual outreach" programs (Fatehi et al. 2015b).

26.6   Establishing a Sustainable Telehealth Network: A Case Study from Brazil

The Telehealth Network, which has been established by the state government of Minas Gerais in Brazil, is a partnership of seven public universities. It connects primary care units, most of them located in remote and rural areas, to secondary care as well as to emergency care units in the state's capital city. This network covers more than 95% of the cities in this state and is funded by both the Brazilian government and the state government of Minas Gerais. In addition to clinical services, this network is used for research and education as well (Alkmim et al. 2012).

Several factors have played a role in the success of this telehealth network (Table 26.1), most notably support of public managers, service provision though a collaborative net-

Table 26.1 Enablers and success factors of the Telehealth Network in Brazil.

Enablers	Success factors
1) Disparity in access to specialized care	1) Top managerial support
2) High turnover of practitioners	2) Systematic monitoring of the service
3) Great economic and cultural divide	3) User-friendly system
	4) Short response time
	5) Government-academia partnership
	6) Research capabilities

work, a government-academia partnership, ease of use of the systems, diversity of tele-health activities, and economic viability.

From 2005 to 2015 more than two million consultations were carried out through this network. It is reported that with an investment of around $9 million USD, the cost saving for the public health system was more than $33 million USD, over this ten-year period (Soriano Marcolino et al. 2016).

26.7 Swinfen Telemedicine: A Case Study of Intercontinental Telemedicine

Swinfen Telemedicine Charitable Trust is a good example of a long-running low-cost telemedicine service. It began as a pilot project at the Centre for Rehabilitation of the Paralyzed in Bangladesh in 1999 and has since served many poor and disabled people in low- and middle-income countries by establishing telemedicine links between local doctors and healthcare providers to a network of specialists around the world. This service was initially an email-based service in which referrers could send their cases to the system operator and the operator decided to which specialist each case should be directed. Each case was typically a brief history of the patient with an image taken by a digital camera. The specialist who received the case read it and sent a response back to the referrer. After a few years an email auto-router was developed to automate the process of referral. Later the system utilized a web-based platform for both referrers and specialists and recently the charitable trust has introduced a mobile app to facilitate the process of remote consultation. The number of participating hospitals, centers and specialists has grown over time so that as of April 2017 there was a total of 347 hospital/clinic links with this system. Currently the countries with the most referring hospitals or health centers are Afghanistan, Iraq, Nepal and Pakistan with roughly about 30 cases in total per month.

26.8 Telemedicine in Natural Disaster Responses

Natural disasters occur with little effective prediction and LMICs often suffer larger losses and damage due to lower awareness and preparedness compared to countries in the developed world. During the past several decades telemedicine has been used both for preplanning and post-disaster operations. The American Red Cross divides disaster recovery into four phases:

1) Heroic phase (immediately after a disaster strikes).
2) Honeymoon phase (the first few days of the disaster).
3) Disillusionment phase (a few weeks after the disaster).
4) Reconstruction phase (the long-term phase of a disaster). Telemedicine can provide solutions for a range of medical and public health needs after a disaster, especially during the Heroic phase when there is an immediate need for situation analysis and planning, and the Honeymoon phase when demand for medical treatment and rehabilitation of the injured is increasing.

26.8.1 Role of Telemedicine in Earthquake Relief Operations: A Case Study of the Nepal Earthquake in 2015

On 25th April 2015 Nepal was hit by an 8-magnitude earthquake that killed nearly 9000 people and injured 22 000. Total damage of this earthquake was estimated at approximately $1 billion USD and left 3.5 million people homeless with many villages destroyed. Around 1100 healthcare facilities were destroyed and infrastructure was heavily damaged.

During the Heroic phase a huge amount of medical and health services were required for rescue operations and transporting and treating patients. Members of *Doctors for You*, a humanitarian Indian non-governmental organization (NGO) for medical support in crisis and non-crisis situations used the *WhatsApp* social networking platform to communicate, exchange information and plan for relief operations. This platform helped doctors to get information more quickly, better analyze the situation and allocate their resources more efficiently (Debnath et al. 2016).

Telemedicine was used for treating injured people. Over a period of four weeks after the earthquake, 81 people with spinal injuries were admitted to the Spinal Injury Rehabilitation Centre with complicated cases managed by teleconsultation. Teleconsultations were carried out with both national and international specialists including the Nepalese Spinal Cord Injury Collaboration (SpiNepal), a charitable group within the University of British Columbia in Canada and other specialists such as obstetricians and pediatricians. Various means of communication including *Skype, Viber*, email and phone calls were used for teleconsultation.

A hybrid model of telemedicine was used in which required patient information such as medical history, physical examination, laboratory reports and imaging were sent in advance of real-time teleconsultation. Despite some technical problems such as low bandwidth, poor quality of sound and/or video and disconnections, the recipient clinicians acknowledged the teleconsultations helped provide better patient care and up-skilling of the treating doctors (Dhakal et al. 2015). Further advantages and disadvantages of using telemedicine for Nepal earthquake spinal cord injury patients are shown in Table 26.2.

26.9 Telemedicine for Remote Training of Healthcare Professionals

Rapid advances in information and communications technology in the past two decades have transformed the way people teach and learn. As in other sectors, various

Table 26.2 Advantages and disadvantages of using telemedicine for remote consultation of the Nepal earthquake patients with spinal cord injuries.

Advantages	Disadvantages
1) Rapid assistance to national and international expertise	1) Limited number of available doctors
2) Avoidance of travel costs and issues such as visas	2) Inability to perform physical examination by consulting doctors
3) No threat to consulting doctors	3) Problems with record keeping and hand-over
4) No burden on already limited resources	4) Only included doctors for consultations
5) Fitted well with other relief operations	5) No time for overall rehabilitation

innovative education and training solutions have been introduced for healthcare professionals. Several universities and institutes offer a range of online courses for both their students and the public. Massive open online courses (MOOCs) are delivered via a number of platforms such as *Coursera*. MOOCs have provided an unprecedented opportunity, especially for people in low- and middle-income countries, to get free access to high quality training courses from top ranking universities in developed countries. In many cases this opportunity would not otherwise be accessible to them. Moreover, online forums and social networking facilitate collaborative learning and supportive supervision for healthcare professionals regardless of their physical location. Two case studies of widely implemented and successful applications of telemedicine for remote education and mentoring are presented below.

26.9.1 ECHO (Extension for Community Healthcare Outcomes): A Case Study of an Online Community of Practice

Telemedicine can improve access to healthcare for people in underserved areas and address the issue of maldistribution of specialists, but alone cannot solve the problem of a workforce shortage. Therefore, to provide the required care for more people there is a need to increase the workforce capacity. ECHO aims to up-skill primary care clinicians to provide specialty care services. Project ECHO is a non-profit initiative for medical education and care delivery that links several primary care teams in local communities to a group of specialists in academic medical centers or tertiary hospitals. It is based on hub-and-spoke knowledge-sharing networks, led by specialist teams. The primary care teams (spoke sites) may include doctors, nurses, pharmacists, behavioral health specialists or community health workers from remote and rural areas. The hub is an interdisciplinary team of specialists. This project uses multipoint videoconferencing systems. This allows remote participants to attend virtual rounds or teleECHO clinics hosted by the hub on a weekly basis, and to discuss the management of complex cases with the specialist team (Figure 26.1). In these videoconference sessions primary

Figure 26.1 A group of specialists at the hub of a TeleECHO clinic with remote participants on videoconference.

care teams can present their cases and seek advice from the specialists, or other participants. The topics can be de-identified patient cases, population health issues or health system quality improvement problems.

The ECHO project launched its first virtual clinic for hepatitis C in 2003 in New Mexico and since has expanded to 23 countries around the world, covering more than 70 complex medical conditions and health problems. The ECHO model emphasizes peer-to-peer sharing and learning. Mentoring and coaching by specialists is combined with live discussion of real cases and problem solving in a way that is not accessible to many participants using conventional models. By participating in teleECHO clinics, the primary care clinicians find themselves in a learning community that provides them with support and expertise in their chosen area. Although no patients are seen by the doctors in teleECHO clinics, primary care teams can manage their patients based on the knowledge and skills they acquire in those clinics and provide best practice and evidence-based care. Investigations show the quality and effectiveness of the treatments provided by primary care clinicians who were trained and mentored through ECHO are equal to those provided by university-based specialists.

26.9.2 The RAFT Network: A Case Study of Humanitarian Telemedicine in Africa

Réseau en Afrique Francophone pour la Télémédecine (RAFT) (Network in Francophone Africa for Telemedicine) is a telemedicine network that was established in 2000 in Mali by the Geneva University Hospitals and the University of Geneva.

In 2008 it extended to several other African countries and is becoming more global with approximately 60 active sites connecting hundreds of health professional from all over the world. It aims to provide clinical, public health and educational services to member countries. The management of the network is decentralized and only general coordination is provided by the central team based in Geneva. Each country is responsible for the implementation and management of the activities at a national level.

Apart from distant continuing medical education for healthcare professionals and postgraduate medical education for residents, a tele-expertise service is provided for physicians in remote and rural areas to help them manage complicated cases and to avoid unnecessary referrals. A secure web-based platform called Bogou is used for tele-expertise. Requesting physicians can log into the system and post complicated cases in a wide range of specialties including hematology, endocrinology, cardiology, gynecology and radiology. A community of recognized specialists will respond to the case postings. In some countries this network has been equipped with tele-ultrasonography or telecardiology (Bediang et al. 2014).

26.10   Telemedicine for Mental Health (and see Chapter 29)

Mental disorders account for 32% of years lived with disability (YLDs) globally and are the conditions that hold the greatest proportion of such years, far outdistancing those due to cardiovascular diseases or cancer. However, the proportion of national health budgets allocated to these disorders in many countries, especially the LMICs, is remarkably less than required. Despite robust evidence on the effectiveness of psychosocial interventions transforming the lives of people suffering from mental disorders, the gap between the knowledge and delivery of such healthcare is huge. Thus, innovative strategies are required to facilitate the provision of mental healthcare and improve access to the required services. Telemedicine and eHealth interventions can play a role in mental healthcare by increasing the availability of speciality care (psychiatry and psychology), enhancing treatment adherence and follow-up, and promoting patients' self-care. Interventions of this nature can also reduce the costs of mental healthcare (mainly for the patients) by avoiding unnecessary travel. Provision of mental healthcare at a distance using telecommunication technology is called telemental health.

Similar to telemedicine services for somatic disorders, telemental health has two main modalities: synchronous (real-time) and asynchronous (store-and-forward). A range of telemental health services have been reported from both developed nations and LMICs for prevention, remote consultation, treatment, education, patient self-care, training and research purposes.

26.10.1 Telemedicine for Suicide Prevention: A Case Study from Sri Lanka

Suicide is a preventable mental health problem that takes around one million lives each year. It is estimated that the number of people who attempt suicide is 10–20 times higher in LMICs compared to their developed counterparts. In Sri Lanka the rate of suicide deaths is twice as high as the global average. Preventative mental health programs and support are limited due to insufficient numbers of healthcare staff to follow-up after discharge from hospital people treated for unsuccessful suicide attempts. Research suggests that telemedicine can be effectively used for follow-up care to prevent re-attempts at suicide.

In a randomized controlled trial in Colombo, Sri Lanka, Marasinghe et al. (2012) studied the effectiveness of a brief mobile treatment for the reduction of suicidal ideation, self-harm and depression in 68 participants who were admitted to a university teaching hospital after a suicide attempt. The intervention included two phases:

1) A face-to-face component in which an assessment of mental health was carried out and several meditation and training sessions were provided for problem solving, social support, and alcohol/drug abuse reduction.
2) Distance (mobile/web) follow-up including 10 telephone calls of 10–15 minutes duration, continuous access to 5 minute audio phone messages, and weekly reminders sent via Short Message Service (SMS). The aims of telephone calls were to assess the mood, enhance problem-solving skills, improve social support, and reduce alcohol/drug abuse. The intervention was designed for a duration of six months after discharge from the hospital. The results of this study showed that the brief mobile treatment significantly reduced suicidal ideation and depression compared with usual care. However, its effect was not significant on actual self-harm and substance abuse. Further research is needed to determine the best composition of telemental health for preventing suicide, and undoubtedly such interventions should be customized for each country or community based on their sociocultural status and values (Marasinghe et al. 2012).

26.10.2 Telemedicine for Mental Health in a Conflict Setting: A Case Study from Syria

Syria has been struggling with a humanitarian conflict for a number of years. This situation has both increased the need for mental healthcare and, at the same time, decreased the availability and accessibility of traditional mental health services. Using a store-and-forward telemedicine model, the Syrian Telemental Health Network was established to provide training and education for healthcare workers and make available clinical consultations in the conflict setting of Syria. Rather than providing a direct clinical service to patients, this network aimed to build the capacity of managing mental health for healthcare workers in the affected regions. This global network uses *Collegium Telemedicus*, which is a store-and-forward platform based in the UK and used by non-governmental organizations such as Médecins sans Frontières (MSF). This platform helps physicians, nurses, and other healthcare workers obtain specialized consultation about their cases from many Arabic mental health specialists working in academic centers all around the world.

The staff in the participating clinics in this network receive an introductory training on how to use the system, either web-based or in real-time by videoconferencing. The requesting healthcare workers can log into the system by a secure connection via any web browser and make a new referral by describing the clinical manifestation of the patient or uploading audio-visual clinical information such as an interview with the patient, and then ask one or more questions on the management of the patient.

A trained coordinator reviews the submitted case and assigns it to the best available specialist to provide advice. The consulting specialist will see the case and give advice on the clinical management of the patient, and will be contacted if further information or supervision is required. The specialist does not interact with the patient directly.

Through a pilot trial in 2014, and based on the feedback from the users, the guidelines and referral process were refined. In addition, the procedural instructions were improved to make the provided specialist advice as useful as possible, and a mental health history template was developed to allow the requesting healthcare worker to provide a complete and consistent set of information on the mental health history and examination of the patient. As of September 2015 the Syrian Telemental Health Network connected 35 referring healthcare providers working at 14 refugee clinics in Syria, Turkey, Lebanon and Jordan to 16 Arabic-speaking mental health specialists from North America, Europe and the Middle East. In addition to the provision of clinical consultations, this network has contributed to capacity building of mental healthcare workers through effective supervision and clinical mentoring of healthcare providers in conflict-affected regions (Jefee-Bahloul et al. 2016).

26.11   The Rise of Mobile Health (mHealth)

It is estimated that five billion people around the globe are mobile phone users and this number is on the rise, especially in LMICs. In fact, the number of people who have a mobile phone is higher than those who have access to clean water or electricity worldwide. Such a high availability of mobile phones provides many opportunities for health promotion and improving access to healthcare. Early interventions of mobile health (mHealth), which is the use of mobile phones and other wireless digital devices for achievement of health objectives, took advantage of the ubiquity of mobile phones for communicating health-related messages, enhancing adherence to management plans and reminding of appointments mainly via (automated) voice calls or text messaging. Moreover, personal digital assistants (PDAs) were used for health data collection and communication. With the advent of smartphones, generally marked with the introduction of the Apple iPhone in 2007, a new era in mHealth started and smartphone software applications (apps) for health and medical purposes were developed. The number of healthcare apps in the major app stores is continuously increasing and with remarkable growth (57%) in 2016. It now exceeds 250 000 apps. Despite existing concerns about regulation and approval of apps, especially for clinical purposes, researchers and entrepreneurs in many countries have reported successful utilization of smartphones by healthcare providers and consumers for a range of purposes including screening, prevention, management and follow-up of many medical conditions.

26.11.1 A Low Cost Mobile-Based Eye Care System: A Case Study of Kenya (and see Chapter 25)

Peek (Portable Eye Examination Kit) is a smartphone-based eye care system that has been successfully used in rural areas of Kenya. It uses a cheap clip-on piece of hardware to transform a smartphone into a portable and user-friendly ophthalmic imaging device (Figure 26.2).

Using Peek Acuity, a mobile app running on Peek, community health workers are able to perform vision tests for community screening. The information is automatically sent to a local health center for assessment and identification of people who may need to be referred to a specialist. The results of eye tests conducted by this system have been shown to be comparable with those of the traditional tests and the system was acceptable to both the patients and healthcare workers (Bolster et al. 2015; Lodhia et al. 2016).

26.12 Social Networking for Clinical Purposes

The increasing rate of ownership of mobile phones and availability of cellular Internet has decreased the technological gap between developed counties and LMICs in terms of telecommunications. The processing power and software applications (apps) of smartphones have extended their capabilities far beyond the basic functions of mobile phones (i.e. phone calls and short messaging services) and offered a range of innovative uses in our daily lives. Among other applications, mobile-based social networking has

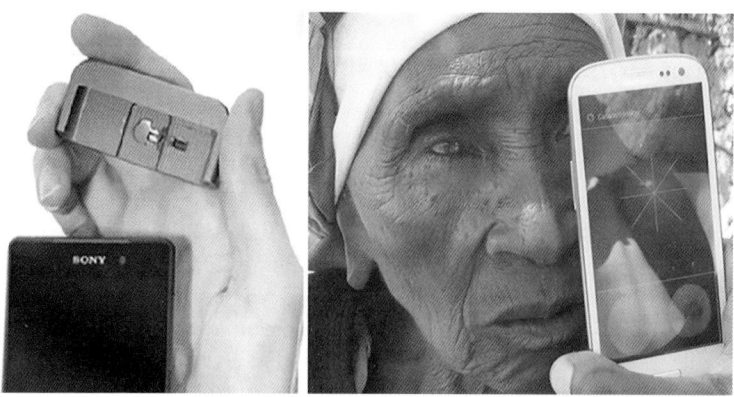

Figure 26.2 Using a smartphone for eye care in Kenya showing the peripheral device connection (Peek project).

gained unprecedented popularity especially among adolescents and young adults. Instant messaging platforms such as *WhatsApp, Line, Viber,* and *WeChat* have drastically changed the communication pattern among smartphone users in terms of ease-of-use and timeliness. While these platforms were initially used for informal social interactions between people, their use has gradually extended to formal business and professional purposes, including health and medical care.

26.12.1   Using Telegram Instant Messaging for Clinical Consultation: A Case Study from Iran

Telegram is the most popular instant messaging service in Iran. It is a cross-platform service that allows users to send text messages to their contacts, or share files, multimedia content (voice, image and video) and location information with them. Moreover, Telegram supports group chats for participation in a conversation with up to 20 000 users, and broadcasting channels for disseminating information and multimedia contents to unlimited users instantly. It is estimated that around 60% of the Iranian population are Telegram users and consume a high proportion of the national bandwidth for communication, sharing photos and videos, listening to music and watching movies on Telegram. Software developers can create bots to automate some tasks and interact with users. (An Internet bot, also known as web robot, WWW robot or simply bot, is a software application that runs automated tasks over the Internet. Typically, bots perform tasks that are both simple and structurally repetitive, at a much higher rate than would be possible for a human alone.)

TeleVisit is a Telegram bot that connects users with healthcare providers (HCPs) including specialist doctors, general practitioners, nurses, midwifes, physiotherapists, and nutritionists. HCPs need to register with the company that runs this service to be listed as a provider. Currently there are around 70 HCPs collaborating with this service. Users can see a list of available providers, choose the best one based on the specialty of the provider, and direct their questions to them.

The questions can be sent in text, voice, or video format, along with accompanying information such as photos, radiographic images, lab results, and prescriptions. To use the service, users need to buy credit in advance to pay the consultation fee that is variable and set by each provider. Users can hide their identity (real name and telephone number) by changing their privacy settings in the control panel of Telegram, so the consultation may be anonymized from the patient side (Hashtarkhani et al. 2017). Further advantages as well as the disadvantages of TeleVisit are listed in Table 26.3.

26.12.2   Short Message Service (SMS) in Healthcare

Telecommunication technology has advanced rapidly over the past two decades. The emergence and development of mobile technology has transformed the way people live, work and interact worldwide. In 2002 the number of mobile subscribers exceeded the number of landline subscribers and the expansion of mobile phone ownership has been most dominant in LMICs. In fact, the number of mobile phones is greater than landline phones in many LMICs including sub-Saharan African countries, and the use of mobile phones in Asian countries is rapidly growing.

Short Message Service (SMS) is one of the basic functions of mobile phones and is based on the Global System for Mobile Communication. It is available on both cheap

Table 26.3 Advantages and disadvantages of TeleVisit (a messenger-based clinical consultation platform) in Iran.

Advantages	Disadvantages
1) Instant access to online doctors and other healthcare providers regardless of their geographical location	1) Confidentiality
	2) Lack of real-time communication (videoconferencing)
2) Low cost	3) Not suitable for serious or complex conditions
3) User rating for providers	4) No reimbursement by insurance companies
4) Avoids travel for a face-to-face visit	5) No prescriptions can be issued
5) Ability to send images, sounds, videos in addition to text messaging	6) Legal considerations
6) Anonymous consultation	

"simple" phones as well as smartphones. SMS makes it possible to send a message of up to 160 characters from one mobile phone to another in almost real-time and at a relatively low cost. SMS gateways and computer software applications can be used to send bulk messages or personalized messages at predetermined times to thousands of mobile users, and request delivery/read receipts from the receiver's mobile phone.

Although recent research studies on mobile health have focused on the use of smartphone applications for healthcare, SMS remains a valuable means of rapid, effective, low-cost communication suitable for health promotion, particularly in resource-limited settings such as low-income countries.

26.12.3 Short Message Service for Improving Maternal and Child Health: A Case Study from Rwanda

Pregnant women in Rwanda, as in many other LMICs, are at higher risk of death during pregnancy and after childbirth than women in developed countries. This can be due to poor maternal care, failure to diagnose life threating-conditions and delay in access to emergency services. Rwanda managed to reduce the maternal mortality rate from 910 to 340 maternal deaths per 100 000 live births between 1990 and 2010. However, it became clear that further improvement toward achieving the target mortality rate of 228 maternal deaths per 100 000 live births required innovative actions to enhance maternal care, especially in remote and rural areas. Mobile technology can be used for improving maternal and antenatal care by facilitating the communication between pregnant women, healthcare providers and emergency obstetric care. With the increasing penetration rate of mobile phone coverage (around 50% by 2012), Rwanda has strategically considered mobile health for improving communication between healthcare providers and for improving the flow of information to enable better decision making on pregnancy-related issues.

Through a collaboration between the Ministry of Health, UNICEF, and Management Sciences for Health, a pilot project was launched in Musanze District, which is the most mountainous area of Rwanda and which had the lowest rate of assisted deliveries in healthcare facilities. The aim of this project was to improve maternal and child health by using an SMS-based service, called RapidSMS-MCH, for monitoring pregnancies and enhancing health information exchange between community health workers. The system was developed based on the RapidSMS platform. The RapidSMS system is a free and open-source platform for rapidly building customizable SMS (text messages) with web-based dashboards. It can be used for data collection, streamlining complex work-flows and group coordination using basic mobile phones. Local programmers developed the Rwandan RapidSMS-MCH and deployed further improvements based on feedback from the users and the specific needs of the district through an iterative approach. The main function of the system was to facilitate two-way communication between registered pregnant women, community health workers and health facilities for the duration of each pregnancy (Figure 26.3).

In each village of the district, the community health workers received training on the system and were provided with a mobile phone. They were able to report new pregnancies, request referrals, and notify emergency conditions of pregnant women to their health center or the district hospital by SMS trough the RapidSMS-MCH system. Through a partnership with the mobile operator MTN, the Ministry of Health covered the SMS costs. In the case of normal pregnancies, the system sent automated reminders for upcoming visits. In the case of emergency events, the system alerted the nearest available ambulance, health facility manager, and an officer in the district health office, to streamline the transportation of the patient, while supporting the health worker with initial patient management. The underlying information infrastructure is a national database that connects community health workers, health facilities, and the ambulance service. A web-based user interface allows authorized persons to access individual and aggregated data for clinical, managerial, and auditing purposes. Errors and mistakes are also logged in order to provide feedback to the users and to debug the system. Success factors and challenges of the RapidSMS-MCH system are summarized in Table 26.4.

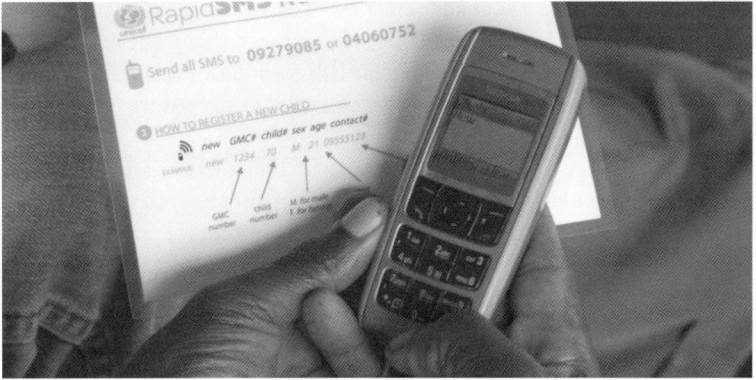

Figure 26.3 Coding a message for RapidSMS system.

Table 26.4 Success factors and challenges of RapidSMS-MCH.

Key success factors	Challenges
1) Government commitment to innovation	1) Maintenance of mobile phones
2) Public-private partnership	2) Limited access to electricity in some areas
3) Local sourcing	
4) Co-creation	
5) Essential role of community health workers	
6) Regular feedback and evaluation	

The RapidSMS-MCH was used from June 2010 to May 2011 in the Musanze District as a pilot project. During this period 432 community health workers registered and monitored 11 502 pregnancies by sending a total of 35 734 SMS, of which approximately 1% (362 SMSs) represented an emergency situation. Compared to a similar period of time prior to the pilot trial, the number of first antenatal care visits increased by 24% and maternal mortality dropped by 25%. Moreover, the percentage of deliveries at a healthcare facility increased from 68% to 95%, while the number of stillbirths and deaths in newborns decreased by 69% and 48%, respectively. Based on the notable improvement to the outcomes of maternal and child health in this pilot project, the Rwandan Ministry of Heath decided to scale-up and implement the RapidSMS-MCH system nationwide, with more than 7000 community health workers serving 18 Districts (Ngabo et al. 2012; WHO 2014).

26.13 The World Health Organization and Telemedicine

In December 2012 the United Nations General Assembly passed a resolution on universal health coverage (UHC), which is the availability of a full range of essential health services (for example, prevention, promotion, treatment, rehabilitation and palliative care) for all people when needed without the risk of financial hardship. The World Health Organization highlighted that UHC is not achievable without the support of eHealth, and encouraged state members to use information and communications technology for the delivery of comprehensive, integrated and people-centered health services. The third global survey on eHealth, which was conducted by the WHO Global Observatory for eHealth in 2015, had a special focus on the use of eHealth in support of UHC and was completed by more than 125 countries. It explored the eHealth developments of the countries in eight thematic subjects: eHealth foundations, mHealth, telehealth, eLearning in health science, electronic health records, legal frameworks, social media and big data (WHO 2016a, b).

The results of this survey documented that more than half of the countries that responded to the survey had an eHealth strategy and 90% of those strategies directly addressed the objectives of UHC. Eighty-three percent of countries reported at least one mHealth initiative; however, very few of them had any assessment of implemented mHealth programs. Limited evaluations of mHealth programs make it difficult to understand what works and why interventions may fail. The most widely implemented telemedicine service is teleradiology (77% of the countries) followed by telepathology, teledermatology and remote patient monitoring. Many countries have attempted to use telehealth for the delivery of healthcare at a distance to improve the equity of their healthcare coverage. Nearly half of the countries had developed national electronic health records (EHRs) that could contribute to the quality and timeliness of the care by improving access to patients' data for healthcare providers at the point-of-care. International standards are essential for interoperability of EHRs. Despite the need for a sound legal framework to fully utilize eHealth, only 54% of countries reported legislation that could protect electronic patient data. With the rapid proliferation of social media in the past few years, healthcare organizations in almost 80% of the countries surveyed had used social media, to various extents, for health promotion. Although a large proportion of the countries had used social media for the purpose of health promotion, the full potential of social media in healthcare is yet to be fully explored.

Big data is a term for data sets that are so voluminous and complex that traditional data processing application software is inadequate to deal with them. Big data challenges include capturing data, data storage, data analysis, search, sharing, transfer, visualization, querying, updating and information privacy. There are five dimensions to big data known as Volume, Variety, Velocity and the recently added Veracity and Value. Lately, the term "big data" tends to refer to the use of predictive analytics, user behavior analytics, or certain other advanced data analytics methods that extract value from data, and seldom to a particular size of data set. This is an emerging field in healthcare. Around 17% of countries have developed their national policy or strategy regarding big data in healthcare.

WHO advocates eHealth playing a key role in the progress toward achieving the UHC that the member states committed to provide. Countries are encouraged to move from silo solutions to integrated people-centered interoperable solutions that reduce the burden of data collection and improve the quality of care. National strategies should be developed based on the specific needs of citizens, healthcare professionals and health managers. Moreover, an enabling environment comprising technical infrastructures, legal frameworks, standards, and skill sets, as well as monitoring and evaluation plans, are required for a successful national eHealth strategy.

26.14   Challenges and Barriers to Implementation

Despite an increasing number of papers published on the role of telemedicine in improving patient care, the overall adoption of telemedicine has been slow. Several important issues need to be addressed before telemedicine can be widely accepted and adopted.

26.14.1   Funding

Based on a global survey by WHO, the most important barrier to implementing telemedicine was a lack of funding to develop and support telemedicine programs. For many healthcare systems, the high cost of capital investment for equipment and infrastructure, as well as maintenance and technical support costs, cannot be justified by the potential savings from implementation of telemedicine programs.

26.14.2   Lack of Evidence of Effectiveness

Although many clinical trials have demonstrated the efficacy of medical and health services provided via telemedicine, the evidence on the scalability of interventions and societal impact of telemedicine is scant and inconclusive. In fact, the results of closely supervised clinical trials under very controlled settings are not often generalizable to real-life settings. Now it is clear that traditional randomized controlled trials are not the best study design for the evaluation of telemedicine. There is a need for more robust evaluation methods with a realistic approach to explain which telemedicine services work best for which group of patients under what circumstances.

26.14.3 Reimbursement

Apart from a few instances in developed countries, reimbursement of telemedicine services remains a major barrier to the expansion of telemedicine services in many countries. In 2012 the Australian Government introduced 11 Medicare Benefit Schedule items for video consultation in eligible areas, provided the patient and specialist are separated by 15 km or more, and specialists still meet all the requirements of traditional consultation such as minimum time spent with the patient. Similarly, in the United States a limited number of telemedicine activities are covered by public health payers. Policy makers are cautious about the potential overutilization of telemedicine services if reimbursement is established.

26.14.4 Sustainability

Several studies have reported the results of telemedicine projects in LMICs, but most of them are funded by research grants and thus less likely to survive beyond the research phase. This a common issue worldwide and only a few large-scale telemedicine services have been reported in the literature. Two examples of successful ongoing telemedicine services using real-time solutions are the Ontario Telemedicine Network (https://en .wikipedia.org/wiki/Ontario_Telemedicine_Network) and Kaiser Permanente Programs (https://healthy.kaiserpermanente.org/health-wellness).

26.14.5 Legal Issues

Since the model of care in telemedicine is different from traditional services, countries should consider legal issues pertaining to the delivery of medical care through information technology-based modalities. Roles and responsibilities of parties involved in a remote consultation, as well as patient privacy and confidentiality are among the issues that require revised or new legislation. When the provider and recipient of the service are located in different jurisdictions, the issues of accreditation and authorization of providers become more complicated.

26.14.6 The Future of Telemedicine

Artificial Intelligence (AI) is one of several fast-growing technologies that holds promise for numerous applications in healthcare. Using machine learning algorithms, which is an approach to AI, public health specialists can understand the spread of communicable diseases and locate where an outbreak is likely to occur next. In addition, AI will enable specialist consultants to predict the progress of chronic conditions and the likelihood of the development of clinical complications. Similarly, healthcare managers could utilize AI-enabled hospital information systems to predict their workload, optimize their resources, and prioritize their services to better serve patients and save more lives. AI has already shown superiority to human specialists in diagnosing rare diseases.

The most advanced form of AI is known as Deep Learning. Deep learning (also known as deep structured learning or hierarchical learning) is part of a broader family of machine learning methods based on learning data representations, as opposed to task-specific algorithms. Deep learning is expected to significantly improve the precision and efficacy of traditional clinical decision support systems. In addition to voice recognition and recommender systems, deep learning is being studied in medical image processing for screening and diagnosis of diseases such as diabetic retinopathy. In several hospitals, AI has been used to aid radiologists in reading x-rays and CT scans, and in identifying suspicious lesions. However, a number of people have expressed concerns that a runaway growth in AI would enable super intelligent systems to upgrade themselves and surpass human intelligence (known as the "technical singularity").

The Internet of Things (IoT) and cloud computing will revolutionize the process of patient data collection and remote monitoring (Kvedar et al. 2015). Mobile devices and wearable sensors are attracting more attention, and with their decreasing price, they can play a greater role in healthcare provision. Healthcare providers will have access to timely and quality data, which is collected and transmitted with minimal effort from patients. Intelligent personal assistants such as Siri (Apple Inc.) and Cortana (Microsoft Corp), and smart devices such as Amazon Echo and Google Home already help people to get some jobs done easier and faster. Smart homes will enable closer and more precise monitoring of the inhabitants and allow elderly people to live independently longer.

Robotics technology is also advancing rapidly and gaining ground in health and medical services. We will see more robots in hospitals helping patients and undertaking some routine tasks. Telepresence robots can virtually bring specialists to the bedside of the patients at rural hospitals or carers to patients' homes. Social robots will be programmed to help with the education and treatment of patients, especially for repetitive therapeutic tasks in disorders that require therapist supervision and are currently available only in clinics or hospitals. Intelligent agents and chat bots will be employed to coach patients and interact with them for therapeutic purposes (for example, for people with autism), to monitor chronic conditions, or to collect health data. (A chatbot [also known as a talkbot, chatterbot, Bot, IM bot, interactive agent, or Artificial Conversational Entity] is a computer program which conducts a conversation via auditory or textual methods.)

Drones and Unmanned Aerial Vehicles are transforming transport in healthcare, both in developed nations and LMICs (see also Chapter 35). Medical drone delivery has already been successfully used in parts of Rwanda and Tanzania where ground vehicles cannot reach people in need of healthcare due to lack of roads or poor condition of roads during rainy seasons. In fact, delivery of blood supplies by Zipline drones has saved the lives of many patients in these countries. Drones also are being studied as an alternative means of delivering emergency equipment (such as defibrillators) and videoconference units to cardiovascular patients in metropolitan areas when emergency services are unable to reach patients in time because of congested traffic. A flying ambulance is a good fit for autonomous aerial transport technology and will help airlifting of patients in spaces in which helicopters cannot land.

26.15 Conclusions

Telemedicine presents many opportunities to improve access to quality healthcare for people all around the world and these opportunities are expanding with the advances of technology and the emergence of a digital native generation. The remarkable progress of many countries, including low- and middle-income ones, in implementing eHealth solutions and providing telemedicine services reflect their attitude toward strengthening healthcare systems with innovations and new technologies. However, to mimic the success story of pioneer countries in this field and to unleash the full potential of using ICT in the health sector, countries should consider several fundamental issues that can hinder the adoption of telemedicine. The most important ones are securing funding, enhancing the infrastructure, promoting inter-sectoral collaboration, and developing policy and strategies. Finally, while the aim of many research projects in telemedicine has been to improve access to healthcare for people in LMICs, the outcome of these studies can also be used for service improvements in developed countries – a phenomenon which is known as "reverse innovation."

Bibliography

Alkmim, M.B., Figueira, R.M., Marcolino, M.S. et al. (2012). Improving patient access to specialized health care: the Telehealth Network of Minas Gerais, Brazil. *Bull. World Health Organ.* 90: 373–378.

Bagayoko, C.O., Muller, H., and Geissbuhler, A. (2006). Assessment of Internet-based tele-medicine in Africa (the RAFT project). *Comput. Med. Imaging Graph.* 30 (6–7): 407–416.

Bediang, G., Perrin, C., De Castaneda, R. et al. (2014). The RAFT telemedicine network: lessons learnt and perspectives from a decade of educational and clinical services in low- and middle-incomes countries. *Front. Public Health* 2: 180.

Bolster, N.M., Giardini, M.E., and Bastawrous, A. (2015). The Diabetic Retinopathy Screening Workflow: Potential for Smartphone Imaging. *J. Diabetes Sci. Technol.* 10: 318–324.

Debnath, P., Haque, S., Bandyopadhyay, S. and Roy, S., eds. (2016). Post-disaster Situational Analysis from Whats App Group Chats of Emergency Response Providers. ISCRAM.

Dhakal, R., Gurung, J., Gyanwali, S. and Poudel, M. (2015). Rapid implementation of telemedicine after earthquake disaster 2015 at Spinal Injury Rehabilitation Centre, Nepal. 14th Asian Spinal Cord Network Conference; Kathmandu, Nepal.

Fatehi, F. and Wootton, R. (2012). Telemedicine, telehealth or e-health? A bibliometric analysis of the trends in the use of these terms. *J. Telemed. Telecare* 18: 460–464.

Fatehi, F., Armfield, N.R., Dimitrijevic, M., and Gray, L.C. (2015a). Technical aspects of clinical videoconferencing: a large scale review of the literature. *J. Telemed. Telecare* 21: 160–166.

Fatehi, F., Martin-Khan, M., Smith, A.C. et al. (2015b). Patient satisfaction with video teleconsultation in a virtual diabetes outreach clinic. *Diabetes Technol. Ther.* 17: 43–48.

Hashtarkhani, S., Azimi, A. and Fatehi, F. (2017). Televisit: an innovative mobile-based system for remote medical consultation using Telegram messaging app in Iran. Success and Failures in Telehealth (SFT-17); 30–31 October 2017; Brisbane, Australia.

Jefee-Bahloul, H., Barkil-Oteo, A., Shukair, N. et al. (2016). Using a store-and-forward system to provide global telemental health supervision and training: a case from Syria. *Acad. Psychiatry* 40: 707–709.

Kvedar, J.C., Colman, C., and Cella, G. (2015). *The Internet of Healthy Things*. Boston: Partners Connected Health.

Lodhia, V., Karanja, S., Lees, S., and Bastawrous, A. (2016). Acceptability, usability, and views on deployment of peek, a mobile phone mHealth intervention for eye care in Kenya: qualitative study. *JMIR Mhealth Uhealth* 4: e30.

Marasinghe, R.B., Edirippulige, S., Kavanagh, D. et al. (2012). Effect of mobile phone-based psychotherapy in suicide prevention: a randomized controlled trial in Sri Lanka. *J. Telemed. Telecare* 18: 151–155.

Ngabo, F., Nguimfack, J., Nwaigwe, F. et al. (2012). Designing and Implementing an Innovative SMS-based alert system (RapidSMS-MCH) to monitor pregnancy and reduce maternal and child deaths in Rwanda. *Pan Afr. Med. J.* 13: 31.

Ribeiro, A.L.P., Alkmim, M.B., Cardoso, C.S. et al. (2010). Implementation of a telecardiology system in the state of Minas Gerais: the Minas Telecardio Project. *Arquivos brasileiros de cardiologia* 95 (1): 70–78.

Soriano Marcolino, M., Minelli Figueira, R., Pereira Afonso Dos Santos, J. et al. (2016). The experience of a sustainable large scale Brazilian telehealth network. *Telemed. e-Health* 22: 899–908.

WHO (2010). *Telemedicine: Opportunities and Developments in Member States: Report on the Second Global Survey on eHealth*. Geneva: World Health Organization.

WHO (2014). eHealth and innovation in women's and children's health: a baseline review: based on the findings of the 2013 survey of CoIA countries by the WHO Global Observatory for eHealth. 9241564725.

WHO (2016a). *Global Diffusion of eHealth: Making Universal Health Coverage Achievable: Report of the Third Global Survey on eHealth*. Geneva: WHO.

WHO (2016b). *mHealth; New Horizons for Health Through Mobile Technologies: Second Global Survey on eHealth*. Geneva: WHO.

Webliography

Ontario Telemedicine Network https://en.wikipedia.org/wiki/Ontario_Telemedicine_Network

Kaiser Permanente Programs https://healthy.kaiserpermanente.org/health-wellness

27

Telemedicine for the Delivery of Specialist Pediatric Services

Anthony C. Smith, Monica Taylor, Farhad Fatehi, and Liam J. Caffery

Centre for Online Health, The University of Queensland, Brisbane, Queensland, Australia

Revolutionizing Tropical Medicine: Point-of-Care Tests, New Imaging Technologies and Digital Health,
First Edition. Edited by Kerry Atkinson and David Mabey.

27.1   Introduction

Of the 2.2 billion children in the world, over 86% (1.9 billion) live in low- and middle-income countries (LMICs) (UNICEF 2005). Children in LMICs are 10 times more likely to die before the age of five than children in developed countries WHO (2011). Further, they have far worse health outcomes than children in more developed countries. In almost 40 LMICs the median age of the population is in the teens, highlighting the importance of pediatric care (CIA 2016). The population of least developed countries (LDCs) is expected to more than double over the next 30 years (United Nations 2017). In combination with already limited health resources this population growth will likely further exacerbate access issues with health services, especially for children who make up the largest percentage of the population. Telemedicine presents an opportunity for LMICs to increase access to child health services and to reduce costs in some cases. Furthermore, the benefit in telemedicine of saving patient travel is especially important in pediatrics as it is not just the patient but also a parent who has to bear the time and costs associated with the journey (Smith et al. 2003). Quality pediatric care may not be available in LMICs due to a lack of education in pediatric subjects for medical professionals which could in part be addressed by telemedicine (Nolan et al. 2001). Another benefit of telemedicine is the ability to remotely mentor clinicians.

Broadly speaking, telemedicine is the transmission of healthcare information over a distance using information and communication technologies (ICTs) for the purpose of diagnosis, treatment, or management of a patient. However, there is not a strict, consistent definition of telemedicine. Some commentators believe telemedicine is the delivery of clinical services only, whereas the World Health Organisation (WHO) (among others) sees telemedicine as a method of providing continuing education of healthcare providers (WHO 2010). Many rural areas in LMICs have clinics that are staffed by community health workers who have lesser qualifications than other healthcare providers. Hence, the opportunities for education and remote supervision and mentoring that telemedicine offers is an important consideration.

Another debate is whether teleradiology is a form of telemedicine (Wootton 2014). Further, there is contention as to whether telephone calls and text messages are a form of telemedicine. Mobile health (mHealth) is the use of mobile devices such as tablet computers and smartphones to provide health services. mHealth applications may be considered a form of telemedicine if they facilitate communication between a local and distant healthcare provider or a patient and a distant healthcare provider. Smartphones can also be used as respiratory or pulse rate counters, gestational age date calculators, drug dose calculators, drip rate calculators and drug reminder alarms when installed in mobile phones (Källander et al. 2013) However, as none of these require the transmission of information they would not necessarily be considered as part of telemedicine. For the purposes of this chapter we take a holistic view of telemedicine and include any form of ICT-mediated exchange (including telephone and text messaging) of healthcare information (including teleradiology) for clinical or educational purposes.

Despite the ambiguous definition, the motivation for telemedicine is more consistent. Telemedicine is used to increase accessibility to health services. Poor accessibility may due to a shortage of healthcare providers in a particular geographical area. This scenario is often worsened when the patient lives in a rural or remote area and is further exacerbated by the subspecialty nature of some specialist services, most notably pediatrics. In addition to geographic maldistribution, a simple lack of available workforce is another reason for poor accessibility. The WHO has identified that the most acute shortage of healthcare providers is in LMICs (WHO 2010) and is often the result of population growth outstripping the training capacity of the health workforce (Pacqué-Margolis et al. 2011) or the migration of the health workforce to richer countries (Stilwell et al. 2004). A case in point is that there are only 2.3 health workers per 1000 population in Africa, whereas countries in the Americas have up to 24 health workers per 1000 population (WHO 2006).

This chapter will provide a background for telepediatrics in LMICs, explore the variety of telepediatric models of care that can be utilized and provide case study examples.

27.2 Technical Consideration for Telemedicine in LMICs

27.2.1 Real-Time

Telemedicine can be delivered using different technologies. Videoconferencing is one of the main modalities of telemedicine. Videoconferencing is used to enable a healthcare provider to consult with a patient at a distant location. This process is termed a real-time consultation, videoconsultation, or a teleconsultation. The patient may attend a hub site such as the local hospital or clinic where videoconferencing equipment is available. Alternatively, the videoconsultation can be done directly in the patient's home but this is less common. During a videoconsultation the patient is often accompanied by their local healthcare provider. The local healthcare provider can be used to perform a physical examination (including assessment of vital signs, neurological assessment, abdominal assessment, auscultation and range of motion assessment) and convey this information to the remote clinician. Not all specialties require a physical examination – for example, psychiatric consultation most often relies on conversation as the means of clinical assessment.

27.2.2 Store-and-Forward

Store-and-forward is another modality of telemedicine. Store-and-forward gets its name from the steps in the consultation process. Firstly, some form of digital information is collected from the patient by their local healthcare provider and stored. This information may include patient history, assessment tests (e.g. nutritional assessment), video imaging (e.g. gait), still image (e.g. skin lesion, burns, wound, radiograph or retinal image), sound (e.g. recorded auscultation) and biometric testing (e.g. electrocardiograph). Next, the stored information is forwarded to the remote clinician for diagnosis or management advice.

The simplest way to implement store-and-forward telemedicine is to use email. An alternative is where the patient-end healthcare provider uploads information to a web portal where it can be viewed by the remote clinician. Web portals often require bespoke development but do have the advantage of being able to collect structured information. Another option may be the use of short message services (SMSs) or multi-media messaging services (MMSs) or a smartphone messenger app to transmit information via a cell or mobile phone.

27.2.3 Internet Connectivity

Regardless of the modality used for telemedicine they all rely on ICTs. For this reason, it is important to examine both the availability of internet connectivity and devices to connect to the Internet in LMICs. Additionally, telemedicine requires information and a relatively stable supply of electricity and people to maintain and support the infrastructure (Mars 2013).

Globally 3.2 billion people are currently using the Internet of which 2 billion are from LMICs (International Telecommunication Union [ITU] 2015). LMICs have substantially less access to the Internet than developed countries. The ITU reported that in LMICs 35% of the population has access to broadband Internet (ITU 2015). However, in least developed countries (LDCs), only 10% of the population has access. In comparison, 97% of the population of developed countries has broadband Internet access. In both developed countries and LMICs, the Internet is generally less accessible in rural areas compared to metropolitan areas. In LMICs mobile broadband exceeds fixed broadband services. In the last five years a high growth rate (10%) in subscription to mobile broadband has been seen in LMICs (ITU 2017).

27.2.4 Mobile Devices

Similar proportions of adults own cell or mobile phones in LMICs as in developed countries. Smartphones specifically, however, are less widely used (Pew Research Center 2015). Across African countries, for example, the Pew Research Center showed that the median proportion of adults with a mobile phone was 80% However, only 15% had smartphones (Pew Research Center 2014). Mobile phones can facilitate telemedicine in a variety of ways. Firstly, the device can act as the interface to web portals, email and apps (see Section 27.2.7) and can thus be used for telemedicine. Further, there is a wide range of attachments that can convert a mobile phone into a dedicated imaging or diagnostic device (see Section 27.2.10). Furthermore, the built-in camera can be used to facilitate store-and-forward telemedicine in such applications as dermatology, plastic surgery and wound care. Photographing radiographic images can also be used to implement a low-cost teleradiology system. Concordance between smartphones photos and a gold standard method have shown 100% specificity and 96% sensitivity in past research (Zennaro et al. 2013).

27.2.5 Telemedicine Platforms for Low-Resource Settings

LMICs may have limited funding for ICTs that can be used to facilitate telemedicine. However, the availability of telemedicine options for low-resource settings offer opportunities to participate in telemedicine. This section investigates low-cost platforms that can be used for telemedicine including video conference software, apps and web-based portals. Further, we describe how simple email infrastructure can be leveraged to support telemedicine.

27.2.6 Videoconferencing

In some countries, such as Australia, videoconsultation is the predominant modality of telemedicine. Historically, videoconsultations were performed using dedicated videoconferencing end-points often referred to as room-based systems or platforms. Expensive room-based systems can now be replaced with low-cost or free software-based systems, making them an attractive option for telemedicine services where cost may be a barrier to implementation.

Software-based video consultations still require hardware (for example, a personal computer, tablet computer or smartphone) to run the videoconferencing software. The technical quality of software-based videoconsultation platforms can be enhanced by using peripheral devices such as an external microphone and speaker or a web cam instead of the hardware device's built-in microphone, speaker and web cam. The increased quality will come with a trade-off of increased costs (Liu et al. 2015). Perhaps the best known software-based system is *Skype* which has been described as "feasible and having benefit" for delivery of telemedicine (Armfield et al. 2015). In addition to *Skype* there are a growing number of cloud-based videoconferencing platforms based on the Web Real-Time Communication (WebRTC) protocol (WebRTC 2017). WebRTC allows videoconferencing to be conducted from within a browser and does not require software installation.

Videoconferencing may consume substantial bandwidth and the quality of the videoconsultation is dependent on the reliability of the Internet connection (Taylor et al. 2015). While the infiltration of 3G networks in LMICs is high, 4G networks are better suited to videoconferencing (Caffery and Smith 2015). In LMICs, lack of appropriate and reliable Internet connectivity may be a barrier to videoconsultations and thus less bandwidth-intensive modalities of telemedicine (such as store-and-forward or mHealth) may be the modality of choice for telemedicine.

27.2.7 Apps for Smart Devices

There is growing evidence to support the use of mobile phone apps which can be used to deliver health services in LMICs. Commercial messaging and communication apps are often used to facilitate teleconsultations (Mars and Scott 2016). One such application is *WhatsApp*, which is a unified communication app that can be used for voice, video and text communication as well as the transmission of still and/or video images.

An app that has been developed specifically for telemedicine is the Swinfen Charitable Trust app25, where licensed health professionals working in a LMIC can download the app for free and sign up as a referrer and who can then send in medical questions and get advice or act as a consultant. Health professionals see a "case inbox" on the app requesting consultation advice or with a response to the referral (The Swinfen Charitable Trust 2017).

27.2.8 Email

Email-based telemedicine can facilitate store-and-forward telemedicine consultations in a large number of medical specialties and applications. Email has a niche application in low-bandwidth, image-based specialties such as dermatology, pathology, wound care and ophthalmology where attached digital camera images are used for telediagnosis (Caffery and Smith 2010). Email-based telemedicine can leverage existing infrastructure used for personal communication for clinical purposes. As previously discussed, email-based telemedicine may be appropriate for limited bandwidth applications.

27.2.9 Hosted Services

There are a number of hosted telemedicine services suitable for use in LMICs. There are distinctions between the types of hosted services. Some services provide both the telemedicine platform, hosting of the platform, and medical consulting services. One such example is Orbis Cybersight (Orbis International 2017). Referrers from LMICs can enroll and post a complex ophthalmic case via the web portal. A network of volunteers can provide advice to the referrer. The consulting services are coupled with online training opportunities for healthcare providers in LMICs.

The Swinfen Charitable Trust runs a multi-speciality service using a hosted web application and additional a smart device app (The Swinfen Charitable Trust 2017).

Other services only provide the platform and hosting. One such platform is the *Collegium Telemedicus* which facilitates a store-and-forward network (Collegium Telemedicus 2017) The *Collegium Telemedicus* platform has been adopted by a number of distinct services. Each service effectively has its own virtual instance of *Collegium Telemedicus* ensuring privacy across services. *Collegium Telemedicus* has been adopted by Médecins Sans Frontières to run a teleradiology service of which over half the cases were children (Wootton et al. 2014).

The last variation of hosted services are those providing software as a service. For example, the iPath system developed at the University of Basel in Switzerland. As the name implies, iPath was originally designed for telepathology but has more recently been adapted for other specialty consultations and image sharing in dermatology and radiology. iPath has also been used as the platform for a pediatric store-and-forward telemedicine network that supports 40 small rural hospitals in Tanzania (Kruger and Niemi 2012). The iPath system can be downloaded from the Internet for installation on the healthcare organizations' servers (Brauchli 2017). iPath is currently being used in Africa, Asia and the Pacific (Zolfo et al. 2009).

27.2.10 Smartphone-Attached Devices

There is wide range of devices that can be attached to mobile phones that can be used for diagnostic or monitoring purposes. These include blood pressure monitors, blood glucose monitors, electrocardiography, ultrasound imaging, electronic stethoscopes, dermatoscopes, otoscopes (Figure 27.1a, b) and ophthalmic slip lamp imaging devices.

Images or data collected by these devices immediately appear on the mobile phone and can then be easily sent onward for consultation. Most do not require extensive training to operate. While not extensively tested, there is some evidence that a mobile attached device performs similarly to dedicated medical devices – for example, an *iPhone*-attached otoscope was preferred to a traditional otoscope in a pediatric emergency department (Richards et al. 2015).

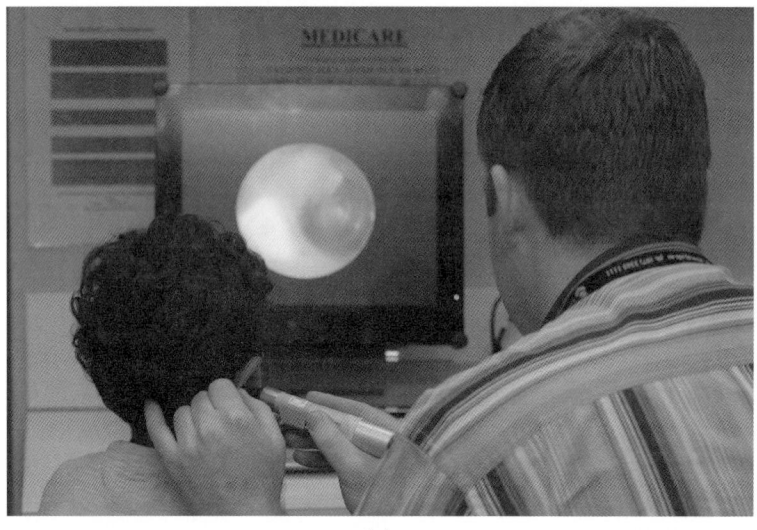

(a)

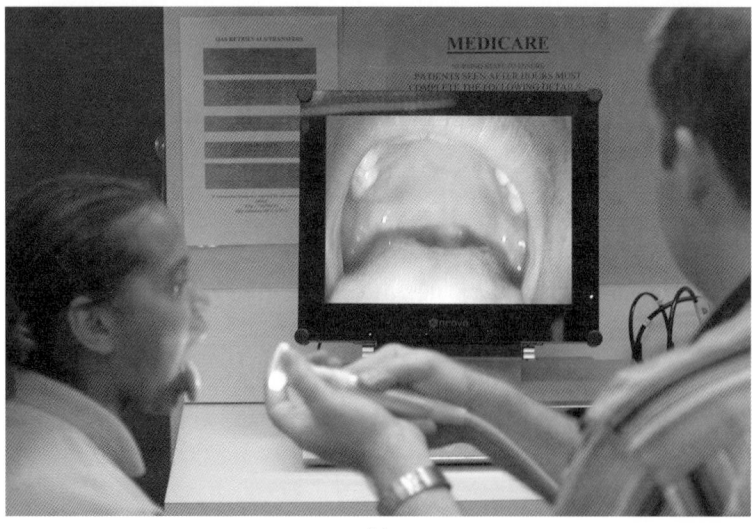

(b)

Figure 27.1 (a) Video-otoscopy used for examining the ears and nose. (b) Video-otoscopy for examining the throat.

27.3   Models of Care in Telepediatrics

Telemedicine can be used to support the delivery of pediatric health services using both asynchronous and synchronous methods. Various models have been demonstrated to be useful for LMICs as a means of providing healthcare advice, training opportunities and lifesaving instructions during emergency situations. From a legislative perspective specialist advice is normally given to clinicians on the understanding that the referring doctor remains responsible for the patient. The specialist is providing an opinion which the primary referrer can choose to either accept or reject, depending on the circumstances. Decisions regarding treatment are often dependent on available resources (medications, surgical supplies, equipment) and the skill mix of the staff involved in the care of the patient. Some telemedicine services provide interim advice before further specialist care can be provided – for example in cases where an emergency retrieval is arranged for a patient who needs to be transported to another facility for treatment.

Models of care which rely on store-and-forward or asynchronous communication methods are most common due to the low technology requirements and the ability to manage cases in a flexible timeframe. Clinical cases may involve a health worker or clinician requesting advice via email, and sharing information such as clinical history, test results or images. If this information is available for review by a specialist, advice can then be provided to assist with assessment, diagnosis, and/or treatment planning. This may be done through email, web-based portals or secure online databases.

Other models of care may utilize real-time (synchronous) communication methods such as telephone and videoconference support. Common examples of real-time telemedicine services include videoconsultations between clinicians to discuss cases (case consultations) or sessions which include the referring person, the advising clinician and the patient (patient consultation). In the context of pediatric services the patient-end support person plays an important role. Video consultations which require the patient to present usually require a parent or caregiver to be present for support. In regards to technological requirements, real-time telemedicine usually requires reliable Internet connections to ensure an effective connection for audio and video communication. This can be challenging in many countries where Internet access remains extremely limited.

27.4   Swinfen Charitable Trust Telemedicine Service

One example of a well-known store-and-forward telemedicine service in LMICs is the Swinfen Charitable Trust (SCT) telemedicine program (The Swinfen Charitable Trust 2017). The Trust has been providing free "Humanitarian Telemedicine" for almost two decades, making it the longest running telemedicine charity. As of 2017 the SCT network comprised 346 referring hospitals and 741 volunteer medical specialists. Even though most referrals sent through the SCT are non-urgent, some cases have warranted an urgent response within hours of sending the request for information (see the case examples below). SCT support is provided to medical centers or health services which have subscribed to the service. Clinicians in LMICs are able to sign on to a secure online server and submit case details which an operator then sends to the appropriate specialist for a response. Specialists are located all around the world. The person at the remote location can then log back on to the server to see the response. Based on referral records held by the SCT, pediatric referrals are the most common (Patterson and Wootton 2013).

27.5 Selected Examples of SCT Referrals

27.5.1 Neurosurgery

Origin: Iraq, 2005
Original email message:
Dear Sir,
This unlucky, and lucky at same time, child exposed to a bullet from above of unknown origin to the left upper part of the frontal bone passed through the frontal lobe as we see in the x-ray of the left orbit behind the globe (Figure 27.2a and b). Conscious, oriented and fully active, the child only complains of heaviness in the left orbit.

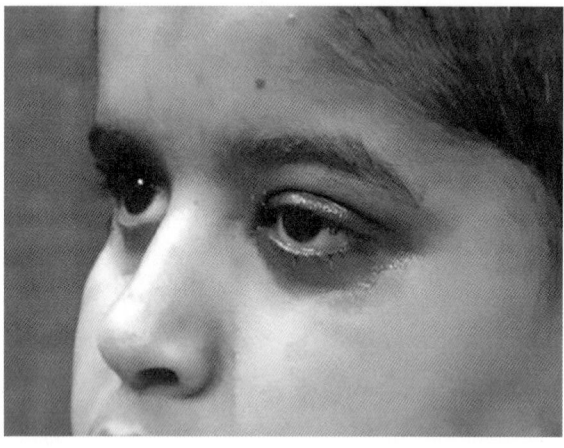

(a)

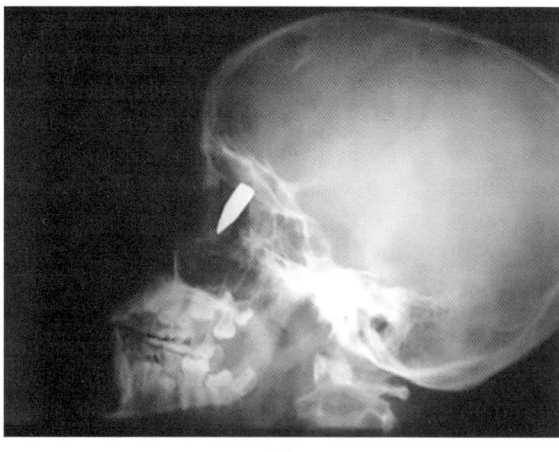

(b)

Figure 27.2 (a) Mild left-sided proptosis in a child with a bullet lodged in the skull. (b) Skull x-ray of this child showing a bullet lodged in the frontal bone.

Examination:

Visual acuity: 6/6 × 6/6. Mild proptosis; no bruit nor thrill.

Mild limitation of ocular motility in lateral gaze.

Good alignment in primary position and no diplopia.

Pupil: round, regular, reacts to light.

Normal fundi and optic disc.

Your opinion please.

With best regards,

Referring clinician (name withheld).

Follow-up:

Exact outcome of case unknown; however, treatment advice was provided by a neurosurgeon including a plan for surgery. The neurosurgeon also liaised with SCT to organize surgery free of charge; however, the parents of the child opted not to leave their country for the surgery. It is believed that the child is still being cared for by the health service in Iraq.

Pat and Roger Swinfen

27.5.2 Dermatology

Origin: Zambia, 2017

Original email message:

16 July 2017 – 22:47 hours

This infant was seen by our doctors on one of our mobile medical clinics in rural Zambia this week. The rash developed one week before attending clinic. Some lesions were crusting and the rash was present on the face, arms, back and buttocks (Figure 27.3a and b). The rash did not appear to be bothering the infant and there was no fever or systemic symptoms.

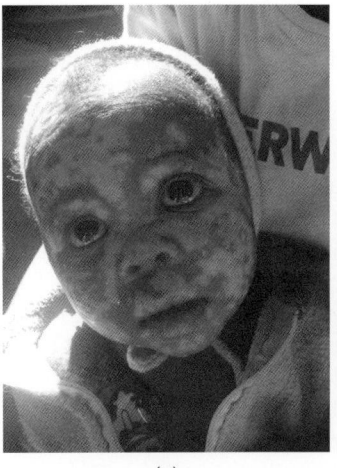
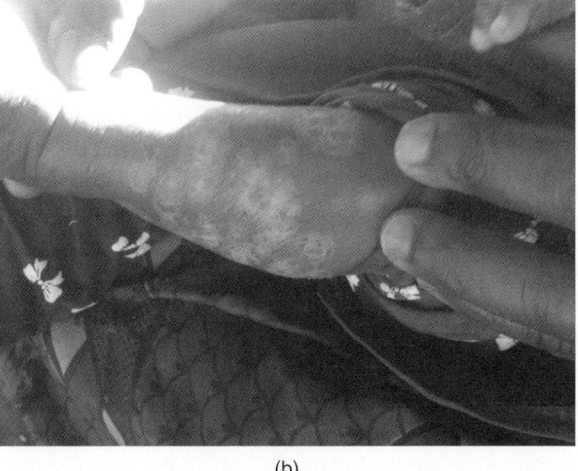

(a)   (b)

Figure 27.3 (a and b) Generalized crusting annular rash most prominent on the face with an overall annular scaly appearance and impressive hypopigmentation in an otherwise well baby. The teledermatologist thought the most likely diagnosis was neonatal lupus erythematosus (see text). (*See color plate section for the color representation of this figure.*)

The infant was otherwise fit and well, HIV negative and had been seen previously by our doctors at which time there was no evident rash.

This rash had been treated by the community health worker with IM benzyl penicillin with some effect.

Referring clinician (name withheld)

Follow up:

17 July 2017 – 04:29 hours

Dear Dr. (name withheld),

Congratulations on taking some of the finest images I have ever seen as a Swinfen consultant. The photo of the infant's face is particularly impressive and has perhaps the most valuable feature for dermatology: tangential light which shows texture. Now, if only I could be certain of the diagnosis! It is 12:30 a.m. as I write this (and the infant you describe is stable and not in crisis) so I will get some sleep and answer fully later today. Thanks in advance for your patience.

Pediatric dermatology consultant.

(Name withheld)

17 July 2017 – 16:14 hours

The more generalized distribution can be seen in neonatal lupus erythematosus (NLE) but the face is always worst, presumably due to sun exposure. Some infants have rash at birth even before any UV exposure. This infant has an eruption that looks very much like the subacute cutaneous lupus that we often see in adult patients.

Certainly there is a differential diagnosis here which would include congenital infections such as rubella and syphilis but I would expect other findings and the rash likely present at birth. Genetic photosensitivity disorders would also have additional findings. I suppose a primary immunodeficiency (and there are many to choose from including new, recently described conditions) should be on the list but I would expect other problems and failure to thrive. My money is still on NLE.

Pediatric dermatology consultant

(Name withheld)

17 July 2017 – 19:03 hours

Hello Referring Doctor,

Thank you for sending this fascinating case. I would like to establish first that (i) the infant had no rash until the recent visit at about six months of age and (ii) the rash is confined to face and one arm? and (iii) there is no family history of skin disease OR autoimmune disease?

This very prominent facial rash with an overall annular scaly appearance and impressive hypopigmentation is characteristic of neonatal lupus erythematosus, a rare disorder in which infants have maternal antibodies such as anti-Sjögren's-syndrome-related antibody A (SS-A), or SS-B sometimes called Ro and La antibodies, and rarely ribonucleoprotein (RNP) antibodies which are photosensitizing, so the rash appears after sun exposure. The rash will fade in time as the infant clears maternal antibodies but sun protection is important until this occurs. Scarring is unlikely. Topical hydrocortisone 1% or 2.5% ointment or cream applied twice daily is safe and should result in improvement, although return of normal pigmentation will likely take many months. Sunscreens and/ or complete sun avoidance is strongly recommended until this occurs. The most IMPORTANT ASPECT of NLE is the possibility of CONGENITAL HEART BLOCK which is often complete and can be fatal. When NLE is suspected an EKG as soon as possible after delivery is key. In this infant who appears to be thriving and has reached the age of six months, heart block is less likely but EKG is still recommended. Other organ systems can be involved causing, among other things, anemia and liver enzyme elevations. Mothers of infants with NLE are more often than not asymptomatic and unaware of the presence of antibodies. There is a high risk of NLE in subsequent pregnancies.

I hope this is helpful. Serologies may not be available for this infant but I hope an EKG can be arranged.

This is an absolutely fascinating case and I am so pleased to review it. I would be very grateful for additional information about this baby and especially grateful for clinical follow-up. Please contact me with any questions or to discuss this case further.

Most sincerely,

Pediatric dermatology consultant.

(Name withheld)

PS. Brilliant case which beautifully illustrates just why Swinfen Telemedicine is making such an important contribution to world health one person at a time. Love this!

27.5.3 Orthopedics

Origin: Centre for Rehabilitation of the Paralyzed (CRP) – Bangladesh, 2000 (CRP 016?)
Original email message:

Dear Sir,

A four year-old boy presented to me at the CRP outpatient department with the complaint of huge swelling of both thighs, shortening of both thighs and inability to walk without support. The boy is quite normal with regards to his other functions and systems. He has very short stature according to his age, height is 70 cm and the diameter of each mid-thigh is 28 cm. It seems that both the thighs are short in comparison to his legs and trunk (Figure 27.4a). X-ray showed apparent dislocation of both hips (Figure 27.4b).

I did not find any other abnormalities in the boy nor any other congenital anomalies in his family.

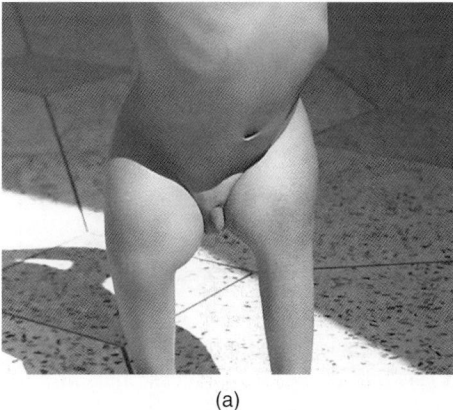 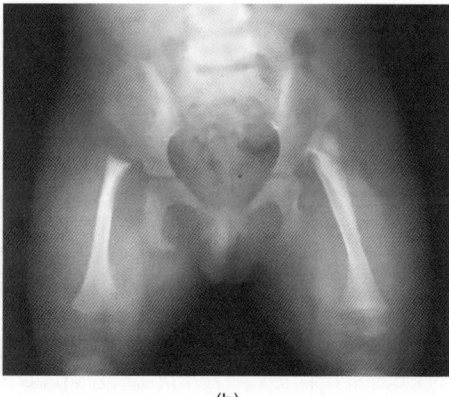

(a)   (b)

Figure 27.4 (a) Photograph showing child's gait thought most likely due to congenital dislocation of the hips. (b) X-ray showing dislocation of the hips in the same child.

Referring clinician
(Name withheld)

Follow up (same day):
I think it is possible he has bilateral congenital dislocation of the hip. That said, on the basis of the photos it is a very severe case. The Canadian experience suggests that leaving them is probably best with late presenters.
Regards,
Orthopedic surgeon
(Name withheld)

Follow up (next day):
I showed these films in Edinburgh today at an orthopedic meeting. No real consensus, but the absence of the femoral head epiphysis was noted … with the acetabular deficiency. I don't feel you have much prospect of getting an anatomical hip joint. My feeling is leave alone for fear of making it worse.
Regards,
Orthopedic surgeon
(Name withheld)

27.6 National and International Telemedicine Services

27.6.1 Telemedicine Services in Tanzania

The Tanzanian Telemedicine Network supports pediatric care at many remote hospitals in Tanzania. They use open-source web-based software called iPath to send patient information and images to volunteer pediatric specialists for advice. Part of the process also includes e-mail alerts to the consultants (Kruger and Niemi 2012). One of the strengths of this model is that it was developed from within a LMIC, creating a sense of ownership and empowerment within the local health centers. Pediatricians were not the only specialists used for telepediatric store-and-forward consultations in this case as obstetricians, dermatologists, radiologists, and surgeons as well as other consultants also provided responses. Over 75% of referrals to the network were responded to within 24 hours (Kruger and Niemi 2012). Store-and-forward methods work especially well in low-bandwidth scenarios since real-time video is difficult to achieve in areas where telecommunications are limited.

27.6.2 Médecins Sans Frontières

Médecins Sans Frontières (MSF), also known as Doctors Without Borders, is an international non-profit medical organization founded in France with associated offices worldwide (Médecins Sans Frontières 2017a). MSF has established a telemedicine network based on the *Collegium Telemedicus* (CT) system of medical professionals providing advice and support to clinicians in remote regions. For example, with the absence of a cardiologist in a South Sudan hospital, excerpts of a patient's heart ultrasound imaging were sent through MSF's 24 hours referral allocation system to a telemedicine doctor in Canada and a cardiologist in the USA. Later, the opinion of an anesthesiologist in

France and an obstetrician in Australia were also called upon using the system, in order to formulate a plan for the patient's treatment (Médecins Sans Frontières 2017b).

Over 40% of all telemedicine cases referred to MSF were pediatric. The median age of the pediatric patients was four years old (Martinez-Garcia et al. 2014). Pediatric cases encountered by MSF have used telemedicine in a number of ways including seeking a second opinion on a rash from a dermatologist located remotely, or sending x-rays to a radiologist in another country for assessment (Médecins Sans Frontières 2017b).

27.6.3 Case Conferencing in LMICs

Pediatric case-conferencing in LMICs has been shown to be a key enabler of better health outcomes for patients and for an increase in detection of pediatric disease. In the case of a real-time telepediatric case-consultation service running from clinicians in Somalia to pediatricians in Nairobi, a significant change to case management was made in 64% of cases and adverse outcomes were reduced by 30% (Zachariah et al. 2012).

Provider to patient video-conferencing allows for the remote examination of pediatric patients and discussion of diagnosis and treatment with the family. Cases of pediatric head trauma were managed in this way in India, with the real-time video also proving very valuable for post-treatment follow up (Ganapathy 2005).

27.6.4 Medical Missions for Children

Medical Missions for Children (MMC) is a USA-based pediatric non-profit organization that uses telemedicine to help diagnose and treat children in LMICs, as well as to educate medical professionals in those areas. In a format similar to the previously mentioned groups, a doctor in need of assistance from one of the 110 countries involved contacts the organization and forwards patients' records. MMC then arranges a video consultation with a medical specialist. MMC installs the Polycom video conferencing equipment required to engage in video consultations for these hospitals in need. They are increasingly pushing toward communication with hospitals beyond just the video conferencing system, looking to use hand-held mobile devices more in the future (Medical Missions for Children 2017). MMC relies on charitable donations and volunteers to carry out their work.

27.7 mHealth Applications for LMICs

While children themselves may be too young to engage in the use of mHealth on mobile applications, this mode of telepediatrics can be very beneficial to parents, caregivers, and health service providers including appointment reminders, treatment tips and health education. Hand-held communication devices are becoming increasingly popular in LMICs, as indicated by a survey of pediatric cases presenting to an emergency department in Kenya which showed nearly 90% of families had mobile phone access (House et al. 2015). Messages can be sent via SMS or using mobile device applications.

While mHealth has been more widely used for chronic diseases such as diabetes mellitus (Martínez-Pérez et al. 2013) LMICs can utilize similar methods to tackle infectious diseases in children as well as maternal health issues. Daily text messaging to parents in LMICs has been shown to significantly improve oral health knowledge and practice in

India and on-time immunizations for children in Thailand and Myanmar (Higgs et al. 2014). Community health workers in LMICs have used mHealth methods to prevent mother-to-child transmission of HIV by coordinating their activities using text messages (Mushamiri et al. 2015). Guidance on how to treat children under 5 with malaria has also been provided to healthcare providers using mHealth in Kenya (Higgs et al. 2014).

ImTeCHO, an abbreviation for Innovative mobile-phone Technology for Community Health Operations, is a mobile phone application for Accredited Social Health Activists (ASHAs) in India to use for assistance in their maternal, newborn, and child health work (ImTeCHO 2013). This program supports a mother from the beginning of pregnancy and her infant up until two years of age. The app allows ASHAs to schedule home visits, record services provided, track high risk cases and record births and deaths. There is also a web-based interface used by primary healthcare staff to manage supplies, track high risk cases, and gather information (ImTeCHO 2013). This application has been found to be acceptable, feasible, and useful (Modi et al. 2015).

27.8   Telemedicine Screening Services

Many childhood health problems can be identified early and subsequently avoided by using screening programs. This applies to both developed countries and LMICs. Telemedicine allows for rapid diagnostics and for information collected during screening in underserved areas to be sent immediately to specialists located in major cities for assessment.

School-based screening presents an opportunity to screen many children at once in a building where equipment can be set up. Hearing screening is one service that has been provided by telehealth both in developed countries and LMICs (Smith et al. 2012; Swanepoel and Hall III 2010). For example, schools in India were set up with a laptop, audiometer, and video-otoscope as well as a web camera for video conferencing, and connected over the Internet to a hospital site with an audiologist to deliver the screening service (Monica et al. 2017). Training was provided to a facilitator in the school so they could assist with the screening. When tele-hearing screening was compared to in-person screening no significant difference in pure tone audiometry or distortion product otoacoustic emissions was found (Monica et al. 2017). In locations with limited Internet access, mobile phone hot spot and dongle devices were found to be useful. (An Internet dongle is usually a universal serial bus-sized gadget that plugs into the USB port on a laptop or computer to provide easy mobile broadband access. A dongle is basically a tiny unpowered modem that uses the computer's power to access the Internet.) When testing hearing in schools, staff should be aware that ambient noise can cause issues.

27.9   Telemedicine Support during Disaster Situations

Telemedicine is useful in disaster situations for many reasons including medical advice and triage as well as planning patient transportation (Simmons et al. 2008). Often in these situations the demand on medical services is very high, while the ability to deploy the necessary specialist resources is often limited. Telemedicine support can provide timely advice during critical situations.

War can destroy a country's infrastructure, including its medical services. Due to the destruction of hospitals as a result of the Syrian civil war, a tele-intensive care unit (ICU) program was set up from the USA using volunteer clinicians, social media applications, satellite Internet, a web-based medical records system and video cameras (Moughrabieh and Weinert 2016). The remote physicians triaged victims of attacks and directed transfers to hospitals, in addition to advising nurses and managing postoperative care.

Disease epidemics such as cholera outbreaks put vulnerable populations such as children at an even higher risk. Online daily reporting of inpatient and outpatient cases using telemedicine from a hospital in India to a referral center showed an emerging risk of a cholera outbreak. Onsite microbiologists at a large religious fair were able to take stool cultures and rectal swabs from pilgrims presenting with diarrhea, confirmed a diagnosis of cholera and then relayed the information to the main hospital in the area and to health authorities, who then recommended increased sanitation via the telemedicine system, thus preventing a widespread outbreak (Ayyagari et al. 2003).

Disaster situations can cause mental health issues. One example was the 54% incidence of depression and suicidal ideation in children and adolescents exposed to radiation from the Chernobyl nuclear powerhouse disaster in Ukraine (Contis and Foley 2015). Another is the 50% of children that tested positive for severe symptoms of post-traumatic stress disorder (PTSD) after the 2010 Haiti earthquake (Blanc et al. 2015) While child and adolescent telepsychiatry has been run in large scale in developed countries (Pignatiello et al. 2011; Wood et al. 2012; Yellowlees et al. 2008), it is one area of disaster relief that is lacking in LMICs (Augusterfer et al. 2015).

With that said, incorporating any form of telepediatrics under these conditions should be considered cautiously as disaster situations can be associated with an interruption in telecommunications, especially if the disaster is environmental or weather-related.

27.10 Challenges Associated with Telemedicine Adoption in LMICs

One way of solving a workforce shortage is not only to look for more local workers, but to bring in workers through international collaborations. However, any increase in service provision, including the use of telemedicine, can put an extra burden of work on to health service administrators and medical staff. Nevertheless, bringing in an international workforce using ICT offers a solution to human resource-low environments often found in LMICs. Many non-profit organizations operate in LMICs in an attempt to provide telepediatric services to underserved regions.

While international telepediatric services provide a great option to increase access to care temporarily, if the LMICs become too dependent on them, there is a serious risk of adverse patient outcomes should the service stop. Most organizations are reliant on donations to operate which is not always a stable source of income. It is for that reason that a useful aspect in many of these services also includes mentoring and case-conference aspects to upskill the local workforce in the event that the program can no longer continue.

International agreements are not without challenges. Organizations offering assistance from overseas must be aware of the local context in which they are working. It is ideal if the clinicians providing services from outside the local area have previously worked in the area on an outreach basis or even visited the region temporarily for work. For example, in the case of the Tanzanian Telemedicine Network, consultants from overseas had work experience as expatriate doctors in Tanzania (Kruger and Niemi 2012).

Other issues that complicate international organizations and governments providing telepediatric services includes taking into account time differences, language barriers and legal or ethical problems. Connectivity and integration of technology at both ends must also be considered.

With multiple charity organizations such as MSF, the SCT and MMC working in isolation of each other, there is possible room for further collaboration and resource sharing.

27.11 Telepediatric Case Studies in LMICs

There are many excellent examples of telemedicine applications that support children and families living in LMICs. Some examples based on clinical specialty are described below.

27.11.1 Pediatric Cardiology

Hospitals in LMICs may struggle to have adequate specialists including cardiologists, let alone a pediatric cardiologist which many developed countries would routinely have as part of their pediatric health service team. A telepediatric cardiology network known as the Royal Portuguese Hospital Circle of the Heart (RCP-CirCor) was established in Brazil to deal with the lack of pediatric cardiologists in rural areas. Neonatologists and nurses were trained by video conference (WebEx by Cisco) to carry out cardiac examination, to perform screening electrocardiograms and to film the screening on a tablet device (Galdino et al. 2016). Other telemedicine tools included an electronic stethoscope, a portable ultrasound machine and a pulse oximeter. The network has now surpassed 100 000 neonatal screenings using these methods. Such telemedicine facilities in remote areas enable local healthcare providers to perform an electrocardiogram which can lead to a diagnosis: the network can then also be used to plan transfer or further treatment. Timing is very important especially for neonates.

A similar service has been set up in India where pediatric cardiologists are confined to the major metropolitan areas: remote hospitals have been connected to a tertiary care cardiac centre using satellite bandwidth to transmit tele-echocardiographs, so that the children may be diagnosed or other pathologies excluded (Sekar and Vilvanathan 2007).

27.11.2 Cancer Care Services

Around 70% of cancer deaths occur in the LMICs, with highly curable cancers harder to treat in LMICs versus more developed regions. This is, at least in part, due to the lack of infrastructure for dealing with conditions that require chronic care (Chan 2010). A tele-oncology service was set up through the SickKids-Caribbean Initiative utilizing specialists from a large Canadian pediatric hospital – The Hospital for Sick Children in Toronto. This program enabled the local healthcare providers the ability to conduct case consultation review rounds with hospitals in six English-speaking Caribbean countries. A secure electronic file transfer system was used to send case reports. Additionally, patient care education rounds were conducted with remote hospital staff to upskill them in caring for children receiving treatment for cancer (Adler et al. 2015). Evaluations from the Caribbean staff showed they were highly satisfied with the program.

The Brazilian National Telemedicine Network in Oncology (ONCONET) used digital technology to combine many services for childhood cancers including a web-based system for electronic patient records, a cancer reporting database for effective health service planning and software for videoconference links between a major children's hospital and remote hospitals in order to provide second medical opinions on cases (Hira et al. 2006). ONCONET started in 2000 and grew to include more hub sites, with much planning on how to embed it into the national infrastructure and how to set up the system server components (web, application, database and storage), so that the network could be easily be scaled up.

27.11.3 Orthopedics

Even in developed countries pediatric orthopedic surgeons are less available than surgeons in other surgical specialities. Clinicians in Djibouti (a French and Arabic speaking country located in the Horn of Africa) that face diagnostic uncertainties in pediatric orthopedics have been able to participate in asynchronous teleconsultations for advice. Radiographs are photographed using a digital camera and sent by e-mail to specialists in France. In emergencies the medical worker located in Djibouti would call the specialist by phone in order to better alert them of the incoming request. It was found that the teleconsultation resolved most diagnostic uncertainties and expert advice changed the way the case was managed in 77% (n = 37) of cases (Bertani et al. 2012).

Pacific Island nations have also incorporated store-and-forward telemedicine in pediatric orthopedic consultations in the USA using e-mail. Referring health providers sent radiographs of a fracture site (pre- and post-reduction) and a consultant was able to provide advice on traction weight and set up. Many families in the Pacific have several children so the parent having to leave the local area with one child or to travel with multiple children is very disruptive to the family (Person et al. 2003). One example involved a young girl who fell from the window of a house. Even though the hospital did not have an orthopedic surgeon, the local staff were able to take an x-ray and e-mail a photograph and receive advice back via e-mail that enabled better patient management.

27.11.4 Pediatric Surgery

A need for pediatric surgery and more pediatric surgical training programs has been identified in LMICs (Bickler et al. 2001). Web-meeting platforms like *Adobe Connect* have been used for pediatric surgery education. An example of this was in South Africa where international speakers have been set up to present on pediatric surgery online so that anyone in a resource-limited setting could connect (Numanoglu 2017). Recordings were made available after the meeting on a website for future viewing, which could also include PowerPoint presentations.

Low-bandwidth telemedicine has also been used in surgical pre-screening to exclude those children who may have a disease too advanced to operate on, or who do not have the local support required for recovery. Preoperative diagnoses performed by telemedicine in Kenya using only a computer and a digital camera, agreed with those made in person (Lee et al. 2003). One major benefit to the system is that it prevented patient and family members from traveling a long distance only to be told that their child was not an appropriate case for surgery.

27.12 Pathology Services

Pathologists, like many medical specialists, are in short supply in LMICs. In Malawi, for example, it has been said that four pathologists were responsible for a population of 14 million, including children (Carey et al. 2014). To deal with this, pediatric telepathology methods were used, including a microscope with a camera set up at distant medical sites and images sent to a receiving site in the United Kingdom where pathologists provided a report. Clinicians at the distant sites simply uploaded the clinical details and photographs to a Dropbox account, which the pathologist could access and then report back, as well sending a copy of the report back via e-mail.

Another model of telepathology in LMICs is the iPATH Telemedicine Platform. iPATH operates as a web-based platform for case-conferencing. The online system, based at the University of Basel in Switzerland, can be used by multiple independent groups (Brauchli and Oberholzer 2005). A pathologist on the server has access to remote controlled microscopes at the patient end to make a diagnosis, or can just refer to what has been previously sent through asynchronously to make a diagnosis. Users are separated into closed groups with moderators. LMICs that have used this system include Cambodia, the Solomon Islands, Bangladesh, Iran, Laos, Ukraine, and India. A country might set up a series of groups for their nation under different specialties, such as "Tanzania Pediatrics."

27.13 Radiographic (Imaging) Services

Sending radiographic images using ICT is not a novel concept, and even in countries with low bandwidth limits, image compression is possible and the process still occurs. An example is the World Federation of Pediatric Imaging (WFPI) Program (Prabhu 2017). WFPI is an international platform of volunteer radiologists who provide pediatricians and hospitals with assistance in reading child radiographs. One South African program utilized this network in a pilot project, converting digital radiographs to JPEG images and e-mailing them along with a referral request to a pediatric radiologist for their opinion (Griggs et al. 2014).

In Liberia, the Island Pediatric Hospital in Monrovia had no radiographers but did have a fully digital X-ray unit that high school graduates with no medical training and limited technical knowledge learned to use. Radiographic images were downloaded and sent to a radiologist in the USA along with clinical and laboratory information. Radiologist responses were sent within a few hours (Andronikou et al. 2011).

27.14   Maternal Health Services

Postnatal maternal and infant health has seen benefits in LMICs from the use of mHealth technologies. In Ecuador new mothers were provided with postnatal education on their mobile phones and text access to a nurse, which increased the likelihood that they attended a check-up and decreased the number of infant illnesses (Maslowsky et al. 2016). In Argentina, Cormick et al. (2015) used SMS messages to mothers who had tested positive for Chagas disease. Seventy percent of mothers had a mobile phone and thus SMS appointment reminders for post-partum mothers were used and were well received.

Of note in regards to maternal health in LMICs is the fact that that globally each year approximately 16 million girls aged 15–19 years of age give birth, and around 1 million girls give birth under the age of 15 (WHO 2014). Most of these girls are in low- and middle-income countries and babies born to adolescent mothers have a greater risk of dying compared to those born to women aged 20–24 (WHO 2017). The fact that these mothers still fall under pediatric care themselves further complicates healthcare provision to them and their babies.

27.15   Conclusions

Telemedicine has a vital role for supporting the health of children and their families in LMICs. The benefits of telemedicine also extend to clinicians working in areas where resources are limited and demand on services may be high. While there are some very good examples of telemedicine programs throughout the world, the issue of equity of access remains. In an ideal world we would like to see every person, irrespective of where they live, have access to the healthcare they need in a fair and equitable manner. In the right circumstances, telemedicine can help improve health access and collaboration throughout the world.

27.16   Acknowledgements

We sincerely thank Lord Roger Swinfen and Lady Pat Swinfen, Swinfen Charitable Trust, Canterbury, United Kingdom, for providing case examples and permission to reproduce selected case studies in this book chapter.

27.17   Useful Websites

Collegium Telemedicus: https://collegiumtelemedicus.org
Cybersight: https://cybersight.org
ImTeCHO: http://www.imtecho.com
iPath Telemedicine Platform: https://sourceforge.net/projects/ipath
Médecins Sans Frontières: http://www.msf.org
Medical Missions for Children: http://www.mmissions.org
Swinfen Charitable Trust: http://www.swinfencharitabletrust.org
UNICEF: https://www.unicef.org

WebRTC: https://webrtc.org
World Federation of Pediatric Imaging: www.wfpiweb.org
World Health Organisation: http://www.who.int/en

Bibliography

Adler, E., Alexis, C., Ali, Z. et al. (2015). Bridging the distance in the Caribbean: telemedicine as a means to build capacity for care in pediatric cancer and blood disorders. *Studies in Health Technology and Informatics* 209: 1–8.

Andronikou, S., McHugh, K., Abdurahman, N. et al. (2011). Pediatric radiology seen from Africa. Part I: providing diagnostic imaging to a young population. *Pediatric Radiology* 41 (7): 811–825.

Armfield, N.R., Bradford, M., and Bradford, N.K. (2015). The clinical use of Skype – for which patients, with which problems and in which settings? A snapshot review of the literature. *International Journal of Medical Informatics* 84 (10): 737–742.

Augusterfer, E.F., Mollica, R.F., and Lavelle, J. (2015). A review of telemental health in international and post-disaster settings. *International Review of Psychiatry* 27 (6): 540–546.

Ayyagari, A., Bhargava, A., Agarwal, R. et al. (2003). Use of telemedicine in evading cholera outbreak in Mahakumbh Mela, Prayag, UP, India: an encouraging experience. *Telemedicine Journal and e-Health* 9 (1): 89–94.

Bertani, A., Launay, F., Candoni, P. et al. (2012). Teleconsultation in pediatric orthopaedics in Djibouti: evaluation of response performance. *Orthopaedics & Traumatology: Surgery & Research (OTSR)* 98 (7): 803–807.

Bickler, S.W., Kyambi, J., and Rode, H. (2001). Pediatric surgery in sub-Saharan Africa. *Pediatric Surgery International* 17 (5): 442–447.

Blanc, J., Bui, E., Mouchenik, Y. et al. (2015). Prevalence of post-traumatic stress disorder and depression in two groups of children one year after the January 2010 earthquake in Haiti. *Journal of Affective Disorders* 172: 121–126.

Brauchli K. (2017). iPath Telemedicine Platform. SourceForge; [Accessed: 26 Oct 2017]. Available from: https://sourceforge.net/projects/ipath

Brauchli, K. and Oberholzer, M. (2005). The iPath telemedicine platform. *Journal of Telemedicine and Telecare* 11 (2 Suppl): 3–7.

Caffery, L.J. and Smith, A.C. (2010). A literature review of email-based telemedicine. *Studies in Health Technology and Informatics* 161: 20–34.

Caffery, L.J. and Smith, A.C. (2015). Investigating the quality of video consultations performed using fourth generation (4G) mobile telecommunications. *Journal of Telemedicine and Telecare* 21 (6): 348–354.

Carey, P., Fudzulani, R., Scholfield, D. et al. (2014). Remote and rapid pathological diagnosis in a resource challenged unit. *Journal of Clinical Pathology* 67 (6): 540–543.

Central Intelligence Agency. (2016). The World Factbook. Washington, DC. [Accessed: 26 Oct 2017]. Available from: https://www.cia.gov/library/publications/the-world-factbook/fields/2177.html

Chan, M. (2010). Cancer in developing countries: facing the challenge. Geneva, Switzerland World Health Organistion [26 Oct 2017]. Available from: http://www.who.int/dg/speeches/2010/iaea_forum_20100921/en

Collegium Telemedicus. (2017). Telemedicine for remote or low resource settings.

Contis, G. and Foley, T.P. (2015). Depression, suicide ideation, and thyroid tumors among Ukrainian adolescents exposed as children to Chernobyl radiation. *Journal of Clinical Medicine Research* 7 (5): 332–338.

Cormick, G., Ciganda, A., Cafferata, M.L. et al. (2015). Text message interventions for follow up of infants born to mothers positive for Chagas disease in Tucuman, Argentina: a feasibility study. *BMC Research Notes* 8: 508.

Galdino, M.M., Hazin, S.M., de Araujo, J.S. et al. (2016). Diagnosis and management of transposition of great arteries within a pediatric cardiology network with the aid of telemedicine: a case report from Brazil. *Journal of Telemedicine and Telecare* 22 (3): 179–182.

Ganapathy, K. (2005). Telemedicine and neurosciences. *Journal of Clinical Neuroscience* 12 (8): 851–862.

Griggs, R., Andronikou, S., Nell, R. et al. (2014). World Federation of Pediatric Imaging (WFPI) volunteer outreach through tele-reading: the pilot project in South Africa. *Pediatric Radiology* 44 (6): 648–654.

Higgs, E.S., Goldberg, A.B., Labrique, A.B. et al. (2014). Understanding the role of mHealth and other media interventions for behavior change to enhance child survival and development in low- and middle-income countries: an evidence review. *Journal of Health Communication* 19 (Suppl 1): 164–189.

Hira, A.Y., Nebel de Mello, A., Faria, R.A., et al. (2006). Development of a telemedicine model for emerging countries: a case study on pediatric oncology in Brazil. Conference proceedings: Annual International Conference of the IEEE Engineering in Medicine and Biology Society IEEE Engineering in Medicine and Biology Society Annual Conference. 1:5252–5256.

House, D.R., Cheptinga, P., and Rusyniak, D.E. (2015). Availability of mobile phones for discharge follow-up of pediatric emergency department patients in western Kenya. *Peer J* 3: e790.

ImTeCHO. India (2013). [Accessed: 26 Oct 2017]. Available from: http://www.imtecho.com

International Telecommunication Union. ICT Facts & Figures. Geneva, Switzerland: 2015. Available from: https://www.itu.int/en/ITU-D/Statistics/Documents/facts/ICTFactsFigures2015.pdf

International Telecommunication Union. (2017). ICT Facts & Figures. Geneva, Switzerland: Available from: http://www.itu.int/en/ITU-D/Statistics/Documents/facts/ICTFactsFigures2017.pdf

Källander, K., Tibenderana, J.K., Akpogheneta, O.J. et al. (2013). Mobile health (mHealth) approaches and lessons for increased performance and retention of community health workers in low- and middle-income countries: a review. *Journal of Medical Internet Research* 15 (1): https://doi.org/10.2196/jmir.2130.

Kruger, C. and Niemi, M. (2012). A telemedicine network to support pediatric care in small hospitals in rural Tanzania. *Journal of Telemedicine and Telecare* 18 (1): 59–62.

Lee, S., Broderick, T.J., Haynes, J. et al. (2003). The role of low-bandwidth telemedicine in surgical prescreening. *Journal of Pediatric Surgery* 38 (9): 1281–1283.

Liu, W.-L., Zhang, K., Locatis, C., and Ackerman, M. (2015). Cloud and traditional videoconferencing technology for telemedicine and distance learning. *Telemedicine Journal and e-Health* 21 (5): 422–426.

Mars, M. (2013). Telemedicine and advances in urban and rural healthcare delivery in Africa. *Progress in Cardiovascular Diseases* 56 (3): 326–335.

Mars, M. and Scott, R.E. (2016). WhatsApp in clinical practice: a literature review. *Studies in Health Technology and Informatics* 231: 82–90.

Martinez-Garcia, D., Bonnardot, L., Olson, D. et al. (2014). A retrospective analysis of pediatric cases handled by the MSF tele-expertise system. *Frontiers Public Health* 2: 266.

Martínez-Pérez, B., de la Torre-Díez, I., and López-Coronado, M. (2013). Mobile health applications for the most prevalent conditions by the World Health Organization: review and analysis. *Journal of Medical Internet Research* 15 (6): e120.

Maslowsky, J., Frost, S., Hendrick, C.E. et al. (2016). Effects of postpartum mobile phone-based education on maternal and infant health in Ecuador. *International Journal of Gynecology & Obstetrics* 134 (1): 93–98.

Médecins Sans Frontières (MSF). (2017a). [Accessed: 26 Oct 2017]. Available from: http://www.msf.org

Médecins Sans Frontières. (2017b). 3 Questions: Telemedicine on the front lines. Australia: Digital Agency. [Accessed: 26 Oct 2017]. Available from: www.msf.org.au/article/project-news/3-questions-telemedicine-front-lines

Medical Missions for Children. (2017). New Jersey, U.S. [Accessed: 26 Oct 2017]. Available from: www.mmissions.org

Modi, D., Gopalan, R., Shah, S. et al. (2015). Development and formative evaluation of an innovative mHealth intervention for improving coverage of community-based maternal, newborn and child health services in rural areas of India. *Global Health Action* 8: 26769.

Monica, S.D., Ramkumar, V., Krumm, M. et al. (2017). School entry level tele-hearing screening in a town in South India – lessons learnt. *International Journal of Pediatric Otorhinolaryngology* 92: 130–135.

Moughrabieh, A. and Weinert, C. (2016). Rapid deployment of international tele–intensive care unit services in war-Torn Syria. *Annals of the American Thoracic Society* 13 (2): 165–172.

Mushamiri, I., Luo, C., Iiams-Hauser, C., and Ben, A.Y. (2015). Evaluation of the impact of a mobile health system on adherence to antenatal and postnatal care and prevention of mother-to-child transmission of HIV programs in Kenya. *BMC Public Health* 15: 102.

Nolan, T., Angos, P., Cunha, A.J. et al. (2001). Quality of hospital care for seriously ill children in less-developed countries. *Lancet* 357 (9250): 106–110.

Numanoglu, A. (2017). Using telemedicine to teach pediatric surgery in resource-limited countries. *Pediatric Surgery International* 33 (4): 471–474.

Orbis International. (2017). Cybersight. [Accessed: 26 Oct 2017]. Available from: https://cybersight.org

Pacqué-Margolis, S., Muntifering, C., Ng, C. and Noronha, S. (2011). Population Growth and the Global Health Workforce Crisis. Technical Brief. CapacityPlus. Available from: http://www.who.int/workforcealliance/knowledge/resources/capacityplus_techbrief_nov2011/en

Patterson, V. and Wootton, R. (2013). A web-based telemedicine system for low-resource settings 13 years on: insights from referrers and specialists. *Global Health Action* 6: https://doi.org/10.3402/gha.v6i0.21465.

Person, D.A., Hedson, J.S., and Gunawardane, K.J. (2003). Telemedicine success in the United States Associated Pacific Islands (USAPI): two illustrative cases. *Telemedicine Journal and e-Health* 9 (1): 95–101.

Pew Research Center. (2014). Global Attitudes & Trends Spring 2014 Survey – Q68&69. [Accessed: 26 Oct 2017]. Available from: http://www.pewglobal.org/2014/06/05/spring-2014-survey-data

Pew Research Center. (2015). Cell Phones in Africa: Communication Lifeline. [Accessed: 26 Oct 2017]. Available from: http://www.pewglobal.org/2015/04/15/cell-phones-in-africa-communication-lifeline

Pignatiello, A., Teshima, J., Boydell, K.M. et al. (2011). Child and youth telepsychiatry in rural and remote primary care. *Child and Adolescent Psychiatric Clinics of North America* 20 (1): 13–28.

Prabhu, S. (2017). The World Federation of Pediatric Imaging. [Accessed: 26 Oct 2017]. Available from: http://www.wfpiweb.org

Richards, J.R., Gaylor, K.A., and Pilgrim, A.J. (2015). Comparison of traditional otoscope to iPhone otoscope in the pediatric ED. *The American Journal of Emergency Medicine* 33 (8): 1089–1092.

Sekar, P. and Vilvanathan, V. (2007). Telecardiology: effective means of delivering cardiac care to rural children. *Asian Cardiovascular and Thoracic Annals* 15 (4): 320–323.

Simmons, S., Alverson, D., Poropatich, R. et al. (2008). Applying telehealth in natural and anthropogenic disasters. *Telemedicine Journal and e-Health* 14 (9): 968–971.

Smith, A.C., Armfield, N.R., Wu, W.-I. et al. (2012). A mobile telemedicine-enabled ear screening service for indigenous children in Queensland: activity and outcomes in the first three years. *Journal of Telemedicine and Telecare* 18 (8): 485–489.

Smith, A.C., Youngberry, K., Christie, F. et al. (2003). The family costs of attending hospital outpatient appointments via videoconference and in person. *Journal of Telemedicine and Telecare* 9 (Suppl 2): S58–S61.

Stilwell, B., Diallo, K., Zurn, P. et al. (2004). Migration of health-care workers from developing countries: strategic approaches to its management. *Bulletin of the World Health Organization* 82: 595–600.

Swanepoel, D.W. and Hall, J.W. III (2010). A systematic review of telehealth applications in audiology. *Telemedicine and e-Health* 16 (2): 181–200.

Taylor, A., Morris, G., Pech, J. et al. (2015). Home telehealth video conferencing: perceptions and performance. *JMIR mHealth and uHealth* 3 (3): e90.

The Swinfen Charitable Trust. (2017). Mobile App. [Accessed: 26 Oct 2017]. Available from: http://www.swinfencharitabletrust.org/mobile-app.php

The Swinfen Charitable Trust. (2017). Swinfen Charitable Trust – Telemedicine Links. U.K. [Accessed: 26 Oct 2017]. Available from: http://www.swinfencharitabletrust.org

UNICEF. (2005). Children Living in Poverty. [Accessed: 26 Oct 2017]. Available from: https://www.unicef.org/sowc05/english/poverty.html

United Nations. (2017). World Population Prospects. [Accessed: 26 Oct 2017]. Available from: https://esa.un.org/unpd/wpp

WebRTC. (2017) [Accessed: 26 Oct 2017]. Available from: https://webrtc.org

Wood, J., Stathis, S., Smith, A., and Krause, J. (2012). E-CYMHS: an expansion of a child and youth telepsychiatry model in Queensland. *Australasian Psychiatry* 20 (4): 333–337.

Wootton, R. (2014). Twenty years with the JTT. *Journal of Telemedicine and Telecare* 20 (8): 425–426.

Wootton, R., Wu, W., and Bonnardot, L. (2014). Store-and-forward teleradiology in the developing world--the Collegium Telemedicus system. *Pediatric Radiology* 44 (6): 695–696.

World Health Organisation. (2006). The World Health Report 2006 – working together for health. [Accessed: 26 Oct 2017]. Available from http://www.who.int/whr/2006/en

World Health Organisation. (2010). Telemedicine: opportunities and developments in Member States: report on the second global survey on eHealth 2009. Geneva, Switzerland. Available from http://www.who.int/goe/publications/goe_telemedicine_2010.pdf

World Health Organisation. (2011). Child mortality. [Accessed: 26 Oct 2017]. Available from http://www.who.int/pmnch/media/press_materials/fs/fs_mdg4_childmortality/en

World Health Organisation. (2014). Adolescent pregnancy. [Accessed: 26 Oct 2017]. Available from http://www.who.int/mediacentre/factsheets/fs364/en

World Health Organisation. (2017). Adolescent pregnancy. Available from http://www.who.int/news-room/fact-sheets/details/adolescent-pregnancy

Yellowlees, P.M., Hilty, D.M., Marks, S.L. et al. (2008). A retrospective analysis of a child and adolescent eMental Health program. *Journal of the American Academy of Child and Adolescent Psychiatry* 47 (1): 103–107.

Zachariah, R., Bienvenue, B., Ayada, L. et al. (2012). Practicing medicine without borders: tele-consultations and tele-mentoring for improving pediatric care in a conflict setting in Somalia? *Tropical Medicine and International Health* 17 (9): 1156–1162.

Zennaro, F., Bava, M., Casalino, A. et al. (2013). Smartphones for mobile teleradiology in pediatric imaging: evaluation of diagnostic accuracy. *Egyptian Computer Science Journal* 37 (5).

Zolfo, M., Renggli, V., Koole, O., and Lynen, L. (2009). Telemedicine in low-resource settings: experience with a telemedicine service HIV/AIDS care. In: *Telehealth in Developing Countries* (ed. R. Wootton, R.E. Scott, N.G. Patil and K. Ho), 91–100. London: Royal Society of Medicine Press.

Webliography

https://www.unicef.org/sowc05/english/poverty.html. UNICEF. Children Living in Poverty. 2005.

http://www.who.int/pmnch/media/press_materials/fs/fs_mdg4_childmortality/en. World Health Organisation. Child mortality. 2011. [26 Oct 2017].

https://www.cia.gov/library/publications/the-world-factbook/fields/2177.html. Central Intelligence Agency. The World Factbook. Washington, DC: 2016. [26 Oct 2017].

https://esa.un.org/unpd/wpp. United Nations. World Population Prospects. 2017. [26 Oct 2017].

http://www.who.int/goe/publications/goe_telemedicine_2010.pdf. World Health Organisation. Telemedicine: opportunities and developments in Member States: report on the second global survey on eHealth 2009. Geneva, Switzerland: 2010.

http://www.who.int/workforcealliance/knowledge/resources/capacityplus_techbrief_nov2011/en. Pacqué-Margolis, S., Muntifering, C., Ng, C. and Noronha, S. Population Growth and the Global Health Workforce Crisis. Technical Brief. CapacityPlus. 2011.

http://www.who.int/whr/2006/en. World Health Organistion. The World Health Report 2006 – working together for health. 2006. [26 Oct 2017].

https://www.itu.int/en/ITU-D/Statistics/Documents/facts/ICTFactsFigures2015.pdf. International Telecommunication Union. ICT Facts & Figures. Geneva, Switzerland: 2015.

http://www.itu.int/en/ITU-D/Statistics/Documents/facts/ICTFactsFigures2017.pdf. International Telecommunication Union. ICT Facts & Figures. Geneva, Switzerland: 2017.

http://www.pewglobal.org/2014/06/05/spring-2014-survey-data. Pew Research Center. Global Attitudes & Trends Spring 2014 Survey – Q68&69. 2014. [26 Oct 2017].

http://www.pewglobal.org/2015/04/15/cell-phones-in-africa-communication-lifeline. Pew Research Center. Cell Phones in Africa: Communication Lifeline. 2015. [26 Oct 2017].

https://webrtc.org. WebRTC. [26 Oct 2017].

http://www.swinfencharitabletrust.org/mobile-app.php. The Swinfen Charitable Trust. Mobile App. 2017. [26 Oct 2017].

https://cybersight.org. Orbis International. Cybersight. 2017. [26 Oct 2017].

http://www.swinfencharitabletrust.org. The Swinfen Charitable Trust. Swinfen Charitable Trust – Telemedicine Links. U.K. 2017. [26 Oct 2017].

https://sourceforge.net/projects/ipath. Brauchli K. iPath Telemedicine Platform. SourceForge; 2017. [26 Oct 2017].

http://www.msf.org. Médecins Sans Frontières (MSF). 2017a. [26 Oct 2017].

www.msf.org.au/article/project-news/3-questions-telemedicine-front-lines. Médecins Sans Frontières. 3 Questions: Telemedicine on the front lines. Australia: Digital Agency. 2017b. [26 Oct 2017].

www.mmissions.org. Medical Missions for Children. New Jersey, U.S. 2017. [26 Oct 2017].

http://www.imtecho.com. ImTeCHO. India. 2013. [26 Oct 2017].

http://www.who.int/dg/speeches/2010/iaea_forum_20100921/en. Chan, M. Cancer in developing countries: facing the challenge. Geneva, Switzerland World Health Organistion; 2010. [26 Oct 2017].

http://www.wfpiweb.org. Prabhu, S. The World Federation of Pediatric Imaging. 2017. [26 Oct 2017].

http://www.who.int/mediacentre/factsheets/fs364/en. World Health Organisation. Adolescent pregnancy. 2014. [26 Oct 2017].

28

Telemedicine in the Diagnosis and Management of Skin Diseases

Giselle Prado[1], Odinaka Anyanwu[2], and Carrie Kovarik[3]

[1] *Orange Park Medical Center, Jacksonville, FL, USA*
[2] *Ross University School of Medicine and University of Texas Soutwestern, Dallas, TX, USA*
[3] *Dermatology, Dermatopathology, and Infectious Diseases, University of Pennsylvania, Philadelphia, PA, USA*

Revolutionizing Tropical Medicine: Point-of-Care Tests, New Imaging Technologies and Digital Health,
First Edition. Edited by Kerry Atkinson and David Mabey.
© 2019 John Wiley & Sons, Inc. Published 2019 by John Wiley & Sons, Inc.

28.1   Introduction

Skin disease may be associated with underlying systemic illnesses and can be the cause of significant morbidity, particularly in countries with limited resources, where access to specialty medical care may be challenging. Telemedicine is a potential method of providing healthcare to such populations worldwide. By using a telemedicine platform, whether through a computer or mobile device, specialized care and education can be delivered to healthcare providers and patients in need. Teledermatology has been shown to be an effective method for providing care to underserved populations (Norton et al. 1997).

Many low- and middle-income countries (LMICs) have a limited number of trained physicians and even fewer dermatologists. Over half of all physicians live in places with less than a fifth of the world's population. The 20% of the world's population that is most economically challenged is served by only 2% of the world's physicians (Desai et al. 2010). In sub-Saharan Africa, for example, there can be as few as 10 physicians per 100 000 inhabitants with no dermatologists in many areas (Schmid-Grendelmeier et al. 2003). This is in comparison to 3.4 dermatologists per 100 000 in the United States (Glazer et al. 2017). This vast difference portrays the disparity in dermatological care that exists in most of the developing parts of the world. Exacerbating these numbers is the fact that most dermatologists reside in urban areas. In India, almost all of the country's dermatologists live in cities, whereas 72% of the population live in villages (Desai et al. 2010). In Peru 80% of the country's dermatologists live in Lima which has 33% of the country's population (Desai et al. 2010).

Thus, providing dermatologic care to all people in need becomes difficult to impossible. Cutaneous disease is very common in LMICs due to the rural environment, low socioeconomic status, close living quarters and inadequate living conditions. These factors lead to high rates of cutaneous infections (Bobbs et al. 2016). Depending on the region of the world, patients may be at risk for systemic diseases that may manifest with cutaneous symptoms, including HIV/AIDS infection, tuberculosis, and leprosy for which consultation with a dermatologist may be critical.

The relatively recent rise and scale-up of cellular telephone networks, as well as the adoption of mobile telephone technology in the LMICs, has greatly improved the ability to reach underserved patients through telemedicine. Dermatology is especially suited for telemedicine due to its inherently visual nature. It is possible to capture quality data on a dermatologic condition through photography. Additionally, the healthcare provider at the referring end can provide important information to the teledermatologist through direct palpation and measurement of the skin lesion. Teledermatology can not only provide assistance with diagnosis and treatment plans, but it can also be used for triaging patients to referral centers, as well as for education of both providers and patients.

The use of teledermatology can also have significant economic impact on middle- and low-income countries. Several studies have shown that teledermatology leads to lower healthcare costs (Ferrandiz et al. 2008; Moreno-Ramírez et al. 2009). Due to the shortage of specialists, most patients must travel long distances and pay the accompanying transportation costs as well as missing time from work.

28.2 Methods of Delivering Teledermatology: Store and Forward Versus Live Interactive Methods

There are two main techniques for conducting teledermatology: live interactive (LI) and store and forward (SAF), the latter also known as asynchronous technique. LI allows a real time visit to occur between the patient and provider through the use of an audio and video connection. With SAF images of the patient's lesions are captured and sent to the consulting teledermatologist together with a clinical history. Whereas LI visits are only separated by space, SAF visits are separated by both time and space. In the LMICs, SAF is typically the most convenient and readily accessible method since it does not require a high level of technological sophistication or expensive equipment such as video cameras, audio hardware, dedicated spaces for conducting the live visit, video conferencing software or high-speed internet connections (American Telemedicine Association 2016). Additionally, SAF visits allow consulting clinicians to review cases when they are available, allowing for differences in time zones.

28.3 The History of Teledermatology

The history of teledermatology in the LMICs, dates back to the late 1990s. Some of the pioneers of teledermatology include the Pacific Island Healthcare system and the African Teledermatology Project. With support from the American Academy of Dermatology, the Commission for Development Studies, the Austrian Academy of Sciences and e-derm-consult, the African Teledermatology Project has been providing store-and-forward dermatologic services to 15 sites in 13 countries since 2007.

Informal teledermatology networks have been in place for many years. Physician colleagues in different countries often share cases and receive informal consultations. One of these larger networks is the telederm.org project sponsored by the International Society of Teledermatology, which hosts a free online community for physicians who want to consult with other doctors. This network was started in 2002 and seeks to raise the knowledge of physicians and dermatologists worldwide through its forums.

28.4 Global Teledermatology Programs

28.4.1 Swinfen Charitable Trust

Operating since 1999 the Swinfen Charitable Trust provides second opinions for physicians practicing in LMICs (Wootton et al. 2012). Healthcare providers in hospitals in LMICs have the opportunity to submit cases to specialists worldwide. A referring provider can submit a case via an email telemedicine system using an automatic email messaging service. This service then triages the case to the appropriate specialist. Consultations are provided free of charge. Remote hospitals are provided with imaging equipment and training to conduct a telemedicine consult. The Swinfen Charitable Trust operates in over 50 countries and delivers specialty care in many fields including dermatology, pediatrics and neurology (Wootton et al. 2012). (Email details: swinfencharitabletrust.org; info@swinfentrust.org).

28.4.2 Médecins Sans Frontières

In 2010 Médecins Sans Frontières (MSF) launched a SAF telemedicine system to provide expert clinical support for healthcare workers in the field (Delaigue et al. 2014). MSF offers tele-expertise in many areas including radiology, pediatrics, infectious disease, and dermatology. In the first four years after its inception this program handled 65 dermatology cases including infectious diseases, inflammatory conditions and genetic diseases and with a rapid increase in uptake from 2012 to 2014. Healthcare workers living in low resource settings can securely message distant specialists with cases using the MSF platform. Most cases have been referred from African countries with South Sudan, Ethiopia, and the Democratic Republic of Congo making the majority of submissions. A study of these first cases highlighted several areas requiring improvement including increasing the quality of referrals by standardizing the clinical examination and providing clear photographs, increasing awareness of the service among other MSF member countries, and establishing better ways to obtain follow-up of cases (Delaigue et al. 2014).

28.5 Teledermatology in Africa

Most of Africa's 1.2 billion inhabitants do not have a computer with Internet access (Nyirenda-Jere and Biru 2015). However, in a survey of 7052 inhabitants in Ghana, Kenya, Nigeria, Senegal, South Africa, Tanzania and Uganda, approximately two thirds of respondents reported that they owned a cellphone (Pew Research Center 2015). Among these cellphone users 53% reported using their cellphone to take photographs or videos (Pew Research Center 2015). Smartphone ownership rates are steadily increasing. This will open up access to a panacea of applications and internet resources aimed at improving healthcare (see Chapters 31 and 31).

28.5.1 The Africa Teledermatology Project

The Africa Teledermatology Project (ATP) was established in 2007 as a SAF teledermatology consultation network between nine hospitals in sub-Saharan Africa and dermatologists in Austria, Australia and the USA (Kaddu et al. 2009b). In the first 15 months over 140 cases were submitted for evaluation and diagnoses included cutaneous infections, skin tumors, papulosquamous lesions and drug reactions (Kaddu et al. 2009a). The majority of consults were resolved within two days of submission, greatly reducing the time to treatment in these complicated patients. The major limiting factor was access to telecommunications networks at the originating sites.

A follow-up study analyzed all 1229 cases submitted to the ATP from 2007 to 2013 (Lipoff et al. 2015). The consulting clinician agreed with the originating clinician's diagnosis 60% of the time. There was a steady increase in the number of consults submitted during the study period, likely due to increased awareness of the ATP and widespread access to the Internet and cellphones. Mobile phone users accounted for 13% of all submitted cases suggesting again that mobile teledermatology is the most efficient way to deliver care (Lipoff et al. 2015).

Poor follow-up remains a problem in this network and has been reported with other teledermatology systems. Unfortunately the ATP submission system lacks the features necessary to track returning patients (Lipoff et al. 2015). The system was designed to be as simple as possible in order to accommodate the slow Internet speeds encountered by the originating providers. This low tech system also led to major differences in patient documentation among providers. Some referring clinicians' input included detailed patient histories while others did not provide any history at all (Lipoff et al. 2015). Future iterations of the ATP submission system will work toward standardizing documentation while maintaining simplicity.

The ATP also offers *dermatopathology* services for cases in which a definitive diagnosis is necessary to guide patient management and where dermatopathology services are not available locally. Samples are shipped with appropriate approval from the African country of origin to dermatopathologists in the USA. This is of particular importance in the high numbers of immunosuppressed patients with HIV/AIDS who can have atypical presentations of skin disease. Definitive histopathologic diagnosis in this population guides treatment strategy and avoids incorrect empiric therapies which can worsen the patient's condition. A retrospective study of 55 cases sent for pathology revealed 15 cases of Kaposi sarcoma, 12 cases of infection and 8 cases of dermatitis (Tsang and Kovarik 2011). Only 5% of biopsies were nondiagnostic, mainly due to poor sampling of the dermis and absence of organisms on staining (Tsang and Kovarik 2011). The diagnostic concordance rate between the submitting clinician and pathologist was 58% (Tsang and Kovarik 2011).

The ability to obtain gold standard histopathologic diagnosis in LMICs is limited, but when it is available for complex dermatologic disease, it can greatly augment the quality of care. The method used in this study required physically shipping specimens over long distances (Tsang and Kovarik 2011). One solution for the limited access to dermatopathologists in LMICs is to process the tissue locally and submit this for teledermatopathology.

28.5.2   *The Réseau en Afrique Francophone pour la Télémédecine (RAFT) network* (The Network in Francophone Africa for Telemedicine)

RAFT was created in 2001 to provide support to healthcare professionals living in isolated areas of French-speaking African countries (Bediang et al. 2014). It has since been extended to 60 active sites worldwide including Bolivia and Nepal. RAFT is managed locally by a team in each country that is responsible for the strategy, implementation and coordination of activities with oversight from a small team based in Geneva. The RAFT network provides continuing medical education lectures, virtual patients for training, help with implementation of e-health (healthcare practice supported by electronic processes and communication) and mHealth (the provision of healthcare via mobile technology) as well as telemedicine consultation opportunities for remote providers (Bediang et al. 2014).

28.5.3   The Institute of Tropical Medicine (ITM) Telemedicine Website

This program was developed in 2003 by the Department of Clinical Sciences at the Institute of Tropical Medicine in Antwerp, Belgium (Desai et al. 2010). The main purpose of the ITM's telemedicine initiative is to provide advice for healthcare providers who are treating HIV in Africa (Wootton et al. 2012). The website features training modules, distance support, and other educational materials. A specialist provides advice to the healthcare provider by email. The specialist can then post the clinical case to a forum for additional discussion.

28.5.4   Mobile Teledermatology in Africa

HIV patients often present with unique skin diseases or more severe presentations of common dermatologic complaints. With the high incidence of HIV in Africa, it is imperative that these patients have access to dermatologists and mobile teledermatology may be the only solution for many of these patients. HIV patients must deal with the additional challenge of negative stigmas associated with their disease and as such may have heightened privacy concerns.

A survey of 75 HIV-positive dermatology clinic patients in Botswana found that 82% of respondents had no concerns with mobile teledermatology consults, and 91% believed they would get the same quality of care via teleconsultation as they would face-to-face (FTF) (Azfar et al. 2011). The initial concern that HIV patients would be more hesitant to use telemedicine due to privacy concerns was not substantiated. In this study patients were more concerned with the time taken, distance traveled, and costs of obtaining care and were willing to use teledermatology in order to mitigate these barriers.

A study in Egypt used mobile teledermatology to determine diagnostic concordance rates between a dermatology resident in Egypt and two remote dermatologists (Tran et al. 2011). A specialized mobile app (ClickDoc) was used to input the patient information and images. Thirty patients were evaluated and the diagnostic concordance was found to be 75% overall between the trainee and senior dermatologists (Tran et al. 2011). The authors believe that the adoption of mobile teledermatology will surpass traditional computer-enabled teledermatology in the LMICs due to decreased hardware demands and the rise in cellular connectivity.

One study in Uganda sought to evaluate the impact of mobile teledermatology on patient outcomes and the perceptions of local healthcare workers regarding the use of this technology. The study, conducted over an eight month period, connected healthcare workers in four health centers in Uganda with dermatologists in Europe, Australia and the USA (Frühauf et al. 2013). A total of 72 patients were evaluated for conditions that included non-infectious skin diseases and inflammatory skin diseases (50%) infectious skin diseases (26%), neoplastic lesions (11%) and autoimmune disease (4%) (Frühauf et al. 2013). Most cases were managed satisfactorily via mobile teledermatology with total remission or improvement in the skin condition. The referring healthcare workers were satisfied with the system and cited portability, ease of operability and instantaneous access to expert dermatologists as benefits.

28.6   BUP: The Botswana – University of Pennsylvania Partnership

In Botswana there are 3.4 physicians per 10 000 people, and specialists are few and far between (Littman-Quinn et al. 2013). As with most LMICs, access to doctors is limited outside of major cities. Mobile phone access, however, is widespread with over 95% of the population having access to mobile phone networks (Littman-Quinn et al. 2013). In 2013 a partnership between the University of Botswana, the Botswana Ministry of

Health and the University of Pennsylvania piloted several mobile initiatives throughout Botswana to provide teledermatology and educational resources to patients and healthcare providers. The use of mobile devices helped to ameliorate issues with low bandwidth internet access and outdated computer equipment.

The Botswana – University of Pennsylvania Partnership (BUP) developed four mobile telemedicine initiatives in the areas of cervical cancer screening, radiology, oral health and dermatology (Littman-Quinn et al. 2013). Figure 28.1 shows local training of clinicians in the use of oral telemedicine.

From 2011 to 2013, 126 teledermatology cases were seen and 5 clinicians were trained in teledermatology. Healthcare workers in Botswana used a smartphone to collect the patient's medical history and skin images. This information was then sent to an in-country remote specialist who either consulted on the case or forwarded the case to an international specialist for further input. Additionally, a national specialty manager trained the referring healthcare workers on how to use these mobile health applications.

BUP has also leveraged the mobile network to create several mobile learning programs for residents and medical students at the University of Botswana School of Medicine. Residents were provided with smartphones that had been preloaded with medical apps such as *Dynamed*, *Medscape*, and *ePocrates* and given additional training on how to access the medical information on the phones during a clinical encounter (Littman-Quinn et al. 2013). The benefits of this strategy include increased clinician empowerment and increased collaboration between primary clinicians and trainees.

BUP's mobile healthcare strategy did not come without challenges during its development, both technical and social. Several internal technical challenges arose such as problems with software, device malfunctions, battery problems, lack of hardware maintenance, and threats to cyber security from hackers. In addition to these problems, issues with power supplies and internet access that are commonly seen in LMICs also hindered the project. Social challenges to the implementation of mHealth included negative perceptions from new users, damage to devices from misuse, misplacements or theft and a high turnover of key personnel (Littman-Quinn et al. 2013).

Learning from this partnership, it became apparent that the key to success is focusing on sustainability and local ownership. Partners in LMICs should be invested in, and driven to, make their programs work by incorporating costs into the existing healthcare budget. BUP has worked to identify the most cost-effective options for sustaining the mobile telemedicine project. For example, software problems proved to be a major difficulty since all the initial software had been encoded outside the country using it, and it was complicated to "debug" and maintain the code (Littman-Quinn et al. 2013). As a result BUP began to use only local information technology (IT) professionals to create the required software and set a "coded in country" requirement for all new mobile telemedicine systems.

28.7 Teledermatopathology in Botswana

Dermatologic diagnoses often hinge on the findings from skin biopsies. Obtaining and analyzing a skin biopsy requires technical expertise in acquiring the specimen, a method for preserving the tissue, the ability to transfer the tissue to a pathology laboratory where it can be further processed prior to being examined by an expert in

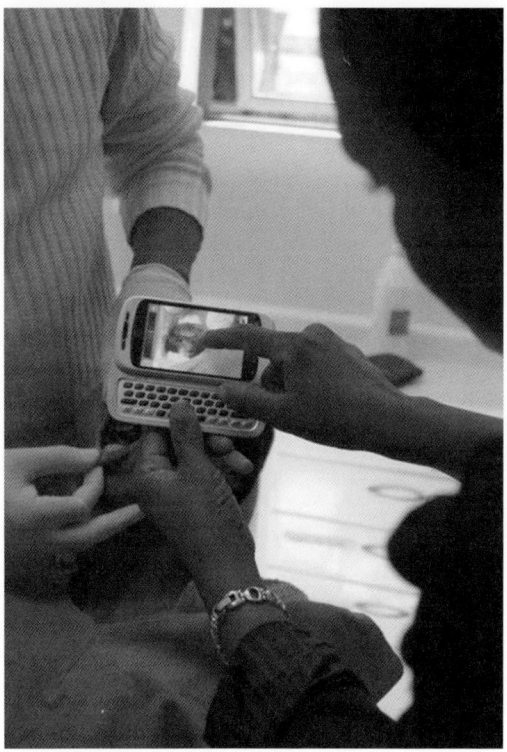

Figure 28.1 Local training of clinicians in the use of mobile oral telemedicine in Botswana (*Source: photo credit Ryan Littman-Quinn*).

dermatopathology. Additionally, a reliable method for following up with the patient is essential. However, this process can be near to impossible in low resource countries, and dermatologists often rely exclusively on history and clinical presentation to make a diagnosis.

In Africa physicians often practice in professional isolation. Telepathology could allow one pathologist to meet the demands of many referring providers (Figures 28.2 and 28.3).

One study in telepathology in Botswana used a motorized microscope that could be controlled by the remote referrer (Fischer et al. 2011), although the authors highlighted that the use of telepathology does not preclude the need for tissue processing capabilities at the originating site, high speed Internet access, and knowledgeable personnel to maintain equipment (Fischer et al. 2011). These barriers can make the implementation of telepathology difficult.

28.8 Diagnostic Concordance

The use of mobile teledermatology was evaluated in Ghana to determine the diagnostic concordance rate between three FTF dermatologists, a remote Ghanaian dermatologist using a mobile platform and a United States teledermatologist using a desktop computer (Osei-tutu et al. 2013). FTF visits were conducted with 34 patients by Ghanaian

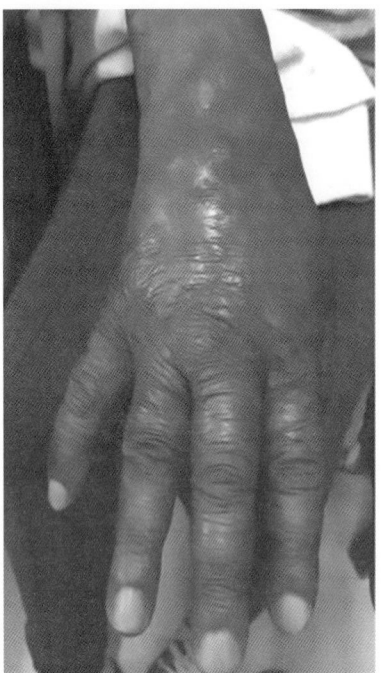

Figure 28.2 Teledermatology consultation of a patient with leprosy in Botswana (*Source:* photo credit Dr. Tori Williams).

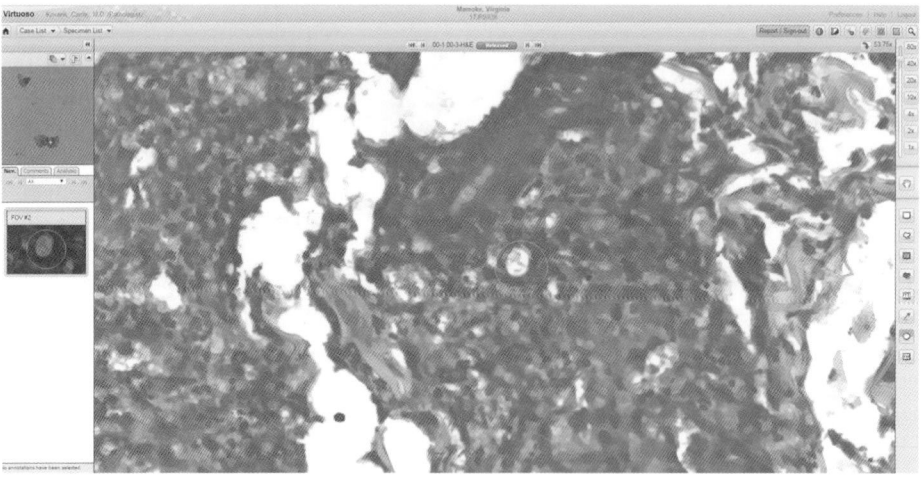

Figure 28.3 Teledermatopathology consultation of the patient with leprosy in Botswana, confirming the diagnosis and demonstrating organisms staining positively with Fite's acid fast stain.

dermatologists who made a primary diagnosis and then sent the histories and images to the consulting teledermatologists (Osei-tutu et al. 2013). Interestingly, this study sought to compare the use of a mobile-only platform to the use of a desktop computer for making the telediagnosis. The diagnostic concordance was 79.4% for the Ghanian teledermatologist (mobile-only) and 78.8% for the US teledermatologist (desktop computer) (Osei-tutu et al. 2013). Thus in this study there was no difference between the use of a mobile device or a desktop computer for determining a diagnosis. The use of mobile-only solutions for teledermatology will likely prove more cost-effective and can help increase access to dermatologic expertise.

Another study conducted at a hospital in Egypt on 600 patients with dermatologic complaints found diagnostic concurrency rates of 87% between an in-person dermatology resident and two off-site dermatologists (Saleh et al. 2017). The authors emphasized that the simplicity and low cost of the teledermatology system was an advantage over other traditional telemedicine systems. It is important that the software used for such interactions be secure in order to maintain patient privacy and confidentiality.

The use of teledermatology has been shown to increase the diagnostic acumen of referring providers. A study conducted in South Africa found that six referring healthcare providers became more adept at diagnosing the correct dermatologic condition and had decreased referrals the more they used a teledermatology service (Colven et al. 2011). This is likely due to primary care providers' recall of previous training in dermatology and greater abilities in pattern recognition. This South African report noted that sufficient clinical history was provided most of the time, but that, despite having a referral template available, most referring physicians did not use it to standardize documentation and preferred an informal reporting style.

28.9 Teledermatology in Asia

Primary care physicians consulted teledermatologists regarding 206 patients with skin conditions in rural India (Patro et al. 2015). The referring providers gave the patients an initial diagnosis and advised management. The patient cases and images were then sent to a dermatology department for evaluation. This study found a diagnostic concordance rate of 56% with the greatest differences occurring in cases of eczema and psoriasis (Patro et al. 2015). A major limitation of the teledermatology program instituted in this study is that the images were only reviewed once a month. This delayed time to consultation supports the fact that teledermatology is still an academic exercise at many institutions, not a trusted avenue of care.

28.10 Teledermatology in Latin America

28.10.1 Teledermatology in Mexico

A teledermatology program was created in Mexico in 2010 through a partnership between Community Dermatology Mexico (CDG) and the State of Guerrero (Figure 28.4).

The aim of this program is to teach dermatology to healthcare workers, conduct epidemiologic research and to provide specialist services to healthcare personnel working in remote areas. The project cites many benefits to using a teledermatology service including lower costs, the ability to order medications on demand, and better distribution of resources to patients with complicated illnesses living in these areas. In 2016 CDG was able to connect 22 health centers employing over 10 000 healthcare personnel with teledermatology services (Pasquali 2017).

Figure 28.4 Clinic in Atzacoaloya, Guerrero, Mexico, where hands-on clinical dermatology training for general healthcare providers was given after the first Teledermatology Training Course was completed in 2010.

28.10.2   Teledermatology in Belize

The Medical College of Wisconsin (MCW) and Hillside Healthcare International (HHCI) in Belize created a partnership to serve the residents of a rural district in Belize (Bobbs et al. 2016). The partnership has been found to be mutually beneficial. HHCI identified a need for specialist dermatology care in this community that could be served by dermatologists at MCW, and the dermatology residents at MCW were exposed to tropical diseases and telemedicine. Referring clinicians in Belize are now able to send dermatologic cases to consultants in the USA and receive recommendations within one week.

This project faced many of the same challenges as other new teledermatology initiatives. The physician-to-physician interactions were limited by unreliable Internet connections and an inability to ask follow-up questions in real time (Bobbs et al. 2016). The program's value was also limited by the engagement of the physicians at the referring end. Additionally, the clinic was located in a low-resource setting where skin biopsies and laboratory monitoring could not be obtained.

28.11   Barriers

It is evident that telemedicine may provide an opportunity for LMICs to expand services within their health systems, increase access to advanced medical education and provide specialized care to patients in need. The question now is what prevents short-term telemedicine pilot exercises from becoming long-term institutional programs that can be implemented on a wider scale. There are several reasons to explain why some programs do not succeed and have been phased out over time.

28.12   Costs

Funding is a critical aspect of maintaining a successful telemedicine initiative. There must be sufficient funding to purchase equipment, acquire broadband Internet or cellular phone access, and pay healthcare workers. Many of resource-poor countries do not have the funds to sustain telemedicine programs over the longterm. However, several studies have shown that teledermatology is a cost-effective method for delivering dermatology care (Ferrandiz et al. 2008; Moreno-Ramírez et al. 2009). One study of cost in teledermatology in Uganda and Guatemala found that the total cost for supplies and technology was \$42 per consult and \$64 per biopsy, but due to upfront costs of equipment this cost decreased as additional cases were seen (Greisman et al. 2015).

Cost is becoming more feasible now that there is an increase in the number of healthcare providers with personal cellular phones, cheaper mobile plans, and apps that can both take quality photos and transmit secure and encrypted data. The key for sustainability in the future will be to use the technology that the local providers already have available.

28.13   Education and Training

Competent teleproviders are vital to the success of telemedicine programs. Special training and education is required to ensure that teleproviders are able to properly utilize the available technologies in order to perform a teleconsultation. Teleproviders should strive to complete a standardized documentation of the patient encounter. Limited training on the use of teledevices and poor technique standards may compromise the reliability of the presented information.

Additionally a high turnover of staff trained to perform the teleconsultations in LMICs can be problematic.

28.14   Equipment and Internet Access

Nowhere is the digital divide more apparent than in LMICs. Telemedicine has the potential to reduce healthcare disparities but there can be problems with maintaining a steady power supply, limits on international bandwidth and poor connectivity outside of the larger cities (Bagayoko et al. 2006). Ideally, designated trained personnel will be responsible for the installation and maintenance of the telemedicine technology in order to alleviate such issues. Remote power supplies and new technology, such as Television White Space Internet (Chavez et al. 2016) will make connectivity to remote areas more feasible in the future. (White Space Internet derives from a frequency range that is unused within the wireless spectrum know more commonly as White spaces [radio]). This frequency range is created when there are gaps between television channels. These spaces can provide broadband Internet access that is similar to that of 4G mobile. Figure 28.5 shows live video telemedicine training with local physicians during the Microsoft TV White Program in Botswana.

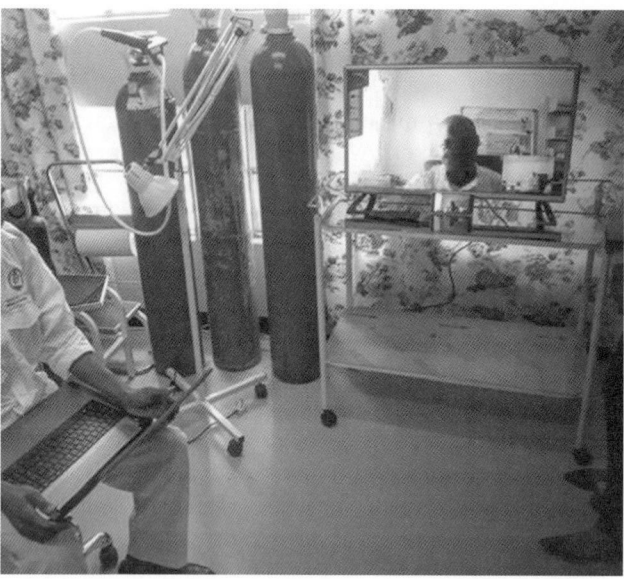

Figure 28.5 Live video telemedicine training with local clinicians during the Microsoft TV White Space Program in Botswana (*Source:* photo credit Ryan Littman-Quinn).

28.15 Privacy Concerns

Telemedicine provided in LMICs should be executed with the same high level of security precautions undertaken in developed countries, and any local additional standards or exceptions should be discussed with the local governments. The Health Insurance Portability and Accountability Act and the Health Information Technology for Economic and Clinical Health Act in the U.S. set minimum standards for security and encryption of medical information. Telemedicine networks launched in countries without these legal protections should still strive to avoid and deter security leaks. Indeed one study from Botswana recounted hacker issues on telemedicine servers and sealing security breaches (Littman-Quinn et al. 2013).

28.16 Cultural Hesitancy

Although telemedicine offers many convenient advantages, from a cultural perspective it is not the conventional method of delivering healthcare. Some patients may not be comfortable receiving care from a provider across a computer screen and may prefer in-person services. Some patients worry that their privacy can be compromised by usage of teleconsultations. Other patients may not want photos taken of their face or genitals, and each patient and culture needs to be approached individually.

In addition, the service provided to both the patient and provider needs to be seen as valuable and relevant. The telemedicine service needs to fit seamlessly into the busy daily workflow, and the local providers should have input into how the service works.

28.17 Language Barriers

As with a FTF visit, the clinician should keep in mind language fluency when talking to patients. Luckily, most teledermatology visits are conducted in a SAF format, precluding the need for instant translation. If LI telemedicine is being conducted, certified medical translators must be involved to insure proper communication between patient and teleproviders or referring physicians and teleproviders if the primary spoken language is not the same.

28.18 Availability of Treatments

One decisive limitation of telemedicine in LMICs is the ability of patients to follow through with treatment recommendations. It may be simple to make a diagnosis online but acquiring the medications or topical agents necessary to complete a treatment course may be impossible. Teleconsultants should be familiar with the local medical care, formularies, social issues, patient culture, and be able to develop a treatment plan that is locally relevant and useful.

28.19 Legal Issues

Local physicians are ultimately responsible for the care of their patients. The remote expert dermatologist provides advice from a distance, and the normal standards of care and skill should apply. In order for the teledermatologist to perform at his or her best ability, the referring provider should provide accurate clinical history and high quality images. Many LMICs have not adopted official rules and regulations that address the legal aspects of telemedicine to safeguard patients.

28.20 Follow-up

The gold standard for diagnosis is histopathologic analysis of a skin biopsy. This is the most accurate way to confirm that a teleconsultation diagnosis is correct. However, only a small percentage of patients will need a biopsy for definitive diagnosis, and in many cases, biopsies are not available. In resource limited settings, work with the local providers to come up with the best possible care plan and continue to collaborate and follow up as needed.

From a purely economic perspective, teledermatology is an effective means to reduce transportation costs and time away from work for those patients seeking specialist care. Removing the travel obstacle for patients who live in remote areas has the potential to reduce the time to initial presentation of skin disease and greatly accelerate diagnosis and management.

28.21 Ensuring Success of a New Teledermatology Initiative

In order to institute a successful teledermatology program, there are several practical steps that should be followed (Kaddu et al. 2009a). The first step is to research and review any prior telemedicine projects conducted in the region. A realistic appraisal of the available technologies and telecommunications access should be completed. Then, one should assess how teledermatology can meet the needs of the healthcare professionals and hospitals in the area of interest. The legal policies in place must be permissive of telemedicine. Extensive information should be provided to all parties potentially involved in the project.

Additionally, a sustainable partnership must be established between referring and consulting institutions. Both of these stakeholders should have mutually beneficial reasons for entering into the partnership. The institutions should work together to target a specific need in the local community and agree on an information sharing practice that meets the standard of both parties (Bobbs et al. 2016). They should also work to identify the costs associated with starting a teledermatology program and obtain appropriate funding before starting.

After this preliminary analysis, one should identify which teledermatology technology solutions would best fit the clinical needs, technical requirements, and security of the referral centers. Once a software and equipment vendor has been chosen, one should work with the vendor to test and implement the chosen option. A team should be assembled to help with planning, implementing the initial phase, and evaluating the project. This team can also educate and train users of the telemedicine system.

Once the project has launched, there should be a quality assurance program in place to evaluate the design, technical performance, and staff and patient satisfaction of the teledermatology system. In order to increase awareness and engage future potential sites, the developers should create ongoing education programs directed at individuals who are not part of the original program.

28.22 Conclusions

Telemedicine has revolutionized access to care in LMICs. There are long-running established teledermatology programs in many countries, and as access to internet and mobile phones continues to rise these programs will continue to grow in reach and influence. Programs in African countries especially have served as models for delivering quality dermatologic care partnerships through telemedicine.

Bibliography

American Telemedicine Association. (2016). Practical Guidelines for Dermatology. April 2016.

Azfar, R.S., Weinberg, J.L., Cavric, G. et al. (2011). HIV-positive patients in Botswana state that mobile teledermatology is an acceptable method for receiving dermatology care. *J. Telemed. Telecare.* 17 (6): 338–340.

Bagayoko, C.O., Müller, H., and Geissbuhler, A. (2006). Assessment of Internet-based tele-medicine in Africa (the RAFT project). *Comput. Med. Imaging Graph.* 30 (6): 407–416.

Bediang, G., Perrin, C., de castañeda, R. et al. (2014). The RAFT telemedicine network: lessons learnt and perspectives from a decade of educational and clinical services in low- and middle-incomes countries. *Front. Public Health* 2: 180.

Bobbs, M., Bayer, M., Frazer, T. et al. (2016). Building a global teledermatology collaboration. *Int. J. Dermatol.* 55 (4): 446–449.

Chavez, A., Littman-Quinn, R., Ndlovu, K., and Kovarik, C.L. (2016). Using TV white space spectrum to practise telemedicine: A promising technology to enhance broadband internet connectivity within healthcare facilities in rural regions of developing countries. *J. Telemed. Telecare.* 22 (4): 260–263.

Colven, R., Shim, M.H., Brock, D., and Todd, G. (2011). Dermatological diagnostic acumen improves with use of a simple telemedicine system for underserved areas of South Africa. *Telemed. J. E Health.* 17 (5): 363–369.

Delaigue, S., Morand, J.J., Olson, D. et al. (2014). Teledermatology in low-resource settings: the MSF experience with a multilingual tele-expertise platform. *Front. Public Health.* 2: 233.

Desai, B., Mckoy, K., and Kovarik, C. (2010). Overview of international teledermatology. *Pan Afr. Med. J.* 6: 3.

Ferrándiz, L., Moreno-Ramírez, D., Ruiz-de-Casas, A. et al. (2008). An economic analysis of presurgical teledermatology in patients with nonmelanoma skin cancer [in Spanish]. *Actas Dermosifiliogr.* 99 (10): 795–802.

Fischer, M.K., Kayembe, M.K., Scheer, A.J. et al. (2011). Establishing telepathology in Africa: lessons from Botswana. *J. Am. Acad. Dermatol.* 64 (5): 986–987.

Frühauf, J., Hofman-wellenhof, R., Kovarik, C. et al. (2013). Mobile teledermatology in sub-Saharan Africa: a useful tool in supporting health workers in low-resource centres. *Acta Derm. Venereol.* 93 (1): 122–123.

Glazer, A.M., Holyoak, K., Cheever, E. et al. (2017). Analysis of US dermatology physician assistant density. *J. Am. Acad. Dermatol.* 76: 1200–1202.

Greisman, L., Nguyen, T.M., Mann, R.E. et al. (2015). Feasibility and cost of a medical student proxy-based mobile teledermatology consult service with Kisoro, Uganda, and Lake Atitlán, Guatemala. *Int. J. Dermatol.* 54 (6): 685–692.

Institute of Tropical Medicine (2017). Telemedicine. http://telemedicine.itg.be. Accessed June 5, 2017.

Kaddu, S., Kovarik, C., Gabler, G., and Soyer, H.P. (2009a). Teledermatology in developing countries. In: *Telehealth in the Developing World*, 1e (ed. R. Wootton, N.G. Patil, R.E. Scott and K. Ho), 121–134. London, UK: IDRC.

Kaddu, S., Soyer, H.P., Gabler, G., and Kovarik, C. (2009b). The Africa Teledermatology Project: preliminary experience with a sub-Saharan teledermatology and e-learning program. *J. Am. Acad. Dermatol.* 61 (1): 155–157.

Lipoff, J.B., Cobos, G., Kaddu, S., and Kovarik, C.L. (2015). The Africa teledermatology project: a retrospective case review of 1229 consultations from sub-Saharan Africa. *J. Am. Acad. Dermatol.* 72 (6): 1084–1085.

Littman-Quinn, R., Mibenge, C., Antwi, C. et al. (2013). Implementation of m-health applications in Botswana: telemedicine and education on mobile devices in a low resource setting. *J. Telemed. Telecare.* 19 (2): 120–125.

Moreno-Ramírez, D., Ferrándiz, L., Ruiz-de-Casas, A. et al. (2009). Economic evaluation of a store-and-forward teledermatology system for skin cancer patients. *J. Telemed. Telecare.* 15 (1): 40–45.

Norton, S.A., Burdick, A.E., Phillips, C.M. et al. (1997). Teledermatology and underserved populations. *Arch. Dermatol.* 133: 197–200.

Nyirenda-Jere, T. and Biru, T. (2015). Internet development and Internet governance in Africa. Internet Society. http://www.internetsociety.org/sites/default/files/Internet%20development%20and%20Internet%20governance%20in%20Africa.pdf

Osei-tutu, A., Shih, T., Rosen, A. et al. (2013). Mobile teledermatology in Ghana: sending and answering consults via mobile platform. *J. Am. Acad. Dermatol.* 69 (2): e90–e91.

Pasquali, P. (2017). Teledermatology in Mexico. Accessed Jun 8, 2017. http:// teledermatology-society.org/teledermatology-in-mexico

Patro, B.K., Tripathy, J.P., De, D. et al. (2015). Diagnostic agreement between a primary care physician and a teledermatologist for common dermatological conditions in North India. *Indian Dermatol. Online J.* 6 (1): 21–26.

Pew Research Center (2015). Cell Phones in Africa: Communication Lifeline. http://www .pewglobal.org/2015/04/15/cell-phones-in-africa-communication-lifeline

Saleh, N., Abdel Hay, R., Hegazy, R. et al. (2017). Can teledermatology be a useful diagnostic tool in dermatology practice in remote areas? An Egyptian experience with 600 patients. *J. Telemed. Telecare.* 23 (2): 233–238.

Schmid-Grendelmeier, P., Doe, P., and Pakenham-Walsh, N. (2003). Teledermatology in Sub-Saharan Africa. *Telemed. Teledermatol.* 32: 233–246.

Tran, K., Ayad, M., Weinberg, J. et al. (2011). Mobile teledermatology in the developing world: implications of a feasibility study on 30 Egyptian patients with common skin diseases. *J. Am. Acad. Dermatol.* 64 (2): 302–309.

Tsang, M.W. and Kovarik, C.L. (2011). The role of dermatopathology in conjunction with teledermatology in resource-limited settings: lessons from the African teledermatology project. *Int. J. Dermatol.* 50 (2): 150–156.

Wootton, R., Geissbuhler, A., Jethwani, K. et al. (2012). Long-running telemedicine networks delivering humanitarian services: experience, performance and scientific output. *Bull. World Health Organ.* 90: 341–347D.

29

Digital Technology, Including Telemedicine, in the Management of Mental Illness

John A Naslund[1], Sophia M. Bartels[2], and Lisa A. Marsch[2]

[1] Harvard Medical School, Boston, MA, USA
[2] Dartmouth College, Hanover, NH, USA

CHAPTER MENU

29.1 Introduction and Background

In this chapter we explore the potential to substantially expand the reach and quality of mental health services within low- and middle-income countries (LIMCs) with the use of digital technologies. Throughout this chapter we define digital technologies as mobile devices including cellphones or smartphones, as well as telepsychiatry applications and web-enabled interventions delivered through online platforms for diagnosis, screening, treatment and providing education or facilitating self-management of mental disorders across diverse low-resource settings. We also broadly define "mental disorder," and consider how digital technology interventions could support efforts targeting depression or major depressive disorder, serious mental illnesses such as schizophrenia spectrum disorders or bipolar disorder, anxiety disorders, post-traumatic stress disorder, epilepsy, and problematic alcohol abuse or other substance abuse disorders.

The global growth in telecommunications may yield new opportunities to improve mental healthcare for those at greatest risk. Therefore, in this chapter we specifically concentrate on the following five key objectives: first, we demonstrate why the treatment

Revolutionizing Tropical Medicine: Point-of-Care Tests, New Imaging Technologies and Digital Health,
First Edition. Edited by Kerry Atkinson and David Mabey.
© 2019 John Wiley & Sons, Inc. Published 2019 by John Wiley & Sons, Inc.

and prevention of mental disorders is a global public health priority, and why it is an especially pressing concern in LMICs. Second, we consider how the recent availability and widespread access to digital technologies throughout many LMICs may afford new opportunities to advance efforts to treat mental disorders globally. Third, we summarize promising examples where digital technologies have been successfully used to support care for common mental disorders in lower income settings. Fourth, we critically assess the risks and potential limitations with using digital technologies for supporting mental healthcare in LMICs. Fifth, and lastly, we highlight key areas and future research directions where emerging technologies could substantively advance efforts to alleviate the global burden of mental disorders in the years ahead.

29.2 Why Mental Disorders?

Worldwide, mental disorders represent the leading cause of years lived with disability (Patel et al. 2016). It is estimated that mental disorders account for over eight million deaths each year (Walker et al. 2015). The burden of mental disorders is especially severe in LMICs, where health systems are underdeveloped and lack the necessary resources to meet the needs of the millions of individuals living with mental disorders without access to adequate mental healthcare (Saxena et al. 2007). Illicit drug abuse and alcohol abuse disorders also represent a serious global public health concern, and specifically account for over 20% of the years lived with disability caused by mental and substance abuse disorders (Whiteford et al. 2013).

One of the main barriers to the improvement of mental health services is that mental healthcare is not recognized as a national priority within many countries, nor is it ranked highly on national or international development agendas.

For example, it is estimated that less than 1% of development assistance for healthcare in LMICs is dedicated to the treatment or prevention of mental disorders (Patel et al. 2016). Furthermore, mental healthcare was not mentioned as part of United Nations/ WHO Millennium Development Goals (MDGs), even though there is a clear link between mental illness and many of the other development targets given that mental disorders are known to have severe consequences for the prosperity of communities and entire economies (Patel et al. 2016; Saraceno et al. 2007). The low-degree of importance directed toward mental healthcare impedes progress, and means that fewer funds arc allocated to preventing and treating mental illness (Patel et al. 2016). The World Health Organization Mental Health Atlas survey from 2014 revealed that in many LMICs less than US$2 per person per year is spent on the prevention and treatment of mental illness (Chisholm et al. 2016). In many countries there are often misconceptions about the cost effectiveness and societal benefits of investing in mental health services, including the notion that insufficient gains can be attained from investment in mental healthcare (Saraceno et al. 2007). Other factors responsible for the low levels of mental health funding include fragmentation of advocacy efforts among stakeholders and difficulties effectively communicating the importance of mental healthcare to government officials and policymakers, given the wide variation in the type and complexity of mental disorders (Saraceno et al. 2007). Additionally, those affected by mental illness are often socially marginalized and as a result face difficulty advocating for themselves or generating sufficient political influence to produce meaningful change (Saraceno et al. 2007).

Figure 29.1 Efforts to advocate for mental healthcare. Most people affected by mental disorders do not have access to the treatment that they need. The World Health Organization launched the Mental Health Gap Action Programme (mhGAP) to support the scale up of mental health services in low- and middle-income countries (http://www.who.int/mental_health/mhgap/en). This photo shows a parade in Monrovia, Liberia on World Mental Health Day advocating for better mental healthcare and treatment. *Source:* Photo retrieved from http://www.afro.who.int/news/liberia-plans-strengthen-mental-health

Another major barrier to the provision of mental healthcare is the limited availability of mental health providers (Figure 29.1).

It is estimated that in low-resource settings as many as 90% of individuals living with mental illness do not have access to mental healthcare (Patel et al. 2010). Specifically, it is estimated that in LMICs there is a sizeable human resource shortage of roughly one million mental health workers (Kakuma et al. 2011). This shortage represents a significant challenge to delivering even basic mental health services to the millions of individuals in need in these countries (Becker and Kleinman 2013). In many LMICs the low status and poor pay of mental health workers means that few individuals choose to enter this profession (Saraceno et al. 2007). Moreover, mental health services are typically centralized in more developed urban settings, and professionals are provided with little incentive to work in rural areas, leading to further gaps and shortages in care (Saraceno et al. 2007). To mediate this dramatic shortage of mental health workers, local and national governments will need to make a commitment to prioritizing mental healthcare and strengthening workforce capacity through training non-specialist health workers and making services available in remote areas (Becker and Kleinman 2013). It is clear that addressing this immense gap in available mental healthcare is an urgent global public health priority. Worldwide there is also substantial economic burden attributable to mental disorders.

Multiple studies have documented the immense and wide-ranging societal costs of mental disorders through reduced employee productivity, decreased rates of labor participation, poorer overall health and functioning, fewer tax contributions and greater welfare expenses (Chisholm et al. 2016). Mental disorders also contribute additional costs due to increased health expenditures (Chisholm et al. 2016). In 2010 it was estimated that the global costs of substance abuse, mental, and neurological disorders was nearly US$9 trillion as a result of lost output (Chisholm et al. 2016). Furthermore, one study estimated that making basic treatment for anxiety and depression available worldwide would contribute to a return on investment of nearly US$150 billion in direct costs, over 40 million additional years of healthy life and economic benefits exceeding $310 billion (Chisholm et al. 2016).

Addressing mental disorders worldwide can have many measurable benefits; however, there are currently numerous barriers that restrict access to these services, particularly in resource-poor settings. Digital technologies hold potential to overcome several of these barriers and increase access to mental healthcare worldwide.

29.3   Growing Access to Digital Technology and New Opportunities

As highlighted in previous chapters, low-cost and widely available digital technologies afford many promising opportunities to transform the delivery of a broad range of healthcare services in LMICs. In the context of mental illness, numerous studies have demonstrated that digital technologies, ranging from mobile phones or telepsychiatry applications to online programs and remote sensors, have had a substantial impact on the delivery of mental healthcare in high-income countries (Naslund et al. 2015). However, in most lower income countries these emerging technologies are not commonly used for mental health (Farrington et al. 2014).

Digital technologies hold immense promise for transforming mental healthcare globally, an emerging phenomenon driven largely by the growing access to, and use of, these technologies across most settings. By 2015 the World Bank estimated that there were nearly 96 mobile cellular subscriptions for every 100 people in LMICs (The World Bank 2016). Figure 29.2 shows the proportion of the world's mobile cellular subscriptions, Internet connections, and population by region.

As illustrated in this figure, it is clear that many gaps in access to mobile and online connectivity remain, especially in Africa and South Asia. This is largely because many individuals residing in rural areas or highly impoverished settings do not yet have access to mobile devices. Overall, however, access has spread rapidly to a large percentage of the global population.

It is important to be aware that an increasing proportion of the world's Internet traffic is now coming from mobile devices, which suggests that a greater number of individuals from lower income regions may be finding new ways to access the Internet through their mobile devices (Internet Society 2015). As of 2017 over half of the world's web traffic came from mobile phones, an increase attributed to rapidly growing access in developing economies (We Are Social 2017). This remarkable increase in access to, and affordability of, mobile telecommunications and mobile Internet allows for new opportunities to improve the reach and quality of services for mental healthcare and to begin bridging the mental health treatment gap in lower resource settings (Fairburn and Patel 2017).

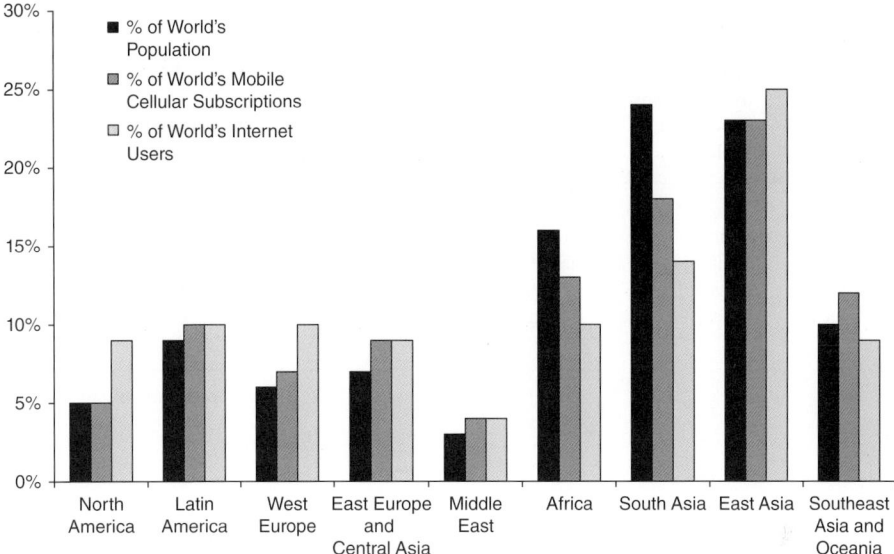

Figure 29.2 Share of global mobile cellular telephone subscriptions and Internet users by region. Source: The mobile cellular telephone subscribers, Internet users, and population data were obtained from the Central Intelligence Agency World Factbook from 2014 (Central Intelligence Agency 2014, 2015).

29.4 Promising Examples from Low- and Middle-Income Countries

In recent years there has been an emergence of studies from LMICs describing the use of digital technology interventions for supporting the treatment and prevention of mental disorders. A recent review of the literature identified 49 studies of interventions specifically for the treatment, prevention, diagnosis and management of various mental disorders from over 20 LMICs (Naslund et al. 2017a). By drawing from the findings from this review as well as other recent studies and relevant publications in global mental health, we highlight four key areas where digital technology holds promise for transforming the delivery of mental healthcare in low-resource settings. We present these potential applications of digital technology informed by the existing literature in Table 29.1, and describe each in detail in the sections that follow.

29.4.1 Prevention and Promotion

There are numerous opportunities to leverage digital technology to support efforts aimed at preventing mental illness or promoting health and wellbeing among individuals living with mental illness. For example, technology could be used to help with educating the public and disseminating information about common mental disorders through anti-stigma campaigns, substance-abuse prevention messaging, or efforts to promote awareness using short message service (SMS) text message blasts or social media pages. For instance, in one study conducted in Bangalore, India, 40 young women

Table 29.1 Opportunities for digital technologies to support global mental healthcare.

Application of technology	Countries where interventions have been evaluated	Types of mental disorders that have been targeted	Types of digital technology that has been utilized for this purpose	Main outcomes achieved
1) Prevention and promotion	India, Mexico, China, Malaysia, Russia, Iraq, Brazil, Uruguay	Promoting positive mental health; depression; social anxiety; suicide prevention; obsessive compulsive disorder; post-traumatic stress disorder	SMS text messaging; web-based programs; email support	Feasibility and acceptability; reduction in symptoms; reduction of problematic alcohol use; promotion of program engagement
2) Detection and assessment	South Africa, India	Depression	Web-based program; SMS text messaging	Acceptable validity; reliability; feasibility; overall participant satisfaction
3) Treatment and management	Somaliland, Colombia, South Africa, Chile, Brazil, Niger, Nigeria, Sri Lanka, Romania, Thailand, Turkey	Depression; serious mental illness; schizophrenia; alcohol abuse	Telepsychiatry; SMS text messaging; telephone follow-up calls/ coaching; mobile applications	Feasibility; high levels of satisfaction; increased adherence to treatment and medications; reduction in relapses; reduction in alcohol consumption; improved depressive symptoms; increased weight-loss
4) Training and supervision	Pakistan, India, Brazil, Somaliland	Alcohol abuse	Asynchronous telepsychiatry; e-learning (online platforms)	Effectively used to diagnose patients; highly rated

from urban slums and with access to a mobile phone were sent daily SMS text messages containing mental health tips, positive messages, or helpline information (Chandra et al. 2014). This approach appeared feasible and acceptable for promoting positive mental health among these young women, as reflected by their high response rates and their expressed satisfaction with the program (Chandra et al. 2014).

Online programs may also make it possible to reach individuals without access to care, or who may be reluctant to seek services due to stigma, long travel distances or out-of-pocket expenses. For example, online self-help programs could provide on-demand access to resources and supportive online communities, while offering a discrete and anonymous method for seeking support. Online communities could be especially important for promoting mental wellbeing by allowing individuals with mental illness to feel less alone and to find support from others with shared experiences of living with mental illness (Naslund et al. 2016b). It may be possible for family members to also access important resources such as social support, recommended coping strategies and self-help programs through online communities. There is compelling evidence from high-income countries showing the promise of online programs for supporting prevention and promotion efforts (Andersson and Cuijpers 2009; Naslund et al. 2015; Spek et al. 2007). There is also growing evidence supporting these types of online self-help programs for mental health in lower income settings.

In one example from Mexico a pilot feasibility study tested an online cognitive behavioral intervention among eight participants with depression (Plata et al. 2014). This online intervention consisted of weekly sessions focused on problem solving, cognitive reframing, emotional expression and self-esteem, and contributed to improvement in symptoms of anxiety and depressive symptoms following treatment and after six-months of follow-up (Plata et al. 2014). In another pilot study from Mexico, 66 participants with social anxiety were randomly assigned to receive an online psychoeducation program or to a waitlist control condition (Plata et al. 2014). The program in this study emerged as potentially effective and feasible for treating social anxiety using cognitive behavioral therapy techniques when compared to the control group (Plata et al. 2014).

In another example an online Chinese-language depression and suicide screening program reached over 4700 Chinese-speakers who searched for depression and suicide information online during a seven-month period (Liu et al. 2015). This study demonstrated the feasibility of employing an online program for identifying and reaching individuals at-risk of suicide and depression (Liu et al. 2015). There is also potential to deliver cognitive behavioral therapy self-help interventions in low-resource settings through online programs, as highlighted in a four-week randomized pilot study of an online Russian-language program (Moritz and Russu 2013). In total, 72 participants with obsessive–compulsive disorder were enrolled in the study and those who received the online cognitive behavioral therapy intervention appeared to show improvement in obsessions and depression over time compared with the control group (Moritz and Russu 2013).

There may also be ways for online self-help programs to reach individuals who have access to the Internet across many countries. This was demonstrated in a study of an online program for preventing post-partum depression by teaching women how to create a healthy lifestyle for themselves and their newborn child through a series of eight weekly sessions (Barrera et al. 2015). This program enrolled 852 mostly Spanish-speaking (83%) pregnant women from 23 different countries, and included many lower income countries in Latin America (Barrera et al. 2015). In China a series of studies of the My Trauma Recovery online self-help program, which includes modules with videos and educational materials on trauma, coping skills, social support and relaxation techniques, demonstrated effectiveness by reducing symptoms of post-traumatic stress disorder among victims of trauma (Wang et al. 2013, 2014; Xu et al. 2015). In a randomized controlled trial of the My Trauma Recovery program, 197 trauma survivors were enrolled, consisting of 103 from urban areas and 94 from rural areas (Wang et al. 2013). In this trial, both urban and rural participants in the My Trauma Recovery program showed significant reduction in post-traumatic symptom severity; these benefits persisted at three-month follow-up assessments (Wang et al. 2013).

Online self-help programs for the prevention and promotion of mental disorders could also yield opportunities to reach individuals in conflict zones or disaster areas where it is likely not possible to access a mental health provider. In surveying the literature, we discovered two studies of an online program designed to support victims of trauma living in Iraq; one of the studies enrolled 159 participants and the other enrolled 40 participants (Knaevelsrud et al. 2015; Wagner et al. 2012). These studies demonstrated the feasibility of reaching individuals in a conflict zone and providing evidence-based cognitive behavioral therapy as well as for contributing to significant reductions in post-traumatic stress symptoms (Knaevelsrud et al. 2015; Wagner et al. 2012).

There is also potential for online programs to support efforts aimed at the prevention of problematic alcohol abuse and other substance abuse disorders. For example, the online "Bebermenos" program from Brazil was developed with support from the World Health Organization and consisted of alcohol use self-monitoring, goal setting with automated feedback, exercises to manage relapse and risky situations, weekly reminders, progress reports and access to discussion forums to connect with peers (Andrade et al. 2016). A 6-week study of this program enrolled 929 participants and demonstrated the potential effectiveness of the program for reducing alcohol consumption among individuals who self-identified as being harmful or hazardous users (Andrade et al. 2016).

In a separate study 333 college students in Brazil participated in an online motivational intervention for preventing general substance abuse (Christoff and Boerngen-Lacerda 2015). The three-month program appeared as effective at reducing abuse of alcohol, cannabis and other drugs when compared to a face-to-face version of the same program (Christoff and Boerngen-Lacerda 2015). In another three-month study from Uruguay an online program called COLOKT was implemented in 10 schools in Montevideo and tested with 359 adolescents in the 9th and 10th grades (Balsa et al. 2014). COLOKT was developed to prevent substance abuse among adolescents, and email and SMS text message reminders appeared to support participation and engagement in the program (Balsa et al. 2014).

Each of these examples highlight the potential for digital technology, especially online self-help programs, to support the prevention of mental health and substance abuse disorders, and to achieve widespread reach toward promoting mental wellbeing. However, it is clear that further efforts are needed to rigorously evaluate the effectiveness of such online self-help or prevention programs using randomized designs and inclusion of appropriate comparison conditions.

29.4.2 Detection and Assessment

In both high-income and lower-income countries studies have demonstrated that digital tools have the potential to improve the detection and diagnosis of mental disorders (Naslund et al. 2015, 2017a). For example, it may be possible to access effective screening tools for depression or anxiety through online platforms, SMS text messages and smartphone applications. These forms of screening tools can potentially be powerful in the detection and assessment of various mental illnesses, especially in settings where access to mental health providers is limited.

In countries such as South Africa and India, web-based screening tools designed to be delivered on mobile phones and used by non-medical health workers and community health providers have appeared effective for reliably diagnosing depression and common psychiatric disorders. As an example, in India a 6-month study was conducted with 100 outpatients with a range of mental disorders to evaluate an Internet-based psychiatric screening and diagnosis application (Malhotra et al. 2015a). This application was created for use in remote settings by non-specialists, and was found to have acceptable validity, reliability and feasibility (Malhotra et al. 2015a). Similarly, another study in India enrolled 123 patients with varying levels of depression and pilot tested the use of web-based screening tools delivered by non-medical healthcare workers (Chattopadhyay 2012). The screening tool tested in this one year trial appeared affective for supporting diagnosis of depression (Chattopadhyay 2012).

In another study, non-medical health workers, volunteers and trainees used an Android smartphone application to diagnose epilepsy among 132 outpatients with possible epilepsy within primary care settings in India and Nepal (Patterson et al. 2015). The mobile application in this study required minimal training and achieved comparable diagnostic agreement with diagnoses provided by medical specialists such as neurologists (Patterson et al. 2015). These studies demonstrate how technology can be important in shifting the tasks that are generally performed by specialist providers to non-medical health workers as a means of meeting the needs of people with mental health disorders who have limited access to care.

Technology has also been used to screen for depression among refugees in South Africa. In this study a mobile SMS-based depression screener was implemented to test for symptoms of depression among 153 refugees from Zimbabwe and the Democratic Republic of Congo living in a social service setting (Tomita et al. 2016). In general, participants in this 10-month pilot study were satisfied with the SMS screening intervention, and there was agreement between the SMS intervention and a face-to-face depression screening intervention (Tomita et al. 2016).

The studies described here demonstrate the ways in which simple mobile technologies can provide vulnerable populations with access to mental healthcare, and can allow non-medical healthcare workers to consistently screen for, and diagnose, mental disorders in remote and low-resource areas. As online platforms increasingly become more popular and big data analytic techniques continue to develop, new opportunities may emerge to analyze patterns of online interactions and behaviors, and to detect individuals who are potentially at-risk for depression, psychosis, suicide or substance abuse before their conditions progress and become more severe.

29.4.3 Treatment and Management

Studies in high-income countries have demonstrated the role of digital technologies for supporting self-management of various disorders, promoting mental wellbeing and improving adherence to treatment for people living with mental disorders (Alvarez-Jimenez et al. 2014; Donker et al. 2013; Hilty et al. 2013; Naslund et al. 2015). Increasingly throughout many LMICs mobile and online technologies such as SMS text reminders, smartphone sensors, mobile apps, and telephones are being used to support telepsychiatry efforts and to deliver treatment for mental disorders (Arjadi et al. 2015; Naslund et al. 2017a; Ruzek and Yeager 2017).

Telepsychiatry involves using telephones, email or videoconferencing to allow speciality mental health clinicians to consult with patients, or to support primary care clinicians or non-medical providers in remote locations that may not have access to specialty providers (Hilty et al. 2002). Telepsychiatry is especially useful because it has the potential to allow mental health providers to connect with patients for clinical consultations for diagnosis, follow-up care, or long-term support (Chipps et al. 2012; Naslund et al. 2017a).

Telepsychiatry has been successfully utilized in multiple countries around the world, including Somaliland (Abdi and Elmi 2011), Colombia (Barrera-Valencia et al. 2014), South Africa (Chipps et al. 2012), Chile (Rojas et al. 2014) and Brazil (Hungerbuehler et al. 2015), to support care for various mental disorders such as depression and other serious mental illnesses. For example, in Somaliland, *Skype*-based online telepsychiatry consultations were delivered over an 18-month period from psychiatrists in Scandinavia to 132 patients with a range of diagnoses, including schizophrenia and mood disorders, using a computer located in a central mental health clinic (Abdi and Elmi 2011). This program emerged as feasible, and clinic staff, community members, and local health authorities reported high levels of satisfaction with this service (Abdi and Elmi 2011).

Several studies have also demonstrated that SMS text message reminders appear to be an effective tool for delivering mental health treatment, promoting engagement and adherence and supporting self-management of various mental disorders. For example, SMS text messages have been used to encourage participation in treatment as reported in randomized controlled trials conducted in Niger (Maiga 2011) and Nigeria (Thomas et al. 2017). In the trial from Niger, 100 patients with psychosis were randomized to receive text message reminders two days before their monthly follow-up appointments compared to a control group who did not receive such reminders (Maiga 2011). Participants who received the text message reminders attended significantly more follow-up visits compared to the control group (Maiga 2011). Similarly, in Nigeria, a randomized controlled trial conducted over a six-month period enrolled 200 patients with first episode psychosis (Thomas et al. 2017). In this study SMS reminders were sent five and three days before patients' appointments with their mental health provider (Thomas et al. 2017). These text message reminders appeared effective for promoting participants' attendance at their clinical appointments as reflected by significantly greater attendance among participants who received the SMS reminders compared to participants in the control group (Thomas et al. 2017).

In another example, from China, SMS text messaging was used to support clinical care for schizophrenia (Fang et al. 2011). In this trial 91 patients with schizophrenia were randomized either to an intervention consisting of daily SMS text messages in combination with telephone follow-up calls and antipsychotic treatment, or to a control condition consisting of antipsychotic treatment and telephone follow-up only (Fang et al. 2011). Compared to the control condition, the intervention involving the addition of daily SMS text messages appeared effective by contributing to a greater reduction in relapse among patients with schizophrenia (Fang et al. 2011). In an additional example from Sri Lanka, SMS text messaging was found to be useful for supporting an intervention aimed at reducing suicidal ideation and depression among patients undergoing treatment following a suicide attempt (Marasinghe et al. 2012). In this study 68 outpatients were randomly assigned to a brief mobile treatment consisting of face-to-face training in problem-solving and mental health, telephone follow-up calls to assess suicidality and weekly supportive SMS text messages, or to a waitlist control group receiving usual care (Marasinghe et al. 2012). This program contributed to significant reductions in suicidal ideation and depression when compared to the control group (Marasinghe et al. 2012).

Technology has also been harnessed for supporting the treatment and diagnosis of mental health conditions through the use of websites and mobile applications. These platforms can be used to deliver evidence-based programs, such as psychoeducation, to patients who have limited access to care, and can potentially afford opportunities to reach individuals who have difficulty seeking care due to stigma, out-of-pocket expenses or inability to travel long distances. For example, a six-month pilot study enrolled 27 patients with major depressive disorder or depressive symptoms from Romania, Spain, and Scotland to test the Help4Mood mobile application for depression treatment (Burton et al. 2016). This application was used to support depression treatment between appointments using guided mood checks, relaxation exercises, and activities to track progress (Burton et al. 2016). While the intervention appeared feasible, no significant improvement in depressive symptoms was observed (Burton et al. 2016).

Telephone interventions also hold promise for encouraging treatment adherence and promoting positive clinical outcomes among patients with mental health or substance abuse disorders. In Brazil 637 self-reported alcohol abusers who had called a national helpline were enrolled in a randomized controlled trial where a motivational telephone-based intervention significantly reduced alcohol consumption compared to standard telephone follow-up and printed materials (Signor et al. 2013). Similarly, in Thailand 60 participants were randomized to receive a six-week telephone-based motivational support program or printed self-help materials (Wongpakaran et al. 2011). In this study the telephone intervention significantly reduced alcohol consumption compared with the control group, with these effects sustained at follow-up assessments (Wongpakaran et al. 2011).

In Chile, a randomized controlled trial enrolled 345 women with depression in primary care settings and evaluated a telephone monitoring program involving contact from non-professionals to provide education and medication monitoring compared to usual care (Fritsch et al. 2007). In this study the telephone monitoring program was more effective for alleviating depressive symptoms at three-months and six-months follow-up compared to the control group (Fritsch et al. 2007). In South Africa a 12-month telephone-based lifestyle coaching intervention enrolled 761 participants with serious mental illness, and contributed to significant weight loss and improved health ratings over time (Temmingh et al. 2013).

Telephone interventions could also be used to support the caregivers of persons with mental illness. For example, in a study from Turkey, 62 participants with schizophrenia and their caregivers were provided weekly telephone-based psychoeducation follow-up over a six month period (Özkan et al. 2013). The patients with schizophrenia who received the intervention showed increased treatment adherence and emotional expression over time, while their caregivers also benefited by exhibiting reduced depressive symptoms and family burden (Özkan et al. 2013).

As a whole, this collection of diverse studies demonstrates that digital technology is both feasible and acceptable as a means of supporting the treatment and self-monitoring of patients with mental disorders. These studies also highlight how technology, such as SMS text messaging, telepsychiatry, telephones and mobile applications can be used to improve adherence and retention in care, reduce symptom severity and promote mental wellbeing among patients living with mental disorders and their caregivers. As this field advances new opportunities may arise to leverage wearable sensors or smartphone-based location, time, or activity data to track patients in LMICs and to enable real-time alerts to their family or caregivers.

29.4.4 Training and Supervision

A fourth important function of digital technology is the potential to increase the capacity of the limited number of clinical providers in low-resource settings. Opportunities include facilitating contact between the limited number of available mental health providers and community health workers or non-medical providers at a distance, thereby improving their capacity to screen for mental illness, provide treatment recommendations and support medical education or training activities (Fairburn and Patel 2017). In this way digital technology could extend the reach of mental health specialists by allowing them to supervise lay providers and non-medical health workers in the delivery of mental health services in primary care or other community-based settings.

For example, through mobile platforms, psychiatrists can provide remote supervision and assistance with consultations to community providers. Community healthcare workers with limited mental health training could also benefit from mobile applications through opportunities to improve the accuracy of the diagnoses that they make and to facilitate screening procedures for common mental disorders in community care settings. Thus, mobile platforms could be used to provide non-medical health workers with critical decision support tools informed by the mhGAP (WHO Mental Health Gap Action Programme) guidelines (World Health Organization 2017). The mhGAP training manuals were developed as part of an initiative launched by the World Health Organization to provide guidance for the assessment and management of mental, neurological and substance abuse disorders in non-specialized health settings (World Health Organization 2017). Digital technology interventions could help distribute and facilitate care for both common and more complex mental disorders, thereby reducing workforce burnout and high turnover among frontline health workers (Fairburn and Patel 2017).

In one possible example, asynchronous telepsychiatry, involving the use of email by expert providers to support consultations to community health workers at a distance (Hilty et al. 2013), has been used in several LMICs to support delivery of psychiatric care in remote areas. A study in Pakistan implemented asynchronous telepsychiatry as a means of supporting clinical care for child or adolescent outpatients with various mental disorders by sending information records on complex cases via email to psychiatrists in the UK for clinical guidance and feedback (Rahman et al. 2006). In another study from India, over a six-year duration, asynchronous telepsychiatry methods appeared feasible for supporting the diagnosis of patients with a variety of mental illnesses (Balasinorwala et al. 2014). In this study, primary care physicians in district hospitals in India emailed notes, patient records and investigation reports to psychiatrists at a central hospital to determine diagnoses and to obtain treatment recommendations for patients with a range of diagnoses (Balasinorwala et al. 2014).

Another area where digital technologies could be especially impactful is in supporting educational opportunities for healthcare workers. In LIMCs many mental healthcare providers and community health workers lack access to adequate educational opportunities (Frehywot et al. 2013). The limited availability of training and professional development opportunities further contributes to challenges in developing the mental health workforce capacity in LMICs (Saraceno et al. 2007).

Digital learning programs represent one promising way that healthcare workers could access basic education and training for providing mental health services (Fairburn and Patel 2017). Digital learning programs, also referred to as e-learning, utilize online platforms to provide educational content to health workers (Bollinger et al. 2013). In LIMCs access to online education programs could serve an essential role in addressing gaps in medical education (Frehywot et al. 2013). There have been several promising e-learning initiatives related to mental health and substance abuse disorders developed and evaluated in Brazil. These included education programs about child and adolescent mental health (Lowenthal et al. 2012) and general mental healthcare (Novaes et al. 2012), as well as an online program aimed at enhancing primary care workers and doctors' knowledge about alcohol abuse (Pereira et al. 2015) and to educate school teachers about childhood mental disorders (Pereira et al. 2015).

In one example of an e-learning initiative from Somaliland, the Aqoon program involved a four-month online psychiatry education program where medical students from Somaliland paired with medical students from the United Kingdom to support cross-cultural medical education (Keynejad et al. 2016). In an evaluation of this program, 48 students (half from Somaliland and half from the United Kingdom) sent text messages to each other through the MedicineAfrica web platform (Keynejad et al. 2016). The Aqoon peer-education program covered core psychiatry curriculum and was highly rated by the students who used it (Keynejad et al. 2016).

This selection of promising studies showcases how digital technologies can potentially be used as a means of supporting training efforts and education about mental healthcare for non-medical health workers and community providers. Ongoing research is needed to evaluate whether technology-supported education efforts are effective for increasing workforce capacity and the reach of mental health services and improving patient outcomes in LMICs.

29.5 Critical Assessment of the Risks and Limitations

Although there are many areas where digital technology holds potential to improve access to, and quality of, mental healthcare, there are also risks and limitations associated with the use of these technologies that require careful consideration. One likely concern related to the use of digital technologies is lack of a clear understanding about how these technologies will be paid for, and what economic and business models will ensure that digital mental health programs can be sustained over the long-term. Cost represents a prominent barrier to accessing digital technologies; therefore, this question is particularly pertinent for LMICs keen to implement these technologies, and for governments, health systems, and local authorities that will require a better understanding of the costs of using digital technologies for improving mental healthcare services.

Lower income countries are also disproportionately affected by poor infrastructure, unreliable access to electricity and lower literacy rates, which can act as significant barriers to accessing and using digital technology (Cullen 2001). For instance, many low-resource settings lack robust telecommunications infrastructure with sufficient and reliable bandwidth for supporting Internet connections with continuous access necessary for enabling the use of digital technology in health interventions (Cullen 2001). Additionally, many people in LIMCs may be unable to readily use emerging digital technologies because of low levels of computing and technology skills and low levels of literacy, especially in rural areas, for those who are most marginalized, and among older individuals (Cullen 2001).

Cultural and behavioral attitudes about using technology may also be a factor that could impact use and impede adoption of digital technology interventions in many settings. For example, behavioral perceptions toward technology might include the misperception that technology use is limited to young people, males or more affluent individuals (Cullen 2001). Cultures that value oral and personal communication may also be less inclined to use computers or mobile devices (Cullen 2001). Strict governmental regulation of the telecommunications industry can also act as a further challenge for accessing digital technologies in several low-resource settings (Howard and Mazaheri 2009).

Use of digital technologies could also have the unintended consequence of worsening inequalities in access to mental healthcare. Many inequalities already exist between individuals who have access to Internet or mobile technologies and those who do not. These include inequalities based on gender, age, education, ethnicity, social class and rural/urban divide. For example, older adults, women, marginalized ethnic groups, impoverished individuals, and individuals living in rural settings have low access to mobile devices. In particular, women are much less likely to own a mobile phone than men; globally across LMICs it is estimated that there are 200 million fewer women than men who own a mobile phone (sGroupe Spéciale Mobile Association 2015). In utilizing digital technologies to support the delivery of mental healthcare, it is essential to address these systematic inequalities, particularly in low-resource settings, at the patient, family, community, policy, and government level.

A more direct risk associated with digital technologies for mental healthcare relates to the security of using online or mobile platforms. Concerns include ethical risks, such as privacy, confidentiality and the potential for intrusion or coercion. Currently, among available smartphone applications for mental health, there is limited evidence of robust privacy policies, or standards for ensuring safe storage and handling of private patient information collected through these applications (Bhugra et al. 2017). For example, while doctors are bound by privacy guidelines, mobile applications can collect aggregate data from users, such as information imputed by users, as well as passive information, such location and purchasing history, and this personal information may be sold by digital technology companies (Torous and Roberts 2017). In another example, the 2014 Samaritans Radar project, a service that scanned social media for negative posts and alerted individuals' friends saying that a person might need support, was quickly ended after the public expressed strong fears about confidentiality and consent (Bhugra et al. 2017). Developers and healthcare providers will also need to ensure that governments or authorities will not be able to use these technologies as a tool for monitoring individuals with mental illness to discriminate against them. These concerns will need to be addressed to ensure that the information that is shared through these platforms is reliable, safe and trustworthy. Another potential harm of digital technologies is that their use could lose the "human" element that forms a critical part of the provision of mental health services, potentially reducing the effectiveness of the care provided.

In order to address these concerns, a set of regulations for the use of digital technologies for supporting the provision of mental healthcare should be developed to ensure their safe and effective use in community and clinical settings. Digital technologies that do not fall under the category of medical devices but could be used for mental healthcare are largely unregulated in many countries. To ensure their safe usage, healthcare professionals should partner with technology developers and regulators to create a set of safety and efficacy standards for the use of digital technologies for providing mental healthcare. These regulations could include standards to protect against predatory companies and guidelines to prevent use of interventions that are not supported by evidence. These regulations should also account for the need to implement and scale up digital technologies in LMICs. Limitations and risks will need to be considered to ensure the safe and effective implementation of digital technology interventions for mental health.

29.6 Future Directions and Implications

In this chapter we have described some of the ways in which digital technologies have been used to treat and prevent mental disorders, and described studies that support the feasibility and acceptability of many recent efforts. These studies have demonstrated that online, text messaging, and telephone support interventions have the potential to be safe and effective for treating mental illness. However, many of the pilot studies described in this chapter only discuss initial results related to the feasibility or acceptability of technological interventions in community or clinical settings, and are not rigorously designed trials with adequate comparison conditions. To truly evaluate the effectiveness of technology interventions, researchers will need to undertake large-scale randomized controlled trials. Greater research efforts are especially needed in LMICs to determine how digital technologies can be used to scale-up mental health treatment. In the paragraphs that follow, we describe a selection of areas that we consider highly important for exploring in greater detail in order to inform efforts aimed at fully utilizing digital technologies for the prevention, detection, treatment and management of mental disorders worldwide. We summarize these key areas in Table 29.2.

First, digital technology has the potential to increase the capacity and effectiveness of the mental health workforce in multiple innovative ways. Digital technologies can be used both to facilitate treatment in clinical settings (Chen et al. 2007; Fang et al. 2011; Fritsch et al. 2007; Lua and Neni 2013; Maiga 2011; Özkan et al. 2013; Thomas et al. 2017) and to allow non-medical providers to screen and diagnose various mental disorders (Chattopadhyay 2012; Malhotra et al. 2015b, c; Patterson et al. 2015). For example, digital technology could support the integration of mental healthcare into primary care settings in community clinics (Figure 29.3).

Technology has also been used to connect patients or community providers with specialists (Abdi and Elmi 2011; Balasinorwala et al. 2014; Chipps et al. 2012; Hungerbuehler et al. 2015; Rahman et al. 2006; Rojas et al. 2014), and has been used as a way to provide training to community health workers and non-medical specialist providers in regions with limited access to specialty training programs (Lowenthal et al. 2012; Novaes et al. 2012; Pereira et al. 2015). As technology can be used to facilitate mental health training, it may also be used to extend the reach of mental health specialists by allowing them to supervise and support community health workers, lay providers and non-medical workers in delivering mental healthcare at a distance (Figure 29.4).

Task sharing is an important area where digital technology holds promise for improving the reach and capacity of the limited number of available mental health specialists (Fairburn and Patel 2017), particularly in rural areas. Fewer than 10% of the mental health workforce is located in rural areas, greatly limiting access to necessary mental healthcare in these regions (Hoeft et al. 2017). Task sharing can involve the sharing of care between rural and urban providers through the shifting of tasks from healthcare workers with higher levels of training to those with lower levels of training, and can allow scarce specialty mental health providers to reach a greater number of patients in need of care (Hoeft et al. 2017). Task sharing can also support community health workers and non-medical health workers in becoming more involved in the delivery of

Table 29.2 Future directions for digital technologies to support the global scale up of mental healthcare.

Potential for digital technology	Key need	Challenges	Potential solutions and impact
1) Expand workforce capacity	Increased reach of mental health providers, particularly in remote areas and resource-poor settings	Low numbers of mental health specialists; lack of mental health specialists in rural areas; lack of formal training programs for mental health providers in rural areas	Task sharing; using technology to: facilitate treatment in clinics; allow non-medical providers to screen and diagnose mental disorders; connect patients and community health providers with specialists; facilitate mental health training
2) Implementation of digital technology interventions	Understanding how to implement digital technology mental health interventions in clinical settings	The majority of mental health treatment technology has been designed outside of clinical settings; outdated research	Conduct research in settings where technologies will be used; use of implementation science to speed up studies and learn how to better implement interventions
3) Economic and cost benefit of digital technology interventions	Understanding the costs of using technology in the treatment of mental disorders	Few studies report on the costs and resources required for implementation or the sustainability of these interventions; decision makers rely on costs in deciding to adopt models of care	Conduct more studies on the cost-effectiveness and sustainability of the implementation of digital technologies for the treatment of mental disorders in a variety of settings

mental healthcare, through technology such as videoconferencing (Hoeft et al. 2017). Use of digital technology for scaling up mental health interventions in lower income settings is a nascent field, though ongoing research efforts across many different regions of the world are poised to substantially advance our understanding of best practices and strategies for improving mental healthcare globally.

It is also important to consider the role of behavioral theory for guiding the development, evaluation, and implementation of digital technology interventions for mental health in lower income settings. Studies have shown that interventions that are developed using theories of human behavior are generally more effective than those that are not guided by theory (Glanz and Bishop 2010). Although theory can be influential in guiding interpretations of study findings, keeping research focused and demonstrating causal processes of behavior change, few digital technology interventions have been developed using behavioral theory (Riley et al. 2011). Widely studied and applied behavior theories and models include the Health Belief Model, the Theory of Planned Behavior, the Transtheoretical Model and Social Cognitive Theory (Glanz et al. 2008).

Figure 29.3 A rural community clinic located in Tanzania. Digital technology holds promise for supporting task shifting of mental healthcare to community providers working at clinics such as these. With the use of digital tools, it may be possible to integrate routine mental health screening and treatment into primary care services offered in community clinics. *Source:* Photo: US Army Africa. Retrieved from https://yaleglobalhealthreview.com/2015/05/16/depression-in-developing-countries

Figure 29.4 Digital technology for supporting mental healthcare in rural India. In this photograph a rural health worker uses a tablet with a specially designed mobile application to support screening for common mental disorders such as depression or suicidal risk in community settings. This project is supported by The George Institute for Global Health in India with funding from the Wellcome Trust/DBT India Alliance and Grand Challenges Canada (Maulik et al. 2017). *Source:* Photo: The George Institute for Global Health. Retrieved from https://www.georgeinstitute.org/our-impact/using-mobile-technology-to-treat-mental-health-in-rural-india

Although these behavior theories are useful, they also face some important limitations in the context of digital technology interventions because they view human behavior as static, focus on between-person differences in behavior, and have remained largely unchanged since they were first proposed (Naslund et al. 2017b). With the emergence of digital technologies that can track individual-level data in real time, behavior theories will need to shift to understanding within-person differences in behavior (Naslund et al. 2017b). In this way, more advanced behavior theories will be needed as new digital technology delivery platforms become widely available to effectively guide the development and delivery of intervention strategies aimed at achieving better health and mental health outcomes (Naslund et al. 2017b). For example, there have been recent advances in behavior theory including dynamic models of behavior, persuasive system design, the behavioral intervention technology model and behavior models for just-in-time adaptive interventions (Naslund et al. 2017b). It will be especially important to apply behavior theories for guiding the development of digital technology interventions for mental health in LMICs in order to maximize the success of these efforts and to support implementation and sustained use of digital technology interventions (Figure 29.5).

Second, there is immense need to better understand how to effectively implement digital technology interventions for mental health into routine clinical and community

Figure 29.5 Youth mental health public engagement program in India. Digital technology can potentially support efforts to raise awareness about mental health and promote public engagement among young people. *It's Ok To Talk* combines public events and social media to engage young people in discussions about mental health and to increase awareness about mental health and wellbeing. This program is delivered by Sangath in India with support from the Wellcome Trust, UK (Gonsalves and Pal 2017). *Source:* Photo: It's Ok to Talk Program, Sangath, India. Retrieved from §1 http://itsoktotalk.in

settings in lower income countries. The majority of mental health treatment technology has been designed and created in highly controlled laboratory settings or by commercial developers with limited consideration of the needs for supporting implementation and sustained use of these interventions in real world settings (Mohr et al. 2017). Translation of digital technology interventions into clinical and community settings is often more complex and costly than in the laboratory-based context (Damschroder et al. 2009). Several important considerations for healthcare providers and researchers include the adaptability of the digital technology intervention to fit local needs, the complexity of the intervention, such as duration, scope, and disruptiveness, the cost of the intervention, and the characteristics such as the size, age and culture of the organization where the digital technology intervention will be implemented (Damschroder et al. 2009). Therefore, greater research focused on the implementation of digital technology interventions for mental health in the real world settings where they will be used is essential, as this will make it possible to determine their true effectiveness and sustainability (Mohr et al. 2017). Consideration of key constructs of implementation science throughout the development, evaluation and delivery process can help ensure that digital technologies support clinical, administrative and patient workflows, rather than impeding them, thereby facilitating successful implementation (Glasgow et al. 2014).

One of the concerns with research on digital technology interventions is that often by the time the research has been published, the technology has advanced such that the findings may no longer be relevant. Implementation science can help inform the research process, so that interventions can be more readily disseminated and integrated into the target settings while ensuring that the intervention is delivered as intended (Glasgow et al. 2014). Of particular importance in the context of LMICs is the potential for digital technologies to support the effective implementation of mental health programs by ensuring that programs are delivered in a more consistent and reliable way than could be possible when implemented by providers alone.

Third, it is important to understand the costs of using technology in the treatment of mental disorders. Several studies have demonstrated the effectiveness and cost-effectiveness of mental health interventions delivered using digital technology (Mohr et al. 2017); however, further research is needed to determine if digital technology interventions can be implemented in a cost-effective way into clinical and community settings, and if these programs can also contribute to cost savings for patients (Mohr et al. 2017). For example, technology could benefit patients with mental disorders by allowing them to avoid lengthy and expensive travel to mental health clinics located in central areas, and by providing them with care closer to their homes. Furthermore, by accessing mental health services through a personal mobile phone, there is the potential to reduce out-of-pocket expenses typically paid by patients and their families. These out-of-pocket savings for individuals are especially important in low-resource settings where financial resources are scarce and even small expenses can be prohibitive or have serious consequences on individuals and families.

To date, few studies have reported on the costs and resources that are required for the implementation of digital technology interventions (Glasgow et al. 2014), yet the cost of implementation is generally one of the first questions that decision-makers ask when deciding whether to adopt a new program or service. Costs are important for informing health policies, as governments are less likely to support the integration of costly digital technology interventions for mental healthcare into routine health services without robust findings to demonstrate the benefits and associated cost savings. Moreover, cost-effectiveness studies generally examine only the implementation costs of a specific intervention and frequently do not recognize the wider social and economic benefits of investing in mental healthcare, such as gains from increased work participation and productivity (Chisholm et al. 2016). Further research is necessary to determine the cost-effectiveness of digital technologies for mental healthcare as a way to inform the long-term sustainability of these programs in LMICs.

29.7 Conclusions

As technologies such as mobile phones become increasingly available and form an integral part of most societies worldwide (Hampshire et al. 2015), they may be used to advance the delivery of mental healthcare services (Jones et al. 2014). In LMICs it may be possible to use digital technology to improve access to specialty mental health providers and to offer mental healthcare services through mobile devices, as a means of providing direct-to-consumer care. These technologies have the potential to decrease inpatient services and costly psychiatric care practices, and increase access to quality mental healthcare for those who lack access to even the most basic services (Eaton et al. 2011). Technology could act as an important driver of advances in mental healthcare delivery, as well as a facilitator of increased access.

Further research and controlled effectiveness studies are required to determine the ways in which these emerging technologies can most effectively be implemented, as well as their potential social and economic benefits, such as increases in productivity, employment and enabling individuals to fulfill family or community responsibilities. In this chapter, we considered several of the ways in which digital technologies have contributed to important clinical benefits, such as improved functioning and symptom remission, in LMICS. As mentioned in previous sections, ongoing efforts are needed to explore indicators such as effectiveness and safety related to the use of mobile or online approaches for preventing and treating mental disorders, as well as the potential risks and limitations with the use of these technologies. Lastly, it is important to carefully consider segments of the population who may be excluded from access to these technologies, as these individuals may have an elevated need for mental health services. Women who have less access to mobile devices than men, the impoverished, the elderly, individuals living in rural areas without access to electricity or reliable mobile coverage and individuals living in regions where mobile phones or the Internet are restricted due to governmental policy (Howard and Mazaheri 2009) are among those who may experience challenges accessing mental healthcare made available through digital technologies. Technology developers, mental health providers, community health workers, and policymakers will need to work together to develop innovative solutions to overcome these persistent challenges of unequal access.

Digital technologies have the potential to greatly improve access to interventions for the prevention, detection and treatment of mental disorders worldwide. There may also be exciting opportunities to leverage digital technologies to support the training of healthcare workers and community health workers in order to develop capacity and skills in delivering mental healthcare across a wide range of settings. Ongoing efforts are needed to examine the safety, effectiveness, costs and risks associated with the use of digital technology for mental disorders, particularly in LMICs and among diverse cultural and ethnic groups and among those who are most vulnerable and socially marginalized.

Bibliography

Abdi, Y.A. and Elmi, J.Y. (2011). Internet based telepsychiatry: a pilot case in Somaliland. *Med Confl Surviv* 27 (3): 145–150.

Alvarez-Jimenez, M., Alcazar-Corcoles, M., Gonzalez-Blanch, C. et al. (2014). Online, social media and mobile technologies for psychosis treatment: a systematic review on novel user-led interventions. *Schizophr Res* 156 (1): 96–106.

Andersson, G. and Cuijpers, P. (2009). Internet-based and other computerized psychological treatments for adult depression: a meta-analysis. *Cogn Behav Ther* 38 (4): 196–205.

Andrade, A.L., de Lacerda, R.B., Gomide, H.P. et al. (2016). Web-based self-help intervention reduces alcohol consumption in both heavy-drinking and dependent alcohol users: a pilot study. *Addict Behav* 63: 63–71.

Arjadi, R., Nauta, M., Chowdhary, N., and Bockting, C. (2015). A systematic review of online interventions for mental health in low and middle income countries: a neglected field. *Glob Ment Health* 2: e12.

Balasinorwala, V.P., Shah, N.B., Chatterjee, S.D. et al. (2014). Asynchronous telepsychiatry in Maharashtra, India: study of feasibility and referral pattern. *Indian J Psychol Med* 36 (3): 299–301.

Balsa, A.I., Gandelman, N., and Lamé, D. (2014). Lessons from participation in a web-based substance use preventive program in Uruguay. *J Child & Adolesc Subst Abuse* 23 (2): 91–100.

Barrera, A.Z., Wickham, R.E., and Muñoz, R.F. (2015). Online prevention of postpartum depression for Spanish-and English-speaking pregnant women: a pilot randomized controlled trial. *Internet Interv.* 2 (3): 257–265.

Barrera-Valencia, C., Benito-Devia, A.V., Vélez-Álvarez, C. et al. (2014). Cost-effectiveness of synchronous vs. asynchronous telepsychiatry in prison inmates with depression. *Rev. Colomb. Psiquiatr.* 46 (2): 65–73. https://doi.org/10.1016/j.rcp.2016.04.008.

Becker, A.E. and Kleinman, A. (2013). Mental health and the global agenda. *N Engl J Me.* 369 (1): 66–73.

Bhugra, D., Tasman, A., Pathare, S. et al. (2017). The WPA-lancet psychiatry commission on the future of psychiatry. *The Lancet Psychiatry* 4 (10): 775–818. https://doi.org/10.1016/S2215-0366(17)30333-4.

Bollinger, R., Chang, L., Jafari, R. et al. (2013). Leveraging information technology to bridge the health workforce gap. *Bull World Health Organ.* 91 (11): 890–892.

Burton, C., Szentagotai Tatar, A., McKinstry, B. et al. (2016). Pilot randomised controlled trial of Help4Mood, an embodied virtual agent-based system to support treatment of depression. *J Telemed Telecare* 22 (6): 348–355.

Central Intelligence Agency. (2014). The World Factbook: Internet Users. Retrieved from https://www.cia.gov/library/publications/the-world-factbook/rankorder/2153rank.html

Central Intelligence Agency. (2015). The World Factbook: Population. Retrieved from https://www.cia.gov/library/publications/the-world-factbook/rankorder/2119rank.html

Chandra, P.S., Sowmya, H.R., Mehrotra, S., and Duggal, M. (2014). "SMS" for mental health – feasibility and acceptability of using text messages for mental health promotion among young women from urban low income settings in India. *Asian J Psychiatr* 11: 59–64.

Chattopadhyay, S. (2012). A prototype depression screening tool for rural healthcare: a step towards e-health informatics. *J Med Imaging Health Inform* 2 (3): 244–249.

Chen, L., Li, Z., and Chen, Y. (2007). Comparative study of relapse prevention in schizophrenia by network. *J Clin Psych Med* 17 (4): 234–236.

Chipps, J., Ramlall, S., Madigoe, T. et al. (2012). Developing telepsychiatry services in KwaZulu-Natal – an action research study. *Afr J Psychiatry (Johannesburg)* 15 (4): 255–263.

Chisholm, D., Sweeny, K., Sheehan, P. et al. (2016). Scaling-up treatment of depression and anxiety: a global return on investment analysis. *Lancet Psych* 3 (5): 415–424.

Christoff, A.O. and Boerngen-Lacerda, R. (2015). Reducing substance involvement in college students: a three-arm parallel-group randomized controlled trial of a computer-based intervention. *Addict Behav* 45: 164–171.

Cullen, R. (2001). Addressing the digital divide. *Online Information Review* 25 (5): 311–320.

Damschroder, L.J., Aron, D.C., Keith, R.E. et al. (2009). Fostering implementation of health services research findings into practice: a consolidated framework for advancing implementation science. *Implementation Science* 4 (1): 50.

Donker, T., Petrie, K., Proudfoot, J. et al. (2013). Smartphones for smarter delivery of mental health programs: a systematic review. *J Med Internet Res* 15 (11): e247. https://doi.org/10.2196/jmir.2791.

Eaton, J., McCay, L., Semrau, M. et al. (2011). Scale up of services for mental health in low-income and middle-income countries. *Lancet* 378 (9802): 1592–1603.

Fairburn, C.G. and Patel, V. (2017). The impact of digital technology on psychological treatments and their dissemination. *Behav Res Therapy* 88: 19–25.

Fang, C.-X., Ye, M.-J., Yang, Y.-F., and Chen, S.-L. (2011). Effects of daily short message reminder for preventing schizophrenia recrudescence. *Nursing J Chinese People's Lib Army* 7: 74–76.

Farrington, C., Aristidou, A., and Ruggeri, K. (2014). mHealth and global mental health: still waiting for the mH 2 wedding? *Globalization and Health* 10 (1): 1.

Frehywot, S., Vovides, Y., Talib, Z. et al. (2013). E-learning in medical education in resource constrained low-and middle-income countries. *Hum Resour Health* 11 (4): https://doi.org/10.1186/1478-4491-11-4.

Fritsch, R., Araya, R. et al. (2007). Un ensayo clínico aleatorizado de farmacoterapia con monitorización telefónica para mejorar el tratamiento de la depresión en la atención primaria en Santiago, Chile. *Revista médica de Chile* 135 (5): 587–595.

Glanz, K. and Bishop, D.B. (2010). The role of behavioral science theory in development and implementation of public health interventions. *Annu Rev Pub Health* 31: 399–418.

Glanz, K., Rimer, B.K., and Viswanath, K. (2008). *Health Behavior and Health Education: Theory, Research, and Practice*. San Francisco, CA: Jossey-Bass.

Glasgow, R.E., Phillips, S.M., and Sanchez, M.A. (2014). Implementation science approaches for integrating eHealth research into practice and policy. *Int J Med Inform* 83 (7): e1–e11.

Gonsalves, P. and Pal, S. (2017). It's OK To Talk, Annual Report 2016-17, Insights from a youth mental health public engagement program in India, Sangath, India. http://www.sangath.in/wp-content/uploads/2017/02/Public-Engagement-Annual-Report-2016-2017.pdf Retrieved from: http://itsoktotalk.in

Hampshire, K., Porter, G., Owusu, S.A. et al. (2015). Informal m-health: how are young people using mobile phones to bridge healthcare gaps in sub-Saharan Africa? *Soc Sci Med* 142: 90–99.

Hilty, D.M., Ferrer, D.C., Parish, M.B. et al. (2013). The effectiveness of telemental health: a 2013 review. *Telemed e-Health* 19 (6): 444–454.

Hilty, D.M., Luo, J.S., Morache, C. et al. (2002). Telepsychiatry. *CNS Drugs* 16 (8): 527–548.

Hoeft, T.J., Fortney, J.C., Patel, V., and Unutzer, J. (2017). Task-sharing approaches to improve mental health care in rural and other low-resource settings: a systematic review. *The J Rural Health.* 34 (1): 48–62.

Howard, P.N. and Mazaheri, N. (2009). Telecommunications reform, internet use and mobile phone adoption in the developing world. *World Development* 37 (7): 1159–1169.

Hungerbuehler, I., Leite, R.F.M., Bilt, M.T., and Gattaz, W.F. (2015). A randomized clinical trial of home-based telepsychiatric outpatient care via videoconferencing: design, methodology, and implementation. *Arch Clin Psych* 42 (3): 76–78.

Internet Society. (2015). Global Internet Report 2015: Mobile Evolution and Development of the Internet. Retrieved from http://www.internetsociety.org/globalinternetreport/2015/assets/download/IS_web.pdf

Jones, S.P., Patel, V., Saxena, S. et al. (2014). How Google's 'ten things we know to be true'could guide the development of mental health mobile apps. *Health Affairs* 33 (9): 1603–1611.

Kakuma, R., Minas, H., van Ginneken, N. et al. (2011). Human resources for mental health care: current situation and strategies for action. *Lancet* 378 (9803): 1654–1663.

Keynejad, R., Garratt, E., Adem, G. et al. (2016). Improved attitudes to psychiatry: a global mental health peer-to-peer E-learning partnership. *Acad Psychiatry* 40 (4): 659–666.

Knaevelsrud, C., Brand, J., Lange, A. et al. (2015). Web-based psychotherapy for posttraumatic stress disorder in war-traumatized Arab patients: randomized controlled trial. *J Med Internet Res* 17 (3): e71.

Liu, N.H., Contreras, O., Muñoz, R.F., and Leykin, Y. (2015). Assessing suicide attempts and depression among Chinese speakers over the internet. *Crisis* 35 (5): 322–329.

Lowenthal, R., Wen, C.L., and Paula, C.S. (2012). Child mental health training for primary care providers in Brazil by telemedicine. *Neuropsychiatr Enfance Adolesc* 60 (5): S30–S31.

Lua, P.L. and Neni, W.S. (2013). A randomised controlled trial of an SMS-based mobile epilepsy education system. *J Telemed Telecare* 19 (1): 23–28.

Maiga, D.D. (2011). Intérêt de l'utilisation du téléphone mobile dans la réponse aux rendez-vous des patients atteints de psychoses aiguës fonctionnelles au service de psychiatrie de l'Hôpital national de Niamey. *Inf Psychiatr* 87 (2): 127–132.

Malhotra, S., Chakrabarti, S., Shah, R. et al. (2015a). A novel screening and diagnostic tool for child and adolescent psychiatric disorders for Telepsychiatry. *Indian J Psychol Med* 37 (3): 288–298.

Malhotra, S., Chakrabarti, S., Shah, R. et al. (2015b). Validity and reliability of diagnostic module of telepsychiatry software applications used by non-specialists. *Indian J Psychiatry* 57 (5): S4.

Malhotra, S., Chakrabarti, S., Shah, R. et al. (2015c). Diagnostic accuracy and feasibility of a net-based application for diagnosing common psychiatric disorders. *Psychiatry Re* 230 (2): 369–376.

Marasinghe, R.B., Edirippulige, S., Kavanagh, D. et al. (2012). Effect of mobile phone-based psychotherapy in suicide prevention: a randomized controlled trial in Sri Lanka. *J Telemed Telecare* 18 (3): 151–155.

Mohr, D.C., Lyon, A.R., Lattie, E.G. et al. (2017). Accelerating digital mental Health Research from early design and creation to successful implementation and sustainment. *J. Med. Internet Res* 19 (5): e153.

Moritz, S. and Russu, R. (2013). Further evidence for the efficacy of association splitting in obsessive-compulsive disorder: an Internet study in a Russian-speaking sample. *J Obsessive-Compuls Relat Disord* 2 (2): 91–98.

Naslund, J.A., Aschbrenner, K.A., Araya, R. et al. (2017a). Digital technology for treating and preventing mental disorders in low-income and middle-income countries: a narrative review of the literature. *Lancet Psychiatr* 4 (6): 486–500.

Naslund, J.A., Aschbrenner, K.A., Kim, S.J. et al. (2017b). Health behavior models for informing digital technology interventions for individuals with mental illness. *Psychiatr Rehabil J* 40 (3): 325–335.

Naslund, J.A., Marsch, L.A., McHugo, G.J., and Bartels, S.J. (2015). Emerging mHealth and eHealth interventions for serious mental illness: a review of the literature. *J of Ment. Health* 24 (5): 321–332.

Naslund, J.A., Aschbrenner, K.A., Marsch, L.A., and Bartels, S.J. (2016). The future of mental health care: peer-to-peer support and social media. *Epidemio Psychiatr Sci* 25 (2): 113–122.

Novaes, M.N., Machiavelli, J.L., Verde, F.C.V. et al. (2012). Tele-educação para educação continuada das equipes de saúde da família em saúde mental: a experiência de Pernambuco, Brasil. *Interface – Comunicação, Saúde, Educação* 16 (43): 1095–1106.

Özkan, B., Erdem, E., Özsoy, S.D., and Zararsiz, G. (2013). Şizofreni hastalarına verilen ruhsal eğitim ve telepsiklyatrik izlemenin hasta işlevselliği ve ilaç uyumuna etkisi. *Anadolu Psikiyatri Dergisi* 14 (3): 192–199.

Patel, V., Chisholm, D., Parikh, R. et al. (2016). Addressing the burden of mental, neurological, and substance use disorders: key messages from disease control priorities. *Lancet* 387 (10028): 1672–1685.

Patel, V., MAJ, M., Flisher, A.J. et al. (2010). Reducing the treatment gap for mental disorders: a WPA survey. *World Psychiatry* 9 (3): 169–176.

Patterson, V., Singh, M., Rajbhandari, H., and Vishnubhatla, S. (2015). Validation of a phone app for epilepsy diagnosis in India and Nepal. *Seizure* 30: 46–49.

Pereira, C.A., Wen, C.L., and Tavares, H. (2015). Alcohol abuse management in primary care: an e-learning course. *Telemed J e-Health* 21 (3): 200–206.

Plata, L.A.F., López, G.C., Baca, X.D., and Gómez, A.D.L.R. (2014). Psicoterapia vía Internet: Aplicación de un programa de intervención cognitivo-conductual para pacientes con depresión. *Psicología Iberoamericana* 22 (1): 7–15.

Rahman, A., Nizami, A., Minhas, A. et al. (2006). E-mental health in Pakistan: a pilot study of training and supervision in child psychiatry using the internet. *Psychiatr Bull* 30 (4): 149–152.

Riley, W.T., Rivera, D.E., Atienza, A.A. et al. (2011). Health behavior models in the age of mobile interventions: are our theories up to the task? *Translat Behav Med* 1 (1): 53–71.

Rojas, G., Castro, A., Guajardo, V. et al. (2014). Programa colaborativo a distancia para el tratamiento de la enfermedad depresiva. *Revista médica de Chile* 142 (9): 1142–1149.

Ruzek, J. and Yeager, C. (2017). Internet and mobile technologies: addressing the mental health of trauma survivors in less resourced communities. *Glob Mental Health* 4: e16.

Saraceno, B., van Ommeren, M., Batniji, R. et al. (2007). Barriers to improvement of mental health services in low-income and middle-income countries. *Lancet* 370 (9593): 1164–1174.

Saxena, S., Thornicroft, G., Knapp, M., and Whiteford, H. (2007). Resources for mental health: scarcity, inequity, and inefficiency. *Lancet* 370 (9590): 878–889.

sGroupe Spéciale Mobile Association. (2015). Connected Women 2015 – Bridging the gap: mobile access and usage in low- and middle-income countries. Retrieved from http://www.gsma.com/mobilefordevelopment/wp-content/uploads/2016/02/Connected-Women-Gender-Gap.pdf

Signor, L., Pierozan, P., Ferigolo, M. et al. (2013). Efficacy of the telephone-based brief motivational intervention for alcohol problems in Brazil. *Rev. Bras. Psiquiatr.* Available from http://dx.doi.org/10.1590/1516-44462011-07.

Spek, V., Cuijpers, P., Nyklícek, I. et al. (2007). Internet-based cognitive behaviour therapy for symptoms of depression and anxiety: a meta-analysis. *Psychol Med* 37 (3): 319–328.

Temmingh, H., Claassen, A., van Zyl, S. et al. (2013). The evaluation of a telephonic wellness coaching intervention for weight reduction and wellness improvement in a community-based cohort of persons with serious mental illness. *J Nerv Mental Dis* 201 (11): 977–986.

The World Bank. (2016). Mobile cellular subscriptions (per 100 people): International Telecommunication Union, World Telecommunication/ICT Development Report and database. Retrieved from http://data.worldbank.org/indicator/IT.CEL.SETS.P2?contextual=default&end=2015&locations=XL-1W-XO-F1&start=2000&view=chart

Thomas, I.F., Lawani, A.O., and James, B.O. (2017). Effect of short message service reminders on clinic attendance among outpatients with psychosis at a psychiatric Hospital in Nigeria. *Psychiatr Serv* 68 (1): 75–80.

Tomita, A., Kandolo, K.M., Susser, E., and Burns, J.K. (2016). Use of short messaging services to assess depressive symptoms among refugees in South Africa: implications for social services providing mental health care in resource-poor settings. *J Telemed Telecare* 22 (6): 369–377.

Torous, J. and Roberts, L.W. (2017). The ethical use of mobile health Technology in Clinical Psychiatry. *J Nerv Mental Dis* 205 (1): 4–8.

Wagner, B., Brand, J., Schulz, W., and Knaevelsrud, C. (2012). Online working alliance predicts treatment outcome for posttraumatic stress symptoms in Arab war-traumatized patients. *Depress Anxiety* 29 (7): 646–651.

Walker, E.R., McGee, R.E., and Druss, B.G. (2015). Mortality in mental disorders and global disease buranden implications: a systematic review and meta-analysis. *JAMA Psychiatry* 72 (4): 334–341.

Wang, Z., Küffer, A., Wang, J., and Maercker, A. (2014). Nutzung eines webbasierten Selbsthilfeprogramms für Traumaopfer: Auswertung von Dropout-Raten und ihren Prädiktoren. *Verhaltenstherapie* 24 (1): 6–14. https://doi.org/10.1159/000358472.

Wang, Z., Wang, J., and Maercker, A. (2013). Chinese my trauma recovery, a web-based intervention for traumatized persons in two parallel samples: randomized controlled trial. *J Med Internet Res* 15 (9): e213. https://doi.org/10.2196/jmir.2690.

We Are Social. 2017. Digital in 2017: Global Overview. Retrieved from https://wearesocial.com/special-reports/digital-in-2017-global-overview

Whiteford, H.A., Degenhardt, L., Rehm, J. et al. (2013). Global burden of disease attributable to mental and substance use disorders: findings from the global burden of disease study 2010. *Lancet* 382 (9904): 1575–1586.

Wongpakaran, T., Petcharaj, K., Wongpakaran, N. et al. (2011). The effect of telephone-based intervention (TBI) in alcohol abusers: a pilot study. *J Med Assn Thailand* 94 (7): 849–856.

World Health Organization. (2017). mhGAP training manuals for the mhGAP Intervention Guide for mental, neurological and substance use disorders in non-specialized health settings version 2.0 (for eld testing). (W. H. Organization Ed.). Geneva, Switzerland: World Health Organization.

Xu, W., Wang, J., Wang, Z. et al. (2015). Web-based intervention improves social acknowledgement and disclosure of trauma, leading to a reduction in posttraumatic stress disorder symptoms. *J Health Psychol* 21 (11): 2695–2708.

Webliography

http://data.worldbank.org/indicator/IT.CEL.SETS.P2?contextual=default&end=2015&locations=XL-1W-XO-F1&start=2000&view=chart. The World Bank 2016. Mobile cellular subscriptions (per 100 people): International Telecommunication Union, World Telecommunication/ICT Development Report and database.

Wang, Z., Küffer, A., Wang, J., and Maercker, A. (2014). Nutzung eines webbasierten Selbsthilfeprogramms für Traumaopfer: Auswertung von Dropout-Raten und ihren Prädiktoren. *Verhaltenstherapie* 24 (1): 6–14. http://dx.doi.org/10.1159/000358472.

http://itsoktotalk.in

http://www.afro.who.int/news/liberia-plans-strengthen-mental-health

http://www.gsma.com/mobilefordevelopment/wp-content/uploads/2016/02/Connected-Women-Gender-Gap.pdf. Groupe Spéciale Mobile Association 2015. Connected Women 2015 – Bridging the gap: mobile access and usage in low- and middle-income countries.

http://www.internetsociety.org/globalinternetreport/2015/assets/download/IS_web.pdf. Evolution and Development of the Internet.

http://www.who.int/mental_health/mhgap/en

https://msf.exposure.co/a-year-in-pictures

https://wearesocial.com/special-reports/digital-in-2017-global-overview We Are Social 2017. Digital in 2017: Global Overview.

https://www.cia.gov/library/publications/the-world-factbook/rankorder/2153rank.html Central Intelligence Agency 2014. The World Factbook: Internet Users.

https://www.cia.gov/library/publications/the-world-factbook/rankorder/2119rank.html Central Intelligence Agency 2015. The World Factbook: Population.

https://www.georgeinstitute.org/our-impact/using-mobile-technology-to-treat-mental-health-in-rural-india

https://www.hrw.org/news/2016/03/20/indonesia-treating-mental-health-shackles

https://yaleglobalhealthreview.com/2015/05/16/depression-in-developing-countries

30

The Use of Mobile Chest X-Rays for Tuberculosis Telemedicine

Meghan L. Jardon, Kelsey L. Pomykala, Ishita Desai, and Kara-Lee Pool

Department of Radiology, University of California, Los Angeles, CA, USA

30.1 Background

Tuberculosis (TB) is an ancient infectious disease caused by *Mycobacterium tuberculosis*. Yet while the oldest recorded cases of TB can be traced back to fossils over 10 000 years old, TB continues to have worldwide significance today (Lee et al. 2012 and see Chapter 1). The World Health Organization (WHO) estimates that TB was responsible for 1.7 million deaths in 2016, making it the leading infectious cause of death worldwide (WHO 2017). The low- and middle-income countries (LMICs) bears a disproportionate volume of TB cases, however, with 64% of new worldwide cases in 2016 occurring in India, India, Indonesia, China, Philippines, Pakistan, Nigeria, and South Africa (WHO 2017). Additionally, the HIV epidemic has increased the spread of TB, and TB became the leading cause of death in patients with HIV in 2017 (WHO 2017). Furthermore, partial treatment of TB worldwide has led to the development of multi-drug resistant TB, increasing the importance of early detection and complete treatment of all tuberculosis cases.

Revolutionizing Tropical Medicine: Point-of-Care Tests, New Imaging Technologies and Digital Health,
First Edition. Edited by Kerry Atkinson and David Mabey.
© 2019 John Wiley & Sons, Inc. Published 2019 by John Wiley & Sons, Inc.

30.2   Lack of Access to Radiology

Establishing a diagnosis of TB, whether in its latent or active form, depends on a varying combination of a nucleic acid amplification test (GeneXpert MTB/RIF), purified protein derivative (PPD) testing, sputum acid-fast staining/culture and chest radiography (Jiménez, Medlen, and Fleitas-Estévez 2014). Historically, despite its inclusion in the WHO's building blocks essential for all health systems as well as the WHO-recommended algorithms for initial workup for TB, access to chest radiography in the LMICs has been poor. The WHO estimates that although x-ray technology was initially developed in the late 1800's, as much as two thirds of the world's population does not have access to x-ray technology today (Shah 2014; Waters 2011). When new diagnostic imaging technology first expands to low resource countries, it typically reaches larger regional hospital centers first, leaving the healthcare facilities in rural communities the most noticeably lacking in radiology services. The medical equipment that does reach these regional facilities is often unreliable, and if the staff at the receiving facility are not trained, this equipment may not be used safely or effectively. In many resource-poor countries, there is a lack of centralized distribution for imaging technology, which combined with a lack of understanding of each site's individual needs, leads to fragmentation of services provided to patients (Jiménez et al. 2014; WHO 2012). As the developed world's use of radiology in diagnosis and treatment continues to expand rapidly, the LMICs access to basic imaging technologies crucial for the detection and treatment of tuberculosis remains inadequate.

30.3   Implementation

As the current statistics on lack of access to radiology demonstrate, the complex implementation processes required to establish these services often serve as barriers to patient care. In 2012 the WHO released a statement on barriers to access to medical devices in low-income countries. Among the reasons they cited for this disparity were: inadequate data gathering systems to address the need for specific medical devices, insufficient infrastructure to support the use of medical devices in delivering care, lack of development of local industry to develop technologies needed, inadequate financing and insufficiently trained workforce to operate and maintain diagnostic imaging devices (WHO 2012).

With regards to the institution of x-ray capabilities in a medical facility to aid in the diagnosis of tuberculosis, there are many required infrastructure specifications. Standard stationary x-ray technology requires ample space to align the x-ray generator and detector equipment, as well as a reliable power source (Malkin and Teninty 2014). While these specifications are commonplace in hospitals in resource-rich countries, unreliable power grids in underdeveloped countries may require the use of a generator to provide an adequate and consistent power source, and special rooms may need to be constructed to house traditional x-ray equipment (KNCV Tuberculosis Foundation 2008). (KNCV is the Koninklijke Nederlandse Centrale Vereniging tot bestrijding der Tuberculose [Dutch Tuberculosis Foundation]). One way to overcome the infrastructure limitations, as well as underdeveloped local industry, is through the use of mobile x-ray units produced abroad and distributed globally. One such company is MinXray which specializes in producing lightweight, portable x-ray machines that produce digital x-ray images (MinXray, Inc. 2015). These machines can be set up in any room quickly and easily, with two-wheel carts that allow for easy transport within a medical center or from site to site (Figure 30.1).

In addition to mobile x-ray machines that can be transported within an existing medical center, portable x-ray units can be housed in a mobile van. Delft imaging systems has developed a "OneStopTB" clinic, which combines x-ray technology and Xpert testing, and is currently being utilized in multiple African countries (Delft Imaging Systems) (Figures 30.2–30.4).

These clinics help overcome barriers associated with medical facility infrastructure, as well as the difficulties of widespread distribution to rural sites. Delft's mobile clinics assist in mobile diagnosis, allowing for computer-aided diagnostic interpretation of chest x-ray with Bayesian applications that can detect cavities and nodules with pattern recognition (Shen, Cheng, and Basu 2010) and computer-aided diagnostic systems (CAD) (Jaeger et al. 2014; Shen et al., 2010). This CAD for TB (computer-aided detection for tuberculosis – CAD4TB) technology has been shown in multiple studies to be an effective tool in analyzing chest x-rays for the presence of TB (Breuninger et al. 2014;

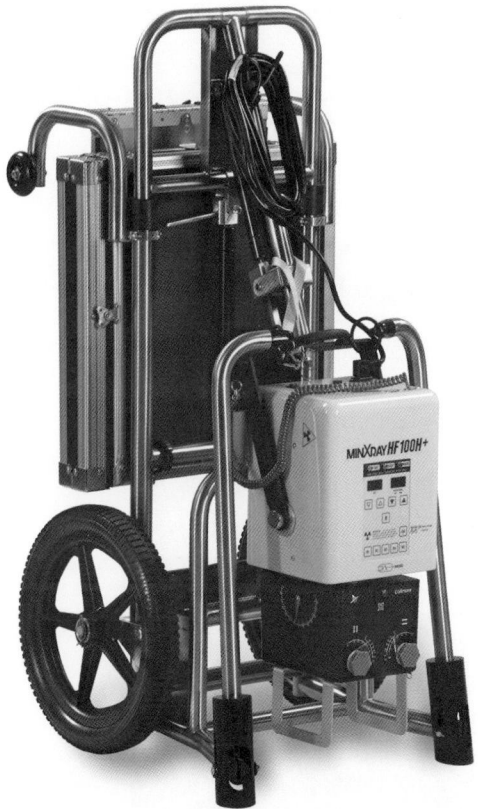

Figure 30.1 MinXray model CMDR.PE.120.60.S portable digital radiography system.

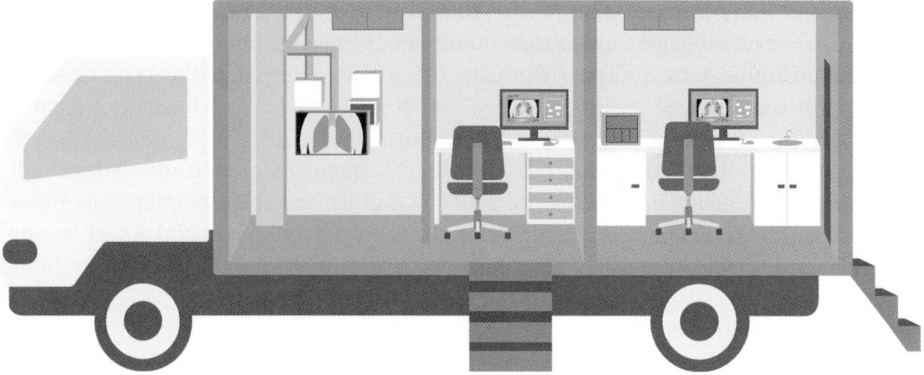

Figure 30.2 Cartoon Diagram of Delft OneStopTB clinic.

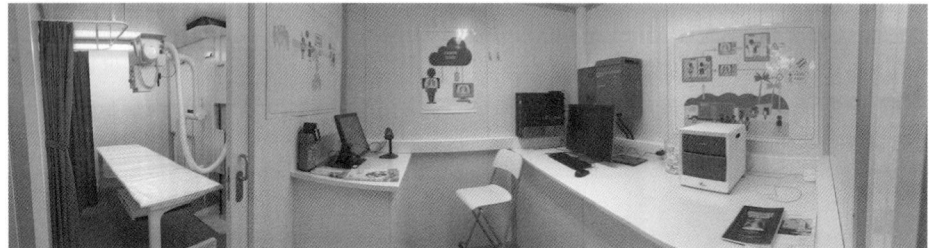

Figure 30.3 Panorama view inside Delft OneStopTB clinic.

Figure 30.4 Mobile Delft OneStop Lung Health clinic in Ghana.

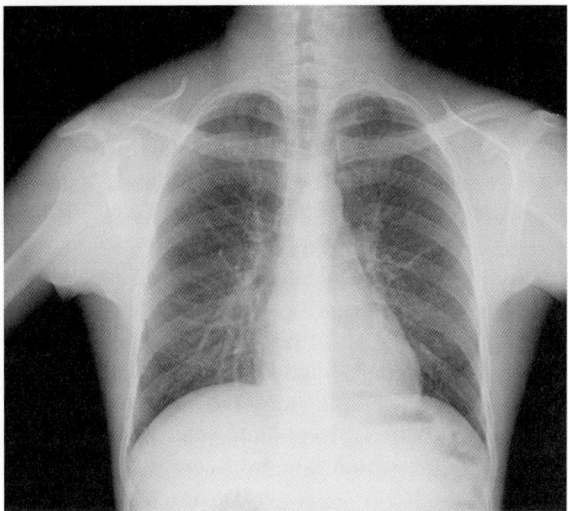

Figure 30.5 Normal posterior–anterior (PA) chest radiograph.

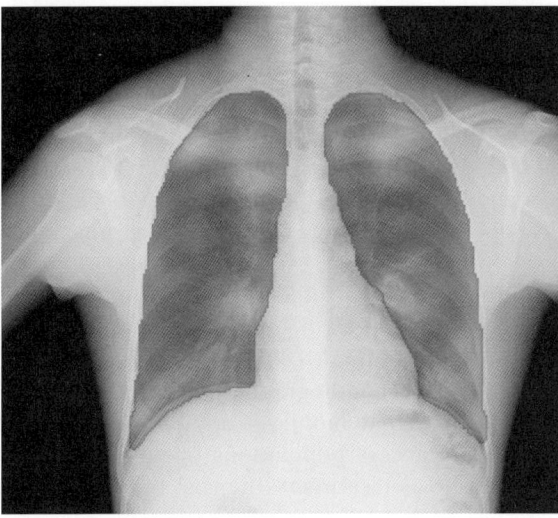

Figure 30.6 Heat map score 19. (*See color plate section for the color representation of this figure.*)

Maduskar et al. 2013; Muyoyeta et al. 2014). CAD4TB analyzes digital chest x-rays and provides a score from 0 to 100 on relevant TB findings, with heat maps indicating the areas of abnormalities (Figures 30.5–30.10).

The mobile medical groups that provide digital x-ray services are not only a convenience to rural providers but have been shown to be an effective use of medical resources and to reduce the time to TB detection (Morishita et al. 2017; Steiner et al. 2015). These van-based mobile clinics can reach remote areas, providing both technical imaging and interpretation of chest x-rays to those who often do not typically initially benefit from developing technology.

30.4 Cost

The cost of implementing new diagnostic imaging technologies is high at multiple levels. First, there is the initial purchase cost of the radiography system, which varies greatly depending on whether older analogue/film systems are purchased or newer digital technology. Analogue systems can cost US$30000 to 50000, while the costs of systems using digital detectors is over three times higher (Garra 2014; KNCV Tuberculosis Foundation 2008). While it is possible for some hospitals to receive radiography systems as donations, this is rarely without some expense to the receiver, whether in shipping costs or repairs needed on damaged or non-functional equipment (Malkin and Teninty 2014). In addition to the initial purchase price, there is cost associated with each image that is obtained. This cost is substantially higher with analogue systems when taking into consideration the price of film, developing materials and physical storage of films (Eijkman and van Doren Geneva 2009). Despite its higher initial purchase price, the price of continuing to operate a digital system remains lower than that of an analogue system.

In addition, there are personnel-related costs at every step of image acquisition. Staff must be hired and/or trained to acquire images and radiologists must be hired to interpret these images. While these costs will vary greatly depending on the site and the approach to expanding staff capabilities, hiring of new employees or additional training for existing employees has associated cost. Furthermore, there is the cost of continued use and maintenance of the equipment, which will be discussed in detail in a later section of this chapter. However, maintenance costs of radiographic equipment can frequently exceed twice the original purchase price. For example, a US$50000 x-ray machine can cost over US$100000 to maintain over its lifetime (Hockel and Hamilton 2011; Malkin and Teninty 2014). In resource poor settings, consistent access to funds required to operate and maintain portable x-ray machines can prove difficult.

Given the high upfront and continuing costs associated with chest radiography, acquiring the ability to obtain chest x-rays to aid in the diagnosis of tuberculosis can be a large financial burden for medical institutions in LMICs to take on. Despite this, multiple studies have concluded that chest radiography improves screening coverage and tuberculosis identification, and that mobile x-ray units are a cost-effective method of reaching these hard-to-access populations (Heuvelings et al. 2017). Compared with a course of tuberculosis treatment, the cost of which often exceeds US$7000, implementing chest radiography on portable systems that can reach the highest risk populations ultimately proves to be worth the cost (Heuvelings et al. 2017).

30.5 Sustainability

In addition to ensuring sufficient funding and personnel to obtain and operate x-ray technology, the machines must be regularly serviced to function reliably. This has proven to be an additional barrier to access, and it is estimated that in some countries in LMICs nearly 50% of all x-ray equipment is out of service (Perry and Malkin 2011). Typically, the best way to ensure that equipment is well maintained

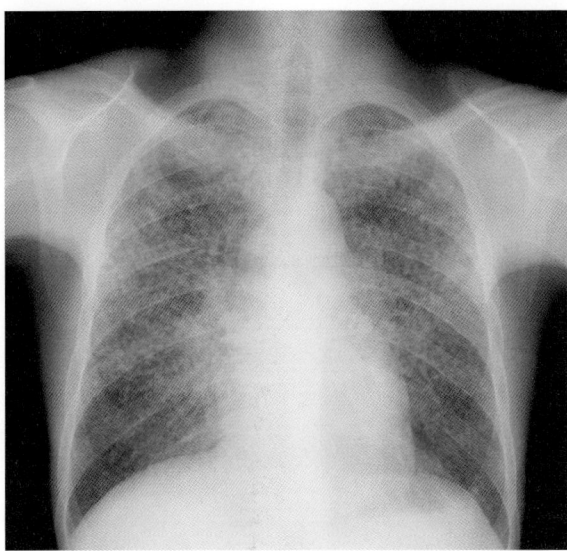

Figure 30.7 PA chest radiograph with innumerable randomly distributed micronodules, a miliary pattern.

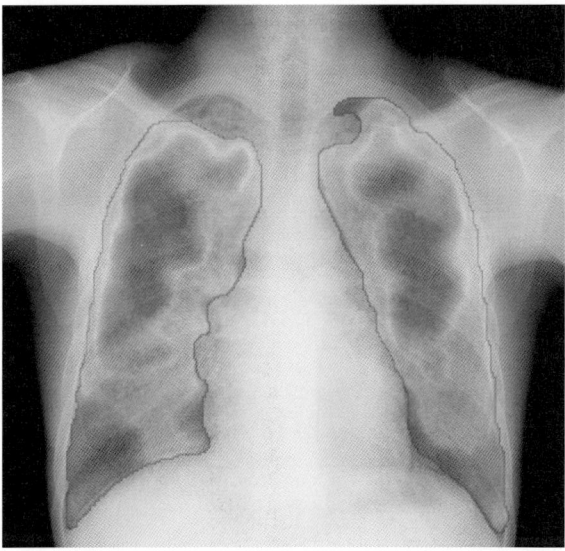

Figure 30.8 Heat map score 90. (*See color plate section for the color representation of this figure.*)

is through service contracts, where a knowledgeable and licensed technician services the equipment at least yearly to ensure that it functions properly. Establishing these service contracts, however, can be costly upfront, and it is not uncommon for service contractors in LMICs to take the initial payment and not fulfill their service obligations (Malkin and Teninty 2014). Failure to maintain equipment properly may ultimately lead to more costly repairs in the future and unreliable access to imaging.

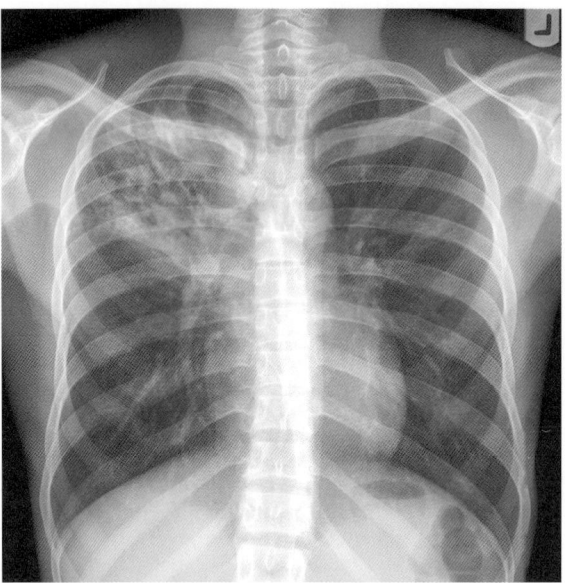

Figure 30.9 PA chest radiograph demonstrating a thick-walled right upper lobe cavitary mass consistent with reactivation of tuberculosis.

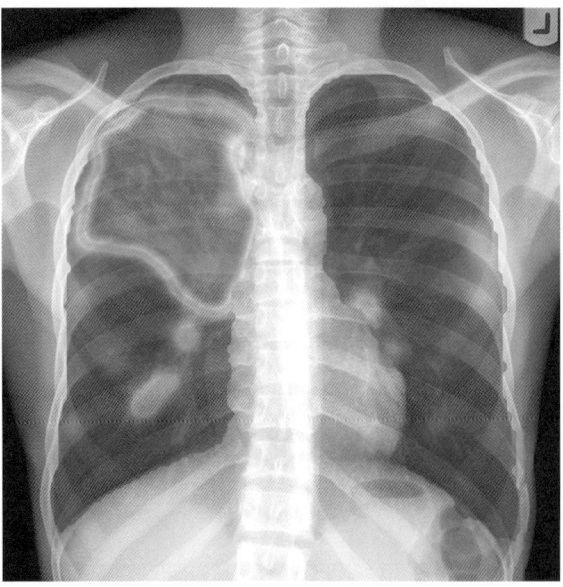

Figure 30.10 Heat map score 100. (*See color plate section for the color representation of this figure.*)

30.6 Chest X-Ray Information Technology (IT)

30.6.1 Imaging Device Systems

When choosing an x-ray imaging device system in the global health setting, hospitals and clinics must consider how the images will be obtained and transferred. As for the way images are obtained, chest x-ray information technology has been advancing over the years and most systems now use digital detectors instead of film. While conventional film radiography does have the benefits of good image quality and high spatial resolution, it does have disadvantages that include limited exposure range, inflexibility of image display and film management, as well as a high retake rate (Bacher et al. 2003). In addition, due to less demand, fewer companies are manufacturing the equipment needed for film images, causing equipment prices to increase.

Digital radiography systems, on the other hand, can offer instant image display with a wide dynamic range and linear signal response, as well as flexibility in image processing and archiving. Digital radiography systems originally used phosphor plates due to their compatibility with existing radiography equipment (Sonoda et al. 1983). This method is called computed radiography, where the phosphor plate is scanned with a helium neon laser, the emitted light is captured by a photomultiplier tube and then converted to an analogue electrical system which is digitized (Bansal 2006). However, there were concerns with image quality and radiation doses. Newer systems include direct radiography where a semiconductor-based sensor directly converts x-ray energy into electrical signals. There are solid state detectors (selenium drum) and flat panel detectors (selenium and cesium iodide), which convert x-ray photons to light, which are then converted to electrons by amorphous silica. These systems have been proven to have exceptional image quality (Aufrichtig 1999; Fink et al. 2002; Floyd et al. 2001; Rong et al. 2001).

Importantly, before obtaining the images, healthcare systems must consider how the images will be transferred so that compatibility can be ensured. Image transfer will be discussed in detail in the next section. Briefly, images should ideally be transferred using the digital imaging and communications in medicine (DICOM) system, which contains protocols that facilitate communication between different medical machines. This communication includes the transfer of patient information, medical history, image creation method and the images themselves. Although DICOM is the standard for medical imaging, other file formats including the Joint Photographic Experts Group (JPEG) format and the Audio Video Interleaved (AVI) format can be used. These, however, will not include patient information and may lead to mislabeled or missing images (Kahn et al. 2007). Therefore, an imaging device system compatible with a DICOM system is preferred (Garra 2014).

30.6.2 Transfer of Images

The electronic transfer of digital images is particularly significant for resource-poor regions due to the increasing cost of x-ray film, the lack of image interpretation experts in these regions and poor transportation infrastructure such as unreliable roads and expensive fuel. Moreover, image acquisition, interpretation monitoring, maintenance and training can all be better accomplished digitally (Garra 2014).

Key factors for the efficient transfer of images include limiting communication bandwidth use by compressing images so that less data has to be transferred, avoiding the use of cloud-hosted applications that require a continuous network connection, using software that can continue to function with intermittent connectivity and variable data transfer rates and a possibly unreliable power supply. To overcome these barriers it is required to have a back-up generator, and to keep software and hardware costs low by obtaining donations and discounts or sharing software with other low budget companies and benefiting from multi-user licenses (Garra et al. 2008; Garra 2014; Jones 2016).

As discussed previously, the first step is to acquire the images in a file format that can be transferred reliably (for example, DICOM), then physically transfer the images via Ethernet cord, USB, Wi-Fi or cellular network. Most imaging device systems include an Ethernet and USB connection, but newer systems also have Wi-Fi capabilities. Wired data is usually preferred due to higher speed, but when working with mobile devices, the use of Wi-Fi or cellular network is preferred. Exporting data to a USB drive and then uploading to a workstation is an antiquated form of transfer but highly reliable and can be used as a backup method if wireless means are unavailable (Garra 2014).

In order to transfer the images, the data need to be compressed to conserve bandwidth. If the images are being transferred locally over Wi-Fi, images may be sent without compression or lossless compression. When data are transferred over greater distances, however, increasing compression is necessitated in order to reduce speed and cost (Garra 2014).

30.6.3 Image Interpretation and Quality Assurance IT

Next, images must be able to reach qualified interpreters. When the interpreters are local, transfer via an Ethernet connection from the imaging source to a work-station less than 100 m away is the most economical and efficient option. Reading stations must have DICOM reading software and a DICOM server to support reception and storage of the images. Reading stations can consist of a simple laptop or a larger station with many monitors. If a laptop is to serve as the reading station, however, the addition of at least one high resolution monitor is recommended. When sending images to interpreters in other regions or countries, electronic transfer is necessary. Images can be uploaded to a reliable wide area network (WAN) or to the Internet and then transferred to a DICOM server (Garra 2014).

Reports can be sent to the provider via written note, phone, email or text message. There are reservations when communicating via text message due to security concerns, but there are options to text message a link to a secure server that contains the complete report and some sample images. Because email and text message are vulnerable to network outages, phone and hand-written messages can be used as a backup. A PACS (picture archiving and communication system)-based reporting system is a second, more expensive option (Choplin et al. 1992). PACS systems often include voice recognition and structured report templates that may be translated (Garra 2014).

Additionally, quality assurance should to be included. Quality assurance programs monitor performance of image system operators, image readers and the equipment itself. Many PACS systems have built in computerized quality assurance modules. Finally, there should be network maintenance and redundancy. Networks in LMICs are not reliable, thus making back-up transfer methods for images and reports highly recommended. It is also important to have technicians available who can monitor the performance of the network and troubleshoot or repair such problems as they arise (Garra 2014).

30.7 Mobile Devices

Modern means of communication, including mobile phones and the Internet, can greatly improve communication within healthcare systems. Although Internet reliability remains low in remote areas, mobile networks are expanding and the number of mobile subscribers is surpassing the number of people using a fixed line connection (Asangansi and Braa 2010; Krohn 2010; Teltscher et al. n.d.; Tuijn et al. 2011). Additionally, a survey of Internet users found that greater than 50% use instant messaging services on their mobile devices, with the most popular messaging application worldwide being *WhatsApp* with 1.3 billion active monthly users (Stahl et al. 2017; Statistica n.d.; WhatsApp n.d.).

Mobile devices can address barriers in accessibility, quality, effectiveness, efficacy and cost of transferring and interpreting images. These devices can expedite and facilitate access to expert advice, contributing to effective diagnosis and treatment, while also reducing cost for both healthcare systems and patients. Recent studies demonstrated that experts can accurately diagnose images on tablets and cell phones (Caffery et al. 2015; Choi et al. 2011; Goost et al. 2012; Scheuermeyer et al. 2016; Schwartz et al. 2014). Research has also found images on smartphones and tablets to be of comparable quality to those viewed on computer screens (Boissin et al. 2017; Toomey et al. 2014).

Despite the great potential of using mobile devices to transport and interpret images, there are limitations and concerns that should be addressed. For example, medical images are often obtained in large sets and may overwhelm a mobile device. Additionally, downloaded images on a mobile device are vulnerable to breach of patient confidentiality, including lost and stolen data. Regulation and quality control are needed to ensure that the images are safe and can be visualized and manipulated in a clinically acceptable manner (Hirschorn et al. 2014). Most important, however, is the potential limitation of lost spacial resolution and poor image quality which can lead to false negative interpretations.

30.8   Education to Ensure Sustainability

The use of mobile x-ray units and telemedicine can greatly improve the quality of radiological services that are available to resource-poor countries. These countries face many challenges when providing radiological services, but a large burden is the lack of human resources. Often, nurses without training acquire the images, general practitioners have to make the interpretations and untrained physicists and engineers are responsible for quality, safety and maintenance (Jiménez et al. 2014). Proper initial and continuing education will be essential for the success of the mobile x-ray unit/ telemedicine combination.

Training for physicians in underdeveloped regions may come in different forms. This can be in the form of a service from aid groups, non-profit organizations and international residency programs. Recently an agreement was made between the International Atomic Energy Agency and the European Society of Radiology to provide state of the art training in Europe for radiologists in LMICs (Gudkova n.d.) Furthermore, RAD-AID provides many resources to training health workers in China, Nepal, India, Vietnam, Laos, Bhutan, Guyana, Nicaragua, Haiti, Jamaica and more (RAD-AID n.d.). [RAD-AID is a not-for-profit organization focused on delivering radiology services to LMICs.] The WHO established a Global Steering Group for Education and Training in Diagnostic Imaging in 1999, whose sole purpose is to coordinate various training activities organized by international and regional societies in order to improve quality, quantity and equity of diagnostic imaging services worldwide (WHO n.d.).

Alongside improved education, there is a push for increasing research into the development of more reliable computer-assisted diagnostics as evidenced by new research regarding artificial intelligence-assisted diagnosis of pneumonia on chest x-rays. These developments may serve as a major step forward for underserved regions within and outside resource-rich regions (Rajpurkar et al. 2017).

30.9 Conclusions

As evidenced by the information presented here, there remains an obvious need for the delivery of radiological services in resource-poor regions for the diagnosis of TB, as well as numerous other common and uncommon ailments. Sustainable high quality radiologic services for such diagnoses is difficult and expensive to implement. This is, however, a barrier that can be overcome with persistence, continued education and innovative research.

Bibliography

Asangansi, I. and Braa, K. (2010). The emergence of mobile-supported National Health Information Systems in developing countries. *Stud Health Tech Inform* 160 (Pt 1): 540–544.

Aufrichtig, R. (1999). Comparison of low contrast detectability between a digital amorphous silicon and a screen-film based imaging system for thoracic radiography. *Med Physics* 26 (7): 1349–1358. https://doi.org/10.1118/1.598630.

Bacher, K., Smeets, P., Bonnarens, K. et al. (2003). Dose reduction in patients undergoing chest imaging: digital amorphous silicon flat-panel detector radiography versus conventional film-screen radiography and phosphor-based computed radiography. *Am J Roentgenology* 181 (4): 923–929. https://doi.org/10.2214/ajr.181.4.1810923.

Bansal, G.J. (2006). Digital radiography: a comparison with modern conventional imaging. *Postgrad Med J.* 82 (969): 425–428. https://doi.org/10.1136/pgmj.2005.038448.

Boissin, C., Blom, L., Wallis, L., and Laflamme, L. (2017). Image-based teleconsultation using smartphones or tablets: qualitative assessment of medical experts. *Emerg Med J.* 34 (2): 95–99. https://doi.org/10.1136/emermed-2015-205258.

Breuninger, M., van Ginneken, B., Philipsen, R.H.H.M. et al. (2014). Diagnostic accuracy of computer-aided detection of pulmonary tuberculosis in chest radiographs: a validation study from sub-Saharan Africa. *PLOS One* 9 (9): e106381. https://doi.org/10.1371/journal.pone.0106381.

Caffery, L.J., Armfield, N.R., and Smith, A.C. (2015). Radiological interpretation of images displayed on tablet computers: a systematic review. *Brit J Radiology* 88 (1050): 20150191. https://doi.org/10.1259/bjr.20150191.

Choi, B.G., Mukherjee, M., Dala, P. et al. (2011). Interpretation of remotely downloaded pocket-size cardiac ultrasound images on a web-enabled smartphone: validation against workstation evaluation. *J Am Soc Echocardiogr.* 24 (12): 1325–1330. https://doi.org/10.1016/j.echo.2011.08.007.

Choplin, R.H., Boehme, J.M., and Douglas Maynard, C. (1992). PACS mini refresher course. *Radiographics* 12: 127–129. https://doi.org/10.1148/radiographics.12.1.1734458.

Delft Imaging Systems. n.d. "One Stop TB Clinics. Accessed October 20, 2017. http://www .delft.care/onestoptb-clinics

DICOM: Digital Imaging and Communications in Medicine. n.d. Accessed October 30, 2017. http://dicom.nema.org/dicom/about-DICOM.html

Eijkman, W. and van Doren Geneva, F. (2009). Innovative Chest X-Ray Solutions Supporting TB Prevalence Studies. http://www.who.int/tb/advisory_bodies/impact_ measurement_taskforce/meetings/prevalence_survey/chest_x_ray_solutions.pdf

Fink, C., Hallscheidt, P.J., Noeldge, G. et al. (2002). Clinical comparative study with a large-area amorphous silicon flat-panel detector: image quality and visibility of anatomic structures on chest radiography. *Am J Roentgenol* 178 (2): 481–486. https://doi. org/10.2214/ajr.178.2.1780481.

Floyd, C.E., Warp, R.J., Dobbins, J.T. et al. (2001). Imaging characteristics of an amorphous silicon flat-panel detector for digital chest radiography. *Radiology* 218 (3): 683–688. https://doi.org/10.1148/radiology.218.3.r01fe45683.

Garra, B.S. (2014). Information technology in global health radiology. In: *Radiology in Global Health.* (ed. D.J. Mollura and M.P. Lungren), 61–74. New York, NY: Springer doi: 10.1007/978-1-4614-0604-4_8.

Garra, B.S., DeStigter, K., Maguire, S., et al. (2008). Nov 30. Data compression for transmission of volume imaging from developing countries. Presented at the 2008 Convention of the Radiological Society of North America; Chicago, IL, USA.

Goost, H., Witten, J., Heck, A. et al. (2012). Image and diagnosis quality of X-ray image transmission via cell phone camera: a project study evaluating quality and reliability." Edited by H. Peter Soyer. *PLOS One* 7 (10): e43402. https://doi.org/10.1371/journal. pone.0043402.

Gudkova, O. n.d. Training Opportunities Increase for Developing Country Radiologists. Accessed October 30, 2017. https://www.iaea.org/newscenter/news/training-opportunities-increase-for-developing-country-radiologists

Heuvelings, C.C., de Vries, S.G., Greve, P.F. et al. (2017). Effectiveness of interventions for diagnosis and treatment of tuberculosis in hard-to-reach populations in countries of low and medium tuberculosis incidence: a systematic review. *Lancet Infect Dis.* 17 (5): e144–e158. https://doi.org/10.1016/S1473-3099(16)30532-1.

Hirschorn, D.S., Choudhri, A.F., Shih, G., and Kim, W. (2014). Use of mobile devices for medical imaging. *J Am Col Radiology.* 11 (12): 1277–1285. https://doi.org/10.1016/ j.jacr.2014.09.015.

Hockel, D. and Hamilton, T. (2011). Understanding Total Cost of Ownership. https://www .hpnonline.com/inside/2011-09/1109-equipplan-tco.html

Jaeger, S., Karargyris, A., Candemir, S. et al. (2014). Automatic tuberculosis screening using chest radiographs. *IEEE Transact Med Imag* 33 (2): 233–245. https://doi.org/10.1109/ TMI.2013.2284099.

Jiménez, P., Medlen, K.P., and Fleitas-Estévez, I. (2014). Diagnostic imaging for Global Health: implementation and optimization of radiology in the developing world. In: *Radiology in Global Health* (ed. D.J. Mollura and M.P. Lungren), 127–137. New York, NY: Springer doi: 10.1007/978-1-4614-0604-4_13.

Jones, J. (2016). Imaging 3.0 Case Study: Direct Image Transfer. https://www.acr.org/ Advocacy/Economics-Health Policy/Imaging-3/Case-Studies/IT/Direct-Image-Transfer

Kahn, C., Carrino, J., Flynn, M. et al. (2007). DICOM and radiology: past, present and future. *J Am Col Radiology* 4 (9): 652–657.

KNCV Tuberculosis Foundation. (2008). Working Document on Chest X-Ray Equipment for Use in TB Prevalence Surveys. http://www.who.int/tb/advisory_bodies/impact_measurement_taskforce/meetings/prevalence_survey/chest_x_ray_eqpt.pdf

Krohn, R. (2010). There's an app for that – MHealth takes center stage. *J Healthcare Info Manag.* 24 (3): 9–10.

Lee, O.Y.-C., Houdini, H.T.W., Donoghue, H.D. et al. (2012). Mycobacterium tuberculosis complex lipid virulence factors preserved in the 17,000-year-old skeleton of an extinct bison, bison Antiquus. *PLOS One* 7 (7): e41923. https://doi.org/10.1371/journal.pone.0041923.

Maduskar, P., Muyoyeta, M., Ayles, H. et al. (2013). Detection of tuberculosis using digital chest radiography: automated reading vs. interpretation by clinical officers. *Int J Tubercul Lung Dis* 17 (12): 1613–1620. https://doi.org/10.5588/ijtld.13.0325.

Malkin, R. and Teninty, B. (2014). Medical imaging in the global public health: donation, procurement, installation, and maintenance. In: *Radiology in Global Health* (ed. D.J. Mollura and M.P. Lungren), 33–39. New York, NY: Springer doi: 10.1007/978-1-4614-0604-4_6.

MinXray, Inc. (2015). MinXray: Medical. http://www.minxray.com/products/medical

Morishita, F., Garfin, A.M.C.G., Lew, W. et al. (2017). Bringing state-of-the-art diagnostics to vulnerable populations: the use of a mobile screening unit in active case finding for tuberculosis in Palawan, the Philippines. Edited by Madhukar Pai. *PLOS One* 12 (2): e0171310. https://doi.org/10.1371/journal.pone.0171310.

Muyoyeta, M., Maduskar, P., Moyo, M. et al. (2014). The sensitivity and specificity of using a computer aided diagnosis program for automatically scoring chest X-rays of presumptive TB patients compared with Xpert MTB/RIF in Lusaka Zambia. Edited by Robert J. Wilkinson. *PLOS One* 9 (4): e93757. https://doi.org/10.1371/journal.pone.0093757.

Perry, L. and Malkin, R. (2011). Effectiveness of medical equipment donations to improve health systems: how much medical equipment is broken in the developing world? *Med Bio Engineer Comp.* 49 (7): 719–722. https://doi.org/10.1007/s11517-011-0786-3.

RAD-AID. n.d. A Nonprofit Public Service RAD-AID.Org Radiology Serving the Wolrd. Accessed October 30, 2017. https://www.rad-aid.org/programs

Rajpurkar, P., Irvin, J., Zhu, K., et al. (2017). CheXNet: Radiologist-Level Pneumonia Detection on Chest X-Rays with Deep Learning. https://arxiv.org/pdf/1711.05225.pdf

Rong, X.J., Shaw, C.C., Liu, X. et al. (2001). Comparison of an amorphous silicon/cesium iodide flat-panel digital chest radiography system with screen/film and computed radiography systems--a contrast-detail phantom study. *MedPhysics* 28 (11): 2328–2335. https://doi.org/10.1118/1.1408620.

Scheuermeyer, F., Grunau, B., Cheyne, J. et al. (2016). Speed and accuracy of mobile BlackBerry messenger to transmit chest radiography images from a small community emergency department to a geographically remote referral center. *J Telemed Telecare* 22 (4): 244–251. https://doi.org/10.1177/1357633X15595734.

Schwartz, A.B., Siddiqui, G., Barbieri, J.S. et al. (2014). The accuracy of mobile Teleradiology in the evaluation of chest X-rays. *J Telemed Telecare* 20 (8): 460–463. https://doi.org/10.1177/1357633X14555639.

Shah, N. (2014). Access to imaging Technology in the Developing World. In: *Radiology in Global Health* (ed. D.J. Mollura and M.P. Lungren), 13–17. New York, NY: Springer doi: 10.1007/978-1-4614-0604-4_3.

Shen, R., Cheng, I., and Basu, A. (2010). A hybrid knowledge-guided detection technique for screening of infectious pulmonary tuberculosis from chest radiographs. *IEEE Trans Biomed Engineering* 57 (11): 2646–2656. https://doi.org/10.1109/TBME.2010.2057509.

Sonoda, M., Takano, M., Miyahara, J., and Kato, H. (1983). Computed radiography utilizing scanning laser stimulated luminescence. *Radiology* 148 (3): 833–838. https://doi.org/10.1148/radiology.148.3.6878707.

Stahl, I., Dreyfuss, D., Ofir, D. et al. (2017). Reliability of smartphone-based Teleradiology for evaluating thoracolumbar spine fractures. *Spine J* 17 (2): 161–167. https://doi.org/10.1016/j.spinee.2016.08.021.

Statistica. n.d. Mobile Messenger Apps – Statistics & Facts. Accessed October 30, 2017. https://www.statista.com/topics/1523/mobile-messenger-apps

Steiner, A., Mangu, C., van den Hombergh, J. et al. (2015). Screening for pulmonary tuberculosis in a Tanzanian prison and computer-aided interpretation of chest X-rays. *Pub Health Action* 5 (4): 249–254. https://doi.org/10.5588/pha.15.0037.

Teltscher, S., Gray, V., van Welsum, D., Biggs, P. and Magpantay, E. n.d. World Telecommunication/ICT Development Report 2010. Accessed October 30, 2017. http://www.itu.int/ITU-D/ict/publications/wtdr_10/material/WTDR2010_e.pdf

Toomey, R.J., Rainford, L.A., Leong, D.L. et al. (2014). Is the iPad suitable for image display at American Board of Radiology Examinations? *Am J Roentgenol.* 203 (5): 1028–1033. https://doi.org/10.2214/AJR.13.12274.

Tuijn, C.J., Hoefman, B.J., van Beijma, H. et al. (2011). Data and image transfer using mobile phones to strengthen microscopy-based diagnostic Services in low and Middle Income Country Laboratories. Edited by Charles Jonathan Woodrow. *PLOS One* 6 (12): e28348. https://doi.org/10.1371/journal.pone.0028348.

Waters, H. (2011). The First X-Ray, 1895. https://www.the-scientist.com/?articles.view/articleNo/30693/title/The-First-X-ray--1895

WhatsApp. n.d. www.whatsapp.com

World Health Organization (2012). Local Production and Technology Transfer to Increase Access to Medical Devices: Addressing the Barriers and Challenges in Low- and Middle-Income Countries. http://www.who.int/medical_devices/1240EHT_final.pdf

World Health Organization (2017). Tuberculosis: Global Tuberculosis Report 2017. http://www.who.int/tb/publications/factsheet_global.pdf?ua=1

World Health Organization (2017). Diagnostic Imaging: Global Collaboration. Accessed October 30, 2017. http://www.who.int/diagnostic_imaging/collaboration/en

Webliography

https://doi.org/10.1118/1.598630Aufrichtig, R. (1999). Comparison of low contrast detectability between a digital amorphous silicon and a screen-film based imaging system for thoracic radiography. *Med Physic.s* 26 (7): 1349–1358.

https://doi.org/10.2214/ajr.181.4.1810923Bacher, K., Smeets, P., Bonnarens, K. et al. (2003). Dose reduction in patients undergoing chest imaging: digital amorphous silicon flat-panel detector radiography versus conventional film-screen radiography and phosphor-based computed radiography. *Am J Roentgenol* 181 (4): 923–929.

https://doi.org/10.1136/pgmj.2005.038448Bansal, G.J. (2006). Digital radiography: a comparison with modern conventional imaging. *Postgrad Med J.* 82 (969): 425–428.

https://doi.org/10.1136/emermed-2015-205258Boissin, C., Blom, L., Wallis, L., and Laflamme, L. (2017). Image-based teleconsultation using smartphones or tablets: qualitative assessment of medical experts. *Emerg Med J.* 34 (2): 95–99.

https://doi.org/10.1371/journal.pone.0106381Breuninger, M., van Ginneken, B., Philipsen, R.H.H.M. et al. (2014). Diagnostic accuracy of computer-aided detection of pulmonary tuberculosis in chest radiographs: a validation study from sub-Saharan Africa. *PLOS One* 9 (9): e106381.

https://doi.org/10.1259/bjr.20150191Caffery, L.J., Armfield, N.R., and Smith, A.C. (2015). Radiological interpretation of images displayed on tablet computers: a systematic review. *Brit J Radiol* 88 (1050): 20150191.

https://doi.org/10.1016/j.echo.2011.08.007Choi, B.G., Mukherjee, M., Dala, P. et al. (2011). Interpretation of remotely downloaded pocket-size cardiac ultrasound images on a web-enabled smartphone: validation against workstation evaluation. *J Am Soc Echocardiogra.* 24 (12): 1325–1330.

https://doi.org/10.1148/radiographics.12.1.1734458Choplin, R.H., Boehme, J.M., and Douglas Maynard, C. (1992). PACS mini refresher course. *Radiographics* 12: 127–129.

http://www.delft.care/onestoptb-clinics Delft Imaging Systems. n.d. "One Stop TB Clinics. Accessed October 20, 2017. http://www.delft.care/onestoptb-clinics

http://dicom.nema.org/dicom/about-DICOM.html DICOM: Digital Imaging and Communications in Medicine. n.d. Accessed October 30, 2017.

http://www.who.int/tb/advisory_bodies/impact_measurement_taskforce/meetings/prevalence_survey/chest_x_ray_solutions.pdf Eijkman, W. and van Doren Geneva, F. 2009. Innovative Chest X-Ray Solutions Supporting TB Prevalence Studies.

https://doi.org/10.2214/ajr.178.2.1780481Fink, C., Hallscheidt, P.J., Noeldge, G. et al. (2002). Clinical comparative study with a large-area amorphous silicon flat-panel detector: image quality and visibility of anatomic structures on chest radiography. *Am J Roentgenol.* 178 (2): 481–486.

https://doi.org/10.1148/radiology.218.3.r01fe45683Floyd, C.E., Warp, R.J., Dobbins, J.T. et al. (2001). Imaging characteristics of an amorphous silicon flat-panel detector for digital chest radiography. *Radiology* 218 (3): 683–688.

https://doi.org/10.1007/978-1-4614-0604-4_8Garra, B.S. (2014). Information technology in global health radiology. In: *Radiology in Global Health* (ed. D.J. Mollura and M.P. Lungren), 61–74. New York, NY: Springer.

https://doi.org/10.1371/journal.pone.0043402Goost, H., Witten, J., Heck, J. et al. (2012). Image and diagnosis quality of X-ray image transmission via cell phone camera: a project study evaluating quality and reliability. Edited by H. Peter Soyer. *PLoS One.* 7 (10): e43402.

https://www.iaea.org/newscenter/news/training-opportunities-increase-for-developing-country-radiologists Gudkova, O. n.d. Training Opportunities Increase for Developing Country Radiologists. Accessed October 30, 2017.

https://doi.org/10.1016/S1473-3099(16)30532-1Heuvelings, C.C., de Vries, S.G., Greve, P.F. et al. (2017). Effectiveness of interventions for diagnosis and treatment of tuberculosis in hard-to-reach populations in countries of low and medium tuberculosis incidence: a systematic review. *Lancet Infect Dis.* 17 (5): e144–e158.

https://doi.org/10.1016/j.jacr.2014.09.015Hirschorn, D.S., Choudhri, A.F., Shih, G., and Kim, W. (2014). Use of mobile devices for medical imaging. *J Am Col Radiol.* 11 (12): 1277–1285.

https://www.hpnonline.com/inside/2011-09/1109-equipplan-tco.html Hockel, D. and Hamilton, T. 2011. "Understanding Total Cost of Ownership.

https://doi.org/10.1109/TMI.2013.2284099Jaeger, S., Karargyris, A., Candemir, S. et al. (2014). Automatic tuberculosis screening using chest radiographs. *IEEE Transact Med Imaging.* 33 (2): 233–245.

https://doi.org/10.1007/978-1-4614-0604-4_13Jiménez, P., Medlen, K.P., and Fleitas-Estévez, I. (2014). Diagnostic imaging for Global Health: implementation and optimization of radiology in the developing world. In: *Radiology in Global Health* (ed. D.J. Mollura and M.P. Lungren), 127–137. New York, NY: Springer.

https://www.acr.org/Advocacy/Economics-Health Policy/Imaging-3/Case-Studies/IT/Direct-Image-Transfer Jones, J. 2016. Imaging 3.0 Case Study: Direct Image Transfer.

http://www.who.int/tb/advisory_bodies/impact_measurement_taskforce/meetings/prevalence_survey/chest_x_ray_eqpt.pdfKahn, C., Carrino, J., Flynn, M. et al. (September 2007). DICOM and radiology: past, present and future. *J Am Col Radiol* 4 (9): 652–657.

https://doi.org/10.1371/journal.pone.0041923Lee, O.Y.-C., Houdini, H.T.W., Donoghue, H.D. et al. (2012). Mycobacterium tuberculosis complex lipid virulence factors preserved in the 17,000-year-old skeleton of an extinct bison, bison Antiquus. *PLoS One* 7 (7): e41923.

https://doi.org/10.5588/ijtld.13.0325Maduskar, P., Muyoyeta, M., Ayles, H. et al. (2013). Detection of tuberculosis using digital chest radiography: automated reading vs. interpretation by clinical officers. *Int J Tubercu Lung Dis.* 17 (12): 1613–1620.

https://doi.org/10.1007/978-1-4614-0604-4_6Malkin, R. and Teninty, B. (2014). Medical imaging in the global public health: donation, procurement, installation, and maintenance. In: *Radiology in Global Health* (ed. D.J. Mollura and M.P. Lungren), 33–39. New York, NY: Springer.

http://www.minxray.com/products/medical MinXray, Inc. 2015. MinXray: Medical.

https://doi.org/10.1371/journal.pone.0171310Morishita, F., Garfin, A.M.C.G., Lew, W. et al. (2017). Bringing state-of-the-art diagnostics to vulnerable populations: the use of a mobile screening unit in active case finding for tuberculosis in Palawan, the Philippines. Edited by Madhukar Pai. *PLOS One* 12 (2): e0171310.

https://doi.org/10.1371/journal.pone.0093757Muyoyeta, M., Maduskar, P., Moyo, M. et al. (2014). The sensitivity and specificity of using a computer aided diagnosis program for automatically scoring chest X-rays of presumptive TB patients compared with Xpert MTB/RIF in Lusaka Zambia. Edited by Robert J. Wilkinson. *PLOS One* 9 (4): e93757.

https://doi.org/10.1007/s11517-011-0786-3Perry, L. and Malkin, R. (2011). Effectiveness of medical equipment donations to improve health systems: how much medical equipment is broken in the developing world? *Med Bio Engineer Comp.* 49 (7): 719–722.

https://www.rad-aid.org/programs RAD-AID. n.d. A Nonprofit Public Service RAD-AID. Org Radiology Serving the Wolrd. Accessed October 30, 2017.

https://arxiv.org/pdf/1711.05225.pdf Rajpurkar, P., Irvin, J., Zhu, K., et al. 2017. CheXNet: Radiologist-Level Pneumonia Detection on Chest X-Rays with Deep Learning.

https://doi.org/10.1118/1.1408620Rong, X.J., Shaw, C.C., Liu, X. et al. (2001). Comparison of an amorphous silicon/cesium iodide flat-panel digital chest radiography system with screen/film and computed radiography systems--a contrast-detail phantom study. *Med Physics.* 28 (11): 2328–2335.

https://doi.org/10.1177/1357633X15595734Scheuermeyer, F., Grunau, B., Cheyne, J. et al. (2016). Speed and accuracy of mobile BlackBerry messenger to transmit chest radiography images from a small community emergency department to a geographically remote referral center. *J Telemed Telecare.* 22 (4): 244–251.

https://doi.org/10.1177/1357633X14555639Schwartz, A.B., Siddiqui, G., Barbieri, J.S. et al. (2014). The accuracy of mobile Teleradiology in the evaluation of chest X-rays. *J Telemed Telecare* 20 (8): 460–463.

https://doi.org/10.1007/978-1-4614-0604-4_3Shah, N. (2014). Access to imaging Technology in the Developing World. In: *Radiology in Global Health* (ed. D.J. Mollura and M.P. Lungren), 13–17. New York, NY: Springer.

https://doi.org/10.1109/TBME.2010.2057509Shen, R., Cheng, I., and Basu, A. (2010). A hybrid knowledge-guided detection technique for screening of infectious pulmonary tuberculosis from chest radiographs. *IEEE Transact Biomed Engineer.* 57 (11): 2646–2656.

https://doi.org/10.1148/radiology.148.3.6878707Sonoda, M., Takano, M., Miyahara, J., and Kato, H. (1983). Computed radiography utilizing scanning laser stimulated luminescence. *Radiology* 148 (3): 833–838.

https://doi.org/10.1016/j.spinee.2016.08.021Stahl, I., Dreyfuss, D., Ofir, D. et al. (2017). Reliability of smartphone-based Teleradiology for evaluating thoracolumbar spine fractures. *Spine J.* 17 (2): 161–167.

https://www.statista.com/topics/1523/mobile-messenger-apps Statistica. n.d. Mobile Messenger Apps – Statistics & Facts. Accessed October 30, 2017.

https://doi.org/10.5588/pha.15.0037Steiner, A., Mangu, C., van den Hombergh, J. et al. (2015). Screening for pulmonary tuberculosis in a Tanzanian prison and computer-aided interpretation of chest X-rays. *Pub Health Act.* 5 (4): 249–254.

http://www.itu.int/ITU-D/ict/publications/wtdr_10/material/WTDR2010_e.pdf Teltscher, S., Gray, V., van Welsum, D., Biggs, P. and Magpantay, E. n.d. "World Telecommunication/ICT Development Report 2010. Accessed October 30, 2017.

https://doi.org/10.2214/AJR.13.12274Toomey, R.J., Rainford, L.A., Leong, D.L. et al. (2014). Is the IPad suitable for image display at American Board of Radiology Examinations? *Am J Roentgenol.* 203 (5): 1028–1033.

https://doi.org/10.1371/journal.pone.0028348Tuijn, C.J., Hoefman, B.J., van Beijma, H. et al. (2011). Data and image transfer using mobile phones to strengthen microscopy-based diagnostic Services in low and Middle Income Country Laboratories. Edited by Charles Jonathan Woodrow. *PLoS One* 6 (12): e28348.

https://www.the scientist.com/?articles.view/articleNo/30693/title/The-First-X-ray--1895 Waters, H. 2011. The First X-Ray, 1895.

www.whatsapp.com. WhatsApp. n.d.

http://www.who.int/medical_devices/1240EHT_final.pdf World Health Organization. 2012. Local Production and Technology Transfer to Increase Access to Medical Devices: Addressing the Barriers and Challenges in Low- and Middle-Income Countries.

http://www.who.int/tb/publications/factsheet_global.pdf?ua=1 World Health Organization. 2017. Tuberculosis: Global Tuberculosis Report 2017.

http://www.who.int/diagnostic_imaging/collaboration/en World Health Organization. 2017. n.d. Diagnostic Imaging: Global Collaboration. Accessed October 30, 2017.

Part VI

The Future

31

An Introduction to Digital Health

Kerry Atkinson[1, 2]

[1] *University of Queensland Centre for Clinical Research, Brisbane, Queensland, Australia*
[2] *The University of Technology/Institute of Health and Biomedical Innovation, Brisbane, Queensland, Australia*

Ever tried. Ever failed. No matter. Try Again. Fail again. Fail better.
From Worstwood Ho (1984) by Samuel Beckett. Nobel Laureate in Literature, 1969.

There's no success like failure, and failure's no success at all.
From Love Minus Zero/No Limit from the album by Bob Dylan Bringing it all back home (1965). Nobel Laureate in Literature, 2016.

CHAPTER MENU

Revolutionizing Tropical Medicine: Point-of-Care Tests, New Imaging Technologies and Digital Health,
First Edition. Edited by Kerry Atkinson and David Mabey.
© 2019 John Wiley & Sons, Inc. Published 2019 by John Wiley & Sons, Inc.

31.1 Introduction

Digital health is a collective term for eHealth and mHealth technologies. eHealth is the cost-effective and secure use of information and communication technology for health and health-related fields. mHealth (mobile health) is a component of eHealth involving the provision of health services and information via mobile technologies such as mobile phones, tablet computers and personal digital assistants. eLearning is a term for the use of electronic technology in learning and teaching.

Cognitive computers have been developing rapidly over the last few years following three technological breakthroughs. First, cheap parallel computation due to a new kind of chip called a graphics processing unit (GPU). Secondly, accessible big data due to massive databases, web cookies, wearable devices and decades of search results. Thirdly, the building of better algorithms by companies such as Google, Amazon and Microsoft, often in conjunction with academic partners or other organizations.

Artificial intelligence, wearable sensors, virtual reality and medical robots are disruptive technologies which are completely changing the way people think and act about healthcare.

This chapter describes the tools available for digital healthcare delivery. The following chapter delves into this in much greater detail.

31.2 The Pillars and Components of Digital Health for Use in LMICs

A good example of the use of digital health in the low- and middle-income countries (LMICs) is a project organized by the World Health Organization in order to attain the goal of ending tuberculosis globally (Falzon et al. 2016). The authors of the program identified three policy pillars, each with a set of components. The three pillars are:

1) Integrated patient-centered care and prevention
2) Bold policies and supportive systems
3) Intensified research and innovation

The components to support the first pillar include automated laboratory results, video-observed therapy, short message services (SMS) and e-learning for staff.

The components to support the second pillar include e-learning for patients, mobile telephone credit as an enabler, e-notification of disease and provision of a unique identifier for patients.

The components to support the third pillar include add-on hardware to smartphones to permit clinical assessment. Examples of this are already in use. They include PEEK Vision for measuring visual acuity, fields of vision and retinal photography in remote rural areas (see Chapter 25). Another example is the U-HEAR app on a smartphone for assessing deafness: it was found to be a feasible screening test to exclude significant hearing loss above a pure-tone average (PTA) of 40 dB. It was highly sensitive for detecting threshold changes at high frequencies, making it reasonably well suited for detecting presbycusis (age-related hearing loss) and ototoxic hearing loss from HIV, tuberculosis therapy or chemotherapy (Peer and Fagan 2015). In 2016 Clarius Mobile Health introduced the world's first handheld ultrasound scanner with a mobile application; the images can be displayed on an iOS or android smartphone. It is possible to carry around this ultrasound device for quick examinations and to guide procedures such as nerve blocks and targeted injections (Clarius Mobile Health: https://www.clarius.me (Figures 31.1 and 31.2).

Falzon and colleagues identified three stages for realizing their goals. The first was

1) Diagnostic connectivity by connecting medical diagnostic devices for digitization and transmission of data. This area is the current focus.
2) The second stage is data repository and gateway, requiring secure data storage and routing. This stage has not yet begun.
3) The third future goal is data presentation and application, involving support for timely and actionable data use such as notification to patient records and to a medical records system.

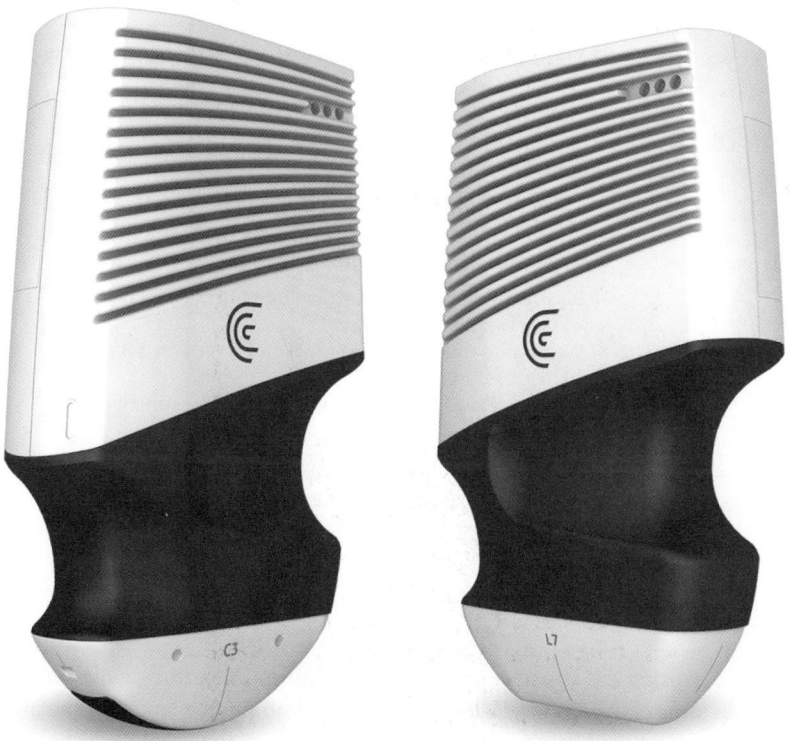

Figure 31.1 The Clarius handheld ultrasound scanner with a mobile application.

Figure 31.2 Ultrasound images captured by the Clarius handheld ultrasound scanner and displayed on iOS or android smartphones or tablets.

The characteristics of the desired digital health products were described using target product profiles (TPPs) (Table 31.1).

Table 31.1 Summary of target product profiles (TPPs) for the End TB Strategy (as of February 2016).

Function	Target product profiles
Patient care	1) Video treatment support/video-observed therapy (VOT) for TB patients via mobile telephones. 2) eHealth portal to improve TB and tobacco abuse.
Surveillance and monitoring	3) Digital dashboard for TB indicators and epidemiological trends. 4) Digital notification of TB cases. 5) Digital application for active TB drug safety monitoring.
Programme management	6) Diagnostic device connectivity for TB.
eLearning	7) Information resources platform for patients on TB and smoking cessation programs. 8) Web-based training for health professionals on TB and smoking cessation programs. 9) Clinical decision support systems for TB treatment and smoking cessation programs.

Source: Adapted with permission from Falzon et al. (2016).
Abbreviations used: eLearning, electronic learning; VOT, video (virtually) observed therapy;
TB, tuberculosis; eHealth, electronic health

31.3 Smartphones and Internet Access

As in the developed world, the most commonly used communication device in the LMICs is the mobile phone. By 2015 half the world's population had a smartphone and 40% could access the Internet. Morgan Stanley analysts predict that there will be 266 million Apple iPhones sold in 2018 with a current user base of 600 million users worldwide. Table 31.2 shows the uptake of mobile phones in sub-Saharan Africa (The Australian, 11 September 2017; theaustralian.com.au/wsj).

Table 31.2 The uptake of mobile phones in sub-Saharan Africa in September 2017.

Country	Mobile phone subscribers (millions)	Penetration (%)
Nigeria	84	45
South Africa	37	68
Ethiopia	35	34
Kenya	28	59
Tanzania	22	42
Democratic Republic of Congo	20	26
Ghana	18	67
Uganda	17	41
Mozambique	13	53

The Red Cross in Haiti used mobile phones to send important health information to millions of Haitians. It used mobile phones for a nationwide malaria prevention campaign, sending over 3.5 million SMS messages which included information on how malaria is transmitted, how to recognize its symptoms, treatment options and simple steps for prevention.

The Apple iPhone 8, 8s and the iPhone X were released on 12 September 2017 and are able to non-invasively measure blood sugar and HbA1c levels. This can be done by downloading the free app "Sugar Sense Diabetes" from the Apple App Store™ to any device using the iOS platform.

G Medical Innovations in Australia has its product Prisma approved by the FDA, the EU and China. (China is the world's biggest smartphone market.) Prisma is a sleeve that fits over a smartphone and, once there, continually consolidates and analyses medical data from its wearer. The sleeve can measure heart rate, heart rhythms, body temperature and blood oxygen and perform electrocardiograms and stress tests via touchpads on the phone once the Prisma app has been downloaded to the phone from the Apple App Store or Google Play. These data can then be sent to a health professional for interpretation and advice. The company is working on expanding its range of applications including measurement of blood glucose, uric acid and cholesterol (http://gmedinnovations.com).

31.4   Wearables

Wearable sensors and devices can stream data to a doctor's smartphone, notifying them whenever vital signs become abnormal. These will make it possible to offer treatment advice remotely. In turn, this will increase the time health providers have to treat and advise patients locally. Smart algorithms will ensure that health providers can tap expert advice on rare diseases and act as a gatekeeper for specialist treatment. Wearable devices that monitor the mother's vital signs and those of her unborn baby will ensure that, in an emergency, delivering care will no longer depend on being close to a healthcare facility.

Molly, developed by the biotechnology company Sens.ly, is the world's first virtual nurse (https://www.crunchbase.com/organization/sense-ly). On a smartphone or portable computer screen she has a smiling, amiable face and a pleasant voice. The goal is to help people monitor their condition and treatment. The interface uses machine learning to support patients with chronic conditions in between doctor's visits. It provides customized monitoring and follow-up care and has a strong focus on chronic diseases.

31.4.1 The Apple Watch Series 3 and 4

The Apple Watch Series 3 (https://www.apple.com/au/apple-watch-series-3) can monitor the heart rate and the heart rhythm of the person wearing it (Figure 31.3).

A study conducted using the heartbeat measurement app "Cardiogram" downloaded to an Apple Watch Series 3 found the Apple Watch to be 97% accurate in detecting the most common abnormal heart rhythm when paired with an AI-based algorithm. The study involved 6158 participants who were recruited through the Cardiogram app on the Apple Watch. Two hundred of these people had been diagnosed previously as having atrial fibrillation. Engineers then trained a deep neural network to identify abnormal heart rhythms from Apple Watch heart rate data.

The Apple Watch Series 4 was released in the USA in 2018. It can produce the wearer's electrocardiogram on demand. Both the Series 3 and the Series 4 Watches have been approved by the USA Food and Drug Administration for these features.

More people, including older populations most prone to stroke risk, are starting to use wearable technology such as Fitbit or the Apple Watch, both of which can double as heart monitors.

31.4.2 AliveCor's ECG Reader

The Mayo Clinic conducted a study involving artificial intelligence (AI) and Alive-Cor's version of an electrocardiogram (ECG) reader, which is attached to the back of

Figure 31.3 The Apple Watch Series 3.

a smartphone and which uses the "Kardia" app to detect abnormal heart rhythms (http://www.alivetec.com/alivecor-heart-monitor). It was found that this device was as good as other ECG devices used in the doctors' offices.

31.4.3 The MOCAheart Device

The MOCAheart device pairs with a matching app running on a tablet or smartphone via Bluetooth (MOCAheart Consumer Cardiac Monitor: A Medgadget Review. https://www.medgadget.com/2015/10/mocaheart-consumer-cardiac-monitor-medgadget-review.html).

Pressing the only button on the device activates it. If one thumb covers the button and the other thumb covers the flat metal part, the app recognizes that the device is activated and that readings are being captured (Figure 31.4). The app displays a countdown timer that ensures that the time required to measure the average heart rate and blood oxygen saturation is sufficient. The app then displays the results on the smartphone screen. It retails on Amazon.com for US$99.00.

31.4.4 Digital Contact Lens for Measuring Glucose Levels

The digital contact lens patented by Google and Novartis aims to change the course of diabetes management by measuring blood glucose levels from tears. Google and Novartis said the lens would contain a tiny and ultra slim microchip embedded in one of its thin concave sides. Through its equally tiny antenna, it would send data about glucose measurements from the user's tears to his or her paired smartphone via installed software.

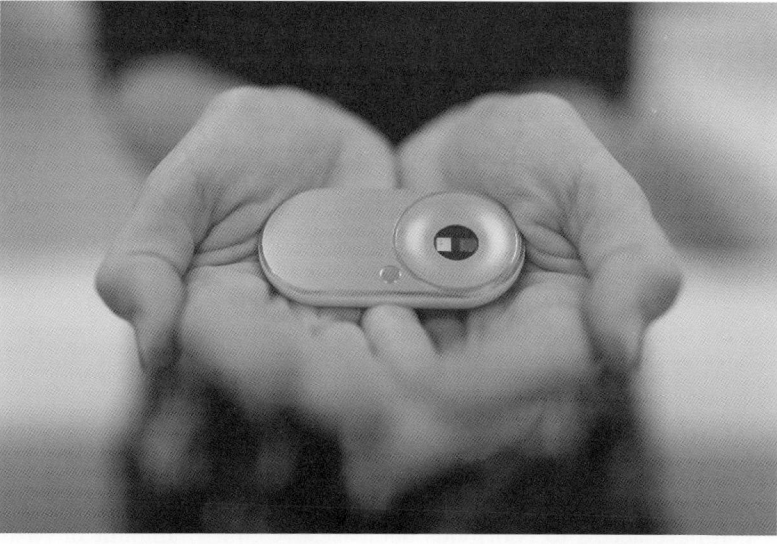

Figure 31.4 The MOCAheart device.

31.5   Personal Digital Assistants and Chatbots

Personal digital health assistants and medical chatbots may significantly ease the burden on doctors. A chatbot (also known as a talkbot, chatterbot, Bot, IM bot, interactive agent, or Artificial Conversational Entity) is a computer program which conducts a conversation by auditory or textual methods. It is likely that patients will turn to chatbots with simple questions about their health or specific drugs. The United Kingdom's National Health Service (NHS) has already recognized the potential of chatbots. The NHS started to use a chatbot app for dispensing medical advice in 2017, in order to mitigate the pressure on its 111 non-emergency helplines. The NHS also developed an app with Babylon Health, a paid, doctor-on-demand service. Bots such as HealthTap (www.healthtap.com) or Your.Md (www.your.md) aim to help patients find a solution to the most common symptoms through AI.

31.6   Augmented Reality

Augmented reality (AR) creates the illusion that a virtual object is in the physical world. One of the first example of this was by the furniture manufacturer IKEA in which you could pick an object from their catalog and then see how it looked in your living room – all using a smartphone. The Apple iPhones, released since September 2017, have AR software.

A clinic in Germany started using AR during operations; surgeons could see through anatomical structures such as blood vessels in the liver without opening organs. This enabled them to perform more precise excisions.

Another example of AR in medicine is a pill bottle that glows blue when a medication dose should be taken and red when a dose is missed.

The intelligent surgical knife, iKnife, works by using an old technology by which an electrical current heats tissue to make incisions with minimal blood loss (https://en .wikipedia.org/wiki/Iknife). With the iKnife the vaporized smoke is analyzed by a mass spectrometer to detect the chemicals in the biological sample. This means it can identify whether the tissue is malignant or not in real time.

31.7   Big Data

With the evolution of digital capacity, more and more data are produced and stored digitally. The amount of available digital data is doubling every two years. In 2013 it encompassed 4.4 zettabytes. However, by 2020 it is speculated to be 44 zettabytes (44 trillion gigabytes). Big data is so huge that AI is required to keep track of it.

31.8   Artificial Intelligence (AI)

Artificial intelligence is a term used for advanced computers that learn from their mistakes and which have access to enormous databases. A limited amount of AI is currently in general use. Examples include Google searches and Amazon suggestions.

Apple's Siri, Microsoft's Cortana, Google's OK Google and Amazon's Echo. These services are able to extract questions from speech using natural-language processing and then perform a limited set of tasks such as getting driving directions, finding an open slot for a meeting, or running a simple web search.

31.8.1   The Application of AI to Data Management

The most obvious application of AI in healthcare is data management. Collecting it, storing it, normalizing it and tracing its lineage. It is the first step in revolutionizing existing healthcare systems. Recently, the AI research branch of Google launched its Google DeepMind Health project. This is used to mine medical records data in order to provide better and faster health services. The project is in its initial phase. At present Google is also collaborating with Moorfields Eye Hospital NHS Foundation Trust, London, to improve treatment for eye diseases.

The company Enlitic (www.enlitic.com) aims to couple deep learning with huge stores of medical data to improve diagnostics and clinical outcomes. Using deep learning their system can readily handle a broad spectrum of diseases throughout the body as well as all imaging modalities including x-rays and CT scans.

31.8.2   The Increasing Complexity of Algorithms

Advances in imaging analysis are making algorithms increasingly capable of performing interpretations currently done by radiologists. Algorithms can analyze the pixel and other bits and bytes of data contained within the image to detect patterns associated with a specific pathology. The outcome of the algorithmic analysis is a metric. In the current early stage of imaging analytics, these metrics complement the analysis of the images made by radiologists, and help them render a more accurate or faster diagnosis.

Another example is calculation of bone mineral density by applying an algorithm on any CT image of a bone. The resulting number is then compared with a threshold metric to determine whether the patient is at risk of fracture. In the near future radiologists will be equipped with thousands of predictive algorithms to automatically detect the patterns of the most common diseases. This application of advanced data analysis also holds the prospect of preventing diseases. Evolving to this level of preventative care requires not only expertise in imaging analytics to develop the algorithms, but also access to huge libraries of images to refine the algorithms. Like other AI tools, the algorithms will evolve, becoming more knowledgeable and more accurate as they analyze more cases.

31.8.3   The Advent of Intelligent Machines

Analysts predict that intelligent machines, programmed to think and reason like the human mind, will revolutionize healthcare in the near future. Supercomputers can speed up the reading and interpretation of results from radiographs, electrocardiograms, ultrasound and CT scans, and even the analysis of blood samples. Given the significant shortage of health workers, the application of AI to healthcare could potentially reduce some of the burden on overloaded health staff. Emerging AI technologies are shown in Table 31.3.

Table 31.3 Emerging AI technologies.

Function	Components	Examples
Sense	Computer vision Audioprocessing	Virtual agents Identity analytics
Comprehension	Natural language processing Knowledge representation	Cognitive robotics Speech analytics
Action	Machine learning Expert systems	Data visualization

AI can mine medical records or medical images in order to discover previously unknown implications or signals, design treatment plans for cancer patients or create drugs from existing pills, or re-use old drugs for new purposes.

IBM's supercomputer is called Watson (https://www.ibm.com/watson). Watson for Oncology has an advanced ability to analyze the meaning and context of structured and unstructured data in clinical notes and reports that may be critical to selecting a treatment pathway. Thus, AI is not making the decision per se but offers the doctor the most rational options.

31.8.4 The Increasing Use of AI

The increasing use of AI has been enabled by

1) Unlimited access to computing power. Public cloud computing was estimated to realize almost US$70 billion in 2015 worldwide.
2) Data storage has also become abundant.
3) Growth in big data. Global data has seen a compound annual growth rate (CAGR) of more than 50% since 2010 as more devices have become connected.

AI can analyze entire healthcare systems. For example, 97% of healthcare invoices in the Netherlands are digital and contain data regarding the treatment, the doctor and the hospital used. These invoices can be easily retrieved. The company Zorgprisma Publiek analyzes the invoices and uses IBM Watson to mine the data. They can tell if a doctor, clinic or hospital makes mistakes repetitively in treating a certain type of condition.

IBM has also launched an algorithm called Medical Sieve qualified to assist in clinical decision making in radiology and cardiology (https://researcher.watson.ibm.com/researcher/view_group.php?id=4384). The "cognitive health assistant" is able to analyze radiology images to detect problems faster and more reliably than the doctor. After IBM purchased Merge Health in 2015, Watson got access to millions of radiology studies and a vast amount of existing medical record data. These were used to help train Watson in evaluating patient data and improving image interpretation.

In order to obtain some estimation of when machine learning might be introduced on a wider scale, it is useful to look at how machine learning takes place in radiology. Firstly, the algorithm should be fed by thousands or millions of images and will learn to detect differences in tissues. If the algorithm makes a mistake, the researcher notices it and adjusts the code. Thus, it is a lengthy process. The algorithm will probably create a structured, minable, preliminary report. Thus, it will do the quantification and will likely do it very well.

IBM is grooming Watson Health to help physicians make diagnoses. Tapping into streams of imaging data, Watson may be able to look for signs of disease and adjust scan parameters to optimize data acquisition. The USA FDA approved the first cloud-based deep learning algorithm for cardiac imaging developed by Arterys in 2017.

Another way of using AI in medicine is to turn groups of human experts into super experts at diagnosis. The algorithm creates a form of AI, called swarm AI, that helps radiologists form a consensus. Swarm AI has already proven effective in radiological applications. In one study a collective intelligence of radiologists reduced false positives and false negatives when interpreting mammograms. The study demonstrated that this swarm intelligence could improve mammography screening and had the potential to improve many other types of medical decision-making.

In another study a dozen radiologists increased their ability to diagnose skeletal abnormalities correctly. The researchers concluded that the algorithm's accuracy in distinguishing normal from abnormal images was significantly higher than the radiologists' mean accuracy. In developing the algorithm, the authors took note of examples in the natural world where a given species can accomplish more by participating in a flock, colony or swarm than they can individually.

31.8.5   Current Uses of AI

31.8.5.1   Athletes
The first generation of activity trackers focused on people who exercise regularly, but only provided basic insight into how they were performing. A second generation of devices were tailored to professional athletes – for example, Fitbit Blaze, GymWatch and Wahoo. With detailed insight into movement patterns and force output in any movement, sports medicine physicians will be able to acquire data to measure how athletes are improving. Additionally, video consoles from XBox to Microsoft Kinect offer a way of remotely monitoring the athlete.

31.8.5.2   Radiology
AI algorithms can improve radiologists' efficiency by prioritizing cases during the examination that require immediate medical attention. Built into the scanners themselves, these algorithms may improve the efficiency by determining which examinations should be performed, shortening exam times by ending data acquisition when certain thresholds – set by the radiologist – are met.

In the future it is likely that deep learning (DL) algorithms will be trained to recognize disease patterns, identifying and outlining masses, measuring them, and transposing the measurements into the radiology report. It is possible that DL/AI will direct the attention of radiologists to suspicious areas in images in order to make efficient use of radiologists' time, while reducing the possibility that disease might be overlooked. Additionally, DL algorithms, fashioned into clinical decision-making tools, may help doctors identify and narrow the choices of different scans based on clinical observations in the patient's electronic medical record (EMR).

31.8.5.3   Oncology
This specialty is currently paving the way for precision medicine and targeted treatments. Oncologists already customize therapies based on the patients' genetic background and the molecular signature of certain tumors. Cheaper genome sequencing

and measuring blood biomarkers are speeding up this process. Companies like GRAIL are developing fluid biopsies. These blood tests are able to detect all types of cancer from a very early stage.

31.8.5.4 Dermatology

With the cdevice idoc24.com, patients are able to photograph skin lesions and dermatologists can provide medical advice in a 24 hours online service (www.idoc24.com).

IBM's supercomputer Watson uses machine learning algorithms in conjunction with image recognition technology to detect patterns in skin moles. It has been partnered with a skin cancer detection program MoleMap and the Melanoma Institute of Australia to teach the computer how to recognize cancerous skin lesions. So far 41 000 melanoma images have been fed into the system with accompanying clinician notes. It had a 91% accuracy at detecting skin cancers. Using a DermaScope a trained clinician can identify a skin cancer with 80% accuracy.

31.8.5.5 Cardiology

The AliveCor device (www.alivecor.com) can record the user's heart rate and ECG recording through their fingertips and transmit the results to an app. If placed on a person's chest it can diagnose an acute myocardial infarction in real time.

Ng and colleagues from Stanford University developed a deep learning algorithm to detect 14 types of arrhythmia from ECG signals (Rajpurkar et al. submitted 2017). It was able to diagnose arrhythmias approximately as accurately as cardiologists and to outperform them in most cases. The group trained their algorithm on data collected from a wearable ECG monitor. Patients wore a small chest patch for two weeks and carried out their normal day-to-day activities while the device recorded each heartbeat. The group took approximately 30 000 30-second clips from various patients that represented a variety of arrhythmias. To test the accuracy of the algorithm, researchers gave a group of six expert cardiologists 300 undiagnosed clips and asked them to reach a consensus on any arrhythmias present in the recordings. The algorithm could then predict how those cardiologists would label every second of other ECGs with which it was presented, in essence, giving a diagnosis. The algorithm could be a step toward expert-level arrhythmia diagnosis for people who do not have access to a cardiologist, such as in many parts of LMICs. More immediately, the algorithm could be part of a wearable device that at-risk people keep on at all times that would alert emergency services to potentially fatal arrhythmias as they occur.

31.8.5.6 Gastroenterology

Capsule endoscopy is a new technology designed to examine the small bowel. The test involves swallowing a camera capsule (Pillcam) which is the size of a jelly bean (Figure 31.5) (http://www.givenimaging.com/en-int/Innovative-Solutions/Capsule-Endoscopy/Pages/default.aspx). The Pillcam takes multiple digital photos of the small bowel. The images are transmitted via sensors attached to the abdomen with adhesive stickers.

More than 1.5 million patients have already had this capsule endoscopy. Physicians can visualize the small bowel, enabling them to monitor and diagnose disorders of the upper gastrointestinal tract without sedation or invasive endoscopic procedures. The small bowel is about 6 m long and the capsule takes approximately eight hours to travel through it. It is normally passed per rectum without any side effects.

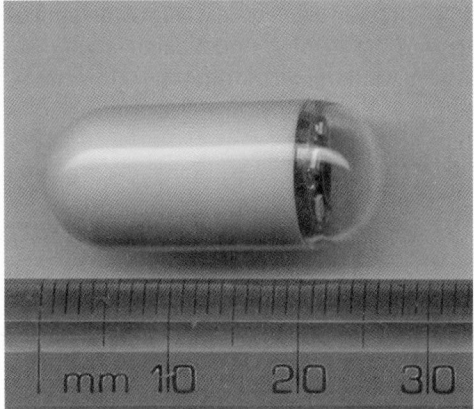

Figure 31.5 The Pillcam device.

31.8.5.7 Surgery

By 2020 sales of surgical robotics are expected to almost double to $6.4 billion. The most commonly known surgical robot is the da Vinci Surgical System featuring a magnified 3D high-definition vision system and tiny wristed instruments that can bend and rotate far more greatly than the human hand. The surgeon is 100% in control of the robotic system at all times; he or she is able to carry out more precise operations than previously thought possible.

31.8.5.8 Genetics and Genomics

Artificial intelligence will have a huge impact on genetics and genomics. Deep Genomics aims at identifying patterns in huge data sets of genetic information and medical records, looking for mutations and linkages to disease. A new generation of computational technologies is being developed that can tell doctors what will happen within a cell when DNA is altered by genetic variation, whether natural or therapeutic.

31.8.5.9 Medications

Developing pharmaceutical agents through clinical trials take sometimes more than a decade and costs billions of dollars. Speeding this up and making it more cost-effective would have a significant effect on healthcare and how innovations reach everyday medicine. Atomwise uses supercomputers that root out therapies from a database of molecular structures (www.atomwise.com; https://www.crunchbase.com/organization/atomwise). Last year Atomwise launched a virtual search for safe, existing medicines that could be redesigned to treat the Ebola virus. They found two drugs predicted by the company's AI technology which may significantly reduce Ebola infectivity. This analysis, which typically would have taken months or years, was completed in less than one day.

31.9 The Game Changer – A Smartphone with AI Access

At the 2017 Internationale Funkausstellung (IFA) Berlin Technology Show Huawei discussed their Kirin 970 smartphone which could access AI data on the phone itself (without using other computers). The chip will be 8 CPU and 12-core GPU with 5.5 billion transistors packed into $1\,\text{cm}^2$. It will need a large battery. Such a neural processing unit will be part of Huawei's goals for its next generation of high-end smartphones (The Australian, 7 September 2017, p. 14. http://theaustralian.com.au/life). Image recognition of photographs will be possible. Two other smartphones have similar specifics: Samsung's Galaxy 9 and Apple's iPhone X.

31.10   Conclusions

Given the power of super computers and superchips to mine and organize huge amounts of data, it is easy to envision a number of applications in the health sector. Machines can aid the work of doctors, organize, rationalize, and streamline the processes leading to a diagnosis or other medical decision, but AI is unlikely to replace doctors and nurses in their interactions with patients. Overall, however, the potential of AI in healthcare is enormous.

Bibliography

Clarius Mobile Health n.d. https://www.clarius.me

Falzon, D., Timimi, H., Kurosinski, P. et al. (2016). Digital health for the end TB strategy: developing priority products and making them work. *Eur. Respir. J.* 48: 29–45.

Peer, S. and Fagan, J.J. (2015). Hearing loss in the developing world: evaluating the iPhone mobile device as a screening tool. *S. Afr. Med. J.* 105: 35–39.

Rajpurkar, P, Hannun, A.Y., Haghpanahi, M., Bourn, C. and Ng, A.Y. Cardiologist-Level Arrhythmia Detection with Convolutional Neural Networks. arXiv: 1707.01836v1 [cs.CV]. Submitted 2017.

The Apple Watch Series 3 n.d. https://www.apple.com/au/apple-watch-series-3

Webliography

http://gmedinnovations.com A manufacturer of a product that enables multiple health parameters to be measured.

http://www.alivetec.com/alivecor-heart-monitor A portable cardiac monitor.

www.atomwise.com This company uses supercomputers to develop therapies from a database of molecular structures.

http://www.givenimaging.com/en-int/Innovative-Solutions/Capsule-Endoscopy/Pages/default.aspx This company makes a small device that, when swallowed, sends back images of the small intestine.

https://arterys.com This company has a platform for changing medical imaging and healthcare through ultra-fast cloud computing, advanced visualization and deep learning.

https://en.wikipedia.org/wiki/Iknife An intelligent surgical knife.

www.alivecor.com This company makes a small device for measuring heart rate and taking electrocardiograms.

https://www.apple.com/au/apple-watch-series-3 The Apple Watch Series 3.

www.clarius.me Clarius Mobile Health – portable ultrasound scanner.

https://www.crunchbase.com/organization/atomwise Crunchbase was founded to be the master record of data on the world's most innovative companies using data collection leveraging a strong community of contributors, the largest venture partner network, and in-house data teams armed with powerful machine learning.

https://www.crunchbase.com/organization/sense-ly Sensely is an avatar-based, empathy-driven clinical platform that helps clinicians and patients better monitor and manage their health.

www.enlitic.com A company which aims to couple deep learning with huge stores of medical data.

www.healthtap.com A bot to help in clinical decision making.

https://www.ibm.com/watson IBM's supercomputer called Watson uses AI in both health and business decision making.

www.idoc24.com A company that answers dermatology questions online.

https://www.medgadget.com/2015/10/mocaheart-consumer-cardiac-monitor-medgadget-review.html MOCAheart Consumer Cardiac Monitor: A Medgadget Review.

www.your.md A bot that uses AI to help in clinical decision making.

https://researcher.watson.ibm.com/researcher/view_group.php?id=4384&mhq=medicalsieve&mhsrc=ibmsearch_a Medical Sieve is a long-term exploratory grand challenge project to build a next generation cognitive assistant with advanced multimodal analytics, clinical knowledge and reasoning capabilities that is qualified to assist in clinical decision making in radiology and cardiology.

http://theaustralian.com.au/wsj Projected uptake of mobile phones. The Australian 11 September 2017.

32

Digital Health in Low- and Middle-Income Countries

Martin Seneviratne[1] and David Peiris[2]

[1] Department of Biomedical Informatics, Stanford School of Medicine, Stanford, CA, USA
[2] George Institute for Global Health, Sydney, Australia and the University of New South Wales, Sydney, Australia

---

CHAPTER MENU

Revolutionizing Tropical Medicine: Point-of-Care Tests, New Imaging Technologies and Digital Health,
First Edition. Edited by Kerry Atkinson and David Mabey.
© 2019 John Wiley & Sons, Inc. Published 2019 by John Wiley & Sons, Inc.

32.1 Introduction – The Digital Health Revolution

Arguably one of the greatest drivers of economic participation and prosperity in low- and middle-income countries (LMICs) over the last generation has been the advent of mobile banking. A decade since its founding in 2007, the Kenyan service M-PESA now enables over 20 million people in Africa to transfer money and receive micro-credit using their mobile phones (Harford 2017). (M-Pesa [M for mobile, pesa is Swahili for money] is a mobile phone-based money transfer, financing and microfinancing service, launched in 2007 by Vodafone for Safaricom and Vodacom, the largest mobile network operators in Kenya and Tanzania. It has since expanded to Afghanistan, South Africa, and India.) M-Pesa allows users to deposit, withdraw, transfer money and pay for goods and services easily with a mobile device. The downstream effects are clear – digital banking has made it possible for millions of people to receive a stable income, has reduced the risk of corruption, and opened new channels of commerce. The BBC has called it the "M-PESA Revolution." Mobile banking was revolutionary not because the technology was advanced, but because it was simple. This was a way to leverage existing mobile network infrastructure to solve the burning economic need for simple, secure payments.

The same trend is playing out in healthcare, where simple technologies are being co-opted to deliver critical health services in low resource environments in Africa and other areas. The need is even more pressing than for financial transactions, with 400 million people still lacking access to essential health services based on a 2015 World Health Organization (WHO) and World Bank Group report (Joint WHO/World Bank Group Report 2015). Meanwhile, the burden of chronic disease in low- and middle-income countries continues to rise precipitously. Over 85% of the premature deaths attributed to chronic diseases, including cardiovascular disease, diabetes, cancers and chronic obstructive pulmonary disease, occur in low- and middle-income countries (WHO 2013). The economic burden of this unchecked rise of non-communicable diseases (NCDs) in the LMICs has been estimated at US$7 trillion over a 15-year timespan. The United Nations Sustainable Development Goals (SDGs) have set a 2030 target to reduce premature deaths from NCDs by one third; to curb the infectious epidemics of HIV/AIDS, malaria, tuberculosis (TB) and other tropical diseases; and to reduce maternal mortality by providing universal access to reproductive health services. However, the global medical workforce looks to be overwhelmed by these looming public health trends, with the WHO estimating a shortage of almost 13 million healthcare workers internationally by 2035 (Global Health Workforce Alliance and World Health Organization 2013). The bold vision of the SDGs will only be achieved by rapidly upscaling the capacity of an already overburdened healthcare infrastructure using digital technologies.

Accompanying the rising burden of chronic disease is the increasing digital literacy of the LMICs. According to the most recent estimates from the International Telecommunication Union, mobile phone penetration in LMICs is approximately 85% in the 15–74 age group, with a rising rate of mobile broadband subscriptions, at 44 per 100 persons in 2017 (International Telecommunication Union 2017). Meanwhile, plummeting costs of computers, tablets, and wearable devices are creating a rich digital infrastructure even in low resource settings.

The last decade has seen tremendous innovation in the use of mobile technologies for healthcare delivery (mHealth) and the development of clinical software and data platforms (eHealth). From Nicaragua to Zimbabwe, there has been a wide array of digital health interventions, from large-scale public health efforts based around short message services (SMS) and smartphone apps, through to grassroots initiatives built by community health workers to streamline their own workflows. Some of the most successful examples have come from maternal health – delivering ante- and postnatal care via SMS and telemedicine services. However, the applications are far reaching, from point-of-care TB screening, to gamified platforms for self-management of chronic diseases, to inventory management and capacity planning services for health administrators. (Gamification software is any tool or platform used for applying game mechanics to non-game contexts in order to boost engagement and successful end-results. Common use cases include customer loyalty, e-learning, employee engagement, and performance management.) The value proposition is clear: digital technologies stand to increase healthcare access, improve clinical outcomes and reduce cost.

Traditionally, mHealth has involved SMS reminders and telemedicine consults; however, the proliferation of smartphones presents a range of new possibilities. Smartphone penetration in the LMICs rose from 21% to 37% between 2013 and 2015 (Poushter 2016), although East Africa and South Asia continue to lag behind global averages, with 25 and 30% penetration in 2017 respectively (GSMA Intelligence 2017a). Nevertheless, an increasing proportion of the world's population stands to carry with them a computer thousands of times more powerful than NASA's Apollo navigation systems, with a suite of sensors such as a gyroscope, camera and global positioning system (GPS) poised to collect new streams of physiological and behavioral data. Illiteracy and language barriers continue to be challenges. However, in the words of one MobileMedic user: "Even though I am not able to read and write, I am able to work with this mobile phone" (MedicMobile 2018). Capitalizing on this ambient technology in an engaged user base will open the door to a new mHealth revolution far beyond the traditional SMS reminder services.

This Chapter presents a high level overview of mHealth and eHealth technologies in the LMICs, highlighting key emerging trends, critical challenges and presenting a vision for the future. While we refer to these technologies collectively as "digital health," we recognize that the "broad church" of digital health contains further diversity still, including technologies such as wearables, virtual reality, and predictive analytics. While these technologies certainly have potential in low resource settings, the majority of case studies to date have related to mHealth and eHealth.

For this Chapter we have cataloged 120 such applications that were developed for, and typically in, low-income countries. This is by no means a comprehensive list. However, it is broadly representative of the digital health solutions identified by searching PubMed, the WHO Digital Health Atlas, the GSMA 2017 review of digital health trends in developing markets and a range of local and international mobile health news outlets. Sixty-eight per cent had significant operations in Africa, 28% in Asia and 13% in Latin America, with several solutions having inter-continental footprints and counted in each region. Although many solutions have multiple functionalities, 17% were categorized as being predominantly focused on patient education and awareness campaigns, 23% on self-management and remote monitoring, 35% on empowering frontline healthcare workers and managing electronic medical records, 4% on powering remote diagnostic devices, 13% on disease-tracking and epidemiology, 5% on financing and insurance and 3% on inventory management and supply chain.

32.2   The Current Landscape

To set the scene I begin by reviewing three clinical use-cases where digital health has been applied across the continuum of care. Specifically, I focus on one communicable disease (HIV), one non-communicable disease (diabetes mellitus) and maternal health.

32.3   HIV/AIDS

According to WHO estimates there were almost 37 million people living with HIV in 2016, of which over 25 million lived in Africa (World Health Organization 2017). Despite the development of effective anti-retroviral therapies (ART), an estimated one million people die each year from AIDS-related causes. The ongoing AIDS epidemic is multifactorial, related to awareness, cultural stigma and treatment access and adherence.

One problem space where digital health is making a particular impact is ensuring adherence to ART regimens for HIV-positive subjects. WelTel is an international platform specializing in SMS support programs for HIV. In a landmark randomized study in Kenya WelTel sent patients in the treatment group a weekly text message starting with a simple "mambo?" ("how are you?") (Lester et al. 2010). This weekly check-in was associated with improved adherence to ART (relative risk [RR] of non-adherence 0.81) and reduced viral load (RR 0.84). Other solutions are attempting to assist care providers. iDART (intelligent Dispensing of Anti-Retroviral Treatment) is an open-source software platform developed by the South African non-governmental organization (NGO) Cell-Life, which helps pharmacies in the distribution of ART by offering inventory management, patient tracking and efficient barcoding so that medication packages can be reliably sent to remote distribution sites.

While medication adherence is an issue for those on treatment, there is also the upstream issue that many subjects are not screened for HIV. An estimated 24% of HIV-positive individuals in South Africa remained undiagnosed as of 2012, although this number is declining (Johnson et al. 2015). In order to reach the UNAIDS target of 90% diagnosis rate by 2020, various mobile health solutions have attempted to increase awareness of testing services. An early example was Project Masiluleke, which used the clever strategy of inserting screening advice into the blank space of "Please call me" SMS messages indicating that HIV screening were being offered for free in South Africa. In their first year of deployment in 2008/2009, they saw a tripling in the number of calls made to South Africa's AIDS Helpline (Project Masiluleke 2009).

More recent initiatives are even more comprehensive. The Moyo project in Lesotho also aims to improve the access of remote communities to HIV screening and ART. Mobile clinics visit these communities offering free HIV testing and simultaneously register patients into a surveillance database via a smartphone application (app). If a patient needs medical attention or is unable to access ART, they are sent money for transportation to the nearest health facility using third-party mobile banking tools. In a country where 40% of pregnancy-related mortality is attributed to HIV, this holistic solution for remote communities is extremely promising.

32.4 Diabetes Mellitus

The prevailing pattern of morbidity in the LMICs has shifted from infectious diseases to non-communicable diseases, of which diabetes represents a significant burden. Eight and one half per cent of the adult population has diabetes internationally, with prevalence rising fastest in low- and middle-income countries (WHO 2016). The penetration of mobile phones into the population, and into the daily life of the individual, makes them a natural tool to effect behavioral change in chronic diseases. mHealth solutions have been deployed throughout the continuum of care for diabetes, from preventative medicine through to screening, diagnosis and longitudinal support for both patient and clinician.

With the International Diabetes Federation estimating that two thirds of diabetics in Africa remain undiagnosed (International Diabetes Federation 2017), screening is a particular priority. A variety of simple risk scoring tools exist, which have been packaged into mobile apps for use by patients and community health workers. The Diabetes Online Risk Assessment (DORA) initiative was launched across 12 African nations to provide an accessible web-based tool for diabetes risk stratification. Moderate and high-risk patients were given an electronic voucher for a blood sugar level (BSL) test which could be performed at local pharmacies, and those found to have high sugar readings received a diabetes management app with educational resources and symptom tracking, as well as a home BSL monitor.

Mobile technologies have also been used to screen for specific diabetic complications such as diabetic retinopathy. "Fundus on the Phone" is one example of a smartphone-enabled device for capturing high-quality retinal scans, currently deployed in Bangalore, India. Additionally, the PEEK Retina is a portable ophthalmoscope that connects to a smartphone, allowing remote ophthalmologists to make a diagnosis and suggest treatment strategies (see Chapter 25).

A new class of self-management platforms for diabetes has recently emerged throughout the LMICs. The Indian company JanaCare is one of the leaders in this space, having developed a suite of tools including a smartphone test-strip able to perform blood glucose levels, HBA1c levels, a lipid profile and hemoglobin levels, as well as behavioral coaching through an app called Habits with personalized advice and goal-setting, and a physician dashboard that allows patients to be monitored remotely. In Senegal an app is helping diabetic patients better manage their sugar levels during the Ramadan fast. mRamadan is an SMS reminder service that provides advice for patients and care-providers around diet and medication titration in the context of fasting.

Evidence is emerging to suggest that engagement with diabetes management apps is associated with improved clinical outcomes. A systematic review of mHealth diabetes interventions in mostly high-resource settings found a significant reduction in HbA1c in patients "treated" with a mobile app (Kitsiou et al. 2017). This has led to an emerging view of diabetes management apps known as "digital therapeutics."

32.5 Maternal Health

Each day approximately 830 women die during pregnancy or childbirth from preventable causes. Ninety-nine per cent of these women live in LMICs (WHO 2016). The Sustainable Development Goals call for a reduction in maternal mortality to less than

70 per 100 000 live births by 2030 with universal access to reproductive health services. This is a particular challenge in low-income countries because of significant resource shortages, especially in the maternal health workforce. In 2010, the UN launched the *Every Woman Every Child* initiative – a global movement advocating for major maternal health reform – which pioneered mHealth initiatives through a targeted grant program and support resources. Digital health is seen as an opportunity to upscale the capacity of overburdened health services by empowering mothers, training community health workers and better distributing specialized resources.

The Wired Mothers initiative in Zanzibar is a classical mHealth intervention involving SMS-based reminders during the antenatal and postnatal periods, as well as teleconsults with a midwife. It was an early example of rigorous evaluation with a cluster randomized trial – a trial which ultimately demonstrated an increase in the number of prenatal visits and a reduction in overall maternal mortality (from 3.6 to 1.9%) (GSMA Intelligence 2017; Sondaal et al. 2016). There are many similar solutions: the Mobile Alliance for Maternal Action (MAMA) has developed SMS-based pregnancy quizzes and an interactive question-and-answer service. The *MayMay* app from Myanmar provides regular antenatal reminders, educational tips and a listing of local doctors that women can access directly.

Alongside the wealth of patient-facing pregnancy apps is a growing class of tools designed for healthcare workers, which attempt to address the severe workforce shortage by upskilling existing staff. MedicMobile is a San Francisco based not-for-profit organization that has implemented a number of open-source solutions for health workers in low-resource settings across Africa, Asia, and Latin America. Their maternal health package enables community health workers to register pregnancies, track antenatal care visits and screen for high-risk pregnancies and red flags. A systematic review of mobile solutions for maternal health workers found widespread applications of mobile technologies for providing education, tracking patients, collecting epidemiological data, and communicating between care providers. However, more evidence around the impact on maternal and neonatal outcomes must be gathered (Sondaal et al. 2016).

32.6 Core Functionalities

The previous section reviewed the breadth of digital health solutions within three clinical verticals. We now draw out the common functionalities that appear across clinical domains. Figure 32.1 shows a breakdown of these core functionalities, categorized into patient-facing and clinician-facing functions.

These functions may be viewed as the building blocks of digital health tools – a useful frame-of-reference in an ecosystem that is increasingly modular, with open-source packages for various functions that may be stitched together into a custom app. The sections below briefly review each of these functions and provide relevant examples.

32.7 Patient-facing Functions

32.7.1 Patient Education

Patient education is among the most fundamental functions of digital health tools in low-resource settings. A number of systematic reviews in the maternal and child health

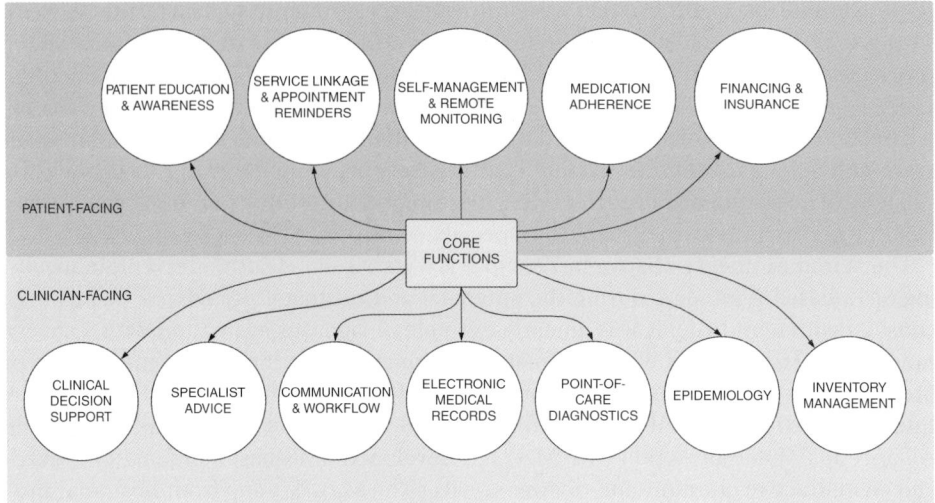

Figure 32.1 Core functions of digital health applications categorized into patient-facing and clinician-facing functionalities.

space found that the provision of educational resources was the most common function (Chen et al. 2018; Sondaal et al. 2016). Patient education can be provided either passively or actively. The former refers to automated content delivered via mobile or web; the latter refers to interactive dialog with the patient via telemedicine or chatbots. Key challenges across the spectrum include

1) Ensuring content is accurate and up-to-date,
2) Making content engaging to users. A range of engagement strategies have proved effective, from simple SMS-based quizzes for HIV education (for example, using *Text to Change* in Uganda) to more sophisticated gamification such as the *Fooya* app, which aims to promote healthy eating habits and targets childhood metabolic syndrome in India and beyond.

32.7.2 Service Linkage and Reminders

mHealth tools can assist patients both in finding relevant health programs and in reminding patients of their appointment schedule. Numerous examples of SMS- and app-based reminders exist in the management of HIV, TB, antenatal care, diabetes, asthma and mental health. One systematic review across low- and high-resource settings showed improved follow-up adherence in 77% of studies using text reminders (Kannisto et al. 2014). Reminder services have been particularly effective in vaccination uptake, with one study in Zimbabwe showing a significant increase in childhood immunizations with simple SMS reminders (Bangure et al. 2015).

32.7.3 Self-management and Remote Monitoring

Self-management refers to solutions that empower patients to be in control of their own care, from assisted diagnosis through to longitudinal disease management. The related concept of remote monitoring refers to any two-way communication between patients and care providers that enables remote care delivery. I have discussed some of the applications in diabetes and antenatal care above, largely based around interactive mobile software and telemedicine consults with remote healthcare workers. However, the rise of sophisticated home sensors is creating a new dimension to self-management. One example is *UChek*, a low-cost urine dipstick reader powered by a smartphone. The associated app gives readings directly to patients about the presence of urinary leukocytes or high glucose, enabling assisted triage for conditions such as urinary tract infections as well as better home management of diabetes.

32.7.4 Medication Adherence

Medication non-adherence is known to be a significant contributor to morbidity in both communicable diseases such as HIV and TB, and in NCDs such as congestive heart failure. mHealth approaches to medication adherence range from simple SMS reminder services, video messages, regular phone calls by healthcare workers, and gamified medication journals powered by smartphone apps. In TB, where directly observed therapy (DOT) is recommended, medication adherence tools have needed to become more rigorous, with creative techniques such as video observation of therapy (VOT), embedded sensors in tablets, smart pill bottles, and low-cost urine tests which detect relevant drug metabolites (DiStefano and Schmidt 2016).

32.7.5 Financing and Insurance

mHealth is helping to streamline the finances of healthcare in the same way as mobile banking has streamlined general commerce. With over 75% of Kenyans uninsured, the Grameen Foundation pioneered the Uzima Project – a low-cost health insurance product interlinked with a smartphone app that provides patient education and reminders (Grameen Foundation 2018). Meanwhile, another Kenyan innovation, Changamka Microhealth, is allowing patients to set aside money for medical treatments with predefined costings for procedures. mHealth platforms have also begun to offer financial incentives for positive behavioral change. A systematic review of 10 such programs offering conditional cash transfers for healthy behaviors in low-resource environments showed increased utilization of preventative health services and, in some cases, improved clinical outcomes (Lagarde et al. 2007).

32.8 Clinician-facing Functions

32.8.1 Clinical Decision Support

Clinical decision support (CDS) systems are being used to upskill community health workers with minimal training to deliver high-quality evidence-based care. This decision support can be in the form of clinician education about consensus guidelines. An example is the sending of regular text messages with guidance about malaria management (which improved adherence when trialed in Kenya) (Zurovac et al. 2011). CDS may also involve point-of-care advice, personalized to the patient at hand. The George Institute for Global Health developed the SMARTHealth India system – a plain-language mobile CDS tool to conduct cardiovascular screening and basic management in rural India (Praveen et al. 2014). This enabled task-shifting from physicians to community health workers and allowed more intelligent triage of patients requiring specialist review. One CDS system in Pakistan even empowered laypeople from the community to screen others for TB, with financial incentives for case reporting (Khan et al. 2012). A review of CDS platforms deployed in sub-Saharan Africa for conditions including pneumonia, antenatal care, pediatric triage and hypertension, found significant potential in the use of such mHealth tools by health workers, although evidence around outcome improvement was still lacking (Adepoju et al. 2017). More work is needed on translating high level treatment guidelines into practical advice at the point-of-care. This ecosystem would likely benefit from a library of clinically-validated decision support tools accessible via application programming interfaces (APIs).

32.8.2   Specialist Advice

Myanmar was ranked last in a WHO ranking of 190 international health systems in 2000, and has faced ongoing challenges since. The first Myanmar-based telemedicine solution was implemented in 2014 for the purpose of providing specialist advice to remote clinicians (Leroux et al. 2016). Rural health workers were able to send x-rays, ECGs and ultrasound images to specialists in Yangon, as well as requesting support via video consultation. The service had promising uptake, with over 4500 specialist consults provided in the 16 month trial period. In Tanzania *First Derm* is being used by local physicians for dermatology support, with images being sent via the app to consultants in Dar es Salaam to select which patients should be referred for specialist investigation (Chung and Hyunsoo 2017). In Botswana similar technology is being used to transmit cervical photographs for review by specialist gynecologists, providing effective cervical cancer screening to women in remote communities (Quinley et al. 2011).

32.8.3   Communication and Workflow

mHealth technologies are helping clinicians to communicate with one another in order to streamline various components of their daily workflows. In Ghana and Liberia, the *MDNet* Program allows free physician-to-physician communication nationwide to promote education and collaboration. mHealth also accelerates feedback loops so that test results are communicated rapidly and not overlooked. An example is in Zambia where SMS is used for delivering infant HIV screening results back to the referring clinician more rapidly (Seidenberg et al. 2012).

32.9   Electronic Medical Record Management

Even in the highest resource settings electronic medical records (EMRs) are challenging to implement and configure, with significant shortcomings around user interface and interoperability. A number of simple modular EMR architectures have been developed for low resource settings, notably *OpenMRS* – an open-source EMR suite built and supported by a distributed community of developers. Even basic EMRs can help streamline patient care, capacity planning and observational research, especially if data starts to be collected longitudinally and shared between care providers (Williams and Boren 2008).

32.10   Point-of-Care Diagnostic Tests

mHealth solutions provide the backbone for the emerging sector of point-of-care diagnostics. *Swasthya Slate* is a mobile platform paired with a number of low-cost sensors, which allows a healthcare worker to perform over 30 diagnostic tests from vital signs through to rapid tests for HIV and syphilis (Wunker 2014). The platform, which also offers patient record management and decision support for the clinician, has now been deployed widely across India. A number of tools are leveraging the higher-performing cameras in smartphones to analyze blood samples for malaria (for example, *Lifelens*) and screen for anemia based on the color of the sclera (for example, *Eyenaemia*).

32.11   Epidemiology

During the recent Ebola outbreak in West Africa, mHealth apps such as the Ebola Care App were used to provide real-time information from the field about new exposures. This allowed health authorities and NGOs to track the epidemic, better allocate resources and facilitate contact-tracing. The open-source tool *GeoChat* is being used to track disease outbreaks such as cholera and avian influenza in Cambodia via crowd-sourced updates (http://instedd.org/technologies/geochat). Similar systems have been deployed for malnutrition, vaccine coverage and dengue fever. The same principle of mobile-powered data collection can be extended beyond the acute setting of outbreak tracking to epidemiological surveys and academic research.

32.12   Inventory Management and Supply Chain

A survey of 12 national AIDS programs in 2011 found that 67% of countries experienced episodes of insufficient antiretroviral stock lasting an average of 40 days, largely due to maldistribution and forecasting problems (Sued et al. 2011). mHealth can be used to build supply chain resilience by allowing frontline healthcare workers to monitor inventory and by enabling citizens to report counterfeit medications that can arise in times of undersupply (Agarwal et al. 1996). In Malawi, for example, *cStock* was developed to track medication inventory, providing dashboards for monitoring stock levels across the community and sending text messages to health workers when new stock became available.

32.13   Challenges to Scale

Despite the volume of new mHealth initiatives, few projects manage to achieve scale beyond the pilot stage. The WHO has recently published their MAPS (mHealth Assessment and Planning for Scale) toolkit as a framework for digital health solutions to cross that bridge from pilot to established initiative (WHO 2015). MAPS outline six "axes of scale": groundwork, partnerships, financial sustainability, technology architecture, operations, and monitoring/evaluation. The toolkit provides self-assessment tools to strengthen an mHealth program along each axis. In the following section, I will zoom out one level and present four common challenges that the digital health ecosystem at large is facing as it matures and scales.

32.13.1   Evidence-based Practice

Evidence-based clinical content is critical for scalability, but is often overlooked in digital health designs. A review of apps for TB prevention found numerous examples with poorly-substantiated and, in some cases, incorrect information, with only 2 of 24 apps citing peer-reviewed publications (Iribarren et al. 2016). Although the US Food and Drug Administration (FDA) is bolstering oversight around digital health (US Food and Drug Administration 2017), mHealth solutions in the LMICs continue to operate without any clinical vetting process. *Txt2MEDLINE* in Botswana aims to tackle this problem by allowing physicians to question MEDLINE literature via SMS. The ecosystem would benefit from more open-source documents, specifically clinically approved educational resources with multi-language support that could be readily customized and integrated into new tools. Ideally this would include multimedia options such as educational graphics and videos. Furthermore, it is becoming increasingly important to gather evidence around clinical outcomes following mHealth interventions. Although many of the systematic reviews in this space lament the shortage of high-quality evidence beyond feasibility studies (Chib et al. 2015; Oliver-Williams et al. 2017), others argue that the rigor of mHealth studies is increasing rapidly (Labrique et al. 2013). Digital health innovators are realizing that they must improve clinical endpoints or reduce costs in order to gain widespread adoption.

32.13.2   Integration into Clinical Workflows

A common pitfall of digital health interventions is a lack of integration with the surrounding health infrastructure, resulting in multiple, fragmented point solutions. Successful initiatives, such as the deployment of the *CommCare* app by the *m4Change2* project in Nigeria, have been deeply embedded within existing care pathways. This was a simple case-management and decision support tool for Nokia and Android devices, deployed across 10 rural primary health centers. However, the healthcare workers received good quality training and were able to use the app alongside their pre-existing workflows. Additionally, the system tied into a cash-transfer scheme supported by the federal health authorities to provide a small incentive payment to women accessing antenatal care. The same integration mantra applies for patient-facing tools – they are most valuable if they are interlinked with the health ecosystem, for example by connecting patients to relevant facilities or providing telemedicine support from local physicians.

32.13.3   Financial Sustainability

It can be difficult for digital health solutions to draw revenue from the end user, hence creative funding models must be considered. Many of the most successful initiatives to date have been funded by governments or NGOs. However, large technology companies such as Philips and GE, along with many of the cellular networks such as Telefonica and Telenor are contributing significantly to the ecosystem. There are also several recent examples of public–private partnerships being used as a successful model for building digital health infrastructure in the LMICs. An example is the one between Philips and Telkom Indonesia for mobile obstetric monitoring (GMSA Intelligence 2017; Latif et al. 2017). A recent set of recommendations from the United States National Academies of Sciences, Engineering, and Medicine called for USAID to provide seed funding to local governments in order to facilitate public–private collaborations (Dzau et al. 2017).

As health systems internationally move toward bundled or capitated payment schemes, there may also be more scope for digital health programs to provide preventative medicine in return for a share of cost savings. *AxisMed* is a chronic disease management platform in Brazil that has created a business case around reducing overall expenditure through decreased emergency room visits and hospitalizations. In Bangladesh Telenor and Grameenphone have developed *Tonic* – a health management platform offering education, service linkage, telemedicine consults and emergency insurance packages to over two million users – based around a low-cost subscription model where users pay a monthly access fee (US$1–4). Digital health innovators must consider creative reimbursement models to support their initiatives.

32.13.4 Technical Scalability

There are inherent limitations associated with mobile phone technologies in extremely low-resource environments. Mobile Internet continues to be lacking in some regions, and up to 10% of the world's population (predominantly in sub-Saharan Africa and rural Asia) still live without any cellular connectivity (GSMA Intelligence 2015). There can be issues around electricity coverage, phone battery life and the cost of mobile devices, making access to the most vulnerable populations an ongoing challenge (Latif et al. 2017).

Another aspect of achieving technical scale is data interoperability. It is important for digital health tools to be able to push and pull information from other health IT systems; however, to date there has been poor adherence to standards such as HL7 and FHIR. (Health Level-7 or HL7 refers to a set of international standards for transfer of clinical and administrative data between software applications used by various healthcare providers. These standards focus on the application layer, which is "layer 7" in the OSI model. The OSI model is The Open Systems Interconnection model, a conceptual model that characterizes and standardizes the communication functions of a telecommunication or computing system without regard to its underlying internal structure and technology. Its goal is the interoperability of diverse communication systems with standard protocols. The HL7 standards are produced by the Health Level Seven International, an international standards organization, and are adopted by other standards issuing bodies such as American National Standards Institute and International Organization for Standardization. Fast Healthcare Interoperability Resources (FHIR, pronounced "fire") is a draft standard describing data formats and elements (known as "resources") and an application programming interface (API) for exchanging electronic health records).

Digital health tools that support interoperability will help to power the emerging paradigm of data sharing and large-scale analytics, which stands to improve efficiency at a systems level (Latif et al. 2017). In Malaysia, the 2016–2020 eHealth strategy emphasizes interoperability as a key priority, aiming toward a "one person, one record" health system with an associated data warehouse (World Health Organisation, Malaysian Ministry of Health 2016).

Finally, data security and privacy must be a priority for technical scale, especially when dealing with vulnerable populations and highly sensitive information such as HIV status. The ecosystem would benefit from a simplified version of HIPAA (Health Insurance Portability and Accountability Act, US) and GDPR (General Data Protection Regulation, EU), the prevailing health data privacy guidelines for the western world, tailored to low-resource settings with practical guidance around identity management and encryption.

32.14   Emerging Trends and Future Vision

As the digital health ecosystem overcomes the challenges listed above, it can be predicted that a number of trends will emerge in moving from the "SMS era" into one of interactive, personalized digital health. Some of these trends have been pioneered in high-resource settings, fueled by the massive injections of talent and capital into digital health in recent years. However, others germinated in low-resource settings and are rapidly gaining momentum. I briefly highlight some of the most promising trends.

32.14.1   Modularity and Application Programming Interfaces (APIs)

The OpenMRS example is an open source, community-supported suite of EMR modules that can be analogized across digital health. The *Be He@lthy Be Mobile* campaign from the WHO aims to scale up national digital health initiatives by providing reusable mHealth toolkits, streamlined service models and collaboration platforms for digital health globally (WHO 2014). Imagine if local health organizations could drag and drop functionalities into templated software products with easy-to-use software development kits (SDKs), plug into the latest diagnostic and treatment guidelines using application programming interfaces (APIs), and connect with third-party services such as telemedicine consults. Technical modularity is helping to power the revolution toward human-centered design, where solutions are created in close consultation with end-users, and thus reshaping digital health more broadly (Pym 2015).

32.14.2   Personal Health Clouds

As mobile internet access expands, we can expect digital health tools to increasingly move into the cloud, allowing patients to store and manage their personal health data. (Cloud computing is an information technology paradigm that enables ubiquitous access to shared pools of configurable system resources and higher-level services that can be rapidly provisioned with minimal management effort, often over the Internet. Cloud computing relies on sharing of resources to achieve coherence and economies of scale, similar to a public utility.) The vision of an electronic "health passport" that travels with a patient over time (regardless of how fragmented health facilities may be on the ground) would empower patients, streamline diagnosis and treatment and enable better population health tracking.

32.14.3 Social Care Models

Platforms such as *Omada* and *WealLife* (both San Francisco-based) have recognized that strong social networks, between similar patients or even within a patient's own circle of family and friends, can assist in managing chronic disease. This principle is already in action organically in communities across the LMICs. However, technology can support this process by explicitly forming patients into "care teams," potentially built around different stages of chronic disease management.

32.14.4 Home-based Care

In the words of the UK's National Health Service director Sir Bruce Keogh, "the hospital of the future is in the home." Digital health will likely be an enabler of home-based care, providing tailored advice to patients and their carers, and allowing remote health workers to manage dozens of cases remotely.

32.14.5 Crowdsourced Medical Advice

The applications shown in Figure 32.1 summarize the mission to democratize medical knowledge by allowing physicians to upload clinical photographs and discuss cases with a network of specialists from around the globe. This technology is being used by Médecins Sans Frontières and a number of other healthcare NGOs to crowdsource medical advice. Given the impending workforce shortages and maldistribution of specialist talent between rural and urban areas, we anticipate that crowdsourcing for clinical education and decision support will be a growing trend. (Crowdsourcing is a sourcing model in which individuals or organizations obtain goods and services, including ideas and finances, from a large, relatively open and often rapidly-evolving group of Internet users. It divides work between participants to achieve a cumulative result.)

32.14.6 Artificial Intelligence (AI)

AI can assist with diagnosis, management, and population health predictions. One startup in Nigeria has applied AI to analyze the acoustic patterns of the cry of newborn babies in order to screen for neonatal asphyxia (*freelanceqz* 2017). Another project is using automated image analysis to screen for otitis media in tympanic membrane photographs with a view to deploying low-cost otoscopes and automating the diagnostic step (Myburgh et al 2016). The Artificial Intelligence and Medical Epidemiology (AIME) platform in Malaysia aims to predict disease outbreaks by aggregating public health and climate data, with promising results already demonstrated in tracking dengue epidemics.

32.14.7 Blockchain Technologies

A flurry of organizations are currently trying to adapt blockchain technology to store and manage health data securely. (Blockchain is a continuously growing list of records, called blocks, which are linked and secured using cryptography.) The advantages of this decentralized digital ledger includes an auditable ground-truth for important health data, identity management with custom access levels and smart contracts.

Blockchain-based solutions have helped to decrease corruption and improve transparency in the LMICs, such as in managing property rights in Ghana, and there are likely some transferable lessons for health systems, especially in settings where unauthorized practitioners or corrupt payment systems reduce access for the most vulnerable.

32.15 Conclusions

Digital health has a bright future in helping to improve access, quality, and cost for health systems around the globe. The benefits will arguably be felt most strongly in the LMICs, in the face of severe workforce shortages and the rising burden of chronic disease. The digital health ecosystem faces challenges such as a shortage of clinical evidence, technical limitations, a lack of sustainable financial models, and interfacing barriers with the broader health system. A number of upstream factors can accelerate the maturity of digital health in the LMICs: a supportive policy environment, more flexible funding (from governments, aid organizations and industry), and better educational resources for digital health innovators. Bidirectional knowledge and talent transfer between low- and high-resource settings is particularly valuable. The future will likely see a combination of tried-and-tested technologies such as SMS better embedded into care pathways, as well as some of the emerging technologies such as artificial intelligence powering new solutions altogether. We have already seen the "M-PESA" revolution for mobile banking – the next chapter will be the digital health revolution where these technologies blend into the fabric of healthcare and society.

Bibliography

Adepoju, I.-O.O., Albersen, B.J.A., De Brouwere, V. et al. (2017). mHealth for clinical decision-making in Sub-Saharan Africa: a scoping review. *JMIR Mhealth Uhealth.* 5: e38.

Agarwal, S., Tamrat, T., Fønhus, M.S. et al. (1996). Tracking health commodity inventory and notifying stock levels via mobile devices. In: *Cochrane Database of Systematic Reviews.* Wiley Blackwell.

Bangure, D., Chirundu, D., Gombe, N. et al. (2015). Effectiveness of short message services reminder on childhood immunization programme in Kadoma, Zimbabwe – a randomized controlled trial, 2013. *BMC Pub Health.* 15: 137.

Chen, H., Chai, Y., Dong, L. et al. (2018). Effectiveness and appropriateness of mHealth interventions for maternal and child health: systematic review. *JMIR Mhealth Uhealth* 6: e7.

Chib, A., van Velthoven, M.H., and Car, J. (2015). mHealth adoption in low-resource environments: a review of the use of mobile healthcare in developing countries. *J. Health Commun.* 20: 4–34.

Chung & Hyunsoo (2017). How smartphone technology is changing healthcare in developing countries. *J. Glob. Health.*

Digital Health Technology Vision (2017). Technology for People. Accenture, 2017.

DiStefano, M.J. and Schmidt, H. (2016). mHealth for tuberculosis treatment adherence: a framework to guide ethical planning, implementation, and evaluation. *Glob. Health Sci. Pract.* 4: 211–221.

Dzau, V., Fuster, V., Frazer, J., and Snair, M. (2017). Investing in global health for our future. *N. Engl. J. Med.* 377: 1292–1296.

freelanceqz. (2017). This Nigerian AI health startup wants to save thousands of babies' lives with a simple app. Quartz. https://qz.com/1158185/nigerian-ai-health-startup-ubenwa-hopes-to-save-thousands-of-babies-lives-every-year (Accessed 17 January 2018).

Global Health Workforce Alliance and World Health Organization. (2013). A Universal Truth: No Health Without a Workforce: Third Global Forum on Human Resources for Health Report. http://www.who.int/workforcealliance/knowledge/resources/hrhreport2013/en

Grameen Foundation. (2018). Grameen Foundation, MicroEnsure, Penda Health and Clifford Chance Launch New Health Insurance Initiative for Low-Income Kenyans. https://www.grameenfoundation.org/press-releases/grameen-foundation-microensure-penda-health-and-clifford-chance-launch-new-health (Accessed 17 January 2018)

GSMA Intelligence. (2015). Rural coverage: strategies for sustainability. https://www.gsmaintelligence.com/research/2015/07/rural-coverage-strategies-for-sustainability/512/

GSMA Intelligence. (2017a). Accelerating affordable smartphone ownership in emerging markets. https://www.gsma.com/mobilefordevelopment/programme/connected-women/accelerating-affordable-smartphone-ownership-in-emerging-markets/

GSMA Intelligence. (2017b). Scaling digital health in developing markets: Opportunities and recommendations for mobile operators and other stakeholders. https://www.gsmaintelligence.com/research/2017/06/scaling-digital-health-in-developing-markets/628/

Harford, T. (2017). Money via mobile: The M-Pesa revolution. http://www.bbc.com/news/business-38667475 (Accessed 17 January 2018)

International Diabetes Federation. (2017). Diabetes Atlas - 8. http://www.diabetesatlas.org/resources/2017-atlas.html

International Telecommunication Union. (2017). ICT Facts and Figures 2017.

Iribarren, S.J., Schnall, R., Stone, P.W., and Carballo-Diéguez, A. (2016). Smartphone applications to support tuberculosis prevention and treatment: review and evaluation. *JMIR Mhealth Uhealth* 4: e25.

Johnson, L.F., Rehle, T.M., Jooste, S., and Bekker, L.-G. (2015). Rates of HIV testing and diagnosis in South Africa: successes and challenges. *AIDS* 29: 1401–1409.

Joint WHO/World Bank Group report. (2015). Tracking universal health coverage: First global monitoring report.

Kannisto, K.A., Koivunen, M.H., and Välimäki, M.A. (2012). Use of mobile phone text message reminders in health care services: a narrative literature review. *J. Med. Internet Res.* 16: e222.

Khan, A.J., Khowaja, S., Khan, F.S. et al. (2012). Engaging the private sector to increase tuberculosis case detection: an impact evaluation study. *Lancet Infect. Dis.* 12: 608–616.

Kitsiou, S., Paré, G., Jaana, M., and Gerber, B. (2017). Effectiveness of mHealth interventions for patients with diabetes: an overview of systematic reviews. *PLoS One* 12: e0173160.

Labrique, A., Vasudevan, L., Chang, L.W., and Mehl, G. (2013). H_pe for mHealth: more 'y' or 'o' on the horizon? *Int. J. Med. Inform.* 82: 467–469.

Lagarde, M., Haines, A., and Palmer, N. (2007). Conditional cash transfers for improving uptake of health interventions in low- and middle-income countries: a systematic review. *JAMA* 298: 1900–1910.

Latif, S., Rana, R., Qadir, J. et al. (2017). Mobile health in the LMICs: review of literature and lessons from a case study. *IEEE Access* 5: 11540–11556.

Leroux, E.J., Ohn, T., Lwin, P.M. et al. (2016). The first Myanmar-based telemedicine solution for the people of Myanmar: a pilot study at 3 diverse facilities. *Ann. Glob. Health* 82: 458.

Lester, R.T., Ritvo, P., Mills, E.J. et al. (2010). Effects of a mobile phone short message service on antiretroviral treatment adherence in Kenya (WelTel Kenya): a randomised trial. *Lancet* 376: 1838–1845.

MedicMobile. (2016). Empowering Health Workers in Nepal with mHealth. https://medicmobile.org/blog/empowering-health-workers-in-nepal-with-mhealth (Accessed 17 January 2018)

Myburgh, H.C., van Zijl, W.H., Swanepoel, D. et al. (2016). Otitis media diagnosis for developing countries using tympanic membrane image-analysis. *EBioMedicine* 5: 156–160.

Oliver-Williams, C., Brown, E., Devereux, S. et al. (2017). Using mobile phones to improve vaccination uptake in 21 low- and middle-income countries: systematic review. *JMIR Mhealth Uhealth* 5: e148.

Poushter, J. (2016). *Smartphone Ownership and Internet Usage Continues to Climb in Emerging Economies*. Pew Research Center.

Praveen, D., Patel, A., Raghu, A. et al. (2014). SMARTHealth India: development and field evaluation of a mobile clinical decision support system for cardiovascular diseases in rural India. *JMIR Mhealth Uhealth* 2: e54.

Project Masiluleke. (2009). PopTech.

Pym, H. (2015). Technology is changing healthcare. BBC News http://www.bbc.com/news/health-30997413 (Accessed 17 January 2018).

Quinley, K.E., Gormley, R.H., Ratcliffe, S.J. et al. (2011). Use of mobile telemedicine for cervical cancer screening. *J. Telemed. Telecare* 17: 203–209.

Seidenberg, P., Nicholson II, N., Schaefer, M. et al. (2012). Early infant diagnosis of HIV infection in Zambia through mobile phone texting of blood test results. *Bull. World Health Organ.* 90: 348–356.

Sondaal, S.F.V., Browne, J.L., Amoakoh-Coleman, M. et al. (2016). Assessing the effect of mHealth interventions in improving maternal and neonatal care in low- and middle-income countries: a systematic review. *PLoS One* 11: e0154664.

Sued, O., Schreiber, C., Girón, N., and Ghidinelli, M. (2011). HIV drug and supply stock-outs in Latin America. *Lancet Infect. Dis.* 11: 810–811.

US Food and Drug Administration, 2017. Digital Health Innovation Action Plan.

Williams, F. and Boren, S.A. (2008). The role of the electronic medical record (EMR) in care delivery development in developing countries: a systematic review. *Inform. Prim. Care* 16: 139–145.

World Health Organisation, Malaysian Ministry of Health (2016). Malaysia–WHO Country Cooperation Strategy 2016–2020.

World Health Organization (2013). Global Action Plan for the Prevention and Control of NCDs 2013–2020.

World Health Organization (2014). Be he@lthy, be mobile.

World Health Organization (2015). The MAPS Toolkit: mHealth Assessment and Planning for Scale.

World Health Organization (2016a). Global Report on Diabetes.

World Health Organization (2016b). Maternal Mortality Fact Sheet.

World Health Organization (2017). HIV/AIDS Fact Sheet.

Wunker, S. (2014). How the swasthya slate is revolutionizing healthcare, and why it steers clear of the US. https://www.forbes.com/sites/stephenwunker/2014/11/22/how-the-swasthya-slate-is-revolutionizing-healthcare-and-why-it-steers-clear-of-the-united-states/#2a4c6d9a477e

Zurovac, D., Sudoi, R.K., Akhwale, W.S. et al. (2011). The effect of mobile phone text-message reminders on Kenyan health workers' adherence to malaria treatment guidelines: a cluster randomised trial. *Lancet* 378: 795–803.

Webliography

https://qz.com/1158185/nigerian-ai-health-startup-ubenwa-hopes-to-save-thousands-of-babies-lives-every-year Accessed 17 January 2018. *freelanceqz*, 2017. This Nigerian AI health startup wants to save thousands of babies' lives with a simple app. Quartz

http://www.who.int/workforcealliance/knowledge/resources/hrhreport2013/en Global Health Workforce Alliance and World Health Organization. 2013. A Universal Truth: No Health Without a Workforce: Third Global Forum on Human Resources for Health Report.

https://www.grameenfoundation.org/press-releases/grameen-foundation-microensure-penda-health-and-clifford-chance-launch-new-health Accessed 17 January 2018. Grameen Foundation. Grameen. Foundation, MicroEnsure, Penda Health and Clifford Chance Launch New Health Insurance Initiative for Low-Income Kenyans.

https://www.gsmaintelligence.com/research/2015/07/rural-coverage-strategies-for-sustainability/512 GSMA Intelligence, 2015. Rural coverage: strategies for sustainability.

http://www.bbc.com/news/business-38667475 Accessed 17 January 2018. Harford, T. 2017. Money via mobile: The M-Pesa revolution.

https://medicmobile.org/blog/empowering-health-workers-in-nepal-with-mhealth Accessed 17 January 2018. MedicMobile. 2018. Empowering Health Workers in Nepal with mHealth.

https://www.bbc.com/news/health-30997413 Accessed 17 January 2018. Pym, H. Technology is changing healthcare. BBC News, 2015.

33

Nucleic Acid Detection of Tuberculosis Via Innovative Point-of-Care Nanotechnologies Targeted for Low Resource Settings

Benjamin Y.C. Ng[1,2], Eugene J.H. Wee[1], Nicholas P. West[2], and Matt Trau[1,2]

[1] *Centre for Personalized NanoMedicine, Australian Institute for Bioengineering and Nanotechnology, The University of Queensland, Brisbane, Queensland, Australia*
[2] *School of Chemistry and Molecular Biosciences, The University of Queensland, Brisbane, Queensland, Australia*

CHAPTER MENU

33.1   Introduction

Tuberculosis (TB) is an ancient and highly successful scourge that is thought to be as old as the origin of mankind. The pathogen ranks as one of the most deadly infectious diseases in the world, causing the deaths of 1.1 million people in 2015, and afflicting a further 10 million. It is proving difficult to eradicate. Commonly perceived as a disease of poverty, TB still remains a major threat to the developed world due to its ability to mutate into drug-resistant forms which are challenging to treat successfully. A large

Revolutionizing Tropical Medicine: Point-of-Care Tests, New Imaging Technologies and Digital Health,
First Edition. Edited by Kerry Atkinson and David Mabey.
© 2019 John Wiley & Sons, Inc. Published 2019 by John Wiley & Sons, Inc.

proportion of new TB cases and TB deaths are concentrated in LMICs that face an uphill challenge in keeping TB under control with scant resources and infrastructure (Bjune 2008). In vulnerable populations such as these, rapid diagnostic tests that can be deployed at the point-of-care are urgently sought in order to hasten initiation of treatment and minimize transmission in the community. Unlike conventional TB diagnostic methods used at the healthcare periphery (such as sputum smear microscopy and the tuberculin skin test), nucleic acid detection methods have advantages such as greater specificity and sensitivity, as well as a short time to obtain results. This will aid efforts to screen for the disease in vulnerable populations, or conduct active case finding when tracing contacts. In addition, nucleic acid detection assays can be interpreted with a variety of readout methods to suit a broad range of applications, and potentially provide a solution to empower healthcare agencies in low resource settings with the tools to make headway in the global effort against TB.

Details on the global burden of tuberculosis are described in Chapter 1, the currently available diagnostic tests in Chapter 5, and the use of chest x-rays for its diagnosis in Chapter 30.

This chapter focuses on current and future methods of nucleic acid testing to diagnose TB early at the primary point-of-care, *but is translatable to the detection of any microbial agent, the genome of which has been sequenced.*

33.2 Nucleic Acid Detection of Tuberculosis

Rapid diagnostic tests for TB have come under the spotlight recently with the endorsement of the the Xpert MTB/RIF (*Mycobacterium tuberculosis*/rifampicin-resistant) assay by the WHO in 2010 (WHO 2017). As described in Chapter 5, The Xpert MTB/RIF is a real-time polymerase chain reaction (PCR) system that samples and interprets results of real-time PCR reactions performed on sputum samples collected in proprietary cartridges (Lawn and Nicol 2011). In addition to determining whether TB is present in a sample, it is also able to detect most mutations associated with rifampicin resistance using fluorescent probes that bind and amplify at a slower rate when a mutation is present (Lawn and Nicol 2011; Piatek et al. 1998). A standard test run using the Xpert system takes less than two hours (Helb et al. 2010). Since the introduction of the Xpert assay, an abundance of other nucleic acid amplification tests based on real-time PCR technology have been announced by various companies and are in various phases along the development pipeline (WHO 2017).

33.3 The Availability of Rapid Diagnostic Tests at the Peripheral Healthcare Level

It is useful to consider TB diagnostic tests by their accessibility and availability. These are categorized into three tiers (Parsons et al. 2011). Peripheral laboratory tests are ones that can be conducted at the most basic level, typically a community or district clinic, and include sputum smear microscopy and skin tests (Parsons et al. 2011). These tests are simple to perform and do not consume many resources – features ideal for screening and active case finding. Occasionally Xpert MTB/RIF platforms are also available at the peripheral level for rapid TB and rifampicin resistance testing (Helb et al. 2010).

The second, or intermediate, tier is at the regional level and tests usually include the Xpert MTB/RIF platform, conventional culture and first-line drug susceptibility testing (DST). The highest level, central, is occupied by national reference laboratories that have the necessary tools and data for speciation, rapid culture, phenotypic DST and molecular tests. Sample isolates may also be referred to central laboratories in other countries for confirmation of results (Parsons et al. 2011).

Peripheral healthcare level tests in particular are in great demand in countries with a high TB incidence and burden of disease (WHO 2017). One of the main priorities in TB management is the need for early diagnosis through systematic screening of high risk groups such as those with HIV/AIDS. This is most feasible when conducted at the peripheral level in order to stop transmission in a community at an early stage. Necessary measures such as quarantine processes and treatment plans can then be activated without delay, as opposed to triaging the patient to a hospital for results and treatment (McNerney et al. 2012; Parsons et al. 2011). Drug susceptibility and speciation can be performed at higher laboratory tiers through stepwise referral after the initial diagnosis has been made. The WHO targets a minimum of 90% of all new and relapsed TB cases to be tested with a WHO-recommended rapid diagnostic (WRD), highlighting the use of rapid diagnostic testing to combat and control the spread of TB (WHO 2017).

The Xpert MTB/RIF is one such WRD, and while it is available at some peripheral settings and fulfills some of the criteria of a rapid test, it has a number of limitations. Firstly, the Xpert system is expensive compared to conventional diagnostics: it costs approximately US$17000 for a four module platform and each test requires the consumption of a single disposable cartridge that can cost more than US$15 each (McNerney et al. 2012). Moreover, the Xpert platform performs thermal cycling for quantitative (q) PCR and requires a continuous power supply which may not be available on-site in remote areas. It is a complex device that requires routine maintenance and calibration, which further increases its cost. These limitations can hinder the Xpert MTB/RIF in delivering rapid point-of-care (POC) TB diagnosis in low resource settings (Boehme et al. 2011; McNerney et al. 2012).

In the rest of this chapter we present and examine technologies that could potentially revolutionize diagnosis at the peripheral healthcare level. We think that new POC tests developed for this purpose will only be useful if they are able to fill the diagnostic gap between conventional peripheral healthcare tests and the Xpert MTB/RIF assay. A new POC test can have selective advantages by being more sensitive and specific than conventional tests, produce results as rapidly as the Xpert MTB/RIF assay, but do not depend on specialized laboratory-based equipment or electricity. A list of such POC requirements has been compiled by the Médecins Sans Frontières (Table 33.1) summarizing the criteria for such a POC test (MSF 2009).

Apart from high sensitivity and specificity, criteria include a time to result of less than three hours for effective case management, minimal reliance on equipment and instrumentation, an unambiguous yes or no result, and a cost of less than US$10/test after scaling up (MSF 2009). We believe that innovative nanotechnologies can be used to enable high throughput, effective, simple and affordable tests.

Table 33.1 Requirements for a point-of-care TB test as recommended by an expert meeting organized by Médecins Sans Frontières, 2009.

Specification	Requirement
Sensitivity	• Smear positive cases: 95% • Smear negative cases: 60–80%*
Specificity	• 95% compared to culture
Time to results	• Maximum: 3 hr (patient must receive result in the same day)
Throughput	• Minimum: 20 tests day^{-1} by a single operator
Number of samples	• One sample per test (no need for multiplexing)
Storage	• Stable at 30 °C and high humidity
Instrumentation	• If needed, no maintenance required • Fits in backpack • Acceptable replacement value
Power requirement	• Can be battery powered
Cost	• Less than US$10 after scale up
Training	• Minimal training and can be performed by any health worker
Waste disposal	• Environmentally friendly • Simple burning or sharps disposal acceptable

* Consensus cannot be reached on a minimum sensitivity for smear-negative cases.

33.4 Leveraging Innovative Nanotechnologies for Point-of-Care TB Diagnosis

Leveraging innovative nanotechnologies can bypass restrictions on tests due to lack of technology or equipment in resource-poor areas. Using isothermal nucleic acid amplification tests (NAATs) as part of the diagnostic assay is one such innovation, as it enables laboratories to perform DNA amplification with simple heating devices capable of maintaining a single temperature, instead of complex, programmable thermal cyclers. We will specifically examine some innovations that are amenable at the POC for a TB diagnostic assay workflow.

33.4.1 Nucleic Acid Amplification Tests (NAATs)

NAATs, which generally refer to PCR-based methods, are a critical component in next generation TB diagnostics as they have several advantages over conventional methods (Eisenach et al. 1990; Smith et al. 1996). NAATs have the potential to be highly specific to their target gene, and additional controls are included in the test to invalidate results due to non-specific amplification. NAATs can also be very sensitive depending on the purity of sample used as the DNA template, possessing a theoretical limit of detection (LOD) of a single DNA copy. It is common for NAATs to take under two hours to complete, a significant improvement over conventional methods. Finally, NAATs are amenable to various analytical methods, which expand their potential in POC diagnostics. It is this versatility that sees the use of NAATs in broad translational applications in various fields of work.

Central to the design of NAAT assays are the DNA oligonucleotides, or primers, used for the amplification reaction. As DNA amplification only occurs if the primers are able to bind to their complementary sequences in a potential target, NAATs are only as specific as the divergence of a primer set's sequence of nucleotides from all present DNA templates in a sample. NAATs used in TB detection often employ primers that target highly conserved gene regions in the *M. tuberculosis* complex, a group of mycobacteria that is known to cause tuberculosis in humans or other living organisms. Popular targets include the insertion elements IS1081 and IS6110 (Eisenach et al. 1990), which have been chosen because they are present in multiple copies in the *M. tuberculosis* genome and translate to greater sensitivity in a NAAT, as well as the beta subunit of the DNA-dependent RNA polymerase (*rpoB*) gene, used in the Xpert MTB/RIF (Lawn and Nicol 2011). However, the specificity of these insertion elements has come under scrutiny in recent studies as they were noted to exist in other similar organisms, and may not be exclusive to *M. tuberculosis* (Coros et al. 2008; Gillespie et al. 1997). Alternative target regions that recent methods have employed are the genes encoding ESAT-6 and CFP-10, proteins used in interferon-γ release assays as they are deleted from the attenuated BCG strain (Maartens et al. 2007). In particular, the rifampicin-resistance determining region (RRDR), an 81 base pair sequence within the *rpoB* gene, is frequently selected because of its role in identifying rifampicin-resistant TB strains (Piatek et al. 1998; Ramaswamy and Musser 1998; Telenti et al. 1993).

In recent years a subset of NAATs known as isothermal amplification reactions has generated significant interest. These reactions differ from the traditional PCR in that they do not employ thermal cycling, a core process that allows for the controlled denaturation, annealing and extension of the template and primers for DNA replication. Essentially, isothermal amplification utilizes special strand displacement enzymes that circumvent the use of high temperatures for DNA denaturation. This allows the DNA amplification process to occur at much lower and constant temperatures, which is a key advantage for diagnosis in peripheral healthcare centers. Isothermal NAATs have been employed in the detection of TB in recent years, with loop-mediated isothermal amplification (LAMP) and recombinase polymerase amplification (RPA) reactions being the more popular techniques used in TB NAATs. These two methods will be described in the following sections. Notably, a commercial molecular assay using LAMP was recommended by the WHO as a replacement test for sputum smear microscopy in 2013 (Iwamoto et al. 2003; WHO 2017), highlighting the significant niche that isothermal DNA amplification technologies can fill in the diagnostic landscape for TB.

33.4.2 Loop Mediated Isothermal Amplification (LAMP)

LAMP was developed in 2000 (Notomi et al. 2000) and has since been demonstrated in a wide variety of applications requiring DNA amplification for the detection of human, animal and plant pathogens (Das et al. 2012; Iwamoto et al. 2003; Tomlinson et al. 2010). LAMP utilizes two sets of primers, an inner and outer primer set that recognizes six distinct target sequences in DNA, and produces DNA loop structures. Because of its multiple checks on site recognition before DNA synthesis, LAMP is a highly specific isothermal amplification reaction that is capable of producing a large amount of amplified DNA product, or amplicons, of varying sizes (Notomi et al. 2000). This is ideal for

DNA readouts that require a large amount of product for the visualization of results. Due to the single enzyme used in LAMP and the formation of magnesium pyrophosphate as a by-product, successful amplification can occasionally be assessed based on the presence or absence of a white precipitate in the reaction mixture (Mori et al. 2001). However, the atypical double primer set adds a layer of complexity in primer design and assay optimization. DNA amplicons of varying sizes could also be a double-edged sword when DNA products of specific sizes are required downstream, such as for identification or cloning. Nevertheless, Connelly and colleagues (2015) detected *Escherichia coli* using LAMP together with end-point detection using a handheld UV source and a camera phone

33.4.3 Recombinase Polymerase Amplification

Recombinase polymerase amplification (RPA) was developed in 2010 and has since been shown capable of TB detection in several studies (Boyle et al. 2014; Piepenburg et al. 2006). Unlike LAMP, RPA uses two enzymes in DNA amplification – the eponymous recombinase forms complexes with primers and searches for complementary target sequences in the template for binding, and a strand-displacing polymerase is used to generate copies while releasing a template strand for further amplification (Piepenburg et al. 2006). RPA is generally considered a rapid test as it only takes around 30 minutes for completion, and a reaction temperature of between 37 and 42 °C may be perceived as an advantage over LAMP, which reacts at temperatures between 60 and 65 °C. In one study body heat was shown able to substitute for a conventional heating device during the RPA process (Crannell et al. 2014).

33.5 Sample Preparation Workflow

Sample preparation is one of the more challenging aspects of TB diagnosis at the point-of-care. In a laboratory setting that follows up with bacteriological culture, sputum samples are digested and decontaminated following a widely accepted protocol with N-acetyl-L-cysteine and sodium hydroxide (NALC-NaOH) (Whittier et al. 1993). This kills off contaminants in the sample such as pseudomonad bacteria but recovers viable *M. tuberculosis* bacilli for subsequent culture. However, not being able to recover viable bacteria is of little consequence in a NAAT: on the contrary it is beneficial to kill *M. tuberculosis* bacilli through cell lysis to facilitate primer access to the DNA template. Due to the thick mycobacterial cell wall of *M. tuberculosis*, cell lysis is conventionally achieved through mechanical disruption techniques such as bead beating. This is not ideal in a peripheral healthcare setting that seeks to minimize reliance on equipment use. An alternative for cell lysis is the use of strong chaotropic reagents, such as guanidinium hydrochloride, which has been shown to be effective in lysing *M. tuberculosis*, and even tougher plant cell walls, for DNA amplification purposes (Ng et al. 2015a; Wee et al. 2015). (A chaotropic agent is a molecule in water solution that can disrupt the hydrogen bonding network between water molecules. This has an effect on the stability of the native state of other molecules in the solution, mainly macromolecules, such as proteins and nucleic acids, by weakening the hydrophobic effect.) Such a chemical lysis can take place at room temperature and does not require additional laboratory devices to provide shaking or heating.

The extraction and purification of *M. tuberculosis* DNA is another crucial step after cell lysis, as the yield and purity of DNA directly influences the sensitivity of the assay. In addition, some chemical agents used in extraction can impede enzyme activity in DNA amplification and must be completely removed from the purified DNA sample. *M. tuberculosis* DNA extraction and purification is typically achieved with a standard phenol-chloroform extraction protocol in the laboratory, which presents several challenges for implementation at the point-of-care (Sambrook and Russell 2006). Chief among these issues is the hazardous properties of the chemicals that mandate the use of chemical fume hoods, as well as the lengthy and tedious nature of the protocol. Instead of the above, the well-established technique of solid phase reversible immobilization (SPRI) can be used to perform DNA extraction and purification with good yield and purity (Deangelis et al. 1995; Hawkins et al. 1994). SPRI can be used in conjunction with commercially available carboxyl-coated magnetic nanoparticles that have an affinity for nucleic acids, enabling a simple DNA purification system based on magnetic separation. By precipitating DNA on to the surface of these magnetic nanoparticles, buffer exchange and elution can be performed using a magnet in a single tube, which reduces the cost and complexity associated with consumables and reagents. This method has been used with isothermal DNA amplification reactions with good efficiency (Ng et al. 2015a; Wee et al. 2015).

33.6 Nanotechnologies for TB DNA Sensing and Readouts

Traditionally the outcome of a DNA amplification reaction is determined using gel electrophoresis, where charged DNA molecules migrate within agarose gel under the influence of an electric field applied to a suitable buffer (Southern 1975). The speed at which DNA amplicons migrate correlates with their mass, which, in this case, is the relative size of the amplicon. A standard known as a ladder is also run together with the DNA amplicons to provide a basis for size comparison. The stained DNA molecules, including unused primers, can be visualized under ultraviolet light. Clearly, this is an unsuitable method for interpreting the result of DNA amplification in a peripheral healthcare setting due to the sheer amount of technical manipulation and apparatus required. Therefore, developing a simple, appropriate, and accurate sensor for the presence or absence of DNA amplicons is very relevant to the development of a point-of-care TB diagnostic test. Readouts for DNA amplification have very much come into their own as a field of important research for all types of DNA-based tests (Koo et al. 2017). These range from the measurement of glucose levels with a glucose meter through the hydrolysis of sucrose by DNA-conjugated invertase (Xiang and Lu 2012) to quantitative measurements using surface-enhanced Raman spectroscopy (Wang et al. 2017). (Surface-enhanced Raman spectroscopy [SERS] is a surface-sensitive technique that enhances Raman scattering by molecules adsorbed on rough metal surfaces or by nanostructures such as plasmonic-magnetic silica nanotubes. Raman scattering or the Raman effect is the inelastic scattering of a photon by molecules which are excited to higher vibrational or rotational energy levels. It was discovered by C.V. Raman and K.S. Krishnan in liquids, and independently by Grigory Landsberg and Leonid Mandelstam in crystals.)

Some suitable DNA readouts for point-of-care TB diagnostic tests will be examined in the following sections.

33.6.1 Naked Eye Detection of TB DNA

The major benefit of naked eye readouts is that the interpretation of DNA amplification results can be performed without a measurement device. This is valuable in two ways: firstly, it reduces dependence on equipment which brings distinct utility to point-of-care diagnosis in resource-poor settings, and secondly, it does not require a technician skilled in this field to operate the measurement device. This introduces the possibility of having staff, or even the patients themselves, being able to perform and interpret DNA test results. This reduces the burden of expertise required. A key disadvantage of naked eye detection lies in indeterminate results, where it is difficult to call a result positive or negative due to minute changes in state or coloration. This is generally an issue of low DNA product yield and can be minimized by ensuring that the DNA reaction is optimized to be sensitive and robust.

33.6.2 Bridging Flocculation

Bridging flocculation is a phenomenon in colloidal chemistry that was first described in 1953 and established as a well-known practice for colloidal separation processes (Healy and Lamer 1964), such as those used in water purification applications. It occurs due to the surface adsorption of particles on to long polymers via cross-linking that bring these particles out of solution (Lamer 1966). The same separation technique can be applied to DNA detection where DNA amplicons function as long polymers that flocculate particles in a positive amplification reaction, whereas shorter primers present in a negative amplification reaction will fail to do so (Figure 33.1). This technique has been demonstrated recently for DNA amplicons and carboxyl-coated magnetic beads, as used in SPRI purification. This is convenient because it reduces the variety of reagents in a single test (Ng et al. 2015b; Wee et al. 2015).

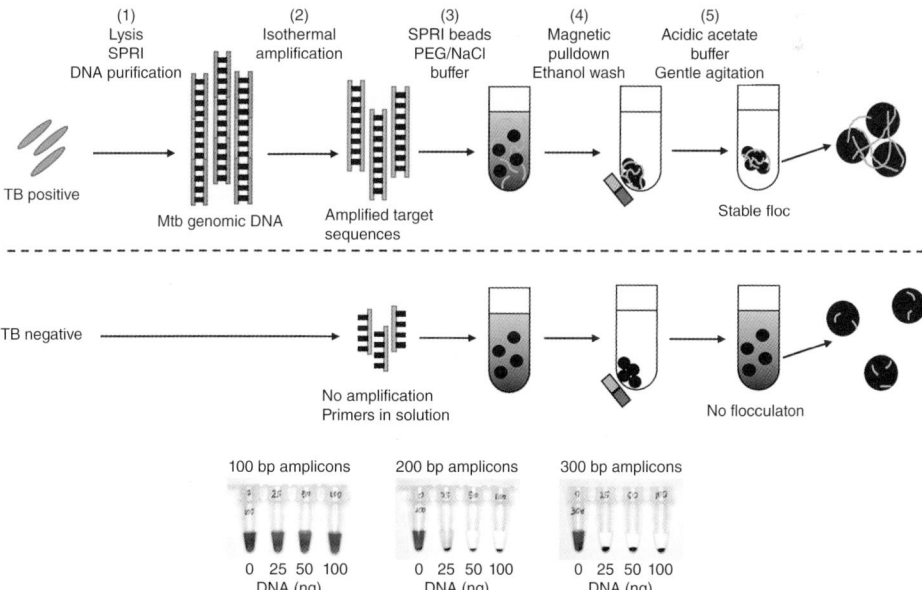

Figure 33.1 Bridging flocculation assay for nucleic acid detection. *(See color plate section for the color representation of this figure.)*

A positive result is determined when there are stable flocculates in a clear solution that are not disrupted through minor mechanical agitation, and a negative result presents as a suspension of magnetic particles. The two main factors affecting the sensitivity of the test are the length of DNA amplicons, which was determined to be at least 200 base pairs, and the amount of DNA produced after amplification. This naked eye approach has been used successfully for TB detection with close to single cell sensitivity and excellent specificity using the primers discussed earlier (Ng et al. 2015b).

33.6.3 Colorimetric Readouts

Colorimetric readouts are another type of naked eye readouts that are simple to interpret and do not require additional quantitative devices for measurement. A prominent example is the oxidation of tetramethylbenzidine (TMB) with horseradish peroxidase (HRP), which turns the colorless TMB a distinct blue (Josephy et al. 1982). One simple method is to incorporate HRP into the double stranded DNA product, making use of the well understood affinity between biotin and streptavidin (Figure 33.2). When biotinylated nucleotides are added into a nucleotide mix, the resulting DNA product possesses biotin handles that act as suitable binding sites for streptavidin-conjugated substrates. Examples are HRP-streptavidin and HRP-conjugated magnetic beads.

Subsequent washing with a buffer using a simple magnetic pulldown of DNA amplicons allows excess HRP to be removed from the product. In such an assay specificity for the target template is of paramount importance to ensure that there are no false positive results. Extreme care needs to be taken to prevent the formation of artifacts such as primer-dimers, as these double stranded products could also incorporate HRP and produce a false positive result. Additionally, it is vital to include a control, usually the absence of any DNA template, to rule this out. This method has been applied to the detection of TB, and shows excellent sensitivity and specificity for *M. tuberculosis* species (Figure 33.2) (Ng et al. 2016).

33.7 Quantitative DNA Detection Methodologies

Quantitative DNA readouts offer greater sensitivity and finer control over assay design at the expense of cost and overall complexity. Quantitative measurements provide feedback during the assay design process that allow critical thresholds to be established for

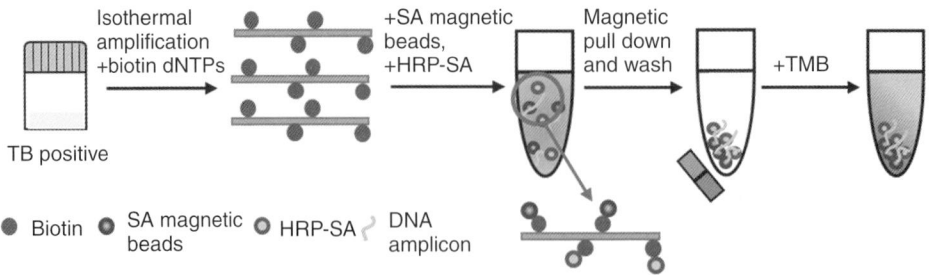

Figure 33.2 Naked eye colorimetric assay using HRP in amplified TB DNA to catalyze TMB. *Source:* Adapted with permission from The American Chemical Society. *(See color plate section for the color representation of this figure.)*

positive and negative results, thus eliminating visual bias with naked eye readouts and minimizing indeterminate outcomes. As a result, much lower limits of detection can be achieved with quantitative readouts that are not possible with simple visual confirmation. Measurement devices are integral to these assay readouts and must be selected based on their cost and amenability to point-of-care requirements. In particular, the broad field of electrochemical sensors exhibit good potential for quantitative DNA measurements and have been used for other translational applications in the past. Electrochemical detection encompasses all types of detection methodologies applied to a wide variety of sample analytes, and essentially refers to a system comprising molecular recognition, signal transduction and a device capable of measuring the signal output. The signal output is usually in the form of changes in electric current or potential. Electrochemical sensors remain popular as readouts due to their excellent sensitivity with small analyte input, in large part due to the amplification effect through signal transduction (Leng et al. 2010; Xu et al. 2012). Two different electrochemical techniques applied to TB detection will be described in detail in the following sections.

33.7.1 Differential Pulse Voltammetry with Gold Nanoparticles

The usage of gold nanoparticles (AuNPs) as labels is a cornerstone of nanoparticle-based electrochemical detection strategies as they are relatively easy to synthesize within a monodispersed range of nanoparticle sizes, and can be functionalized for a wide range of biological applications (Low et al. 2015; Pinijsuwan et al. 2008) (Figure 33.3a). Gold (Au) is oxidized by HCl to $AuCl_4^-$, and the electrochemically active species can be detected with methods such as differential pulse voltammetry (DPV) (Leng et al. 2010; Xu et al. 2012). For DNA detection double-stranded DNA product is labeled with AuNPs and immobilized on electrode surfaces for follow-up electrochemical measurement.

For the detection of TB in resource-poor areas the format is modified slightly to make use of much cheaper screen-printed carbon electrodes (SPCEs) and a handheld potentiostat instead of a workstation electrochemical sensor in the laboratory (Ng et al. 2015b). An added benefit of SPCEs is their disposability after each run, which simplifies the process by

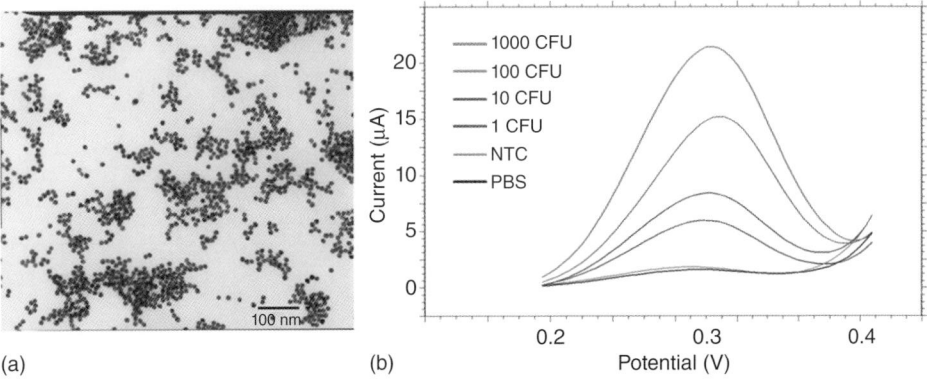

(a) (b)

Figure 33.3 Detection of TB DNA using Differential Pulse Voltammetry. (a) TEM image of as-prepared colloidal gold nanoparticles showing monodispersity. (b) DPV responses for AuNP-SA labeled DNA duplexes immobilized on SPCE from 1000, 100, 10, 1 CFU of *Mtb*, no template and PBS controls. *Source: Adapted with permission from The American Chemical Society. (See color plate section for the color representation of this figure.)*

removing electrode washing and potential contamination. DPV has been demonstrated to be able to detect as little as one colony-forming unit of *M. tuberculosis*, with excellent specificity (Figure 33.3b). This can also be used on a handheld potentiostat, which shows promise as a method that can be further developed for low resource settings (Ng et al. 2015b).

33.7.2 Ampometry and Spectrophotometry with TMB

Another simple electrochemical detection strategy is to employ ampometry to measure changes in the electric current that is the result of a redox reaction. Ampometry can be applied to oxidized TMB, as discussed above, to produce a secondary quantitative measurement that is used to validate the naked eye readout (Figure 33.4a). A current measurement is obtained when TMB is reduced back to its base oxidation state (Fanjul-Bolado et al. 2005). As with DPV, ampometry can be performed on SPCEs but does not require that the analyte be immobilized on the electrode surface (Ng et al. 2016).

An alternative to ampometry is to quantify the absorbance at 650 nm using a spectrophotometer, of which there are handheld versions. When used to detect the presence of TB, both ampometry and spectrophotometry produced highly concordant results in sensitivity and specificity and are helpful for establishing a conclusive result when the visual readout is indeterminate (Figure 33.4) (Ng et al. 2016).

33.8 Drug-resistant Tuberculosis

As discussed in Chapter 1, drug-resistant tuberculosis (MDR-TB) represents a particular challenge to public health systems because its treatment requires a much longer course of expensive and toxic drugs than does drug-susceptible TB (Tanimura et al.

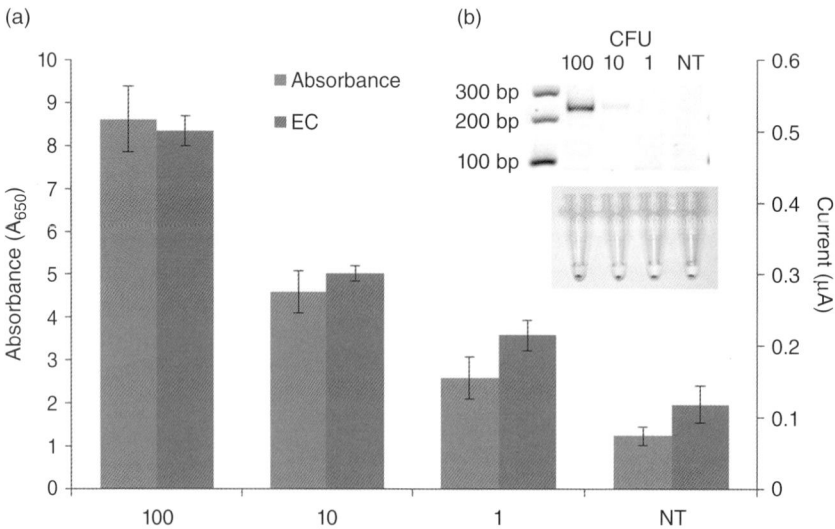

Figure 33.4 Quantitative analysis of colorimetric TB DNA assay. (a) Comparison of relative absorbance (light gray) and relative current (dark gray) scores between (L-R) 100, 10, 1 CFU of input gDNA and no template control. (b) Gel electrophoresis and photo of colorimetric assay of the experiments in (a). *Source:* Adapted with permission from The American Chemical Society.

2014). Multidrug-resistant TB is defined as *M. tuberculosis* resistant to both rifampicin (RIF) and isoniazid (INH), the two frontline drugs used to treat drug-susceptible TB. In 2015, 580 000 people were newly diagnosed with MDR-TB, and 250 000 patients died from the disease, making it responsible for more than 10% of the total deaths caused by TB in that year (WHO 2017). The gold standard for MDR-TB diagnosis is drug susceptibility testing with bacterial cultures, but these can take a number of weeks to obtain a result. In many resource-poor settings, a rapid screening diagnostic test is urgently needed to quarantine and treat such patients.

33.8.1 Specific Detection of Point Mutations in *M. tuberculosis* DNA

The basis for RIF resistance has been well-characterized, with 95% of all resistance shown to be caused by mutations within an 81-base pair (bp) region in the RpoB gene of *M. tuberculosis* (Ramaswamy et al. 1998; Telenti et al. 1993). More significantly, a comprehensive study of clinical isolates has determined that point mutations form the majority of rifampicin-resistant *M. tuberculosis* (RRTB) strains, with two of these point mutations accounting for more than 75% of all RRTB strains, namely, the $C \rightarrow T$ point mutation at codon 526 (Tyr mutant) and the $C \rightarrow T$ point mutation at codon 531 (Leu mutant) (Ramaswamy et al. 1998; Telenti et al. 1993). Past studies have shown that almost 90% of RRTB is also resistant to isoniazid, which led to rifampicin resistance being used extensively as a surrogate marker of choice for MDR-TB (Cavusoglu et al. 2002; Morgan et al. 2005).

Unfortunately, isothermal DNA amplification methods such as RPA lack temperature and cycle controls and are considered poorly suited for discriminating between point mutants and the wild type strain of *M. tuberculosis* at point-of-care testing. However, recent studies have reported that RPA amplifies targets with up to nine primer-template mismatches, and this has spurred the investigation of highly polymorphic primers for TB detection (Boyle et al. 2013; Daher et al. 2015).

As part of a solution toward specific detection of MDR-TB in low resource settings, a nested RPA assay consisting of a first round for enrichment of the target DNA followed by a second round using primers mismatched to the wild type yielded excellent specificity for mutant variants of *M. tuberculosis* (Figure 33.5a) (Ng et al. 2017). Multiple assay conditions, such as, but not limited to, primer length and concentration, type and position of base mismatches and temperature were investigated and optimized to improve assay specificity and sensitivity. Intercalating fluorescent dye specific for double-stranded DNA was used to determine the result of DNA amplification.

A useful application of this assay is a twostep workflow system (Figure 33.5b). The first round allows potential TB cases to be triaged, treatment to be initiated, and healthcare systems to be notified upon receipt of a positive result. The second round of amplification can then be performed to establish the presence of drug-resistant strains. In order to demonstrate its translational potential for point-of-care use, this assay has been tested on a simple heating device assembled from 3D printed parts and store-bought electronics that cost less than US$30 (Ng et al. 2017).

33.9 Conclusions

As part of the push to end TB, the rapid and simple detection of TB at the initial point-of-care in the peripheral healthcare setting remains an urgent priority in the LMICs.

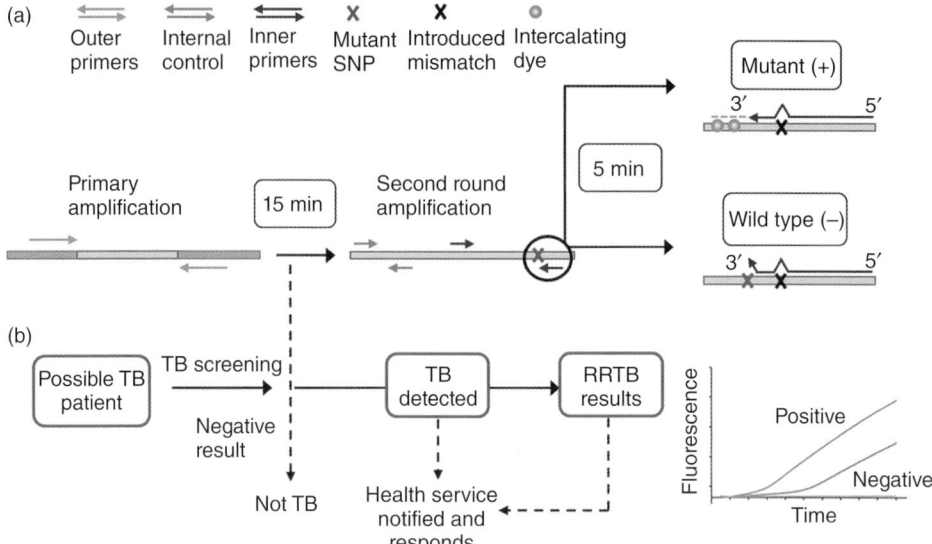

Figure 33.5 Point mutation detection for MDR TB screening. (a) Assay scheme using mismatched primers for specific point mutation discrimination. (b) Two step workflow for TB screening at the point-of-care. *Source:* Adapted with permission from The American Chemical Society. *(See color plate section for the color representation of this figure.)*

This chapter covers recent nanotechnologies that have been applied to tuberculosis, but they are translatable to other infectious agents whose genome has been sequenced. We believe that such tests may become reality very soon.

References

Bjune, G. (2008). A systematic review of delay in the diagnosis and treatment of tuberculosis. *BMC Public Health* 8: 15.

Boehme, C.C., Nicol, M.P., Nabeta, P. et al. (2011). Feasibility, diagnostic accuracy, and effectiveness of decentralised use of the Xpert MTB/RIF test for diagnosis of tuberculosis and multidrug resistance: a multicentre implementation study. *Lancet* 377 (9776): 1495–1505. https://doi.org/10.1016/s0140-6736(11)60438-8.

Boyle, D.S., Lehman, D.A., Lillis, L. et al. (2013). Rapid detection of HIV-1 proviral DNA for early infant diagnosis using recombinase polymerase amplification. *Mbio* 4 (2): https://doi.org/10.1128/mBio.00135-13.

Cavusoglu, C., Hilmioglu, S., Guneri, S., and Bilgic, A. (2002). Characterization of rpoB mutations in rifampin-resistant clinical isolates of *Mycobacterium tuberculosis* from Turkey by DNA sequencing and line probe assay. *J. Clin. Microbiol.* 40 (12): 4435–4438. https://doi.org/10.1128/jcm.40.12.4435-4438.2002.

Crannell, Z.A., Rohrman, B., and Richards-Kortum, R. (2014). Equipment-Free Incubation of Recombinase Polymerase Amplification Reactions Using Body Heat. *PLoS One* 9 (11): https://doi.org/10.1371/journal.pone.0112146.

Connelly, J.T., Rolland, J.P., and Whitesides, G.M. (2015). "Paper Machine" for molecular diagnostics. *Analytical Chemistry* 87 (15): 7595–7601. https://doi.org/10.1021/acs.analchem.5b00411.

Coros, A., DeConno, E., and Derbyshire, K.M. (2008). IS6110, a Mycobacterium tuberculosis complex-specific insertion sequence, is also present in the genome of Mycobacterium smegmatis, suggestive of lateral gene transfer among mycobacterial species. *J. Bacteriol.* 190 (9): 3408–3410. https://doi.org/10.1128/jb.00009-08.

Daher, R.K., Stewart, G., Boissinot, M. et al. (2015). Influence of sequence mismatches on the specificity of recombinase polymerase amplification technology. *Mol. Cell. Probes* 29 (2): 116–121. https://doi.org/10.1016/j.mcp.2014.11.005.

Das, A., Babiuk, S., and McIntosh, M.T. (2012). Development of a Loop-Mediated Isothermal Amplification Assay for Rapid Detection of Capripoxviruses. *J. Clin. Microbiol.* 50 (5): 1613. https://doi.org/10.1128/JCM.06796-11.

Deangelis, M.M., Wang, D.G., and Hawkins, T.L. (1995). Solid-phase reversible immobilization for the isolation of PCR products. *Nucleic Acids Res.* 23 (22): 4742–4743. https://doi.org/10.1093/nar/23.22.4742.

Eisenach, K., Cave, M.D., Bates, J., and Crawford, J. (1990). Polymerase chain reaction amplification of a repetitive DNA sequence specific for Mycobacterium tuberculosis. *J. Infect. Dis.* 16 (5): 977–981.

Fanjul-Bolado, P., Gonzalez-Garia, M.B., and Costa-Garcia, A. (2005). Amperometric detection in TMB/HRP-based assays. *Anal. Bioanal. Chem.* 382 (2): 297–302. https://doi.org/10.1007/s00216-005-3084-9.

Gillespie, S.H., McHugh, T.D., and Newport, L.E. (1997). Specificity of IS6110-based amplification assays for Mycobacterium tuberculosis complex. *J. Clin. Microbiol.* 35 (3): 799.

Hawkins, T.L., Oconnormorin, T., Roy, A., and Santillan, C. (1994). DNA purification and isolation using a solid-phase. *Nucleic Acids Res.* 22 (21): 4543–4544. https://doi.org/10.1093/nar/22.21.4543.

Healy, T.W. and Lamer, V.K. (1964). Energetics of flocculation and redispersion by polymers. *J. Colloid Sci.* 19 (4): 323–332. https://doi.org/10.1016/0095-8522(64)90034-0.

Helb, D., Jones, M., Story, E. et al. (2010). Rapid Detection of Mycobacterium tuberculosis and Rifampin Resistance by Use of On-Demand, Near-Patient Technology. *J. Clin. Microbiol.* 48 (1): 229–237. https://doi.org/10.1128/jcm.01463-09.

Iwamoto, T., Sonobe, T., and Hayashi, K. (2003). Loop-Mediated Isothermal Amplification for Direct Detection of Mycobacterium tuberculosis Complex, M. avium, and M. intracellulare in Sputum Samples. *J. Clin. Microbiol.* 41 (6): 2616–2622. https://doi.org/10.1128/JCM.41.6.2616-2622.2003.

Josephy, P.D., Eling, T., and Mason, R.P. (1982). The horseradish peroxidase-catalyzed oxidation of 3,5,3',5'-tetramethylbenzidine – Free-radical and charge-transfer complex intermediates. *J. Biol. Chem.* 257 (7): 3669–3675.

Koo, K.M., Wee, E.J.H., Wang, Y., and Trau, M. (2017). Enabling miniaturised personalised diagnostics: from lab-on-a-chip to lab-in-a-drop. *Lab on a chip*. https://doi.org/10.1039/c7lc00587c.

Lawn, S.D. and Nicol, M.P. (2011). Xpert (R) MTB/RIF assay: development, evaluation and implementation of a new rapid molecular diagnostic for tuberculosis and rifampicin resistance. *Future Microbiol.* 6 (9): 1067–1082. https://doi.org/10.2217/fmb.11.84.

Low, K.F., Rijiravanich, P., Singh, K.K.B. et al. (2015). An electrochemical genosensing assay based on magnetic beads and gold nanoparticle-loaded latex microspheres for vibrio cholerae detection. *J. Biomed. Nanotech.* 11 (4): 702–710. https://doi.org/10.1166/jbn.2015.1956.

McNerney, R., Maeurer, M., Abubakar, I. et al. (2012). Tuberculosis Diagnostics and Biomarkers: Needs, Challenges, Recent Advances, and Opportunities. *J. Infect. Dis.* 205: S147–S158. https://doi.org/10.1093/infdis/jir860.

Morgan, M., Kalantri, S., Flores, L., and Pai, M. (2005). A commercial line probe assay for the rapid detection of rifampicin resistance in *Mycobacterium tuberculosis*: a systematic review and meta-analysis. *BMC Infect. Dis.* 5 (1): 62. https://doi.org/10.1186/1471-2334-5-62.

Mori, Y., Nagamine, K., Tomita, N., and Notomi, T. (2001). Detection of Loop-Mediated Isothermal Amplification Reaction by Turbidity Derived from Magnesium Pyrophosphate Formation. *Biochem. Biophys. Res. Commun.* 289 (1): 150–154. https://doi.org/10.1006/bbrc.2001.5921.

MSF (2009). *Defining Specifications for a TB Point-of-Care Test*. Paris, France: Medicins Sans Frontieres.

Ng, B.Y.C., Wee, E.J.H., West, N.P., and Trau, M. (2015a). Naked-eye colorimetric and electrochemical detection of Mycobacterium tuberculosis – towards rapid screening for active case finding. *ACS Sens.* 1 (2): 173–178.

Ng, B.Y.C., Xiao, W., West, N.P. et al. (2015b). Rapid, single-cell electrochemical detection of mycobacterium tuberculosis using colloidal gold nanoparticles. *Anal. Chem.* 87 (20): 10613–10618.

Ng, B.Y.C., Wee, E.J.H., West, N.P., and Trau, M. (2016). Naked-eye colorimetric and electrochemical detection of Mycobacterium tuberculosis – towards rapid screening for active case finding. *ACS Sensors* 1 (2), 173–178.

Ng, B.Y.C., Wee, E.J.H., Woods, K. et al. (2017). Isothermal point mutation detection: toward a first-pass screening strategy for multidrug-resistant tuberculosis. *Anal. Chem.* 89 (17): 9017–9022.

Notomi, T., Okayama, H., Masubuchi, H. et al. (2000). Loop-mediated isothermal amplification of DNA. *Nucleic Acids Res.* 28 (12): e63–e63. https://doi.org/10.1093/nar/28.12.e63.

Parsons, L.M., Somoskövi, Á., Gutierrez, C. et al. (2011). Laboratory Diagnosis of Tuberculosis in Resource-Poor Countries: Challenges and Opportunities. *Clin. Microbiol. Rev.* 24 (2): 314–350. https://doi.org/10.1128/cmr.00059-10.

Piatek, A.S., Tyagi, S., Pol, A.C. et al. (1998). Molecular beacon sequence analysis for detecting drug resistance in Mycobacterium tuberculosis. *Nat. Biotechnol.* 16 (4): 359–363. https://doi.org/10.1038/nbt0498-359.

Piepenburg, O., Williams, C.H., Stemple, D.L., and Armes, N.A. (2006). DNA detection using recombination proteins. *PLoS Biol.* 4 (7): 1115–1121. https://doi.org/10.1371/journal.pbio.0040204/.

Pinijsuwan, S., Rijiravanich, P., Somasundrum, M., and Surareungchai, W. (2008). Sub-femtomolar electrochemical detection of DNA hybridization based on latex/gold nanoparticle-assisted signal amplification. *Anal. Chem.* 80 (17): 6779–6784. https://doi.org/10.1021/ac800566d.

Ramaswamy, S. and Musser, J.M. (1998). Molecular genetic basis of antimicrobial agent resistance in *Mycobacterium tuberculosis*: 1998 update. *Tubercle and lung disease : the official journal of the International Union against Tuberculosis and Lung Disease* 79 (1): https://doi.org/10.1054/tuld.1998.0002.

Sambrook, J. and Russell, D.W. (2006). Purification of nucleic acids by extraction with phenol:chloroform. *CSH protocols* 2006 (1): https://doi.org/10.1101/pdb.prot4455.

Smith, K.C., Starke, J.R., Eisenach, K. et al. (1996). Detection of Mycobacterium tuberculosis in clinical specimens from children using a polymerase chain reaction. *Pediatrics* 97 (2): 155.

Southern, E.M. (1975). Detection of specific sequences among DNA fragments separated by gel electrophoresis. *J. Mol. Biol.* 98 (3): 503–517. https://doi.org/10.1016/S0022-2836(75)80083-0.

Tanimura, T., Jaramillo, E., Weil, D. et al. (2014). Financial burden for tuberculosis patients in low- and middle-income countries: a systematic review. *Eur. Respir. J.* 43 (6): 1763.

Telenti, A., Imboden, P., Marchesi, F. et al. (1993). Detection of rifampicin-resistance mutations in mycobacterium-tuberculosis. *Lancet* 341 (8846): 647–650. https://doi.org/10.1016/0140-6736(93)90417-f.

Tomlinson, J.A., Boonham, N., and Dickinson, M. (2010). Development and evaluation of a one-hour DNA extraction and loop-mediated isothermal amplification assay for rapid detection of phytoplasmas. *Plant Pathol.* 59 (3): 465–471. https://doi.org/10.1111/j.1365-3059.2009.02233.x.

Wang, J., Koo, K.M., Wee, E.J.H. et al. (2017). A nanoplasmonic label-free surface-enhanced Raman scattering strategy for non-invasive cancer genetic subtyping in patient samples. *Nanoscale* 9 (10): 3496–3503. https://doi.org/10.1039/c6nr09928a.

Wee, E.J.H., Lau, H.Y., Botella, J.R., and Trau, M. (2015). Re-purposing bridging flocculation for on-site, rapid, qualitative DNA detection in resource-poor settings. *Chem. Commun.* 51 (27): 5828–5831. https://doi.org/10.1039/c4cc10068a.

Whittier, S., Hopfer, R.L., Knowles, M.R., and Gilligan, P.H. (1993). Improved recovery of mycobacteria from respiratory secretions of patients with cystic fibrosis. *J. Clin. Microbiol.* 31 (4): 861–864.

WHO. (2017). Global Tuberculosis Report 2016. http://apps.who.int/medicinedocs/en/d/Js23098en.

Xiang, Y. and Lu, Y. (2012). Using commercially available personal glucose meters for portable quantification of DNA. *Anal. Chem.* 84 (4): 1975–1980. https://doi.org/10.1021/ac203014s.

Xu, Q., Yan, F., Lei, J. et al. (2012). Disposable electrochemical immunosensor by using carbon sphere/gold nanoparticle composites as labels for signal amplification. *Chem. Eur. J.* 18 (16): 4994–4998. https://doi.org/10.1002/chem.201200171.

34

The Use of Functional Nanoparticles for Water Purification

Jing Zhang[1,2], Chuanping Feng[2], and Chengzhong Yu[1]

[1] Australian Institute for Bioengineering and Nanotechnology, The University of Queensland, Brisbane, Queensland, Australia
[2] School of Water Resources and Environment, China University of Geosciences (Beijing), Beijing, China

---

CHAPTER MENU

34.1 Introduction

Clean water is vital to human health and food security. However, water contamination has become a major problem with rapidly developing industry and agriculture as well as population growth in recent decades. The risks in safe drinking water supply mainly come from pathogenic microorganisms and chemical contamination. Pathogenic microorganisms presenting in drinking water lead to waterborne diseases such as cholera and typhoid (Hoyer et al. 2015). Some of the protozoal, bacterial and viral agents that can cause water-borne diseases are shown in Table 34.1.

Escherichia coli, widely recognized as an indicator of fecal contamination, is used to assess the microbiological quality of drinking water (McQuaig et al. 2006). It is estimated that 1.8 billion people drink water contaminated with *E. coli* (Bain et al. 2014). There are about 600000 deaths annually caused by water-borne disease, mainly African children under age five (WHO 2016). Moreover, 300–500 million tons of water contaminated by heavy metal ions, solvents and other toxic waste are released into water environment by industrial activity annually (Tesh and Scott 2014). Therefore, disinfection and contamination removal technologies are in urgent demand to provide safe drinking water.

Revolutionizing Tropical Medicine: Point-of-Care Tests, New Imaging Technologies and Digital Health,
First Edition. Edited by Kerry Atkinson and David Mabey.
© 2019 John Wiley & Sons, Inc. Published 2019 by John Wiley & Sons, Inc.

Table 34.1 Examples of diseases caused by water-borne microbial agents.

Disease	Microbial agent	Sources of microbial agent in water supply
1. Protozoal diseases		
Amoebiasis	*Entamoeba histolytica*	Sewage, non-treated drinking water, flies in water supply
Cryptosporidiosis	*Cryptosporidium parvum*	Collects on water filters and membranes that cannot be disinfected, animal manure, seasonal runoff of water.
Giardiasis	*Giardia lamblia*	Untreated water, poor disinfection, pipe breaks, leaks, groundwater contamination, camp-grounds where humans and wildlife use same source of water
2. Bacterial diseases		
Campylobacter infection	Most commonly *Campylobacter jejuni*	Water contaminated with feces
Cholera	*Vibrio cholerae*	Water contaminated with the bacterium
Escherichia coli infection	Certain strains of *E. coli*	Water contaminated with the bacterium
Dysentery	A number of species in the genera Shigella and Salmonella with the most common being *Shigella dysenteriae*	Water contaminated with the bacterium
Leptospirosis	Bacteria of genus Leptospira	Water contaminated by the urine of animal carrying the bacteria
Salmonellosis	Caused by many bacteria of the genus Salmonella	Water contaminated with the bacteria. (More commonly as a food borne illness)
Typhoid fever	*Salmonella typhi*	Water contaminated with feces of an infected person
3. Viral diseases		
Hepatitis A	Hepatitis A virus	Water or food contaminated with infected feces

Various techniques have been developed in recent decades for advanced treatment of drinking water, generally including disinfection (Sorlini et al. 2015), adsorption (Zietzschmann et al. 2016), membrane technology (Oka et al. 2017), and electrochemistry (Li et al. 2015). Despite their advantages, toxic byproducts can be generated during the disinfection process when using chlorine, chloramines, chlorine dioxide and ozone as disinfectants (Pan et al. 2016). Moreover, membrane fouling predominantly caused by deposition and growth of microorganisms is one of the reasons for the low efficiency in membrane-based water purification (Iorhemen et al. 2016). Traditional adsorption and electrochemistry technology are inefficient due to the low specific surface areas and pore volume of adsorbents and electrodes (Wan et al. 2012; Zhang et al. 2015). To address these challenges, it is necessary to develop new disinfectants, anti-fouling membranes and advanced adsorbents with integrated multi-functions and electrodes with huge specific surface areas.

Nanomaterials can meet the above needs for water treatment because of their unique nanostructures and high specific surface areas. Many excellent reviews have been published on nanomaterials or nanocomposites for water decontamination focused on the morphology (e.g. membrane, beads and porous structures) (Tesh and Scott 2014) or the functionalities (e.g. adsorption, photocatalysis and disinfection) (Santhosh et al. 2016) of nanomaterials, and each functionality has also been reviewed in details categorized by active substances (Chong et al. 2010; Le Ouay and Stellacci 2015; Spasiano et al. 2015; Lofrano et al. 2016; Yadav et al. 2016; Tan et al. 2016). The aim of this review is to highlight why nanomaterials play an important role in water purification. Thus, this review will introduce nanomaterials used for various water treatment technology, including disinfection, adsorption and electrochemistry. In the disinfection section, several disinfectants and anti-fouling membranes will be reviewed from three aspects: advantages compared with traditional ones, parameters that affect the disinfection efficacies, and selected products commercialized or patented. In the adsorption and electrochemistry sections, some advanced adsorbents with multi-functions and three-dimensional nanostructured electrodes used for capacitive deionization will be introduced, respectively.

34.2 Disinfection

Nanomaterials are widely used for disinfection as disinfectants and for membrane technology to prevent bacterial fouling of membrane filters in water purification. Among them, silver, some transition metal oxides such as titanium dioxide (TiO_2) and carbon-based nanomaterials exhibit excellent antibacterial performance (Xiao et al. 2015; Dhanalakshmi and Palanimurugan 2017; Xia et al. 2017). We will choose silver and TiO_2 as typical examples to explain why nanostructures have excellent disinfection efficacy.

34.2.1 Disinfectants

Silver ions have excellent performance on inhibition of *E. coli* because silver ions deactivate cellular enzymes and disrupt membrane permeability of bacteria (Akhigbe et al. 2016). However, the toxicity of silver is not obviously observed in wastewater treatment system because silver ions easily form complexes with various coexisting ligands including chloride, sulphide and dissolved organic carbon species (Wang et al. 2003). Moreover, silver chloride colloids have limited inhibition efficacy due to their big size which can reach microns in diameter. Compared to silver ions and colloids, silver nanoparticles are a more effective disinfectant toward autotrophic nitrifying bacteria (Choi et al. 2008) (Figure 34.1).

This is because nanoparticles are able to penetrate inside the bacteria and then cause damage through interaction with sulfur- and phosphorus-containing substances such as proteins and DNA (Morones et al. 2005). Importantly, silver nanoparticles with sizes larger than 10 nm have low toxicity toward mammalian cells (Ivask et al. 2014).

The size and morphology of silver nanoparticles have a significant impact on their disinfection efficacy (Camacho et al. 2015). Lu et al. (2013) synthesized silver nanoparticles of different sizes using a simple reduction or hydrothermal method. Compared with silver nanoparticles of larger sizes of 15 and 55 nm, the 5 nm ones exhibited the best antibacterial activity against *E. coli* and *Fusobacterium nucleatum*.

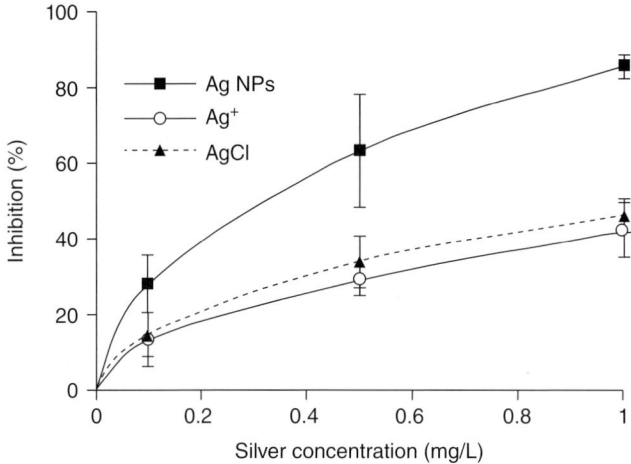

Figure 34.1 Nitrification inhibition as a function of the concentrations of silver (Ag) in the form of Ag nanoparticles (NPs), Ag^+ ions, and AgCl colloids. *Source:* Copyright 2008, Elsevier. Reproduced by permission.

In addition, two different shapes of silver nanoparticles with similar sizes, namely truncated octahedral silver nanoparticles (AgNOct) enriched in (111) facets and spherical silver nanoparticles (AgNS) enriched in (100) facets, were synthesized and compared for their antibacterial activity (Alshareef et al. 2017). AgNOct were found to be more active than AgNS, suggesting that the reactivity of silver is also dependent on the crystalline structure of exposed facets. Taking the influences of sizes and morphologies of silver nanoparticles on their disinfection efficacy altogether, silver nanoparticles with sizes smaller than 15 nm and (111) dominant facets are the most promising disinfectants.

Despite their advantages, nanoparticles applied in water disinfection have major limitations including aggregation and retention in liquid media (Das et al. 2012; Meier et al. 2017). To address the challenges and to achieve easy separation, immobilization of silver nanoparticles on natural substances, such as the fungus *Rhizopus oryzae*, wood and zeolite, were conducted (Das et al. 2012; Lin et al. 2014a; Akhigbe et al. 2016). The results showed that the size and distribution of silver nanoparticles can be controlled by changing the synthesis conditions, such as initial silver ion concentration and reaction temperature. However, a common phenomenon observed when using these natural materials as support material is the limited disinfection efficacy accompanied by a large amount of silver ion leaching. The leaching of silver ions is relieved by in situ reducing silver ions and depositing silver nanoparticles on the surface of graphene oxide, multi-walled nanotubes, polyurethane and other macromolecular compounds (Gunawan et al. 2011; Ma et al. 2015; Borse et al. 2016), but the effectiveness is limited and transitory. More stable structures are obtained by combining the silver nanoparticle with macromolecular compounds through chemical bonding or chelation. For example, silver ions are reduced by diatomite to generate silver nanoparticles which are conjugated by the silanol groups on diatomite; consequently, the silver concentration in filtered water in most pH environments is less than 100 ppb, the maximum limit recommended by the World Health Organization (WHO) (Xia et al. 2017).

Many silver nanoparticles as disinfection products have been commercialized for application medical, sanitary, and environment fields. Zibo Xingze Environmental Protection Technology Co., Ltd. provides a product composed of ceramic beads modified with silver nanoparticles for bathing water disinfection, which work effectively against a wide spectrum of bacteria and fungi, including *E. coli, Staphylococcus aureus* and *Candida albicans*, with disinfection efficacy of more than 90%. Several other water filters containing silver nanoparticles used for water purification have also been commercialized. However, with the increasing number of products containing silver nanoparticles being manufactured, researchers have considered the impact of nanotechnology on human health because small-sized silver nanoparticles with enhanced disinfection efficacy have increased health risks (Soares et al. 2016). Therefore, in the future, the long-term stability of products containing silver nanoparticles should be monitored and regulated.

TiO_2, a semiconducting material with a wide band gap of 3.2eV, has also been considered as one of the most promising photocatalysts and a disinfectant for widespread environmental applications due to its high reactivity, high physical and chemical stability, biocompatibility and commercial availability (Pathakoti et al. 2013). Under UV irradiation when the energy is higher than the bandgap energy of TiO_2, electrons at the valence band are excited to the conducting band, leaving positively charged holes in the valence band. The electron and hole pairs then migrate to the surface of TiO_2 where they react with water and oxygen to produce reactive oxygen species (ROS). These ROS come in contact with organic molecules, such as organic pollutants and microorganisms, and ultimately mineralize the organics into CO_2 and H_2O (Leong et al. 2014; Yadav et al. 2016).

The photocatalytic property of TiO_2 has made it widely used in water sterilization, which is also controlled by the structure of TiO_2 materials. To study the size influence of TiO_2 on their potential toxicity, the cytotoxicity of different-sized TiO_2 (10, 20, and 100 nm) toward mouse macrophage cells was tested (Xiong et al. 2013). The results showed that under photoactivation, the toxicity of TiO_2 nanoparticles significantly increased with decreasing particle size. This observation is consistent with the findings that smaller particles tend to have a higher rate of hydroxyl radical generation under photoactivation, because the surface area increases with decreased particle size, providing more photogenerated electrons. Similar results were also obtained by Lin et al. (2014b) on the size effect of TiO_2 on toxicity toward *E. coli*. Their results showed that the 50% mortality (LC50) of TiO_2-NP 10A (anatase TiO_2 with particle size of 10 nm) was 17.0 mg L^{-1} while that of TiO_2-NP 50A (anatase TiO_2 with particle size of 50 nm) was 304 mg L^{-1}. In addition, malondialdehyde was detected in the TiO_2 nanoparticle treated *E. coli*, indicating the nanoparticles induced the peroxidation of cell membrane lipid. The amount of malondialdehyde generated also depended on the TiO_2 sizes, exhibiting the same tendency found with ROS production (Figure 34.2).

UV light is the energy source used to activate TiO_2 in order to generate electrons and holes; however, it accounts for less than 4% of full solar spectrum, limiting the photocatalytic performance and broad applications of TiO_2 materials (Paul et al. 2007; Liu et al. 2012). In this regard, incorporation of Ag/AgCl, Au, Fe_2O_3, or transition metal ions such as Cr (VI) as electron donors is a useful strategy to make doped TiO_2 with photocatalytic activity under visible light irradiation (Liu et al. 2011; Wang et al. 2014; Wang et al 2016b; Xiao et al. 2015). Alternatively, Wang et al. (2016a) fabricated an ultrathin photocatalytic film using two dimensional (2D) titanium oxide nanosheets (TONs) via

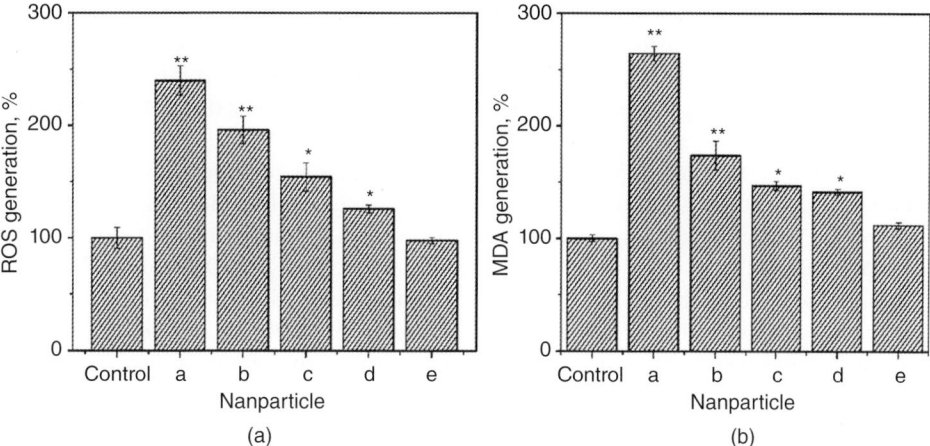

Figure 34.2 Relative contents of (a) intracellular ROS and (b) malondialdehyde (MDA) in the bacterial cells after 3 hours exposure to the TiO$_2$ nanoparticles (NPs) (50 mg L^{-1}). (a–e) stands for TiO$_2$-NP 10A, TiO$_2$-NP 25A, TiO$_2$-NP 25AR (Rutile TiO$_2$), TiO$_2$-NP 50A, and TiO$_2$-NP 50R, respectively; A and R stand for anatase and rutile, respectively, and AR stands for the mixture of anatase and rutile TiO$_2$. Asterisks indicate a significant difference relative to the control (*, $p < 0.05$; **, $p < 0.01$) based on the Student's t test. Error bars represent standard deviation (n = 3). *Source:* Courtesy Lin et al. (2014b). Reproduced by permission.

a layer-by-layer method. Interestingly, it was demonstrated that the TONs can store electrons under UV irradiation, and then discharge the stored electrons to produce antibacterial radicals in the dark due to the reduction/oxidation of Ti^{4+}/Ti^{3+} on the 2D TONs. The as-prepared film showed bactericidal activity toward both Gram-positive (*Enterococcus faecalis*) and Gram-negative (*E. coli*) bacteria in the dark, which further widens the application of TiO$_2$ as a photocatalyst.

Besides silver nanoparticles which have metal toxicity toward bacterial and semiconducting nanoparticles (e.g. TiO$_2$, ZnO and CuO) which produce ROS as disinfectants, some carbon-based materials also present antibacterial activity via carbon nanotubes, graphene oxide and mesoporous carbon nanoparticles. These nanomaterials possess unique structural and electronic properties: the nanoscale edges of the carbon-based materials can damage the integrity of bacteria by physical piercing, and the formation of superoxide anions cause further bacterial damage (Kang et al. 2007; Orooji et al. 2017). While carbon-based materials have advantages in water disinfection without causing secondary contamination (such as metal ion leaching) and toxicity to humans, their antibacterial efficacy in general cannot compete with silver or TiO$_2$ materials.

34.2.2 Anti-fouling Membranes to Improve the Performance of Water Filters

Bacterial growth control is a challenging issue in environmental applications because bacteria can use either inorganic or organic sources as nutrients. Long-term growth of bacteria may cause the formation of biofilms on the membrane surface, resulting in membrane fouling, decreased filter efficiency and potential risks to human health if the membrane is used for drinking water purification. For example, microcystins, which are often present in water supply systems, release cyanotoxins which cause hepatic and neuromuscular lesions and tumors (Teixeira and Sousa 2013).

Silver and TiO_2 nanoparticles are commonly added into membranes for desalination and ultrafiltration to address the membrane fouling issue for two reasons: they can change the surface property of hydrophobic membranes into hydrophilic to increase the flux (Tang et al. 2016), and they have antibacterial activities as discussed in the previous section.

Polyvinylidene fluoride (PVDF), as one of the excellent polymers widely used for membrane synthesis, has outstanding chemical and thermal stability as well as good mechanical properties (Zhang et al. 2008). However, its strong hydrophobic nature has become a barrier to its application in separation processes. A TiO_2 deposited PVDF/ sulfonated polyethersulfone (SPES) membrane was developed via self-assembly of TiO_2 nanoparticles along with UV irradiation (Rahimpour et al. 2012). The results showed that the hydrophilicity of the membrane was significantly improved by TiO_2 deposition and UV irradiation, and the antifouling property and long-term flux stability were also significantly enhanced. Meier et al. (2017) switched initially hydrophobic silicone nanofilaments (SNF) into hydrophilic ones by oxygen plasma treatment and then used them as substrates to load silver nanoparticles (AgNPs).

After coating the SNF or AgNP-SNF on the glass beads (GB), the amount of silver loading on AgNP-SNF-GB increased to as much as 30-fold of that on AgNP-GB. Consequently, no bacterial clogging on the filamentous structure of the SNF coating was observed. Furthermore, with columns packed by AgNP-SNF-GB, a contact time of about three seconds was sufficient to observe strong antibacterial activity and 91% reduction in colony-forming units (CFU, a unit used to estimate the number of viable bacteria cells in a sample) was achieved in comparison to pristine glass beads. The above findings suggest that doping membranes with silver and TiO_2 nanoparticles is a promising strategy to make functional membranes with potent anti-fouling activities.

Another method to make anti-fouling membranes is to add disinfection active substances into membranes. Gunawan et al. (2011) developed silver nanoparticle/multiwalled carbon nanotubes (Ag/MWNTs) coated on a polyacrylonitrile (PAN) hollow fiber membrane (Ag/MWNTs/PAN). This was achieved by coating silver nanoparticles with controlled sizes on polyethylene glycol-grafted MWNTs and then coating Ag/ MWNTs on to the external surface of a chemically modified PAN hollow fiber membrane to act as a disinfection barrier. The sizes of Ag nanoparticles synthesized were 2–5 nm (Figure 34.3).

It was found that the Ag/MWNT coating significantly enhanced the antimicrobial activities and anti-fouling properties of the functionalized membrane. From an

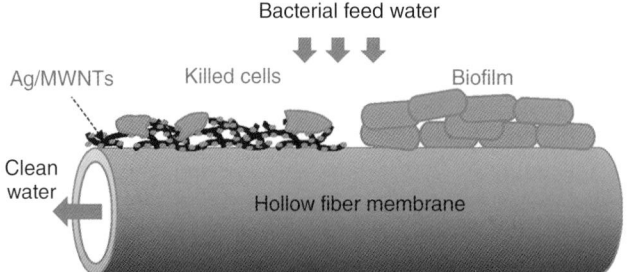

Figure 34.3 Ag/MWNT hollow fiber membrane for water disinfection. *Source:* Copyright 2011, American Chemical Society. Reproduced by permission.

engineering aspect, it is suggested that outer surface modification of membranes could be a more convenient and more efficient method than making blended membranes (mixing the active substances with the polymer precursors before making membranes) to ensure anti-fouling properties.

34.3 Adsorption

Nanomaterials have been widely used for the removal of heavy metal ions, organics, and inorganic anions from contaminated water because of their huge surface areas and rapid adsorption kinetics. Recently, multi-functional nanoadsorbents have drawn much attention because they can streamline the water treatment process. A few selected examples are introduced below to highlight recent progress in this direction.

34.3.1 Membranous Adsorbent Filter

Boron nitride (BN) ultrathin fibrous nanonets were developed by Lian et al. (2013) through a one-step solvothermal process. The average diameter of BN nanofibers was only approximately 8nm. The nanonets displayed outstanding performance for water purification with the maximum adsorption capacity for methylene blue (MB) of $327.8\,mg\,g^{-1}$. Moreover, the nanonets presented an ultrafast adsorption process for MB taking only one minute to achieve the adsorption equilibrium. In addition, a filtration membrane consisting of these nanonets also displayed a sieving performance for nano-particles with different sizes via a filtration process (Figure 34.4).

Incorporating nanostructured adsorbents into macroscopic membranes can enhance the filtration efficiency and achieve easy separation and recovery of adsorbents after the adsorption process (Liang et al. 2011). Usually, polysulfone, polyvinylidene fluoride and poly-acrylonitrile are employed to form blend membranes with nanoparticles (Zheng et al. 2013; He et al. 2014; Venkatesh et al. 2016; Yurekli 2016). The adsorption is inefficient because some of the adsorbents are trapped by the polymers so that they cannot contact with the adsorbate and thus cannot be used. A novel concept of dual-functional ultrafiltration (DFUF) membranes was recently developed by entrapping hollow porous Zr(OH)x nanospheres (HPZNs) as adsorbents into the fingerlike pores of poly (ether sulfone) (PES) ultrafiltration (UF) membranes followed by polydopa-mine sealing (Figure 34.5) (Pan et al. 2017).

DFUF exhibited improved removal efficiencies for low molecular weight and ionic contaminants than traditional UF membrane, allowing for the effective removal of mul-tiple contaminants including colloidal golds, polyethylene glycol and Pb (II) with an adsorption capacity of about $230\,mg\,g^{-1}$. Importantly, the DFUF membranes showed negligible leakage of nanoadsorbents during testing, and the membrane can be easily regenerated and reused.

34.3.2 Other Multifunctional Adsorbents

Multi-functional adsorbents with antibacterial or catalytic activities have also been investigated (Bazyari et al. 2016; Groiss et al. 2017). Purwajanti et al. (2015) introduced an easy and economical synthesis of MgO hierarchical microspheres. It was shown that

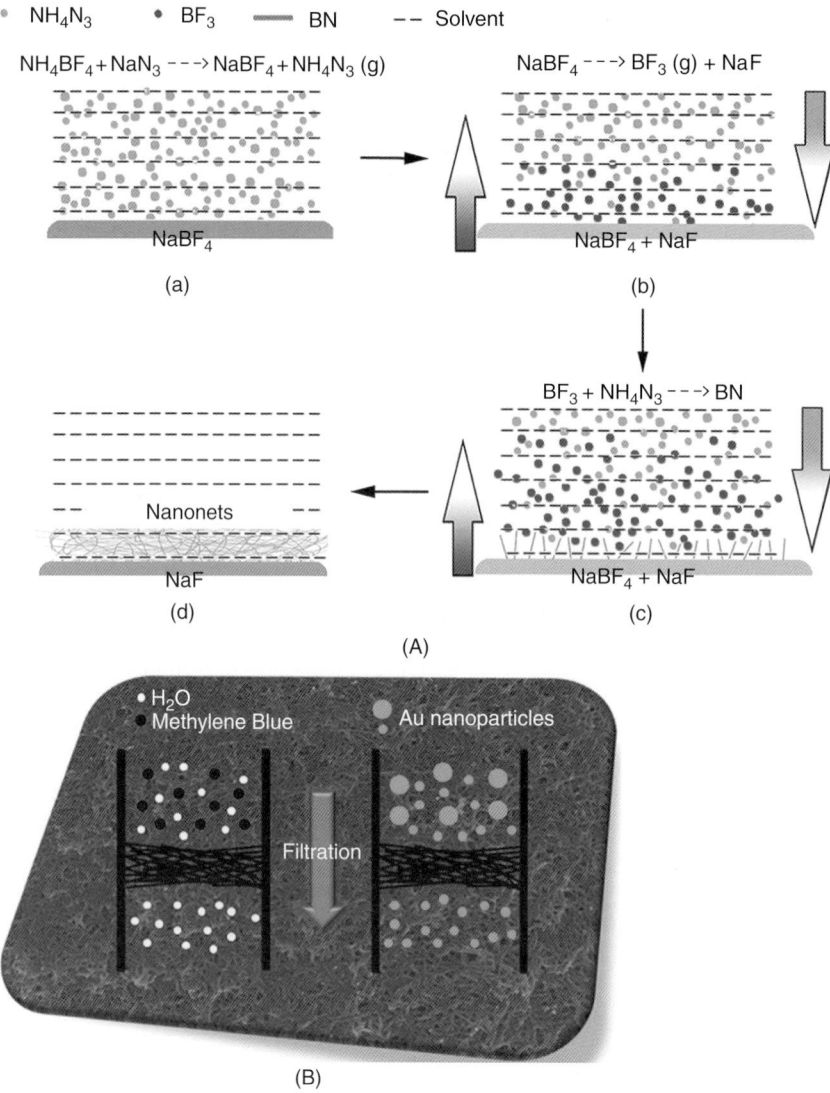

Figure 34.4 (A) Schematic illustration of the formation process of BN ultrathin fibrous nanonets; (B) The nanonets used for water purification. *Source:* Copyright 2013, American Chemical Society. Reproduced by permission. (*See color plate section for the color representation of this figure.*)

hierarchical MgO microspheres calcined at 500 °C exhibited the best trade-off between As (III) adsorption ($502 \, mg \, g^{-1}$) and antibacterial performance (complete elimination at $700 \, \mu g \, mL^{-1}$). In another example, CeO_2 NPs were used to develop a catalytic-sorbent with antibacterial property for the adsorption of lignin and photodegradation of organic dye (Shuhailath et al. 2016). With a high lignin adsorption efficacy of 97.53% and methyl orange removal efficiency up to 95% within 40 minutes, CeO_2@AlOOH/polyethylene imine (PEI) showed excellent antibacterial efficacy toward *E. coli*, *Klebsiella pneumoniae* and *S. aureus* bacteria. The ROS generated by CeO_2 NPs including hydroxyl, super-oxide radicals, and H_2O_2 which were mainly responsible for the antibacterial performance.

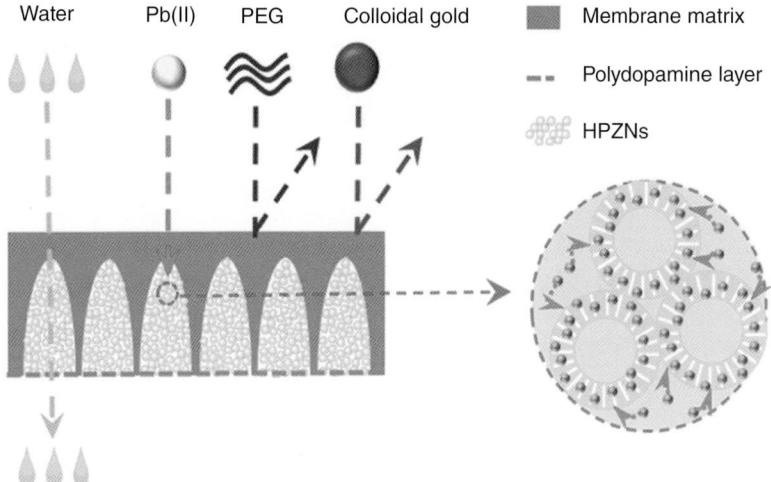

Water	Pb(II)	PEG	Colloidal gold		Membrane matrix
				--	Polydopamine layer
					HPZNs

Figure 34.5 The purification process for multiple water pollutants by DFUF membrane. *Source: Copyright 2017, American Chemical Society. Reproduced by permission. (See color plate section for the color representation of this figure.)*

34.4 Electrochemistry

The desalination efficiency of capacitive de-ionization (CDI) depends on the surface area, the conductivity and surface wettability of selected electrode materials. Kong et al. (2016) reported a three-dimensional holey graphene hydrogel (HGH) electrode with abundant in-plane pores. With this unique structure, the electrode material exhibited a large specific surface area for electrosorption of ions as well as interconnected channels for electron transport. As a result, the electrosorption capacity of the electrode achieved $26.8\,mg\,g^{-1}$ under the conditions of $5000\,mg\,L^{-1}$ NaCl feeding concentration and $1.2\,V$ applied voltage, showing excellent potential in CDI desalination applications. Zhang et al. (2016) introduced Li-ion treated graphene/carbon nanofiber (Li$^+$/G/CNF) architectures prepared by electrospinning for desalination applications. The abundant and self-supporting porous structure with a large surface area of $393\,m^2\,g^{-1}$ contributed to a high electroadsorption capacity as well as a fast adsorption rate. The results showed that Li$^+$/G/CNF exhibited a higher desalination efficiency of 84% and better conductivity compared to conventional active carbon particles (ACP) plates (Figure 34.6).

34.5 Conclusions and Future Perspectives

We have briefly reviewed typical nanoparticles used for water purification, including their applications in disinfection, anti-fouling membranes, multi-functional adsorbents and electrodes. Compared with bulk silver and silver ions, silver nanoparticles have demonstrated their advantages by carefully adjusting the particle size and morphology. The high effectiveness of nano-sized TiO$_2$ and carbon-based materials is also showcased. Control over the nanostructure and composition has significant impact on the generation of ROS and consequently the antibacterial performance. Furthermore, adding these active nanoparticles into membranes is a useful strategy in the preparation of multifunctional devices in water treatment.

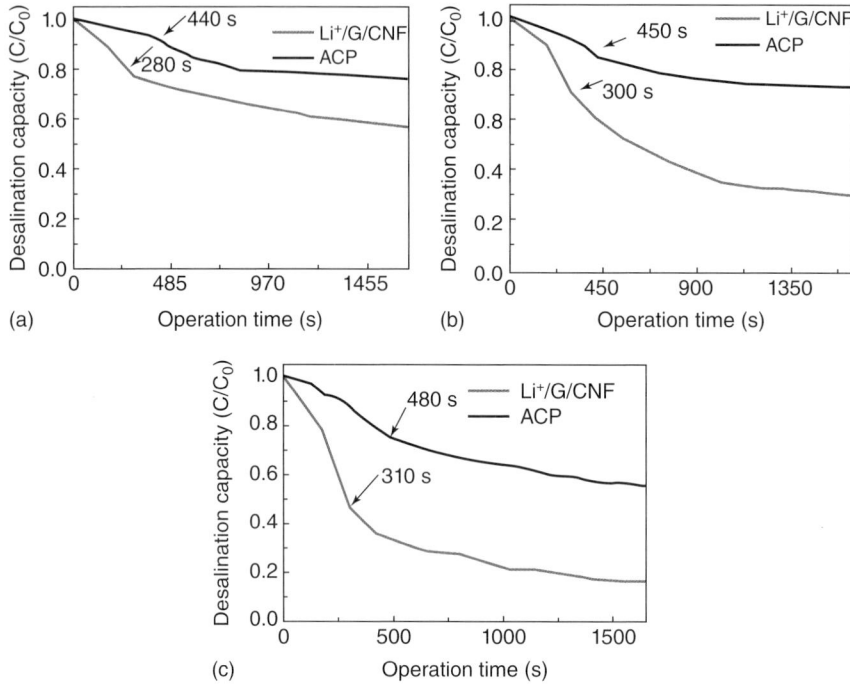

Figure 34.6 The desalination characteristics of Li$^+$/G/CNF architectures and conventional ACP plates in a CDI test with different initial sodium chloride (NaCl) concentrations: (a) 0.5 ‰, (b) 1 ‰ and (c) 3.5%. *Source:* Copyright 2015, Elsevier B.V. Reproduced by permission.

To date, only a small number of nano products are applied to water purification, because little is known about the final fate of released nanoparticles in the environment and their long-term toxicity to living creatures. To take advantage of the promising properties of nanomaterials and use them in water treatment, scale-up synthesis of advanced nanoparticles and their engineering into functional products need tremendous inputs from various sectors including industries. Simultaneously, regulatory concerns of safe manufacturing and environmental risks of nanomaterials/nano-products should be addressed. The rapid progress in nanoscience and nanotechnology is expected to add a significant contribution in controlling water-borne diseases and chemical threats to public health.

References

Akhigbe, L., Ouki, S., and Saroj, D (2016). Disinfection and removal performance for *Escherichia coli* and heavy metals by silver-modified zeolite in a fixed bed column. *Chem. Eng. J.* 295: 92–98.

Alshareef, A., Laird, K., and Cross, R.B.M (2017). Shape-dependent antibacterial activity of silver nanoparticles on *Escherichia coli* and *Enterococcus faecium* bacterium. *Appl. Surf. Sci.* https://doi.org/10.1016/j.apsusc.2017.03.176.

Bain, R., Cronk, R., Hossain, R. et al. (2014). Global assessment of exposure to faecal contamination through drinking water based on a systematic review. *Trop. Med. Int. Health* 19: 917–927.

Bazyari, A., Khodadadi, A.A., Mamaghani, A.H. et al. (2016). Microporous titania–silica nanocomposite catalyst-adsorbent for ultra-deep oxidative desulfurization. *Appl. Catal. B Environ.* 180: 65–77.

Borse, S., Temgire, M., Khana, A., and Joshi, S (2016). Photochemically assisted one-pot synthesis of PMMA embedded silver nanoparticles: antibacterial efficacy and water treatment. *RSC Adv.* 6: 56674–56683.

Camacho, A., Holt, K., Kouri, J.B., Ramirez, J.T. and Yacaman, M.J. (2005). The bactericidal effect of silver nanoparticles. *Nanotechnology* 16, 2346–2353.

Choi, O., Deng, K.K., Nam-Jung Kim Jr., L.R., Surampalli, R.Y. and Hu, Z. (2008). The inhibitory effects of silver nanoparticles, silver ions, and silver chloride colloids on microbial growth. *Water Res.* 42, 3066–3074.

Chong, M.N., Jin, B., Chow, C.W.K., and Saint, C. (2010). Recent developments in photocatalytic water treatment technology: a review. *Water Res.* 44: 2997–3027.

Das, S.K., Khan, M.R., Guha, A.K. et al. (2012). Silver-nano biohybride material: synthesis, characterization and application in water purification. *Bioresour. Technol.* 124: 495–499.

Dhanalakshmi, A. and Palanimurugan, B.N. (2017). Enhanced antibacterial effect using carbohydrates biotemplate of ZnO nano thin films. *Carbohydr. Polym.* 168: 191–200.

Groiss, S., Selvaraj, R., Varadavenkatesan, T., and Vinayagam, R (2017). Structural characterization, antibacterial and catalytic effect of iron oxide nanoparticles synthesised using the leaf extract of *Cynometra ramiflora. J. Mol. Struct.* 1128: 572–578.

Gunawan, P., Guan, C., Song, X. et al. (2011). Hollow fiber membrane decorated with Ag/MWNTs: toward effective water disinfection and biofouling control. *ACS Nano* 5: 10033–10040.

He, J., Takeshi, M., and Paul Chen, J (2014). A novel Zr-based nanoparticle-embedded PSF blend hollow fiber membrane for treatment of arsenate contaminated water: material development, adsorption and filtration studies, and characterization. *J. Membr. Sci.* 452: 433–445.

Hoyer, A.B., Schladow, S.G., and Rueda, F.J (2015). A hydrodynamics-based approach to evaluating the risk of waterborne pathogens entering drinking water intakes in a large, stratified lake. *Water Res.* 83: 227–236.

Iorhemen, O.T., Hamza, R.A., and Tay, J.H (2016). Membrane bioreactor (MBR) technology for wastewater treatment and reclamation: membrane fouling. *Membranes* 6 (2): 33. https://doi.org/10.3390/membranes6020033.

Ivask, A., Kurvet, I, Kasemets, K. et al. (2014). Size-dependent toxicity of silver nanoparticles to bacteria, yeast, algae, crustaceans and mammalian cells in vitro. *PLoS One* 9: 102108.

Kang, S., Pinault, M., Pfefferle, L.D., and Elimelech, M (2007). Single-walled carbon nanotubes exhibit strong antimicrobial activity. *Langmuir* 23: 8670–8673.

Kong, W., Duan, X., Ge, Y. et al. (2016). Holey graphene hydrogel with in-plane pores for high-performance capacitive desalination. *Nano Res.* 9: 2458–2466.

Le Ouay, B. and Stellacci, F (2015). Antibacterial activity of silver nanoparticles: a surface science insight. *Nano Today* 10: 339–354.

Leong, S., AmirRazmjou, K.W., Hapgood, K. et al. (2014). TiO$_2$ based photocatalytic membranes: a review. *J. Membr. Sci.* 472: 167–184.

Li, Y., Shen, W., Shujie, F. et al. (2015). Inhibition of bromate formation during drinking water treatment by adapting ozonation to electro-peroxone process. *Chem. Eng. J.* 264: 322–328.

Lian, G., Zhang, X., Si, H. et al. (2013). Boron nitride ultrathin fibrous nanonets: one-step synthesis and applications for ultrafast adsorption for water treatment and selective filtration of nanoparticles. *ACS Appl. Mater. Interfaces* 5: 12773–12778.

Liang, H., Cao, X., Zhang, W. et al. (2011). Robust and highly efficient free-standing carbonaceous nanofiber membranes for water purification. *Adv. Funct. Mater.* 21: 3851–3858.

Lin, L., Li, J., Ma, S. et al. (2014a). Toxicity of TiO_2 nanoparticles to *Escherichia coli*: effects of particle size, crystal phase and water chemistry. *PLoS One* 9 (10): e110247.

Lin, X., Wang, F., Kuga, S. et al. (2014b). Eco-friendly synthesis and antibacterial activity of silver nanoparticles reduced by nano-wood materials. *Cellulose* 21: 2489–2496.

Liu, H., Shon, H.K., Sun, X., Vigneswaran, S. and Nan, H. (2011). Preparation and characterization of visible light responsive Fe_2O_3–TiO_2 composites. *Appl. Surf. Sci.* 257, 5813–5819.

Liu, L., Liu, Z., Bai, H., and Sun, D.D (2012). Concurrent filtration and solar photocatalytic disinfection/degradation using high-performance Ag/TiO_2 nanofiber membrane. *Water Res.* 46: 1101–1112.

Lofrano, G., Carotenuto, M., Libralato, G. et al. (2016). Polymer functionalized nanocomposites for metals removal from water and wastewater: an overview. *Water Res.* 92: 22–37.

Lu, Z., Rong, K., Li, J. et al. (2013). Size-dependent antibacterial activities of silver nanoparticles against oral anaerobic pathogenic bacteria. *J. Mater. Sci. Mater. Med.* 24: 1465–1471.

Ma, S., Zhan, S, Jia, Y., and Zhou, Q (2015). Highly efficient antibacterial and Pb(II) removal effects of $AgCoFe_2O_4$-GO nanocomposite. *ACS Appl. Mater. Interfaces* 7: 10576–10586.

McQuaig, S.M., Scott, T.M., Harwood, V.J. et al. (2006). Detection of human-derived fecal pollution in environmental waters by use of a PCR-based human polyomavirus assay. *Appl. Environ. Microbiol.* 72: 7567–7574.

Meier, M., Suppiger, A., Eberl, L., and Seeger, S (2017). Functional silver-silicone-nanofilament-composite material for water disinfection. *Small* 13: 1601072.

Morones, J.R., Elechiguerra, J.L., Comacho, O. et al. (2005). The bactericidal effect of silver nanoparticles. *Nanotech.* 16 (10): 2346–2353.

Oka, P.A., Khadem, N., and Berube, P.R (2017). Operation of passive membrane systems for drinking water treatment. *Water Res.* 115: 287–296.

Orooji, Y., Faghih, M., Razmjou, A. et al. (2017). Nanostructured mesoporous carbon polyethersulfone composite ultrafiltration membrane with significantly low protein adsorption and bacterial adhesion. *Carbon* 111: 689–704.

Pan, S., Li, J., Noonan, O. et al. (2017). Dual-functional ultrafiltration membrane for simultaneous removal of multiple pollutants with high performance. *Environ. Sci. Technol.* 51: 5098–5107.

Pan, Y., Li, W.B., Li, A., et al. (2016). A new group of disinfection byproducts in drinking water: trihalo-hydroxy-cyclopentene-diones. *Environ. Sci. Technol.* 50, 7344–7352.

Pathakoti, K., Morrow, S., Han, C. et al. (2013). Photoinactivation of *Escherichia coli* by sulfur-doped and nitrogen–fluorine-codoped TiO_2 nanoparticles under solar simulated light and visible light irradiation. *Environ. Sci. Technol.* 47: 9988–9996.

Paul, T., Miller, P.L., and Strathmann, T.J (2007). Visible-light-mediated TiO_2 photocatalysis of fluoroquinolone antibacterial agents. *Environ. Sci. Technol.* 41: 4720–4727.

Purwajanti, S., Liang, Z., Nor, Y.A. et al. (2015). Synthesis of magnesium oxide hierarchical microspheres: a dualfunctional material for water remediation. *ACS Appl. Mater. Interfaces* 7: 21278–21286.

Rahimpour, A., Jahanshahi, M., Mollahosseini, A., and Rajaeian, B (2012). Structural and performance properties of UV-assisted TiO_2, deposited nano-composite PVDF/SPES membranes. *Desalination* 285: 31–38.

Santhosh, C., Velmurugan, V., Jacob, G. et al. (2016). Role of nanomaterials in water treatment applications: a review. *Chem. Eng. J.* 306: 1116–1137.

Shuhailath, K.A., Linsha, V., Kumar, S.N. et al. (2016). Photoactive, antimicrobial CeO_2 decorated AlOOH/ PEI hybrid nanocomposite: a multifunctional catalytic-sorbent for lignin and organic dye. *RSC Adv.* 6: 54357–54370.

Soares, T., Ribeiro, D., Proença, C. et al. (2016). Size-dependent cytotoxicity ofsilver nanoparticles in human neutrophils assessed by multiple analytical approaches. *Life Sci.* 145: 247–254.

Sorlini, S., Biasibetti, M., Collivignarelli, M.C., and Crotti, B.M (2015). Reducing the chlorine dioxide demand in final disinfection of drinking water treatment plants using activated carbon. *Environ. Technol.* 36: 1499–1509.

Spasiano, D., Marotta, R., Malato, S., Fernandez-Ibañez, P. abnd Di Somma, I. (2015). Solar photocatalysis: materials, reactors, some commercial, and pre-industrialized applications. A comprehensive approach. *Appl. Catal. B Environ.* 170–171, 90–123.

Tan, X., Liu, Y., Yanling, G. et al. (2016). Biochar-based nano-composites for the decontamination of wastewater: a review. *Bioresour. Technol.* 212: 318–333.

Tang, C., Bai, H., Liu, L. et al. (2016). A green approach assembled multifunctional Ag/ AgBr/TNF membrane for clean water production and disinfection of bacteria through utilizing visible light. *Appl. Catal. B Environ.* 196: 57–67.

Teixeira, M.R. and Sousa, V.S. (2013). Fouling of nanofiltration membrane: effects of NOM molecular weight and microcystins. *Desalination* 315: 149–155.

Tesh, S.J. and Scott, T.B (2014). Nano-composites for water remediation: a review. *Adv. Mater.* 26: 6056–6068.

Venkatesh, K., Arthanareeswaran, G., and Chandra Bose, A (2016). PVDF mixed matrix nano-filtration membranes integrated with 1D-PANI/TiO_2 NFs for oil–water emulsion separation. *RSC Adv.* 6: 18899–18908.

Wang, G., Zheng, X., Zeng, X. et al. (2016a). Ultrathin titanium oxide nanosheets film with memory bactericidal activity. *Nanoscale* 8: 18808–18809.

Wang, J., Huang, C.P., and Pirestani, D (2003). Interactions of silver with wastewater constituents. *Water Res.* 37: 4444–4452.

Wang, Y., Xiong, D., Wang, Z. et al. (2016b). Surface plasmon resonance of gold nanocrystals coupled with slow-photon-effect of biomorphic TiO_2 photonic crystals for enhanced photocatalysis under visible-light. *Catal. Today* 274: 15–21.

Wang, Y.-S., Jyun-Hong, S., and Horng, J.-J (2014). Chromate enhanced visible light driven TiO_2 photocatalytic mechanism on Acid Orange 7 photodegradation. *J. Hazard. Mater.* 274: 420–427.

Wang, Z., Dou, B., Lu, Z. et al. (2012). Effective desalination by capacitive deionization with functional graphene nanocomposite as novel electrode material. *Desalination* 299: 96–102.

WHO (2016). *Climate Change and Health*. World Health Organisation.

Xia, Y., Jiang, X., Zhang, J. et al. (2017). Synthesis and characterization of antimicrobial nanosilver/diatomite nanocomposites and its water treatment application. *Appl. Surf. Sci.* 396: 1760–1764.

Xiao, G., Zhang, X., Zhang, W. et al. (2015). Visible-light-mediated synergistic photocatalytic antimicrobial effects and mechanism of Ag-nanoparticles@chitosan–TiO$_2$ organic–inorganic composites for water disinfection. *Appl. Catal. B Environ.* 170-171: 255–262.

Xiong, S., George, S., Ji, Z. et al. (2013). Robert Damoiseaux, Bryan France, Kee Woei Ng, Say Chye Joachim Loo, size of TiO$_2$ nanoparticles influences their phototoxicity: an in vitro investigation. *Arch. Toxicol.* 87: 99–109.

Yadav, H.M., Kim, J.-S. and Pawar, S.H. (2016). Developments in photocatalytic antibacterial activity of nano TiO$_2$: a review. *Korean J. Chem. Eng.* 33(7): 1989-1998.

Yurekli, Y. (2016). Removal of heavy metals in wastewater by using zeolite nano-particles impregnated polysulfone membranes. *J. Hazard. Mater.* 309: 53–64.

Zhang, J., Chen, N., Zheng, T. et al. (2015). A study of the mechanism of fluoride adsorption from aqueous solutions onto Fe-impregnated chitosan. *Phys. Chem. Chem. Phys.* 17: 12041–12050.

Zhang, M., Zhang, A.-Q., Zhu, B.-K. et al. (2008). Polymorphism in porous poly(vinylidene fluoride) membranes formed via immersion precipitation process. *J. Membr. Sci.* 319: 169–175.

Zhang, T., Zhao, H., Huang, X.X., and Wen, G (2016). Li-ion doped graphene/carbon nanofiber porous architectures for electrochemical capacitive desalination. *Desalination* 379: 118–125.

Zheng, X., Chen, D., Wang, z., Lei Y. and Cheng R. (2013). Nano-TiO$_2$ membrane adsorption reactor (MAR) for virus removal in drinking water. *Chem. Eng. J.* 230, 180–187.

Zietzschmann, F., Stutzer, C., and Jekel, M. (2016). Granular activated carbon adsorption of organic micro-pollutants in drinking water and treated wastewater - aligning breakthrough curves and capacities. *Water Res.* 92: 180–187.

35

The Use of Drones in the Delivery of Rural Healthcare

Debrah I. Boeras[1,2], Blanche C. Collins[3], and Rosanna W. Peeling[1]

[1] *International Diagnostics Centre, London School of Hygiene and Tropical Medicine, London, UK*
[2] *Global Health Impact Group, Atlanta, GA, USA*
[3] *Centers for Disease Control and Prevention; Center for Surveillance, Epidemiology, and Laboratory Services, Atlanta, GA, USA*

This book chapter was co-written by Blanche C. Collins in her private capacity. No official support or endorsement by the Centers for Disease Control and Prevention, Department of Health and Human Services is intended, nor should be inferred.

Revolutionizing Tropical Medicine: Point-of-Care Tests, New Imaging Technologies and Digital Health,
First Edition. Edited by Kerry Atkinson and David Mabey.
© 2019 John Wiley & Sons, Inc. Published 2019 by John Wiley & Sons, Inc.

35.1 Challenges in Healthcare Delivery – Opportunities for Innovation

Innovation in healthcare represents an opportunity for the implementation of new or improved products, services, or processes (Omachonu and Einspruch 2010; Haughom n.d.). It can also support new organizational models and service delivery models, including patient services.

Traditional healthcare involves the encounter-based care delivery model where a patient visits a clinic to see a healthcare provider. A specimen may be obtained and sent off for diagnostic testing to a laboratory with results returned to the healthcare provider to guide patient management. In emergency situations, access to a hospital or clinic with emergency supplies, life-saving blood, critical diagnostic tests and medicines is possible in most urban settings. But what happens in rural areas where there is often a critical shortage of healthcare personnel and basic facilities, supplies and where laboratory medicine is lacking? Or what happens in disease outbreak situations, often hitting hard-to-reach areas with very limited resources, and time becomes the most critical factor? These are some of the many challenges driving the need for innovations in rural healthcare. In this Chapter we review the ongoing challenges in healthcare delivery and explore the use of drones as an innovation in the delivery of rural healthcare.

35.2 The Need for Disruptive Solutions for Healthcare Delivery in Rural Areas

Resource-limited countries already struggle with weak infrastructures that prohibit equitable access to healthcare. An estimated one billion people live in rural areas that lack access to an all-weather road and hence have limited access to essential supplies such as food, healthcare supplies and other life-saving interventions otherwise easily accessible by people living in urban centers (The World Bank 2017). It is estimated that between 26 400 and 61 000 people die annually from rabies, with most of those deaths occurring in rural areas (WHO 2013). Rabies has the highest fatality rate of any infectious disease and in rural areas availability of the vaccine is limited. Because of this, children bitten by rabid animals have to wait hours or days for the necessary vaccine. The highest rate of mortality in childbirth is seen in women living in rural areas where the drug to stop postpartum hemorrhage is not often available.

With the commitment of countries to the Sustainable Development Goals advocating for universal healthcare and leaving no one behind, it is time to question existing systems and re-imagine rural healthcare using disruptive transport systems that will create new markets and new opportunities to deliver services directly to rural communities (United Nations 2015).

For early infant diagnosis of the human immunodeficiency virus (HIV) disease, we have seen innovations at work to enable specimen collection in remote and rural areas through the use of dried blood spots (DBS) where, after a simple finger prick, a small drop of blood is collected on to filter paper (Smit et al. 2014). These specimens are air dried, stored and transported in batches to a laboratory for testing. Though DBS is an innovative method for collecting specimens, results often take months to return to the provider, by which time, some of the infants have already become seriously ill or died.

In an effort to overcome some of the delay and other challenges associated with DBS, point-of-care (POC) diagnostic technology has been introduced (Meggi et al. 2017). While bringing diagnostic testing closer to the patient provides a solution, it remains expensive and cannot be introduced into every rural community. An efficient, alternative specimen transport solution is needed.

35.3 The Use of Drones for Healthcare Delivery

One of the most radical and revolutionary innovations in healthcare systems has been the use of unmanned aerial vehicles (UAVs) or drones These were initially developed for military use in (surprisingly) the late 1800s and early 1900s, but they are now also being used in varying and growing roles such as public safety, commercial uses and healthcare delivery (Choi-Fitzpatrick et al. 2016; Unitaid 2015).

The first non-military uses for drones were for humanitarian purposes – to provide aid and assess damage in areas affected by major disasters (Balasingam 2017). Examples are given in this section and in greater detail in Section 35.4. of this chapter.

Drones were successfully used in the 2010 earthquake in Haiti, the 2012 hurricane Sandy, that affected the north-eastern United States, Canada and the Caribbean, the 2015 category five cyclone 'Pam' that struck the islands of Vanuatu, and the 2015 earthquake Gorkha in Nepal (Meier 2015; Howard 2015; Sharma 2016). Commercially, drones are being used to deliver packages and food (McFarland 2017). The healthcare industry is also exploring the varying using of drones and their capabilities to transform patient care (Global Health Supply Chain Summit 2017; Drogolea n.d.).

Drone batteries need to be replaced after flights. Since drones do not operate on easy-to-access batteries, they need to be ordered and are more expensive than traditional batteries. One way to help reduce drone operating costs would be to work with drone battery manufacturers to create rechargeable batteries and solar powered battery rechargers.

35.3.1 Modes of Operation for Drones

Drones come in various shapes and sizes with numerous capacities. Small drones can even be miniaturized to insect-sized devices used for military situational awareness while much larger commercial drones are being used in the movie industry to carry cameras. In general, the larger the drone, the longer its endurance, speed, tolerance for higher altitudes, and ability to carry larger payloads (cargo). Figure 35.1 shows the relationship between payload and range for various types of drones.

Drones can vary in payload from <0.5 kg up to 100 kg for commercial use (McFarland 2017). Their range extends up to approximately 500 km, switching from rotary wings to fixed wings, accordingly. They can be launched in the air vertically like a helicopter, hand thrown, or catapulted like a rocket (Fahlstrom and Gleason 2012). Some can even perform wheeled take-offs and landings using a runway. Landings can also occur by various methods, but usually do not require much room. Some are capable of two-way launching and landing with cargo while others are primarily one-way systems which launch with cargo but instead of landing to deliver, they hover and drop their cargo. They can be radio-controlled remotely, relying on global positioning system (GPS) and radio frequencies and can live stream using a camera controller. They can be manually

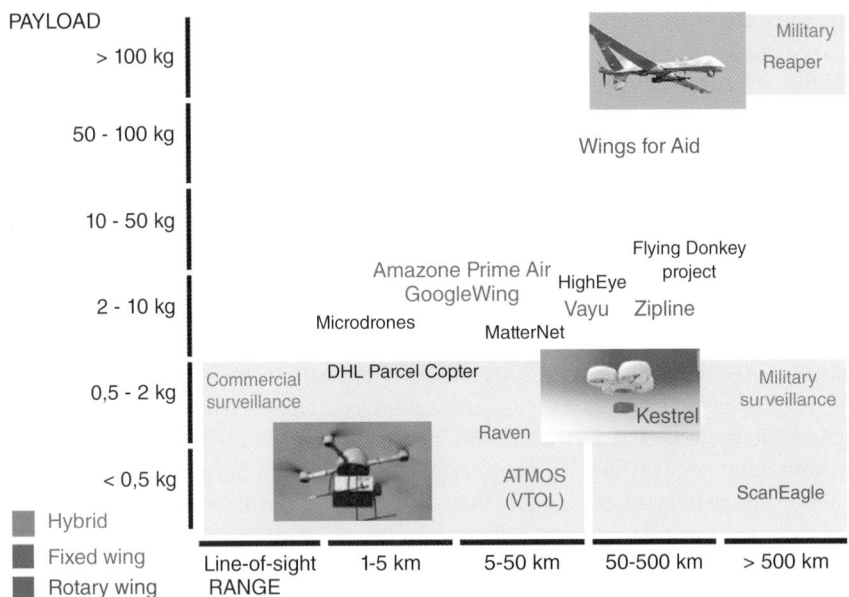

Figure 35.1 The relationship between payload and range for various types of drones. (*See color plate section for the color representation of this figure.*)

operated or preprogrammed to fly specific routes. Drones are navigated using GPS and the cellular network of the country in which they are operating. Drones usually can sense and avoid obstacles, but this is an area of ongoing development.

In February 2017, United States Agency for International Development performed a landscape analysis of drones in the development context and compared merits of different drone technologies used for the delivery of cargo (Table 35.1) (USAID 2017). The fastest drones such as Zipline can operate at 90 miles per hour (mph) while the drones used by the commercial companies Matternet and DHL operate at lower speeds of 25–40 mph.

The initial use of drones for health-related purposes includes delivering aid packages to disaster areas as well as delivering blood and specimens, medicines and other medical supplies, vaccines, anti-venom and diagnostic test kits to hard-to-reach areas.

35.3.2 Delivery in Disasters and Health Emergencies

Drones have been used during disasters on a trial basis to deliver food and medical supplies. In 2012 small aid packages were delivered to camps set up in Haiti and Nepal after the 2010 earthquakes (Gaitan 2014; Sharma 2016; USAID 2017). While helpful, authorities in Nepal complained that drones were an added burden to the situation since there were no restrictions or regulations, allowing them to fly anywhere and offering no feedback to the government. Google has launched Project Wing, an effort to build delivery drones that can be used for disaster relief by delivering aid, including water and medical supplies to affected areas (Stewart 2014). Communication equipment and mobile technologies could be delivered by drones to disaster areas to provide critical support. Portable shelters could also be delivered by drones in a rapid fashion to areas where critical infrastructure is needed to support emergency and healthcare systems.

Table 35.1 USAID landscape analysis and specifications of drones for potential delivery of cargo.

	Fixed wing	Multi-rotor	Hybrid
Range	Up to 160 km	About 20 km	About 80 km
Payload	Up to 5 kg	Up to 2 kg	Up to 5 kg
Launch	Catapult	Vertical	Vertical
Variations	Gas or electric	Gas or electric	Gas or electric
Advantages	• Long range • More efficient • Heavier payloads than multi-rotor • More stable flying • Well established concept with the weight of aerospace engineering behind it	• Maneuverability in small spaces • Vertical takeoff and landing • Generally cheaper • Can fly with a minimum of two rotors	• Vertical take-off and landing but with comparable range to fixed wing • More options for landing and takeoff sites • Heavier payloads than multi-rotor • Easier for "safe" emergency landings
Disadvantages	• Large space required for take-off and landing (no VTOL) • Limited maneuverability in small spaces • Emergency landings are generally less easy to control	• Low payload limit • Generally more complex designs (high software requirements to keep in the air) requiring expert maintenance and trained staff at health centers • Limited range • Inefficient in some settings	• Generally more expensive • Neither as long range as fixed wing nor as maneuverable as multi-rotor
Manufacturers	• Zipline • Wings for Aid • UAVaid	• Matternet • Flirtey • Microdrones	• Amazon • Google • DHL • Drones for Development - Dr.One. • Quantum Systems • Vayu
Example of Users	• Government of Rwanda • MOAS	• MSF • World Bank • UNICEF • Swiss Post	• MSF (planned) • We Robotics (planned)

Abbreviation used: VTOL, vertical take-off and landing.

35.3.3 Transporting Blood and Specimens

Drones can be used as part of the routine healthcare system to supplement and strengthen existing systems and networks that support healthcare, focusing on healthcare logistics (Amukele et al. 2017a).

In the USA drones recently set a new distance record with a three hour flight carrying human blood samples across 161 miles of desert (Johns Hopkins Medicine 2017).

In October 2016, **Rwanda** started delivering lifesaving blood products for eight million Rwandans across 21 hospitals (Zipline n.d.). Blood products could be dropped onsite in less than 15 minutes compared to hours by road transportation. Given this capacity, drones had the ability to complete about 150 blood deliveries a day regionally to transfusing facilities. A hospital could order blood and medicines via text messages and receive the supplies within 30 minutes (Toor 2016). Specific blood volumes and types can be easily moved across the country, reaching the areas where they are most needed (Amukele et al. 2017b).

Drones are also being used to deliver samples to laboratories for testing. The laboratory may be distant or unreachable due to geography or impassable roads due to flooding. For example, **Madagascar** is using drones to collect medical samples from remote villages in the most rural areas of the country to carry blood and stool samples for testing at the country's central laboratory (Stony Brook University 2016; Drogolea n.d.).

Malawi has successfully used motorcycle transport to deliver blood samples to the testing facilities, thus improving the turnaround time in obtaining results. Ten percent of the population are HIV-positive and there are only eight laboratories equipped for HIV testing. Drones can thus support the sample transport system to provide additional outreach, particularly in hard-to-reach areas during rainy seasons (Reuters 2016).

In **Papua New Guinea**, Médecins Sans Frontières (MSF) International used drones to transport tuberculosis test samples from a remote village to a large city for testing (Médecins Sans Frontières 2014).

35.3.4 Transporting Medicines and Medical Supplies

The delivery of medication is also being explored. Both Amazon and DHL are using weather- and waterproof containers to deliver medicines directly to individuals or to local pharmacies (Hern 2014).

In **sub-Saharan Africa** condoms and contraceptives are being transported by drones to regions where women lack access to basic services (Blau 2017).

In the **USA** the National Aeronautics and Space Administration (NASA) piloted a drone to deliver medical supplies to a rural clinic in a hard-to-reach rural area (DeAmicis 2015). The supplies included medications for asthma, hypertension and diabetes mellitus. It was shown to be safe and to dramatically reduce delivery time.

35.3.5 Vaccine Delivery

Vaccine delivery has been an area of strong interest for drone use. Rapid delivery of vaccines to poor and inaccessible areas could prevent outbreaks of life-threatening communicable diseases. Apart from improving accessibility, the use of drones could potentially reduce the costs of vaccines, and, in turn, improve vaccination rates (Johns Hopkins Bloomberg School of Public Health 2016).

The price of routine immunizations is expected to rise by 80% between 2010 and 2020 with supply chain logistics being a major driver of cost. In resource-limited settings, one third of the cost of a vaccine can be attributed to supply chain logistics (WHO, UNICEF,

World Bank 2009). Using drones for transportation could offset these costs and ensure vaccines remain affordable. Development of temperature-controlled chambers for vaccine delivery is ongoing.

The government of **Vanuatu** and United Nations International Children's Fund (UNICEF) are testing drones in the delivery of vaccines to health centers in hard-to-reach Vanuatu islands (Hackett 2017) to increase childhood vaccination rates and reduce vaccine-preventable diseases.

35.3.6 Anti-venom Delivery

In some areas of the Amazon, such as **Peru**, snakebites are common but quick access to anti-venom is not. In a simulated snake-bite scenario WeRobotics and the Peruvian Ministry of Health field-tested drones to deliver anti-venom to a remote community in the Amazon and reduced the delivery time from 6 hours to 35 minutes (Meier and Bergelund 2017).

35.3.7 Diagnostic Test Kit Delivery

Supplying HIV diagnostic test kits to hard-to-reach areas is a challenge. In countries where HIV prevalence rates are high, overcoming this challenge is essential for controlling the spread of disease. In **Malawi** UNICEF has tested the use of drones to deliver HIV test kits to rural areas (McNeish 2016).

35.4 Further Focus on Uptake of Drone Technology by Different Countries

35.4.1 Malawi

In Malawi, as indicated above, drones are increasing access and coverage for early infant diagnosis of HIV and are reducing the test result turn-around time. Like many other countries in resource-limited settings, motorbikes have been used to transport specimens along rural roads, bringing dried blood spots (DBS) to testing facilities in urban centers. However, the country's varied topography and severe flooding events can make motorbike transport challenging. The rough roads and unstable bridges often become impassable or completely destroyed and it can take months for infants to receive their results and to start treatment. Malawi also has one of the highest rates of HIV infection in the world but with only eight laboratories offering testing for early infant diagnosis. Drones have been viewed as a desirable option to support the healthcare sample transport system. UNICEF is demonstrating how drones can deliver HIV testing to infants in these hard-to-reach areas (Allen 2016). The drones being piloted for this project are small, weighing 5–15 kg, with payloads up to 5 kg, which means they could deliver up to 60 DBS cards per flight. VillageReach, a non-government organization (NGO) working with UNICEF, examined the use of drones for this purpose (Phillips et al. 2016). One 5 kg UAV could travel a distance no greater than 25 km between service delivery points. At an average speed of 40.5 km per hour, the drone could fly 6 hours a day, visiting each service delivery point at least twice per week.

35.4.2 Rwanda

In Rwanda drones are providing medical supplies to hospitals using an on-demand ordering system and expansively delivering vaccines to facilities. Currently 5 of the 45 country hospitals are using the drone-on-demand innovative ordering system to improve their healthcare system, with plans for servicing nearly half of all hospitals within one year. A healthcare worker places an order by text message and within minutes a drone is launched. The medical products are dropped off, landing gently and accurately in an open area about the size of a few car parking spaces. These same drones are also delivering vaccines to healthcare facilities through a partnership between Zipline and the Global AIDS Vaccine Alliance (GAVI 2016). From a single base 15 drones are able to serve 21 clinics. The drones can reach speeds up to $100\,km\,h^{-1}$ for 150 deliveries a day.

35.4.3 Tanzania

Tanzania is planning the largest drone healthcare delivery operation in the world. Four bases across the country will deploy up to 120 drones daily, serving more than 10 million people at 1000 clinics. These 30-lb electric drones will travel up to 50 miles from their base, delivering blood, vaccines, essential medications and medical supplies such as sutures and intravenous tubing (National Public Radio 2017). The drone on-demand delivery system will be used to service the entire healthcare system. Clearly such a large operation requires detailed planning of logistics, integration with the existing laboratory system, supply chain, human resources, and information technology.

35.5 Models of Potential Public-Private Collaboration

There are numerous drone companies supporting these efforts in resource-limited settings, but one that has been receiving much attention lately from the healthcare community is the San Francisco based company, Zipline (http://flyzipline.com/product). Founded in 2011, Zipline has established drone networks using a multitude of 25 lb (11.3 kg) drones that can fly up to 75 miles before needing recharging. The drones are electric and can carry 3 lb (1.3 kg) of essential medicines and supplies across countries. After a hospital places an order, a worker packs the products into a shoebox-sized container and loads it onto a drone, which flies to the hospital, drops the box by parachute and flies back to the distribution center. Here one of its batteries is replaced and a new package loaded. In five minutes, the drone can take off again, in any weather, day-in and day-out.

Their innovative approach is to work with governments to fully integrate their drone system within the national healthcare system, focusing on human resources and logistics to successfully maintain 500 flights a day. With this approach, the company is building country capacity and strengthening national systems and infrastructures. The company has trained local workers to operate and manage the drones and to run the logistics and supporting systems. This cadre of logistical human resources run the distribution centers, stock products and medical supplies, and work with country supply chain management.

35.6 Promises and Challenges of the Use of Drones in Healthcare Delivery

From all the examples included in this chapter, it is clear that drones have the potential to dramatically improve rural healthcare delivery and services with numerous benefits including:

- *Increasing access and routine healthcare coverage* – Deliver laboratory specimens to central laboratories for testing and bring vaccines and medical supplies to rural areas.
- *Delivering life-saving products* – Reach individuals in rural and remote areas within minutes with emergency medical deliveries, such as blood and blood products, snake anti-venom and post-exposure prophylaxis (PEP) treatment for rabies, often making the difference between life and death.
- *Increasing system efficiencies* – Decrease the turn-round time for delivery of diagnostic tests, samples, and test results.

However, significant challenges remain that need to be addressed:

- *Cost* - A major consideration is the cost of using drones. Drones are able to extend life-saving services or potentially decrease the costs associated with emergency medical care. Studies evaluating costs associated with routine services showed that drones have the potential – in some instances to strengthen health systems by increasing access and availability in a cost-efficient way. However, no studies thus far have measured health impact parameters, total cost of country-based ownership, nor have any yet contributed to building the evidence-base to support the expansion of this innovation at scale (Haidari et al. 2016; Phillips et al. 2016).
  The largest cost–benefit of this innovation may exist relative to the existing ground-based logistics system. Recently published evaluations have demonstrated the impact that modifications to transportation routes can have on the cost-effectiveness of supply chains. For example, Assi et al. (2013) found that removing the regional level and restructuring the transportation network in **Niger** so that districts collect vaccines directly from the central level resulted in improved vaccine availability by 30% as well as reduced total cost per dose of vaccine delivered. Other studies of similar transportation network changes have demonstrated comparable impacts (Assi et al. 2013; Lee et al. 2015, 2016).
- *Regulatory considerations* – Country regulatory legislations need to be in place before drones can be used for healthcare delivery. These include establishing predetermined flight corridors along which drones fly for routine deliveries as well as on-call systems for drone deliveries. For medical emergencies the fastest route needs to be used and the air corridor cleared. Regulatory restrictions can set back the widespread use of drones in healthcare. Federal, state and local regulatory bodies must all agree on operational procedures (Thiels et al. 2015).
- *Logistics* – Storage and transportation of all medications, vaccines, medical supplies, and specimens via drones need to be carefully implemented and monitored for the integrity of the products and the safety of the persons involved. This includes storage temperature and transportation duration. For example, in the case of blood transport, coolers would be required to keep blood products at the correct temperature. Blood would need to be packaged appropriately to prevent inadvertent exposure during transit. Additionally, the range of drones, their payload, climate zones and topogra-

phy, and the resulting network will need to be carefully mapped. The smaller more affordable drones can carry a 5 lb (1.8 kg) payload for 30–60 minutes of flight time and have a range of approximately 20–60 miles (31–97 kms).

- *Sustainability* – As with most innovations, the sustainability of drones will depend on many factors, including cost, uptake, policies regarding their use, and logistical and technical support. Once drones are in place, they could potentially be used for other activities, but the longevity and maintenance of drones still needs to be determined. Drones use batteries so battery life and battery supplies must be considered. Consumer demand is an often-overlooked factor. New technologies see a swift surge as there is interest around the novelty, but often programs then see a plateau and lack of continued uptake as challenges surface and novelty turns to nuisance. Community attitudes and acceptance will need to be evaluated and communities will need to become aware of, and comfortable with, drones.

Ultimately, for this technology to succeed at a national scale and serve the communities at need, stakeholders will need to work with governments to perform the critical mapping and costing exercises, drive necessary policy changes, and ensure this technology and system is integrated into their national health system (Thornton 2017).

35.7 Outlook for the Future

Is it possible to make drones even smarter? Could countries and programs benefit if drones were automated across mapped networks for routine and emergency deliveries, able to expand cellular data networks, and support two-way services, as well as serve multiple functions while in use (for example, health systems surveillance for potential disease outbreaks?).

- *Two-way services*
  If drones are already delivering specimens, can the same drones be used to pick up the necessary emergency supplies, medications and other treatments for that patient as a result of their diagnostic test? This type of service is what is actually needed for true clinical care and response during outbreaks. Currently, not all delivery drones make pickups, but if a drone can make pickups, it can also perform deliveries (Landhuis 2016).

35.7.1 Manual Versus Automatic Control of Drones

There are significant concerns regarding having drones flying in air space without control and beyond eyesight (Popper 2016). Perhaps one way to deal with flight regulation issues may be to have drones fully automated and flying along pre-defined flight corridors.

35.7.2 Delivering Within Hospitals

Large medical centers and hospitals could utilize drones to transport blood and medications within the facility. Intra-hospital drone use may provide more cost-effective floor-to-floor transport or courier service between buildings. However, it could also pose challenges related to GPS or radiofrequency communication interference and equipment size and cost constraints.

- *Ambulance drones*

Google has patented a device that can call for a drone in emergency situations to fly in with life-saving medical equipment on board (Murphy 2016). With the push of a button an ambulance drone equipped with specific lifesaving medical equipment would quickly appear. Approximately 30% of the time, lifesaving technologies such as automatic external defibrillators (AEDs) and cardiopulmonary resuscitation (CPR) aids are inaccessible. There are also specific drones being developed that can deliver AEDs for use on individuals undergoing an acute myocardial infarction (Engineering strategic communications 2016; Van de Voorde et al. 2017). In addition, ambulance drones also have the capability of instructing bystanders on how to perform CPR and can instruct them on the use of AEDs until emergency services arrive (Momont n.d.).

- *Telemedical drones*

Perhaps one of the most revolutionary drone concepts for healthcare delivery is a drone equipped with video-conferencing capabilities. Hologram technology could provide an immersive physician-patient virtual interaction. The Health Integrated Rescue Operations (HIRO) drone system delivers a case that includes medical supplies as well as cellular-connected Google Glass smart glasses. A person near the sick patient is expected to put on the glasses, which then send a video of what is in front of them to a remote physician. The physician can then see what is happening and lead the deputized civilian through the necessary treatment steps that utilize the supplies in the case (Tesser 2017). This would allow doctors to communicate with victims of natural disasters in situations where they may not be reachable in-person.

This also expands the potential impact for drones delivering specialized medical equipment such as portable ultrasound scanners to remote areas. The delivery of such equipment in concert with telemedicine via the drone has the ability to increase access to care and provide people living in rural areas with earlier diagnosis and treatment recommendations (Balasingam 2017).

- *Transporting humans and organs*

It is thought that a large drone or the coordination of a fleet of pilotless drones could potentially transport humans. While still early, there have been a few pilot experiments on this ambitious goal. One drone could potentially carry a single passenger weighing up to 220 lb (100 kg) and a small suitcase for 30 minutes. Drones could also be used to transport lifesaving organs for organ transplantation. In the EHang, a company in the USA has created the Manufactured Organ Transport Helicopter (MOTH) system to help save lives by swift delivery of organs to people in need rather than waiting for a helicopter or courier (EHang 2016).

- *Detection of methicillin-resistant Staphylococcus aureus*

Hospital-acquired methicillin-resistant *Staphylococcus aureus* (MRSA) is a growing concern for healthcare settings. IBM has patented a drone-based microbial analysis system in which drones could potentially monitor and map bacterial infections (Kozloski 2016). Drones can be trained to detect and analyze microbes in an environment, to recommend treatment and predict their spread.

- *Lab-on-a-drone*

Polymerase chain reaction (PCR) is considered the laboratory gold standard for diagnostic tests but is often limited to large centralized pathology laboratories with significant infrastructure. More recently molecular point-of-care instruments have been developed. Priye et al. (2016) are investigating smartphone-enabled PCR diagnostic tests for mobile healthcare using drones. Drones would enable the mobility of very complex diagnostic tools to be deployed exactly where they are needed most in a timely manner.

- *Health system surveillance*

Drones have the potential to detect disease outbreaks. Researchers are developing drones to deliver mosquito traps to remote locations and to collect them later and deliver them to a laboratory for testing (The Seattle Times 2017). Test results can alert public health officials of potential mosquito-borne disease outbreaks such as Zika, dengue, malaria and chikungunya (NBC News 2016).

Sterile male mosquitoes have been released as part of a mosquito control program in parts of Florida. USAID has funded a project to test and develop the use of drones to release sterile mosquitos over areas that are affected by mosquito-borne illnesses such as Zika (NBC News 2016).

In Guatemala the company RTI International is exploring the use of drones to identify and map potential mosquito breeding sites (specifically for Zika and dengue) by taking aerial photographs of items that collect stagnant water such as old tires. The mapping of possible breeding sites could be used by public health officials to inform communities about risk for mosquito-borne disease and recommended prevention methods. These maps can also be used to target areas for larvicide application (RTI International 2018).

35.8 Conclusions

Innovations have become vital to rural and resource-limited areas to meet the challenges of implementing universal healthcare systems. The majority of these innovations has centered on the development of new technologies such as point-of-care diagnostics to bring testing closer to the patient. However, these have been slow and incremental in the face of increasing healthcare needs due to emerging and re-emerging disease outbreaks, population expansion and mobility and the increasing importance of non-communicable diseases in the LMICs.

Drones are proving to be very exciting for healthcare delivery to rural areas. The possibilities seem to be endless and only limited by current logistics such as payload capacities, flight regulation ranges and potential costs. Countries need to collect data and model the true cost of this technology for the potential impact that it may have on models of healthcare service delivery that leaves no one behind.

Bibliography

Allen, K. (2016). Using drones to save lives in Malawi. BBC News. March 2016. Available from http://www.bbc.com/news/world-africa-35810153

Amukele, T.K., Hernandez, J., Snozek, C.L.H. et al. (2017a). Drone transport of chemistry and hematology samples over long distances. *Am. J. Clin. Pathol.* 148 (5): 427–435.

Amukele, T., Ness, P.M., Tobian, A.A. et al. (2017b). Drone transportation of blood products. *Transfusion* 57 (3): 582–588.

Andrews, R. (2016). Drone Capable Of Carrying A Human At 100KM Per Hour Unveiled. http://www.iflscience.com/technology/self-flying-human-carrying-chinese-drone-unveiled-0

Assi, T.M., Brown, S.T., Kone, S. et al. (2013). Removing the regional level from the Niger vaccine supply chain. *Vaccine* 31 (26): 2828–2834.

Balasingam, M. (2017). Drones in medicine-The rise of the machines. *Int. J. Clin. Pract.* 71 (9): https://doi.org/10.1111/ijcp.12989.

Blau, M. (2017). 6 Ways Drones Could Change Health Care. Scientific American. June 2017. Available from https://www.scientificamerican.com/article/6-ways-drones-could-change-health-care

Choi-Fitzpatrick, A., Chavarria, D., Cychosz, E., et al. (2016). Up in the Air: A Global Estimate of Non-Violent Drone Use 2009–-2015. https://digital.sandiego.edu/cgi/viewcontent.cgi?article=1000&context=gdl2016report

DeAmicis C. (2015). Watch the first government approved drone delivery. Jul 2015. Available from http://www.recode.net/2015/7/18/11614838/watch-the-first-government-approved-drone-delivery

Drogolea N. (n.d.). Doctor Preneurs. 9 Drones That Will Revolutionise Healthcare. Available from http://www.doctorpreneurs.com/9-drones-that-will-revolutionise-healthcare

Drone Use 2009-2015, 2016. Available from http://digital.sandiego.edu/gdl2016report/1

EHang. (2016). EHang announces development agreement with lung biotechnology to enable drone delivery of manufactured organs for transplant. http://www.ehang.com/news/135.html

Engineering Strategic Communications. (2016). Many life-saving defibrillators behind locked doors during off-hours, study finds. August 2016. Available from http://news.engineering.utoronto.ca/many-life-saving-defibrillators-behind-locked-doors-off-hours-study-finds

Fahlstrom, P. and Gleason, T. (2012). *Introduction to UAV Systems*, 4e. Oxford: Wiley http://site.ebrary.com/id/10580221?ppg.

Gaitan D. (2014). Drones being developed to deliver medical aid, not bombs. Reuters. August 2014. Available from https://www.reuters.com/article/us-medical-drones/drones-being-developed-to-deliver-medical-aid-not-bombs-idUSKBN0GF17I20140815

GAVI (2016). The Vaccine Alliance. Rwanda Launches World's First National Drone Delivery Service Powered by Zipline. October 2016. Available from Document 8. http://www.gavi.org/library/news/gavi-features/2016/rwanda-launches-world-s-first-national-drone-delivery-service-powered-by-zipline

Global Health Supply Chain Summit. (2017). Available from http://ghscs.com/wp-content/uploads/2017/01/T2-4_Turning-Innovation-into-Impact_The-role-of-collaboration-in-leveraging-UAVs-for-payload-delivery-.pdf

Hackett, D.W. (2017). Drones Delivering Vaccines: Vanuatu embracing high tech for vaccine delivery Precision Vaccinations. June 2017. Available from http://www.precisionvaccinations.com/vanuatu-embracing-high-tech-vaccine-delivery

Haidari, L.A., Brown, S.T., Ferguson, M. et al. (2016). The economic and operational value of using drones to transport vaccines. *Vaccine* 34 (34): 4062–4067.

Haughom J. n.d. Innovation in Healthcare: Why It's Needed and Where It's Going. Available from https://www.healthcatalyst.com/innovation-in-healthcare-why-needed-where-going

Hern, A. (2014). DHL launches first commercial drone 'parcelcopter' delivery service, The Guardian. September 2014. Available from https://www.theguardian.com/technology/2014/sep/25/german-dhl-launches-first-commercial-drone-delivery-service

Howard, B. (2015). Vanuatu puts drones in the sky to see cyclone damage. *National Geographic. April 2015*. Available from http://news.nationalgeographic.com/&/2015/04/150406-vanuatu-cyclone-pam-relief-drones-uavs-crisis-mapping

Johns Hopkins Bloomberg School of Public Health. (2016). Drones Could Be Cheaper Alternative To Delivering Vaccines in Developing World. June 2016. Available from https://www.jhsph.edu/news/news-releases/2016/drones-could-be-cheaper-alternative-to-delivering-vaccines-in-developing-world.html

Johns Hopkins Medicine. (2017) Study sets new distance record for medical drone transport. September 2017. Available from https://www.sciencedaily.com/releases/2017/09/170912093108.htm

Kozloski, J. (2016). Drone-based antimicrobial analysis system. https://www.ibm.com/blogs/research/2017/01/drones-to-reduce-outbreaks

Landhuis, E. (2016). Doctors Test Drones To Speed Up Delivery Of Lab Tests. Shots Health News from NPR. September 2016. Available from https://www.npr.org/sections/health-shots/2016/09/13/493289511/doctors-test-drones-to-speed-up-delivery-of-lab-tests

Lee, B.Y., Connor, D.L., Wateska, A.R. et al. (2015). Landscaping the structures of GAVI country vaccine supply chains and testing the effects of radical redesign. *Vaccine* 33 (36): 4451–4458.

Lee, B.Y., Haidari, L.A., Prosser, W. et al. (2016). Re-designing the Mozambique vaccine supply chain to improve access to vaccines. *Vaccine* 34 (41): 4998–5004.

McFarland, M. (2017). Amazon's delivery drones may drop packages via parachute. CNN Tech. February 2017. Available from http://money.cnn.com/2017/02/14/technology/amazon-drone-patent/index.html

McNeish H. (2016). The First HIV-Fighting Drones Have Been Deployed in Africa. Vice. March 2016. Available from https://www.vice.com/en_ca/article/8gkpbk/unicef-just-launched-the-first-hiv-fighting-drones-in-africa

Médecins Sans Frontières (MSF). (2014). Papua New Guinea: Innovating to reach remote TB patients and improve access to treatment. November 2014. Available from http://www.msf.org/en/article/papua-new-guinea-innovating-reach-remote-tb-patients-and-improve-access-treatment

Meggi, B., Bollinger, T., Mabunda, N. et al. (2017). Point-of-care p24 infant testing for HIV may increase patient identification despite low sensitivity. *PLoS One* 12 (1): e0169497.

Meier, P. (2015). Chapter 6: UAVs and Humanitarian Response. Drones and Aerial Observation: New Technologies for Property Rights, Human Rights, and Global Development. A Primer. Washington DC: New America. Available from http://drones.newamerica.org/primer.

Meier, P. and Bergelund, J. (2017). Field-Testing the First Cargo Drone Deliveries in the Amazon Rainforest. We Robotics. February 2017. Available from http://werobotics.org/wp-content/uploads/2017/02/WeRobotics-Amazon-Rainforest-Cargo-Drones-Report.pdf

Momont A. n.d. Ambulance Drone. TU Delft. Available from https://www.tudelft.nl/en/ide/research/research-labs/applied-labs/ambulance-drone

Murphy M. (2016). Google wants to call drones to medical emergencies at the push of a button. Quartz. August 2016. Available from https://qz.com/655314/google-wants-to-call-drones-to-medical-emergencies-at-the-push-of-a-button

National Public Radio, USA (NPR). (2017). Tanzania Gears Up To Become A Nation Of Medical Drones. August 2017. Available from https://www.npr.org/sections/goatsandsoda/2017/08/24/545589328/tanzania-gears-up-to-become-a-nation-of-medical-drones

NBC News. (2016). Feds to Fund Plans to Take on Zika Virus With Drones. October 2016. Available from https://www.nbcnews.com/storyline/zika-virus-outbreak/feds-fund-plans-take-zika-virus-drones-n665241

Omachonu, V.K. and Einspruch, N.G. (2010). Innovation in healthcare delivery systems: a conceptual framework. *Public Sect. Innov. J.* 15 (1): 2.

Phillips, N., Blauvelt, C., Ziba, M., et al. (2016). Costs Associated with the Use of Unmanned Aerial Vehicles for Transportation of Laboratory Samples in Malawi. Seattle: VillageReach. Available from http://www.villagereach.org/wp-content/uploads/2017/06/Malawi-UAS-Report_MOH-Draft_-FINAL_14_07_16.pdf

Popper, B. (2016). Amazon's drone delivery launches in the UK. The Verge. December 2016. Available from https://www.theverge.com/2016/12/14/13952240/amazon-drone-delivery-launch-uk

Priye, A., Wong, S., Bi, Y. et al. (2016). Lab-on-a-drone: toward pinpoint deployment of smartphone-enabled nucleic acid-based diagnostics for mobile health care. *Anal. Chem.* 88 (9): 4651–4660. https://doi.org/10.1021/acs.analchem.5b04153.

Reuters. (2016). Drones could speed up HIV tests in remote areas. April 2016. Available from https://www.reuters.com/article/us-malawi-hiv-drones/drones-could-speed-up-hiv-tests-in-remote-areas-idUSKCN0XH1ZN

RTI International, (2018). Using Drones for Vector Control and Surveillance of Aedes Mosquitoes in Guatemala. Available from https://www.rti.org/impact/using-drones-vector-control-and-surveillance-aedes-mosquitoes-guatemala

Sharma G. (2016). Armed with drones, aid workers seek faster response to earthquakes, floods. Reuters. May 2016. Available from https://www.reuters.com/article/us-humanitarian-summit-nepal-drones/armed-with-drones-aid-workers-seek-faster-response-to-earthquakes-floods-idUSKCN0Y7003

Smit, P.W., Sollis, K.A., Fiscus, S. et al. (2014). Systematic review of the use of dried blood spots for monitoring HIV viral load and for early infant diagnosis. *PLoS One* 9 (3): e86461.

Stewart J. (2014). Google tests drone deliveries in Project Wing trials. BBC. August 2014. Available from http://www.bbc.com/news/technology-28964260

Stony Brook, (2016). Drones used to improve healthcare delivery in Madagascar. August 2016. Available from http://sb.cc.stonybrook.edu/news/general/2016_08_05_DronesInMadagascar.php

Tesser, M. (2017). The HiRO Drone project: a DIY rescue system for first responders in wilderness. http://www.emergency-live.com/en/equipment/the-hiro-drone-critical-injuries-are-no-more-problem

The Seattle Times. (2017). Microsoft helps build a mosquito trap with brains. February 2017. Available from https://www.seattletimes.com/business/technology/microsoft-helps-build-a-mosquito-trap-with-brains

The World Bank. (2017). Transport. September 2017. Available from http://www.worldbank.org/en/topic/transport/overview

Thiels, C.A., Aho, J.M., Zietlow, S.P., and Jenkins, D.H. (2015). Use of unmanned aerial vehicles for medical product transport. *Air. Med. J.* 34 (2): 104–108.

Thornton J. (2017). Drones and phones: how mobile tech is fighting global diseases. August 2017. London School of Hygiene and Tropical Medicine.

Toor A. (2016). The Verge. This startup is using drones to deliver medicine in Rwanda. April 2016. Available from 2017

Unitaid, (2015) HIV/AIDS Diagnostics Technology Landscape. 5th edition. October 2015. Available from http://unitaid.eu/assets/UNITAID_HIV_Nov_2015_Dx_Landscape.pdf

United Nations. (2015). Sustainable Development Goals. https://sustainabledevelopment.un.org

USAID. (2017). Global Health Supply Chain Program-Procurement and Supply Management, Unmanned Aerial Vehicles Landscape Analysis: Applications in the

Development Context, February 2017, Washington, DC: Chemonics International Inc. Available from https://www.ghsupplychain.org/sites/default/files/2017-06/GHSC_PSM_UAV%20Analysis_final.pdf

Van de Voorde, P., Gautama, S., Momont, A. et al. (2017). The drone ambulance [A-UAS]: golden bullet or just a blank? *Resuscitation* 116: 46–48. https://doi.org/10.1016/j.resuscitation.2017.04.037.

WHO, UNICEF, World Bank (2009). *State of the World's Vaccines and Immunization*, 3e. Geneva: World Health Organization.

World Health Organization. (2013). WHO Expert Consultation on Rabies: second report. Available from http://apps.who.int/iris/bitstream/10665/85346/1/9789240690943_eng.pdf

Zipline. n.d. Making Instant Deliveries Across Rwanda. Available from http://flyzipline.com/now-serving

Webliography

http://digital.sandiego.edu/gdl2016report/1 Drone Use 2009-2015, 2016

http://drones.newamerica.org/primer Meier, P. Chapter 6: UAVs and Humanitarian Response. Drones and Aerial Observation: New Technologies for Property Rights, Human Rights, and Global Development. A Primer. Washington DC: New America; 2015.

http://ghscs.com/wp-content/uploads/2017/01/T2-4_Turning-Innovation-into-Impact_The-role-of-collaboration-in-leveraging-UAVs-for-payload-delivery-.pdf Global Health Supply Chain Summit.

http://money.cnn.com/2017/02/14/technology/amazon-drone-patent/index.html McFarland, M. Amazon's delivery drones may drop packages via parachute. CNN Tech. February 2017.

http://news.engineering.utoronto.ca/many-life-saving-defibrillators-behind-locked-doors-off-hours-study-finds Engineering Strategic Communications. Many life-saving defibrillators behind locked doors during off-hours, study finds. August 2016.

http://news.nationalgeographic.com/&/2015/04/150406-vanuatu-cyclone-pam-relief-drones-uavs-crisis-mapping Howard, B. Vanuatu puts drones in the sky to see cyclone damage. National Geographic. April 2015.

http://sb.cc.stonybrook.edu/news/general/2016_08_05_DronesinMadagascar.php Stony Brook. Drones used to improve healthcare delivery in Madagascar. August 2016.

http://site.ebrary.com/id/10580221?ppg Fahlstrom, P. and Gleason, T. Introduction to UAV Systems. Fourth Edition, 2012.

http://werobotics.org/wp-content/uploads/2017/02/WeRobotics-Amazon-Rainforest-Cargo-Drones-Report.pdf Meier, P. and Bergelund, J. Field-Testing the First Cargo Drone Deliveries in the Amazon Rainforest. We Robotics. February 2017.

http://www.bbc.com/news/technology-28964260 Stewart, J. Google tests drone deliveries in Project Wing trials. BBC. August 2014.

http://www.bbc.com/news/world-africa-35810153 Allen, K. Using drones to save lives in Malawi. BBC News. March 2016.

http://www.doctorpreneurs.com/9-drones-that-will-revolutionise-healthcare/ Drogolea, N. Doctor Preneurs. 9 Drones That Will Revolutionise Healthcare.

http://www.ehang.com/news/135.html EHang, 2016. EHang announces development agreement with lung biotechnology to enable drone delivery of manufactured organs for transplant.

http://www.gavi.org/library/news/gavi-features/2016/rwanda-launches-world-s-first-national-drone-delivery-service-powered-by-zipline GAVI. The Vaccine Alliance. Rwanda Launches World's First National Drone Delivery Service Powered by Zipline. October 2016. Document 8.

http://www.iflscience.com/technology/self-flying-human-carrying-chinese-drone-unveiled-0 Andrews, R. 2016. Drone Capable Of Carrying A Human At 100KM Per Hour Unveiled.

http://www.msf.org/en/article/papua-new-guinea-innovating-reach-remote-tb-patients-and-improve-access-treatment Médecins Sans Frontières (MSF). Papua New Guinea: Innovating to reach remote TB patients and improve access to treatment. November 2014.

http://www.precisionvaccinations.com/vanuatu-embracing-high-tech-vaccine-delivery Hackett, D.W. Drones Delivering Vaccines: Vanuatu embracing high tech for vaccine delivery Precision Vaccinations. June 2017.

http://www.recode.net/2015/7/18/11614838/watch-the-first-government-approved-drone-delivery DeAmicis, C. Watch the first government approved drone delivery. Jul 2015.

http://www.reuters.com/article/us-humanitarian-summit-nepal-drones-idUSKCN0Y7003 Sharma, G. Armed with drones, aid workers seek faster response to earthquakes, floods. May 2016.

http://www.villagereach.org/wp-content/uploads/2017/06/Malawi-UAS-Report_MOH-Draft_-FINAL_14_07_16.pdf Phillips, N., Blauvelt, C., Ziba, M., Sherman, J., Saka, E., Bancroft, E. and Wilcox, A. Costs Associated with the Use of Unmanned Aerial Vehicles for Transportation of Laboratory Samples in Malawi. Seattle: VillageReach 2016.

https://digital.sandiego.edu/cgi/viewcontent.cgi?article=1000&context=gdl2016report Choi-Fitzpatrick, A., Chavarria, D., Cychosz, E., et al. Up in the Air: A Global Estimate of Non-Violent Drone Use 2009–2015.

https://qz.com/655314/google-wants-to-call-drones-to-medical-emergencies-at-the-push-of-a-button Murphy, M. Google wants to call drones to medical emergencies at the push of a button. Quartz. August 2016.

https://www.healthcatalyst.com/innovation-in-healthcare-why-needed-where-going Haughom, J. Innovation in Healthcare: Why It's Needed and Where It's Going.

https://www.ibm.com/blogs/research/2017/01/drones-to-reduce-outbreaks Kozloski, J. 2016. Drone-based antimicrobial analysis system.

https://www.jhsph.edu/news/news-releases/2016/drones-could-be-cheaper-alternative-to-delivering-vaccines-in-developing-world.html Johns Hopkins Bloomberg School of Public Health. Drones Could Be Cheaper Alternative To Delivering Vaccines in Developing World. June 2016.

https://www.nbcnews.com/storyline/zika-virus-outbreak/feds-fund-plans-take-zika-virus-drones-n665241 NBC News. Feds to Fund Plans to Take on Zika Virus With Drones. October 2016.

https://www.npr.org/sections/goatsandsoda/2017/08/24/545589328/tanzania-gears-up-to-become-a-nation-of-medical-drones National Public Radio, USA (NPR). Tanzania Gears Up To Become A Nation Of Medical Drones. August 2017.

https://www.npr.org/sections/health-shots/2016/09/13/493289511/doctors-test-drones-to-speed-up-delivery-of-lab-tests Landhuis, E. Doctors Test Drones To Speed Up Delivery Of Lab Tests. Shots Health News from NPR. September 2016.

https://www.reuters.com/article/us-humanitarian-summit-nepal-drones/armed-with-drones-aid-workers-seek-faster-response-to-earthquakes-floods-idUSKCN0Y7003

Sharma, G. Armed with drones, aid workers seek faster response to earthquakes, floods. Reuters. May 2016.

https://www.reuters.com/article/us-malawi-hiv-drones/drones-could-speed-up-hiv-tests-in-remote-areas-idUSKCN0XH1ZN Reuters. Drones could speed up HIV tests in remote areas. April 2016.

https://www.reuters.com/article/us-medical-drones/drones-being-developed-to-deliver-medical-aid-not-bombs-idUSKBN0GF17I20140815 Gaitan, D. Drones being developed to deliver medical aid, not bombs. Reuters. August 2014.

https://www.rti.org/impact/using-drones-vector-control-and-surveillance-aedes-mosquitoes-guatemala RTI International, 2018. Using Drones for Vector Control and Surveillance of *Aedes* Mosquitoes in Guatemala.

https://www.sciencedaily.com/releases/2017/09/170912093108.html Johns Hopkins Medicine. Study sets new distance record for medical drone transport. September 2017.

https://www.scientificamerican.com/article/6-ways-drones-could-change-health-care Blau, M. 6 Ways Drones Could Change Health Care. Scientific American. June 2017.

https://www.theguardian.com/technology/2014/sep/25/german-dhl-launches-first-commercial-drone-delivery-service Hern, A. DHL launches first commercial drone 'parcelcopter' delivery service, The Guardian. September 2014.

https://www.theverge.com/2016/12/14/13952240/amazon-drone-delivery-launch-uk Popper, B. Amazon's drone delivery launches in the UK. The Verge. December 2016.

https://www.tudelft.nl/en/ide/research/research-labs/applied-labs/ambulance-drone Momont, A. Ambulance Drone. TU Delft.

https://www.vice.com/en_ca/article/8gkpbk/unicef-just-launched-the-first-hiv-fighting-drones-in-africa McNeish, H. The First HIV-Fighting Drones Have Been Deployed in Africa. Vice. March 2016.

36

Implementation of Point-of-Care Tests: Lessons Learnt

Rosanna W. Peeling and Debrah I. Boeras

International Diagnostics Centre and Clinical Research Department, London School of Hygiene and Tropical Medicine, London, UK

36.1   Synopsis

Advances in point-of-care (POC) technologies to ensure universal access to affordable quality-assured diagnostic tests have the potential to transform patient management, surveillance programs and control of infectious (communicable) diseases. Lessons learnt from implementation of POC tests showed that decentralization of testing can put tremendous stresses on fragile health systems if POC testing is not introduced within the context of an appropriate architecture. Implementation science is needed to understand the political, cultural, economic, and behavioral context for technology introduction. The new paradigm should include

Revolutionizing Tropical Medicine: Point-of-Care Tests, New Imaging Technologies and Digital Health,
First Edition. Edited by Kerry Atkinson and David Mabey.
© 2019 John Wiley & Sons, Inc. Published 2019 by John Wiley & Sons, Inc.

1) Building a connected "diagnostic system" which consists of a comprehensive system of laboratories and POC testing sites to provide quality-assured diagnostic services with good laboratory-clinic interface to build trust in test results and linkage to care.
2) Building more human resources for data connectivity, training through machine learning, and coordinating a comprehensive national surveillance and communication system for disease control and global health emergencies.
3) Conducting research to monitor the impact of new tools and interventions on improving patient care.

36.2 Healthcare Needs in Low- and Middle-Income Countries

The year 2018 marks the 40th anniversary of the Declaration of Alma-Ata at the International Conference of Primary Health Care, during which United Nations member states expressed the need for urgent action to strengthen primary healthcare (Declaration of Alma-Ata 1978). This major milestone in the field of public health identified primary healthcare as the key to the attainment of the goal of Health for All, allowing people to lead socially and economically productive lives. The Declaration urges "WHO and UNICEF, and other international organizations, as well as multilateral and bilateral agencies, nongovernmental organizations, funding agencies, all health workers and the whole world community to support national and international commitment to primary health care and to channel increased technical and financial support to it, particularly in developing countries/low- to middle-income countries."

Ill health is often more than a health issue. It can both be caused by poverty and be a cause of poverty (The World Bank 2014). Poverty is a barrier to directly accessing healthcare as well as health information. The 2004 World Health Report found that the poorest 5th quintile of the population have to travel 5–27 km to reach the nearest health facility (WHO 2004). This is a particular hardship for the sick, young children and pregnant women and for the poor; additional hardships include transport costs and loss of income from missed days at work. Lack of access to health information in resource-limited settings also contributes to increasing inequalities and a cascade of poverty-related health issues.

In low- and middle-income countries (LMICs) infectious/communicable diseases remain the leading cause of morbidity and mortality, and The Big Three – HIV/AIDS, malaria and tuberculosis – continue to be the major causes of morbidity and mortality (see Chapter 1). Diagnosis of infectious diseases is often based on clinical presentation, which can be non-specific. The quality of care is dependent on whether the clinical diagnosis is supported by diagnostic testing. Most countries in the LMICs do not have an extensive or robust laboratory infrastructure to support complex and centralized diagnostic testing. Therefore, the development and deployment of simple rapid diagnostic tests that can be used by health providers at the point-of-care (POC) is recognized as an urgent priority (Mabey et al. 2004).

In 2003 the acronym ASSURED was coined at a WHO Special Program for Research and Training in Tropical Diseases (WHO/TDR) meeting, for the ideal characteristics of a test that can be accessed at all levels of the healthcare system (Kettler et al. 2004; Mabey et al. 2004) (Table 36.1). This acronym embodies the key attributes of affordability, accuracy and accessibility that address the inequity and needs of patients in the LMICs.

Table 36.1 The attributes of the ideal POC test.

A: Affordable
S: Sensitive
S: Specific
U: User-friendly
R: Rapid and robust
E: Equipment-free or minimal equipment
D: Deliverable to those who need them

Unfortunately, development of simple rapid tests is by no means "simple" (Senior 2009) and trade-offs need to be made between a test being accurate and accessible to the target population. A study in the USA showed that a POC test with 65% sensitivity to detect genital chlamydial infection, compared to a molecular assay, led to more patients being treated than those undergoing a molecular assay which required patients to return for test results (Gift et al. 1999). The important question that needs to be addressed is what is an acceptable level of accuracy that one can trade for accessibility of a POC test?

The Bill and Melinda Gates Foundation convened a Global Health Diagnostics Forum to model the attributable benefits of improved diagnostics for global health for the diseases associated with the largest mortality in LMICs (Burgess et al. 2006; Urdea et al. 2006). Recognizing that laboratory infrastructure supporting diagnostic testing for these diseases is currently limited, the models estimated the potential impact of POC tests requiring no laboratory infrastructure and acceptable levels of accuracy. These models showed that investment in increasing access to diagnostic tests would have a greater impact than improvements in test performance. A summary of the performance for POC tests from the Gates Forum (Nature Supplements 2006) is shown in Table 36.2 (Aledort et al. 2006; Lim et al. 2006).

Donors such as the Bill and Melinda Gates Foundation and the Wellcome Trust, and technical agencies such as Grand Challenges Canada, the US National Institutes of Health, the UK Department for International Development and the European Commission responded to this challenge and invested in the development of appropriate diagnostic tests for LMICs.

TDR, the Special Program for Research and Training in Tropical Diseases, is a global program of scientific collaboration that helps facilitate, support and influence efforts to combat diseases of poverty. It is hosted at the World Health Organization (WHO), and is sponsored by the United Nations Children's Fund (UNICEF), the United Nations Development Program (UNDP), the World Bank and WHO. The TDR and product development partnerships, such as the Foundation for Innovative New Diagnostics (FIND) among others, supported public-private efforts to develop POC tests for which there is no viable commercial market and which require procurement through donor agencies such as the Global Fund for HIV, Tuberculosis and Malaria, The President's Emergency Program for AIDS Relief (PEPFAR) and the President's Malaria Initiative.

In the last two decades a range of simple rapid diagnostic tests that can be deployed at the point-of-care have been developed, with an emphasis on appropriate technology that is affordable, relevant to the needs of the population and scientifically sound. More

Table 36.2 Potential impact of POC tests (extracted from Nature Supplement, December 2006).

Disease	Population	Sensitivity/ specificity	Potential impact
ALRI	Children, <5 years	95/85	Save ~405 000 lives
HIV	Infants, <12 months	90/90	Save 2.5 million DALYs if 100% access to treatment
Malaria	Children, <5 years	90/90	Save 2.2 million DALYS and prevent 447 million unnecessary treatments
Tuberculosis	Symptomatic	85/97	Save ~400 000 lives
Syphilis	Prenatal	86/72	Save 201 000 DALYS, avert 215 000 stillbirths
Ct/Ng	Sex workers	85/90	Save ~4 million DALYs, avert 16.5 million new cases and prevent 212 000 HIV cases

Abbreviations used: ALRI, acute lower respiratory tract infections; Ct/Ng: *Chlamydia trachomatis/Neisseria gonorrhoeae*

recently, a call for countries to reach elimination targets (90-90-90 for HIV), dual elimination of mother-to-child transmission (MTCT) of HIV and syphilis and the UN/WHO Millennium Development Goals (MDGs) and Sustainable Development Goals (SDGs) have further incentivized industry to develop POC tests that increase access to the lowest levels of the healthcare system and "leave no one behind" (Jani et al. 2013; Fitzpatrick and Engels 2016).

To date few POC tests fulfill all of the ASSURED criteria. Table 36.3 scores POC tests for diseases listed in Table 36.2 that are currently available against the ASSURED criteria.

36.3 Rapid Diagnostic Tests for Human Immunodeficiency Virus (HIV) Disease (and See Chapter 4)

HIV antibody detection RDTs using finger prick whole blood specimens have shown acceptable performance (Pai et al. 2007). Implementation of blood-based rapid HIV tests has been a great success, enabling HIV control programs to identify infections early and prevent onward transmission. In 2014 as many as 150 million children and adults in 129 low- and middle-income countries received HIV testing services (WHO 2014). A major challenge is training and quality assurance to ensure accurate results, especially in health facilities where there is high turnover of staff and training can be limited or often inadequate. A study in South Africa found that only 3% of HIV RDTs were done correctly (Ghani et al. 2015). Although not every error will result in a wrong result, the alarming reality is that with 150 million tests being performed annually worldwide, assuming a 99% accuracy rate, as many as one million wrong results per year could potentially be generated. Quality assurance efforts are ongoing that include the groundwork for developing key policy and quality documents for the implementation of HIV-related POC testing (Fonjungo et al. 2016).

Table 36.3 ASSURED POC tests available and unmet needs.

Disease/ syndrome	POC tests needed	POC Tests		
		Available	Accurate	Affordable
HIV	RDT for antibody detection, blood based;	+	+++	+
	RDT for antibody detection, oral fluid	+	++	No
	Devices for viral load and EID	+	+++	No
Malaria	Rapid antigen tests	+	++	+
TB	POC NAT and resistance detection	+	+++	No
Respiratory infections	RDTs to distinguish between bacterial and viral infections	–	–	–
Sexually transmitted infections	*Chlamydia trachomatis*	+	–	No
	Gonorrhea	+	–	No
	Syphilis	+	++	+

+++ highly accurate; + affordable defined as 0.5–1.0 US$.

Abbreviations used: POC, point-of-care; RDT, rapid diagnostic test; EID, early infant diagnosis; NAT, nucleic acid amplification test.

36.4 Rapid Diagnostic Tests for Syphilis (and See Chapter 6)

Syphilis, caused by the spirochete *Treponema pallidum*, has a long latent period during which patients have no signs or symptoms, but can remain infectious. Syphilis in pregnancy can lead to adverse outcomes such as stillbirths, miscarriages and babies born with congenital syphilis. In LMICs babies born with congenital syphilis have only a 50% chance of survival during the first two years of life (Peeling et al. 2018; WHO 2007). Despite the availability of simple diagnostic tests for antenatal screening and the effectiveness of treatment with a single dose of long-acting penicillin, syphilis is re-emerging as a global public health problem. Screening of pregnant women for syphilis is recommended policy in most countries, yet it is estimated that 500 000 babies die each year as a result of stillbirths and congenital syphilis. This in turn is due to lack of access to antenatal screening (Gomez et al. 2013; Newman et al. 2013). Congenital syphilis is preventable if a single dose of penicillin is given to an infected pregnant woman during the first two trimesters of pregnancy. Two systematic reviews reported that RDTs for syphilis have acceptable performance (Tucker et al. 2010; Jafari et al. 2013). Although the introduction of RDTs was acceptable to patients and healthcare providers, ensuring adequate training for healthcare workers and supplies of commodities were cited as key barriers to implementation (Sweeney et al. 2014; Ansbro et al. 2015; Shelley et al. 2015). A systematic review showed that RDTs for antenatal syphilis screening contribute to the improvement of antenatal care in low resource settings (Swartzendruber et al. 2015).

Dual HIV and syphilis RDT testing in countries was prioritized by WHO for the dual elimination of MTCT of HIV and syphilis by 2020 (Peeling et al. 2004; Taylor et al. 2017). Elimination MTCT programs for HIV has resulted in a dramatic decrease in HIV-positive infants and yet in sub-Saharan Africa the rate of syphilis screening of pregnant women has remained at approximately 30% (Wijesooriya et al. 2016). Disparity in funding and lack of political will continue to neglect antenatal screening for syphilis

(Peeling et al. 2016). With the availability of acceptable dual rapid HIV-syphilis RDTs, there are opportunities for women to be screened for both infections using a single drop of blood in a single visit to a healthcare facility (Gliddon et al. 2017).

36.5   Rapid Diagnostic Tests for Tuberculosis (TB) (and See Chapter 5)

In 2016 10.4 million people developed tuberculosis and 1.7 million died from the disease, including 0.4 million among people with HIV disease. Over 95% of TB deaths occur in low- and middle-income countries. Multidrug-resistant TB (MDR-TB) remains a public health crisis and a health security threat. WHO estimates that there were 600 000 new cases with resistance to rifampicin – the most effective first-line drug, of which 490 000 had MDR-TB (WHO Tuberculosis Fact Sheet 2018).

The development pipeline has shown progress for the evaluation and implementation of new technologies. Assays such as the Xpert MTB/RIF allow for simultaneous detection of *Mycobacterium tuberculosis* (MTB) and rifampicin (RIF) resistance. Detection of tuberculosis and rifampicin resistance in 1 hour and 45 minutes using a POC molecular TB assay should improve case detection and decrease transmission. As of June 2012 two-thirds of countries with a high tuberculosis burden and half of countries with a high multidrug-resistant tuberculosis burden had incorporated the assay into their national tuberculosis program guidelines. Challenges still remain with implementation, uptake and sustainability that will need to be addressed (Abubakar et al. 2013).

Point-of-care technologies such as the Xpert MTB/RIF, if introduced appropriately, can improve patient outcomes and provide an opportunity to strengthen health systems. (Pathmanathan et al. 2017). In clinics where the assay was introduced without addressing patient pathways, the impact was not fully realized (Lawn et al. 2013; Schito et al. 2015). Evidence-based recommendations are needed to incorporate Xpert MTB/RIF into clinical and laboratory guidelines and policies that will strengthen coordination between laboratory systems, laboratory-program interfaces and tuberculosis-HIV program interfaces.

If successful, the benefits of this tool could extend progress toward the goals of the global End TB Strategy, to improve system-wide capacity for global disease detection and control (WHO 2015). Globally, TB incidence is falling at about 2% per year. An estimated 53 million lives were saved through TB diagnosis and treatment between 2000 and 2016. This needs to accelerate to a 4–5% annual decline to reach the 2020 milestones of the End TB Strategy. Ending the TB epidemic by 2030 is among the health targets of the Sustainable Development Goals (https://sustainabledevelopment.un.org/?menu=1300).

36.6   Rapid Diagnostic Tests for Malaria (and See Chapter 7)

An estimated 40% of the world population today is at risk of malaria infection (WHO 2016). The WHO estimates that each year there are more than 300 million episodes of acute illness and at least a million deaths due to malaria worldwide. More than 90% of the global burden is in sub-Saharan Africa.

Malaria RDTs can provide a rapid, accurate diagnosis in circumstances where demonstration of parasitemia has previously been impossible or where microscopy-based diagnosis may be unreliable. There are over 120 brands of malaria RDTs made by approximately 60 companies selling RDTs of varying quality in the absence of regulatory oversight. In most LMICs the lack of requirement for regulatory approval before malaria RDTs are marketed means that the quality of tests sold in shops are not guaranteed. Many lack thermostability at temperatures above 30 °C making them unsuitable for settings where malaria is endemic.

The price of most RDT brands is between US$0.65 and US$2.50 per test (Frost and Reich 2008). Pan-specific tests are usually about 40% more expensive than products detecting *Plasmodium falciparum* only. Commercial demand for this product has been fueled by new external financing for purchasing malaria diagnostics, especially from the Global Fund, leading to a rapid expansion of the RDT market. A major challenge for countries that have adopted malaria RDTs is future affordability. It is not clear how testing will be sustainable if funding sources decrease as malaria is eliminated in different countries.

36.7 Lessons Learnt from the Implementation of POC Tests

"The delivery of health care to the poor often fails on the last lap. While more fortunate individuals and families have access to life-saving technologies and health-improving services, the poor often have to do without essential health care" – Adetokunbo O. Lucas, Former Professor of Preventive and Social Medicine, University of Ibadan, Nigeria (Frost and Reich 2008).

Despite the successful rollout of drugs and vaccines in the LMICs, technological innovations remain out of reach of its intended population in the absence of political will, leadership, strategic planning and funding. Access to diagnostic technologies is complex and challenging in any environment, but particularly in the LMICs. The focus on cost, pricing and patents has obscured other critical barriers to access to diagnostics, such as distribution, delivery and adoption.

36.7.1 Policy Barriers

Implementation starts with a clearly defined and implementable policy based on estimation of the most cost-effective strategies to ensure sustainability. The policy needs to demonstrate that POC technologies are needed, affordable and can lead to a clear benefit in health outcomes for the intended population. However, many countries do not have policies for POC testing, or the decisions are made ad hoc for each technology depending on specific disease control programs and partner demands. The policy needs to have a clear plan of action that specifies who makes the decisions on test adoption, test selection for procurement, who can use them, how the results should be interpreted and reported, how the quality of tests and testing can be assured and, of course, how to deal with the test results and linkage to care. Most importantly, and often overlooked, is the requirement that countries must ask who pays for implementing the policy.

36.7.2 Implementation Barriers

A major obstacle to access is the cost of validating novel technologies as appropriate and the cost of sustainable implementation. In LMICs much needed health technologies are often unaffordable to governments and individuals that need them most.

Another important barrier to accessibility for novel POC tests in resource-limited settings is the limited capacity of public health systems to roll out and support decentralized testing, including a significant increase in training, supply chain management and quality assurance. Many health systems already struggle with critical staff shortages, high rates of staff turnover, persistent stockouts of essential supplies of health commodities, including vaccines, drugs, and diagnostics (UN Commission on Life Saving Commodities). Introduction of POC testing can exacerbate the situation. In 1981 the WHO published an Essential Drugs list to advocate that countries make medicines on the list available to those who need them. However, problems of access to these essential drugs remain unsolved to date.

36.7.3 Architecture

Frost and Reich (2008) conducted a systematic analysis of the dimensions, distribution and determinants of inequitable access to healthcare and developed a framework for developing feasible solutions to improving access for the poor, which they labeled as Architecture. As illustrated by Brooks et al. in their 2012 publication, *Implementing new health interventions in developing countries: Why do we lose a decade or more?* architecture is the organizational structures and relationships established with the purpose of coordinating and steering the availability, affordability and adoption activities. Availability refers to the logistics of making, ordering, shipping, storing, distributing, and delivering a new health technology to ensure it reaches the end-user. Affordability is ensuring that health technologies and related services are not too costly for the people who need them. Adoption is gaining acceptance and creating demand for a new health technology from global organizations, government actors, providers and dispensers and individual patients.

36.8 Lessons Learnt from the Implementation of POC Tests for Three Diseases

Lessons learnt from the implementation of RDTs for HIV, malaria and syphilis illustrate the importance of "Architecture" as defined by Frost and Reich and the role of partnerships within this Architecture. RDTs for all three infections are available, accessible and affordable but the success of their implementation differs mainly because funding for procurement of RDTs for HIV and malaria is financially and technically supported by global programs such as the Global Fund, PEPFAR and the President's Malaria Initiative. In addition, each country has a national control program for HIV and malaria which provide the architecture needed to implement the RDTs through global programs which help countries implement their control strategy. In contrast, the lack of funding and limited architecture available for syphilis screening are in large part responsible for the lack of progress in introduction and scale up of prenatal syphilis screening for the dual elimination of MTCT of HIV and syphilis (Table 36.4).

Table 36.4 Lessons learnt from implementation of POC tests.

POC test	Availability	Accessibility	Affordability and procurement	Architecture		
				Supply chain	Training	Quality assurance	Monitoring and evaluation
HIV	Yes	Finger prick; 2–3 easy steps; results in 15–20 minutes	~1 US$; Global Fund, PEPFAR and others	Provided by national HIV programmes, most of which are well resourced and well organized		
Malaria	Yes		~0.5–0.6 US$; Global Fund, PMI and NGOs	Provided by national malaria programmes, most of which are well resourced and well organized		
Syphilis	Yes		~0.5–1 US$; none	Frequent stockouts	Not organized	Not consistent	Not coordinated

Abbreviations used: PEPFAR: President's Emergency Program for AIDS Relief; PMI, President's Malaria Initiative; NGOs: non-governmental organizations

Technological advancement is only one side of the coin (Pang and Peeling 2007). Even when a POC test with acceptable performance is available, there are considerable challenges and difficulties in introducing new tests in LMICs where healthcare infrastructures are weak or fragile. Test introduction and sustainable adoption depend on a robust healthcare system and many other factors, including supply chain management, training and quality assurance. Supply chain management is needed to avoid frequent stockouts of diagnostics and/or drugs. The demand for training can be amplified as much as a 1000-fold when testing is decentralized and quality assurance for the expanded testing at thousands of POC testing sites are stresses that POC testing puts on often already fragile healthcare systems.

In implementing rapid syphilis tests countries have found that, although there are challenges in the provision of adequate training, supply chain management and quality assurance, the introduction of POC tests increased the efficiency of the healthcare system (Mabey et al. 2012; Garcia et al. 2013). With many countries having a severe shortage of healthcare personnel, having testing and treatment delivered at a single visit instead of several visits reduces burden on both the patient and the healthcare system.

36.9 The Way Forward

36.9.1 Delivering Interventions

Global health initiatives, such as the Global Fund for AIDS, TB, and Malaria, the President's Emergency Program for AIDS Relief, and the President's Malaria Initiative, have made significant progress in terms of improving the capacity of LMICs to offer healthcare that includes quality diagnostic services. Other non-governmental organizations and universities have also emulated and increased their efforts at building capacity for diagnostic testing and quality management of laboratories.

Bill Gates made a very important point in his speech before the World Health Assembly on May 16, 2005, when he said, "The world has to devote more thinking and funding to delivering interventions – not just discovering them" (Gates 2005).

36.9.2 Access as a Human Right

Apart from having global, national and regional coordinating bodies for implementation, availability and affordability of technologies, awareness has to be added to the Architecture. There are mechanisms built in to assure that the highest quality of the technology is available. Systems and partnerships are available for securing the continued financing and affordability that can ensure ongoing access to technological innovation. However, there is still a lack of awareness and education that demands access to diagnostics and medicines as a human right. Frost and Reich (2008) suggested that Access can be approached in several ways:

1) Rights-based approach: Access to a health technology depends on providing the "right" product at the "right" place with the "right" protocol at the "right" time. How that is accomplished varies depending on public policies and social values.
2) Cost-effectiveness approach: One common way of defining what is right for access is through a cost-effectiveness perspective (getting the "biggest bang for the buck," based on utilitarian principles), in which a government seeks to maximize health for a particular population under certain resource constraints.

3) Market-based approach: A different normative perspective on how to provide access would be a market-based approach that makes products available for sale to people who can pay the prices set by producers or governments.
4) Egalitarian values: These would preferentially provide financially subsidized access to effective health technologies for those groups that are worst off within a population.
5) Rule-of-rescue approach: This approach would provide free access to life-saving health technologies for those individuals within a population who need these technologies but cannot afford them.

36.9.3 A Package of Care

Access to an essential package of diagnostic tests would significantly improve care and prevent mortality and morbidity due to many and multiple diseases. This Essential Diagnostic Package could include rapid tests for HIV-syphilis, anemia, diabetes and a test to predict and diagnose pre-eclampsia. There are no limitations to the concept and development of the package – only barriers that need to be lowered to service patients with a more holistic approach. This Package should be considered in the list of the UN Commission for Life Saving Technologies.

36.10 The New Paradigm for Technological Innovation and Implementation

Moving forward, there are several ways in which we can improve the implementation of POC tests so that their true potential can be realized. Countries should leverage social media and other means of communication to increase awareness of diseases and to enable a rights-based approach. Further social and anthropologic studies can inform us of how to engage healthcare users and providers so that the design of healthcare services and linkage to care and treatment can be improved and be appropriate for the populations they are to serve.

The rapid pace of technological innovation in recent years has outstripped regulatory science and the traditional ways of assessing risk and benefit. A new paradigm in which policy makers, regulators and subject area experts get together to assess acceptable risk for incremental clinical benefit of novel technologies is required. Securing agreements among countries to reduce regulatory bureaucracy by adopting harmonized regulatory standards with a multinational or regional platform would accelerate approval and access to new technologies (McNerney et al. 2014). This approach could also reduce the cost for companies seeking to register their products in multiple countries. Those savings can be passed on to the customers, while ensuring purchasing options and competition to optimize the choice of proven diagnostic technologies. There should be sustained efforts to curtail counterfeit products entering the healthcare system. Regulation of Internet services that promote and sell unproven diagnostic tests is also required. Education for professionals on technological innovations will allow providers to be updated on new tests and their performance and will incentivize uptake.

The parallel advances in information technology and data connectivity can be leveraged by the diagnostic industry and national programs (Cheng et al. 2016). An information system that liaises between the laboratory and the individual or caregiver can ensure transparency and optimal treatment within a public health control program. The use of mobile phones to drive a diagnostic reaction or to read a test result has been explored (Laksanasopin et al. 2015). Since mobile phones are now widely available and used in the LMICs, there is an opportunity to build on data connectivity to diagnostic tests or for devices to revolutionize the delivery of healthcare services in primary healthcare settings. While there is the need to respect data governance issues of privacy and confidentiality, the timely availability of data and testing trends in a central database, allows emergency programs to optimize control strategies and immediately respond with control interventions. At a national level these systems can support the supply chain systems and immediately address critical stockouts (Baker et al. 2015). Connectivity is also a component of a quality assurance program that allows proficiency testing to be linked to testing (Boeras and Peeling 2016). Alerts of failure in proficiency testing will trigger investigation and corrective action.

Finally, connectivity will allow surveillance of diseases of epidemic potential in order to identify and limit outbreaks, inform policy changes, recognize treatment failures and evaluate the effectiveness of new interventions. Connectivity can bring together a "diagnostic system" that is comprehensive of laboratories and POC testing sites and provides quality-assured diagnostic services and epidemic surveillance (Boeras et al. 2017). This new paradigm will improve and accelerate access to novel diagnostic technologies in the future and strengthen health systems instead of weakening them (Mabey et al. 2012; Garcia et al. 2013; Pathmanathan et al. 2017).

36.11 Conclusions

There have been many lessons learnt in the implementation of POC tests in resource-limited settings. Decentralization of testing can put tremendous stresses on fragile health systems if POC testing is not introduced within an architecture that can provide the framework for implementation. Countries should leverage social media and other means of educating the public and professionals on novel technologies. Implementation science is needed to understand the political, cultural, economic and behavioral context for technology introduction. The new paradigm should include:

1) Building a "diagnostic system" representing a comprehensive system of laboratories and POC testing sites to provide quality-assured diagnostic services with good laboratory-clinic interfaces in order to build trust in test results and linkage to care.
2) Building more human resources for information and communication technology, machine learning and coordinating a comprehensive national surveillance and communication system for disease control and global health emergencies.
3) Conducting research to monitor the impact of new tools and interventions on improving patient care.

Bibliography

Abubakar, I., Zignol, M., Falzon, D. et al. (2013). Drug-resistant tuberculosis: time for visionary political leadership. *Lancet Infect. Dis.* 13 (6): 529–539. https://doi. org/10.1016/S1473-3099(13)70030-6.

Aledort, J.E., Ronald, A., Le Blanq, S.M. et al. (2006). Reducing the burden of HIV/ AIDS in infants: the contribution of improved diagnostics. *Nature* 444 (suppl 1): 19–28.

Aledort, J.E., Ronald, A., Rafael, M. et al. (2006). Reducing the burden of sexually transmitted infections in resource-limited settings: the role of improved diagnostics. *Nature* 444 (suppl 1): 59–72.

Ansbro, É.M., Gill, M.M., Reynolds, J. et al. (2015). Introduction of syphilis point-of-care tests, from pilot study to National Programme Implementation in Zambia: a qualitative study of healthcare workers' perspectives on testing, training and quality assurance. *PLoS One* 10 (6): e0127728. https://doi.org/10.1371/journal.pone.0127728.

Baker, U., Okuga, M., Waiswa, P. et al. (2015). Bottlenecks in the implementation of essential screening tests in antenatal care: syphilis, HIV, and anemia testing in rural Tanzania and Uganda. *Int. J. Gynecol. Obstet.* 130: S43–S50.

Boeras, D.I., Nkengasong, J.N., and Peeling, R.W. (2017). Implementation science: the laboratory as a command centre. *Curr. Opin. HIV AIDS* 12 (2): 171–174.

Boeras, D.I. and Peeling, R.W. (2016). External quality assurance for HIV point-of-care testing in Africa: a collaborative country-partner approach to strengthen diagnostic services. *Afr. J. Lab. Med.* 5 (2): 556.

Brooks, A., Smith, T., de Savigny, D., and Lengeler, C. (2012). Implementing new health interventions in developing countries: Why do we lose a decade or more? *BMC Public Health* 12: 683. https://doi.org/10.1186/1471-2458-12-683.

Burgess, D.C., Wasserman, J., and Dahl, C.A. (2006). Global health diagnostics. *Nature* 444 (suppl 1): S1–S2.

Cheng, B., Cunningham, B., Boeras, D.I. et al. (2016). Data connectivity: a critical tool for external quality assessment. *Afr. J. Lab. Med.* 5 (2): 535.

Declaration of Alma-Ata. (1978). International Conference on Primary Health Care, Alma-Ata, USSR, 6–12 Sep 1978. http://www.who.int/publications/almaata_declaration_en.pdf

FIND (2014 Sep). Product development partnerships. https://www.finddx.org/wp-content/uploads/2016/03/PDP_Brief_ENG_final_Sep_2014.pdf

FIND, Foundation for Innovative New Diagnostics: https://www.finddx.org.

Fitzpatrick, C. and Engels, D. (2016). Leaving no one behind: a neglected tropical disease indicator and tracers for the sustainable development goals. *Int. Health* 8 (Suppl 1): i15–i18. https://doi.org/10.1093/inthealth/ihw002.

Fonjungo, P.N., Osmanov, S., Kuritsky, J. et al. (2016). Ensuring quality: a key consideration in scaling-up HIV-related point-of-care testing programs. *AIDS* 30 (8): 1317–1323.

Frost, L.J. and Reich, M.R. (2008). Access: How do good health technologies get to poor people in poor countries? published by Harvard Center for Population and Development Studies, ISBN 9780674032156.

Garcia, P.J., Carcamo, C.P., Chiappe, M. et al. (2013). Rapid syphilis tests as catalysts for health systems strengthening: a case study from Peru. *PLoS One* 8 (6): e66905.

Gates, B. (2005). Remarks of Mr. Bill Gates, Co-founder of the Bill & Melinda Gates Foundation, at the World Health Assembly, Fifty-eighth World Health Assembly, Geneva, Switzerland, 16 May 2005, http://www.who.int/mediacentre/events/2005/wha58/gates/en/index.html (retrieved January 2, 2008).

Ghani, A.C., Burgess, D.H., Reynolds, A., and Rousseau, C. (2015). Expanding the role of diagnostic and prognostic tools for infectious diseases in resource-poor settings. *Nature* 528 (7580): S50–S52. https://doi.org/10.1038/nature16038.

Gift, T.L., Pate, M.S., Hook, E.W., and Kassler, W.J. (1999). The rapid test paradox: when fewer cases detected led to more cases treated. *Sex. Transm. Dis.* 26: 232–240.

Gliddon, H.D., Peeling, R.W., Kamb, M.L. et al. (2017). A systematic review and meta-analysis of studies evaluating the performance and operational characteristics of dual point-of-care tests for HIV and syphilis. *Sex. Transm. Infect.* 93 (S4): S3–S15. https://doi.org/10.1136/sextrans-2016-053069.

Global Fund for HIV, Tuberculosis and Malaria. www.theglobalfund.org

Gomez, G.B., Kamb, M.L., Newman, L.M. et al. (2013). Untreated maternal syphilis and adverse outcomes of pregnancy: a systematic review and meta-analysis. *Bull. World Health Organ.* 91: 217–226. https://doi.org/10.2471/BLT.12.107623.

Jafari, Y., Peeling, R.W., Shivkumar, S. et al. (2013). Are Treponema pallidum specific rapid and point-of-care tests for syphilis accurate enough for screening in resource limited settings? Evidence from a meta-analysis. *PLoS One* 8 (2): e54695. https://doi.org/10.1371/journal.pone.0054695.

Jani, I.V. and Peter, T.F. (2013). How point-of-care testing could drive innovation in global health. *N. Engl. J. Med.* 368 (2324): 2319–2324.

Kettler, H., White, K., and Hawkes, S. (2004). *Mapping the Landscape of Diagnostics for Sexually Transmitted Infections*. Geneva: WHO/TDR.

Laksanasopin, T., Guo, T.W., Nayak, S. et al. (2015). A smartphone dongle for diagnosis of infectious diseases at the point of care. *Sci. Transl. Med.* 7 (273): 273re1. https://doi.org/10.1126/scitranslmed.aaa0056.

Lawn, S.D., Mwaba, P., Bates, M. et al. (2013). Advances in tuberculosis diagnostics: the Xpert MTB/RIF assay and future prospects for a point-of-care test. *Lancet Infect. Dis.* 13 (4): 349–361. https://doi.org/10.1016/S1473-3099(13)70008-2.

Lim, Y.W., Steinhoff, M., Girosi, F. et al. (2006). Reducing the global burden of acute lower respiratory infections in children: the contribution of new diagnostics. *Nature* 444 (suppl 1): 9–18.

Mabey, D., Peeling, R.W., Ustianowski, A., and Perkins, M. (2004). Diagnostics for the developing world. *Nat. Rev. Microbiol.* 2: 231–240.

Mabey, D.C., Sollis, K.A., Kelly, H.A. et al. (2012). Point-of-care tests to strengthen health systems and save newborn lives: the case of syphilis. *PLoS Med.* 9 (6): e1001233.

McNerney, R., Sollis, K., and Peeling, R.W. (2014). Improving access to new diagnostics through harmonised regulation: priorities for action. *Afr. J. Lab. Med.* 3 (1): 123.

Newman, L., Kamb, M., Hawkes, S. et al. (2013). Global estimates of syphilis in pregnancy and associated adverse outcomes: analysis of multinational antenatal surveillance data. *PLoS Med.* 10: e1001396.

Pai, N.P., Tulsky, J.P., Cohan, D. et al. (2007). Rapid point-of-care HIV testing in pregnant women: a systematic review and meta-analysis. *Trop. Med. Int. Health* 12: 1–12.

Pang, T. and Peeling, R.W. (2007). Diagnostic tests for infectious diseases in the developing world: two sides of the coin. *Trans. R. Soc. Trop. Med. Hyg.* 101: 856–857.

Pathmanathan, I., Date, A., Coggin, W.L. et al. (2017). Rolling out Xpert MTB/RIF* for tuberculosis detection in HIV-positive populations: an opportunity for systems strengthening. *Afr. J. Lab. Med.* 6 (2): pii: a460. https://doi.org/10.4102/ajlm.v6i2.460.

Peeling, R.W. and Mabey, D. (2016). Celebrating the decline in syphilis in pregnancy: a sobering reminder of what's left to do. *Lancet Glob. Health* 4 (8): e503–e504.

Peeling, R.W., Mabey, D., Fitzgerald, D.W., and Watson-Jones, D. (2004). Avoiding HIV and dying of syphilis. *Lancet* 364: 1561–1563.

Peeling, R.A.W., Mabey, D., Kamb, M.L. et al. (2017). Syphilis. *Nat. Rev. Dis. Prim.* 3: 17073. https://doi.org/10.1038/nrdp.2017.73.

Schito, M., Migliori, G.B., Fletcher, H.A. et al. (2015). Perspectives on advances in tuberculosis diagnostics, drugs, and vaccines. *Clin. Infect. Dis.* 61 (Suppl 3): S102–S118. https://doi.org/10.1093/cid/civ609.

Senior, K. (2009). The complex art of making diagnostics simple. *Lancet* 9: 467.

Shelley, K.D., Ansbro, É.M., Ncube, A.T. et al. (2015). Scaling down to scale up: a health economic analysis of integrating point-of-care syphilis testing into antenatal Care in Zambia during pilot and national rollout implementation. *PLoS One* 10 (5): e0125675. https://doi.org/10.1371/journal.pone.0125675.

Nature Supplements. (2006). Improved Diagnostic Technologies for the Developing World. https://www.nature.com/nature/archive/supplements.html

Sustainable Development Goals. https://sustainabledevelopment.un.org/?menu=1300

Swartzendruber, A., Steiner, R.J., Adler, M.R. et al. (2015). Introduction of rapid syphilis testing in antenatal care: a systematic review of the impact on HIV and syphilis testing uptake and coverage. *Int. J. Gynaecol. Obstet.* 130 (Suppl 1): S15–S21. https://doi.org/10.1016/j.ijgo.2015.04.008.

Sweeney, S., Mosha, J.F., Terris-Prestholt, F. et al. (2014). The costs of accessible quality assured syphilis diagnostics: informing quality systems for rapid syphilis tests in a Tanzanian setting. *Health Policy Plan.* 29 (5): 633–641. https://doi.org/10.1093/heapol/czt049.

Taylor, M., Newman, L., Ishikawa, N. et al. (2017). Elimination of mother-to-child transmission of HIV and syphilis (EMTCT): process, progress, and program integration. *PLoS Med.* 14: e1002329.

The President's Emergency Program for AIDS Relief (PEPFAR). www.pepfar.gov

The President's Malaria Initiative. www.pmi.gov

The World Bank. (2014). Poverty and Health. Brief. http://www.worldbank.org/en/topic/health/brief/poverty-health

Tucker, J.D., Bu, J., Brown, L.B. et al. (2010). Accelerating worldwide syphilis screening through rapid testing: a systematic review. *Lancet Infect. Dis.* 10 (6): 381–386. https://doi.org/10.1016/S1473-3099(10)70092-X.

UN Commission on Life Saving Commodities. https://www.unfpa.org/publications/un-commission-life-saving-commodities-women-and-children

United Nations Commission on Life-Saving Commodities. http://www.who.int/medical_devices/unclsc/en (Accessed March 2018)

Urdea, M., Penny, L.A., Olmsted, S.S. et al. (2006). Requirements for high impact diagnostics in the developing world. Nature supplement: determining the global health impact of improved diagnostic technologies for the developing world. *Nature* 444 (suppl 1): 73–79.

WHO. (2007). Sexual and reproductive health. The global elimination of congenital syphilis: rationale and strategy for action. http://www.who.int/reproductivehealth/publications/rtis/9789241595858/en

WHO. (2016). Fact Sheet: World Malaria Report. 13 December 2016. http://www.who.int/malaria/media/world-malaria-report-2016/en

WHO. WHO End TB Strategy. http://www.who.int/tb/post2015_strategy/en

WHO. (2004). Global Health Observatory Data. World Health Statistics. http://www.who.int/gho/publications/world_health_statistics/2014/en

WHO. (2014). Global Health Observatory Data. World Health Statistics. Available at http://www.who.int/gho/hiv/en

WHO. Tuberculosis Fact Sheet. (2018). http://www.who.int/mediacentre/factsheets/fs104/en

Wijesooriya, N.S., Rochat, R.W., Kamb, M.L. et al. (2016). Global burden of maternal and congenital syphilis in 2008 and 2012: a health systems modelling study. *Lancet Glob. Health* 4: e525–e533.

Webliography

http://www.who.int/publications/almaata_declaration_en.pdf Declaration of Alma-Ata. International Conference on Primary Health Care, Alma-Ata, USSR, 6–12 September 1978.

https://www.finddx.org/wp-content/uploads/2016/03/PDP_Brief_ENG_final_Sep_2014. pdf FIND. Product development partnerships. 2014.

https://www.finddx.org. FIND (Foundation for Innovative New Diagnostics)

http://www.who.int/mediacentre/events/2005/wha58/gates/en/index.html (retrieved January 2, 2008). Gates, B. 2005. "Remarks of Mr. Bill Gates, Co-founder of the Bill & Melinda Gates Foundation, at the World Health Assembly," Fifty-eighth World Health Assembly, Geneva, Switzerland.

www.theglobalfund.org. Global Fund for HIV, Tuberculosis and Malaria.

http://www.who.int/gho/publications/world_health_statistics/2014/en/ WHO Global Health Observatory Data. World Health Statistics 2004.

https://www.nature.com/nature/archive/supplements.html Nature Supplements, 2006. Improved Diagnostic Technologies for the Developing World.

https://sustainabledevelopment.un.org/?menu=1300 Sustainable Development Goals.

www.pepfar.gov. The President's Emergency Program for AIDS Relief (PEPFAR).

www.pmi.gov. The President's Malaria Initiative

http://www.worldbank.org/en/topic/health/brief/poverty-health. The World Bank. Poverty and Health Brief. 2014.

https://www.unfpa.org/publications/un-commission-life-saving-commodities-women-and-children. UN Commission on Life Saving Commodities.

http://www.who.int/medical_devices/unclsc/en (accessed March 2018) United Nations Commission on Life-Saving Commodities.

http://www.who.int/malaria/media/world-malaria-report-2016/en WHO. 2016. Fact Sheet: World Malaria Report 2016.

http://www.who.int/mediacentre/factsheets/fs104/en WHO. 2018. Tuberculosis Fact Sheet.

http://www.who.int/mediacentre/factsheets/fs104/en WHO. 2018. Tuberculosis Fact Sheet.

http://www.who.int/tb/post2015_strategy/en WHO. WHO End TB Strategy.

http://www.who.int/gho/hiv/en WHO. Global Health Observatory Data. World Health Statistics 2014.

http://www.who.int/reproductivehealth/publications/rtis/9789241595858/en WHO. Sexual and reproductive health. 2007. The global elimination of congenital syphilis: rationale and strategy for action.

37

Useful Electronic Healthcare Resources Available for Those Working in Remote Settings

Tyler Evans

AIDS HealthCare Foundation, Los Angeles, CA, USA

CHAPTER MENU

37.1   Introduction

Accessing information remotely has become increasingly easier for a number of professional fields, as traditional brick and mortar repositories have become obsolete and rapidly replaced with online databases. Healthcare is certainly no exception. Indeed, the need for biomedical research to be housed within a universal cloud has finally been met and our rapidly expanding multipolar world has precipitated a recent explosion of biomedical research. This relative increase in data, and access to it, has been coupled with a paradigm shift of clinicians, irrespective of global hemisphere, to follow a uniform

Revolutionizing Tropical Medicine: Point-of-Care Tests, New Imaging Technologies and Digital Health,
First Edition. Edited by Kerry Atkinson and David Mabey.
© 2019 John Wiley & Sons, Inc. Published 2019 by John Wiley & Sons, Inc.

set of evidence-based guidelines for both the prevention and treatment of disease. As this information exchange requires transparent and easily accessible platforms, clinicians working in remote settings in the global south, including low- and middle-income countries (LMICs), seem to preserve relative parity to literature and data as do those based in academic centers in the developed world in the global north.

Based on a number of laudable collaborative efforts from various academic institutions, global funds, non-governmental organizations (NGOs), and state-funded agencies, this information is particularly abundant for those working in the fields of tropical medicine/infectious disease. Notwithstanding the importance of effective communicable disease management in rural settings, the importance of education to manage non-communicable diseases (NCDs) has also started to gain prominence.

The reasoning behind the need for this access will vary considerably, and this Chapter aims to target a versatile environment. As clinicians begin traveling more into unfamiliar territories, a number of unknowns begin to emerge. As a case in point, a well-educated and trained western family medicine physician may find himself or herself identifying a salient pattern of splenomegaly in children/adolescents along the Nile River. Alternatively, a western trained orthopedic surgeon may find himself or herself delivering a baby at sea and having to respond in emergency mode to a postpartum hemorrhage. Of course, this chapter would be remiss if it did not consider the well-trained physician in an LMIC in the global south (erudite in tropical medicine) and now with grant-funded access to an academic center with the capacity for screening of many diseases including cancer.

As these complex migratory webs start to unravel, clinicians find themselves needing answers in fields they were not trained in. Fortunately, information is readily available, but the poisoned chalice of general access can be overwhelming and even daunting. As such, this chapter serves to organize these resources so that the reader can navigate them with relative ease. Lastly, in our contemporary world of evidence-based medicine, we find ourselves frequently searching for any research updates and discoveries that could fundamentally change our practice. A chapter such as this provides one of the most effective platforms to access these updates.

37.2 General Web-Based Resources

The following sites are very helpful in terms of general medical information, and are arguably the most replete with comprehensive, updated and evidence-based guidelines (EBGs). While UpToDate is the gold standard for most clinical scenarios nowadays in the USA, it may be less helpful when managing tropical diseases with limited resources. However, it is certainly helpful when managing NCDs and obviates the need in many scenarios to consult with a respective specialist, particularly if such specialists have limited access. An added benefit is that it also provides American Medical Association (AMA) established continuing medical education (CME) credit. The app is free to download but a financial subscription is required to log in and use the program. Further details are available at https://www.uptodate.com.

Alternatively, the National Health Service (NHS, UK) provides an impressive repository of succinct medical information, but seems to rely less on updated EBGs. Importantly, NHS is a free resource, whereas, as indicated above, UpToDate requires a subscription (often provided by one's home institution, if based in the USA and some other countries). Further information on the NHS UK website can be obtained at National Health Service (UK). Its website address is:

https://www.nhs.uk/Conditions/Pages/hub.aspx

The following sites are also free (and mostly government-sponsored, with excellent access to medical information. However, they focus more on research publications, which can be less helpful when looking for a quick, evidence-based solution to a specific clinical issue. However, these sites are easy to navigate and highly germane to the implementation of sustainable public health systems, if looking to impact larger scale systems. They include

- PubMed

This is a data repository within the US-funded National Institutes of Health (NIH), with a simple navigational platform. However, while it is free to use, most of the publications are in abstract form only, requiring either a fee or a subscription for information beyond this.

- Google Scholar

This site seems to employ a similar algorithm to that of PubMed, but may be easier to access in certain settings because of the power of Google's search engines.

- The Medical Research Council (MRC, UK)
  This site essentially offers the UK equivalent to PubMed.
- Hinari

This site is intended for a very different audience (LMICs) and is the result of collaboration between the World Health Organization (WHO) and several major publishers to provide open (free) access to a number of biomedical journals. While the search engine is not as elegant as the others noted in this section, its unrestricted access is a definite advantage.

The website addresses for these organizations are as follows:

Google Scholar
https://scholar.google.com

Medical Research Council (MRC, UK)
www.mrc.ac.uk/publications/browse

Hinari
http://www.who.int/hinari/en

Pubmed
https://www.ncbi.nlm.nih.gov/pubmed

37.3 Travel Medicine

The National Travel Health Network and Centre (NaTHNaC) provides an excellent dashboard with options to link either with Travel Health Pro, an excellent travel medicine resource providing updated publications in the field, as well as with country health profiles. It also houses a platform for online training courses, and information on global access to yellow fever distribution. Importantly, it also provides an opportunity to connect with a UK-based specialist referral desk with any more detailed queries.

The Centers for Disease Control and Prevention (CDC) is a well-known USA government agency focusing mostly on communicable diseases in the USA and beyond.

As such, it is an excellent (and complementary) resource for those working in remote settings who want to obtain a better understanding of disease transmission. Its well-designed platform is probably best known for its succinct schematic microbial (and associative vector, if relevant) life cycles. The CDC is also an excellent resource for travel medicine specialists, although many believe their recommendations overstate disease risk, so their statements may be best considered along with other recommendations (for example, the UK-based Travel Health Pro). Notably, it also has a useful section on HIV/AIDS and tuberculosis, with an impressive gateway to a wealth of information (discussed further below).

Lastly, the International Society of Travel Medicine (ISTM) remains the ultimate authority on travel medicine and has a wealth of publications and recommendations on their website. Many of the world's most prominent scholars in travel medicine remain active within the Society and communication with them provides clinicians with a strong sense of engagement and support. The ISTM also provides a training course, which ultimately leads to a Certificate of Knowledge examination (see Online Courses below).

The website addresses of these organizations are as follows:

National Travel Health Network and Centre (UK)
http://nathnac.net
https://travelhealthpro.org.uk
http://nathnactrainingportal.org.uk
https://travelhealthpro.org.uk/contact

Centers for Disease Control and Prevention (CDC, US)
https://www.cdc.gov/globalhealth/index.html
https://wwwnc.cdc.gov/travel/page/clinician-information-center
https://wwwnc.cdc.gov/travel
https://www.cdc.gov/globalhivtb

International Society of Travel Medicine
http://www.istm.org
http://www.istm.org/certificateofknowledge

37.4 The Big Three Communicable Diseases in Low- and Middle-Income Countries (LMICs)

While many people erroneously conflate the fields of rural and tropical medicine, it is important to treat them as separate entities. This having been said, depending on which part of the world you may be working in, access to knowledge of tropical diseases is a critical resource. Based on the burden of disease, most clinicians are aware that the overwhelming majority of communicable diseases influencing morbidity and mortality are malaria, HIV/AIDS and tuberculosis (see Chapter 1).

37.4.1 Malaria

One of the most efficient resources for the laboratory identification of infectious diseases is the CDC's DPDx, a website developed and maintained by the CDC's Division of Parasitic Diseases and Malaria (DPDM). It not only provides a reference and training function, but also assists with diagnoses. The CDC also offers a CME course on basic malaria management. All resources are complementary.

With funding from the Bill and Melinda Gates Foundation, the Global Health Training Centre has been able to provide a number of helpful free online courses, including one on malaria microscopy, which is strongly recommended. In addition to a predominant focus on malaria, the United Kingdom National External Quality Assessment Service (UK NEQAS) and the American Society of Tropical Medicine and Hygiene (ASTMH) provide an extensive repository of microscopic images to help in the detection of a number of infectious diseases using a number of different techniques. There is also an impressive collection of microscopic images published by a Czech Republic-based microbiology faculty.

Lastly, PLoS Neglected Tropical Diseases (NTDs) is an open access (free) journal with some of the most important research being published on NTDs. This journal's website has excellent animation and is simple to navigate. It provides a platform for publishing important research results on a range of neglected communicable diseases, as well as providing an opportunity for policy action and advocacy for improved systems to address these important diseases.

Website addresses for these organizations are as follows:

Centers for Disease Control and Prevention DPDx
https://www.cdc.gov/dpdx/index.html
https://www2a.cdc.gov/TCEOnline/registration/detailpage.asp?res_id=2942
Global Health Training Centre
https://globalhealthtrainingcentre.tghn.org/elearning/basic-malaria-microscopy

UK NEQAS
www.ukneqasmicro.org.uk/parasitology/index.php/parasitology-teaching-website
American Society of Tropical Medicine and Hygiene (ASTMH)
http://www.astmh.org/education-resources/zaiman-slide-library

Faculty Teaching Hospital Kralovske Vinohrady (Czech Republic)
http://mikrobiologie.lf3.cuni.cz/parazitologie/#Diphylidium
PLOS Neglected Tropical Diseases
http://journals.plos.org/plosntds

37.4.2 HIV/AIDS

The largest pandemic of the twenty-first century thus far is HIV/AIDS. Organizations including UNAIDS, the CDC, the Joint United Nations Program on HIV and AIDS (UNAIDS), The National Institute of Allergy and Infectious Diseases (NIAID), the AIDS HealthCare Foundation, Johns Hopkins Medical Institute, the University of California San Francisco, the London School of Hygiene and Tropical Medicine and a number of other institutions and facilities across the globe have led to remarkable achievements in terms of awareness and access to testing and effective treatment. This was notably encapsulated in the well-branded "90-90-90" campaign. The targets of this campaign are that by 2020 90% of all people living with HIV will know their HIV status, that by 2020 90% of all people with diagnosed HIV infection will receive sustained antiretroviral therapy and that by 2020 90% of all people receiving antiretroviral therapy will have viral suppression. This disease continues to affect variable populations across the globe. Its synergistic effect on other diseases makes it an even more important disease to control. However, as the disease contains such a complex web of opportunistic infections, chronic comorbidities, and complex psychosocial conditions, it has gradually become its own singular field of medicine. Consequently, the training for HIV medicine is extensive, although much of the training can be fractionated based on the context in which a trainee is working.

It is important to note that HIV/AIDS management in the developed nations and in the LMICs is diametrically different and requires very different lenses. The LMICs suffer more from resource shortages and awareness, thereby leading to a higher incidence of co-infections and sequelae. Alternatively, the developed world tends to be affected more by concurrent complex psychosocial conditions and antiretroviral (ARV) resistance. As such, the science and availability of emerging highly active antiviral treatment (HAART) becomes more germane to the practice of medicine globally. In LMICs the WHO clinical guidelines are the gold standard of care and are free. For the developed world UpToDate provides a very robust and efficient platform for advice on the most effective management practices.

Much like the WHO guidelines, the US Department of Health and Human Services provides a complementary updated collection of EBGs. Many other western governments follow similar guidelines (depending on the availability of HAART). The HIV Medical Association (HIVMA), an affiliate of the Infectious Disease Society of America (IDSA), also provides a number of cutting edge publications in an elegant platform. Unfortunately, there is a subscription cost associated with this.

In addition to these core guidelines, the International Antiviral Society-USA (IAS-USA), and the American Academy of HIV Medicine (AAHIVM) are two well-known educational organizations providing a number of educational opportunities, including internationally renowned conferences. AAHIVM publishes an impressive online textbook, which serves as the gold standard for many HIV clinicians based in the USA and leads to the conferring of specialist certification. IAS-USA is the organization responsible for the Conference on Retroviruses and Opportunistic Infections (CROI), which is a major conference for HIV/AIDS biomedical researchers, scientists and clinicians. The impressive faculties associated with these organizations provide a series of high yield and impressive webinars, which are all free of charge. While the AAHIVM Fundamentals of HIV Medicine has an associated cost, their core curriculum is available free of charge.

The AIDS Education and Treatment Center (AETC) programs are a series of educational sessions provided by the US government-funded Health Resources and Service Administration (HRSA). While the primary focus of these programs is geared toward onsite training, they also provide a number of free and highly engaging webinars, which are both live-streamed and archived.

Website addresses for these organizations are shown below:

World Health Organization (WHO)
http://www.who.int/hiv/pub/guidelines/en

AIDS Info, US Department of Health and Human Services
https://aidsinfo.nih.gov/guidelines

HIV Medicine Association (HIVMA, US)
http://www.hivma.org/Guidelines_and_Patient_Care.aspx

International Antiviral Society-USA (IAS-USA)
https://www.iasusa.org/webinars

American Academy of HIV Medicine
https://aahivm.org/fundamentals-of-hiv-medicine
https://aahivm.org/core-curriculum
https://aahivm.org/clinical-information

AIDS Education and Treatment Center Programs (AETC, USA)

https://www.aidsetc.org/resource-type/webinars

UptoDate
www.uptodate.com

37.4.3 Tuberculosis

Tuberculosis (TB) and HIV/AIDS are intricately intertwined in the LMICs. While over 37 million are living with HIV/AIDS, nearly 2 billion are living with TB. The global epidemiologic overlap is striking (see Chapter 1), and consequently has led to TB being the most common opportunistic infection in those living with HIV/AIDS. While TB has maintained evolutionary fitness over several centuries, its adverse impact on health has been variable. The common denominator of TB-associated morbidity and mortality in the pre-Koch[1] days was poor hygiene and nutrition and a basic misunderstanding of the disease process. While living conditions in many resource-poor parts of the world have not substantially changed (and in certain ways, may have even deteriorated), HIV/AIDS has added a new set of concerns to those living with TB, as illness has been increased by failing immunity. Thus, understanding the scope of TB, and ways to prevent and control it, are germane to public health efforts in remote settings.

Additionally, a new set of unique challenges, largely precipitated by management within resource poor settings, is multi-drug resistant TB (MDR-TB). The incidence of MDR-TB is rapidly increasing (see Chapter 1). Additionally, the pervasive course of disease, often presenting as extra pulmonary TB, leaves clinicians without effective detection systems. Awareness of these challenges and ways to treat and manage MDR-TB is germane to working effectively in remote settings.

Fortunately, there are a number of organizations that have invested heavily in providing free electronic resources to clinicians who may need assistance. The CDC has provided a number of helpful resources for clinicians, including a comprehensive core curriculum, which is available online, as well as animated web courses, which also providing CME credit. In addition, the Curry TB Center at the University of California (UCSF) also provides a number of on demand and archived webinars, which continue to have excellent feedback.

The International Union Against TB and Lung Disease is currently the gold standard for academic publications, information, material resource support, expert consultation, and advocacy for the prevention and management of TB. It is a global collaborative organization with strong links to the WHO and can be an important resource for those working in remote settings. They have published a number of engaging and free online courses, which are strongly recommended.

Lastly, both Médecins Sans Frontières (MSF, Doctors without Borders) and Partners in Health (PIH), a Boston-based nonprofit healthcare organization, have collaborated on a publication, which is available to all MSF clinicians in the field and which is considered the gold standard while on mission. This publication is replete with evidence in terms of treating MDR-TB and should be a part of any rural clinician's armamentarium. The book is now available online and free.

Website addresses for organizations mentioned in this section are as follows:

Centers for Disease Control and Prevention (CDC, US)
https://www.cdc.gov/tb/education/corecurr/index.htm

---

[1] Robert Heinrich Hermann Koch (1843–1910) was a German physician and microbiologist. He identified the specific causative agents of tuberculosis, cholera, and anthrax.

https://www.cdc.gov/tb/webcourses/tb101/default.htm
https://www.cdc.gov/globalhivtb/index.html

Curry TB Center, University of California, San Francisco (UCSF)
http://www.currytbcenter.ucsf.edu/trainings/on-demand-webinars
http://www.currytbcenter.ucsf.edu/trainings/webinar-archive
The International Union Against TB and Lung Disease
http://www.unioncourses.org/category/tb-and-lung-health-courses

Tuberculosis (2014), Médecins Sans Frontières (Doctors without Borders)/Partners in Health
http://refbooks.msf.org/msf_docs/en/tuberculosis/tuberculosis_en.pdf

37.5   Hepatitis C

The global impact of hepatitis C (HCV) is arguably greater than that of the other chronic viral disease, including HIV, with a prevalence nearly twofold higher (approximately 71 million, and likely underreported). While this high prevalence has always been an important public health factor, the ability of clinicians in developed nations to treat this disease was largely anchored by a poorly tolerated and ineffectual interferon alpha-based regime. Consequently, public health organizations in LMICs seemed to largely ignore HCV. However, a major breakthrough in molecular pharmaceuticals established an effective cure requiring an 8–12 weeks course of once daily dosing with specific direct antiviral agents (DAAs).

Consequently, while management has changed dramatically, the appropriate selection of DAAs (including the most cost- effective) has become the most germane area of discussion. HCV was classically managed by hepatologists, gastroenterologists, and infectious disease specialists. However, as this pandemic now has the potential for elimination, a number of academic institutions and organizations have collaborated to provide free education to clinicians without prior experience in HCV management, in order to increase access points for treatment, and thus assist in elimination efforts.

The hepatology department of Mount Sinai Hospital in New York City has created an impressive series of both live and archived webinars that provide a platform for bilateral communication. This platform allows clinicians in remote areas to upload patient data and submit related queries to the hepatology faculty at this hospital. The platform will then suggest the most appropriate DAA regimens and also provide the referring clinicians with an opportunity to track their own data.

The AAHIVM has recently created an Institute for HCV, because of the high prevalence (23% in the USA) of co-infection with HCV among those who are HIV-positive. Its electronic dashboard provides simple navigational tools to access quick facts, epidemiological data and clinical guidelines. Finally, the American Association for the Study of Liver Diseases (AASLD) and the Infectious Diseases Society of America (IDSA) have collaborated to create an updated evidence-based collection of guidelines and recommendations (HCV Guidelines) – including a section on "unique populations." There is also an app available. Importantly, all of these HCV resources are free of charge.

Website addresses for organizations mentioned in this section are as follows:

HepCure, Icahn School of Medicine at Mt. Sinai (NY, NY)

http://hepcure.org

American Academy of HIV Medicine, Institute for Hepatitis C
https://aahivm.org/institute-for-hepatitis-c

HCV Guidelines, American Association for the Study of Liver Diseases (AASLD)/
Infectious Disease Society of America (IDSA)
https://www.hcvguidelines.org

37.6   Other Infectious Diseases (IDs)

While a disproportionate amount of ID literature is provided in the sections described above, there are always general ID questions which may not be covered within the tropical medicine bibliography and webliography. Because of this, UpToDate is recommended for many such questions. In addition, the portals provided by IDSA and its affiliated organizations (the HIVMA as noted above, the Society for HealthCare Epidemiology of America [SHEA], and the Pediatric Infectious Disease Society [PIDS]) are also recommended. The publications from these groups provide results from some of the most preeminent studies in infectious disease. The Journal of Infectious Diseases tends to focus more on global public health issues, whereas Clinical Infectious Diseases tends to have more of a clinical focus mostly on research conducted in the developed nations. Lastly, Open Forum Infectious Disease (OFID) provides open access (free) to a number of publications that reflect the intersection between biomedical science and clinical practice, with a particular focus on translational research.

Website addresses for organizations mentioned in this section are as follows:

UptoDate
www.uptodate.com

Infectious Disease Society of America (IDSA)
http://www.idsociety.org/News_and_Publications
http://www.idsociety.org/OFID

HIV Medicine Association (HIVMA, US)
http://www.hivma.org/Guidelines_and_Patient_Care.aspx

Society for HealthCare Epidemiology of America (SHEA)
https://www.shea-online.org/index.php/practice-resources/research/publications

Pediatric Infectious Disease Society (PIDS, US)
https://www.pids.org
https://www.hopkinsmedicine.org/apps/all-apps/johns-hopkins-hiv-guide

37.7   Dermatology

The site below allows individuals to upload dermatology case studies and obtain a rapid response from a panel of USA dermatologists for a fee starting at US$29.00. This will put it out of reach for many individuals living in LMICs. The website has photographs of many skin disorders https://www.firstderm.com.

DermNet NZ is a charitable trust based in New Zealand and also has an extensive photographic library available online without charge. It does not offer online consultations. Its website address is https://www.dermnetnz.org/image-library.

37.8 Obstetrics and Gynecology

UpToDate, again, is the preferred resource for specialists in women's health. However, the portal for the American College of Obstetrics and Gynecology (ACOG) collection of clinical guidelines and publications is useful. While some of these are free of charge, many of the publications require a paid subscription. There are also book recommendations listed below.

UptoDate
www.uptodate.com

American College of Obstetrics and Gynecology (ACOG)
https://www.acog.org/Clinical-Guidance-and-Publications/Search-Clinical-Guidance
https://www.acog.org/About-ACOG/News-Room/ACOG-App

37.9 Pediatrics

Many of the conditions affecting children in LMICs are referenced in the above sections. However, for general pediatrics, UpToDate is, again, the preferred resource. Also included below is the American Academy of Pediatrics (AAP), which has a very user-friendly dashboard to access ebooks, publications and apps. Unfortunately, there is a cost associated with many of these.

UptoDate
www.uptodate.com

American Academy of Pediatrics (AAP)
http://ebooks.aappublications.org
http://hesperian.org/books-and-resources/safe-pregnancy-and-birth-mobile-app
https://www.hopkinsmedicine.org/apps/all-apps/hopkins-gyn-ob-unbound

37.10 Psychiatry

Mental health is an area that is all too often forgotten – especially in remote settings, where it is, ironically, often needed the most. The problem with accessing information that can help clinicians in rural settings is that the approach to many mental health issues has to be culturally bound, and therefore remains highly variable between different nations or between different areas of the same nation. Most (USA) psychiatrists surveyed agreed that UpToDate was their preferred resource for evidence-based psychiatric care, and their database is recommended for most psychiatric disorders.

In addition, the portal to the American Journal of Psychiatry (the official journal of the American Psychiatric Association [APA]), can be very helpful when looking for a more in-depth analysis. Unfortunately, there is a cost to access this publication. Also included below are MSF's Guidelines on psychosocial and mental health interventions in areas of mass violence, which is more helpful when it comes to the general approach of implementing a mental health program. However, it is not as helpful when it comes to the actual treatment and management of certain conditions. Finally, as Syria currently represents the largest sociopolitical conflict causing mass migration and a number of mental health sequelae, there is a telepsychiatry network that can help clinicians treating this specific population – Syrian Telemental Health Network.

Website addresses of organizations discussed in this section are as follows:

UptoDate
www.uptodate.com
https://www.appi.org/products/dsm-mobile-app
https://www.hopkinsmedicine.org/apps/all-apps/psychiatry-2015-poc-it-guide

American Journal of Psychiatry, American Psychiatric Association (APA)
https://ajp.psychiatryonline.org

Psychosocial and Mental Health Interventions in Areas of Mass Violence, A Community Base Approach (2011), Médecins Sans Frontières (MSF)/Doctors without Borders
http://www.msf.org/sites/msf.org/files/old-cms/source/mentalhealth/guidelines/MSF_mentalhealthguidelines.pdf

Syrian Telemental Health Network
http://www.stmh.net

37.11 Emergency Medicine (EM)

UpToDate is currently also the preferred resource for EM physicians. However, a number of helpful journals can also be accessed at the American College of Emergency Physicians website. Access Emergency Medicine is also included and can provide a number of selected chapters and articles from their list of publications. Electronic textbooks and atlases can be purchased through this portal. Unfortunately, most of the information provided by these two sites is also associated with a cost.

The website addresses of these sites are as follows:

UptoDate
www.uptodate.com

American College of Emergency Physicians (ACEP)
https://www.acep.org

Access Emergency Medicine (McGraw Hill)
http://accessemergencymedicine.mhmedical.com

37.12 Preventive Health

Guidelines on preventive medicine practices are ultimately determined by organizations that regularly study health-related issues and outcomes and use modeling and meta-analyses to determine recommendations for screening and prevention of a wide range of diseases including cancer screening and immunizations. While the guidelines may vary worldwide, many nations tend to follow USA guidelines, if feasible. The US Preventive Services Task Force (USPSTF) is comprised of a panel of experts in primary care and preventive medicine and is advised by the research findings of the US-funded Agency for HealthCare Research and Quality (AHRQ). The USPSTF recommendations are mostly adopted into USA policy. Their guidelines are given lettered rankings, identifying the quality of the evidence to make such recommendations. Access is free of charge and an app is available.

In terms of immunizations only, the CDC and WHO have published sets of gold standard recommendations for both children and adults. While the CDC recommendations are primarily intended for USA populations, many other countries follow similar recommendations, and these respective adoptions are largely dependent on cost and availability.

The website addresses of these sites are as follows:

US Preventive Services Task Force (USPSTF)
https://www.uspreventiveservicestaskforce.org

Centers for Disease Control and Prevention (CDC, US)
https://www.cdc.gov/vaccines/schedules/index.html

World Health Organization (WHO)
http://www.who.int/ith/vaccines/en

37.13   Disease Mapping

Most of those working in remote settings will likely have background knowledge in epidemiology in order to better understand the pattern and distribution of diseases they are dealing with. Three useful disease mapping instruments provide important updated information on current outbreaks and antimicrobial resistance worldwide. Both HealthMap and the Global Infectious Disease and Epidemiology Network (GIDEON) provide updated information on current outbreaks located around the world and represented geographically. The main difference is that the former is free to access and based on local reports (powered by Google and largely unvetted). However, the latter requires a paid subscription, but filters and vets the data being provided. The latter also benefits from a more elegant interface. However, they are both very useful and available as apps.

The Resistance Map summarizes national/subnational data on antibiotic usage and associated resistance around the world. This map is particularly important in selecting the appropriate antibiotic regimen in certain areas, and helps mitigate one of the most formidable public health emergencies currently faced. It is free of charge.

The website addresses of these sites are as follows:

Health Map, Boston Children's Hospital (Cambridge, MA)
http://www.healthmap.org/en

Global Infectious Disease and Epidemiology Network (GIDEON)
https://www.gideononline.com/update-map

Resistance Map, Center for Disease Dynamics Economics and Policy (Washington, DC)
https://resistancemap.cddep.org

37.14   Pharmaceuticals

Information on pharmaceutical agents by IBM's Watson Health are either free of charge or available for a small subscription (currently US$2.99 per year) at
https://truvenhealth.com/products/micromedex/quick-reference/mobile

Epocrates is a publisher of mobile device software applications, designed to provide information about drugs to doctors and other healthcare professionals. Among the

software functions is a check for drug interactions, news feeds for product announce-
ments, medical news, journal article lookups as well as a mobile guide diagnosis code.
The application is widely available and sometimes free. It is possible to download the
application easily to an iPhone or Android device, so it is easily accessible.

http://www.epocrates.com

The official Johns Hopkins ABX (Antibiotic) Guide from Johns Hopkins Medicine
features frequently updated, authoritative, evidenced-based information on the treatment of infectious diseases to help make decisions at the point-of-care. This comprehensive web and mobile resource organizes details of diagnosis, drug indications,
dosing, pharmacokinetics, side effects and interactions, pathogens, management and
vaccines into easily accessible, quick-read entries.

https://www.hopkinsmedicine.org/apps/all-apps/johns-hopkins-abx-guide-2017

37.15   Online Courses

The resources mentioned above may not be sufficient for someone who is planning on
spending an extended period of time in remote settings, and who would like to procure
a defined, sustainable and certifiable skill set. A few recommended courses, many of
which can be completed online, are listed below.

The University of Minnesota is largely a feeder course for those planning on taking
the ASTMH CTropMed Certificate of Knowledge examination. This has been organized by a faculty which provides an excellent foundation into a number of important
topics – including refugee and asylum care, disaster response, tropical medicine and
travel medicine. There is a significant cost associated with the exam.

West Virginia University also provides an online module. However, this course
is only given during certain times and may require on site attendance for certain
components.

As listed above, the National Travel Health Network and Centre, ISTM, AAHIVM,
and the International Union Against TB and Lung Disease all offer comprehensive
courses, with some programs leading to certification. There may be some cost associated with some of these online courses. Each year Dr. Philip Gothard and his colleagues
at the London School of Hygiene and Tropical Medicine organize a three month
diploma course in tropical medicine held for six weeks in Tanzania and six weeks in
Uganda. Attendance requires a fee. There is an examination at the end of the course and
those who are successful are awarded the East African Diploma of Tropical Medicine
and Hygiene (DTM&H). Up to two-dozen DTM&H scholarships are open to students
from low- and middle-income countries with an emphasis on applicants from partner
institutions in Africa.

www.lshtm.ac.uk/study/courses/short-courses/DTMH-east-africa

37.16   Recommended Books

While most publications are currently accessible either through frequently updated
databases or other platforms, there are certain general texts, mostly requiring purchase,
which are considered to be either essential (or at least, highly recommended) for working in remote settings. The first is the Clinical Guidelines text published by MSF

and required for all clinicians on mission. This is a high yield collection of well-evidenced guidelines that is perfectly suited for low resource settings. The text is now available free online.

MSF also makes available free online the following books:

Essential drugs, 2016; Refugee health, 1997; Rapid assessment of refugees or displaced populations; Tuberculosis, 2014; Essential obstetric and newborn care; Public health engineering; Measles, 2017; Clinical care, 2017. These books can be accessed online at: http://www.refbooks.msf.org/msf_docs/en/MSFdocMenu_en.htm

Also highly recommended are the Hesperian Health Guides. They are easy to read with a focus on low resource settings, predominantly in Latin America. Two books are specifically recommended (see below), one of which focuses on women's health. They are available for a nominal fee.

Lastly, for anyone working in remote rural or frontier/wilderness settings, Auerbach's Wilderness Medicine (now in its 7th edition), is strongly recommended. It is the current gold standard for dealing with medical emergencies in austere environments.

Website addresses of organizations or publication described in this section are as follows:

Clinical Guidelines (2016), Médecins Sans Frontières (MSF)/Doctors without Borders
http://refbooks.msf.org/msf_docs/en/clinical_guide/cg_en.pdf
Hesperian Health Guides
https://store.hesperian.org/prod/Where_There_Is_No_Doctor.html
https://store.hesperian.org/prod/Where_Women_Have_No_Doctor.html
Auerbach P. Wilderness Medicine, 7th Edition
https://www.us.elsevierhealth.com/auerbachs-wilderness-medicine-2-volume-set-9780323359429.html
Global Health Course, University of Minnesota School of Medicine
https://www.dom.umn.edu/global-medicine/education-training/courses/online/global-health-course
Global Health Program, West Virginia University School of Medicine
http://medicine.hsc.wvu.edu/tropmed/tropical-medicine-course

37.17 Institutions, Societies and Books

Access Emergency Medicine (McGraw Hill)
 AIDS Education and Treatment Center Programs (AETC, USA)
 AIDS Info, US Department of Health and Human Services
 American Academy of HIV Medicine (AAHIVM)
 American Academy of HIV Medicine, Institute for Hepatitis C
 American Academy of Pediatrics (AAP)
 American College of Emergency Physicians (ACEP)
 American College of Obstetrics and Gynecology (ACOG)
 American Journal of Psychiatry, American Psychiatric Association (APA)
 American Society of Tropical Medicine and Hygiene (ASTMH)
 Auerbach P. Wilderness Medicine, 7th Edition
 Centers for Disease Control and Prevention (CDC, US)
 Centers for Disease Control and Prevention DPDx
 Clinical Guidelines (2016), Médecins Sans Frontières (MSF)/Doctors without Borders

Curry TB Center, University of California, San Francisco (UCSF)
Faculty Teaching Hospital Kralovske Vinohrady (Czech Republic)
Global Health Course, University of Minnesota School of Medicine
Global Health Program, West Virginia University School of Medicine
Global Health Training Centre
Global Infectious Disease and Epidemiology Network (GIDEON)
Google Scholar
HCV Guidelines, American Association for the Study of Liver Diseases (AASLD)/
Infectious Disease Society of America (IDSA)
Health Map, Boston Children's Hospital (Cambridge, MA)
HepCure, Icahn School of Medicine at Mt. Sinai (NY, NY)
Hesperian Health Guides
Hinari
HIV Medicine Association (HIVMA, US)
Infectious Disease Society of America (IDSA)
International Antiviral Society-USA (IAS-USA)
International Society of Travel Medicine
International Society of Travel Medicine (ISTM)
Medical Research Council (MRC, UK)
National Health Service (UK)
National Travel Health Network and Centre (UK)
National Travel Health Network and Centre (UK)
Pediatric Infectious Disease Society (PIDS, US)
PLOS Neglected Tropical Diseases
Psychosocial and Mental Health Interventions in Areas of Mass Violence, A
Community Base Approach (2011), Médecins Sans Frontières (MSF)/Doctors without
Borders
Resistance Map, Center for Disease Dynamics Economics and Policy (Washington, DC)
Society for HealthCare Epidemiology of America (SHEA)
Syrian Telemental Health Network
The International Union Against TB and Lung Disease
The International Union Against TB and Lung Disease
Tuberculosis (2014), Médecins Sans Frontières (Doctors without Borders)/Partners
in Health
UK NEQAS
UptoDate
US Preventive Services Task Force (USPSTF)
World Health Organization (WHO)

Webliography

http://accessemergencymedicine.mhmedical.com
http://ebooks.aappublications.org
http://hepcure.org
http://hesperian.org/books-and-resources/safe-pregnancy-and-birth-mobile-app
http://journals.plos.org/plosntds
http://mikrobiologie.lf3.cuni.cz/parazitologie/#Diphylidium

http://nathnac.net
http://nathnactrainingportal.org.uk
http://nathnactrainingportal.org.uk/library
http://refbooks.msf.org/msf_docs/en/clinical_guide/cg_en.pdf
http://refbooks.msf.org/msf_docs/en/tuberculosis/tuberculosis_en.pdf
http://www.astmh.org/education-resources/zaiman-slide-library
http://www.currytbcenter.ucsf.edu/trainings/on-demand-webinars
http://www.currytbcenter.ucsf.edu/trainings/webinar-archive
http://www.epocrates.com
http://www.healthmap.org/en
http://www.hivma.org/Guidelines_and_Patient_Care.aspx
http://www.idsociety.org/News_and_Publications
http://www.idsociety.org/OFID
http://www.istm.org
http://www.istm.org/certificateofknowledge
http://www.istm.org/onlinelearningprogram
http://www.msf.org/sites/msf.org/files/old-cms/source/mentalhealth/guidelines/MSF_
    mentalhealthguidelines.pdf
http://www.pepid.com
http://www.refbooks.msf.org/msf_docs/en/MSFdocMenu_en.htm
http://www.stmh.net
www.ukneqasmicro.org.uk/parasitology/index.php/parasitology-teaching-website
http://www.unioncourses.org/category/tb-and-lung-health-courses
http://www.who.int/hinari/en
http://www.who.int/hiv/pub/guidelines/en
http://www.who.int/ith/vaccines/en
https://aahivm.org/clinical-information
https://aahivm.org/core-curriculum
https://aahivm.org/credentialing
https://aahivm.org/fundamentals-of-hiv-medicine
https://aahivm.org/institute-for-hepatitis-c
https://aidsinfo.nih.gov/guidelines
https://ajp.psychiatryonline.org
https://globalhealthtrainingcentre.tghn.org/elearning/basic-malaria-microscopy
https://itunes.apple.com/us/app/nejm-this-week/id373156254?mt=8
https://resistancemap.cddep.org
https://scholar.google.com
https://store.hesperian.org/prod/Where_There_Is_No_Doctor.html
https://store.hesperian.org/prod/Where_Women_Have_No_Doctor.html
https://travelhealthpro.org.uk
https://travelhealthpro.org.uk/contact
https://truvenhealth.com/products/micromedex/quick-reference/mobile
https://www.acep.org
https://www.acog.org/About-ACOG/News-Room/ACOG-App
https://www.acog.org/Clinical-Guidance-and-Publications/Search-Clinical-Guidance
https://www.aidsetc.org/resource-type/webinars
https://www.appi.org/products/dsm-mobile-app
https://www.cdc.gov/dpdx/index.html

https://www.cdc.gov/globalhealth/index.html

https://www.cdc.gov/globalhivtb

https://www.cdc.gov/globalhivtb/index.html

https://www.cdc.gov/tb/education/corecurr/index.htm

https://www.cdc.gov/tb/webcourses/tb101/default.htm

https://www.cdc.gov/vaccines/schedules/index.html

https://www.dermnetnz.org/image-library

https://www.dom.umn.edu/global-medicine/education-training/courses/online/global-health-course

https://www.firstderm.com

https://www.gideononline.com/update-map

https://www.hcvguidelines.org

https://www.hopkinsmedicine.org/apps/all-apps/hopkins-gyn-ob-unbound

https://www.hopkinsmedicine.org/apps/all-apps/johns-hopkins-abx-guide-2017

https://www.hopkinsmedicine.org/apps/all-apps/johns-hopkins-hiv-guide

https://www.hopkinsmedicine.org/apps/all-apps/psychiatry-2015-poc-it-guide

https://www.iasusa.org/webinars

www.lshtm.ac.uk/study/courses/short-courses/DTMH-east-africa

https://www.medscape.com/public/applanding

www.mrc.ac.uk/publications/browse

https://www.nhs.uk/Conditions/Pages/hub.aspx

https://www.pids.org

https://www.shea-online.org/index.php/practice-resources/research/publications

www.uptodate.com

https://www.uptodate.com/home/help-manual-install

https://www.us.elsevierhealth.com/auerbachs-wilderness-medicine-2-volume-set-9780323359429.html

https://www.uspreventiveservicestaskforce.org

https://www2a.cdc.gov/TCEOnline/registration/detailpage.asp?res_id=2942

https://wwwnc.cdc.gov/travel

https://wwwnc.cdc.gov/travel/page/clinician-information-center

38

The Future – How Do We Get from Here to There?

Kerry Atkinson[1,2] and David Mabey[3]

[1] *University of Queensland Centre for Clinical Research, Brisbane, Queensland, Australia*
[2] *The University of Technology/Institute of Health and Biomedical Innovation, Brisbane, Queensland, Australia*
[3] *Clinical Research Department, London School of Hygiene and Tropical Medicine, London, UK*

Revolutionizing Tropical Medicine: Point-of-Care Tests, New Imaging Technologies and Digital Health,
First Edition. Edited by Kerry Atkinson and David Mabey.
© 2019 John Wiley & Sons, Inc. Published 2019 by John Wiley & Sons, Inc.

38.1 Progress to Date

38.1.1 Report on The United Nations Millenium Development Goals (MDGs)

The eight millennium development goals were

1) Eradication of extreme poverty and hunger
2) Achievement of universal primary education
3) Promotion of gender equality and the empowerment of women
4) Reduction in child mortality
5) Improvement of maternal health
6) Combatting HIV/AIDS, malaria and other diseases
7) Ensuring environmental sustainability
8) Development of a global partnership for development

While tremendous progress was made on each of these goals, and is described in detail in Chapters 1 and 2, the MDG Report 2015 acknowledged that inequalities persist and that progress had been uneven globally, that the world's poor remain overwhelmingly concentrated in some parts of the world and that in 2011 nearly 60% of the world's one billion extremely poor people lived in just five countries. Additionally, it was acknowledged that too many women continued to die during pregnancy or from child-birth-related complications. Further, that progress tended to bypass women and those lowest on the economic ladder or who are disadvantaged because of their age, disability or ethnicity. Finally, it was accepted that health disparities between rural and urban areas remain pronounced.

(http://www.un.org/millenniumgoals/2015_MDG_Report/pdf/MDG%202015%20rev%20July%201.pdf).

38.1.2 The Bill and Melinda Gates Foundation Letter, 2017

The Bill and Melinda Gates Foundation was launched in 2000. The primary aims of the foundation are, globally, to enhance healthcare and reduce extreme poverty, and in America, to expand educational opportunities and access to information technology. In 2006 Warren Buffett became a major donor to the Foundation. The Gates' first became aware of poverty and disease while on a wild life safari in Africa. They were astonished to learn about the number of children dying of diarrheal diseases, pneumonia and malaria. In their Annual Letter 2017 they noted the following

- More children survived in 2015 than in 2014. More survived in 2014 than in 2013, and so on. If you add it all up, 122 million children under age five have been saved over the past 25 years. These are children who would have died if mortality rates had stayed where they were in 1990.
- When a mother can choose how many children to have, her children are healthier, they are better nourished, their mental capacities are higher and parents have more time and money to spend on each child's health and schooling. That is how families and countries get out of poverty. This link between saving lives, a lower birth rate, and ending poverty was the most important early lesson we learned about global health.
- Coverage for the basic package of childhood vaccines is now the highest it has ever been, at 86%. And the gap between the richest and the poorest countries is the *lowest* it has ever been. Vaccines are the biggest reason for the drop in childhood deaths.
- That led to a partnership between the Gates Foundation, business and government to set up Gavi, the Vaccine Alliance (initially known as the Global Alliance for Vaccines and Immunization), with the goal of getting vaccines to every child in the world. Gavi connects companies who develop vaccines with wealthy governments that help with funding and LMICs that get the vaccines to their people. Since 2000 Gavi has helped immunize 580 million children around the world. However, 19 million children, many of them living in conflict zones or remote areas, are still not fully immunized. It is crucial to the goal of cutting childhood deaths in half again – down below three million by 2030.
- Last year about one million infants died on the day they were born. A total of more than 2.5 million died in their first month of life. As the total number of childhood deaths has dropped, the proportion of newborn deaths has gone up. Newborn deaths now represent 45% of all childhood deaths, up from 40% in 1990.
- Some countries, though, are making big improvements. From 2008 through 2015, Rwanda, one of the poorest countries in Africa, cut its newborn mortality by 30%, down to 19 deaths per 1000 births. By comparison, Mali with a comparable gross domestic product has a newborn mortality rate of 38 deaths per 1000, twice as high as Rwanda. What were they doing differently in Rwanda? A few things so cheap that any government can support them: breastfeeding in the first hour and exclusively for the first six months. Cutting the umbilical cord in a hygienic way. And kangaroo care: skin-to-skin contact between mother and baby to raise the baby's body temperature.
- Malnutrition is partly responsible for 45% of childhood deaths.
- For the first time in history more than 300 million women in LMICs are using modern methods of contraception. It took decades to reach 200 million women. It has taken only another 13 years to reach 300 million and the impact in saving lives is fantastic.
- Extreme poverty has been cut in half over the last 25 years.
- In 1988 when the global vaccination campaign was launched to end polio, there were 350 000 new cases each year. Last year, there were 37. Those 37 cases were confined to northern Nigeria and parts of Afghanistan and Pakistan.

38.1.3   Other Issues

A large proportion of infectious diseases in low and middle income countries (LMICs) are entirely avoidable or treatable with existing medicines or interventions which are also highly cost effective. However, their delivery to affected populations has proven difficult due to

- weak health systems and infrastructures
- access to and utilization of health services
- gender discrimination
- low levels of female literacy
- lack of empowerment of women

These prevent women from seeking care for themselves and their children. Healthcare is also unaffordable for many families due to

- formal and informal healthcare fees
- cost of medicines and tests
- cost of not working during hospitalization, travel, food and accommodation

Effective delivery of proven interventions requires a variety of components including

- training health workers
- effective use of epidemiological data
- proper delivery of safe medicines and commodities
- accurate monitoring and evaluation
- community feedback

Successful implementation requires a positive interrelation between programs for disease control and the health system at large.

Community-based interventions (CBIs) have the potential to overcome barriers of access and availability and, if adequately equipped and supported by parallel structures, can make a significant impact on reducing the burden of the infectious diseases of poverty – malaria, HIV/AIDS, tuberculosis and the neglected tropical diseases (Das et al. 2014).

- A major issue is the availability of a trained health force to scale-up these interventions in population settings. According to a 2006 report by the WHO, 57 countries from Africa and Asia are facing shortages of a healthcare workforce and a total estimate of 4 250 000 workers are needed to fill the gap.
- Many of the interventions targeting infectious diseases have been administered via community platforms through community health workers (CHWs) who have received basic training. Successful examples exist, for example, in Brazil, where CHWs provide coverage to over 60 million people. Ethiopia is training about 30 000 workers with an emphasis on maternal and child health, HIV, and malaria. Other similar programs are also being considered in countries such as India, Ghana, and South Africa. Apart from providing antimicrobial chemotherapy, CHWs can also play a major role in imparting health education regarding general hygiene and sanitation and intervene for vector control measures within household and community settings. These community delivery strategies are not only effective but are also cost efficient, and by training teachers and other school personnel to administer anthelmintic drugs, costs could be reduced by "piggy-backing" on existing programs in the educational sector. In Ghana and Tanzania delivery of school-based targeted anthelmintic treatment costs as little as US $0.03 per child, which is as low as one-tenth of the estimated costs for vertical delivery.

38.2 Major Factors Adversely Affecting Global Health

38.2.1 Global Population Growth

The world's current population is 7.5 billion (http://www.worldometers.info/world-population). Projections of population growth established in 2015 predict that the human population will keep growing until at least 2050, reaching an estimated eight billion people in 2024 and nine billion by 2040 ("World Population Forecast" Worldometers) (Figure 38.1).

By 2050, the bulk of the world's population growth will take place in Africa. Of the additional 2.4 billion people projected between 2015 and 2050, 1.3 billion will be added in Africa, 0.9 billion in Asia and only 0.2 billion in the rest of the world. Africa's share of global population is projected to grow from 16% in 2015 to 25% in 2050 and 39% by 2100, while the share of Asia will fall from 60% in 2015 to 54% in 2050 and 44% in 2100 (World Population Prospects: The 2015 Revision – Key Findings and Advance Tables [PDF]. United Nations Department of Economic and Social Affairs, *Population Division. July 2015*).

India's population is anticipated grow to 1.5 billion by 2100 and it will replace China as the world's most populous nation.

This increase in population will increase the demand for all resources including food and water. One strategist, Bernard Salt, thinks that the key to slowing the rate of increase in the world's population is the emancipation of women through education. This will would likely decrease the fertility rate – currently, for example, 5.4 births per woman in Nigeria compared to 1.9 in Australia (bernard@thedemographicsgroup.co.au)

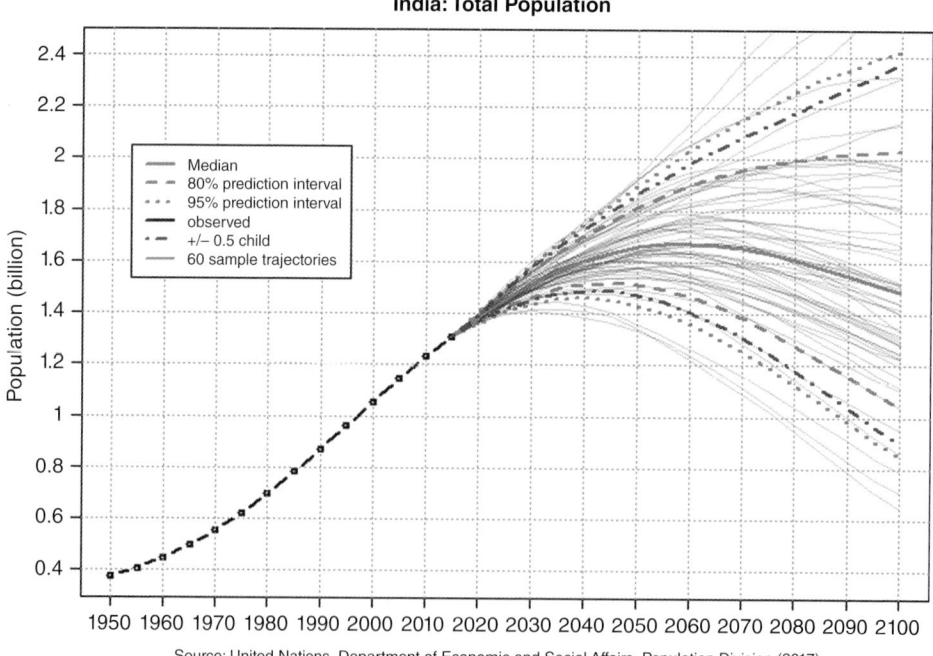

Source: United Nations, Department of Economic and Social Affairs, Population Division (2017).
World Population Prospects: The 2017 Revision. http://esa.un.org/unpd/wpp/

Figure 38.1 Probabilistic changes to the Indian population from 2020 till 2100 (https://esa.un.org/unpd/wpp/Graphs/Probabilistic/POP/TOT: *Source:* permission obtained). (*See color plate section for the color representation of this figure.*)

38.2.2 Global Warming

Multiple lines of scientific evidence show that the Earth's climate is warming (https://www.ipcc.ch/pdf/assessment-report/ar5/syr/AR5_SYR_FINAL_SPM.pdf) (Figure 38.2).

In 2013 the Intergovernmental Panel on Climate Change (IPCC) Fifth Assessment Report concluded that "It is extremely likely that human influence has been the dominant cause of the observed warming since the mid-twentieth century" (IPCC, Climate Change 2013). The largest human influence has been the emission of greenhouse gases such as carbon dioxide (Figure 38.3), methane and nitrous oxide.

Future climate change and associated impacts will differ from region to region around the globe. Anticipated effects include increasing global temperatures, rising sea levels, changing rainfall and expansion of deserts in the subtropics. Other likely changes include more frequent extreme weather events such as heat waves, droughts, heavy rainfall with floods and heavy snowfall and ocean acidification. Effects on humans include the threat to food security from decreasing crop yields and the abandonment of populated areas due to rising sea levels. The rise in temperature will not only exacerbate drought conditions but will make physical labor by humans difficult or impossible. The highest temperature ever experienced in the USA was 127 °F (52.3 °C) in Death Valley, California, in 2017. The highest temperature in India in 2016 was 123 °F (51 °C) in Phalodi, a town in the western state of Rajasthan.

Clearly the world's warming will adversely affect human health everywhere.

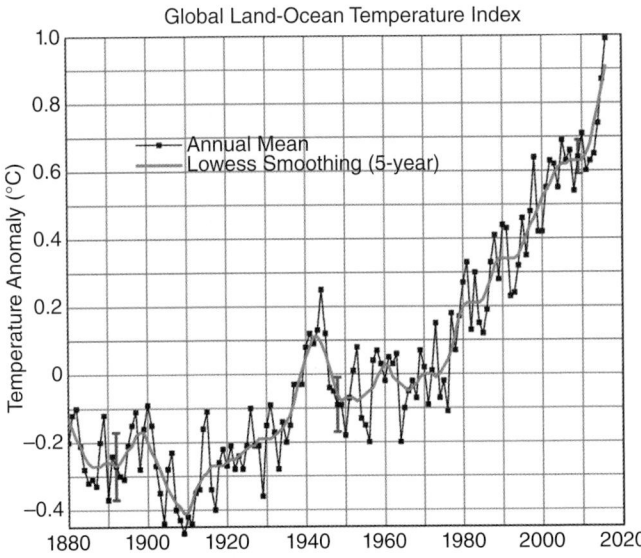

Figure 38.2 Global mean surface-temperature change from 1880 to 2016, relative to the 1951–1980 mean. The black line is the global annual mean, and the red line is the five-year local regression line. The blue uncertainty bars show a 95% confidence interval. This image, reproduced here from Wikimedia Commons, is the work of a USA Federal Government employee and, as such, it is in the public domain. *Source:* Credit: By NASA Goddard Institute for Space Studies – (http://data.giss.nasa.gov/gistemp/graphs, Public Domain, https://commons.wikimedia.org/w/index.php?curid=24363898). (*See color plate section for the color representation of this figure.*)

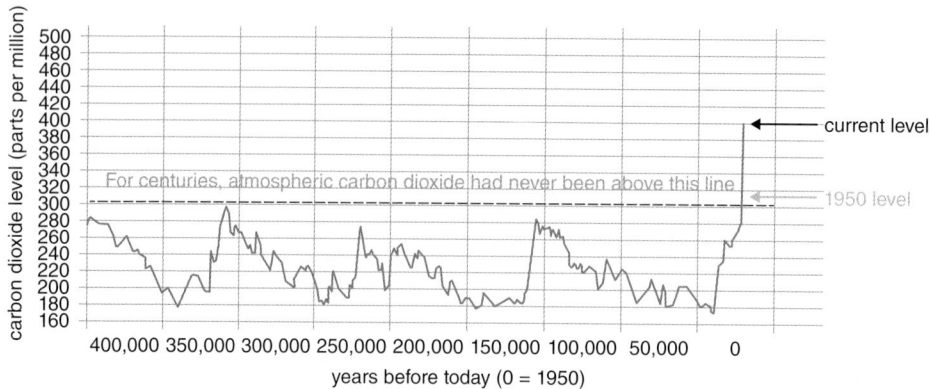

Figure 38.3 The increase in atmospheric CO_2 since the Industrial Revolution. *Source:* Credit: Vostok ice core data/J.R. Petit et al.; NOAA Mauna Loa CO_2 record. This image was created by a USA Government employee in the course of their job with the National Oceanic and Atmospheric Administration (NOAA), USA, and, as such, it is the public domain. The Industrial Revolution was the transition to new manufacturing processes in the period from about 1760 to sometime between 1820 and 1840. (*See color plate section for the color representation of this figure.*)

38.2.3 Famine and Food

Since the early 1990s global undernourishment has fallen by around 27% and currently 10.8% of the global population is undernourished, equivalent to around 795 million people (Gould 2017). Although this is the fewest number of people since records began in 1990, with the current rate of progress, eradicating malnutrition and achieving global food security and meeting the United Nations' Sustainable Development Goal of ending global hunger by 2030 is likely to be missed (Hodson 2017).

Famine is defined when 1 in 5 households in a region face extreme shortage, or when more than 30% of the population are acutely malnourished or when at least 2 people per 10 000 die each day from lack of food.

Currently more than 20 million people face starvation as famine takes hold again in Africa (Wade, M. Crisis point. Sydney Morning Herald, April 8–9th, 2017). According to this report, half of Somalia's population of 12 million people are suffering hunger emergency. It had a similar famine in 2011. Other countries facing the same threat are South Sudan, Nigeria, Yemen and parts of Kenya and Ethiopia. Accompanying drought exacerbates the situation, particularly through the death of livestock. Additional contributors are armed conflict, poor government infrastructure, and rising temperatures. Distress migration occurs with people seeking areas where food is more readily available

38.2.4 Drought and Drinkable Water

38.2.4.1 Drought

Drought is common in many parts of the world and will likely become worse as the planet continues to warm up, producing another obstacle to health in the LMICs (Figure 38.4)

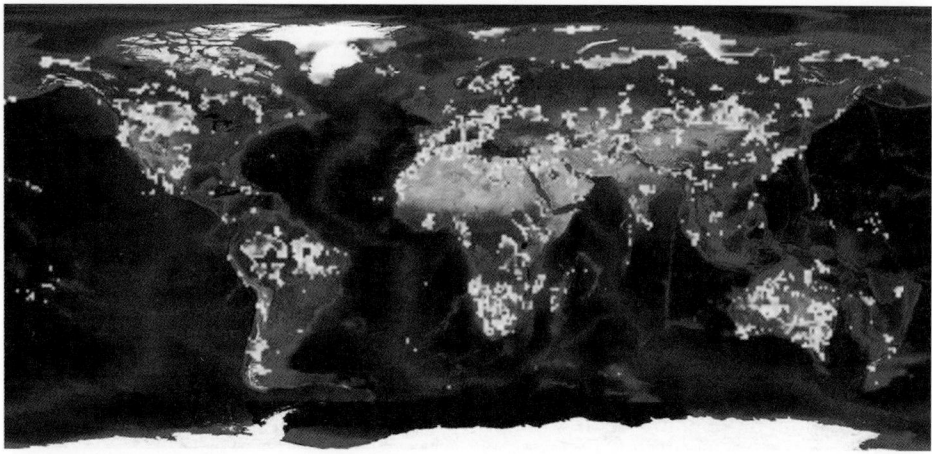

Figure 38.4 Global drought map. Global Drought Information System. 24 month standardized precipitation index (SPI). High SPI value (closer to 3): heavy precipitation (rainfall) event over time period specified. Medium SPI value (approximately = 0): normal precipitation event over time period specified. Low SPI value (closer to – 3): low precipitation event over time period specified (drought?) https://www.drought.gov/gdm/current-conditions. (*See color plate section for the color representation of this figure.*)

38.2.4.2 Water-Borne Diseases

Polluted drinking water plays a role in a large number of communicable diseases including cholera and typhoid. In addition, the infectious agents of many other tropical diseases spend part of their life-cycle in water: malaria and schistosomiasis are examples.

38.2.4.3 Provision of Clean Drinkable Water

If no clean natural water source is available, then installation of water remediation or purification processes should be introduced. The two commonest ways for disinfecting water of pathogenic organisms are ultraviolet irradiation and oxidation with chemicals such as chlorine.

38.2.5 Sanitation

Forty percent of the world's population – 2.5 billion people – practice open defecation or lack adequate sanitation facilities. Even in urban areas, where household and communal toilets are more prevalent, over two billion people use toilets connected to septic tanks that are not safely emptied or use other systems that discharge raw sewage into open drains or surface waters.

Education needs to mandate no open-air defecation or micturition.

The toilets, sewers, and wastewater treatment systems used in the developed world require vast amounts of land, energy, and water, and they are expensive to build, maintain and operate. Novel solutions are required for the LMICs (see Section 38.4).

38.2.6 War

War is currently prevalent in many African countries, the Middle East and parts of Asia including India and Afghanistan. The war in the Middle East has created millions of refugees. These people have inadequate shelter, food, water, and access to medical facilities. Thus, war has a marked adverse effect on the health of people in war zones, either by the measures mentioned above or by direct trauma. A graphic account of the horrors of war imposed on civilian populations is portrayed in the film "Blood Diamond" (2006).

38.3 Continue Doing What Works

This will include

- total vaccination cover for infants and children
- continuation of ongoing efforts in mass drug administration
- continuation of insecticide spraying of mosquito breeding sites
- continuation of ongoing efforts to provide insecticide-impregnated bed nets in malaria, dengue and Zika endemic areas
- wide-spread uptake of Smartphones for delivery of clinical care via telemedicine
- relentless, graphic and clear public advertising especially about HIV/AIDS, malaria and tuberculosis
- the elimination of poverty, since there is a close correlation between poverty and communicable tropical diseases
- provision of shoes to minimize infection with hookworm and other soil-transmitted helminthiases, podoconiosis and snake bite

38.4 New Measures for Improving Remote Rural Healthcare

38.4.1 The Success of Rapid Diagnostic Tests (RDTs) at the Primary Point-of-Care and Their Further Development

Over the last decade the development of RDTs for HIV/AIDS, tuberculosis, syphilis, malaria, human African trypanosomiasis, visceral leishmaniasis, dengue fever, influenza, ebola virus disease and yaws have made it infinitely easier to make a fast and accurate diagnosis of these diseases, enabling more rapid initiation of targeted treatment. Many of them can be performed at the primary point-of-care (POC), thus avoiding the need for transport to clinics or hospitals, another factor in speeding up the initiation of treatment. RDTs for lymphatic filariasis and for onchocerciasis are in development. An ever-increasing number of blood tests can also now be performed at the POC. The current challenge is to develop new, simpler and cheaper POC RDTs for *all* infectious diseases.

38.4.2 The Development of Cheap Innovative Imaging and Other Diagnostic Tests for Provision of Long Distance Medical Advice and Clinical Management

Over the last several years a system involving a smartphone with an additional small piece of hardware and specific software has been developed for measuring visual acuity, visual fields and retinoscopy. These tests can be performed at the POC by anyone suitably trained who has a smartphone with Internet access so that the results can be emailed to an ophthalmologist for interpretation and management (see Chapter 25). A similar approach has been used for assessing deafness.

Portable ultrasound scanning is now feasible at a POC where no radiological facilities exist and can be applied in a wide variety of communicable and non-communicable diseases (see Chapters 23 and 24).

Other POC tests include RDTs for sickle cell disease (Chapter 14), mid upper arm circumference strips for assessing severe acute malnutrition (Chapter 19) and spirometry for chronic obstructive pulmonary disease due to smoke inhalation from indoor fires used for cooking and heating (Chapter 20).

The only requirement for exploiting all these developments to the maximum is training of an individual in their use and Internet access to an appropriate specialist.

38.4.3   The Use of Telemedicine

The combination of smartphones and Internet access is revolutionizing the provision of specialist medical services to remote rural areas and is described in Chapters 26–30.

38.4.4   Novel Approaches to Sanitation

Solving the sanitation challenge in the LMICs will require radically new innovations that are deployable on a large scale. Innovation is especially needed in densely populated areas, where billions of people are only capturing and storing their waste, with no sustainable way to handle it once their on-site storage (such as a septic tank or latrine pit) is full.

Ground breaking improvements in toilet design, pit emptying, and sludge treatment, as well as new ways to reuse waste, can help governments and their partners meet the enormous challenge of providing quality public sanitation services.

This requires understanding issues across the entire sanitation service chain, including waste containment (toilets), emptying (of pits and septic tanks), transportation (to sewage treatment facilities), waste treatment, and disposal/reuse.

The Gates Foundation has a "Reinvent the Toilet Challenge" (RTTC) scheme. These will be next-generation toilets that do not require a sewer or water connection or electricity, cost less than five cents per user per day, and are designed to meet people's needs.

In parallel to the RTTC scheme, the Gates Foundation is developing market-driven ways to stop the dumping of fecal sludge into the environment. The Omni-Ingestor program is developing technologies to make the servicing and maintenance of existing sanitation infrastructure (including latrine pits and septic tanks) easier and more affordable for private companies, public utilities and municipalities. The goal is to develop processors that are smaller than traditional treatment plants, with each processor supporting some 100 000 residents.

38.4.5   A Paradigm Change in the Work Force at the Primary Point-of-Care Delivery

A person trained in performance of physical examination, rapid diagnostic testing and portable ultrasound scanning should be present at every primary point-of-care, with a mobile phone and Internet access.

38.4.6   The Use of Digital Technology at the Primary Point-of-Care

The conventional roll-out of healthcare systems involves building clinics and hospitals and populating them with doctors, nurses, pharmacists and other allied professionals. The rapid development and uptake of smartphones may allow this expensive and slow process to be leapfrogged. A person at the primary point-of-care could be trained in the use of rapid diagnostic tests and portable ultrasound scanning. They will have a smartphone equipped with add-on hardware for assessing sight and hearing and they will have Internet access to physicians and surgeons. They will thus be able to diagnose and treat some illnesses, while being able to triage people with difficult clinical situations to the nearest clinic or hospital. Access to artificial intelligence data banks using smartphones that have greatly enhanced computing power will further complement this approach.

38.4.7   The Supply and Quality of Health Professionals in Remote Rural Areas

As indicated above, it is unrealistic to expect that a doctor, nurse or pharmacist will be widely available for POC healthcare diagnosis and treatment in the near to medium future. For those health professionals who are working at the primary point-of-care in impoverished remote rural areas it is important to have resilience, realism and resourcefulness, since the work environment is challenging.

38.4.8   New Ways Forward for Neglected Tropical Diseases

38.4.8.1   WHO's Overall Approach

The WHO's objectives for neglected tropical diseases (NTDs) are to accelerate the reduction of the disease burden through the control, elimination and eradication of targeted NTDs and contribute to poverty alleviation, increased productivity and better quality of life of the affected people in the African Region. The two most effective approaches for preventing and eliminating NTDs are

- mass administration of medicines (preventive chemotherapy) for diseases such as lymphatic filariasis, onchocerciasis, trachoma, schistosomiasis and soil-transmitted helminthiasis together with post-elimination surveillance
- early case finding and decentralized case management for Buruli ulcer, human African trypanosomiasis, leprosy, leishmaniasis and yaws

Control of lymphatic filariasis and trachoma is targeted by the WHO for 2020, advanced control of schistosomiasis and soil-transmitted helminthiasis by 2020 and onchocerciasis by 2025 (http://www.afro.who.int/en/neglected-tropical-diseases/overview.html).

Additionally, better POC diagnostic tests with increased sensitivity and specificity for *all* infectious diseases need to be developed.

The elimination of blinding trachoma can be managed effectively using the SAFE approach (Surgery, Antibiotics, Facial cleanliness and Environmental changes).

38.4.8.2   Control of Zoonoses

Diseases transmissible from animals to humans through direct contact or though food, water, and the environment, are known as zoonoses. Four neglected tropical diseases have prominent zoonotic components: echinococcosis, foodborne trematodiases, rabies and tenia/cysticercosis.

WHO's strategies for combatting zoonoses are:

- Fostering cross-sectoral collaboration at the human-animal-environment interface among the different relevant sectors at international, regional, and national levels
- Developing capacity and promoting practical, evidence-based, and cost-effective tools and mechanisms for zoonoses' prevention, surveillance and detection, reporting, epidemiological and laboratory investigation, risk assessment, and control, and assisting countries in their implementation.
- Supporting the development of relevant policies, strategies and sustainable programs to prevent and reduce risks and manage outbreaks, and by facilitating their implementation.

In 2015 a global meeting on rabies called for its elimination by 2030.

38.4.8.3 The WASH Program

There are around 2.4 billion people who do not use improved sanitation and 663 million who do not have access to improved water sources (https://www.unicef.org/wash). In 2015 the WHO developed planning, resourcing and delivery of safe water, sanitation and hygiene (the WASH Program) in order to accelerate the control of NTDs and other tropical diseases.

38.4.8.4 Snake Bite

Snake bite is an underestimated public health problem in many tropical LMICs. It is largely an occupational disease of agricultural workers in impoverished rural areas. A study in Bangladesh suggested that 600 000 bites and 6000 snake bite deaths occurred each year (Raman et al. 2010). A study in India in 2001–2003 estimated 45 900 annual snakebite deaths nationally (Mohapatra et al. 2011). Thus, snakebite remains an underestimated cause of accidental death in LMICs. Effective interventions involving education and anti-venom provision would reduce snakebite deaths.

38.5 The UN 2015 Sustainable Development Goals for 2016–2030

38.5.1 The 17 Sustainable Development Goals

In 2015 the United Nations announced 17 new goals to be achieved between 2016 and 2030. These were called the Sustainable Development Goals. Only one of these 17 goals was directly involved in the improvement of health outcomes and was called "Good health and well-being" (Goal 3) (http://www.un.org/sustainabledevelopment/health), although clearly no poverty, zero hunger, quality education, gender equality, clean water and sanitation, affordable and cheap energy and climate action, if achieved, would all have a major positive impact on global health. The 17 Goals are

1) No poverty
2) Zero hunger
3) Good health and well-being

4) Quality education
5) Gender equality
6) Clean water and sanitation
7) Affordable and clean energy
8) Decent work and economic growth
9) Industry, innovation and infrastructure
10) Reduced inequalities
11) Sustainable cities and communities
12) Responsible consumption and production
13) Climate action
14) Life below water
15) Life on land
16) Peace, justice and strong institutions
17) Partnerships for the Goals

Within Goal 3 (Improvement of health outcomes) there are 13 specific sub-goals. They are

1) By 2030 reduce the global maternal mortality ratio to less than 70 per 100 000 live births.
2) By 2030 end preventable deaths of newborns and children under 5 years of age, with all countries aiming to reduce neonatal mortality to at least as low as 12 per 1000 live births and under 5 mortality to at least as low as 25 per 1000 live births.
3) By 2030 end the epidemics of AIDS, tuberculosis, malaria and neglected tropical diseases and combat hepatitis, water-borne diseases and other communicable diseases.
4) By 2030 reduce by one third premature mortality from non-communicable diseases through prevention and treatment and promote mental health and well-being.
5) Strengthen the prevention and treatment of substance abuse, including narcotic drug abuse and harmful use of alcohol.
6) By 2020 halve the number of global deaths and injuries from road traffic accidents.
7) By 2030 ensure universal access to sexual and reproductive healthcare services, including family planning, information and education and the integration of reproductive health into national strategies and programs.
8) Achieve universal health coverage, including financial risk protection, access to quality essential healthcare services and access to safe, effective, quality and affordable essential medicines and vaccines for all.
9) By 2030 substantially reduce the number of deaths and illnesses from hazardous chemicals and air, water and soil pollution and contamination.
10) Strengthen the implementation of the World Health Organization Framework Convention on Tobacco Control in all countries as appropriate.
11) Support the research and development of vaccines and medicines for the communicable and noncommunicable diseases that primarily affect LMICs, provide access to affordable essential medicines and vaccines, in accordance with the Doha Declaration on the Trade-Related Aspects of Intellectual Property Rights (TRIPS) Agreement and Public Health, which affirmed the right of LMICs to use to the full the provisions in the Agreement on Trade Related Aspects of Intellectual Property Rights regarding flexibilities to protect public health, and, in particular, to provide access to medicines for all.

12) Substantially increase health financing and the recruitment, development, training and retention of the health workforce in developing countries/LIMCs, especially in least developed countries and small island developing states.
13) Strengthen the capacity of all countries, in particular LMICs, for early warning, risk reduction and management of national and global health risks.

38.5.2 Details of the Goals Directly or Indirectly Benefiting Health Outcomes

38.5.2.1 No Poverty

Extreme poverty has been cut by more than half since 1990. However, more than one in five people live on less than $1.25 a day. Poverty is more than lack of income or resources: it includes lack of basic services, such as education, hunger, social discrimination and exclusion, and lack or participation in decision making. Gender inequality plays a large role in the perpetuation of poverty and its risks. Pregnant women face potentially life-threatening risks from early pregnancy, and often lose hope for an education and a better income. Age groups are affected differently when struck with poverty. Its most devastating effects are on children, to whom it poses a great threat. It affects their education, health, nutrition, and security. It also negatively affects the emotional, spiritual and emotional development of children through the environment it creates (Data from http://Wikipedia.org: https://en.wikipedia.org/wiki/Sustainable_Development_Goals).

38.5.2.2 Zero Hunger

Globally, one in nine people are undernourished; the vast majority of these people live in LMICs. Agriculture is the single largest employer in the world, providing livelihoods for 40% of today's global population. It is the largest source of income and jobs for poor rural households. Women comprise on average 43% of the agricultural labor force in LMICs, and over 50% in parts of Asia and Africa, yet they only own 20% of the land. Poor nutrition causes nearly half (45%) of deaths in children under five: 3.1 million children each year. (Data from Wikipedia: https://en.wikipedia.org/wiki/Sustainable_Development_Goals).

38.5.2.3 Good Health and Well-Being

Ensure healthy lives and promote well-being for all at all ages. Significant strides have been made in increasing life expectancy and reducing some of the common fatal diseases associated with child and maternal mortality, and major progress has been made on increasing access to clean water and sanitation, reducing malaria, tuberculosis, polio and the spread of HIV/AIDS.

However, only half of women in LMICs have received the healthcare they need, and the need for family planning in increasing exponentially, while the need met is growing slowly: more than 225 million women have an unmet need for contraception. An important target is to substantially reduce the number of deaths and illnesses from pollution-related diseases. (Data from Wikipedia: https://en.wikipedia.org/wiki/Sustainable_Development_Goals).

38.5.2.4 Quality Education

Ensure inclusive and equitable quality education and promote lifelong learning opportunities for all. Major progress has been made for education access, specifically at the primary school level, for both boys and girls. However, access does not always mean quality of education, or completion of primary school. Currently, 103 million youth worldwide still lack basic literacy skills, and more than 60% of them are women. (Data from Wikipedia: https://en.wikipedia.org/wiki/Sustainable_Development_Goals).

38.5.2.5 Gender Equality

Achieve gender equality and empower all women and girls. Providing women and girls with equal access to education, healthcare, decent work, and representation in political and economic decision-making processes. While a record 143 countries guaranteed equality between men and women in their Constitutions by 2014, another 52 had not taken this step. In many nations, gender discrimination is still woven through legal and social norms. (Data from Wikipedia: https://en.wikipedia.org/wiki/Sustainable_Development_Goals).

38.5.3 Reactions to the Goals: Are The Sustainable Development Goals Feasible?

All the goals on the above list would have a direct or indirect beneficial income on global health. However, reaction to the goals has been mixed with some views that the goals do not go far enough and others that they are not specific enough. There has been concern that the dramatic change from the 2005 Millennium Development Goals, which were all health-related, to the 2015 Sustainable Development Goals, in which there is only one health-specific goal, will slow the momentum that has been achieved in the improvement of healthcare galvanized by the 2005 Goals.

A report by the International Food Policy Research Institute (IFPRI) in 2013 criticized the efforts of the SDGs as not being ambitious enough. Instead of aiming for an end to poverty by 2030, the report "An Ambitious Development Goal: Ending Hunger and Undernutrition by 2025" called for a greater emphasis on eliminating hunger and undernutrition and achieving that in five years less, by 2025. It based its claims on an analysis of the experiences from China, Vietnam, Brazil and Thailand and identified three pathways to achieve this: agriculture-led, social protection– and nutrition intervention–led, or a combination of both of these approaches (Fan and Polman 2014).

The SDGs have also been criticized for being contradictory, because in seeking high levels of global GDP growth, they will undermine their own ecological objectives. It has been noted, in relation to the headline goal of eliminating extreme poverty, that $1.25 is actually not adequate for human subsistence and the poverty line should be revised to as high as $5.0.

A commentary in The Economist argued that the 169 targets for the SDGs are too many, calling them "sprawling," "misconceived," and "a mess" compared to the Millennium Development Goals. It also criticized the goals for ignoring local context and promoting "cookie-cutter development policies." They claimed that all other sustainable development goals are founded on achieving SDG number one (No poverty). The Economist estimated that trying to alleviate poverty and achieving the other sustainable development goals will require about US$2–3 trillion per annum for the next 15 years, which critics do not see as being feasible.

In analyzing the SDGs from a medical perspective Murray looked at what worked and what did not work with the MDGs (Murray 2015). He noted that after the Millennium Declaration was signed, the World focused on measuring progress, analyzing roadblocks and celebrating countries that were on track to meet the MDG targets. Similarly, academic research on development focused heavily on the MDGs. In terms of health, the Millennium Declaration and what followed it represented a remarkable success for global collective action. New funding mechanisms emerged, including Gavi (the Vaccine Alliance) and the Global Fund to Fight AIDS, Tuberculosis and Malaria. The U.S. government launched two major funding programs, the President's Emergency Plan for AIDS Relief and the President's Malaria Initiative. Development assistance for health increased from $11.6 billion in 2000 to $33.1 billion in 2012 and has remained steady since then. He felt that the good news was that the inclusive process of defining the SDGs created a new consensus on development to replace the MDGs, but the bad news was that the SDGs are broad, with many aspirational or vague targets and that health clearly does not occupy the central role in the SDGs that it did in the MDGs. Thus health's lower profile in the goals will mean less national level political attention beyond the health sector. He concluded that perhaps the most important aspect of the MDG-to-SDG transition will be the reactions of major donors, and in particular those of the three leading donors that accounted for 61.3% of the funding increase from 2000 to 2014 (the U.S. Government, the UK Government and the Bill and Melinda Gates Foundation). These donors have been, and are likely to remain, strongly committed to the MDG health agenda in the poorest countries.

Water, sanitation and hygiene (WASH) experts have stated that without progress on SDG 6 (Clean water and sanitation) the other goals and targets will not be able to be achieved (Batty 2015; Gupta 2015).

Sridhar (2016) concluded that the SDGs were "mostly vague, largely immeasurable, somewhat attainable, and definitely relevant." He also thought that once the data were collected and put together "the smartest minds and resources would be required to communicate their importance."

38.6 Conclusions

The traditional method of creating a healthcare system involves the building of clinics and hospitals and populating them with doctors and nurses. This model will not work in the poorer LMICs, because it is too expensive. However, it could be replaced by a much cheaper system which would provide a person at the primary-point-care trained to perform rapid diagnostic tests, able to perform portable ultrasound scanning, equipped with a smartphone with vision and hearing hardware and software, and access to medical personnel via the Internet. Either local treatment or triage to the nearest clinic or hospital would be accelerated.

The UN Sustainable Development Goals have been described as a Herculean task, requiring collaboration between governments at the local, regional and national levels, local communities together with major philanthropic donors and healthcare providers.

Seventy-five years ago, the discovery of DNA was the beginning of the molecular biology revolution which led to the identification of genes, RNA and proteins and the basic understanding for the first time of many diseases as well as enabling development of specifically targeted drugs.

More than twenty years ago the Internet together with search engines was another revolutionary development making global access to information easy for many individuals.

At the global level at least, some of the SDGs will be achieved and health in the LMICs will improve.

At the local level the next revolution will be from artificial intelligence through smartphones that have the computing power to search enormous data bases, for example of x-rays, ultrasound scans, retinal images, skin rashes and symptom/sign clusters, and provide a diagnosis when fed data from RDTs, ultrasound scans, retinal images, photographs of rashes, or signs and symptom clusters.

Overall there are some reasons for optimism although burgeoning populations and climate change will make universal healthcare in the LMICs hard to attain.

Bibliography

Batty, M. (2015). Beyond the SDGs: How to deliver water and sanitation to everyone, everywhere. https://www.devex.com/news/beyond-the-sdgs-how-to-deliver-water-and-sanitation-to-everyone-everywhere-86975

Das, J.K., Salam, R.A., Arshad, A. et al. (2014). Community based interventions for the prevention and control of Non-Helmintic NTD. *Infect. Dis. Pov.* 3: 24.

Fan, S. and Polman, P. (2014). An Ambitious Development Goal: Ending Hunger and Undernutrition by 2025. In: 2013 Global Food Policy Report, 2014, pp 15–28 from International Food Policy Research Institute (IFPRI). http://www.ifpri.org/sites/default/files/publications/gfpr2013_ch02.pdf (application/pdf)

Gould, J. Outlook Food Security. A world of insecurity. http://www.nature.com/nature/journal/v544/n7651_supp/pdf/544S6a.pdf

Gupta, G.R. (2015). Opinion: Sanitation, Water & Hygiene For All Cannot wait for 2030. InterPressServiceNewsAgency.http://www.ipsnews.net/2015/10/opinion-sanitation-water-hygiene-for-all-cannot-wait-for-2030

Hodson, R. (2017). Food security. *Nature* 544, S5. doi: https://doi.org/10.1038/544S5a. Published online April 26th 2017.

International Panel for Climate Change. (2013). Final Report 2013. http://www.ipcc.ch/report/ar5/wg1

Mohapatra, B., Warrell, D.A., Suraweera, W. et al. (2011). Snakebite mortality in India: a nationally representative mortality survey. *PLoS Neg. Trop. Dis* https://doi.org/10.1371/journal.pntd.0001018.

Murray, C.J.L. (2015). Shifting to sustainable development goals — implications for global health. *NE J. Med.* 373: 1390–1393.

Rahman, R., Faiz, M.A., Selim, S. et al. (2010). Annual incidence of snake bite in rural Bangladesh. *PLoS Neg. Trop. Dis.* https://doi.org/10.1371/journal.pntd.0000860.

Raman, R., Faiz, M.A., Selim, S. et al. (2010). Annual incidence of snake bite in rural Bangladesh. *PLoS Neg. Trop. Dis.* 4: e860.

Sridhar, D. (2016). Making the SDGs useful: a herculean task. *Lancet* 388: 1453–1454. https://doi.org/10.1016/S0140-6736(16)31635-X.

The Economist. (2015). The 169 Commandments. The proposed sustainable development goals would be worse than useless. Leader, March 15th 2015. https://www.economist.com/news/leaders/21647286-proposed-sustainable-development-goals-would-be-worse-useless-169-commandments

http://Wikipedia.org: https://en.wikipedia.org/wiki/Sustainable_Development_Goals) – The SDGs

Webliography

bernard@thedemographicsgroup.co.au – Prepare for population revolution: published in the Australian, July 27th 2017.

Sridhar, D. (2016). Making the SDGs useful: a Herculean task. *Lancet* 388: 1453–1454. https://doi.org/10.1016/S0140-6736(16)31635-X.

http://www.afro.who.int/en/neglected-tropical-diseases/overview.html: WHO's approach to neglected tropical diseases.

http://www.ifpri.org/sites/default/files/publications/gfpr2013_ch02.pdf (application/pdf). Fan, S. and Polman, P. 2014. An Ambitious Development Goal: Ending Hunger and Undernutrition by 2025. In, 2013 Global Food Policy Report, 2014, pp. 15–28 from International Food Policy Research Institute (IFPRI).

http://www.ipcc.ch/report/ar5/wg1 International Panel for Climate Change 2013: Final Report 2013.

http://www.ipsnews.net/2015/10/opinion-sanitation-water-hygiene-for-all-cannot-wait-for-2030 Gupta, G.R. 2015: Opinion: Sanitation, Water & Hygiene For All Cannot wait for 2030. Inter Press Service News Agency.

http://www.nature.com/nature/journal/v544/n7651_supp/pdf/544S6a.pdf Gould, J. Outlook Food Security. A world of insecurity.

http://www.un.org/millenniumgoals/reports.shtml The UN 2015 MDG Report.

http://www.un.org/sustainabledevelopment/health

http://www.un.org/sustainabledevelopment/health - the United Nations Sustainable Development Goals.

Rahman, R., Faiz, M.A., Selim, S. et al. (2010). Annual Incidence of Snake Bite in Rural Bangladesh. *PLoS Neg. Trop. Dis.* https://doi.org/10.1371/journal.pntd.0000860.

Mohapatra, B., Warrell, D.A., Suraweera, W. et al. (2011). Snakebite Mortality in India: A Nationally Representative Mortality Survey. *PLoS Neg. Trop. Dis.* https://doi.org/10.1371/journal.pntd.0001018.

https://en.wikipedia.org/wiki/Sustainable_Development_Goals). Data from http://Wikipedia.org on the Sustainable Development Goals.

https://esa.un.org/unpd/wpp/Graphs/Probabilistic/POP/TOT - probabilistic estimations for the change in the population of India from 2020 till 2100.

https://www.devex.com/news/beyond-the-sdgs-how-to-deliver-water-and-sanitation-to-everyone-everywhere-86975. Batty, M. 2015. Beyond the SDGs: How to deliver water and sanitation to everyone, everywhere.

https://www.drought.gov/gdm/sites/drought.gov.gdm/files/global-3mon-spi-072017.gif

https://www.economist.com/news/leaders/21647286-proposed-sustainable-development-goals-would-be-worse-useless-169-commandments The Economist. 2015. The 169 Commandments. The proposed sustainable development goals would be worse than useless. Leader, March 15th 2015.

https://www.ipcc.ch/pdf/assessment-report/ar5/syr/AR5_SYR_FINAL_SPM.pdf – International Panel on Climate Change – evidence that the Earth is warming.

https://www.unicef.org/wash - UNICEF's WASH Programme.

Glossary

AAHIVM	American Academy of HIV Medicine
AAP	American Academy of Pediatrics
AASLD	American Association for the Study of Liver Diseases
ABP	acid–base profile
ACE	acetyl choline esterase
ACEP	American College of Emergency Physicians
ACOG	American College of Obstetrics and Gynecology
ACP	active carbon particles
ACT	artemisinin-based combination therapy
AED	automatic external defibrillator
AETC	AIDS Education and Treatment Center
AFB	acid fast bacilli
AHRQ	Agency for HealthCare Research and Quality (USA)
AI	artificial intelligence
AIME	Artificial Intelligence and Medical Epidemiology
AMA	American Medical Association
ANC	antenatal clinic
Anti-HBc	hepatitis B core antibody
Anti-HBs	hepatitis B surface antibody
APA	American Psychiatric Association
API	application programming interface
apps	(software) applications
AR	augmented reality
ART	antiretroviral therapy
ASH	accredited social health activist
ASSURED criteria for diagnostic tests	Affordable, Sensitive, Specific, User-friendly, Rapid and robust, Equipment-free and Deliverable to end-users
ASTMH	American Society of Tropical Medicine and Hygiene
AuNP	gold nanoparticle (aurium being the latin name for gold)
AVI	audio video interleaved (format)
BAL	broncho-alveolar lavage
BGP	blood gas profile

Revolutionizing Tropical Medicine: Point-of-Care Tests, New Imaging Technologies and Digital Health, First Edition. Edited by Kerry Atkinson and David Mabey.

BMI	body mass index
BNP	B-type natriuretic peptide
BOLD	blood oxygen level-dependent
bp	base pair
BRICS	Brazil, the Russian Federation, India, China and South Africa
BSL	blood sugar level
βS	β-globin subunit
BT	bench top
CA	contact angle
CAD	computer-aided detection
CAD	computer-aided design
CAD	computer-assisted diagnosis
CAD4TB	computer-aided detection for tuberculosis
CAGR	compound annual growth rate
CBI	community-based intervention
CDC	Centers for Disease Control and Prevention (USA)
CDS	clinical decision support
CE	Conformité Européene (European Community)
CE-IVD	Conformité Européene *In Vitro* Diagnostics
CFU	colony-forming unit
CHW	community health worker
CKD	chronic kidney disease
CK-MB	creatinine kinase muscle and brain (isoenzyme)
CMAM	community management of acute malnutrition
CME	continuing medical education
CMRO$_2$	cerebral metabolic rate of oxygen consumption
COPD	chronic obstructive pulmonary disease
CPR	cardiopulmonary resuscitation
CrAg	cryptococcal antigen
CROI	Conference on Retroviruses and Opportunistic Infections
CRP	Centre for Rehabilitation of the Paralyzed (Bangladesh)
CSF	cerebrospinal fluid
CT	computerized tomogram
CT	*Collegium Telemedicus*
CW	continuous wave (spectroscopy)
CXR	chest x-ray
DAA	direct antiviral agents
DALY	disability-adjusted life year
DAT	direct agglutination test
DBS	dried blood spot
DENV	dengue virus
DHF	dengue hemorrhagic fever
DICOM	digital imaging and communications in medicine
DL	deep learning
DNDi	Drugs for neglected diseases initiative
DORA	Diabetes Online Risk Assessment
DOT	directly observed treatment
DOTS	directly observed treatment, short course

DPDM	Division of Parasitic Diseases and Malaria (CDC, USA)
DPDx	Laboratory Identification of Parasitic Diseases of Public Health Concern (Centers for Disease Control and Prevention, USA)
DPP	dual path platform
DPV	differential pulse voltammetry
DST	drug susceptibility testing
EBG	evidence-based guidelines
EBOV	Ebola virus
ECG/EKG	electrocardiogram
ECHO	Extension for Community Healthcare Outcomes
EEG	electroencephalography
e-health	electronic health
EHR	electronic health record
EHU	Ebola holding unit
EID	early infant diagnosis
e-learning	electronic learning
ELISA	enzyme-linked immunosorbent assay
ELP	electrolyte profile
EMR	electronic medical record
EMTCT	elimination of mother-to-child transmission
EQA	external quality assessment
EU	European Union
EVD	Ebola virus disease
Fc	Fragment, crystallizable region (Fc receptors bind to antibodies that are attached to infected cells or invading pathogens)
FD	frequency domain (spectroscopy)
FDA	Food and Drug Administration (USA)
FHIR	Fast Healthcare Interoperability Resources
FIND	Foundation for Innovative New Diagnostics
fMRI	functional magnetic resonance imaging
fNIRS	functional near-infrared spectroscopy
FST	filariasis test strip
FTA-Abs	fluorescent treponemal antibody-absorbed
Gavi	Global Alliance for Vaccines and Immunization
GDPR	General Data Protection Regulation, EU
GFR	glomerular filtration rate
GHE	government health expenditures
GIDEON	Global Infectious Disease and Epidemiology Network
GIT	gastrointestinal tract
GLI	Global Laboratory Initiative
GP	general practitioner
GPS	global positioning system
GPU	graphics processing unit
GXM	glucoroxylomannan
HAART	highly active anti-retroviral treatment
Hb	hemoglobin
HbA	normal adult hemoglobin A, also known as adult hemoglobin, hemoglobin A1 or $\alpha 2\beta 2$. It is the most common human hemoglobin

	tetramer, comprising over 97% of the total red blood cell hemoglobin in normal individuals
HbAS	indicates carriers of only 1 allele of the sickle hemoglobin mutation (S) together with one normal hemoglobin allele (A): causes sickle cell trait
HbF	hemoglobin F (fetal hemoglobin)
HbO$_2$	oxygenated hemoglobin
HbS	hemoglobin S (sickle hemoglobin)
HbSC	indicates heterozygosity for the sickle mutation; there is less sickling than in sickle cell disease and acute vaso-occlusive events are fewer. However, people with HbSC disease have more significant retinopathy, ischemic necrosis of bone and priapism than those with sickle cell disease
HbSS	indicates homozygosity for the sickle mutation; causes sickle cell disease, also known as sickle cell anemia
HbSβ$^+$	one sickle cell gene, 1 thalassemia gene inherited
HBC	high burden country
HBeAg	hepatitis B envelope antigen
HBsAg	hepatitis B surface antigen
HBV	hepatitis B
HCG	human chorionic gonadotropin
HCP	healthcare provider
HCT	hematocrit
HCV	hepatitis C
HCVcAg	HCV core antigen
HCW	healthcare worker
HEO	health extension officer
HH	hand held
HH$_b$	deoxygenated hemoglobin
HHB	hemoglobin subunit beta
HIC	high income country
HIPAA	Health Insurance Portability and Accountability Act, USA
HIRO	Health Integrated Rescue Operations
HIV/AIDS	human immunodeficiency virus/acquired immunodeficiency syndrome
HIVMA	the HIV Medical Association
HIVST	HIV self-testing
HL7	Health Level-7
HPLC	high performance liquid chromatography
HRP	horseradish peroxidase
HRP2	histidine-rich protein 2
HRSA	Health Resources and Service Administration (USA)
IAS-USA	International Antiviral Society, USA
ICCM	Integrated Community Case Management guidelines (WHO)
ICF	intensified case finding
ICP	intracranial pressure
ICS	immunochromatographic strip
ICT	immunochromatographic test
ICT	immunochromatographic card test
ICT	information and communication technology

ICU	intensive care unit
iDART	intelligent dispensing of anti-retroviral treatment
IDoP	infectious disease of poverty
IeDea	International Epidemiology Databases to Evaluate AIDS
IFA	immunofluorescence assay
IFA	Internationale Funkausstellung (Berlin)
IFAT	indirect immunofluorescence antibody test
IFCC	International Federation of Clinical Chemistry and Laboratory Medicine
IgG	immunoglobulin G
IMAA	Integrated Management of Adult and Adolescent
IMCI	Integrated Management of Childhood Illnesses
IMRA	immunoradiometric assay
ImTeCHO	Innovative mobile-phone Technology for Community Health Operations
INDIA	Infant Neurodevelopment and Dyadic Interaction Assessment
INH	isoniazid
INR	international normalized ratio
iOS	iPhone operating system
IoT	Internet of things
IPFRI	International Food Policy Research Institute
IPTp	intermittent preventive therapy during pregnancy
IQR	interquartile range
IRS	indoor residual spraying
ISDA	Infectious Diseases Society of America
ISO	International Organization for Standardization
ISTM	International Society of Travel Medicine
IT	information technology
ITN	insecticide-treated net
ITU	International Telecommunication Union
IV	intravenous
JPEG	Joint Photographic Experts Group (format)
KNCV	Koninklijke Nederlandse Centrale Vereniging tot bestrijding der Tuberculose (Dutch Tuberculosis Foundation)
LA	latex agglutination
LAM	lipoarabinomannan
LAMP	loop-mediated isothermal amplification
LDC	least developed countries
LED	light-emitting diode
LF	lymphatic filariasis
LF	lateral flow
LFA	lateral flow assay
Li$^+$/G/CNF	lithium-ion doped graphene/carbon nanofiber
LIPS	luciferase immunoprecipitation system
LMIC	low- and middle-income country
LOD	limit of detection
LOOP	loop-mediated isothermal amplification
LP	lumbar puncture
LPA	line probe assays
MAM	moderate acute malnutrition

MAMA	Mobile Alliance for Maternal Action
MAPS	mHealth Assessment and Planning for Scale (WHO)
MDA	mass drug administration
MDA	malondialdehyde
MDG	Millennium Development Goal
MDR-TB	multi-drug resistant tuberculosis
MEG	magnetoencephalography
mHealth	mobile health
mhGAP	WHO Mental Health Gap Action Programme
MMC	Medical Missions for Children
MMS	multi-media messaging service
MOOC	massive open online course
MOTH	Manufactured Organ Transport Helicopter
mph	miles per hour
MRC	The Medical Research Council (UK)
MRSA	multi-resistant *Staphyloccus aureus*
MSAT	mass screening and treatment
MSF	Médecins Sans Frontières/Doctors Without Borders
MTB	*Mycobacterium tuberculosis*
MTB/RIF	*M.tuberculosis*/rifampicin-resistant (assay)
MTCT	mother-to-child transmission
NAAT	nucleic acid amplification test
NASA	National Aeronautics and Space Administration (USA)
NaTHNaC	The National Travel Health Network and Center (USA)
NGO	non-governmental organization
NHS	National Health Service (UK)
NIAID	The National Institute of Allergy and Infectious Diseases
NIH	National Institutes of Health (USA)
NIRS	near-infrared spectroscopy
NLE	neonatal lupus erythematosus
NOAA	National Oceanic and Atmosphere Administration (USA)
NPR	National Public Radio (USA)
NS1	non-structured protein 1
NT	Northern Territory (of Australia)
NTD	neglected tropical disease
NTM	non-tuberculous mycobacteria
NWS	National Weather Service
OFID	Open Forum Infectious Disease
OPD	outpatient department
OSI	Open Systems Interconnection (model)
PACS	picture archiving and communication system
PCR	polymerase chain reaction
PCW	positive control wells
PDA	personal digital assistant
PEEK	Portable Eye Examination Kit
PEP	post-exposure prophylaxis
PEPFAR	The (USA) President's Emergency Plan For AIDS Relief
PET	positron emission tomography

pfhrp2	*Plasmodium falciparum* histidine-rich protein 2
PFU	plaque forming units
PHC	primary healthcare
PHIL	Public Health Image Library (Centers for Disease Control and Prevention)
PIDS	Pediatric Infectious Disease Society
PIH	Partners in Health
pLDH	*Plasmodium* lactate dehydrogenase
PMI	President's Malaria Initiative
PMTCT	prevention of mother-to-child transmission
POC	point-of-care
POCT	point-of-care testing
PPE	positive protective equipment
PPV	positive predictive value
PTA	pure-tone average
PTSD	post-traumatic stress disorder
QA	quality assurance
qPCR	quantitative polymerized chain reaction
RAFT	Réseau en Afrique Francophone pour la Télémédecine (Network in Francophone Africa for Telemedicine)
RBC	red blood cell
RBM	Roll back malaria
RCT	randomized controlled trial
RDT	rapid diagnostic test
RCP-CirCor	Royal Portuguese Hospital Circle of the Heart
RIF	rifampicin
rK39	recombinant leishmanial antigen K39
RNP	ribonucleoprotein
RPA	recombinase polymerase amplification
RPR	rapid plasma regain
RR	relative risk
RRDR	rifampicin-resistance determining region
RRTB	rifampicin-resistant *M. tuberculosis*
rtPCR	reverse transcription polymerase chain reaction
RTTC	Reinvent the toilet challenge
RT-LAMP	reverse transcription-loop-mediated isothermal amplification
RUTF	ready to use therapeutic feed
SAFE	**s**urgery, **a**ntibiotics, **f**acial cleanliness and **e**nvironmental changes (for control of blinding trachoma)
SARA	Service Availability and Readiness (tool)
SCA	sickle cell anemia
SCD	sickle cell disease
SCT	sickle cell trait
SCT	Swinfen Charitable Trust
SDG	Sustainable Development Goal
SDI	Sexually Transmitted Diseases Diagnostics Initiative (of the WHO)
SDK	software development kit
SERS	surface-enhanced Raman spectroscopy
SHEA	Society for HealthCare Epidemiology of America

SMS	short message service
SPCE	screen-printed carbon electrode
SpO$_2$	peripheral oxygen saturation
SpiNepal	Nepalese Spinal Cord Injury Collaboration
SPRI	solid phase reversible immobilization
SRS	spatially resolved spectroscopy
SSA	sub-Saharan Africa
SS-A	anti-Sjögren's-syndrome-related antibody A[1]
SS-B	anti-Sjögren's-syndrome-related antibody B
ssp	subspecies
STAT	same-day testing and treatment
STH	soil-transmitted helminthiasis
STI	sexually transmitted infection
StO$_2$	tissue oxygen saturation
TB	tuberculosis
TCO$_2$	total carbon dioxide
TDR	Tropical Disease Research (in association with UNICEF, UNDP, the World Bank and the WHO)
TMB	tetramethylbenzidine
TPHA	*T. pallidum* haemagglutination assay
TPPA	*T. pallidum* particle agglutination assay
TPP	target product profiles
TR	time resolved (spectroscopy)
UAV	unmanned aerial vehicle
UK NEQAS	United Kingdom National External Quality Assessment Service
UN	United Nations
UNAIDS	Joint United Nations Programme on HIV and AIDS
UNDP	United Nations Development Programme
UNICEF	United Nations International Children's Fund
UHC	universal health coverage
USB	universal serial bus
USPSTF	US Preventive Services Task Force
VDRL	venereal diseases research laboratory
VL	visceral leishmaniasis
VOC	volatile organic compound
VOT	video-observed therapy
VTOL	vertical take-off and landing
WAN	wide area network
WASH	water, sanitation and hygiene
WAZ	weight for age z-score
WebRTC	web real-time communication
WFPI	World Federation of Pediatric Imaging
WGS	whole genome sequencing
WHA	World Health Assembly

---

[1] Henrik Samuel Conrad Sjögren, 1899–1986, a Swedish ophthalmologist.

WHO	World Health Organization
WHO/TDR	WHO Special Programme for Research and Training in Tropical Diseases
WHZ	weight for height z-score
WRD	WHO-recommended rapid diagnostic (test)
YLD	year lived with disability
XDR-TB	extensively drug-resistant tuberculosis

Index

Revolutionizing Tropical Medicine: Point-of-Care Tests, New Imaging Technologies and Digital Health,
First Edition. Edited by Kerry Atkinson and David Mabey.
© 2019 John Wiley & Sons, Inc. Published 2019 by John Wiley & Sons, Inc.